Animal Brucellosis

Editors
Klaus Nielsen, Ph.D.
J. Robert Duncan, D.V.M., Ph.D.
Animal Diseases Research Institute
Agriculture Canada
Nepean, Ontario

CRC Press
Boca Raton Ann Arbor Boston

Library of Congress Cataloging-in-Publication Data

Animal brucellosis/editors, Klaus Nielsen, J. Robert Duncan.
p. cm.
Includes bibliographical references.
ISBN 0-8493-5878-7
1. Brucellosis. I. Nielsen, Klaus. II. Duncan, J. Robert.
SF809.B8A55 1990
636.089'6957--dc20 89-25248
CIP

This book represents information obtained from authentic and highly regarded sources. Reprinted material is quoted with permission, and sources are indicated. A wide variety of references are listed. Every reasonable effort has been made to give reliable data and information, but the author and the publisher cannot assume responsibility for the validity of all materials or for the consequences of their use.

Direct all inquiries to CRC Press, Inc., 2000 Corporate Blvd., N.W., Boca Raton, Florida 33431.

International Standard Book Number 0-8493-5878-7

Library of Congress Card Number 89-25248
Printed in the United States

PREFACE

Animal brucellosis is a disease that encompasses six bacterial species and principal hosts: *B. abortus* (cattle), *B. melitensis* (goats), *B. suis* (swine), *B. canis* (dogs), *B. ovis* (sheep), and *B. neotoma* (desert rat). As a generalization, the principal manifestations of brucellosis are reproductive failure: that is, abortion or birth of unthrifty offspring in the female and orchitis, epididymitis with frequent sterility in the male. Persistent (lifelong) infection is common with this facultative intracellular parasite with shedding in reproductive and mammary secretions. Man, although considered an end host, often develops a persistent infection characterized by intermittent influenza-like disease termed ''undulant fever''. Because of its economic impact on animal health and the risk of a debilitating human disease, support has been found in most countries for programs to control and eradicate the disease from domestic animal populations. These programs have employed two principal tools: vaccination of young or mature animals and the slaughter of infected/exposed animals, usually on the basis of a reaction in a serological test. Vaccination has been widely used, with the best example being *B. abortus* strain 19 used in cattle. Normally calves would be vaccinated before 8 months of age and would not be tested serologically for brucellosis until 18 months of age when an allowance for elevated agglutination titers would be applied. In the past decade, vaccination of adult cattle with a reduced dosage of *B. abortus* strain 19 has been used as a control measure in heavily infected areas. While vaccination has been demonstrated to protect up to 75% of cattle, serological reactions for reasons that include persistent *B. abortus* strain 19 infection, have interfered with the application of serological tests and therefore the test and cull program.

Definitive diagnosis is by bacteriological culture of the causative organism. However, this is an expensive and time-consuming procedure and as a result, presumptive tests most frequently involving the measurement of antibody in body fluids have been developed. Initially, agglutination tests with a whole cell antigen were used. It was soon realized that nonspecific reactions were frequent, and as a result numerous modifications of the agglutination test were devised. These modifications included acidification of the antigen-serum mixture, heat treatment of serum, treatment with reducing agents such as 2-mercaptoethanol or dithiothreotol, precipitation with Rivanol (6,9-diamino-2-ethoxyacridine lactate) or addition of chelating agents such as ethylene diaminetetraacetic acid. All of these treatments resulted in various levels of inactivation antibody of the IgM class, the main cause of nonspecific reactions. While the agglutination tests have been used successfully in several countries for eradication of brucellosis in animals, these tests have major drawbacks in terms of their very high sensitivity and therefore low specificity and their inability to distinguish antibody resulting from vaccination from that induced by virulent bacteria. A complement fixation test, using a whole-cell antigen and guinea pig serum as a source of complement was also devised and for many years served as the diagnostic standard. This test is only slowly being replaced by primary binding assays in the diagnostic laboratory. While the sensitivity and specificity of the complement fixation test are excellent, its major drawbacks are the technical difficulties in its performance and its inability to distinguish the antibody response of vaccinated from infected animals. Precipitation tests have been developed and were the first serological procedures to employ more purified antigens and in, the case of *B. abortus* strain 19 vaccination, could actually be used for differentiation of vaccine-induced antibody from antibody to field strains. The major problem with precipitin tests is their relative lack of sensitivity. The considerable drawbacks of the classical serological tests have led to the application of primary binding assays to the serodiagnostics. While radioimmunoassay proved very useful in detection of human antibody, the amount of radioisotope required for national animal testing programs would be too difficult to handle and to dispose of. Therefore, primary binding assays utilizing enzymes or fluorochromes are more applicable and a variety of test procedures using both detections systems have been described. These tests are very sensitive, and by selecting the appropriate antigenic components and a suitable specificity for the detection system can be made very specific as well. Other advantages include the ability to distinguish vaccinal antibody from that of true infection, relative ease of automation and data handling, and the ability to manipulate each step of the procedure to suit requirements. The main disadvantages are the expense of the equipment and the standardization requirements.

In this volume we have attempted to include up to date knowledge on brucellosis of animals. Since *B. abortus* has been the topic of most published reports, this species occupies a prominent role in comparison to the other species of *Brucella*. It is clear that a great deal of the research on *B. abortus* is applicable to the other species and it is equally clear that other findings do not apply.

While we currently have most of the tools to efficiently diagnose brucellosis in animals, test and slaughter programs to eradicate this disease are not realistic in large areas of the world and as a result, human brucellosis will continue to be a problem.

A number of notable achievements have occurred over the years through research efforts in brucellosis. These include the development of live attenuated bacterial vaccines for cattle and goats (*B. abortus* strain 19 and *B. melitensis* Rev. 1) and the discovery of poly B with the subsequent demonstration by chemical and immunochemical means of the unique properties of *B. abortus* O polysaccharide. The O polysaccharide of *B. abortus* lipopolysaccharide has been shown to be immunodominant and it induces a readily detectable and discernible antibody response, principally of complement fixing IgG_1. This resulted in the development of a highly specific and sensitive diagnostic complement fixation test that relied on the measurement of IgG_1 antibody rather than IgM, which does not fix guinea pig complement. IgM Fc receptors were demonstrated on this Gram-negative facultative intracellular parasite and these, possibly along with agalactosylated IgG, were shown to be a cause nonspecific agglutination. In a similar vein, it was shown that bovine IgG_1 was capable of agglutinating *B. abortus* cells only at an acid pH, resulting in the development of agglutination tests of increased specificity. *Brucella* was one of the first cases demonstrating reversion from L-forms to intact bacteria. *Brucella* has also been used extensively in basic research, initially as the classical T-independent antigen and to contrast *E. coli* lipopolysaccharide in T and B cell regulation. Subsequently, it has been used in the study of cell-mediated immunity mechanisms and genetic resistance to infection. *Brucella* has also been used as an interferon inducer and in the study of the regulation of polyclonal stimulation, particularly of IgG_{2A} isotype (similar to anti-delta chain antibody) via gamma-interferon. Thus, *Brucella* is a useful model in the delineation of the role of interleukins in the antibody isotype switching/regulation mechanism.

In order to make further inroads into the elimination of this disease, a great deal of further research is required in areas such as diagnostic and protective antigens, production of worthy antigens in a cheap, efficient and safe manner, and possibly most important of all, what constitutes a protective immune response in the host.

THE EDITORS

Klaus H. Nielsen is a Research Scientist in the Immunology section of the Animal Diseases Research Institute at Nepean, Ontario, Canada. Dr. Nielsen was graduated in 1969 and 1971 from the University of Guelph with B.Sc.(Agr.) and M.Sc. degrees, respectively, and in 1974 from the University of Glasgow with a Ph.D. degree. His main areas of study were the mechanisms of homeostasis of antibody production and laboratory diagnosis of human infections such as brucellosis.

Over the past 15 years, Dr. Nielsen has been interested in the immune response of ruminants to infectious disease, using brucellosis and paratuberculosis as models, as well as the interaction of complement in the initiation of the antibody response. Dr. Nielsen is a member of the American Association of Immunologists, the American Association of Veterinary Immunologists, and the British Society for Immunology.

J. Robert Duncan is currently a Research Scientist in the Biologics Evaluation Section, Health of Animals Laboratory of Agriculture Canada in Nepean, Ontario. He is a graduate of the University of Manitoba (B.S.A., 1959), the Ontario Veterinary College (D.V.M., 1963; M.Sc., 1968) and Cornell University (Ph.D, 1971), with special interests in the disciplines of Immunology and Pathology (A.C.V.P., 1972). His research interests have centered on the immunological response of domestic ruminants to infectious diseases which have included bovine brucellosis.

CONTRIBUTORS

Bruce S. Adams
Bega Veterinary Clinic
Bega, New South Wales
Australia

Godfrey G. Alton, D.V.M.
Senior Principal Research Scientist (Retired)
Commonwealth Scientific and Industrial Research Organization
Parkville
Australia

John A. Bantle, Ph.D.
Professor
Department of Zoology
Oklahoma State University
Stillwater, Oklahoma

Jose M. Blasco, D.V.M.
Department of Animal Production
S.I.A./D.G.A.
Zaragoza
Spain

Leland E. Carmichael, D.V.M., Ph.D.
Professor of Virology
Department of Veterinary Microbiology
Cornell University
Ithaca, New York

John W. Cherwonogrodzky, Ph.D.
Research Scientist
Animal Diseases Research Institute
Agriculture Canada
Nepean, Ontario

Richard P. Crawford, D.V.M., Ph.D.
Professor
Department of Veterinary Public Health
Texas A & M University
College Station, Texas

Donald S. Davis, Ph.D.
Associate Research Scientist
Department of Veterinary Pathology
Texas A & M University
College Station, Texas

Gerald Dubray, D.Sc.
Chief, Brucellosis Unit
Department of Pathology of Breeding
INRA
Nouzilly
France

J. Robert Duncan, D.V.M., Ph.D.
Research Scientist
Animal Diseases Research Institute
Agriculture Canada
Nepean, Ontario

Fred M. Enright, D.V.M., Ph.D.
Professor
Department of Veterinary Science
Agricultural Center
Louisiana State University
Baton Rouge, Louisiana

Darla R. Ewalt
Microbiologist
Diagnostic Bacteriology Laboratory
National Veterinary Service Laboratories
U.S. Department of Agriculture
Ames, Iowa

Casimiro García-Carrillo, D.V.M., M.V.M.
Zoonoses Expert
Pan American Zoonoses Center
Pan American Health Organization/World Health Organization
Buenos Aires
Argentina

Jan D. Huber
Domestic Programs Support Staff
Cattle Diseases and Surveillance Staff
APHIS-VS
U.S. Department of Agriculture
Hyattsville, Maryland

W. A. Kelly
Serology Section
Animal Diseases Research Institute
Agriculture Canada
Nepean, Ontario

P. L. McMahon
Research and Development
Idexx Corporation
Portland, Maine

A. P. MacMillan
Central Veterinary Laboratory
Center for Reference and
Research on Brucellosis
Food and Agricultural Organization/
World Health Organization
Weybridge, Surrey
United Kingdom

Herbert Mayer, Ph.D.
Max Planck Institute für
Immunobiologie
Freiburg
Federal Republic of Germany

John E. Mayfield, Ph.D.
Associate Professor
Department of Zoology
Iowa State University
Ames, Iowa

Vincent P. Meador, D.V.M.
Veterinary Medical Officer
National Animal Disease Center
Agriculture Research Service
U.S. Department of Agriculture
Ames, Iowa

Margaret E. Meyer, Ph.D.
Professor of Veterinary Public Health,
Emeritus
School of Veterinary Medicine
University of California
Davis, California

Edgardo Moreno, Ph.D.
Microbiologist
Centro de Investigación en
Biologiá Celular y Molecular
University of Costa Rica
San José
Costa Rica

Paul L. Nicoletti, D.V.M.
Department of Infectious Diseases
College of Veterinary Medicine
University of Florida
Gainesville, Florida

Klaus H. Nielsen, Ph.D.
Research Scientist
Department of Immunology
Animal Diseases Research Institute
Agriculture Canada
Nepean, Ontario

Charlotte E. Rigby, Ph.D.
Research Scientist
Animal Diseases Research Institute
Agriculture Canada
Nepean, Ontario

John E. Searson
Special Veterinary Research Officer
Regional Veterinary Laboratory
N.S.W. Department of Agriculture
and Fisheries
Wagga Wagga, New South Wales
Australia

Michael L. Snyder
Research and Development
Idexx Corporation
Portland, Maine

Shelley S. Sutherland, Ph.D.
Research Officer, Animal Division
W.A. Department of Agriculture
South Perth, Western Australia
Australia

Louisa B. Tabatabai, Ph.D.
Research Chemist
Brucellosis Research Unit
National Animal Disease Center
U.S. Department of Agriculture
Ames, Iowa

Alexander J. Winter, D.V.M.
Department of Veterinary Microbiology,
Immunology, and Parasitology
New York State College of Veterinary
Medicine
Cornell University
Ithaca, New York

Erwin F. Workman, Jr.
Research and Development
Idexx Corporation
Portland, Maine

Peter F. Wright, Ph.D.
Head, Serology Section
Animal Diseases Research Institute
Agriculture Canada
Nepean, Ontario

TABLE OF CONTENTS

Chapter 1

CURRENT CONCEPTS IN THE TAXONOMY OF THE GENUS *BRUCELLA*

Margaret E. Meyer

TABLE OF CONTENTS

I. DEVELOPMENT AND STRUCTURE OF THE GENUS

The isolation of the causal agent of Malta fever by Bruce[1] a century ago had a decisive and permanent impact on clinical medicine as well as on systematic bacteriology. Malta fever had been difficult to differentiate symptomatically and clinically from other fevers, i.e., the so called typhoid-malarial complex, then endemic in countries of the Mediterranean littoral. Thus, to clinicians, the isolation of these organisms separated and defined Malta (undulant) fever as a distinct clinical entity. Since we have the advantage of historical hindsight, we now know that when Bruce[2] later named the causal organisms *Micrococcus melitensis,* he created the first species of *Brucella.*

Interestingly, the realization that undulant fever of man and brucellosis of animals were different manifestations of the same infection led to the creation of the genus *Brucella.* Evans[3] established that *M. melitensis* was, in fact, a small rod (coccobacilli) rather than a coccus and that it was morphologically, culturally, and biochemically essentially indistinguishable from *Brucella abortus.* Because these two organisms shared the distinctive *in vivo* capabilities of producing abortion in animals and undulant fever in man, Meyer and Shaw[4] found unacceptable Evans[1] suggestion that they be classified in the genus *Bacterium,* which included the typhoid-dysentery group of intestinal organisms. To accommodate the distinctive features of *M. melitensis* and *B. abortus* and to commemorate the work of David Bruce, the researchers gave them separate rank as the genus *Brucella.*

From its formation in 1920 to 1963, an additional species, *B. suis* was incorporated into the genus,[5,6] as were several biotypes.[6-11] During these 43 years, there were various critical assessments as to the naming and numbering of biotypes and as to whether these were aberrant, atypical, and/or transitional strains of brucellae.[7,12-17] Nonetheless, the genus membership remained stabilized with the three species of *B. abortus, B. suis,* and *B. melitensis,* now frequently referred to as the three classical species.

Since 1966, three additional species have been added to the genus: *B. neotomae, B. ovis,* and *B. canis,* now appropriately referred to as the three new species. *B. neotomae* was accepted without controversy[18] as it has an essentially smooth colonial morphology, fits other criteria by which *Brucella* organisms can be identified,[19] and also has a distinctive metabolic pattern.[20]

The flow of thought that prevailed concerning both the structure of the genus and the pedigree required for admission into it was abruptly interrupted with the descriptions of *B. ovis*[21] and the accompanying suggestion that it was a *Brucella* organism, and that it should be considered a new species.[22] In fact, for 18 years, doubt and controversy[23] reigned regarding the true identity of *B. ovis* before it was ultimately admitted into the genus *Brucella. B. ovis* had not previously been assigned to a taxonomic niche because it differed markedly from the existing criteria for generic recognition of brucellae and because the manifestations of infection it caused in individual animals, as well as in flocks of sheep, did not fit the classic disease pattern associated with brucellosis. Further, it contradicted the conventional wisdom that only smooth brucellae were virulent and could long maintain themselves in populations of host/reservoir animals. The same circumstances initially clouded the identity of *B. canis.*[24-27]

Hoyer and McCullough[28] ushered in the "high tech" era in this genus in 1968 by being the first to explore species relatedness at the genome level. The results of their DNA-DNA hybridization experiments established that the then four accepted species *(B. abortus, B. suis, B. melitensis,* and *B. neotomae)* had 100% homology among them in their polynucleotide sequences, that *B. ovis* had 94% homology with the other species, and that the base composition of G + C of 56 to 58 mol% was the same in all five species. In a subsequent paper,[29] they established that *B. canis* had DNA homology with the three classical species and by reciprocal DNA-DNA hybridization established that the difference in *B. ovis* was not due to a rearrangement of 6% of the sequences, but that they were actually missing from the genome. On the basis of these results, they concluded that *B. ovis* was a deletion mutant of one of the classical species and that all the species are closely related.

Recently, Verger et al.[30] examined the DNA homologies of the polynucleotide sequences in 51 strains of *Brucella* which included representatives of the six species and several strains of biotypes within each of the classical species (there are no reported biotypes within the three new species). In DNA-DNA reassociation experiments using labeled DNA strands from *B. melitensis* 16M to determine its homology with the other 50 strains, they reported relative binding ratios (percent homology) of from 84 to 100%. In their results on reciprocal DNA-DNA relatedness, they reported percentages ranging 87 to 104%. Even though their 23% range in percentages of binding ratios and 17% range in reciprocal ratios considerably exceeded the reported standard error of 3% in DNA relatedness results,[31,32] they nonetheless denied the validity of Hoyer and McCullough's finding concerning the 6% difference between *B. ovis* and the other species. However, by using a different molecular genetic technique, De Ley et al.[33] established with certainty the genetic similarity of the six species. These investigators previously had found that genome sizes (i.e., molecular complexes) are similar among different strains within a single, well-defined species (standard deviation of a group of averages is less than 14.5%). When the same techniques were applied to the six species of *Brucella,*[34] they found genome molecular complexities of 2.37×10^9, with standard deviation of 8%, indicating an intimate genetic relationship. Their data on DNA ribosomal RNA hybridization also shows, via a similarity map, there to be but little measurable differences among the species. Thus, by all available molecular genetic techniques for ascertaining relatedness at the genome level, it is clear that the relationship among all brucellae is exquisitely close.

II. CURRENT TAXONOMIC STATUS OF THE GENUS

What, then, is the current taxonomic status of this genus? Taxonomy seems to mean different things to different people. However, in its purest sense, it means having a scheme of hierarchical classification that reflects and reveals the evolutionary relatedness of the organisms, ideally at all taxon levels, i.e., biotypes, species, genus, and family. As distinct from the evolutionary relatedness imbued in a taxonomy, the taxonomic process obviously includes a workable identification key and a system of nomenclature.

For many years the genus *Brucella* was sequestered in the family *Brucellaceae,*[35] which also included many other genera (i.e., *Bordetella, Pasteurella,* etc.). Through DNA-DNA hybridization studies and G + C base ratio determinations, the genus *Brucella* was found to be unrelated to the other family members and, in fact, most of the genera in the family were found to be unrelated to each other. In the most recent edition of *Bergey's Manual,*[36] the genus is not subsumed to any family but is free floating in a group of Gram-negative rods and cocci.

However, based on recent results calculated from DNA-ribosomal RNA reciprocal hybridizations, De Ley et al.[34] reported that the genus *Brucella* and plant pathogens in the *Agrobacterium-Rhizobium* complex of organisms have a "rather close phylogenetic origin and they sprang from the same ancestor". These investigators also commented that this unique finding would have to be examined further with other sophisticated genetic techniques. Nonetheless, the genus may now have a family affiliation and a very unexpected one at that.

Because of the exquisite closeness of the genetic relationship among all brucellae, there is no question but what the boundaries of this genus are elegantly defined. Based on all these genetic lines of evidence, the suggestion has been made that all *Brucella* strains are biotypes (biovars) of a single species and should be renamed to reflect this fact.[30,37] This may well be true, but all these sophisticated molecular techniques have revealed only genome similarities, and no attempt has been made to account for the discrete and substantial differences known to exist among these organisms. Additionally, these techniques have revealed only two evolutionary clues, i.e., *B. ovis* is probably a deletion mutant of one of the classical species, and the genus may be descended from some plant pathogens.

III. DIFFERENCES AMONG THE SPECIES

Differences between and among the species that could be useful in ascertaining evolutionary pathways include the following:

Differences in the number of biotypes within a species — Even though the current classification scheme recognizes but 8 biotypes (biovars) within the species *B. abortus,* at least 22 have been reported.[38-41] In the species *B. suis,* there are four recognized biotypes, and a fifth has been reported.[42] In the species *B. melitensis,* there are no reported biotypes, but there are three serotypes. There are no reported biotypes or serotypes within the species *B. neotomae, B. ovis,* or *B. canis.* Thus, the species *B. abortus* is genetically labile, while the others are relatively stable with respect to the numbers of biotypes.

Differences in colonial morphology at time of initial isolation — Each of the three classical species and *B. neotomae* are smooth on initial isolation. *B. ovis* and *B. canis* are nonsmooth (i.e., mucoid), but not fully rough.

Differences in metabolic patterns — Each of the six species has a characteristic and definitive pattern of oxidation on an array of 14 amino acid and carbohydrate substrates. In addition, in the species *B. suis,* the oxidative pattern also is discrete for the biotypes.[14-17,20,27,43,44]

Patterns of growth on appropriate concentrations of the dyes basic fuchsin and thionin — Even though these dyes are man-made products derived from coal tar and are substances the organisms are unlikely to encounter in nature, nonetheless, there is a consistent pattern of growth that recurs throughout the species and their biotypes. Possibly the ring structures of these dyes are mimetic of substrates naturally occurring in mammalian tissues.

Need for CO_2 and serum for growth, especially on initial isolation — While either or both serum and CO_2 are required only by some of the biotypes of *B. abortus* and by *B. ovis,* nevertheless, it is an inherent environmental pabulum unneeded by other genus members and is important in considering lines of descendancy both of the species and of the biotypes.

Susceptibility to *B. abortus* bacteriophage, strain 3, also known as the Tbilisi phage — Several strains of phage have been found that lyse various species of *Brucella,*[45] but strain 3 has consistently provided the most precise results and also is highly correlated to the metabolic patterns.[46] As such, it may offer clues to evolutionary descendancy.

Biological behavior in nature — Each of the three classical species has a preferential reservoir of infection, and each differs in its host range.[47,48] Certainly, each occasionally infects other animals, but they do not perpetuate themselves indefinitely in nonreservoir, nonpreferential hosts. In contrast to the host ranges exhibited by the three classic species, each of the three new species of *B. neotomae, B. ovis,* and *B. canis* has a very limited host range. As far as is known, each is restricted, respectively, to wood rats, sheep (especially rams), and dogs. Also, of probable significance is that all the biotypes of *B. abortus* have the same host range, while in *B. suis,* hosts of the biotypes differ. On the other hand, the host ranges of *B. suis* biotypes may reflect the geographic range of the hosts.

Additionally, no disease in man has been attributed to *B. suis, type 2, B. neotomae,* or *B. ovis.* Also, brucellosis in man is but rarely caused by either *B. canis* or *B. abortus,* type 5.

It is, thus, abundantly clear that there are marked and substantial differences in nature in the biological behavior of the organisms in this genus. Even though classification purists and numerical taxonomists insist that host not be considered when ordering a classification hierarchy, there is experimental evidence to indicate, at least in the genus *Brucella,* that such behavior can provide clues to the lines of evolutionary descent.

MODEL FOR EVOLUTIONARY DERIVATION OF *BRUCELLA* ORGANISMS WITH SUPPORTING EVIDENCE

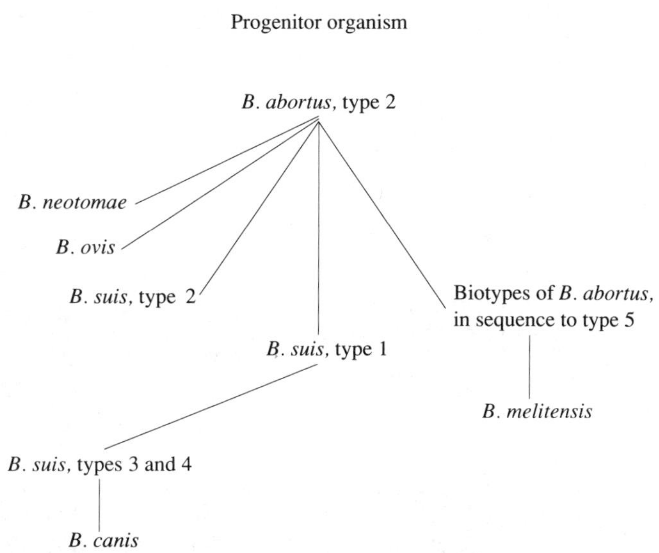

Inasmuch as it has been established by several lines of evidence that all organisms in this genus share an exquisitely close genetic relationship, it is biologically reasonable to assume as a working hypothesis that they arose from a common "stem" or progenitor organism. Also, since "new species" are of recent origin, it is reasonable to assume that the progenitor is an extant organism. The only genus member than can serve in this role is *B. abortus,* type 2. It is sensitive to both basic fuchsin and thionin and requires both CO_2 and serum for growth. Thus, its environmental demands are the strictest among the three classical species, and, except for *B. ovis,* all derivative organisms are less environmentally demanding. In addition, all other *Brucella* organisms either share the pattern of dye sensitivity of *B. abortus,* type 2 or are less sensitive.

There is strong evidence to support considering *B. ovis* a derivative organism of *B. abortus,* type 2.

In fact, an organism essentially indistinguishable from *B. ovis* in its colonial morphology, environmental requirements, and metabolic pattern was derived *in vitro* from a culture of *B. abortus,* type 2. The mechanism of derivation was the induction of L-forms by steroid hormones and the subsequent incomplete reversion of these L-forms to their parental forms. There are substantial reasons to believe that the laboratory lineage of *B. ovis* parallels its natural lineage. *B. abortus,* type 2 is the only extant *Brucella* organism that requires both CO_2 and serum for growth and, thus, is the only species and/or biotype that could transmit these characteristics to a derivative. Hoyer and McCullough[29] hypothesized that the 6% deletion in polynucleotide sequences occurred as a result of multiple, small stepwise losses. Under laboratory conditions, the derivative occurred following a single exposure to the inducing agent, and the resultant derivative was, at the most, two generations removed from its parent — one generation on progesterone and one generation for reversion. The same

TABLE 1
Biotypes of *Brucella abortus* Taxonomically Ordered According to the Sequential Flow of Their Characteristics

Brucella abortus Current Biotype Designation	Line Number	Number of Strains Examined (Total 613)	Characteristics By Conventional Determinative Methods: Serum Required for Growth (■=Yes □=No)	Growth on Dyes (Micrograms Dye/ML Medium): Basic Fuchsin (■=growth) 10	Basic Fuchsin 20	Thionin (□=No growth) 10	Thionin 20	Thionin 50	H_2S Produced (■=Yes □=No)	CO_2 Required for Growth (■=Yes □=No)	Lysis by Bacteriophage (■=Yes □=No; RTD=Routine Test Dilution): RTD	Lysis by Bacteriophage: 10^4xRTD	Agglutination in Monospecific Antiserums (■=Yes □=No): A(abortus)	Agglutination: M(melitensis)
2	1	38	■	□	□	□	□	□	■	■	■	■	■	□
—*	2	35	■	□	□	□	□	□	■	■	■	■	□	■
2	3	18	■	□	□	□	□	□	■	□	■	■	■	□
2	4	11	□	□	□	□	□	□	■	■	■	■	■	□
2	5	9	□	□	□	□	□	□	■	□	■	■	■	□
2	6	8	□	□	□	□	□	□	■	■	■	■	□	■
1	7	163	□	■	■	□	□	□	■	■	■	■	■	□
1	8	108	□	■	■	□	□	□	■	□	■	■	■	□
—*	9	3	□	■	■	□	□	□	■	□	■	■	■	■
4	10	43	□	■	■	□	□	□	■	■	■	■	□	■
3	11	20	□	■	■	■	□	□	■	■	■	■	■	□
3	12	21	□	■	■	■	□	□	■	□	■	■	■	□
9	13	13	□	■	■	■	■	□	■	■	■	■	□	■
9	14	5	□	■	■	■	■	□	■	□	■	■	□	■
—*	15	2	□	■	■	■	■	□	■	■	■	■	■	■
6	16	2	□	■	■	■	■	□	■	□	■	■	■	□
6	17	17	□	■	■	■	■	□	□	□	■	■	■	□
7	18	2	□	■	■	■	■	□	□	□	■	■	■	■
5	19	95	□	■	■	■	■	□	□	□	■	■	□	■

* Strains with combinations of characteristics not previously described.

brevity of time and events must have existed in nature. If the loss of polynucleotide sequences happened over a period of time with many small losses, then strain variants of *B. ovis* should be found. Such variants have not been found. In fact, the characteristics of all reported isolates of *B. ovis* are of such uniformity as to have prompted investigative comment[51] and a separate report on the matter.[52] Further, this uniformity has been confirmed at the genome level by O'Hara et al.,[57] who, after analyzing 33 strains of *B. ovis* DNA with 11 different restriction endonucleases, concluded that *B. ovis* is a very homogenous species and that *B. abortus, B. melitensis,* and *B. canis* are more closely related to each other than each is to *B. ovis*. One lesson learned from this derivation of *B. ovis* is that the order of descendancy from a progenitor is not necessarily linear.

A. DERIVATION OF THE BIOTYPES OF *BRUCELLA ABORTUS*

Table 1 shows the taxonomic ordering of the biotypes of *B. abortus* according to the sequential flow of their characteristics by the conventionally measured characteristics. Table 2 shows those biotypes arranged in six groups according to loss of a characteristic. *B. abortus,* type 2 has the greatest environmental demands (need for serum and CO_2) and is the most sensitive to basic fuchsin and thionin, other dyes,[41] and a variety of antibiotics.[53]

TABLE 2
Biotypes of *Brucella abortus* Divided into 6 Groups

Line Numbers on Table I	Characteristics By Conventional Determinative Methods ■=Yes □=No					
	Needs Serum for Growth	Sensitive to Basic Fuchsin	Sensitive to Thionin	Produces Hydrogen Sulfide	Needs Carbon Dioxide for Growth	Suseptible to Phage
Group I Lines 1,2	■	■	■	■	■	■
Group 2 Lines 3,4,5,6	1st Loss	■	■	■	□■ *	■
Group 3 Lines 7,8,9,10	2nd Loss		■	■	□■	■
Group 4 Lines 11,12	3rd Loss		■	■	□■	■
Group 5 Lines 13,14,15,16	4th Loss			■	□■	■
Group 6 Lines 17,18,19	5th Loss				□ **	■

* Need variable.
** Loss becomes permanent.

The greatest change in biotype characteristics occurs between biotypes 1 and 2, wherein there is a loss of the environmental demands and a concurrent loss in sensitivity to numerous dyes and antibiotics. Since these changes cannot be attributed to a massive occurrence of simultaneous and multiple mutations, this change in sensitivities may well indicate an alteration in permeability and/or structure of the cell surface, wall, or membranes. In any event, the change is dramatic and shared by all subsequent biotypes of *B. abortus*. Even though the metabolic pattern of all biotypes is identical, including type 2, there are changes that are substantial and can be measured by alterations in the sensitivity to basic fuchsin and thionin. When arranged in sequential order, it can be seen (Table 1) that *B. abortus,* type 2 is the only possible progenitor of the biotypes in *B. abortus* and that *B. abortus,* type 5 is, so to speak, the end of the line.

B. POSSIBLE DERIVATION OF *BRUCELLA MELITENSIS*

By all the conventional determinative methods, *B. melitensis* is indistinguishable from *B. abortus,* biotype 5. However, in its metabolic pattern, it differs on three carbohydrate substrates (arabinose, galactose, and ribose). There is evidence, however, that utilization of arabinose and galactose are interdependent, i.e., all species and biotypes oxidize either both or neither of these substrates. Additionally, it is known that one of the by-products of arabinose oxidation is galactose.[54] Thus, alteration in oxidation of one of these two substrates may reflect only one change, the second difference being oxidation of ribose.

Another difference that separates *B. melitensis* from *B. abortus,* biotype 5 is susceptibility to *B. abortus,* phage strain 3. In view of the fact that there is now extant a strain of *B. abortus* that is known to be resistant to this phage and that many strains of *B. melitensis* show lysis from without on exposure to this phage, differences in phage susceptibility may not mark as decisive a difference as previously believed. Nonetheless, there is essentially a permanent change in the phage susceptibility of *B. melitensis,* perhaps reflecting only a change in the phage receptor sites on the cell surface. Differences in cell surface may also be involved in the differing host responses (i.e., the preferential reservoir situation) to *Brucella* organisms and may be related to both host susceptibility and lack of maintenance in nature among the nonpreferential hosts. In any event, according to all available measures of similarity, *B. melitensis* is a close relative of *B. abortus,* type 5. This biotype is rarely pathogenic for man (a characteristic which may be a good indicator of the impending differences between the preferential hosts of *B. abortus* and *B. melitensis*).

C. DERIVATION OF *BRUCELLA SUIS*

The first steps in the derivation of *B. suis* type 2, i.e., loss of need for CO_2 and serum, are identical to the first steps in derivation of *B. neotomae* and of the biotypes of *B. abortus*. Thereafter, all of the biotypes of *B. suis* are derived in the same pattern of sequential flow of characteristics as are the biotypes of *B. abortus,* except the loss of sensitivity to basic fuchsin by the biotypes of *B. suis* is the mirror image of the loss of sensitivity to thionin by the biotypes of *B. abortus*. In the species *B. suis,* type 3 and 4 (type 4 quantitatively has some M antigen) are apparently the "end of the line". The essential difference in metabolic pattern between *B. abortus* and *B. suis* is the oxidation of amino acids in the urea cycle by *B. suis*.

It appears that whatever happens in nature to spark the initiation of altering the characteristics in *B. abortus,* type 2, the critical change can be measured by whether the first loss occurs as a loss of sensitivity to thionin, in which event, the organism becomes characteristic of what we recognize as the species of *B. suis* and its descendant biotypes. On the other hand, if this first loss is measured as a loss of sensitivity to basic fuchsin, the organisms become characteristic of what we recognize as the species *B. abortus* and its descendant biotypes. After this first critical alteration, the biotypes then occur in a sequential and essentially linear descent for each successive biotype.

The change that is measured by the dyes is only mimetic and not duplicative of the changes that occur in nature. By using the techniques of providing selective environments, it is not difficult to obtain mutants with altered dye sensitivities. However, other changes, such as altered sensitivities to other dyes, antibiotics, or alterations in metabolic pattern, do not accompany changes in growth on these two dyes. Obviously, elucidation of the nature of this first critical alteration in *B. abortus,* type 2 is crucial for the recapitulation of the evolution of members in this genus.

D. ORIGIN OF *BRUCELLA CANIS*

There is strong circumstantial evidence to indicate that *B. suis,* type 3 is the progenitor organism of *B. canis,* and that it arose by the same natural phenomenon as did *B. ovis.* By all the conventional determinative methods, it is indistinguishable from *B. suis,* type 3, and it also has the metabolic pattern that characterized the species *B. suis.* Further, it has the identical colonial morphology (i.e., mucoid) to *B. ovis* and, similarly, has a very restricted host range.

E. DERIVATION OF *BRUCELLA NEOTOMAE*

The characteristics of *B. neotomae* indicate that this species also could have been derived from *B. abortus,* type 2, in a fashion similar to *B. suis,* type 2, i.e., loss of the need for CO_2 and serum and loss of sensitivity to thionin. It also is similar to *B. suis* in that it is susceptible only to high concentrations of *B. abortus* bacteriophage, strain 3 and oxidizes one of the major substrates in the urea cycle. As with other of the "new species," it is not pathogenic for man and has a limited host range.

Thus, the model accounts for the known and possible lineages of all organisms in the genus *Brucella* and provides a structural working hypothesis for future investigations.

IV. MECHANISMS THAT COULD ACCOUNT FOR ALTERATIONS IN CHARACTERISTICS OF MEMBERS OF THE GENUS *BRUCELLA*

A. INCOMPLETE REVERSION OF L-FORMS, PRODUCING STRAINS WITH ALTERED CHARACTERISTICS

This phenomenon has already been confirmed experimentally in the case of *B. ovis* and may well occur with other organisms within this genus, especially in consideration of the fact that L-forms can be induced by steroid hormones. Since the preferred habitat of brucellae is the reproductive tract, especially the gravid uterus, no doubt exists about the organisms encountering these hormones *in vivo.*

In fact, a naturally occurring cell wall-defective variant of *B. abortus* has been isolated from bovine tissue.[59] Significantly, this organism required serum for growth, the colonies were nonsmooth, and the animal had a history of having been treated with stilbestrol.[59,60]

The phenomenon of incomplete reversion in the genus *Brucella* was first encountered in a culture of *B. melitensis,*[49] so the mechanism certainly is not confined to the species *B. abortus.* Nor is it confined to this genus. It has been reported to occur in *Streptococcus faecalis*[58] and is responsible for changes in features of *Escherichia coli,*[61] *Salmonella typhimurim,*[62] and β-hemolytic streptococci.[63]

There is no evidence to indicate that all incomplete and/or altered revertants undergo a full 6% loss in polynucleotide sequences, and the possibility certainly exists that nucleotide changes could be more subtle, i.e., less than 3% and thus not detectable by current DNA-DNA techniques.

B. ACTIVITY OF PLASMIDS

There is an increasing body of evidence accumulating to indicate that brucellae do not contain plasmids, or they cannot be liberated by the methods used in other genera and species.

In the brucellosis laboratory at the University of California, Davis, we have examined 600 strains of *Brucella,* which included strains of all species and biotypes. Use of the Kado and Liu,[64] Guerry et al.,[65] and Clewel and Helenski[66] methods, each on essentially equal numbers of strains, found no evidence of plasmids. Simon[67] also reported negative results in his search for plasmids.

C. CONVENTIONAL MUTATIONAL EVENTS INVOLVING DISCRETE CHARACTERISTICS

In view of the repeatedly expressed speculations on the mutability of the *Brucella* species (see review by Meyer in Reference 55), surprisingly little has been reported on the mutation of discrete characteristics. It is known that under laboratory conditions, mutations are responsible for the change from smooth to rough colonial morphology,[68] for loss of CO_2 requirement in *B. abortus* type 1,[69] and for alterations in sensitivities to various dyes,[70,71] erythritol,[72] and penicillin[73] and in hydrogen sulfide production.[74]

However, no mutants of mucoid colonial morphology comparable to the morphology of colonies of *B. ovis* and *B. canis* have been found to occur under laboratory conditions, and we now know that laboratory-selected mutants with altered sensitivities to basic fuchsin and/or thionin are not comparable to the wild-type mutants. However, the essentially minute changes among the species biotypes could easily be accounted for by one step or several one-step sequential mutations.

D. ACTIVITY OF BACTERIOPHAGES

The potential impact that bacteriophages may have on causing alterations in strain characteristics is under active investigation. There are several approaches that can be used. One is to ''cure'' strains of any indigenous phage and check the before and after results. Another is to infect noninfected strains using a variety of phages and ascertain the before and after characteristics. Both these and other avenues for exploring the molecular biology of *Brucella* have been reported by Rigby et al.[75]

Another avenue of approach in the investigation of the relationship of phages to alterations in strain characteristics is to further explore the possible role of sticky white *Brucella* phage carrier colonies. These types of colonies and their associated lytic activity have been reported by Renoux and Suire,[76] McDuff et al.[77] and Meyer,[49] who also commented that carrier phage may be involved in the incomplete reversion process since sticky white colonies and lytic activity invariably accompany this phenomenon.

E. FURTHER USE OF RESTRICTION ENDONUCLEASES

Restriction endonuclease analysis has shown that there is a very close relatedness among all the *Brucella* species and that *B. canis, B. suis,* and *B. melitensis* are more closely related to each other than to *B. ovis*. However, we should now be looking for a technique that will help account for the observed differences.

F. ACTIVITY OF PORINS

The existence of porin channels in the outer membrane of brucellae was first suggested by Verstreate et al.[78] and further elucidated by Douglas et al.,[79] Verstreate and Winter,[80] and Santos et al.[81] The relative sizes of the porin channels differ among the species, being wide in *B. canis,* medium in *B. abortus,* and narrow in *B. melitensis*. These findings certainly may partially account for the differing permeability of the dyes basic fuchsin and thionin and be reflected in the differing patterns of organism sensitivity to these dyes. However, before any conclusions can be drawn about porins, the critical strains remain to be examined, i.e., *B. abortus,* types 2 and 5, *B. neotomae,* and *B. suis,* type 2.

G. USE OF GAS CHROMATOGRAPHY TO ASCERTAIN DIFFERENCES IN CONSTITUENT FATTY ACIDS

Tanaka et al.[82] examined the 16 *Brucella* reference strains and 66 field isolates by gas chromatography for quantitative comparisons of 15 cellular fatty acids. Using the resultant data, they did a numerical taxonomic analysis to determine interspecies similarity matrices and intraspecies matrices on the species *B. abortus, B. melitensis, B. suis,* and *B. canis.* Since only one strain each of *B. ovis* and *B. neotomae* were included in the study, the same type of analysis could not be done on the latter two species. These investigators found that *B. abortus* and *B. melitensis* are quite similar to each other, that *B. ovis* and *B. neotomae* are more similar to *B. abortus* than to the other species, that *B. suis* had overall similarity but was a distinctive group, and that *B. ovis* had greater similarity to *B. suis* than to the other species. Within the species, there was great similarity among biotypes within *B. abortus* and *B. melitensis* and complete homogeneity among strains of *B. canis.* In the *B. suis* species, *B. suis* biotype 2 was distinctly different.

In a later study on the fatty acid content of *B. canis* and *B. suis* by Dees et al.[83] large amounts of 19-carbon cyclopropane acid were present in all strains of *B. suis* and absent in all strains of *B. canis.*

The significance of these findings on the fatty acids lies in the following: (1) if *B. canis* is a derivative of *B. suis* via the route of incomplete reversion of L-forms, then one of the losses must be 19-carbon cyclopropane acid; (2) the marked difference observed in *B. suis* type 2 goes hand in hand with the fact that its behavior in nature is different, i.e., it has a restricted host range and is noninfectious in man. Additionally, *B. suis* type 2 colonies differ subtly, but perhaps significantly, from the classic fully smooth forms. They are neither fully smooth nor fully mucoid. Other investigators have also commented on the colonial morphology of *B. suis,* type 2.[84] This organism may well be from an L-form that has not completely reverted, but has reverted more fully than *B. canis* and *B. ovis.* Since *B. suis,* type 2 is so remarkably different from *B. suis* biotypes 1, 3, and 4, the model shows these biotypes arising independently rather than sequentially.

H. EXPLORATION OF THE RIBOSOMES

One of the newer and significant additions to the elucidation of the process of evolution in prokaryotic cells is the concept that RNA, i.e., ribosides and the ribosome, preceded nucleotides and DNA in the evolutionary development of prokaryotes. Thus, there is not just a duality of genotype and phenotype, but a trinity of genotype, ribotype, and phenotype.[85] Or, as stated by Stanier,[86] "at the level of gene products, information is accumulating which shows that the gene composition of phenotypically similar strains is closely related to ribosomal protein composition." Also, see Stanier[87] and Darnell.[97]

One of the common techniques for ascertaining differences in ribosomal structure and function is to examine ribosomes extracted from disrupted cells or to ascertain the responses of intact cells and mutants thereof with antibiotics known to affect the ribosome, i.e., erythromycin and/or chloramphenicol. This approach has been fruitful in determining discrete ribosomal differences in such organisms as *E. coli*[89,90] and *B. subtilis.*[91-93]

In the brucellosis laboratory at the University of California, Davis, we have examined all the species and biotypes for their sensitivity to erythromycin and chloramphenicol, using both a low and high concentration via antibiotic disks and measuring zones of inhibition in millimeters with calipers. All strains of *B. abortus,* type 2 were sensitive to erythromycin, and all other *B. abortus* biotypes were resistant; all biotypes of *B. suis* were susceptible, as was *B. ovis. B. melitensis, B. canis,* and *B. neotomae* were sensitive only to the higher concentrations. All species and biotypes were equally sensitive to chloramphenicol.

These results indicate that there may be marked ribosomal differences between *B. abortus,* type 2 and all its derivative biotypes. Additionally, there must be subtle but discrete and measurable differences in ribosomal structure and function that ebb and flow through this genus (possibly switched on and off?). Be that as it may, it is known that the differences between sensitivity and resistance to erythromycin are due to changes in the ribosomal structure resulting in a lower binding affinity to this antibiotic.[93] It would appear, therefore, that one of the critical points of inquiry to help account for and to recapitulate the lineage of the species and biotypes of *Brucella* is to further ascertain the importance and occurrence of these ribosomal differences.

V. CURRENT TAXONOMIC STATUS OF THE GENUS *BRUCELLA*

There should be an orderly sequence of processes in establishing the systematics of genus, i.e., strain descriptions, identification key, and a hierarchical ordering of the organisms into a classification scheme based on known and observed evolutionary relationships. Clearly, taxonomy is a dynamic and ongoing rather than a static process.

The present classification scheme that sequesters this genus is actually a determinative identification key, and, as such, it obscures rather than reveals the evolutionary relationships that must exist among the member organisms. Additionally, the species and biotypes have been intercalated into the scheme in the chronological order in which they were isolated and described. Further disorder has been introduced into the scheme by virtue of the fact that there is no consensus on what constitutes a biotype. There is also now the recommendation that the nomenclature be changed to reflect the closeness or DNA-DNA homology[30] and that the genus be considered as one species.

A. WHAT COURSE OF ACTION?

The bottom line is that no one is certain about what constitutes a bacterial species. It is generally and somewhat loosely defined as a group of organisms sharing a mutuality of characteristics as measured phenotypically and that phenotypic discontinuities separates them into species.

In a hierarchical ordering of members of this genus, the great genetic plasticity of the species *B. abortus,* lesser plasticity of *B. suis,* and almost lack of plasticity in *B. melitensis* should be accounted for, as should the phenotypic discontinuities between each species. The working model and submodel include these differences and account for the origin of *B. ovis* and possibly of *B. canis.* Should the genus be considered a single species based on DNA-DNA homology?

DNA-DNA homology of polynucleotide sequences may give the illusion of being more genetically precise than it actually is. For example, human beings and chimpanzees have 98% homology among their nucleotide sequences,[85] and 99% of the amino acids in the proteins are the same.[94] At the bacterial level, among the serovars of *Listeria monocytogenes,* DNA relatedness has been found to be heterogeneous,[95] and the serotypes have been divided into "genomic groups". Intragroup relatedness varies from 90 to 100% homology, while intergroup relatedness falls to 25%.

In view of these disconcerting and somewhat confounding facts and before the species concept in this genus is altered, we should perhaps do what other phylogeneticists and molecular biologists are doing.

First, though overall polynucleotide homology does not separate humans from other primates, a fastidious examination of the DNA base sequences per se revealed altered positions of the protein (amino acid sequences), and these alterations, though very minimal, do separate man from African apes, orangutans and gibbons.[96]

Second is the use of recombinant DNA techniques. As stated by Slatkin,[94] "Even if we were presented with a complete DNA sequence of every individual of every species, we would still not be able to discern what makes species differences. We need to be able to modify genes to see if we can produce new traits and even new species, and to see whether the species act as if they possess some inertia, resisting modification. The tools for this exist in using recombinant DNA."

B. BRINGING ORDER INTO THE TAXONOMY OF THE GENUS *BRUCELLA*

To do this will require consensus and consistency among members of the International Committee on Systematic Bacteriology, Subcommittee on Taxonomy of *Brucella*. Thus far, there has been little consistency and no consensus. In its 1970 meeting, even though it defined *B. suis* by both the conventional determinative methods and metabolic patterns, it assigned *B. canis* as an additional species and simultaneously accepted an organism as a biotype of *B. suis* (type 5) that had not been adequately shown even to be a *Brucella* organism.[98] This organism later was decisively excluded from the genus.[99,100] While there is no uniformity on the definition of a species, Pinigin et al.[101] are still making a plea to have *Brucella suis* type 4 considered a separate species.[101,102]

Further, indecision has been caused by Tolari et al.[103] who suggested that *B. abortus*, biotypes 3 and 6 be considered a single biotype because the only significant difference between them is a sensitivity to thionin. However, in 1984 they suggested strains of *B. suis* with slightly altered resistance to basic fuchsin be considered an additional biotype of this species.[42]

There has also been a suggestion that the species *B. melitensis* can be subdivided into five biovars on minute differences in the oxidative rates of five amino acids.[104] Since it has already been established[105] and observed many times over that age, i.e., number of previous subculturings, and lyophilization tend to cause an increase in utilization rates (but do not alter the basic metabolic pattern), it is really not possible to distinguish metabolic biovars of *B. melitensis* by their oxidative metabolism.

C. CHANGE IN NOMENCLATURE

Verger has proposed that since he considers the genus *Brucella* to be a single species, *B. melitensis*, the names of the constituent organisms reflect this monospecies concept.[30] Thus, *B. abortus* would become *B. melitensis* biovar *abortus*. In view of what we know about the evolutionary paths in this genus, this nomenclature is terribly misleading.

Until more is elucidated about the species and biotype derivations and lineages, it seems to be a good idea to adhere to the advice given in *Bergey's Manual*,[36] i.e., "the advantage of adopting a restrictive species definition must be weighed against its potential impact on well-established and accepted bacteria groups." Thus, it seems the time to consider changing the structure of the genus is when we have more finite and discrete genetic and evolutionary information, especially as to what accounts for the substantial differences among these organisms.

REFERENCES

1. **Bruce, D.,** Note on the discovery of a microorganism in Malta Fever, *Practitioner*, 39, 161, 1887.
2. **Bruce, D.,** Sur une nouvelle forme de fievre, *Ann. Inst. Pasteur Paris*, 7, 289, 1893.
3. **Evans, A. C.,** Further studies on *Bacterium abortus* and related bacteria. II. A comparison *Bacterium abortus* and *Bacterium bronchisepticus* and with the organism which causes Malta Fever, *J. Infect. Dis.*, 22, 580, 1918.
4. **Meyer, K. F. and Shaw, E. B.,** A comparison of the morphological, cultural, and biochemical characteristics of *B. abortus* and *B. melitensis*, *J. Infect. Dis.*, 27, 173, 1920.
5. **Traum, J.,** Report of the Chief of the Bureau of Animal Industry, U.S. Department of Agriculture, Washington, D.C., 1914, 30.

6. **Huddleson, I. F.,** The differentiation of the species in the genus *Brucella,* Tech. Bull. No. 100, Mich. State College of Agriculture Exp. Stn., East Lansing, MI, 1929.
7. **Wilson, G. S.,** The classification of the *Brucella* group: a systematic study, *J. Hyg.,* 33, 516, 1933.
8. **Van der Schaff, A. and Roza, M.,** Brucellosis en onchocerciasis in verband met un chronish gewrichtelijden bij runderen, *Ned. Ind. Bl. Diergenseek,* 52, 1, 1940.
9. **Stableforth, A. W. and Jones, L. M.,** Report of the subcommittee on taxonomy of the genus *Brucella, Int. Bull. Bacteriol. Nomencl. Taxon.,* 13, 145, 1962.
10. **Meyer, M. E. and Cameron, H. S.,** Identification of the causative agents and epidemiology of porcine brucellosis, Bull. W.H.O., 28, 499, 1963.
11. **Meyer, M. E.,** Species identity and epidemiology of *Brucella* strains isolated from Alaskan Eskimos, *J. Infect. Dis.,* 114, 169, 1964.
12. **Renoux, G.,** La notion d'espece dans le genere *Brucella, Ann. Inst. Pasteur Paris,* 94, 179, 1958.
13. **Stableforth, A. W.,** Brucellosis, in *Infectious Diseases of Animals. Vol. 1. Diseases Due to Bacteria,* Stableforth, A. W. and Galloway, I. A., Eds., Butterworth, London, 1959.
14. **Meyer, M. E. and Cameron, H. S.,** Metabolic characterization of the genus *Brucella.* I. Statistical evaluation of the oxidative rates by which type 1 of each species can be identified, *J. Bacteriol.,* 82, 387, 1961.
15. **Meyer, M. E. and Cameron, H. S.,** Metabolic characterization of the genus *Brucella.* II. Oxidative metabolic patterns of the described biotypes, *J. Bacteriol.,* 82, 396, 1961.
16. **Meyer, M. E.,** Metabolic characterization of the genus *Brucella.* III. Oxidative metabolism of strains that show anamolous characteristics by conventional determinative methods, *J. Bacteriol.,* 82, 401, 1961.
17. **Meyer, M. E. and Morgan, W. J. B.,** Metabolic characterization of *Brucella* strains that show conflicting identity by biochemical and serological methods, Bull. W.H.O., 26, 823, 1962.
18. International Committee on Nomenclature of Bacteria, Subcommittee on the Taxonomy of *Brucella,* Minutes of the meeting of August 1970, *Int. J. Syst. Bacteriol.,* 21, 126, 1971.
19. **Stoenner, H. G. and Lackman, D. B.,** A new species of *Brucella* isolated from the desert wood rat, *Neotoma lepida* Thomas, *Am. J. Vet. Res.,* 69, 947, 1957.
20. **Cameron, H. S. and Meyer, M. E.,** Metabolic studies on *Brucella neotomae* (Stoenner and Lackman), *J. Bacteriol.,* 76, 546, 1958.
21. **Buddle, M. B. and Boyes, B. W.,** A *Brucella* mutant causing genital disease in sheep in New Zealand, *Aust. Vet. J.,* 29, 145, 1953.
22. **Buddle, M. B.,** Studies on *Brucella ovis* (n.sp.) a cause of genital disease of sheep in New Zealand and Australia, *J. Hyg.,* 54, 351, 1956.
23. **Meyer, M. E. and Cameron, H. S.,** Studies on the etiological agent of epididymitis in rams, *Am. J. Vet. Res.,* 64, 495, 1968.
24. **Carmichael, L. E. and Bruner, D. W.,** Characteristics of newly recognized species of *Brucella* responsible for canine abortions, *Cornell Vet.,* 48, 579, 1968.
25. **Diaz, R., Jones, L. M., and Wilson, J. B.,** Antigenic relationship of the Gram-negative organism canine abortion to smooth and rough brucellae, *J. Bacteriol.,* 95, 618, 1968.
26. **Jones, L. M., Zanardi, M., Leong, D., and Wilson, J. B.,** Taxonomic position in the genus *Brucella* of the causative agent of canine abortion, *J. Bacteriol.,* 95, 625, 1968.
27. **Meyer, M. E.,** *Brucella* organisms isolated from dogs: comparison of characteristics of members of the genus *Brucella, Am. J. Vet. Res.,* 30, 1751, 1969.
28. **Hoyer, B. H. and McCullough, N. B.,** Polynucleotide homologies of *Brucella* deoxyribonucleic acids, *J. Bacteriol.,* 95, 444, 1968.
29. **Hoyer, B. H. and McCullough, N. B.,** Homologies of deoxyribonucleic acids from *Brucella ovis,* canine abortion organisms, and other *Brucella* species, *J. Bacteriol.,* 96, 1783, 1968.
30. **Verger, J. M., Gremont, F., Gremont, P. A. D., and Grayon, M.,** *Brucella,* a monospecific genus as shown by deoxyribonucleic acid hybridization, *Int. J. Syst. Bacteriol.,* 35, 292, 1985.
31. **Grimont, P. A. D., Popoff, M. Y., Grimont, F., Coynault, C., and Lemelin, M.,** Reproducibility and correlation study of three deoxyribonucleic acid procedures, *Curr. Microbiol.,* 4, 325, 1980.
32. **Grimont, P. A. D. and Popoff, M. Y.,** Use of principal component analysis in interpretation of deoxyribonucleic acid relatedness, *Curr. Microbiol.,* 4, 337, 1980.
33. **De Ley, J., Gillis, M., and de Voss, P.,** Range of the molecular complexities of bacterial genomes within some well established bacterial species, *Zentralbl. Bacteriol. Parasitenkd. Infectionskr. Hyg. Abt. 1 Orig. Reihe C,* 2, 263, 1981.
34. **De Ley, J., Mannheim, W., Segers, P., Lievens, A., Denijin, M., Vanhouke, M., and Gillis, M.,** Ribosomal ribonucleic acid cistron similarities and taxonomic neighborhood of *Brucella* and CDC group Vd, *Int. J. Syst. Bacteriol.,* 37, 35, 1987.
35. **Breed, R. S., Murray, E. G. D., and Smith, N. R., Eds.,** *Bergey's Manual of Determinative Bacteriology,* 7th ed., Williams & Wilkins, Baltimore, 1957.

36. **Buchanan, R. E. and Gibbons, N. E., Eds.,** *Bergey's Manual of Determinative Bacteriology,* 8th ed., Williams & Wilkins, Baltimore, 1974.
37. **Verger, J. M., Grimont, F., Grimont, P. A. D., and Grayson, M.,** Taxonomy of the genus *Brucella, Ann. Inst. Pasteur Microbiol.,* 138, 235, 1987.
38. **Meyer, M. E.,** Evolution and taxonomy in the genus *Brucella:* contemporary evolutionary status of the species *B. abortus, Am. J. Vet. Res.,* 37, 203, 1976.
39. **Harrington, R., Bond, D. R., and Brown, G. M.,** Smooth phage resistant strain of *Brucella abortus* from bovine tissue, *J. Clin. Microbiol.,* 5, 633, 1977.
40. **Ewalt, D. R. and Forbes, L. B.,** Atypical isolates of *Brucella abortus* from bovine tissue, *J. Clin. Microbiol.,* 5, 633, 1977.
41. **Farrell, I. D. and Robertson, L.,** The sensitivity of the biotypes of *Brucella abortus* to three antibiotics used in selective media, and the description of a new biotype, *J. Hyg.,* 65, 165, 1967.
42. **Corbel, M. J.,** Taxonomic studies on some atypical strains of *Brucella suis, Br. Vet. J.,* 140, 34, 1984.
43. **Meyer, M. E.,** Phenotypic comparison of *Brucella ovis* to the DNA-homologous *Brucella* species, *Am. J. Vet. Res.,* 30, 1757, 1969.
44. **Philippon, A.,** Identification de *Brucella abortus:* metabolisme oxidatif et lysotypie, *Ann. Inst. Pasteur,* 115, 367, 1968.
45. **Corbel, M. J.,** Brucella phages: advances in the development of a reliable phage typing system for smooth and non-smooth *Brucella* isolates, in *Brucella* and brucellosis, an update, *Ann. Inst. Pasteur Microbiol.,* 138, 70, 1987.
46. **Meyer, M. E.,** Metabolic characterization of the genus *Brucella.* IV. Correlation of oxidative metabolic patterns and susceptibility to *Brucella* bacteriophage, type abortus, strain 3, *J. Bacteriol.,* 82, 950, 1961.
47. **Meyer, M. E.,** The epizootiology of brucellosis and its relationship to the identification of *Brucella* organisms, *Am. J. Vet. Res.,* 25, 553, 1964.
48. **Meyer, M. E.,** Host-parasite relationships in brucellosis. I. Reservoirs of infection and interhost transmissibility of the parasite, in *Proc. U.S. Livestock Sanitary Assoc.,* 1966, 129.
49. **Meyer, M. E.,** Evolution and taxonomy in the genus *Brucella:* steroid hormone induction of filterable forms with altered characteristics after reversion, *Am. J. Vet. Res.,* 37, 207, 1976.
50. **Meyer, M. E.,** Evolution and taxonomy in the genus *Brucella:* progesterone induction of filterable forms of *Brucella abortus,* type 2 with revertant characteristics essentially indistinguishable in vitro from those of *Brucella ovis, Am. J. Vet. Res.,* 37, 221, 1976.
51. **Todoriu, C. D.,** Position of *Brucella ovis* (Buddle 1956) in bacterial taxonomy (in Bulgarian), *Lucr. Inst. Cercet. Vet. Bioprep. Pasteur,* 3, 199, 1966.
52. **Kaitmazova, E. I., Kurdina, D. S., Dranovskaya, E. A., Grekova, N. A., and Sachnovski, U. G.,** Characteristics of cultures of *B. ovis* (in Russian), *Veterinariya (Moscow),* 48, 44, 1971.
53. **Robertson, L., Farrell, I. D., and Hinchliffe, P. M.,** The sensitivity of *Brucella abortus* to chemotherapeutic agents, *J. Med. Microbiol.,* 6, 549, 1973.
54. **Kaitmazova, I. E. and Ostrovskaya, N. N.,** Concerning the characteristics of *Brucella* isolated in USSR territory (in Russian), *Zh. Mikrobiol.,* 44, 12, 1967.
55. **Meyer, M. E.,** Evolution and taxonomy in the genus *Brucella:* concepts on the origins of the contemporary species, *Am. J. Vet. Res.,* 37, 199, 1976.
56. **Corbel, M. J., Thomas, E. L., and Garcia-Carillo, C.,** Taxonomic studies on some atypical strains of *Brucella suis, Br. Vet. J.,* 140, 34, 1984.
57. **O'Hara, M. J., Collins, D. M., and DeLisle, G. W.,** Restriction endonuclease analysis of *Brucella ovis* and other *Brucella* species, *Vet. Microbiol.,* 10, 425, 1985.
58. **Hoyer, B. H. and King, J. R.,** Desoxyribonuclic acid sequence losses in a stable streptococcal L-form, *J. Bacteriol.,* 97, 1516, 1969.
59. **Corbel, M. J. and Scott, A. C.,** Properties of a cell-wall defective variant of *Brucella abortus* of bovine origin, *J. Hyg. Camb.,* 85, 103, 1980.
60. **Ross, H. M.,** Isolation of a cell-wall defective strain of *Brucella abortus* from bovine tissue, *Vet. Rec.,* 106, 247, 1980.
61. **Rhadakova, E. D.,** Properties of *E. coli* cultures reversed from L-forms (in Russian), *Zh. Mikrobiol.,* 48, 14, 1971.
62. **Levina, G. L.,** Reversion peculiarities of penicillin induced L-forms of causative agent of mouse typhoid fever (in Russian), *Antibiotiki,* 17, 714, 1972.
63. **Kagan, G., Gryzlova, O. N., Mikhailova, V. S., and Levashev, V. S.,** Some peculiarities of the cultures reversed from the L-forms of B. *Hemolyic streptocci* (in Russian), *Zh. Mikrobiol.,* 33, 86, 1961.
64. **Kado, C. I. and Liu, S. T.,** Rapid procedure for detection and isolation of large and small plasmids, *J. Bacteriol.,* 145, 1365, 1981.
65. **Guerry, P., Le Bank, D. J., and Falkow, F.,** General method for the isolation of plasmids and deoxyribonucleic acid, *J. Bacteriol.,* 116, 1064, 1973.

66. **Clewel, D. B. and Helenski, D. R.,** Supercoiled circular DNA protein complex in *Escherichia coli:* purification and induced conversion to an open circular form, *Proc. Natl. Acad. Sci. U.S.A.,* 62, 1159, 1969.
67. **Simon, F.,** Contribution a l'etude des bacteriophages du genera *Brucella* (aspect morphologiques, serologiques et physiologiques (in French), Thesis, University of Paris, 1979.
68. **Braun, W.,** Dissociation in *Brucella abortus:* a demonstration of the role of inherent and environmental factors in bacterial variation, *J. Bacteriol.,* 51, 327, 1946.
69. **Marr, A. G. and Wilson, J. B.,** Genetic aspects of the added carbon dioxide requirements of *Brucella abortus, Proc. Soc. Exp. Biol. Med.,* 75, 438, 1950.
70. **Braun, W. and Ogelsby, G.,** On the problem of naturally occurring aberrant strains of *Brucella, Proc. Soc. Exp. Biol. Med.,* 86, 757, 1954.
71. **Shibata, S., Isayama, Y., and Shimuzu, T.,** A possibility in variation in *Brucella abortus* from type II to Type I, *Natl. Inst. Anim. Health Organ. (Tokyo),* 2, 20, 1962.
72. **Jones, L. M., Montgomery, V., and Wilson, J. B.,** Characteristics of carbon dioxide-independent cultures of *Brucella abortus* isolated from cattle vaccinated with strain 19. I, *Infect. Dis.,* 115, 312, 1965.
73. **Kraft, M. E.,** The identification of *B. abortus* strain 19 by penicillin tolerance, *Am. J. Vet. Res.,* 16, 295, 1955.
74. **Huddleson, I. F.,** Emergence during growth of *Brucella* strains on dye-agar media of cells that show changes in sulfur metabolism, *Bull.* W.H.O., 24, 91, 1961.
75. **Rigby, E. C., Fraser, A. D. F., Garcia, M. M., and Brooks, B. W.,** Studies on the molecular biology of *Brucella,* in Brucellosis Symposium, Animal Disease Research Institute, Nepean, Ontario, Canada, 1984.
76. **Renoux, G. and Suire, A.,** Spontaneous lysis and phage-carrier state in *Brucella* cultures, *J. Bacteriol.,* 86, 642, 1963.
77. **McDuff, C. R., Jones, L. M., and Wilson, J. B.,** Characteristics of *Brucella* phage, *J. Bacteriol.,* 91, 324, 1967.
78. **Verstreate, D. R., Creasy, M. T., Caveney, N. T., Baldwin, C. L., Blab, M. W., and Winter, A. J.,** Outer membrane proteins of *Brucella abortus:* isolation and characterization, *Infect. Immun.,* 35, 979, 1982.
79. **Douglas, J. T., Rosenberg, E. Y., Nikaido, H., Verstreate, D. R., and Winter, A. J.,** Porins of *Brucella* species, *Infect. Immun.,* 44, 16, 1984.
80. **Verstreate, D. R. and Winter, A. J.,** Comparison of sodium dodecyl sulfate-polyacrylamide gel electrophoresis profiles and antigenic relatedness among outer membrane proteins of 49 *Brucella abortus* strains, *Infect. Immun.,* 46, 182, 1984.
81. **Santos, J. M., Verstreate, D. R., Perera, V. Y., and Winter, A. J.,** Outer membrane proteins from rough strains of four *Brucella* species, *Infect. Immun.,* 46, 188, 1984.
82. **Tanaka, S., Suto, T., Isayama, Y., Azuma, R., and Hatakeyama, H.,** Chemotaxonomical studies on fatty acids of *Brucella* species, *Ann. Sclavo,* 19, 67, 1977.
83. **Dees, S. B., Hollis, D. G., Weaver, R. E., and Moss, C. W.,** Cellular fatty acids of *Brucella suis* and *Brucella canis, J. Clin. Microbiol.,* 14, 111, 1981.
84. **Moreira-Jacobs, M.,** New group of virulent bacteriophages showing differential affinity for *Brucella* species, *Nature (London),* 219, 752, 1968.
85. **Barbieri, M.,** *The Semantic Theory of Evolution,* Harwood Academic Publishers, New York, 1985.
86. **Stanier, R. Y.,** Phototropic green and purple bacteria: a comparative systematic survey, *Am. Rev. Microbiol.,* 31, 296, 1977.
87. **Stanier, R. Y.,** Some aspects of the biology of cells and evolutionary significance, in *Organization and Control in Prokaryotic and Eukaryotic Cells,* Charles, H. P. and Knight, B. C., Eds., Cambridge University Press, Cambridge, 1970.
88. Kagan, M. B., *A Microbial Chemotherapy,* W. B. Saunders, Philadelphia, 1980.
89. **Saltzman, L. and Aperion, D.,** Binding of erythromycin to the 50S ribosomal subunit is affected by alterations in the 30S ribosomal subunit, *Mol. Gen. Genet.,* 143, 301, 1976.
90. **Mao, J. C. H. and Robishaw, E. E.,** Erythromycin, a peptidyltransferase effector, *Biochemistry,* 11, 4864, 1972.
91. **Taubman, S. B., Jones, N. R., Young, F. Y., and Corcoran, J. W.,** Sensitivity and resistance to erythromycin in *Bacillus subtilis* 168; the ribosomal binding of erythromycin and chloramphenicol, *Biochim. Biophys. Acta,* 123, 438, 1966.
92. **Oleinick, N., Wilhelm, J. M., and Corcoran, J. W.,** Nonidentity of the site of action of erythromycin A and chloramphenicol on *Bacillus subtilis* ribosomes, *Biochim. Biophys. Acta,* 155, 290, 1968.
93. **Oleinick, N. L. and Corcoran, J. W.,** Two types of binding of erythromycin to ribosomes from antibiotic-sensitive and -resistant *Bacillus subtilis,* 168, *J. Biol. Chem.,* 244, 727, 1969.
94. **Slatkin, M.,** The descent of genes, *Science,* 6, 80, 1985.
95. **Racourt, J., Grimont, F., Grimont, P. A. D., and Seeliger, H. P. R.,** DNA relatedness among serovars of *Listeria monocytogenes sensu lato, Curr. Microbiol.,* 7, 383, 1982.

96. **Wilson, A. C.,** The molecular basis of evolution, *Sci. Am.,* 253, 164, 1985.
97. **Darnell, J. E. D.,** RNA, *Sci. Am.,* 253, 68, 1985.
98. **Renoux, G. and Philippon, A.,** Position taxonomique dans le genre *Brucella* de bacteries isolees de brebis et de vaches, *Ann. Inst. Pasteur,* 117, 524, 1969.
99. **Morris, I. A.,** The use of polyacrylamide gel electrophoresis in taxonomy of *Brucella, J. Gen. Microbiol.,* 761, 231, 1973.
100. **Corbel, M. J.,** Examination of two bacterial strains designated "*Brucella suis* biotype 5", *J. Hyg. Camb.,* 71, 271, 100.
101. **Pinigin, A. F., Petukhova, O. S., and Therinov, S. P.,** Taxonomic status of *Brucella* strains isolated from reindeer, *Zh. Mikrobiol.,* 11, 103, 1983.
102. **Pinigin, A. F., Petukhora, O. S., Merinov, S. P., Verishilova, P. A., and Dranovskaya, E. A.,** A new species, *Brucella rangiferi, Zh. Mikrobiol.,* 6, 98, 1986.
103. **Tolari, F., Thomas, E. L., and Corbel, M. J.,** On the differentiation of *Brucella abortus* biotypes 3 and 6, *Ann. Sclavo,* 23, 320, 1981.
104. **Arnaud-Bosq, C., Brousson-Ialaguir, J., Veron, M., and Roux, J.,** Determination des biovars de *Brucella melitensis* par les caracteres manometriques: interet epidemiologique et relation avec les serovars, *Ann. Inst. Pasteur Microbiol.,* 138, 189, 1987.
105. **Meyer, M. E.,** Characterization of *Brucella abortus* strain 19 isolated from human and bovine tissues and fluids, *Am. J. Vet. Res.,* 46, 902, 1985.

Chapter 2

ANTIGENS OF *BRUCELLA*

John W. Cherwonogrodzky, Gerard Dubray, Edgardo Moreno, and Hubert Mayer*

TABLE OF CONTENTS

* Sections I and IV were the work of John W. Cherwonogrodzky, who served as general editor for the Chapter; Section II was the work of Gevard Dubray; and Section III was contributed by Edgardo Moreno and Hubert Mayer.

I. POLYSACCHARIDES OF *BRUCELLA*

A. INTRODUCTION

The species of *Brucella* have but a few known serotypes. Despite early reports of extensive cross-reaction between the species of *Brucella,*[1-3] distinct antigens between the *abortus* and *melitensis* strains were recognized.[4,5] In 1932, Wilson and Miles designated the A antigen as that which predominates on strains of *B. abortus,* and the M antigen as that which predominates on strains of *B. melitensis.*[6] Two years later, Topping reported that *B. abortus* and *B. melitensis* extracts, treated with heat and dilute acetic acid, yielded polysaccharides which were antigenic.[7] However, she was unable to show any chemical difference between the isolated components of these bacteria. Miles and Pirie[8] later reported that the antigen of *B. melitensis* was a formyl derivative of an amino-polyhydroxy compound (AP) with a specific optical rotation of $[\alpha]_D^{16} + 43°$. This material was released from a larger complex, which contained phospholipids and proteins, by a treatment with acetic acid. Although AP was stable in dilute acetic acid, it lost its antigenicity when treated with 0.1 *M* HCl (100°C/5 min).[9] More recently, Redfearn[10] demonstrated that the phenol-extracted smooth-lipopolysaccharide (SLPS) of *B. abortus* biotype 1 contained the A antigen, whereas Diaz et al.[11] suggested that both the A and M antigens were contained in lipopolysaccharide-protein complexes. Rabbit monospecific anti-A antiserum, but not anti-M nor anti-R antiserum, agglutinated both *B. abortus* and *Yersinia enterocolitica* O:9,[12,13] the cross-reactivity of these bacteria being attributed to O-polysaccharide similarities.[14]

B. COMPOSITION OF THE O-POLYSACCHARIDE OF *BRUCELLA*

It is now well recognized that SLPS, from *Brucella* species as well as from many other organisms of the *Enterobacteriaceae,* consists of an O-polysaccharide attached via the core oligosaccharide to the lipid A which anchors the molecule to the outer membrane. The major antigenic epitopes of SLPS and, indeed, of the cell reside in this O-polysaccharide. Within the core region, one of the 3-deoxy-2-octulosonate (KDO) residues has a ketosidic linkage that is acid labile, and under conditions of mild acid hydrolysis (e.g., 2% acetic acid, 100°C), it is cleaved to release the O-polysaccharide along with attached core sugars. Lipid A, which is insoluble under these conditions, precipitates.[15]

The antigenic O-polysaccharide of *B. abortus* 1119-3 SLPS was subsequently identified as a linear homopolymer of 1,2-linked 4,6-dideoxy-4-formamido-α-D-mannopyranosyl units by Caroff et al.[17] Hot 2 *M* HCl and 1 *M* H_2SO_4 destroyed this polysaccharide but not D-mannose, D-glucose, and quinovosamine, which were obtained in low yields (3.56%) and were probably derived from the terminal core oligosaccharide. The O-polysaccharide was essentially identical in structure to that of the O-polysaccharide of *Y. enterocolitica* O:9.[17] Serological cross-reactivity was confirmed by the use of murine monoclonal antibodies specific for both these antigens.[18] Murine monoclonal antibodies have also been used to show that antibodies specific for the O-polysaccharide are protective in passive immunity to *B. abortus* infection.[19]

The O-polysaccharide structure of *B. melitensis* was found to be related to that of *B. abortus* in that both were homopolymers of 4,6-dideoxy-4-formamido-α-D-mannopyranosyl units.[20,21] Although this similarity in structure provided a rationale for the cross-reaction between these organisms, it did not account for the fact that the A antigen is serologically distinct from the M antigen. Also, the optical rotations of the purified A and M O-polysaccharides differed ($[\alpha]_D + 38°$ [water] for *B. abortus* 1119-3 O-polysaccharide; $[\alpha]_D + 49°$ [water] for *B. melitensis* 16M O-polysaccharide), as did the ^{1}H- and ^{13}C-NMR spectra (*B. melitensis* O-polysaccharide was complex and required extensive analysis).[18,26-28] In combination with classical methods, these showed that while the A antigen is essentially a homopolymer of 1,2-linked 4,6-dideoxy-4-formamido-α-D-mannopyranosyl residues, the M antigen is a linear polymer of a repeating pentasaccharide unit comprising one 1,3-linked and four 1,2-linked 4,6-dideoxy-4-formamido-α-D-mannopyranosyl residues.[25-27]

These similarities and differences may account for the cross-reactivity and different antigenicity of the A and M antigens. That the O-polysaccharides of *B. abortus* (e.g., strain 1119-3) and *B. melitensis* (e.g., strain 16M) are the classic A and M antigens, respectively, is based on the following:

1. Rabbit cross-absorbed serotype-specific antisera for either A or M *Brucella* antigens (from G. M. Brown, National Veterinary Services, Ames, IA; or L. B. Forbes, Health of Animals Laboratory, Saskatoon, Saskatchewan) will selectively precipitate purified *B. abortus* or *B. melitensis* SLPS and will selectively bind to the homologous SLPS coated on polystyrene plates. The latter reaction is inhibited by preincubating the antisera with purified homologous O-polysaccharide.[28]
2. Murine monoclonal antibodies, with specificities for *B. abortus* or *B. melitensis* O-polysaccharides, show similar selectivity to that noted for rabbit antisera.[29] Indeed, monoclonal antibodies can substitute for rabbit cross-absorbed serotype-specific antisera when testing biovars of *Brucella spp.* for the A or M antigen.[261]

Heterologous O-polysaccharide inhibits rabbit anti-A and anti-M antisera from binding to *B. abortus* and *B. melitensis* SLPS coated onto polystyrene plates.[28] Some murine monoclonal antibodies can precipitate the O-polysaccharides and SLPS of *B. abortus* 1119-3, *Y. enterocolitica* O:9, and *B. melitensis* 16M, probably because these antibodies have affinity for common structural entities within the O-polysaccharides.[23]

With regard to the production and use of synthetic antigens, Bundle et al.[30] have linked the aliphatic chain of a fatty acid ester to the O-polysaccharide of either *B. abortus* of *Y. enterocolitica* O:9. This synthetic antigen is comparable to the SLPS of either organism in serodiagnostic tests and has the advantage of giving fewer nonspecific reactions with cattle sera tested in an enzyme-linked immunosorbent assay (ELISA).[31] Peters and Bundle also reported the synthesis of oligosaccharides of 1,2-linked 4,6-dideoxy-4-formamido-α-D-mannopyranosyl residues (i.e., the A antigen) as well as structures containing the α-1,3 linkages (i.e., the M antigen)[32] which should now provide specific diagnostic reagents.

C. CROSS-REACTIONS BETWEEN *B. ABORTUS* AND OTHER ORGANISMS

1. With Other Species of *Brucella*

As evidenced by agglutinations with rabbit cross-absorbed serotype-specific antisera, *B. abortus* cross-reacts with *B. melitensis*. The A and M O-polysaccharides consist of the same carbohydrate with most of the units linked in the same manner (i.e., α-1,2-linkage), and murine monoclonal antibodies confirm that these have common epitopes. However, cross-absorbance should have removed nonspecific antibodies. In 1932, Wilson and Miles[6] suggested that each bacterium has both antigens although in different amounts, *B. abortus* having predominantly the A antigen and *B. melitensis* having predominantly the M antigen. Agglutination of these bacteria with murine monoclonal antibodies specific for the A and M O-polysaccharides and extensive NMR spectra analysis support this conclusion.[262] Also, the species of *Brucella* consist of different biovars, strains which differ in metabolism, sensitivity to dyes, lysis by bacteriophages, and agglutination with serotype-specific antisera.[33] *B. melitensis,* for example, has biovar 1 which expresses the M antigen, biovar 2 which expresses the A antigen, and biovar 3 which expresses both the M and A antigens. It appears that biovar 3 may, indeed, have both antigens on its cell surface, while biovars 1 and 2 may have predominantly the M and A antigens, respectively, and trace amounts of the other.[262]

Despite considerable cross-reaction between *B. abortus* and *B. melitensis,* one antigen should not be used in serodiagnostic tests to detect infections due to more than one *Brucella* species. Rojas et al.[34] reported that approximately 80% of *B. melitensis*-infected human sera could be identified with either *B. abortus* or *B. melitensis* SLPS on a nitrocellulose-ELISA but that the remainder could only be identified with *B. melitensis* SLPS. For *B. ovis* infections in rams, its rough lipoproteins (RLPS)[35] and outer membrane proteins[36] are useful antigens for the detection of serum antibodies.

In considering *B. abortus* 45/20, it has been observed that cattle vaccinated with this bacterium will raise antibodies to the A and M antigens of *B. abortus* and *B. melitensis.*[37] Rabbits immunized with *B. abortus* 45/20 RLPS have antibodies that bind to the SLPS of *B. abortus* 1119-3 or *Y. enterocolitica* O:9 coated onto polystyrene plates.[261] NMR spectra of phenolic extracts of *B. abortus* 45/20 indicate that trace amounts of the O-chain polysaccharide are present on this bacterium and may account for the noted cross-reactions.[262]

2. With Bacteria of Other Genera

Brucella abortus cross-reacts with other bacterial species (for an excellent review, refer to Reference 13), the most widely known being *Y. enterocolitica* O:9. The discovery of this cross-reaction was made in Scandinavia during the 1940s. There, in spite of a successful eradication program against *B. abortus,*[38] human patients still exhibited high antibody titers. *Brucella abortus* could not be isolated from these people, and rather than having ''undulant fever'', their distress was gastroenteritis caused by *Y. enterocolitica* O:9.[39,40] It was subsequently found that these two organisms cross-reacted[41] because of shared antigenic determinants on the O-polysaccharides of their SLPS-protein complexes,[14,42,43] both antigens being identified as linear homopolymers of 1,2-linked 4,6-dideoxy-4-formamido-α-D-mannopyranosyl units.[16,17]

TABLE 1
Comparison of O-Polysaccharides of Cross-Reactive Bacteria

Bacterium	Component glycoside of the SLPS O-polysaccharides	Ref.
Brucella abortus 1119-3	4-Formamido-4,6-dideoxy-D-mannose	16
B. melitensis 16M	4-Formamido-4,6-dideoxy-D-mannose	27
Yersinia enterocolitica O:9	4-Formamido-4,6-dideoxy-D-mannose	17
Vibrio cholerae	*N*-Acylated 4-amino-4,6-dideoxy-D-mannose (where the acylating acid is 3-deoxy-L-glycero-tetronic acid)	44,45
Salmonella landau and *godesburg*	4-Acetamido-4,6-dideoxy-D-mannose (1 part) D-glucose (2 parts) L-fucose (1 part) 2-acetamido-2-deoxy-D-galactose (1 part)	46,47
Escherichia coli O:157H:7	4-Acetamido-4,6-dideoxy-D-mannose (1 part) D-glucose (1 part) L-fucose (1 part) 2-acetamido-2-deoxy-D-galactose (1 part)	48
Pseudomonas maltophilia 555	4-Acetamido-4,6-dideoxy-D-mannose (3 parts) D-rhamnose (1 part) 3-acetamido-3,6-dideoxy-D-galactose (1 part)	49

Other bacteria show serological cross-reactivity with *B. abortus*.[16,17,27,44-49] Structural analysis of the SLPS O-polysaccharides of these cross-reacting Gram-negative bacteria (Table 1) shows that these all contain units of N-acyl derivatives of 4-amino-4,6-α-D-mannopyranose, and it is probable that epitopes involving this aminoglycoside are responsible for the observed cross-reactivity. Some monoclonal antibodies have been raised which agglutinate one bacterium and not other cross-reactive species.[18,29] The potential exists, therefore, for differentiating even cross-reactive bacteria using a panel of monoclonal antibodies of known specificities.

D. OTHER POLYSACCHARIDES OF *BRUCELLA*

1. Polysaccharide B

Polysaccharide B (poly-B, also referred to in the literature as component 1, second component, or PB) is a low molecular weight carbohydrate released from strains of *B. melitensis* or *B. abortus* with 0.2 *M* trichloroacetic acid.[11,50] Interest was given to poly-B since *B. abortus* infected animals were reported to produce antibodies which precipitated poly-B, while vaccinated animals did not.[51,52] The composition of poly-B was controversial, some publications citing its chemical and immunologic difference from both acid hapten (AH) (the O-polysaccharide cleaved from SLPS by mild acid and heat) and native hapten (NH) (a component present in endotoxin preparations from smooth strains of *Brucella* which may be a polysaccharide-protein complex).[53-55] It was shown to have a high glucose content (89%),[54] and its reported isolation from rough, rather than smooth, cells[52] differed from traditional extractions of *Brucella* O-chain polysaccharides. Furthermore, poly-B could be isolated from disrupted cells rather than cell membranes, suggesting that poly-B resides within the cell instead of at the cell surface as does O-polysaccharide.[54]

Bundle et al.[26,56] purified poly-B from an extract of *B. melitensis* 16M. Structural analysis identified the major component as a cyclic polymer of between 17 and 24 1,2-linked β-D-glucopyranosyl residues, being essentially identical to cyclic D-glucans produced by other bacteria.[57] Poly-B appears to be present in all species of *Brucella*.[56] Purified cyclic β-D-glucan does not appear to be antigenic, and the previously described serological reactions attributed to poly-B were therefore probably due to the presence of *Brucella* O-polysaccharide in the preparations.[56,58] This is reinforced by the observation that L'vov et al.[59] noted that the β-D-glucan will bind and form a complex with the SLPS of *B. melitensis* 16M. Therefore, the antigenicity attributed to crude poly-B preparations is probably due to its O-polysaccharide content.

2. Native Hapten

About 50 years ago, Miles and Pirie identified a water-soluble "native antigen" which upon drying or freezing would dissociate into a free lipoid and a formyl derivative of an amino-polyhydroxy compound.[8] The presence of a *Brucella* antigen that is partly O-polysaccharide and yet is distinct from either SLPS or O-polysaccharide remains controversial. Native hapten can be extracted from *Brucella* endotoxin by mild methods unlikely to break covalent bonds in the SLPS,[54] and NH has been characterized as being a large polysaccharide which lacks KDO, does not bind to erythrocytes, and may be associated with protein.[55,60-62] Others have purified NH, found it identical to *Brucella* O-polysaccharide isolated from acid treated SLPS, and have detected core glycoses and a KDO residue, suggesting that NH originates from SLPS.[63]

Moreno et al.[21,62] have identified a *Brucella* polysaccharide which differs from NH by having a small amount of lipid associated with the polysaccharide. Other Gram-negative bacteria have capsules also consisting of polysaccharide associated with lipid,[64,65] and, indeed, an external layer suggesting the presence of a capsule has been reported for *Brucella* species.[66]

E. *BRUCELLA* POLYSACCHARIDE USED TO DISCRIMINATE INFECTED FROM VACCINATED CATTLE

Diaz et al.[11,50,51] reported that animals exposed to *B. abortus* have sera which in an immunodiffusion assay (10% NaCl in the agar) show two precipitin patterns: *B. abortus* field-strain infected cattle sera precipitate both SLPS-protein complex and poly-B, while *B. abortus* S-19 vaccinated cattle sera precipitate only the SLPS-protein complex. Murine monoclonal antibodies (raised to *B. abortus* and *Y. enterocolitica* O:9 and selected for reactivity to both O-polysaccharides) when tested in an immunodiffusion assay (1% NaCl in the agar) show two precipitin patterns: type 1 monoclonal antibodies precipitate both SLPS and O-polysaccharide, while type 2 monoclonal antibodies precipitate only SLPS.[18] Approximately half of the monoclonal antibodies were of each type, and the immunoglobulin class did not appear to be a factor. The similarities between the above precipitation patterns for cattle sera and murine monoclonal antibodies was apparent, and it was found that both *B. abortus* field-strain infected cattle and *B. abortus* S-19 vaccinated cattle sera precipitate *Brucella* SLPS, while only the former cattle sera precipitate *Brucella* O-polysaccharide.[58,67] It appears likely that past reports on the discrimination between infected and vaccinated cattle sera by precipitation with poly-B was due to the *Brucella* O-polysaccharide content.

It may be speculated that the mechanism for this difference in precipitation patterns is due to epitope specificity,[68] type 1 antibodies binding to "length" epitopes along the O-chain and type 2 antibodies binding to "tip" epitopes. Figures 1 and 2 suggest that although type 1 antibodies can form a precipitate with either SLPS or O-polysaccharide, type 2 can form a precipitate only with SLPS, and then because of hydrophobic micelle formation by lipid components. *Brucella abortus* field-strain infected cattle sera may have both types of antibodies, while *B. abortus* S-19 vaccinated cattle sera may have antibodies with the activity of type 2 murine monoclonal antibodies. As the immunodiffusion assay lacks sensitivity and uses relatively large amounts of sera and antigen, an ELISA has been developed which overcomes these limitations.[69]

F. CONCLUSIONS

The composition of *B. abortus* 1119-3 and *B. melitensis* 16M O-polysaccharide has been described as having the O-polysaccharides of bacteria that cross-react with *Brucella*.

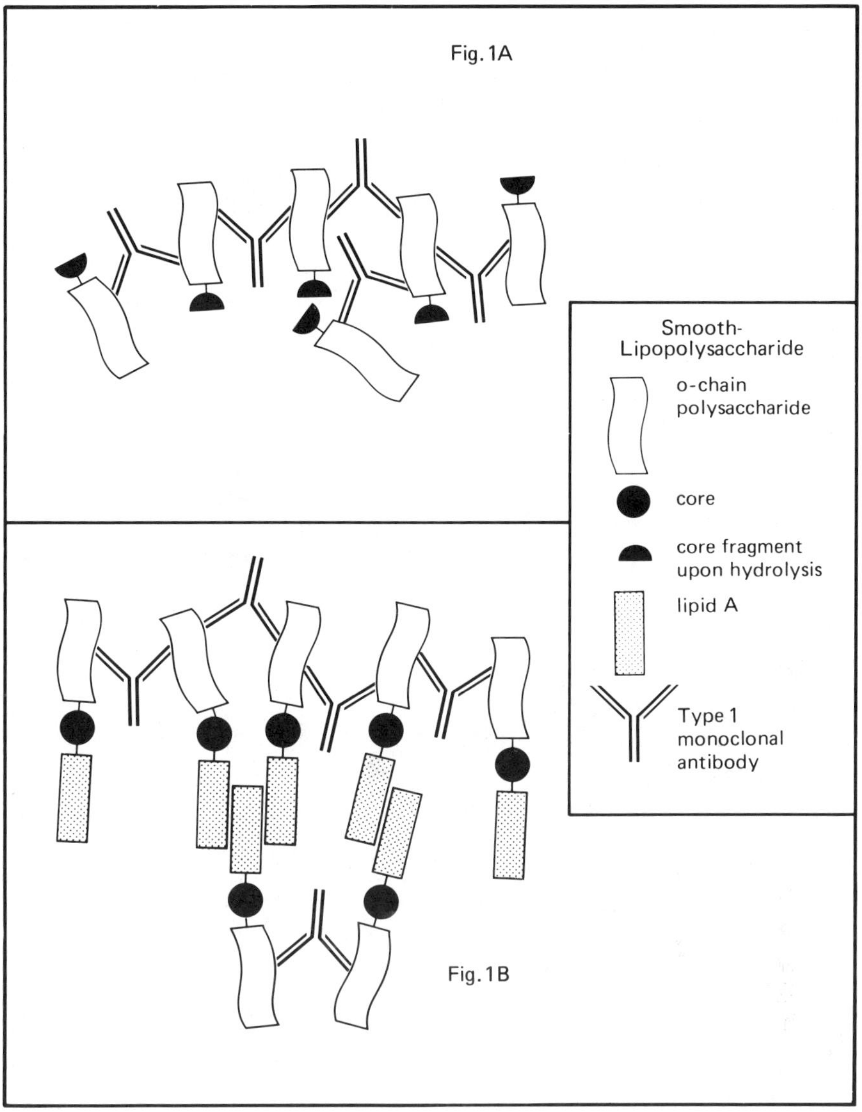

FIGURE 1. (1A) Precipitation of O-chain polysaccharide by type 1 monoclonal antibody. (1B) Precipitation of smooth-lipopolysaccharide by type 1 monoclonal antibody.

These O-polysaccharides have N-acylated derivatives of 4-amino-4,6-dideoxy-α-D-manno-pyranosyl residues in common, and it is likely that this common factor accounts for cross-reaction in serodiagnosis. Monoclonal antibodies have been raised to the A and M antigens, and some can substitute for rabbit cross-absorbed serotype-specific sera, while others can be used in a panel to differentiate *B. abortus* from other cross-reacting bacteria. Murine monoclonal antibodies have given insight into the precipitation patterns of *B. abortus* field-strain infected and *B. abortus* S-19 vaccinated cattle sera against SLPS and O-polysaccharide (crude preparations of poly-B contain *Brucella* O-polysaccharide). Schurig et al.[70] have suggested that immunogens located on the surface of invading microorganisms may be more important than internal immunogens in modulating the host's immune response. This is indirectly supported by the observation that in the production of murine monoclonal antibodies against *B. abortus* and *Y. enterocolitica* O:9 cells between 15 to 18% of the putative hybrids were LPS-specific, and of these, 60% were O-polysaccharide specific[18] (the SLPS of a bacterium is only 1 to 4% of its dry weight).[15]

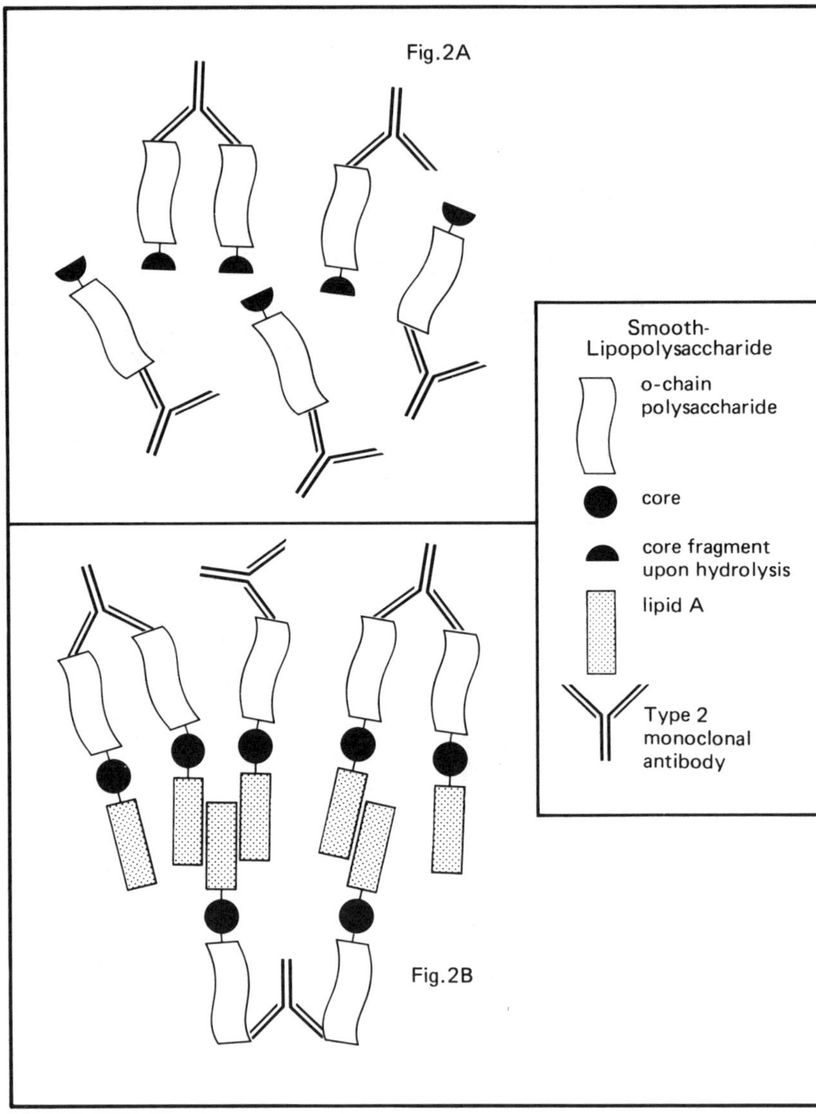

FIGURE 2. (2A) Absence of precipitation of O-chain polysaccharide by type 2 monoclonal antibody. (2B) Precipitation of smooth lipopolysaccharide by type 2 monoclonal antibody.

With the current understanding of the nature of the O-polysaccharide of *Brucella* spp. some questions remain to be answered:

1. If *Brucella* spp. are but biovars of *B. melitensis*,[71] how is the A and/or M antigen genetically regulated, and are these influenced by plasmid or bacteriophage nucleic acid?[72]
2. Although the O-polysaccharide may be immunodominant when compared to other antigens on the bacterium, are there regions of differing immunodominance on the O-polysaccharide (i.e., tip vs. length epitopes)?
3. Can either purified or conjugated antigens be used as protective vaccines, and can the immune status of the inoculated animal be differentiated (i.e., vaccinated vs. infected)?
4. Can screening serodiagnostic tests be improved by having a mixture of *B. abortus* and *B. melitensis* (A + M) antigens?

Although some of these questions are currently being investigated in our laboratories, possibly different ones are of interest to the reader. These are, indeed, interesting times, for with the advances in knowledge of antigen structure and in the development of novel reagents and technologies, the means exist to answer these questions.

II. PROTEIN ANTIGENS OF *BRUCELLA*

A. INTRODUCTION

Protein extracts of *Brucella* cells have been used as antigens, mainly in immunodiffusion and immunoelectrophoretic methods, to study either the antigenic structure of *Brucella* cells with hyperimmune sera or the immune response of experimentally and naturally infected animals and humans.[73] By an anatomic approach (e.g., isolation of cell wall) and by absorption studies of immune sera by whole cells, it has been possible to distinguish between surface (i.e., outer membrane) and internal (i.e., cytoplasmic) protein antigens. This distinction is not definitive because, as yet, little is known about the localization of *Brucella* protein antigens. There is also a need to distinguish antigens revealed by hyperimmune sera, raised by multiple injections of extracts with adjuvant, and antigens revealed by either sera or sensitized immune cells of naturally and experimentally infected hosts, raised by inoculation with live bacteria.

The major advances in the knowledge of *Brucella* protein antigens will be highlighted by a review of the distinction between cell wall and internal antigens in relation to their structure and function. Within this text, further information will be given on the method of obtaining sera from the hyperimmunized and infected host as well as the humoral and cellular immune response of the host.

B. CELL WALL PROTEIN ANTIGENS

The species of *Brucella* are members of the group of Gram-negative bacteria, and as such the cytosol of the cell is surrounded by a complex cell envelope consisting of a cytoplasmic membrane and a cell wall, the latter having a layer of peptidoglycan and an outer membrane.[74] The outer membrane, which forms the physical and functional barrier between the inside of the bacterial cell and its environment, is the first structure in contact with the host's immune cells during the early stages of disease.

1. Structure and Composition of the Cell Wall of *Brucella*

The outer membrane contains lipopolysaccharide (see Section I), proteins, and phospholipids.[75-77] Knowledge about the surface protein molecules (e.g., outer membrane proteins [Omp]) was limited for a long time because the *Brucella* cell wall had structural and biochemical differences with enterobacterial cell walls, notably:

1. The peptidoglycan layer of *Brucella* is strongly associated with the outer membrane and contains far more 1-2 glycols, as revealed by cytochemical reactions, than does the cell wall of *E. coli*.[74]
2. *Brucella* cells do not form spheroplasts under standard conditions.[78]
3. Nonionic detergents are less efficient in extracting Omp of *Brucella* cell wall as compared with *E. coli* Omp.[77,78]
4. Omp can be extracted from *Brucella* by Zwittergent® or sodium dodecyl sulfate (SDS) either without using lysozyme, provided the bacteria are not inactivated before extraction,[78] or by using lysozyme, provided the cells are inactivated by formalin or heat.[77,79]

The major Omp of *Brucella* are porins of group 2 proteins (molecular mass 35 to 40 kDa), group 3 proteins (25 to 30 kDa),[75,76,80] and a lipoprotein (8 kDa) covalently linked to peptidoglycan.[81] Group 1 proteins (88 to 94 kDa) are also found.[77,80] A covalent linkage between these proteins and SLPS has been hypothesized for *B. abortus*.[55,82] This hypothesis has not been confirmed by any structural studies.

The cell wall contains the *Brucella* protective antigens (Bpa).[73,83,84] Bpa are found in either the soluble or the insoluble fractions of extracted cells or cell walls. The soluble fractions, or the supernatants of extracts following centrifugation or ultracentrifugation, contain mainly SLPS and proteins. The insoluble fractions are found in the residues of phenol extraction, done on either cell walls (i.e., the Rh2 fraction)[85] or whole cells (i.e., the phenol insoluble [PI] fraction),[86,87] or in the residues of SDS-boiled cell walls (i.e., the SDS-I or PG fraction).[79]

2. Outer Membrane Proteins

a. Immune Response after Injection of Fractions or Purified Proteins with Adjuvant

i. Humoral Immune Response

Purified group 2 and 3 proteins in Freund's complete adjuvant (FCA) elicited antibodies against Omp and LPS when injected intradermally into cattle.[80] Immunodiffusion and immunoelectrophoretic methods showed two lines of precipitation between LPS absorbed sera and the group 2 and 3 proteins, indicating that these components were immunologically unrelated. Verstreate and Winter showed that porin preparations contained two antigens, "a" and "b", among *B. abortus* strains.[80] Antigen "b" was unique to porins, antigen "a" was common to most group 2 proteins. The common "a" antigenicity could be due to the presence of a murein determinant on both porins and group 3 proteins. These antigens were also present in most strains of *B. melitensis, B. ovis,* and *B. canis*.[80,89]

Porins, derived from *B. abortus* and in oil with immunomodulators such as muramyl dipeptide and trehalose dimycolate, when injected into cattle elicited antibodies that could be detected by ELISA.[88] Protective activity of Zwittergent® purified Omp in laboratory and domestic animals has yet to be reported.

The peptidoglycan-linked lipoprotein cross-reacts with rabbit hyperimmune sera against *E. coli* lipoprotein.[90]

Of the cell wall fractions containing Bpa (pa = protective antigens), only the SDS-I fraction has been characterized.[79] The SDS-I fraction contained two major protein and glycoprotein bands with an apparent molecular mass of 36 to 38 kDa and 25 to 27 kDa, respectively, which are Omp. Minor bands with a molecular mass of 70 and 31 kDa were present. The major isolated bands were not as protective as the entire fraction.[79] The correct identification of Bpa has been impeded by the presence of SLPS. When SLPS-free SDS-I fractions prepared from rough cells were used, these showed varying degrees of protection, but were less effective than the SDS-I fractions prepared from smooth cells. These results favor the hypothesis that probably both SLPS and proteins are involved in protection in the mouse model.

The PI fraction obtained from the phenol extraction of delipidated cells has been used as a vaccine in humans.[91] This fraction contained SLPS as shown by the serological response to SLPS dependent tests in people.[91] Further treatment of the PI fraction with deoxyribonuclease, ribonuclease, pepsin, papain, and pronase hydrolysis led to formation of fraction 4A, which contained peptidoglycan and proteins, plus a small quantity of lipids and carbohydrates. This fraction was about 5% of the original weight of PI and was still highly immunogenic in mice.[86] SDS solubilized about 55% of this material, and the soluble fraction (SF) was slightly protective in mice.[86] This fraction was not characterized in SDS-PAGE, and so it is difficult to determine whether or not it contained the 36 to 38-kDa and 25 to 27-kDa bands.

The identification of Bpa can be done indirectly by passive immunity or by assessing the protection given by the transfer of antibodies or lymphatic cells raised to reagents of rigorously established specificity. Passively transferred polyclonal antibodies with undetermined specificities (mainly against SLPS) were shown to confer protective immunity to *B. abortus* in mice.[92] To define unique specificities, monoclonal antibody technology has been used. Monoclonal antibodies specific to the O-polysaccharide of *B. abortus,* when injected into mice prior to challenge infection, afforded significant protection.[19,93] In contrast, two IgM monoclonal antibodies specific for the porins of *B. abortus* failed to confer protection.[19] Other isotypes of monoclonal antibodies against major protein (36 to 38 kDa) and glycoprotein bands (25 to 27 kDa) have been produced, and their activity will be tested in the protective mouse model.

ii. Cellular Immune Response

Brucella-sensitized immune cells are able to transfer protective immunity in mice.[92,94] Antigens of the SDS-I fraction are involved in this protection, and the two major proteins are good candidates for inducing cell-mediated immunity. Porins, derived from *B. abortus* 45/20 and suspended in oil with immunomodulators (muramyl dipeptide and trehalose dimycolate), elicited blastogenesis and delayed type hypersensitivity when injected into cattle.[88]

b. Immune Response to Antigens during Experimental and Natural Infections

i. Humoral Immune Response

Antibodies against Omp were not detected by immunodiffusion using O-antibody absorbed sera from 11 cattle infected with field strains.[77] The antigenic reactivity of *B. ovis* Omp was further studied by immunoblotting. Antibodies of sera from *B. ovis*-vaccinated rams bound to all Zwittergent®-extracted Omp, while sera from naturally infected rams reacted only with the group 3 proteins.[36] These results have still to be confirmed because the electrophoretic separation of the proteins was poor and the group 3 proteins were located too near the RLPS bands for an accurate assessment.

ii. Cellular Immune Response

Porins were used to probe the cell-mediated immune response of cattle experimentally infected with virulent *B. abortus* field-strains and *B. abortus* strain 19.[95,96] The blastogenic response of peripheral blood lymphocytes of cattle was not indicative of the infection status, as nonpregnant heifers infected with the virulent strain 2308 reacted inconsistently while false-positive responses occurred. As a result of these difficulties, Baldwin et al. "do not recommend the lymphocyte blastogenesis test as a sole method for diagnosis of bovine brucellosis".[96]

The intradermal allergic skin test (AST) and the lymphocyte blastogenesis test (LBT) with cell wall antigens were used to identify infected humans, but no protein antigen was identified.[91] Fraction PI was also used to detect human brucellosis with LBT, but this test does not seem to be any better than the AST when melitine, the PI fraction, is used.[97]

c. Other Cell Surface Protein Antigens

Non-LPS surface antigens were detected with sera of infected and vaccinated animals by the following:

1. The A5 antigen, one of the seven distinct antigens of a soluble extract as revealed by sera from three infected cows, was shown to be present at the surface of *B. abortus* 45/20 cells.[70,98]
2. Unidentified surface antigens, prepared by a hot saline extraction of *B. abortus* 45/20 cells, were precipitated by the sera of *B. abortus* 45/20 vaccinated cattle.[99]
3. Sodium dodecylsulfate extracts of *B. abortus* 544/W were shown to contain three surface antigens: a, b, and X.[100,101] The X antigen, described by Raybould and Chantler, differentiated infected from vaccinated animals by passive hemagglutination,[100] but this assessment was not confirmed by other workers.

4. Salt-extractable protein antigens from *B. abortus* strains developed 1 to 4 immunoprecipitates when analyzed by cross-immunoelectrophoresis using infected cattle sera in the gel. In contrast, 20 to 25 immunoprecipitates developed with hyperimmune rabbit antisera.[102,103] Absorption of infected sera by whole cells did not remove immunoprecipitation against proteins, while the analogous absorption of rabbit hyperimmune sera removed 14 of the 19 immunoprecipitates. These results suggest that protein epitopes, recognized by the immune system of the rabbit after repeated injections in adjuvant, are different from those recognized by the bovine immune system.
5. Nonagglutinating antibodies, obtained by immunizing calves with *B. abortus* S-19 and then cross-absorbing the sera with *B. abortus* 45/20 cells, were able to label two (10 and 12 kDa) or three (10, 12, and 14.3 kDa) bands of a *Brucella* saline extract following SDS-PAGE and electroblotting.[104]

The results of immunoabsorption and surface localization of antigens on rough *B. abortus* 45/20 or other R cells must be interpreted with caution because of the reported nonspecific binding of IgG molecules to their surface.[70]

C. INTERNAL PROTEIN ANTIGENS

Many of the fractions obtained by cell disruption contain internal protein antigens with a variable quantity of cell wall antigens. The numeration and identification of *Brucella* antigens which are able to diffuse through agar and give a precipitate are dependent on:

1. The physical structure of the antigen, whether it is aggregated in vesicles or lamellae
2. The method of extraction of these antigens. Several of the fractions referred to as "soluble antigens" in the literature were crude and poorly characterized, either for physical or chemical properties. These protein extracts have been prepared by numerous methods, from simple extraction with water or saline buffers to cellular disruption by sonication, glass beads, or pressure. These extracts contained nucleic acids, cytoplasmic proteins, lipids, and variable amounts of SLPS or RLPS.
3. Whether the serum is obtained from infected or hyperimmunized animals (the latter may use multiple injections of live, killed, or extracted cells with or without adjuvant) and at what time it is taken
4. The concentration of antigen and antibody which must be optimal to form immunoprecipitates

1. Immune Response after Injection of Cells or Cell Fractions with Adjuvant

The number of immunoprecipitates revealed by hyperimmune sera may vary from 5 to 24.[73,102,103,105-108] A tentative identification of precipitable antigens has been done,[109] but definite relationships between the antigens described can only be established by purification and molecular characterization of the antigens and by exchanges of reagents between research groups for standardization.

2. Enumeration and Identification of Antigens during Experimental and Natural Infections

a. Humoral Immune Response

Numerous complex antigenic fractions have been analyzed by precipitation in gels (e.g., immunodiffusion, immunoelectrophoresis) against infected host sera to study the humoral immune response to *Brucella* (Table 2). As a result, there has also been considerable variation in the enumeration and identification of immunoprecipitates. The number has ranged from 0 to 9 precipitates, depending on the many experimental conditions discussed in the introduction. Only the A2 antigen has been partially characterized. Antibodies against the A2

TABLE 2
Enumeration and Identification of Antigens during Humoral Immune Response of Infected or Experimentally Infected Hosts

Species[a]	Extraction[b] method	Host[c] sera	Test[d]	Number of precipitates	Identified antigens[e]	Ref.
B. m(S)	TCA	r,g,c	ID	1—3		143
	GB	r,h		0—3 or >		144
	Phenol	c		3	SLPS, poly B	113
	Ether					113
(R)	NaCl	r		4		126
(R)		h		1—7		114
(R)		h	CIE	1—3		115
(R)		s		≥1		116
B. m(S,R)	US	r	ID	2—6		113
B. ovis	US	r				113
B. a(S)	Acetone	c	ID	3—6		145
	Ether	c		1—5		146
	US	c	IE	9	SLPS	147
		c	—	11	SLPS	148
		c	ID	5	SLPS	149
		c	IE	3	A2	110
		c		6	A2	98, 117
		c,h	ID	0—3	SLPS?	150
	Saline, GB	c	TDI	3—4		102, 103
B. a(S)	Auto-claved	c	IE	2	SLPS	134
	Heat	c	ID	4		99
	SDS	c	IE	5		109

[a] m = *melitensis*; a = *abortus*; S = smooth; R = rough.
[b] TCA = trichloracetic acid; GB = glass beads; US = ultrasonication; SDS = sodium dodecylsulfate.
[c] c = cattle; g = goat; h = human; r = rabbit; s = sheep.
[d] ID = immunodiffusion; IE = immunoelectrophoresis; CIE = counterimmunoelectrophoresis; TDI = two-dimensional immunoelectrophoresis.
[e] SLPS = lipopolysaccharide from smooth cells; Poly B = polysaccharide B;[113] A2 = A2 antigen.[98]

antigen were found in the sera of infected cows and goats.[98,110,111,117] The A2 antigen is a heat-resistant glycoprotein of high molecular weight which has been partially purified.[112] It detected infected cattle with a moderate sensitivity (55%) and a good specificity (99%), while no A2 precipitins were detected in the sera of calves infected with *Y. enterocolitica* serotype O:9.[112]

One kinetic analysis has shown that for infected rabbits, antibodies arise first against SLPS, then against "polysaccharide B" (which has been characterized as containing a glucan and a polysaccharide related to the O-chain polysaccharide in antigenicity), and last against the surface and inner cell proteins.[113]

A saline extract (refer to the above paragraph on AST) called Brucellin-INRA, which contains at least 20 protein antigens, precipitated with the antibodies in human sera from patients with acute or chronic brucellosis when used in counterimmunoelectrophoresis (CIE).[114,115] The *Brucella* specificity of this antigen mixture eliminates serological cross-reactions due to *Y. enterocolitica* O:9 and *Vibrio cholerae* vaccination.[115] A Brucellin preparation was used in CIE to detect infected sheep and gave a high sensitivity of 82% and a high specificity of 78%.[116]

In conclusion, the host humoral response to *Brucella* is directed against SLPS and other surface or inner cell antigens. One of these antigens, the A2 antigen, was later proposed for diagnosis in immunodiffusion tests with the possibility of differentiating between infected and vaccinated animals. Nevertheless, vaccination of cattle by *B. abortus* strain B19 and of goats by *B. melitensis* strain Rev1 induces anti-A2 antibodies which may persist for 7 months after inoculation.[111,117]

b. Cellular Immune Response

The AST and the LBT have been the main tests used to probe cell-mediated immunity for identifying *Brucella*-infected humans or animals.[118] The following is a review of these assays.

i. Allergic Skin Test

The AST is now used as a diagnostic test in many countries (e.g., U.S.S.R., Italy, Greece, France, and China), and the various allergens have been well described.[119-121] The preparations of allergens are usually mixtures of cell wall and internal antigens and are not physicochemically characterized. Some saline soluble allergenic protein extracts were prepared from smooth strains of *Brucella* with the consequence of being contaminated with SLPS.[122,123] To avoid this problem, saline-soluble allergenic protein extracts (Brucellin) have been prepared for the last 15 years from rough strains of *Brucella* which are free of SLPS.[124-130] The allergenic activity of Brucellin is destroyed by pronase, suggesting that the protein molecules have the biological activity.[122,126] Of these preparations, Brucellin-INRA was the best studied and has the following criteria for being a suitable allergen for detecting delayed hypersensitivity:

1. Its physicochemical and molecular characterizations have been done.
2. It neither sensitizes nor desensitizes the host when several skin tests are made.
3. After infection a true delayed-type hypersensitivity is revealed.
4. It does not induce antibodies reactive in standard diagnostic tests (i.e., this suggests that the preparations are SLPS free).
5. The delayed hypersensitivity is specific for *Brucella*.
6. A titration assay is available, as are reference standards.
7. The allergen is stable for considerable lengths of time (at least 10 years).[125]

Brucellin-INRA contains at least 20 molecular species with no detectable LPS. A commercial preparation for sheep and goats is now available.[127] In cattle and goat herds or sheep flocks, Brucellin-INRA detected more positive infected animals than did standard tests.[131] Usually about 50% of the infected animals react in both serological and allergic tests, while only about 25% are serologically positive, and the remaining 25% are only positive by the skin test. The use of Brucellin to reveal delayed-type hypersensitivity and its use for identifying infected animals has been shown by its high specificity (greater than 95%), though it does have a relatively low sensitivity (about 60%), and by the AST which can detect a significant percentage of infected animals that test serologically as negative. These conclusions show that an AST using Brucellin would be suitable for routine surveillance to identify infected flocks and herds (e.g., beef herds[128] because for dairy herds the milk ring test can be used) and for accelerating an eradication program by using this test in addition to classical serology.

ii. Lymphocyte Blastogenesis Test

The LBT was first developed for chronic human cases of brucellosis often found to be serologically negative[132] and for serologically negative infected cattle.[133] Numerous partially characterized antigens were used with partially purified blood lymphocytes, and several procedures were designed which have been previously discussed.[118] *Brucella abortus* soluble antigen (BASA) containing considerable amounts of SLPS, soluble proteins, and brucellin preparations were tested as antigens.[134,135] BASA was claimed to be superior to brucellin in detecting infected animals and for its ability to discriminate between infected and B19-vaccinated animals.[136] Nevertheless, the validity of using the LBT with SLPS containing antigens, such as BASA, has been seriously questioned because of observations of false-positive reactions and of problems with interpretation.[95,96,137] Also, the lymphocytes used may consist of B- and T-sensitized or immature lymphocytes which may undergo a specific or nonspecific blastogenesis in the presence of a great number of molecules (e.g., LPS and proteins) in the various extracts.[91,138] Additional research is needed on the standardization of LBT procedures as well as the preparation of purified antigens for the correct evaluation of this test.

Nevertheless, Brucellin, although contaminated with small amounts of LPS, as an antigen in LBT was able to differentiate bovine brucellosis from yersiniosis.[139] This result suggests that the *Brucella* protein in the reagent does not cross-react with the protein antigens of *Yersinia*.

D. CONCLUSIONS

Little is yet known about the protein antigens of *Brucella*. Outer membrane proteins which are potentially protective or suitable as diagnostic antigens are the best known. The use of one or a mixture of physicochemically characterized protein antigens in humoral and cellular tests can improve the diagnosis by preventing cross-reaction, since internal proteins differ antigenically from the SLPS of cross-reacting bacteria[139,140] and by detecting hosts infected by R species of *Brucella* (e.g., *B. canis* and *B. ovis*).

Progress in the knowledge of *Brucella* protective antigens, protective immune mechanisms, and diagnostic antigens will depend on the quality of the antigens (i.e., purification and characterization),[141,142] the standardization of operating procedures, and the exchange of reagents between the research groups to define the relationship between the prepared material and the reproducibility of results. The identification of protective antigens, whether SLPS or proteins, will greatly assist the production of an ideal vaccine containing only protective antigens, other than diagnostic antigens, used to differentiate infected and vaccinated hosts. Through the use of technologies such as hybridoma production and gene cloning, purified protein antigens should be available to probe unambiguously the immune response of the host.

III. *BRUCELLA* LIPIDS

A. INTRODUCTION

Members of the genus *Brucella* possess similarities with other Gram-negative bacteria, although they have properties of their own which clearly distinguish them from this group. The cell envelope, morphologically similar to that of other Gram-negative bacteria, consists of a cytoplasmic membrane, the periplasmic space, a peptidoglycan layer, and an outer membrane.[151] However, the chemical composition and physical properties make the cell envelope of *Brucella* unique among the Gram-negatives[152,153] This becomes apparent when the membrane fractions of *Brucella* are isolated and the components analyzed[21,59,62,76-78,82,154]

The role of surface macromolecules in determining the virulence of extracellular parasites has been well established.[155] However, it is not clear which function(s) the surface components play in the virulence of facultative intracellular pathogens such as *Brucella*. It is known that the rough *Brucella* types are more readily killed by polymorphonuclear leukocytes than the smooth counterparts.[154,156] It is likely that the chemistry and macromolecular organization of the surface components are important factors in the virulence of these microorganisms. Therefore, the ability of *Brucella* cells to replicate in phagocytes may be due, at least in part, to the particularities of their outer membrane.[154,157]

The group 1, 2, and 3 proteins of *Brucella* and a lipoprotein covalently linked to the peptidoglycan sacculus seem to be very similar to those present in *Enterobacteriaceae*.[75-77,81,89,152,153,158-160] Thus, it is likely that the major difference in the properties of the outer membrane of *Brucella* is in the proportion, species, and distribution of lipids (free and bound) present in this layer, as well as in their interactions with other molecules of the outer membrane.[21,81,152,161-168] The few reports on the precise composition of the *Brucella* peptidoglycan[169-171] and the cytoplasmic membrane[169-171] suggest that these two layers do not differ significantly from those of *Enterobacteriaceae*.

Brucella cell envelopes are relatively more resistant to the disrupting action of detergents and ethylene diaminetetraacetate (EDTA).[78] In addition, these bacteria are more resistant to the action of polymyxin B than other Gram-negatives,[152,172] polymyxin B is an antibiotic which interacts on membranes by its cationic detergent structure. These properties are probably due to various facts:

1. The *Brucella* LPS present in the outer membrane does not require stabilization by interactions with divalent cations.[78]
2. The properties and quantity of the 2-keto-3-deoxyoctonate (KDO) molecule(s) and the content in phosphate groups linked to the lipid A differ from other Gram-negative bacteria.[82,173]
3. The ionic interactions of some proteins associated to the peptidoglycan layer are weaker than in *Enterobacteriaceae*.[159]
4. The distribution, different proportions, and class of phospholipids in the outer membrane are different from most of the Gram-negatives studied.[152,163]
5. The overall hydrophobic forces operating in *Brucella* envelopes, in comparison with those operating in members of the *Enterobacteriaceae*, seem to be stronger.[78,152,163]

In addition, the cell walls from smooth strains are more resistant to the proteolytic action of enzymes and digestion with lysozyme than strains of rough *Brucella*.[154]

The *Brucella* outer membrane components are the principal molecules interacting with the immune system of the host.[11,54,62,75,79,80] We will devote most of this review to the structure of the "free" and "bound" lipids present in this layer, their implications on pathogenesis, their function in antibiotic susceptibility, the biological activities induced by lipid-bound macromolecules, and their significance for phylogenetic and taxonomic studies.

B. FREE LIPIDS

Free lipids are those which are readily extractable from the bacterial mass with a mixture of ethanol and diethyl ether and which are then separated by gel filtration and repurified by extracting them with chloroform/methanol. The free lipids of *Brucella* (4.5% of bacterial dry weight) include phospholipids (2% of bacterial dry weight) and the neutral lipids (2.5% of bacterial dry weight) which are separated and analyzed by different chromatographic procedures. Thiele and Schwinn[168] found that the amount of total free lipids of *Brucella* is strongly dependent on the culture media. The same seems to be true for the neutral lipid proportions of the total free lipids. Some authors have questioned the idea that the content of neutral lipids and phospholipids vary inversely from *B. abortus* through *B. suis* to *B. melitensis*, i.e., the last species containing the highest proportion of neutral lipids and the lowest of phospholipids.[168]

1. Phospholipids

From the lipid composition of *Brucella* presented in Table 3, several exceptional characteristics can be observed. Phosphatidylcholine (PC) has been found to be the major phospholipid of *Brucella,*[163,166-168] in contrast to *Enterobacteriaceae,* where phosphatidylethanolamine (PE) constitutes the major phospholipid.[174] It is striking that while PC is generally absent in most bacterial species, it is commonly abundant in plant and animal cells.[174] Recently, Moriyon et al.[152] and Gamazo and Moriyon[163] have described the proportion and distribution of phospholipids in the outer membrane (Table 4), taking advantage of the observation that *Brucella* cells release outer membrane fragments devoid of inner membrane and cytoplasmic markers but containing LPS and Omp. Phosphatidylcholine as a membrane constituent has different properties than PE.[175] In most eukaryotic membranes PE is distributed in the inner leaflet of the bilayer, while PC is mostly present in the outer leaflet. It has been established that the polar group of PE is able to form hydrogen bonds with adjacent phosphate groups.[176] Since all other characteristics of PE are very similar to those of PC, the higher viscosity given by PE in the membranes is very likely due to this property. This conclusion has been corroborated by the presence of two enzymes that sequentially methylate PE, transforming it to PC.[177] In eukaryotes, PC moves to the outer leaflet of the bilayer increasing the membrane fluidity. In Gram-negative bacteria containing PC, it is not known whether PC is also present in the outer leaflet of the outer membrane. Thiele and Schwinn[168] detected methylated PE in *Brucella* species, suggesting a similar biosynthetic pathway for PC as has been described for *Agrobacterium tumefaciens.*[178]

The outer membrane of enteric bacteria is highly asymmetric with respect to the distribution of LPS and phospholipids.[179-181] The outer leaflet is devoid of phospholipids and is almost entirely occupied by LPS, glycolipids and the external moiety of proteins. For enteric bacteria, this distribution might be important since they live in the presence of high concentration of detergent cholates in the intestinal tract.[180] In *Brucella* a respective asymmetric distribution in the outer membrane might make the cell less susceptible to the action of phospholipases, proteolytic enzymes, and lysozyme.[154] However, the possibility of having a different distribution of macromolecules in the bilayer of *Brucella,* with significant quantities of phospholipids in the outer leaflet of the outer membrane, cannot be ruled out at the moment.

As proposed by Gamazo and Moriyon,[163] it is likely that the outer membrane fragments released by *Brucella* represent more the average lipid than the average protein composition of the outer membrane. In comparison with the cell envelopes, the outer membrane fragments are enriched in PC (Table 4). Since the total PC content of *Brucella* is similar to the amount of PC found in the respective outer membrane,[163,168] it is feasible to suggest that most, if not all, of the PC is present in this layer.

Phosphatidylglycerol (PG) and diphosphatidylglycerol (DG), also present in *Brucella,* are probably the most commonly occurring bacterial phospholipids. Diphosphatidylglycerol usually occurs together with PG due to their biosynthetic relationship.[182] Diphosphatidylglycerol is the most apolar of the phospholipids, a property resulting from the presence of four acyl residues in the molecule instead of only two.[183] Phosphatidylserine, present in small amounts in *Brucella* as well as in other Gram-negative bacteria, is recognized as an intermediate in the biosynthesis of PE.[183] Phosphatidylinositol, characteristic of the membranes of eukaryotic cells and of some mycobacteria, has not been detected in *Brucella.*[168]

It is known that cyclopropane fatty acids occur frequently in bacterial lipids. The C17:0 cyclic seems to be more frequent in Gram-negative bacteria and the C19:0 cyclic (lactobacillic

TABLE 3
Lipid Composition of *Brucella*[a]

Fatty acid	Bound lipids				Free lipids (4.5% of bacterial dry weight)											
					Phospholipids (2% of bacterial dry weight)[b]								Neutral lipids (2.5% of the bacterial dry weight)[c]			
	LPS (2.5%)[d]		Lipoprotein		DG (20.1%)[e]		PG (9.4%)[e]		PE (32.9%)[e]		PC (37.6%)[e]		Wax-like esters (6%)[f]	1,2- and 1,3-diglycerides, diol monoesters, α-glycol esters, and other acyl esters (28%)[f]	Ornithine-containing lipids (32%)[f]	
	A	E[g]	A	E	1	2[h]	1	2	1	2	1	2			A	E
12:0	—	—	—	—	t[i]	t	1.1[j]	t	1.5	t	t	t	6.4	2.6	t(?)[k]	
14:0	—	—	6.8	12.0	1.1	t	1.6	t	4.1	t	t	t	12.8	3.2	t(?)	
15:0	—	—	—	—	1.3	t	t	t	t	t	t	t	5.4	0.7	—	—
16:0	—	37.0	28.3	20.7	48.9	16.0	72.6	40.4	53.9	10.7	52.7	7.6	34.5	27.3	30.4	—
17:0	—	1.8	—	—	2.5	t	t	t	t	0.2	1.4	0.3	2.0	1.5	0.3(?)	
18:0		4.5	22.6	24.1	6.4	5.2	4.5	6.2	8.4	2.7	4.0	3.6	8.9	9.0	12.2	—
20:0	—	—	4.9	9.9	—	—	—	—	—	—	—	—	—	—	—	—
16:1	—	—	2.9	1.1	39.0	1.9	9.7	4.3	11.1	1.8	31.2	0.7	8.8	4.1	—	1.9
17:1	—	—	—	—	—	—	—	—	—	—	—	—	t	1.3	—	—
18:1	—	—	12.1	3.2	0.9	16.4	t	9.5	10.9	13.9	3.6	8.3	9.8	17.8	—	18.4
3-OH-12:0	7.8	—	—	—	—	—	—	—	—	—	—	—	—	—	—	—
3-OH-13:0	t	—	—	—	—	—	—	—	—	—	—	—	—	—	—	—
3-OH-14:0	15.4	—	—	—	—	—	—	—	—	—	—	—	—	—	—	—
3-OH-15:0	—	—	—	—	—	—	—	—	—	—	—	—	—	—	—	—
3-OH-16:0	15.1	12.4	—	—	—	—	—	—	—	—	—	—	—	—	—	—
3-OH-18:0	—	3.5	—	—	—	—	—	—	—	—	—	—	—	—	—	—
17:Ocyclic	—	—	1.8	3.5	—	—	—	—	—	—	—	—	—	—	—	—
19:Ocyclic	—	—	1.7	0.8	t	57.2	6.8	39.8	7.4	69.6	7.1	79.5	—	31.3	—	29.3
Other	1.3(?)		18.9	23.7	t	3.3	3.7	t	2.7	1.1	—	t	11.4	1.2	7.5(?)	

[a] LPS data are unpublished results obtained with *B. melitensis* by E. Moreno and H. Mayer. Similar results were obtained with *B. abortus*. Fatty acid composition of lipoprotein was obtained from Gomez-Miguel and Moriyon[81] with *B. abortus*. The phospholipid and neutral lipid compositions were taken from Thiele and Schwinn[168] with *B. melitensis*.

[b] Diphosphatidylglycerol (DG), phosphatidylglycerol (PG), phosphatidylethanolamine (PE), phosphatidylcholine (PC). Phosphatidylserine is present in trace quantities in *B. melitensis* and 5.3% in *B. abortus*.

[c] Ubiquinone Q_{10} is the predominant component of neutral lipids, comprising 30% on a dry weight basis.

[d] Expressed as % of the bacterial dry weight.[90]

[e] Expressed as % of dry weight of the total phospholipids.

[f] Expressed as % per dry weight of the total neutral lipids.[154]

[g] Amide- (A) and ester (E)-linked fatty acids. The hydroxylated fatty acids exist as acyl-oxyacyl residues: 3-O-(16:0)12:0, 3-O-(16:0)13:0, 3-O-(16:0)14:0, and 3-O-(18:0)14:0.

[h] Position of the fatty acid in the individual phospholipid.

[i] Trace (t).

[j] In LPS and neutral lipids, numbers express % fatty acid per dry weight of total (A + E) fatty acids. In the lipoprotein and phospholipids, numbers represent % of the corresponding species (A or E linked; 1 or 2 position) of fatty acid.

[k] Linkage is not known (?).

TABLE 4
Lipids in the Cell Envelopes (CE) and Outer Membrane Fragments (OMF) of *Brucella*

	LPS[a]	DG[b]	PE[b]	PC[b]	Other[b]	Ubiquinone Q_{10}	NADH-oxi[c]	SDH[c]	Outer membrane proteins
OMF	48	19	20	46	20	—	0	0	++++
CE	30	25	25	24	25	++++	0.273	0.262	++

[a] % of dry weight of material. Data from Moriyon et al.[152] and Gamazo and Moriyon.[163]
[b] % of total phosphorus.
[c] Enzymatic activity of NADH-oxidase (NADH-oxi) and succinate dehydrogenase (SDH).

acid) in Gram-positives.[167] *Brucella,* as a Gram-negative bacterium, plays an exceptional role in that it contains lactobacillic acid as the main cyclopropane fatty acid besides C18:1 (*cis*-vaccenic acid) as its possible biosynthetic precursor.[168] This is supported by the fact that the amount of lactobacillic acid in *Brucella* increases from 34 to 50% (of cell dry weight) in cells cultured for 24 or 48 h, respectively, accompanied by a concomitant reduction of *cis*-vaccenic acid (31 to 16%, respectively). The data indicate that *cis*-vaccenic acid serves as direct precursor of lactobacillic acid.[161] *B. canis* plays an exceptional role among members of the Brucella because it contains lactobacillic acid only in trace amounts, whereas *cis*-vaccenic acid is the major fatty acid present in this species.[161] It is likely that in *B. canis* the functional enzyme necessary to produce lactobacillic acid from the *cis*-vaccenic acid precursor is lacking or less active and consequently *cis*-vaccenic acid accumulates. In case of the *B. abortus* and *B. melitensis* phospholipids, lactobacillic acid is mainly linked at the C-2 position of glycerol. The C16:0 (palmitic acid) and C16:1 (palmitoleic acid) are also abundant in phospholipids of *Brucella*. However, in contrast to lactobacillic acid, these fatty acids are present mostly at C-1 of the glycerol backbone (Table 3).

1. Neutral Lipids

Thiele and Schwinn[168] isolated five fractions of neutral lipids on a silicic acid-column. Fraction 1 (6% of the neutral lipid weight) formed a single spot in thin layer chromatography. It was identified as a mixture of wax-like esters. This finding deserves further investigation since true wax esters are known to be limited to mycobacteria and corynebacteria. The *Brucella* wax-like esters are substituted by large quantities of palmitic acid (C16:0) and moderate amounts of myristic (C14:0), stearic (C18:0), palmitoleic (C16:1), and *cis*-vaccenic (C18:1) acids. In contrast to the *Brucella* phospholipids, the wax-like esters do not contain lactobacillic acid. Fraction 2, the predominant component (30% of the neutral lipid dry weight), was obtained as orange-red crystals and was identified as ubiquinone Q_{10} (Figure 3). This compound has been found only in a few bacteria so far.[183-185] Gram-negative bacteria usually contain ubiquinone Q with shorter isoprenoid side chain length (Q_7—Q_8), whereas Gram-positive bacteria contain menaquinones instead of ubiquinone Q.[183,184] Since isoprenoid quinones are constituents of the cytoplasmic membrane,[184] they are absent from the outer membranes. Quinones play an important role in the electron transport in the oxidative phosphorylation and also in active transport. Fraction 3 (9% of the neutral lipid weight) contained 1,2- and 1,3-diglycerides (6.7%) as well as small amounts of glycerol, ethylene glycol monoesters and of an alpha-glycol of an unknown structure. Fraction 4 (16% of the neutral lipid weight) contained acyl monoesters of ethylene glycol and another alpha-glycol. Fraction 5 (less than 3% of the neutral lipid weight) also contained acyl esters but lacked glycerol. Fractions 3, 4, and 5 together comprise approximately 28% of the neutral lipids. Their fatty acid compositions are very similar among each other (Table 3).

It might be speculated that ethylene glycol monoesters of *Brucella* are derived from

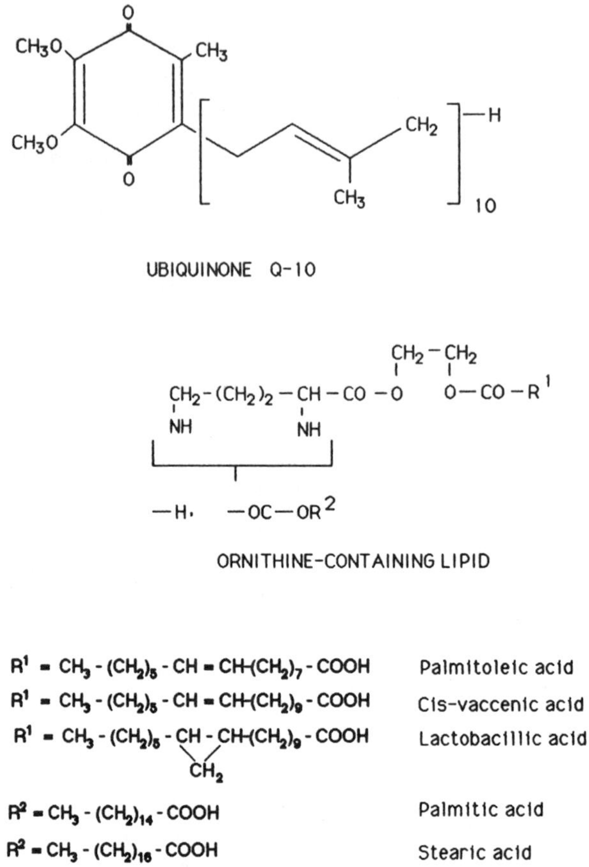

FIGURE 3. Ubiquinone Q_{10} and ornithine-containing lipid of *Brucella*. Based on the work of Thiele and Schwinn.[168]

hydrolytic cleavage of other lipids (see below). They might represent a group of the natural ornithine degradation products.[168] Unlike eukaryotes, glycerides do not occur in large amounts in bacteria and are considered to be precursors of more complex lipids. In accordance with other Gram-negatives, mono- and diglycerides but not triglycerides are found in *Brucella*.[168]

Ornithine-containing lipids (32% of neutral lipids) of *Brucella* represent a major proportion of the total neutral lipids. They might have taxonomic, phylogenetic, and biological significance, since they have a specific distribution among various taxa of bacteria. However, their use as macromolecular markers have received little attention. The *Brucella* ornithine-containing lipid consists of four components:[168] ornithine, ethylene glycol and two different fatty acids (Figure 3). One of the two amino groups of the ornithine is unsubstituted. It is not known whether the alpha- or the delta-amino group is acylated. It has been determined that lactobacillic, *cis*-vaccenic, and palmitoleic acids are present predominantly, if not entirely, in ester linkages, while palmitic and stearic acids are found in amide linkages. Apparently, hydroxylated fatty acids, which are common constituents in other bacteria[186] are absent in the *Brucella* ornithine-containing lipid.

Little is known about the biosynthesis and biological function of ornithine-containing lipids. They are found in various types of bacteria,[186] but their function is not clear. Ornithine-containing lipids might have some special function in the membrane because their structure does not allow hydrolysis by phospholipases. It is generally agreed that ornithine-containing lipids play a structural role within the membrane.[187] Wilkinson[188] suggested that ornithine-containing lipids and PE are functionally interchangeable and that ornithine-containing lipids are synthesized when the phosphate supply of the cell is limited. Thiele et al.[189] found that in *B. abortus* the concentration of ornithine-containing lipids increased severalfold with increasing age of the culture. This is probably due to the gradual consumption of the phosphate supply. Kawai and Yano[190] found that liposomes made of ornithine-containing lipids, in addition to agglutinating cells by hydrophobic interactions between the fatty acid moieties of the aminolipid and the lipids on the surface of the cells, could agglutinate cells by ionic interactions and hydrogen bonding between the lipoamino acid and the lipids or proteins of the membrane.[191] These findings are significant, because *Brucella* cells have the tendency to form rosettes on the surface of macrophages and B lymphocytes.[192] For this to occur, some ornithine-containing lipids would have to be exposed on the surface of the outer membrane of *Brucella*. Since ornithine-containing lipids are not common constituents of vertebrates, and since those present in *Brucella* contain a relatively large polar head group, it would not be unusual for them to be immunogenic for the host.

C. BOUND LIPIDS

Bound lipids are complex molecules such as LPS, glycolipids, and lipoproteins anchored to the outer membrane of Gram-negative bacteria. These lipids are usually not readily removed from the cell by ethanol/ether extraction, and their isolation usually requires stronger conditions. *Brucella* lipoprotein is obtained only after extensive treatment of the cell envelopes with hot sodium dodecyl sulfate (it requires longer periods of extraction than those needed for *E. coli* lipoprotein), followed by proteolytic release from the peptidoglycan. The lipoprotein fragment is precipitated with acetone and purified by gel filtration.[81] Lipopolysaccharides are extracted from the bacteria with a 1:1 mixture of hot phenol and water.[193] In contrast to enterobacterial endotoxin which is extracted into the aqueous phase of phenol/water extracts, *Brucella* S-type LPS is isolated from the phenol phase, together with phenol soluble proteins, phospholipids, neutral lipids, and some polysaccharides.[21,62,82] Further purification is achieved by treatment with nucleases, proteinase K, removal of phospholipids with a mixture of chloroform/methanol/water, ultracentrifugation, and gel filtration in the presence of chaotropic agents.[21,82,194,195] *Brucella* R-type LPS is insoluble in aqueous systems and requires a different protocol for extraction and purification[82,196] On the basis of sodium dodecyl sulfate extraction, SDS-PAGE, as well as laser desorption mass spectroscopy, it is clear that *Brucella* LPS is not a single entity, but rather represents a group of structurally related molecules.[197] The heterogeneity arises from different lengths of the "O"-specific polysaccharide chains, and from different substituents of the amide-linked 3-hydroxy fatty acids in the lipid A backbone.[21] Additional heterogeneity might be caused by differences in the lipid A backbone sugar(s). Like other lipopolysaccharides, *Brucella* lipid A is released from the complete LPS molecule by acid hydrolysis. However, the conditions are stronger than those required to obtain enterobacterial lipid A.[53]

1. Lipoprotein

Gomez-Miguel and Moriyon[81] demonstrated that *Brucella* contains a peptidoglycan-bound lipoprotein, which is similar to that of *E. coli* in molecular weight, behavior in SDS-PAGE, isoelectric point, amino acid composition, and distribution of amide- and ester-linked fatty acids. The data presented by these authors also suggested that the lipoprotein-peptidoglycan linkage might be similar to that of *E. coli*. Furthermore, serological cross-reactivity demonstrated that *Brucella* and *E. coli* lipoproteins share common antigenic determinants.[90] Both the biochemical characterization and the serological analysis indicate a relationship between *E. coli* and *Brucella* lipoproteins closer than that expected from their taxonomic positions. This similarity, probably reflecting a comparable function, stresses the importance of maintaining the integrity of the cell envelope through interactions between the proteins anchored to the peptidoglycan layer and the outer membrane of the bacteria.[74,76,90]

Despite similarities between *Brucella* and *E. coli* lipoproteins, there is a remarkable difference between them with respect to exposure on the cell surface. *E. coli* lipoprotein is not exposed on the surface of smooth cells,[198] while immunoelectron microscopy conclusively demonstrated that *Brucella* lipoprotein is exposed on the cell surface.[90] It must be concluded that, in contrast to *E. coli,* in *Brucella* smooth forms the LPS molecules do not hinder the reaction of antibodies with the lipoprotein molecules. As shown by Dubray and Plommet,[151] the layer corresponding to the polysaccharide chain of the LPS is less thick in smooth *Brucella* than in *E. coli.* Lipoproteins of *Brucella* and *E. coli* also differ in their fatty acid compositions: *Brucella* lipoprotein is very rich in myristic, palmitic, and stearic acids; contains moderate amounts of *cis*-vaccenic and arachidonic (C20:4) acids; contains low amounts of cyclopropane fatty acids (C19:0 cyclic and C17:0 cyclic); and lacks hydroxylated fatty acids. In contrast, *E. coli* lipoprotein is very rich in palmitic acid, unsaturated fatty acids (C16:1 and C18:1), and contains moderate amounts of cyclic fatty acids (C17:0 cyclic) and low amounts of hydroxylated fatty acids (3-OH-C14:0). It is likely that the overall hydrophobic forces operating between *Brucella* lipoprotein and other lipids are stronger than in *E. coli* lipoprotein. Also, the lower amount of cyclopropanic acids, absence of hydroxylated fatty acids, and the presence of fatty acids with longer aliphatic chains may contribute to stronger hydrophobic interactions within the *Brucella* outer membrane.

2. Lipopolysaccharide

As in other Gram-negative bacteria, LPS is the most abundant lipid-containing molecule of the *Brucella* species(2.5 to 3% of the bacterial dry weight). Structurally, LPS is built of an "O"-specific polysaccharide chain, a core polysaccharide, and lipid A. The O-specific chain of *B. abortus* consists of an unbranched linear homopolymer of alpha-(1,2)-linked N-formyl perosaminyl residues,[16] while the structurally related "O" chain of *B. melitensis* is a linear polymer made up of repeating units composed of five *N*-formylperosaminyl residues, four of them being alpha-(1,2) and one alpha-(1,3) linked.[20,25] Mannose, glucose, quinovosamine, and KDO probably form part of the core oligosaccharide, which links the O chain to the lipid A moiety.[16,196] LPS from rough *Brucella* does not contain perosamine and quinovosamine,[196] but has glucose, mannose, and KDO. Heptoses (L-glycero- or D-glyceromannoheptose) have not been found in *Brucella* LPS.[82,196]

The "O" polysaccharide chain and the core oligosaccharide of *Brucella* LPS are of great value in the diagnosis of brucellosis and in the serological identification of *Brucella* species and strains.[53,62,119,196] They also contribute to some of the biological activities induced by the complete LPS molecule.[53] The structure and function of the O-chain of *Brucella* LPS and related polysaccharides have been extensively studied by several groups[16,20,21,25,54,62,206] and are the subject of a separate article in this volume.

Brucella LPS is exposed on the outer leaflet of the outer membrane where it exists either as high molecular weight complexes or associated with other molecules such as proteins, or other lipid-containing molecules.[21,76,78,82,152,159,160,196] This is indicated by coextraction of the *Brucella* smooth LPS with other components of the outer membrane[82] into the phenol phase of the Westphal extraction method.[193] The removal of these components is a difficult task and requires several cycles of purifications.[21,82,194] In *E. coli* and many other Gram-negatives, divalent cations stabilize the outer membrane by interacting with the hydrophilic portion of the LPS.[179,181] Thus, treatment with cationic chelates such as EDTA releases significant amounts of protein and LPS from the outer membrane, allowing the preparation of spheroplasts after lysozyme treatment.[200] However, in *Brucella* the application of EDTA in combination with Tris does not affect the stability of the cell envelopes nor enhance the action of Triton X-100. Thus, it is concluded that the LPS of *Brucella* is not stabilized in the outer membrane by divalent cations.[78] This conclusion was further confirmed by ultracentrifugation of LPS in the presence of detergents and EDTA.[78]

It has been proved that KDO and probably the unsubstituted phosphate groups of both the core and lipid A are the high affinity binding sites for Ca^{++} and polymyxin B in the enterobacterial LPS.[179,201] Therefore, the results obtained by Moriyon and Berman[78] mentioned above are consistent with the presence of lower quantities of KDO in *Brucella* LPS[82,196] and also with our (preliminary) finding of rather low amounts of phosphate in *Brucella* lipid A.[21,263] These data must, however, be taken cautiously since the stronger conditions of hydrolysis to obtain a precipitable *Brucella* lipid A[53] could have led to a release of glycosidically linked phosphate groups linked to the backbone. On the other hand, the relatively high concentration of phosphorus reported by several authors[16,202] could be explained by low quantities of phospholipid or nucleic acid contamination. Moreover, polymyxin B-resistant mutants of *Salmonella* bind less antibiotic to the LPS than the respective wild-type strains.[203] These data are in agreement with the lack of polymyxin B-binding capacity by *Brucella* LPS.[53]

As proposed by Moriyon et al.[152] it seems likely that in contrast to LPS of other Gram-negatives, *Brucella* LPS is stabilized mainly by hydrophobic interactions among their lipid A molecules and by interaction of the lipid A with other lipid and protein molecules. This is also in agreement with the relatively large proportion of long chain fatty acids found in the lipid A of *Brucella* (Table 3). These interactions, which exhibit a remarkably high state of order in the outer membrane as compared to phospholipid arrangements, form a relatively rigid structure.[179] In this respect, Moriyon and Berman[76] reported that the *Brucella* porins remained strongly associated with the LPS. This association could be reestablished *in vitro* and even with heterologous porins from *E. coli,* demonstrating that no species-specific requirements exist for the LPS-porin association. Schindler and Rosenbusch[204] demonstrated that LPS-free porins inserted into PC liposomes do not form stable channels, whose production was found to require the presence of both LPS and the subsequent aggregation of trimers.

Extensive studies on the chemical structure of enterobacterial lipid A have led to a detailed knowledge of the architecture of this molecule.[205,206] The highly toxic *Salmonella* lipid A has been taken as the prototype of the lipid A molecules and, thus, is referred to as the "usual" lipid A type. The backbone structure consists of beta-1,6 linked glucosamine disaccharides, substituted by phosphate on both glucosamine residues in the 1 and 4′ positions, respectively. The amino groups of both the reducing and the nonreducing glucosamine molecules carry amide-linked 3-hydroxy fatty acids. All the hydroxyl groups of the disaccharide, except those in position 4, carry ester-linked 3-hydroxy fatty acids. Amide- and ester-linked 3-hydroxy fatty acids may be further substituted by saturated nonhydroxylated fatty acids which are linked to the 3-hydroxy-groups, thus forming 3-acyl-oxyacyl-residues. KDO, the linkage sugar between the polysaccharide and the lipid A moiety, is ketosidically attached to position 6′ of the nonreducing glucosamine in the lipid A backbone. The two phosphate groups attached to the lipid A backbone are in many cases partially substituted by polar groups (e.g., 4-amino L-arabinose or ethanolamine) linked via phosphodiester bridges.

In contrast to the "usual" lipid A of many Gram-negatives, *Brucella* LPS contains a lipid A molecule of "unusual" structure.[21,165,207,263] As in the case of several photosynthetic bacteria and their nonphotosynthetic relatives, the lipid A of *B. abortus* and *B. melitensis* contains 2,3-diamino 2,3-dideoxy-D-glucose (2,3-diaminoglucose) as backbone sugar. However, in contrast to many of the other "unusual" lipid A molecules described so far, glucosamine is also present in the *Brucella* lipid A preparations which, thus, represents a so-called "mixed" lipid A.[207] Furthermore, *Brucella* lipid A reacts with both rabbit antiserum against 2,3-diaminoglucose containing lipid A of *Rhodopseudomonas viridis* as well as with antiserum against the glucosamine disaccharide lipid A backbone of *E. coli.*[173,264] (See Figure 4.) Similarly the lipid A of *Chromatiaceae* and distinct thiobacilli might be exceptions, since

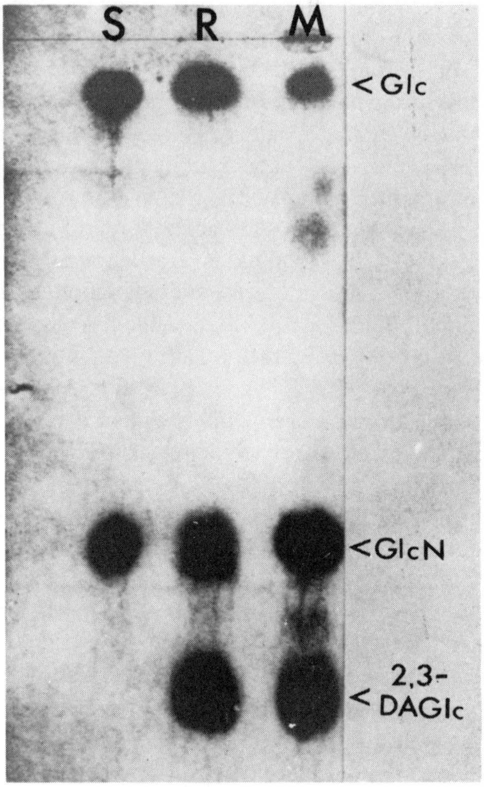

FIGURE 4. High voltage paper electrophoresis (pH 5.3) of hydrolysates (4*M* HCl, 100°C, 14 h) of lipid A from *Brucella melitensis* (M), *Rhodopseudomonas palustris* lipopolysaccharide (R), and glucosamine standard (S). Glucose, Glc; glucosamine, GlcN; and 2,3-diaminoglucose, 2,3-DAGlc.

in addition to the 2,3-diaminoglucose backbone, these bacteria may also contain glucosamine.[207] It is not known whether the 2,3-diaminoglucose and the glucosamine are covalently linked and form a mixed disaccharide or whether they represent two different entities of lipid A backbones. Hydrolysis of *Brucella* lipid A and subsequent preliminary analysis suggest the presence of a disaccharide molecule.[264] According to recent data (Raetz, mentioned in Reference 207), lipid A synthase accepts as the substrate the 2,3-diamino structural analogue of "lipid X", either with or without activation by UDP, making the formation of a "mixed" lipid A possible.

The amide-linked fatty acids, mostly present as acyl-oxyacyl residues, are 3-O-(16:0)12:0 (25% of the diesters), 3-O-(16:0)13:0 (4% of the diesters), 3-O-(16:0)14:0 (64% of the diesters), and 3-O-(18:0)14:0 (7% of the diesters). The amide-linked unsubstituted fatty acids are 3-OH-16:0 (approximately 12% of the total LPS fatty acids). The ester-linked fatty acids are 16:0 (37% of the total fatty acids), 3-OH-16:0 (12.4% total fatty acids), 3-OH-18:0 (3.5% of total fatty acids), and 18:0 (4.5% of the total fatty acids). *Brucella* lipid A seems to contain either low quantities of phosphate or lack it completely. It is probably substituted by KDO at C′-6,[208] but as stated before, this has not been definitely proved. The hydroxylated fatty acids are only present in lipid A, although there is at least one report (Promé et al., cited in Reference 168) claiming their presence also in the ornithine-containing lipid of *Brucella*. Even though lactobacillic acid and unsaturated fatty acids are present in most of

TABLE 5
Biological Activities Induced by *Brucella* and *E. coli* Endotoxins

Activity	μg Necessary to produce maximum biological activity: *Brucella*	μg Necessary to produce maximum biological activity: *E. coli*	Refs.
Toxicity for C3H/HeAU mice[a]	70[b]	9	53, 54
Toxicity for C3H/HeJ mice[c]	70	500(−)[d]	53, 54
Toxicity for macrophages	840	10	237
Lethality for chicken embryos	10	0.003	202
Induction of abortion	500	250	53
Induction of hypoferremia	100	10	238
Leukopenia and leukocytosis	100	10	202
Limulus lysate gelation	0.001	0.002	82
Tolerance to endotoxemia	10	10	202
Nonspecific protection	15	0.2	202
Induction of anaphylaxis	800	ND	239
Anticomplementary activity	50(+)	20(++++)	53
Toxicity for rabbits	3000(−)	500	239
Localized Schwartzman reaction	1000(−)	20	240
Pyrogenicity for rabbits	300(−)	0.15	202
Increase to hystamine susceptibility	16(±)	0.0065	241
Mitogenicity for CBA/N mice[e]	100(±)	20(++)	210
Mitogenicity for BALB/c mice	100(++)	25(++++)	82
Augment of thymocytes mitogenicity	50(++++)	20(++++)	210
Stimulation of hemopoiesis	50(+++)	ND	242
Polyclonal activation	100(++++)	100(++++)	53
Adjuvant activity in C3H/HeJ mice	20(−)	20(−)	196
IgM adjuvant activity in C3H/HeAU	20(++++)	20(++++)	196
IgG adjuvant activity in C3H/HeAU	20(++++)	20(±)	196
Induction of immune IgM	20(++++)	20(++++)	196
Induction of immune IgG3	20(++++)	20(±)	216
Induction of interferon	200(++++)	50(++++)	211
IL-1 induction in C3H/HeJ mice	50(++++)	50(−)	210
IL-1 induction in C3H/HeAU mice	50(++++)	50(++++)	210
TNF[f] production in C3H/HeJ mice	1 × X(++++)[g]	1 × X(−)	194
TNF production in C3H/HeAU mice	5 × X(++++)	1 × X(++++)	194

[a] C3H/HeJ mice are resistant to enterobacterial endotoxin.
[b] Toxicity carried out with carragenan-pretreated mice.
[c] C3H/HeAU mice are sensitive enterobacterial endotoxin and are syngenic with C3H/HeJ mice.
[d] Lack of biological activity (−) and different extents of biological activity (± or ++++).
[e] CBA/N strain of mice have a B cell maturation defect.
[f] TNF: tumor necrosis factor/cachectin.
[g] Unknown quantity (X).

the free lipids and lipoprotein, they are absent in the lipid A of *Brucella*. For practical purposes, the absence of lactobacillic acid and unsaturated fatty acids in the *Brucella* LPS preparations can be taken as one criterion for the purity of these fractions.

Table 5 presents the comparison between the biological activity induced by *Brucella* LPS and by *E. coli* LPS. In general, *Brucella* LPS is less active than *E. coli* LPS, so more material is needed to reach a comparable level of biological activity. In addition to the quantitative differences, there are also qualitative differences between the *Brucella* and *E. coli* lipopolysaccharides. Comparisons of these lipopolysaccharides are used to analyze the overall action of endotoxins. Most of the biological activities of the *Brucella* LPS are associated with the lipid A moiety.

In contrast to many endotoxins, *Brucella* endotoxin (LPS with variable amounts of protein) is practically nonpyrogenic for rabbits, is nonlethal for rabbits and chicken embryos, does not induce localized Shwartzman reaction, and does not increment the susceptibility to histamine. On the other hand, *Brucella* LPS is toxic for mice and macrophages, shows *Limulus* lysate gelation activity, induces leukopenia and leukocytosis, and causes hypoferremia and abortion in mice (references in Table 5). The demonstration that *Brucella* LPS (with less than 1% protein) is toxic for the endotoxin-resistant strain C3H/HeJ mice is one of the strongest pieces of evidence supporting the unique biological activities of this molecule.[53,173,194] It has been recently proposed[194] that lethality of *Brucella* LPS to endotoxin-resistant C3H/HeJ mice is due to the ability of inducing endogenous mediators by macrophages. *Brucella* LPS but not *E. coli* LPS stimulates the production of interleukin-1 (IL-1) in C3H/HeJ adherent cell cultures.[195] Simultaneously, C3H/HeJ mice produce tumor necrosis factor/cachectin in response to *Brucella* LPS. It is important to stress that the lethality for C3H/HeJ mice, induced by *Brucella* LPS preparations containing less than 1% protein, makes it untenable to attribute the toxicity to the contaminant protein.[194] As with other endotoxins, *Brucella* LPS loses its activity on treatment with ionizing irradiation[209] which induces some structural changes in the lipid A.

Enterobacterial LPSs have higher anticomplementary activity than *Brucella* LPS or lipid A. The maximum level of anticomplementarity of *Brucella* LPS is reached at concentrations five times higher than enterobacterial LPS, and the hemolysis inhibition never reaches more than 35% compared with the control. The lower anticomplementary activity of *Brucella* LPS is probably related to the high degree of aggregation and solubility, properties which are associated with the structure of its lipid A moiety.[53]

Brucella LPS and its lipid A are mitogenic for mouse spleen cells and B lymphocytes, but not for T cells.[53,173,210] However, the activity is different from that induced by enterobacterial LPS. *E. coli* LPS usually induces its maximum level of stimulation at concentrations between 10 and 25 μg/ml per 5×10^5 spleen cells. Higher concentrations are toxic for the splenocytes. In contrast, *Brucella* LPS induces its maximum level of mitogenicity at concentrations of 100 μg/ml or higher.[211] It seems likely that above 100 μg/ml, the *Brucella* LPS "saturates" the system without apparent toxicity for the lymphocytes.

Preparations of *Brucella* LPS containing 5 to 6% tightly bound protein are mitogenic for both endotoxin sensitive and resistant mice.[211] After removal of the tightly bound protein from the LPS with proteinase K, *Brucella* LPS maintained a weak mitogenicity for the C3H/HeJ mice as well as for the responsive mice.[194] The level of activity for the C3H/HeJ mice is similar to that of the "lipid A-rich" fraction of detergent dissociated *E. coli* LPS.[212] *Brucella* LPS is not mitogenic for the CBA/N mice strain which possesses a B cell maturation defect.[210] Presently, it is not known if the lipid A-associated protein of *Brucella* is mitogenic by itself or if it only contributes to increase the mitogenic action of the lipid A.

Morrison and Jacobs[213] demonstrated that polymyxin B binds to the lipid A-KDO moiety of enterobacterial LPS, abrogating its mitogenic and polyclonal activities. This antibiotic does not bind to the *Brucella* lipid A and, in consequence, fails to inhibit the mitogenic and polyclonal action of *Brucella* LPS. This observation, together with data on the chemical composition previously cited for *Brucella* LPS, indicates that the lipid A moiety of *Brucella* is different from that of *Enterobacteriaceae*.

In contrast to mitogenicity, the polyclonal activation of murine B lymphocytes by *Brucella* LPS and by its lipid A is similar to that produced by enterobacterial LPS for spleen cells of both the euthymic and athymic mice.[53] Although there is a close biological linkage between mitogenicity and polyclonal activation, the data obtained with *Brucella* LPS suggest that both activities are the result of similar but not identical phenomena.

Similar to other endotoxins, *Brucella* LPS induces the production of endogenous mediators, including IL-1,[195] tumor necrosis factor/cachectin,[194] and interferon.[214] As stated before, *Brucella* LPS but not *E. coli* LPS stimulates the production of IL-1 and tumor necrosis factor in C3H/HeJ mice. On the other hand, twice as much *Brucella* LPS as *E. coli* LPS is required to induce tumor necrosis factor/cachectin in responsive mice. The production of tumor necrosis factor/cachectin in the endotoxin-resistant C3H/HeJ mice requires five times more *Brucella* LPS than in the endotoxin-sensitive C3H/HeAU mice.[194]

Brucella endotoxin induces the same kind of interferon as does the endotoxin of *Enterobacteriaceae*.[214] However, the production is not as effective as with other lipopolysaccharides. The maximum stimulation occurs 2 h after an intravenous injection of the smooth LPS. As with enterobacterial LPS, if the mice are previously treated with cycloheximide, the interferon response is retarded but the titer increased.

Brucella LPS has an unusual adjuvant effect in comparison to other LPS molecules, in that the number of IgG plaque-forming cells is greatly augmented.[215] The lipid A-associated protein is not responsible for the unusual adjuvant activity since the *Brucella* endotoxin does not induce IgG adjuvanticity in athymic or C3H/HeJ mice. It seems likely that the adjuvant effect induced by *Brucella* LPS requires the participation of mature T cells. The unique action of this molecule suggests that C3H/HeJ mice might have a T cell defect, in addition to the well established B cell and macrophage defects.[194]

In contrast to *E. coli, Brucella* endotoxin (containing 5 to 6% tightly bound protein) stimulates primary and secondary responses in which IgM and high levels of IgG are produced by the endotoxin-resistant C3H/HeJ and endotoxin-responsive mice, including athymic and euthymic BALB/c mice.[215,216] Antibodies of all four subclasses of IgG are evoked during the course of the immune responses of mice. The primary response of BALB/c athymic mice consists almost entirely of relatively high levels of IgG_3. Antibody of the IgG_1 subclass appeared following a second injection. The *Brucella* endotoxin evoked strong primary and secondary IgG_3 responses against LPS determinants and strong primary and secondary IgG_1 responses against protein determinants. The primary and secondary responses to protein-free *Brucella* LPS consist mainly of IgG_3 antibody. The demonstration of the unusual ability of *Brucella* LPS to stimulate strong and predominantly IgG_3 antibody levels makes this molecule an attractive probe for studying the mechanisms of the IgM to IgG class switch. While the antigenic determinants of the *Brucella* LPS are mostly located in the polysaccharide moiety, the unique immunogenic activity of this molecule is very probably due to its "unusual" lipid A structure, as well as to the extent of aggregation of the LPS.[196,216]

Leong et al.[202] induced tolerance to the toxic action of homologous and heterologous lipopolysaccharides by testing mice with small doses of *Brucella* LPS prior to a challenge with toxic quantities of *E. coli* endotoxin. The tolerance can be induced reciprocally by treating the mice first with *E. coli* LPS. This result suggests similarities between the toxic action of *Brucella* and *E. coli* LPS. Leong and Wilson[217] suggested that the tolerance to the toxicity by *Brucella* LPS could be due to capture of the endotoxin by mesenteric lymph nodes, avoiding their accumulation in the liver.

Crude preparations of *Brucella* LPS with a high protein content are able to reduce the lytic effect of phages infecting *Brucella* cells.[218]

3. Other Bound Lipids

In addition to LPS, several authors have proposed the presence of different polysaccharide molecules on the surface of *Brucella* cells.[11] However, this has not been definitively proved.[54,62] A polysaccharide hapten, coextracted with LPS and strongly associated with it, has been shown to be difficult to purify, since its physical properties are similar to those of LPS.[11,54] Although this molecule cross-reacts with the "O"-specific chain, it can be distinguished from the LPS by its serological and biological behavior.[54,62] The polysaccharide possesses small quantities of a lipid, it coats erythrocytes, and can be adsorbed to plastic trays. It might be that, as in the case of enterobacterial common antigen,[219] this form of *Brucella* polysaccharide is bound to a lipid different from lipid A (e.g., an L-glycerophosphatidyl

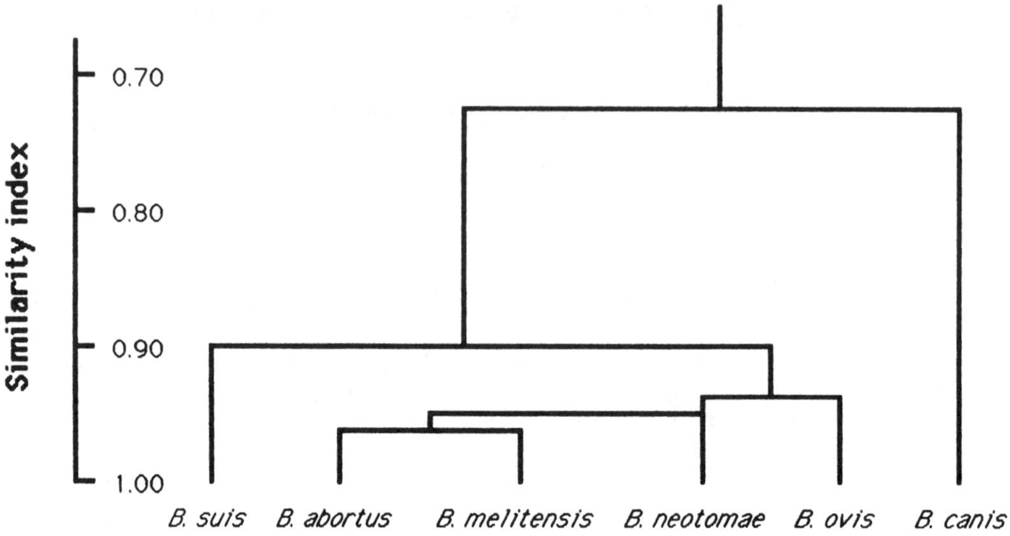

FIGURE 5. Dendogram of the genus *Brucella* based on their fatty acid composition. Modified from Tanaka et al.[224]

residue) and it is present on the surface of the bacteria as originally proposed by Diaz et al.[11]

D. LIPIDS IN TAXONOMIC AND PHYLOGENETIC STUDIES

The identification of bacterial lipids and fatty acids has been established as a powerful approach in diagnostic bacteriology,[161,162] in taxonomic classification,[164,183,185,220] and phylogenetic relationships.[207,221,222] *Brucella* species have been clearly differentiated from other Gram-negatives by their metabolic patterns,[119] guanidine plus cytosine content,[71] DNA hybridization patterns,[71] DNA/rRNA hybridization,[223] lipid composition, and fatty acid profiles.[161,162,168,224] Presently, the features available for the differentiation of the six registered species within the genus *Brucella (B. melitensis, B. abortus, B. suis, B. neotomae, B. ovis,* and *B. canis)* are their individual natural host, their biological and biochemical characteristics, as well as their serological specificities, metabolic patterns, phage susceptibility, and fatty acid and lipopolysaccharide composition.[119,161,215,224]

Studies on the fatty acid methyl esters of the whole cells are useful in determining whether an unknown isolate belongs to the genus *Brucella*.[164,168,224] In limited cases it serves also to differentiate some species from others.[161] Tanaka et al.[224] clustered the *Brucella* species in a dendogram (Figure 5) based on the intra- and interspecies similarities of fatty acid composition. The study used reference strains and a relatively large number of well-characterized field isolates. Among the species, a remarkably low similarity was noted between *B. canis* and other *Brucella* species (level of similarity: 0.40 to 0.74). On the other hand, *B. suis* and four other species formed two smaller clusters and were at the 0.90 level of similarity. The highest fatty acid similarity within species was observed among the *B. canis* strains. Therefore, the homogeneity of *B. canis* seems to be most evident in the genus *Brucella*. The highest interspecies similarity is found between *B. abortus* and *B. melitensis*. The two species are difficult to distinguish between each other on the basis of their fatty acid composition. *Brucella neotomae* shows a high degree of similarity to *B. abortus, B. ovis,* and *B. melitensis*. A close trend of similarity toward the two species mentioned above and *B. neotomae* is presented by *B. ovis*. With respect to the interspecies similarity of *B. suis,* a low and variable similarity index (0.8 to 0.9) is obtained with respect to the other species, with the exception of *B. canis*.

The chemotaxonomic studies on *Brucella* fatty acids have practical applications. Although rough isolates of *B. melitensis* and *B. abortus* can be distinguished from *B. canis* by growth in dilute basic fuchsin and slow hydrolysis of urea, respectively, none of the conventional tests are adequate for the differentiation between rough *B. suis* and naturally occurring rough *B. canis* isolates.[119] However, using the cellular fatty acid composition, Dess et al.[161] clearly differentiated between the two of them. The presence of relatively large amounts of lactobacillic acid in *B. suis* is the principal distinguishing feature, whereas the major fatty acid in *B. canis* is C18:1. Only trace amounts of C16:0 are present in *B. suis*.

Woese[225] could cluster different bacteria in a phylogenetic scheme based on the structural homology of their 16S rRNA. Some nonsulfur purple bacteria and related nonphototrophic organisms have been classified as the alpha-2 subdivision (Table 6), which is in good agreement with the relationships based on cytochrome C_2 structures[226] and lipid A structures.[221] In Table 6, *Brucella* has been included additionally and tentatively, based on the lipid composition similar to members of the alpha-2 subdivision. The alpha-2 subdivision is interesting since it includes Gram-negative bacteria which live in close association with eukaryotic cells, in addition to free-living species. *Brucella* is a highly virulent bacterium which would be an interesting member of this subdivision since it is one of the few facultative intracellular Gram-negatives which infect cells of a wide range of mammals, including humans. Furthermore, close homologies exist between the 16S rRNA of the alpha-2 subdivision and mitochondrial 16S rRNA, suggesting a common possible origin between these bacteria and the endosymbiotic ancestor of mitochondria.[227] The sequences of *Brucella* ribosomal 16S RNA are not known. However, as estimated by DNA-ribosomal RNA hybridization, the genus *Brucella* belongs to the ribosomal RNA superfamily IV, which also includes *Agrobacterium tumefaciens* and *Rhizobium leguminosarum*,[71,223] bacteria which belong to the alpha-2 subdivision. It would be rewarding to sequence the *Brucella* 16S rRNA in order to resolve the question of a possible phylogenetic relationship of *Brucella* to the bacteria of the alpha-2 subdivision. The fact that facultative intracellular bacteria such as *Brucella* have an unusual lipid A moiety containing 2,3-diaminoglucose similar to that of members of the alpha-2 subdivision is of considerable interest and strengthens the hypothesis that *Brucella* belongs to this group.

Although *Brucella* possesses an ''unusual'' lipid A containing 2,3-diaminoglucose related to several photosynthetic bacteria and their nonphotosynthetic relatives, the presence of glucosamine in the isolated lipid A seems to be a distinguishing feature, shared probably also by *Chromatiaceae* and thiobacilli.[207] More detailed analysis concerning the structure of lipid A of these bacteria is necessary in order to prove whether diaminoglucose and glucosamine sugars are covalently linked as a ''mixed'' backbone or if they are present as independent or disaccharidic molecules substituted by different fatty acids in amide and ester linkages.

In addition to the similarity of *Brucella* lipid A with bacteria of the alpha-2 subdivision, *Brucella* shares some common features to the membrane lipid composition from members of this group (Table 6). Phosphatidylcholine is the major component of the phospholipids in several members of the alpha-2 subdivision, whereas in most other Gram-negatives PC is absent and PE is the major phospholipid.[183] In addition, the presence of methylated PE (as possible precursor for PC) in *Brucella* and in several members of the alpha-2 subdivision[168] points to a close biosynthetic pathway of these phospholipids. A similar distribution for PC has been established for ubiquinone Q_{10}. Apart from *Brucella* species, other bacteria, several of which belong to the alpha-2 subdivision, also contain ubiquinone Q_{10}. As outlined previously, most Gram-negatives usually contain ubiquinones Q with shorter isoprenoid side-chain length, whereas Gram-positives usually contain menaquinones instead of ubiquinones. Another interesting characteristic of some of the members of the alpha-2 subdivision, which is shared with *Brucella*, is the presence of ornithine lipids. The taxonomic significance of

TABLE 6
Membrane Lipids from *Brucella* and Members of the Alpha-2 Subdivision[a]

Bacteria	Lipid A backbone composition: GlcN[b]	2,3-Diamino-glucose	Neutral sugars	Heptose	KDO	P[c]	Ubiquinone Q	Ornithine-containing lipid	PC
Rochalimaea quintana	(+)[d]	?[e]	?	+	+	?	?	?	(+)[d]
Rhodomicrobium vannieli	+	−	Mannose	−	+	0.1	10	+	+
Rhodopseudomonas acidophila	+	−	Mannose	−	+	0.1	10	+	+
Agrobacterium tumefaciens	+	−	+	+/−	+	2.5	10	+	+
Rhizobium leguminosarum	+	−	?	+/−	+	2.0	10	+	+
Nitrobacter winogradskyi	−	+	−	+	+	?	?	?	?
Phenylobacterium immobile	−	+	−	+	+	?	?	?	?
Rhodopseudomonas viridis	−	+	−	−	+	0.1	9	?	?
Rhodopseudomonas palustris	−	+	+	−	+	(1.5)[d]	10	+	+
Brucella melitensis	+	+	−	−	+	(0.3)	10	+	+
Brucella abortus	+	+	−	−	+	(0.3)	10	+	+

[a] Alpha-2 subdivision according to Woese.[255] The lipid A data[165,205,208,221,222,243-245,264] and the distribution of ubiquinone, ornithine-containing lipid, and PC were obtained from other works.[183-185,220,228] By DNA-ribosomal RNA hybridization, the genus *Brucella* belongs to the ribosomal-RNA superfamily IV which also includes *A. tumefaciens* and *R. leguminosarum*.[71,223]

[b] Glucosamine (GlcN) in the lipid A of members of the alpha-2 subdivision is present as beta 1,6-linked disaccharide molecule. It is not known whether in *Brucella* lipid A, GlcN is linked to 2,3-diaminoglucose (''mixed'' backbone), to another GlcN, or exists as monosaccharide backbone.

[c] Phosphorus (P) is expressed as percent of dry weight of material.

[d] Preliminary result within ( ).

[e] Unknown (?).

these molecules is not yet known, but they appear to be widely distributed among bacteria.[168,187-191,228,229]

E. CONCLUSIONS

Under the electron microscope, the *Brucella* cell appears similar to cells of *Enterobacteriaceae.*[230] However, a more careful analysis on the composition of the *Brucella* surface indicates that the cell envelope differs considerably. This difference is reflected mainly on the lipid composition: the presence of unusual wax-like esters, the unique ornithine-containing lipid, the presence of relative large quantities of PC, the "unusual" lipid A moiety of the LPS, a lipoprotein exposed at the cell surface, and the long-chain fatty acids substituting these molecules (including lactobacillic acid which is commonly absent from Gram negatives). These traits are the major characteristics which distinguish *Brucella* species from members of *Enterobacteriaceae.* In addition, there are other differentiating features, such as the lack of stability of LPS on the outer membrane due to the presence of divalent cations, the strong interactions among the outer membrane proteins with LPS and phospholipids, and the interactions of the LPS with bacterial polysaccharides. From a practical point of view these characteristics are not trivial, since conventional extraction procedures commonly used to isolate the cell envelopes and to characterize the macromolecules inserted in the layers of *E. coli* must be modified with respect to *Brucella* cells. It is important to take into consideration that the *Brucella* outer membrane is not affected by EDTA; therefore, production of lysozyme spheroplast in order to separate the inner and outer membrane is precluded in these bacteria.

Compared with *E. coli,* nonionic detergents are less efficient in extracting proteins from the cell envelopes of *Brucella.* Likewise, the anionic detergent SDS has little action on *Brucella* cell envelopes. Moreover, *Brucella* outer membrane group 2 proteins retain more fatty acids than *E. coli* OmpC when isolated by the same purification procedures.[78]

The fact that *Brucella* smooth LPS is extracted into the phenol phase rather than into the aqueous phase, using the Westphal method,[82] hampered to some extent the investigations on *Brucella* endotoxin and caused confusion with respect to its chemical characteristics and biological activities.[53,173,194] Several authors, knowing that *E. coli* and *Salmonella* lipopolysaccharides are extracted in the aqueous phase, looked only to this layer and misdefined the respective fraction as *Brucella* "endotoxin".[231-235] Furthermore, coextraction of the *Brucella* LPS with protein, phospholipids, polysaccharides, and free lipids, together with the difficulties encountered in purification of extracted LPS, have created controversy with respect to the chemical and biological characteristics of *Brucella* LPS.[16,53,194,199,205,211] In addition, the difficulties of extracting and purifying polysaccharide(s) from different *Brucella* species and strains have created confusion.[21,25,53,62]

Presently, the importance of the lipid composition of *Brucella* in the specific attachment of the bacteria to immune cells and in the intracellular parasitism is not known. The absorption of *Brucella* cells to B lymphocytes and macrophages might be related to the properties of the LPS and to the ornithine-containing lipids inserted in the outer membrane.[192] In addition, the presence of wax-like esters, rare neutral lipids, an "unusual" lipid A, as well as the high content of the uncommon bacterial PC might have some implications in the resistance of the cell envelope to the action of lysozomal enzymes[154] and in the adaptation to intracellular life.

The recent finding that the surface of *Brucella* cells concentrates a nucleotide-like substance which inhibits the fusion of lysosomes with phago-lysosomes in leukocytes[236] might be related to the lipid composition of the outer membrane of these bacteria.

The overall higher hydrophobicity of *Brucella* cell envelopes and the close association among the macromolecules of the outer membrane are the factors which have implication for the selective penetration of antibiotics. *Brucella* LPS and its lipid A does not bind polymyxin B. In addition, it has been shown that PC liposomes are resistant to the disruptive action of this antibiotic, while PE liposomes are not.[172] Probably these two features account for the greater resistance of *Brucella* to the action of polymyxin B.

Some antibiotics penetrate the Gram-negative cell wall porins.[180,181] Antibiotics with a molecular weight under 600 (e.g., chloramphenicol) usually penetrate the outer membrane through these pores. In *Brucella,* some of the tetracycline and chlorotetracycline antibiotics probably use this route of penetration.[158] However, for some large hydrophobic antibiotics which are acting on *Brucella,* such as erythromycin and rifampicin, it is very probable that the normal route of entrance is through the lipid bilayer. The relatively higher hydrophobicity of the long chain fatty acids (more than 16 carbons) found in the membrane of *Brucella* could present a favorable environment for the penetration of hydrophobic antibiotics. Similarly, the rather hydrophobic properties of the *Brucella* membrane may exclude the penetration of hydrophilic antibiotics with a molecular weight that hampers their penetration through pores.

The bound lipids, such as LPS, antigenic-related glycolipids, and the lipoprotein of *Brucella* are strongly recognized by the immune system. The recognition of these molecules are important in the diagnosis of brucellosis, and very probably in the protection against the disease. It is known that the "O"-polysaccharide chain, the core polysaccharide, and the lipid A backbone of *Brucella* LPS induce a strong antibody response.[173,215,216] Similarly, the lipoprotein is immunogenic, eliciting an antibody response which recognizes the portion of the molecule exposed to the cell surface.[98] It is not known whether the ornithine-containing lipids of *Brucella* are immunogenic. These lipids, which are characteristic of some bacteria, are absent in mammal cells. Therefore, it would not be surprising if they are readily recognized by the immune system.

In general terms, *Brucella* LPS is less biologically active than the LPS of *Enterobacteriaceae*. However, its activity on the endotoxin-resistant mice, its adjuvant effect and the induction of IgG antibodies, makes *Brucella* LPS an interesting molecule to study some aspects of the immune mechanisms and the relationships between the structure and function of endotoxic activity of lipid A.

The lipid composition of *Brucella* can be used for differentiating these bacteria from other Gram negatives. With the exception of *B. canis,* it is not easy to differentiate *Brucella* species by their fatty acid composition. In light of the recent proposition by Verger et al.[71] who proposed the existence of only a single species for the whole genus *Brucella,* the common fatty acid composition of the free lipids may support this hypothesis.

The recent findings, that the lipid A structure and the overall lipid composition of *Brucella* are shared by many members of the alpha-2 subdivision, open new aspects of the phylogenetic relationship of *Brucella* to other species which also live in close association with eukaryotic cells.*

IV. OTHER ANTIGENS OF *BRUCELLA*

A. INTRODUCTION

The preceding sections have dealt with the polysaccharide, protein, and lipid antigens of *Brucella* spp. There is little known about the other antigens of these bacteria, and the following is a brief overview of this topic.

* At the time of writing this article, the 16S RNA of *B. abortus* was sequenced (*Nucleic Acids Res.,* 17, 1765, 1989). It was found that *Brucella* is a member of the alpha-2 subdivision of *Proteobacteria* (purple group of Eubacteria). Within the alpha-2 subgroup, brucellae are specifically related to rickettsiae, agiobacteria, and rhizobiae, organisms that also have the faculty or the obligation of living in close association with eukaryotic cells.

B. NUCLEIC ACIDS

Although the chromosomes of *Brucella* spp. are highly homologous,[71] they differ from those of other bacteria such as *Francisella tularensis* and *E. coli*.[246] Nucleic acids of *B. abortus* have been investigated as immunogens, and it has been found that although the DNA fractions are antigenic[247] the RNA fractions do not appear to be so (false reactions may occur with the latter, for *B. abortus* RNA readily complexes with positively charged proteins, such as lysozyme, to form precipitates).[248] Therefore, although DNA fractions may be immunogenic, RNA does not appear to be so and, indeed, should be removed to avoid misleading results.

C. CELL WALLS

Following cell disruption and enzyme digestion, the cell walls of *B. abortus*[249] and *B. melitensis*[250] have been investigated for immunologic and biological activity. Positive results were observed, but the researchers conceded that contaminants (e.g., proteins, SLPS) may have played a role. However, it has been reported that the peptidoglycan of *E. coli* may initiate an autoimmune disease,[251,252] and autoantibodies have been reported in animals infected with *Brucella* spp.[253-255] Therefore, the peptidoglycan of *Brucella* may have some significance in the immunologic response of the host to the bacterium.

D. EXOTOXINS

With regard to antigens secreted into the exterior environment, brucellacin has been described.[256] Brucellacin is a bacteriocin, probably a protein, produced by one strain of *Brucella* which kills other strains. Also, a water-soluble component from *B. abortus* with bactericidal activity against Gram-positive bacteria has been reported.[257] Evidence of an extracellular protease which initiates or assists the cleavage of IgG1 in whey gives further support for the presence of exotoxins.[258] *Brucella* spp. do not appear to produce hemolysins, as judged by an absence of hemolysis when the bacteria are grown on blood agar.[33]

E. CONCLUSIONS

Although much has been learned about the polysaccharide, protein, and lipid antigens (as well as combinations of these compounds in the form of lipopolysaccharides, lipoproteins, and protein-lipopolysaccharide aggregates), for *Brucella* spp. little is known about the other components. Some studies have been done on their nucleic acids, peptidoglycan, and exotoxins as antigens, but more research is needed, particularly on the latter. The species of *Brucella* do not appear to produce exotoxins that would be of benefit in serodiagnostic tests, but possibly these extracellular components require unusual conditions for their expression (e.g., *Vibrio parahaemolyticus* requires suboptimal pH or amino acid analogues for the derepression of its hemolysins).[259,260]

ACKNOWLEDGMENTS

We thank Dr. Borowiak and Dr. S. Basu (Max-Planck-Institut für Immunbiologie, Freiburg, Federal Republic of Germany) for their help in the lipid A backbone analysis.

Edgardo Moreno received special support from the Universidad Nacional (Heredia, Costa Rica), was a recipient of Costa Rican National Council for Scientific and Technological Research Support Program to Researchers, and Research Fellow from the "Deutscher Akadademischer Austauschdienst" (DAAD), Federal Republic of Germany.

Appendix

PURIFICATION METHODS FOR POLYSACCHARIDES AND LIPOPOLYSACCHARIDES OF *BRUCELLA* SPP.*

ABBREVIATIONS

Acetate-saline	= 2% acetic acid (v/v), 10% sodium chloride (w/v) in water
Aqueous-phenol	= phenol liquified with water to yield 90% phenol (w/v)
DNase	= deoxyribonuclease
ELISA	= enzyme-linked immunosorbant assay
Methanol-acetate	= 1% sodium acetate (w/v) in methanol
Phenol-saline	= 2% phenol (w/v), 0.85% sodium chloride (w/v) in water
PS	= O-chain polysaccharide
Pyridinium-acetate	= 0.05 *M* pyridinium acetate, pH 4.7
	= 0.4% pyridine (v/v), 0.4% acetic acid (v/v) in water
RLPS	= rough lipopolysaccharide
	= core oligosaccharide + lipid A
RNase	= ribonuclease
SLPS	= smooth lipopolysaccharide
	= O-chain polysaccharide + core oligosaccharide + lipid A
TCA	= trichloroacetic acid
tris-saline	= 1% sodium chloride (w/v), 0.02% sodium azide (w/v), 0.01 *M* *tris*(hydroxymethyl)aminomethane, pH 7 with HCl

A. PURIFICATION OF *BRUCELLA ABORTUS* 45/20 RLPS

1. Grow *B. abortus* 45/20 on agar medium, then kill and suspend the cells in phenol-saline (100 g wet wt cells per 500 ml, 2 d at room temperature with occasional stirring). Store the suspension for 1 week at 4°C, during which time test sterility of the culture by plating a loopful onto trypticase soy agar at 37°C.
2. Add an equal volume of aqueous-phenol to the suspension and extract at 70°C for 30 min with constant stirring. Centrifuge the preparation at 20,000 × *g* at room temperature for 30 min, save the aqueous layer, and to it add 5 volumes of methanol-acetate (prechilled at 4°C). Then leave the preparation at 4°C overnight. Centrifuge this at 20,000 × *g* at 4°C for 30 min, discard the decanted supernatant, and resuspend the precipitate in an equivalent volume of methanol-acetate. Resuspension of the precipitate is facilitated using a spatula and then a Polytron® homogenizer (Brinkman Instruments, Rexdale, Ontario). Wash the precipitate twice more with methanol-acetate to remove phenol.
3. Suspend the precipitate in *tris*-saline, then dialyze (Spectrapor® dialysis tubing 1000 mol wt cutoff) twice (4°C for 24 h) against 100-fold volume of *tris*-saline.
4. With constant stirring, digest the preparation by the successive addition of:

   - 25 μg/ml each of RNase and DNase (after the addition of 5 m*M* $MgSO_4$), room temp for 1 h
   - 25 μg/ml lysozyme, room temp for 5 h
   - 50 μg/ml proteinase K, room temp for 48 h

Dialyze the resulting solution against *tris*-saline.

* By John W. Cherwonogrodzky, Nicholaus Rab, and John B. Stevens.

5. Centrifuge the preparation at 20,000 × *g* at 4°C for 30 min to remove any debris and then ultracentrifuge at 120,000 × *g* at 4°C for 18 h to yield a waxy pellet of RLPS. After resuspension of the pellet in *tris*-saline, ultracentrifuge again, dissolve the pellet in water, dialyze against water, and lyophilize.

Yield of RLPS is about 80 mg/100 g wet wt of cells.

B. PURIFICATION OF *BRUCELLA ABORTUS* 413 SLPS

1. Grow, kill, and suspend the cells of *B. abortus* 413 as noted previously for *B. abortus* 45/20. Centrifuge the culture at 20,000 × *g* at 4°C for 30 min and save the cells.
2. Remove contaminating lipids by washing the cells in water, ethanol, acetone, and ethyl ether, after which the cells are immediately dried under vacuum and ground to a fine powder with mortar and pestle. The cells are suspended in *tris*-saline (25 g dry wt of cells per 500 ml).
3. Extract the cells with hot phenol-water as noted for RLPS preparation. Save the phenol layer (N.B.: an equivalent amount of SLPS can be purified if the interphase and insoluble precipitate are pooled and treated as the phenol layer). Add 5 volumes of methanol-acetate (prechilled at 4°C) and leave the mixture overnight at 4°C. Wash the precipitate twice more with methanol-acetate.
4. Suspend and dialyze (Spectrapor® dialysis tubing 3500 m.w. cutoff) against *tris*-saline as noted previously for *B. abortus* 45/20 RLPS.
5. Digest the preparation with enzymes as noted previously for *B. abortus* 45/20 RLPS.
6. Centrifuge the mixture at 20,000 × *g* at 4°C for 30 min (N.B.: save precipitate. If SLPS is not present in the supernatant, this insoluble fraction should be resuspended in *tris*-saline and digested with 100 μg proteinase K at room temp for 48 h). Ultracentrifuge the supernatant at 120,000 × *g* at 4°C for 18 h to yield a waxy pellet of SLPS. After resuspension of the pellet in *tris*-saline, ultracentrifuge again, dissolve the pellet in water, dialyze against water, and lyophilize. If the material has greater than 1% protein, suspend it in *tris*-saline and repeat steps 5 and 6.

Yield of SLPS is about 300 mg/100 g wet wt of cells.

C. PURIFICATION OF *BRUCELLA MELITENSIS* 16M SLPS

1. Grow *B. melitensis* 16M on agar medium, then kill and suspend the cells as noted previously for *B. abortus* 45/20. Centrifuge the culture at 20,000 × *g* at 4°C for 30 min and save the supernatant.
2. To the culture supernatant add 5 volumes of methanol-acetate (prechilled to 4°C) and leave the mixture overnight at 4°C. Wash the precipitate twice more with methanol-acetate.
3. Suspend and dialyze (Spectrapor® dialysis tubing 3500 mol wt cutoff) against *tris*-saline as noted previously for *B. abortus* 45/20 RLPS.
4. Digest the preparation with enzymes as noted previously for *B. abortus* 45/20 RLPS.
5. Extract the preparation with hot phenol-water as noted for *B. abortus* 45/20 RLPS. Remove the phenol layer, add 5 volumes of methanol-acetate (prechilled at 4°C), and leave the mixture overnight at 4°C. Wash the precipitate twice more with methanol-acetate.
6. Centrifuge the preparation at 20,000 × *g* at 4°C for 30 min, resuspend, and dialyze the precipitate against *tris*-saline. Then centrifuge, ultracentrifuge, dialyze, and lyophilize as noted previously for *B. abortus* 413 SLPS.

Yield of SLPS is about 1.0 g from the supernatant of 60 g wet wt of cells.

D. PURIFICATION OF *BRUCELLA ABORTUS* 1119-3 PS (HIGH YIELD METHOD)

1. Grow *B. abortus* 1119-3 in liquid culture, then kill and suspend the cells as noted previously for *B. abortus* 45/20. Centrifuge the culture at 20,000 × *g* at 4°C for 30 min and save the cells.
2. Wash the cells once in acetate-saline, then resuspend the cells in this solution (100 g wet weight of cells per 500 ml). Pour the suspension into propylene centrifuge bottles (filled to $^2/_3$ capacity), tighten the lid, and autoclave the bottle at 15 psi at 121°C for 30 min. Let these stand until cooled to room temp (loosening the caps when the bottles begin to collapse), then pour the suspension into centrifuge tubes, centrifuge at 20,000 × *g* at 4°C for 30 min and save the supernatant. Resuspend the cells in 10% saline (100 ml for 100 g wet weight of cells), centrifuge as before, and combine the supernatants.
3. Neutralize the supernatant with NaOH, add 5 volumes of methanol-acetate (prechilled at 4°C) and leave overnight at 4°C.
4. Centrifuge the preparation at 20,000 × *g* at 4°C for 30 min. Resuspend the precipitate in a small volume of *tris*-saline and dialyze (Spectrapor® dialysis tubing 1000 mol wt cutoff), first against *tris*-saline with a total of 5% NaCl and then against *tris*-saline with no additional salt.
5. Treat the solution with enzymes as noted previously for the purification of *B. abortus* 45/20 RLPS.
6. Extract the preparation with hot phenol-water, centrifuge, save the phenol layer and precipitate, and wash the PS with methanol-acetate as noted for *B. abortus* 45/20 RLPS.
7. Dissolve the precipitate in Tris-saline and dialyze against the same buffer (twice against 100-fold volumes at 4°C over 2 d).
8. Ultracentrifuge the preparation at 120,000 × *g* at 4°C for 18 h to remove debris and contaminating LPS. Save the supernatant, dialyze the liquid against distilled water, and lyophilize. (Yield is about 700 mg of PS per 100 g wet wt of cells.)
9. Dissolve the white powder in pyridinium-acetate (350 mg PS per 3.5 ml) and fractionate the sample by gel filtration on a column of Sephadex® G-50. The PS was recovered from the void volume eluate, elution was monitored by a change of refractive index by an optical refractometer, and the fractions containing the PS were pooled and lyophilized.

Yield of PS was about 500 mg/100 g wet wt of cells.

E. PURIFICATION OF *BRUCELLA ABORTUS* 1119-3 PS (RAPID METHOD)

1. Grow, kill, and suspend *B. abortus* 1119-3 cells as noted in the high yield method.
2. Release the PS from the cells by mild acid hydrolysis as noted in the high yield method.
3. Add 2 *M* TCA to the liquid containing the PS until the final concentration is 0.2 *M* TCA. Chill the flask in icewater and centrifuge the preparation at 20,000 × *g* at 4°C for 30 min. Discard the precipitate and save the liquid.
4. Extract the liquid with an equal volume of aqueous-phenol and heat at 70°C for 30 min with constant stirring. Centrifuge the preparation at 20,000 × *g* at room temp for 30 min, save the phenol layer, and precipitate the PS with methanol-acetate (prechilled at 4°C). Immediately wash twice more with methanol-acetate.

5. Dissolve in and dialyze (Spectrapor® dialysis tubing 1000 mol wt cutoff) against distilled water and lyophilize.

Yield of PS is about 330 mg/100 g wet wt of cells. (N.B.: trace amounts of LPS may be present. 0.1% LPS in PS preparations [w/w] was found to interfere in the ELISA.)

REFERENCES

1. **Orcutt, M. L.,** Agglutination affinities of the *abortus-melitensis* group of bacteria with special reference to two human strains, *J. Exp. Med.,* 43, 225, 1926.
2. **Weigmann, F.,** Experimentelleuntersuchungen uber die Beziehungen zwischen *Bact. abortus* Bang und *Bact. melitense, Zentralbl. Bakteriol.,* 120, 179, 1931.
3. **Kristensen, M.,** Klassification danischer und underer Brucellastamme, *Zentralbl. Bakteriol.,* 120, 179, 1931.
4. **Evans, A. C.,** Further studies on *Bacterium abortus* and related bacteria. II. A comparison of *Bacterium abortus* and *Bacterium bronchisepticus* and with the organism which causes Malta fever, *J. Infect. Dis.,* 22, 580, 1918.
5. **Olitzki, L. and Bromberg, J.,** Verwendungstoffwechsel und Differentialdiagnose in der Bruzellagruppe, *Zentralbl. Bakteriol.,* 120, 347, 1931.
6. **Wilson, G. S. and Miles, A. A.,** The serological differentiation of smooth strains of the *Brucella* group, *Br. J. Exp. Pathol.,* 13, 1, 1932.
7. **Topping, L. E.,** Carbohydrate and nucleoprotein fractions isolated from the *Brucella* group, *J. Pathol. Bacteriol.,* 39, 665, 1934.
8. **Miles, A. A. and Pirie, N. W.,** The properties of antigenic preparations from *Brucella melitensis,* I. Chemical and physical properties of bacterial fractions, *Br. J. Exp. Pathol.,* 20, 83, 1939.
9. **Miles, A. A. and Pirie, N. W.,** The properties of antigenic preparations from *Brucella melitensis.* III. The biological properties of the antigen and the products of gentle hydrolysis, *Br. J. Exp. Pathol.,* 20, 278, 1939.
10. **Redfearn, M. S.,** *An Immunological Study of Antigens of Brucella Extracted by the Westphal Technique,* Ph.D. thesis, University of Wisconsin, Madison, 1960.
11. **Diaz, R., Jones, L. M., Leong, D., and Wilson, J. B.,** Surface antigens of smooth *Brucellae, J. Bacteriol.,* 96, 893, 1968.
12. **Corbel, M. J. and Philip, J. I. H.,** Relationship of *Brucella abortus* agglutinogenic antigens to the receptor sites for Tbilisi phage, *Res. Vet. Sci.,* 13, 91, 1972.
13. **Corbel, M. J.,** Recent advances in the study of *Brucella* antigens and their serological cross-reactions, *Vet. Bull.,* 55, 927, 1985.
14. **Hurvell, B. and Lindberg, A. A.,** Serological cross-reaction between different *Brucella* species and *Yersinia enterocolitica, Acta Pathol. Microbiol. Scand.,* 81B, 113, 1973.
15. **Luderitz, O., Westphal, O., Staub, A. M., and Nikaido, H.,** Isolation and chemical and immunological characterization of bacterial lipopolysaccharides, in *Microbial Toxins,* Weinbaum, G., Kadis, S., and Ajl, S. J., Eds., Academic Press, New York, 4, 145, 1971.
16. **Caroff, M., Bundle, D. R., Perry, M. B., Cherwonogrodzky, J. W., and Duncan, J. R.,** Antigenic S-type lipopolysaccharide of *Brucella abortus* 1119-3, *Infect. Immun.,* 46, 384, 1984.
17. **Caroff, M., Bundle, D. R., and Perry, M. B.,** Structure of the O-chain of the phenol-phase soluble cellular lipopolysaccharide of *Yersinia enterocolitica* serotype O:9, *Eur. J. Biochem.,* 139, 195, 1984.
18. **Bundle, D. R., Gidney, M. A. J., Perry, M. B., Duncan, J. R., and Cherwonogrodzky, J. W.,** Serological confirmation of *Brucella abortus* and *Yersinia enterocolitica* O:9 O-antigens by monoclonal antibodies, *Infect. Immun.,* 46, 389, 1984.
19. **Montaraz, J. A., Winter, A. J., Hunter, D. M., Sowa, B. A., Wu, A. M., and Adams, L. G.,** Protection against *Brucella abortus* in mice with O-polysaccharide-specific monoclonal antibodies, *Infect. Immun.,* 51, 961, 1984.
20. **L'vov, V. L., Mallkov, V. E., Shaskov, A. S., Dranovskaya, E. A., and Dmitriev, B. A.,** Somatic antigens of the genus *Brucella.* Structure of the O-specific polysaccharide chain of the lipopolysaccharide of *Brucella melitensis, Bioorgan. Khim.,* 11, 963, 1985.
21. **Moreno, E., Borowiak, D., and Mayer, H.,** *Brucella* lipopolysaccharides and polysaccharides, *Ann. Inst. Pasteur Microbiol.,* 138, 102, 1987.

22. **Bundle, D. R., Perry, M. B., and Cherwonogrodzky, J. W.,** The identification and structural elucidation of *Brucella melitensis* M antigen, in 13th Int. Symp. Carbohydrates, Cornell University, New York, 1986.
23. **Cherwonogrodzky, J. W., Perry, M. B., and Bundle, D. R.,** Purification and characterization of *Brucella melitensis* 16M lipopolysaccharide, in 86th Annu. Meet. Am. Soc. Microbiol., Washington, D.C., 1986, Abstr. B186.
24. **Perry, M. B., Bundle, D. R., and Cherwonogrodzky, J. W.,** The structures and serology of the A and M antigens of *Brucella abortus* and *Brucella melitensis* and the structure of polysaccharide B of *Brucella* species, in 14th Int. Congr. Microbiology, Manchester, 1986.
25. **Bundle, D. R., Cherwonogrodzky, J. W., Caroff, M., and Perry, M. B.,** The lipopolysaccharides of *Brucella abortus* and *B. melitensis, Ann. Inst. Pasteur Microbiol.,* 138, 92, 1987.
26. **Bundle, D. R., Cherwonogrodzky, J. W., and Perry, M. B.,** The structure of the lipopolysaccharide O-chain (M antigen) and polysaccharide B produced by *Brucella melitensis* 16M, *FEBS Lett.,* 216, 261, 1987.
27. **Bundle, D. R., Cherwonogrodzky, J. W., and Perry, M. B.,** Structural elucidation of the *Brucella melitensis* M antigen by high-resolution NMR at 500 MHz, *Biochemistry,* 26, 8717, 1987.
28. **Cherwonogrodzky, J. W., Perry, M. B., and Bundle, D. R.,** Identification of the A and M antigens of *Brucella* as the O-polysaccharides of smooth lipopolysaccharides, *Can. J. Microbiol.,* 33, 979, 1987.
29. **Bundle, D. R., Cherwonogrodzky, J. W., and Perry, M. B.,** Monoclonal antibodies in the identification and characterization of *Brucella* species, in *Impact of Biotechnology on Microbial Detection, Estimation and Characterization,* Swaminathan, B. and Prikash, G., Eds., Marcel Dekker, New York, 1989, 581.
30. **Bundle, D. R., Perry, M. B., Cherwonogrodzky, J. W., Duncan, J. R., and Caroff, M. G.,** Identification of *Brucella abortus* and brucellosis infection, U.S. Patent No. 1,212,051, 1986.
31. **Wright, P. F., Nielsen, K. H., and Cherwonogrodzky, J. W.,** Comparison of four antigens and two antiglobulins in the detection of bovine anti- *Brucella* antibody by enzyme-linked immunosorbent assay (ELISA), in 37th Annu. Meet. Can. Soc. Microbiol., Saskatoon, 1987, Abstr. VM 5.
32. **Peters, T. and Bundle, D. R.,** Synthesis of 4,6-dideoxy-4-formamido-α-D-mannose containing tri-, tetra-, and penta-saccharides, antigenic determinants of the *Brucella* A and M antigens, *J. Chem. Soc. Chem. Commun.,* p.1648, 1987.
33. **Corbel, M. J. and Brinley-Morgan, W. J.,** *Brucella,* in *Bergey's Manual of Systematic Bacteriology,* Krieg, N. R. and Holt, J. G., Eds., Williams & Wilkins, Baltimore, 1, 377, 1984.
34. **Rojas, X., Guillen, A., and Lopez, A.,** Comparison of serological tests in the detection of *Brucella melitensis* infections, in United Nations University Brucellosis Network Conference, Frank, J., Ed., Chicago, 1986, 18.
35. **Afzal, M., Tengerdy, R. P., Squire, P. G., and Ellis, R. P.,** Characterization of *Brucella ovis* lipopolysaccharide and its use for diagnosis of ram epididymitis by enzyme-linked immunosorbent assay, *J. Clin. Microbiol.,* 20, 1159, 1984.
36. **Afzal, M., Brodie, S. J., Tengerdy, R. P., and Squire, P. G.,** Isolation and antigenic reactivity of *Brucella ovis* outer membrane proteins, *J. Clin. Microbiol.,* 25, 2132, 1987.
37. **Diaz, R. and Jones, L. M.,** The immuno-diffusion method for the identification of cattle vaccinated with *Brucella abortus* strain 45/20, *Vet. Rec.,* 93, 300, 1973.
38. **Huhtala, E.,** L'eradication de la brucellose bovine en Finland, *Bull. Off. Int. Epiz.,* 60, 447, 1963.
39. **Winblad, S.,** Studies on serological typing of *Yersinia enterocolitica,* in 15th Scand. Congr. Pathol. Microbiol., Copenhagen, 1967, 115.
40. **Ahvonen, P. and Sievers, K.,** *Yersinia enterocolitica* infection associated with *Brucella* agglutinins, clinical features of 24 patients, *Acta Med. Scand.,* 185, 121, 1969.
41. **Ahvonen, P., Jansson, E., and Aho, K.,** Marked cross-agglutination between *Brucellae* and a subtype of *Yersinia enterocolitica, Acta Pathol. Microbiol. Scand.,* 75, 291, 1969.
42. **Diaz, R., Lacalle, R., Medrano, M. O., and Leong, D.,** Immunological activities of the endotoxin from *Yersinia enterocolitica* strain M.Y.79, in *Proc. 5th Int. Congr. Infect. Dis.,* Vienna, 2, 119, 1970.
43. **Corbel, M. J. and Cullen, G. A.,** Differentiation of the serological response to *Yersinia enterocolitica* and *Brucella abortus* in cattle, *J. Hyg.,* 68, 519, 1970.
44. **Kenner, L., Lindberg, B., Unger, P., Holme, T., and Holmgren, J.,** Structural studies of *Vibrio cholerae* antigen, *Carbohydr. Res.,* 68, C14, 1979.
45. **Redmond, J. W.,** The structure of the O-antigen side chain of the lipopolysaccharide of *Vibrio cholerae* 569B (INABA), *Biochim. Biophys. Acta,* 584, 346, 1979.
46. **Bundle, D. R., Gerken, M., and Perry, M. B.,** Two-dimensional nuclear magnetic resonance at 500 MHz: the structural elucidation of a *Salmonella* serogroup N polysaccharide antigen, *Can. J. Chem.,* 64, 255, 1986.
47. **Perry, M. B., Bundle, D. R., MacLean, L., Perry, J. A., and Griffith, D. W.,** The structure of the antigenic lipopolysaccharide O-chains produced by *Salmonella urbana* and *Salmonella godesberg, Carbohydr. Res.,* 156, 107, 1986.
48. **Perry, M. B., MacLeann, L., and Griffith, D. W.,** Structure of the O-chain polysaccharide of the phenol-phase soluble lipopolysaccharide of *Escherichia coli* O:157:H7, *Biochem. Cell. Biol.,* 64, 21, 1986.

49. **DiFabio, J. L., Perry, M. B., and Bundle, D. R.,** Analysis of the lipopolysaccharide of *Pseudomonas maltophilia* 555, *Biochem. Cell Biol.*, 65, 968, 1987.
50. **Diaz, R., Toyos, J., Salvo, M. D., and Pardo, M. L.,** A simple method for the extraction of polysaccharide B from *Brucella* cells for use in the radial immunodiffusion test diagnosis of bovine brucellosis, *Ann. Rech. Vet.*, 12, 35, 1981.
51. **Diaz, R., Garatea, P., Jones, L. M., and Moriyon, I.,** Radial immunodiffusion test with a *Brucella* polysaccharide antigen for differentiating infected from vaccinated cattle, *J. Clin. Microbiol.*, 10, 37, 1979.
52. **Jones, L. M., Berman, D. T., Moreno, E., Deyoe, B. L., Gilsdorf, M. J., Huber, J. D., and Nicoletti, P.,** Evaluation of a radial immunodiffusion test with polysaccharide B antigen for diagnosis of bovine brucellosis, *J. Clin. Microbiol.*, 12, 753, 1980.
53. **Moreno, E., Berman, D. T., and Boettcher, L. A.,** Biological activities of *Brucella abortus* lipopolysaccharides, *Infect. Immun.*, 31, 362, 1981.
54. **Moreno, E., Speth, S. L., Jones, L. M., and Berman, D. T.,** Immunochemical characterization of *Brucella* lipopolysaccharides and polysaccharides, *Infect. Immun.*, 31, 214, 1981.
55. **Perera, V. Y., Winter, A. J., and Ganem, B.,** Evidence for covalent bonding of native hapten-protein complexes to smooth lipopolysaccharide of *Brucella abortus, FEMS Microbiol. Lett.*, 21, 263, 1984.
56. **Bundle, D. R., Cherwonogrodzky, J. W., and Perry, M. B.,** Characterization of *Brucella* polysaccharide B, *Infect. Immun.*, 56, 1101, 1988.
57. **Amemura, A. and Cabrera-Crespo, J.,** Extracellular oligosaccharides and low-Mr polysaccharides containing (1 – >2)-β-*D*-glucosidic linkages from strains of *Xanthomonas, Escherichia coli* and *Klebsiella pneumoniae, J. Gen. Microbiol.*, 132, 2443, 1986.
58. **Cherwonogrodzky, J. W. and Nielsen, K. H.,** *Brucella abortus* 1119-3 O-chain polysaccharide to differentiate sera from *B. abortus* S-19-vaccinated and field-strain-infected cattle by agar gel immunodiffusion, *J. Clin. Microbiol.*, 26, 1120, 1988.
59. **L'vov, V. I., Pluzhnikova, G. N., Lapina, E. B., Shashkov, A. S., Askerova, S. A., Malikov, V. E., Dranovskaya, E. A., and Dmitriev, B. A.,** Molecular nature of the polysaccharide B antigen from *Brucella* (poly-B), *Bioorgan. Khim.*, 13, 4070, 1987.
60. **Hoffman, E. M. and Houle, J. J.,** Purification of nonlipopolysaccharide antigen from *Brucella abortus* during preparation of antigen used for indirect hemolysis test, *J. Clin. Microbiol.*, 24, 779, 1986.
61. **Zygmunt, M. S. and Dubray, G.,** Preparation by ultrafiltration and control by high-performance liquid chromatography of the native hapten of *Brucella abortus* for use in radial immunodiffusion diagnostic test, *J. Clin. Microbiol.*, 25, 1860, 1987.
62. **Moreno, E., Mayer, H., and Moriyon, I.,** Characterization of a native polysaccharide hapten from *Brucella melitensis, Infect. Immun.*, 55, 2850, 1987.
63. **Zygmunt, M. S., Dubray, G., Bundle, D. R., and Perry, M. B.,** Purified native hapten of *Brucella abortus* B-19 and *Brucella melitensis* 16M reveal the lipopolysaccharide origin of these antigens, *Ann. Inst. Pasteur Microbiol.*, in press.
64. **Gotschlich, E. C., Fraser, B. A., Nishimura, O., Robbins, J., and Liu, T.-Y.,** Lipid on capsular polysaccharides of gram-negative bacteria, *J. Biol. Chem.*, 256, 8915, 1981.
65. **Schmidt, M. A. and Jann, K.,** Phospholipid substitution of capsular (K) polysaccharide antigens from *Escherichia coli* causing extraintestinal infections, *FEMS Microbiol. Lett.*, 14, 69, 1982.
66. **Oberti, J., Caravano, R., and Roux, J.,** Demonstration of an external layer in several species of the genus *Brucella, Can. J. Microbiol.*, 28, 1300, 1982.
67. **Cherwonogrodzky, J. W., Nielsen, K. H., Duncan, J. R., Wright, P. F., Bundle, D. R., and Perry, M. B.,** Immunoassays for discriminating between brucellosis infections and vaccination, Canadian Patent Appl. No. 519243, 1986.
68. **Schalch, W., Wright, J. K., Rodkey, L. S., and Braun, D. G.,** Distinct functions of monoclonal IgG antibody depend on antigen-site specificities, *J. Exp. Med.*, 149, 923, 1979.
69. **Nielsen, K. H., Cherwonogrodzky, J. W., Duncan, J. R., and Bundle, D. R.,** Differentiation of the antibody response of cattle naturally infected with *Brucella abortus* or vaccinated with *B. abortus* strain 19 by enzyme immunoassay, *Am. J. Vet. Res.*, in press.
70. **Schurig, G. G., Pringle, A. T., and Breese, S. S.,** Localization of *Brucella* antigens that elicit a humoral immune response in *Brucella abortus* infected cattle, *Infect. Immun.*, 34, 1000, 1981.
71. **Verger, J.-M., Grimont, F., Grimont, P. A. D., and Grayon, M.,** *Brucella,* a monospecific genus as shown by deoxyribonucleic acid hybridization, *Int. J. System. Bacteriol.*, 35, 292, 1985.
72. **Fraser, D. E. and Rigby, C. E.,** Conjugation-mediated genetic exchange in *Brucella abortus,* in 36th Annu. Meet. Can. Soc. Microbiol., Toronto, 1986, Abstr. VM4p.
73. **Olitzki, A.,** Immunological methods in brucellosis research. I. In vitro procedures. *The Antigenic Structure of the Brucellae,* S. Karger, Basel, 1970, 1.
74. **Dubray, G.,** Localisation cellulaire des polysides des bacteries des genres *Brucella* et *Escherichia* en phase lisse (S) ou rugueuse (R), *Ann. Microbiol. (Paris),* 127, 133, 1976.

75. **Dubray, G. and Charriaut, C.,** Evidence of three major polypeptide species and two major polysaccharide species in the *Brucella* outer membrane, *Ann. Rech. Vet.*, 14, 311, 1983.
76. **Moriyon, I. and Berman, D. T.,** Isolation, purification, and partial characterization of *Brucella abortus* matrix protein, *Infect. Immun.*, 39, 394, 1983.
77. **Verstreate, D. R., Creasy, M. T., Caveney, N. T., Baldwin, C. L., Blab, M. W., and Winter, A. J.,** Outer membrane proteins of *Brucella abortus:* isolation and characterization, *Infect. Immun.*, 35, 979, 1982.
78. **Moriyon, I. and Berman, D. T.,** Effects of nonionic, ionic, and dipolar ionic detergents and EDTA on the *Brucella* cell envelope, *J. Bacteriol.*, 152, 822, 1982.
79. **Dubray, G. and Bezard, G.,** Isolation of three cell-wall antigens protective in murine experimental brucellosis, *Ann. Rech. Vet.*, 11, 367, 1980.
80. **Verstreate, D. R. and Winter, A. J.,** Comparison of sodium dodecyl sulfate-polyacrylamide gel electrophoresis profiles and antigenic relatedness among outer membrane proteins of 49 *Brucella abortus* strains, *Infect. Immun.*, 46, 182, 1984.
81. **Gomez-Miguel, M. J. and Moriyon, I.,** Demonstration of a peptidoglycan-linked lipoprotein and characterization of its trypsin fragment in the outer membrane of *Brucella* spp., *Infect. Immun.*, 53, 678, 1986.
82. **Moreno, E., Pitt, M. W., Jones, L. M., Schurig, G. G., and Berman, D. T.,** Purification and characterization of smooth and rough polysaccharides from *Brucella abortus, J. Bacteriol.*, 138, 361, 1979.
83. **Dubray, G.,** Protective antigens in brucellosis, *Ann. Inst. Pasteur Microbiol.*, 138, 84, 1987.
84. **Roux, J.,** Les vaccinations dans les Brucelloses humaines et animales, *Bull. Inst. Pasteur,* 70, 145, 1972.
85. **Rasooly, G., Olitzki, A. L., and Sulitzeanu, D.,** Immunization against *Brucella* with killed vaccines. Immunizing activity in mice of *Brucella* cell-wall and of fraction derived from them, *Isr. J. Med. Sci.*, 2, 569, 1966.
86. **Lopez-Merino, A., Asselineau, J., Serre, A., Roux, J., Bascoul, S., and Lacave, C.,** Immunization by an insoluble fraction extracted from *Brucella melitensis:* immunochemical characterization of the active substances, *Infect. Immun.*, 13, 311, 1976.
87. **Roux, J., Asselineau, J., Serre, A., and Lacave, C.,** Proprietes immunologiques d'un extrait phenol-insoluble de *Brucella melitensis* (fraction P.I.), *Ann. Inst. Pasteur,* 113, 411, 1967.
88. **Winter, A. J.,** Outer membrane proteins of *Brucella, Ann. Inst. Pasteur Microbiol.*, 138, 87, 1987.
89. **Santos, J. M., Verstreate, D. R., Perera, V. Y., and Winter, A. J.,** Outer membrane proteins from rough strains of four *Brucella* species, *Infect. Immun.*, 46, 188, 1984.
90. **Gomez-Miguel, M. J., Moriyon, I., and Lopez, J.,** *Brucella* outer membrane lipoprotein shares antigenic determinants with *Escherichia coli* Braun lipoprotein and is exposed on the cell surface, *Infect. Immun.*, 55, 258, 1987.
91. **Serre, A., Bascoul, S., Vendrell, J. P., and Cannat, A.,** Human immune response to *Brucella* infection, *Ann. Inst. Pasteur Microbiol.*, 138, 113, 1987.
92. **Plommet, M.,** Brucellosis and immunity: humoral and cellular components in mice, *Ann. Inst. Pasteur Microbiol.*, 138, 105, 1987.
93. **Limet, J., Plommet, A.-M., Dubray, G., and Plommet, M.,** Immunity conferred upon mice by anti-LPS monoclonal antibodies in murine brucellosis, *Ann. Inst. Pasteur Immunol.*, 138, 417, 1987.
94. **Pavlov, H., Hogarth, M., McKenzie, I. F. C., and Cheers, C.,** In vivo and in vitro effects of monoclonal antibody to Ly antigens on immunity to infection, *Cell. Immunol.*, 71, 127, 1982.
95. **Baldwin, C. L., Antczak, D. F., and Winter, A. J.,** Evaluation of lymphocyte blastogenesis for diagnosis of bovine brucellosis, *Dev. Biol. Stand.*, 56, 357, 1984.
96. **Baldwin, C. L., Verstreate, D. R., and Winter, A. J.,** Immune response of cattle to *Brucella abortus* outer membrane proteins measured by lymphocyte blastogenesis, *Vet. Immunol. Immunopathol.*, 9, 383, 1985.
97. **Bertrand, A., Siffert, M., Bascoul, S., and Mathieu-Daude, J. C.,** Interet pratique de l'exploration de l'immunite cellulaire au cours de la brucellose humaine, *Dev. Biol. Stand.*, 56, 537, 1984.
98. **Schurig, G. G., Jones, L. M., Speth, S. L., and Berman, D. T.,** Antibody response to antigens distinct from smooth lipopolysaccharide complex in *Brucella* infection, *Infect. Immun.*, 21, 994, 1978.
99. **Corbel, M. J. and Redwood, D. W.,** Properties of an immunogenic fraction from *Brucella abortus* 45/20, *Dev. Biol. Stand.*, 56, 159, 1984.
100. **Raybould, T. J. G. and Chantler, S.,** Serological differentiation between infected and vaccinated cattle by using purified soluble antigens from *Brucella abortus* in a hemagglutination system, *Infect. Immun.*, 29, 435, 1980.
101. **Raybould, T. J. G., Beesley, J. E., and Chantler, S.,** Ultrastructural localization of characterized antigens of *Brucella abortus* and distribution among different biotypes, *Infect. Immun.*, 32, 318, 1981.
102. **Tabatabai, L. B. and Deyoe, B. L.,** Characterization of salt-extractable protein antigens from *Brucella abortus* by crossed-immunoelectrophoresis and isoelectrofocusing, *Vet. Microbiol.*, 9, 549, 1984.
103. **Tabatabai, L. B. and Deyoe, B. L.,** Biochemical and biological properties of soluble protein preparations from *Brucella abortus, Dev. Biol. Stand.*, 56, 199, 1984.

104. **Santisteban, C. G., Parma, A. E., Cerone, S. I., and Bowden, R. A.,** Detection of antigenic fractions from *Brucella abortus* S-45/20 which bind to non-agglutinating antibodies using electroblotting and enzyme-linked antibody probes, *Vet. Microbiol.*, 12, 353, 1986.
105. **Atal, P. R., Kulshrestha, R. C., and Wahl, P. N.,** Immuno-electrophoretic characterization of *Brucella abortus* antigenic strain 99. II. Chemical extracts, *Indian J. Med. Res.*, 58, 687, 1970.
106. **Freeman, B. A., McGhee, J. R., and Baughn, R. E.,** Some physical, chemical and taxonomic features of the soluble antigens of the *Brucellae, J. Infect. Dis.*, 121, 522, 1970.
107. **Kulshrestha, R. C., Atal, P. R., and Wahi, P. N.,** Immuno-electrophoretic characterization of *Brucella* species: chemical extract, *Indian J. Med. Res.*, 58, 1662, 1970.
108. **Stone, S. S., Patterson, J. M., and Deyoe, B. L.,** Extraction, separation, and partial characterization of *Brucella abortus* antigens, *Am. J. Vet. Res.*, 43, 149, 1982.
109. **Raybould, T. J. G.,** Antigens of diagnostic significance in *Brucella abortus, Can. J. Microbiol.*, 28, 557, 1982.
110. **Stemshorn, B. and Nielsen, K.,** The bovine immune response to *Brucella abortus*. I. A water soluble antigen precipitated by sera of some naturally infected cattle, *Can. J. Comp. Med.*, 41, 152, 1977.
111. **Stemshorn, B. and Nielsen, K.,** The bovine immune response to *Brucella abortus*. IV. Studies with a double immunodiffusion test for antibody against antigen A2, *Can. J. Comp. Med.*, 45, 147, 1981.
112. **Stemshorn, B., Nielsen, K., and Samagh, B.,** The bovine immune response to *Brucella abortus*. III. Preparation of antisera against a *Brucella* component precipitated by sera of some infected cattle, *Can. J. Comp. Med.*, 45, 77, 1981.
113. **Diaz, R., Jones, L. M., Leong, D., and Wilson, J. B.,** Antigenic relation of *Brucella ovis* and *Brucella melitensis, J. Bacteriol.*, 93, 1262, 1967.
114. **Diaz, R., Maravi-Poma, E., and Rivero, A.,** Comparison of counter-immunoelectrophoresis with other serological tests in the diagnosis of human brucellosis, *Bull. W.H.O.*, 53, 417, 1976.
115. **Laudat, P., Audurier, A., Dubray, G., De Gialluly, C., and Loulergue, J.,** Contre immunoelectrophorese a l'antigene Brucelline dans le diagnostic serologique de la brucellose, *Dev. Biol. Stand.*, 56, 447, 1984.
116. **Muhamed, S. I., Mohammadi, H., and Saadi-Nam, H.,** A comparison of counter-immunoelectrophoresis with the Rose-Bengal and the serum tube agglutination tests in the diagnosis of brucellosis in sheep, *Vet. Microbiol.*, 5, 223, 1980.
117. **Schurig, G. G.,** The immune response of goats vaccinated with low and high doses of *Brucella melitensis* Rev. 1, *Vet. Immunol. Immunopathol.*, 3, 311, 1982.
118. **Stemshorn, B. W.,** Recent progress in the diagnosis of brucellosis, *Dev. Biol. Stand.*, 56, 325, 1984.
119. **Alton, G. G., Jones, L. M., and Pietz, D. E.,** Laboratory techniques in brucellosis, 2nd ed., World Health Organization, Geneva, 1975.
120. **Menasse, I., Skoulas, E., Seimenis, A., and Skyrianos, G.,** Essai sur le mouton d'un nouvel allergene pour le depistage de la brucellose, *Dev. Biol. Stand.*, 31, 171, 1976.
121. **Olitzki, A.,** Allergic reactions in brucellosis, in *Immunological Methods in Brucellosis Research*. II. *In Vivo Procedures,* S. Karger, Basel, 68, 1970.
122. **Bhongbhibhat, N., Elberg, S. S., and Chen, T. H.,** Characterization of *Brucella* skin-test antigens, *J. Infect. Dis.*, 122, 70, 1970.
123. **Cunningham, B., Miler, J. J., Dolan, L., McKeon, F., and O'Meara, M.,** Immunological characteristics in cattle of allergens derived from smooth *Brucella abortus* S99, *Vet. Rec.*, 107, 369, 1980.
124. **Chukwu, C. C.,** Evaluation of brucellin skin test for bovine brucellosis, *Int. J. Zoon.*, 2, 6, 1985.
125. **Fensterbank, R. and Dubray, G.,** Brucellin, an allergen used in diagnosing Brucellosis, *World Health Org.*, 3. 1980.
126. **Jones, L. M., Diaz, R., and Taylor, A. G.,** Characterization of allergens prepared from smooth and rough strains of *Brucella melitensis, Br. J. Exp. Pathol.*, 54, 492, 1973.
127. **Loquerie, R. and Durand, M. P.,** Applications cliniques d'un test cutane d'hypersensibilite retardee pour depistage de la brucellose ovine et caprine au moyen d'un allergene commercial, *Dev. Biol. Stand.*, 56, 407, 1984.
128. **MacDiarmid, S. C. and Hellstrom, J. S.,** An intradermal test for the diagnosis of brucellosis in extensively managed cattle herds, *Prev. Vet. Med.*, 4, 361, 1987.
129. **Nicoletti, P.,** The use of a *Brucella* protein antigen in dermal hypersensitivity as an adjunct method to diagnose bovine brucellosis, *Vet. Immunol. Immunopathol.*, 5, 27, 1983.
130. **Sutherland, A. W.,** Assessment of an intradermal test for the detection of bovine brucellosis, *Vet. Res. Commun.*, 6, 297, 1983.
131. **Fensterbank, R.,** Le diagnostic allergique des Brucelloses animales, *Dev. Biol. Stand.*, 56, 401, 1984.
132. **Renoux, M., Palat, A., Guillaumin, J. M., and Renoux, G.,** Hemagglutination, passive, transformation lymphoblastique et migration des leucocytes appliquees au diagnostic des Brucelloses, *Dev. Biol. Stand.*, 31, 145, 1976.

133. **Kaneene, J. M., Johnson, D. W., Anderson, R. K., and Muscoplat, C. C.,** Utilization of a specific in vitro lymphocyte immunostimulation assay as an aid in detection of *Brucella* infected cattle not detected by serological tests, *J. Clin. Microbiol.*, 8, 512, 1978.
134. **Berman, D. T., Wilson, B. L., Moreno, E., Angus, R. D., and Jones, L. M.,** Characterization of *Brucella abortus* soluble antigen employed in immunoassay, *J. Clin. Microbiol.*, 11, 355, 1980.
135. **Kaneene, J. M., Anderson, R. K., Johnson, D. W., and Muscoplat, C. C.,** *Brucella* antigen preparation for in vitro lymphocyte immunostimulation assays in bovine brucellosis, *Infect. Immun.*, 22, 486, 1978.
136. **Kaneene, J. M., Johnson, D. W., Anderson, R. K., Angus, R. D., Pietz, D. E., and Muscoplat, C.,** Specific lymphocyte stimulation in cattle naturally infected with strains of *Brucella abortus* and cattle vaccinated with *Brucella abortus* strain 19, *Am. J. Vet. Res.*, 39, 585, 1978.
137. **Schultz, R. D.,** The role of cell mediated immunity in infectious diseases of cattle, in *The Ruminant Immune System,* Butler, J. E., Ed., Plenum Press, New York, 1981, 57.
138. **Rasanen, L., Lehto, M., Jokinen, I., and Leinikki, P.,** Polyclonal antibody formation of human lymphocytes to bacterial components, *Immunology,* 58, 577, 1986.
139. **Chukwu, C. C.,** Differentiation of *Brucella abortus* and *Yersinia enterocolitica* serotype O:9 infections: use of lymphocyte transformation test, *Int. J. Zoon.*, 12, 126, 1985.
140. **Diaz, R.,** Estudio de las relaciones antigenicas entre *Yersinia enterocolitica* serotip 9 y otras especies bacterianas gram-negativas, *Microbiol. Esp.*, 27, 1, 1974.
141. **Berman, D. T.,** The importance of being earnest with your antigens, in *The Ruminant Immune System,* Butler, J. E., Ed., Plenum Press, New York, 1981, 217.
142. **Dubray, G.,** Progres recents sur la biochimie et les proprietes biologiques des antigenes de *Brucella, Dev. Biol. Stand.*, 56, 131, 1984.
143. **Bruce, W. and Jones, L. M.,** The use of gel diffusion precipitin plates in the study of *Brucella, Bull. Org. Mond. Sante,* 19(1), 187, 1958.
144. **Glenchur, H., Seal, U. S., Zinneman, H. H., and Hall, W. H.,** Serum precipitins in human and experimental brucellosis, *J. Lab. Clin. Med.*, 59, 220, 1962.
145. **Nagy, L. K.,** Studies on the precipitin content of bovine sera prepared against various biotypes of *Brucella abortus, Immunology,* 12, 463, 1967.
146. **Corbel, M. J.,** Evaluation of an immunodiffusion test for the detection of antibodies to *Brucella abortus* in bovine serum, *J. Med. Microbiol.*, 6, 67, 1973.
147. **Hinsdill, R. D. and Berman, D. T.,** Antigens of *Brucella* abortus. Chemical and immunoelectrophoretic characterization, *J. Bacteriol.*, 93, 544, 1967.
148. **Gaumont, R. and Thorel, M. F.,** Les antigens solubles des *Brucella, Symp. Ser. Immunobiol. Stand.*, 12, 287, 1968.
149. **Corbel, M. J.,** Characterization of antibodies active in the Rose-Bengal plate test, *Vet. Rec.*, 90, 484, 1972.
150. **McMahon, K. J., Renner, E. D., Allmaras, G. W., and Anderson, D. K.,** An agar-gel immunodiffusion test for detection of *Brucella* antibodies in human serum, *Can. J. Microbiol.*, 25, 850, 1979.
151. **Dubray, G. and Plommet, M.,** Structure et constituants des *Brucella.* Caracterization des fractions et proprietes biologiques, *Dev. Biol. Stand.*, 31, 68, 1976.
152. **Moriyon, I., Gamazo, C., and Diaz, R.,** Properties of the outer membrane of *Brucella, Ann. Inst. Pasteur Microbiol.*, 138, 89, 1987.
153. **Winter, A. J.,** Outer membrane proteins of *Brucella, Ann. Inst. Pasteur Microbiol.*, 138, 87, 1987.
154. **Kreutzer, D. L. and Robertson, D. C.,** Surface macromolecules and virulence in intracellular parasitism: comparison of cell envelope components of smooth and rough strains of *Brucella abortus, Infect. Immun.*, 23, 819, 1979.
155. **Smith, H.,** Microbial surfaces in relation to pathogenesis, *Bacteriol. Rev.*, 41, 475, 1977.
156. **Kreutzer, D. L., Buller, C. S., and Robertson, D. C.,** Chemical characterization and biological properties of lipopolysaccharides isolated from smooth and rough strains of *Brucella abortus, Infect. Immun.*, 23, 811, 1979.
157. **Young, E. J., Borchert, M., Kretzerf, L., and Musher, D. M.,** Pagocytosis and killing of *Brucella* by human polymorphonuclear leukocyte, *J. Infect. Dis.*, 151, 682, 1985.
158. **Douglas, J. T., Rosenberg, E. Y., Nikaido, H., Verstreate, D. R., and Winter, A. T.,** Porins of *Brucella* species, *Infect. Immun.*, 44, 16, 1984.
159. **Moriyon, I. and Berman, D. T.,** Interaction of *Escherichia coli* matrix protein with *Brucella abortus* peptidoglycan and chitin, *Dev. Biol. Stand.*, 56, 227, 1984.
160. **Moriyon, I. and Montanes, M. A.,** In vitro interactions between lipopolysaccharides and heterologous outer membrane proteins, *Curr. Microbiol.*, 12, 229, 1985.
161. **Dess, S. B., Hollis, D. G., Weaver, R. E., and Moss, W.,** Cellular fatty acids of *Brucella canis* and *Brucella suis, J. Clin. Microbiol.*, 14, 111, 1981.
162. **Dess, S. B., Thanabalasundrum, S., Moss, C. W., Hollis, D. G., and Weaver, R. E.,** Cellular fatty acid composition of group IV, nonsaccharolytic organisms from clinical sources, *J. Clin. Microbiol.*, 11, 664, 1980.

163. **Gamazo, C. and Moriyon, I.,** Release of outer membrane fragments by exponential growing *Brucella melitensis* cells, *Infect, Immun.*, 55, 609, 1987.
164. **Jantzen, E., Knudsen, E., and Winsnes, R.,** Fatty acid analysis for differentiation of *Bordetella* and *Brucella* species, *Acta Pathol. Microbiol. Scand. B,* 90, 353, 1982.
165. **Mayer, H., Moreno, E., and Weckesser, J.,** Structure of lipid A from photosynthetic and phylogenetically related bacteria, *J. Immunol. Immunopharmacol.*, 6, 35, 1986.
166. **Thiele, O. W. and Keher, W.,** Die frein Lipide aus *Brucella abortus* Bang. Uber die Neutrallipide, *Eur. J. Biochem.*, 9, 167, 1969.
167. **Thiele, O. W., Lacave, C., and Asselineau, J.,** On the fatty acids of *Brucella abortus* and *Brucella melitensis, Eur. J. Biochem.*, 7, 393, 1969.
168. **Thiele, O. W. and Schwinn, G.,** The free lipids of *Brucella melitensis and Bordetella pertusis, Eur. J. Biochem.*, 34, 333, 1973.
169. **Lacave, C. H. and Roux, J.,** Composition chimique des parois de *Brucella,* Glycosaminopeptide de *Brucella abortus* et *Brucella melitensis, C. R. Acad. Sci.*, 260, 1514, 1965.
170. **Mardarowicz, C.,** Isolierung und Charakterisierung das Murein-Sacculus von *Brucella, Z. Naturforsch.*, 21B, 1006, 1966.
171. **Caballero-Castillo, M.,** Isolation of *Brucella abortus* 45/20 peptidoglycan, M.Sc. thesis, University of Wisconsin, Madison, 1984.
172. **Feingold, D. S., Hsuchen, C. C., and Sud, I. J.,** Basis for the selectivity of the action of polymyxin antibiotics on cell membranes, *Ann. N.Y. Acad. Sci.*, 235, 480, 1974.
173. **Moreno, E.,** Caracteristicas y actividades biologicas del lipopolisacarido de *Brucella, Adel. Microbiol. Enf. Infec. (Argentina),* 2, 21, 1983.
174. **Goldfine, N.,** Bacterial membranes and the lipid packing theory, *J. Lipid Res.*, 25, 1501, 1984.
175. **Cullis, P. R. and de-Kruijff, B.,** Lipid polymorphism and functional roles of lipids in biological membranes, *Biochim. Biophys. Acta,* 559, 399, 1979.
176. **Shinitzky, M. and Hankart, P.,** Fluidity of cell membranes: current concepts and trends, *Int. Rev. Cytol.*, 60, 121, 1979.
177. **Hirata, F. and Axelrod, J.,** Phospholipid methylation and biological signal transmission, *Science,* 209, 186, 1980.
178. **Law, J. H., Zalkin, H., and Kaneshiro, T.,** Transmethylation reactions in bacterial lipids, *Biochim. Biophys. Acta,* 70, 143, 1965.
179. **Labischinski, H., Barnickel, G., Bradaczek, H., Naumann, D., Rietschel, E. T., and Giesbrecht, P.,** High state of order of isolated bacterial lipopolysaccharide and its possible contribution to the permeation barrier property of the outer membrane, *J. Bacteriol.*, 162, 9, 1985.
180. **Nikaido, H.,** Outer membrane of *Salmonella typhimurium.* Transmembrane diffusion of some hydrophobic substances, *Biochim. Biophys. Acta,* 433, 118, 1976.
181. **Nakaido, H. and Vaara, M.,** Molecular basis of bacterial outer membrane permeability, *Microbiol. Rev.*, 49, 1, 1985.
182. **Short, S. A. and White, D. C.,** Biosynthesis of cardiolipin from phosphatidylglycerol in *Staphylococcus aureus, J. Bacteriol.*, 109, 820, 1972.
183. **Shaw, N.,** Lipid composition as a guide to the classification of bacteria, *Adv. Appl. Microbiol.*, 17, 63, 1974.
184. **Collins, M. and Jones, D.,** Distribution of isoprenoid quinone structural types in bacteria and their taxonomic implications, *Microbiol. Rev.*, 45, 416, 1981.
185. **Mannheim, W., Stieler, W., Wolf, G., and Zabel, R.,** Taxonomic significance of respiratory quinones and fumarate respiration in *Actinobacillus* and *Pasteurella, Int. J. Syst. Bacteriol.*, 28, 7, 1978.
186. **Thiele, O. W., Biswas, C. J., and Hunneman, D. H.,** Isolation and characterization of an ornithine-containing lipid from *Paracoccus denitrificans, Eur. J. Biochem.*, 105, 267, 1980.
187. **Makula, R. A. and Finnerty, W. R.,** Isolation and characterization of an ornithine-containing lipid of *Desulfovibrio gigas, J. Bacteriol.*, 123, 523, 1975.
188. **Wilkinson, S. G.,** Composition and structure of the ornithine-containing lipid from *Pseudomonas rubescens, Biochim. Biophys. Acta,* 270, 1, 1972.
189. **Thiele, O. W., Dreysel, J., and Hermann, J. D.,** The free lipids of two different strains of hydrogen-oxidizing bacteria in relation to their growth, *Eur. J. Biochem.*, 29, 224, 1972.
190. **Kawai, Y. and Yano, I.,** Ornithine-containing lipid of *Bordetella pertusis,* a new type of hemagglutinin, *Eur. J. Biochem.*, 136, 531, 1983.
191. **Kawai, Y., Suzuki, K., and Hagiwara, T.,** Phosphatidylserine and ornithine-containing lipids of *Bordetella.* Hemagglutinins of lipoamino acid structure, and their control in biomembranes, *Eur. J. Biochem.*, 147, 367, 1985.
192. **Rasanen, L.,** Adherence of bacteria to human lymphocyte subpopulations and the role of monosaccharides in bacterial binding, *Cell. Immunol.*, 58, 19, 1981.

193. **Westphal, O. and Jann, K.,** Bacterial lipopolysaccharides: extraction with phenol-water and further applications of the procedure, in *Methods in Carbohydrate Chemistry,* Vol. 5, Whistler, R. L., Miller, J. N. B., and Wolfrom, M. L., Eds., Academic Press, New York, 1965, 83.
194. **Berman, D. T. and Kurtz, R. S.,** Relationship of biological activities to structures of *Brucella abortus* endotoxin and LPS, *Ann. Inst. Pasteur Microbiol.,* 138, 98, 1987.
195. **Kurtz, R. S. and Berman, D. T.,** Cellular interactions in isotype expression induced by *Brucella abortus* LPS, *Fed. Proc.,* 43 (Abstr. 2886), 1911, 1984.
196. **Moreno, E., Jones, L. M., and Berman, D. T.,** Immunochemical characterization of rough *Brucella* lipopolysaccharides, *Infect. Immun.,* 43, 779, 1984.
197. **Sowa, B. A., Crawford, R. P., Heck, F. C., Williams, J. D., Wu, A. M., Kelly, D. A., and Adams, L. G.,** Size, charge and structural heterogeneity of *Brucella abortus* lipopolysaccharide demonstrated by two-dimensional gel electrophoresis, *Electrophoresis,* 7, 823, 1986.
198. **Lugtemberg, B. and Van-Alphen, L.,** Molecular architecture and functioning of the outer membrane of *Escherichia coli* and other gram-negative bacteria, *Biochim. Biophys. Acta,* 737, 51, 1983.
199. **Marx, A., Ionescu, J., and Pop, A.,** Immunochemical studies on *Brucella abortus* lipopolysaccharides, *Zentralbl. Bakteriol. I. Abt. Orig.,* 253, 544, 1983.
200. **Osborn, M. J., Gandner, J. E., Parisi, E., and Carson, J.,** Mechanism of assembly of the outer membrane of *Salmonella typhimurium.* Isolation and characterization of the cytoplasmic and outer membrane, *J. Biol. Chem.,* 247, 3962, 1972.
201. **Schindler, M. J. and Osborn, J.,** Interaction of divalent cations and polymyxin B with lipopolysaccharide, *Biochemistry,* 18, 369, 1979.
202. **Leong, D., Diaz, R., Milner, K., Rudbach, J., and Wilson, J. B.,** Some structural and biological properties of *Brucella* endotoxin, *Infect. Immun.,* 1, 174, 1970.
203. **Vaara, M.,** Increased outer membrane resistance to ethylenediaminetetraacetate and cations in novel lipid A mutants, *J. Bacteriol.,* 148, 426, 1981.
204. **Schindler, H. and Rosenbusch, J. P.,** Matrix protein in planar membranes: clusters of channels in native environment and their functional reassembly, *PNAS,* 78, 2302, 1981.
205. **Mayer, H. and Weckesser, J.,** "Unusual" lipid As: structures, taxonomical relevance, and potential value for endotoxin research, in *Handbook of Endotoxin. Vol. 1. Chemistry of Endotoxin,* Ritschel, E. T., Ed., Elsevier, Amsterdam, 1984, 221.
206. **Westphal, O., Luderitz, O., Ritschel, E. T., and Galanos, C.,** Bacterial lipopolysaccharide and its lipid A component: some historical and some current aspects, *Biochem. Rev.,* 9, 191, 1981.
207. **Weckesser, J. and Mayer, H.,** Different lipid A types in lipopolysaccharides of phototrophic and related non-phototrophic bacteria, *FEMS Microbiol. Rev.,* 54, 143, 1988.
208. **Wisshar, R., and Lingens, F.,** The lipopolysaccharide of a chloridazon-degrading bacterium, *Eur. J. Biochem.,* 137, 155, 1983.
209. **Dranovskaya, E. A. and Vershilova, P. A.,** The endotoxin of *Brucellae, Ann. Sclavo.,* 19, 109, 1977.
210. **Kurtz, R. S.,** Immune and Adjuvant IgG Responses to *Brucella* Lipopolysaccharide in Mice, Ph.D. thesis, University of Wisconsin, Madison, 1984.
211. **Moreno, E. and Berman, D. T.,** *Brucella abortus* lipopolysaccharide is mitogenic for spleen cells of the endotoxin resistant C3H/HeJ mice, *J. Immunol.,* 123, 2915, 1979.
212. **Vukajlovich, S. W. and Morrison, D. C.,** Conversion of lipopolysaccharides to molecular aggregates with reduced subunit heterogeneity: demonstration of LPS-responsiveness in endotoxin unresponsive C3H/HeJ splenocytes, *J. Immunol.,* 130, 2804, 1983.
213. **Morrison, D. C. and Jacobs, D. M.,** Binding of polymyxin B to the lipid A portion of bacterial lipopolysaccharides, *Immunochemistry,* 13, 813, 1976.
214. **Keleti, G., Feingold, D. S., and Younger, J. S.,** Interferon induction in mice by lipopolysaccharide from *Brucella abortus, Infect. Immun.,* 10, 282, 1974.
215. **Moreno, E., Kurtz, R. S., and Berman, D. T.,** Induction of immune and adjuvant immunoglobulin G responses in mice by *Brucella* lipopolysaccharide, *Infect. Immun.,* 46, 74, 1984.
216. **Kurtz, R. S. and Berman, D. T.,** Influence of endotoxin-protein in immunoglobulin G isotype responses of mice to *Brucella abortus* lipopolysaccharide, *Infect. Immun.,* 54, 728, 1986.
217. **Leong, D. and Wilson, J. B.,** Distribution of chromate-labelled *Brucella abortus* endotoxin in tolerant mice, *J. Bacteriol.,* 97, 32, 1969.
218. **Corbel, M. J.,** Isolation and partial characterization of phage receptor from *Brucella neotomae* 5K33, *Ann. Sclavo.,* 19, 131, 1977.
219. **Huhn, H., Basu, S., and Mayer, H.,** Comparison of enterobacterial common antigen from different species by serological techniques, *Eur. J. Biochem.,* 162, 69, 1987.
220. **Imhoff, J. F., Kushner, D. J., Kushwaha, S. C., and Kates, M.,** Polar lipids in phototrophic bacteria of *Rhodospirillaceae* and *Chromaticeae* families, *J. Bacteriol.,* 150, 1192, 1982.
221. **Mayer, H., Salimath, P. V., Holst, O., and Weckesser, J.,** Unusual lipid A types in phototrophic bacteria and related species, *Rev. Infect. Dis.,* 6, 542, 1984.

222. **Wilkinson, S. G., Galbraith, L., and Lightfoot, G. A.,** Cell walls and lipopolysaccharide of *Pseudomonas* species, *Eur. J. Biochem.*, 33, 158, 1973.
223. **De-Ley, J., Mannheim, W., Segers, P., Kersters, K., Hinz, K. H., and Livens, A.,** Taxonomy of the genus *Bordetella, in Les Bacilles a Gram Negatif d'Interet Medical et en Sante Publique: Taxonomie, Identification, Application,* Institut National de la Sante et de la Recherche Medicale, Paris, 1983, 245.
224. **Tanaka, S., Suto, T., Isayama, Y., Azuma, R., and Hatakeyama, H.,** Chemo-taxonomical studies on fatty acids of *Brucella* species, *Ann. Sclavo.*, 19, 67, 1977.
225. **Woese, C. R.,** Bacterial evolution, *Microbiol. Rev.*, 51, 221, 1977.
226. **Woese, C. R., Gibson, J., and Fox, G. E.,** Do genealogical patterns in purple photosynthetic bacteria reflect interspecific gene transfer?, *Nature (London)*, 283, 212, 1980.
227. **Yang, D., Oyaizu, Y., Oyaizu, H., Olsen, G. J., and Woese, C. R.,** Mitochondrial origins, *PNAS*, 82, 4443, 1985.
228. **Oelze, J. and Drews, G.,** Membranes of photosynthetic bacteria, *Biochim. Biophys. Acta*, 265, 209, 1972.
229. **Thiele, O. W., Oulevey, J., and Hunneman, D. H.,** Ornithine-containing lipids in *Thiobacillus* A2 and *Achromobacter sp.*, *Eur. J. Biochem.*, 139, 131, 1984.
230. **Sanchez, S., Gay, B., Senglat, C., and Roux, J.,** Etude en microscopie electronique de la paroi *Brucella abortus* et de ses fractions antigeniques, *Biol. Cell*, 41, 217, 1981.
231. **Berger, F. M., Fukui, G. M., Ludwing, B., and Rosselet, J. P.,** Increased host resistance to infection elicited by lipopolysaccharide from *Brucella abortus, Proc. Soc. Exp. Biol. Med.*, 131, 1376, 1969.
232. **Kellerman, G. D., Foster, J. W., and Badakhsh, F. F.,** Comparison of chemical components of cell walls of *Brucella abortus* strains of low and high virulence, *Infect. Immun.*, 2, 237, 1970.
233. **Lacave, C., Asselineau, J., Serre, A., and Roux, J.,** Comparison de la composition chimique d'une fraction lipopolysaccharidique et d'une fraction polysaccharidique isolees de *Brucella melitensis, Eur. J. Biochem.*, 9, 189, 198.
234. **Renoux, G., Renoux, M., and Tinelli, R.,** Phenol-water fraction from smooth *Brucella abortus* and *Brucella melitensis*. Immunochemical analysis and biological behavior, *J. Infect. Dis.*, 127, 139, 1973.
235. **Renoux, M., Renoux, G., Palet, A., and Nehemie, C.,** Activite mitogenique des fractions de *Brucella* sur lymphocytes en culture, *Dev. Biol. Stand.*, 31, 230, 1976.
236. **Canning, P. C., Roth, J. A., and Deyoe, B. L.,** Release of 5′-guanosine monophosphate and adenine by *Brucella abortus* and their role in the intracellular survival of the bacteria, *J. Infect. Dis.*, 154, 464, 1986.
237. **Hinsdill, R. D. and Berman, D. T.,** Antigens of *Brucella abortus*. II. Toxicity for macrophages in culture, *J. Infect. Dis.*, 118, 307, 1968.
238. **Baker, P. J. and Wilson, J. B.,** Hypoferremia in mice and its application to the bioassay of endotoxin, *J. Bacteriol.*, 90, 903, 1965.
239. **Diaz, R. and Oyeledun, M. A.,** Studies of some biological activities of *Brucella* endotoxin in normal and infected animals and the role of hypersensitivity factor, *Ann. Sclavo.*, 19, 117, 1977.
240. **Spink, W. W.,** The pathogenesis of brucellosis, in *The Nature of Brucellosis,* University of Minnesota Press, Minnesota, 1956, 112.
241. **Munoz, J. J., Bergman, R. K., and Robbins, K. E.,** Comparison of histamine hypersensitivity and *Limulus* amoebocyte lysate test for endotoxic activity, *Infect. Immun.*, 22, 292, 1978.
242. **Kaulen, D. R., Sanin, A. V., Khorobrykh, V. V., Malikov, V. E., and Dranovskaya, A.,** Stimulating effect of *Brucella melitensis* lipopolysaccharide on hemopoiesis in mice, *Zh. Mikrobiol. Epidemiol. Immunobiol.*, 8, 84, 1983.
243. **Hollingdale, M. R., Vinson, J. W., and Herman, J. E.,** Immunochemical and biological properties of the outer membrane-associated lipopolysaccharide and protein of *Rochalimeae quintana, J. Infect. Dis.*, 141, 672, 1980.
244. **Planque, K., van-Nierop, J. J., and Burgers, A.,** The lipopolysaccharide of free-living and bacteroid forms of *Rhizobium leguminosarum, J. Gen. Microbiol.*, 110, 151, 1979.
245. **Salkinoja-Salonen, M. and Boeck, R.,** Characterization of lipopolysaccharides isolated from *Agrobacterium tumefaciens, J. Gen. Microbiol.*, 105, 119, 1978.
246. **Hoyer, B. H. and McCullough, N. B.,** Polynucleotide homologies of *Brucella* deoxyribonucleic acids, *J. Bacteriol.*, 95, 444, 1968.
247. **Plescia, O. J., Noval, J. J., Palczuk, N. C., and Braun, W.,** Fractionation and immunological properties of a DNA-rich preparation from *Brucella abortus, PNAS*, 106, 748, 1961.
248. **Corbel, M. J. and Brewer, R. A.,** Isolation and properties of an RNA fraction present in *Brucella* culture supernatants, *J. Hyg.*, 84, 223, 1980.
249. **Foster, J. W. and Ribi, E.,** Immunological role of *Brucella abortus* cell walls, *J. Bacteriol.*, 84, 258, 1962.
250. **Ralston, D. J. and Elberg, S. S.,** The cell wall as the antigenic site for antibodies stimulating ingestion (MSF) and inhibition (BIF) of *Brucella* in macrophages from normal and immune animals, *Br. J. Exp. Pathol.*, 52, 130, 1971.

251. **Dziarski, R.,** Preferential induction of autoantibody secretion in polyclonal activation by peptidoglycan and lipopolysaccharide. I. *In vitro* studies, *J. Immunol.*, 128, 1018, 1982.
252. **Dziarski, R.,** Preferential induction of autoantibody secretion in polyclonal activation by peptidoglycan and lipopolysaccharide. II. *In vivo* studies, *J. Immunol.*, 128, 1026, 1982.
253. **Cannat, A., Escande, A., Peraldi, F., and Serre, A.,** Induction of autoantibodies and circulating immune complexes in mice after injection of *Brucella* fraction ''PI'' or inoculation with live *Brucella suis, Ann. Immunol.*, 134C, 43, 1983.
254. **Serikawa, T., Takeda, H., Kondo, Y., Muraguchi, T., and Yamada, J.,** Multiplication of *Brucella canis* in male reproductive organs and detection of autoantibody to spermatozoa in canine Brucellosis, *Dev. Biol. Stand.*, 56, 295, 1984.
255. **George, L. and Carmichael, L.,** Antisperm responses in male dogs with chronic *Brucella canis* infections, *Am. J. Vet. Res.*, 45, 274, 1984.
256. **Tyumentseva, I. S., Taran, I. F., Afansaer, E. N., and Gramotina, L. I.,** Bacteriocinogenicity of *Brucella* isolated in the Caucasus U.S.S.R. and their taxonomic position, *Antibiotiki*, 29, 29, 1984.
257. **Tabatabai, L. B., Schmidt, L. A., and Deyoe, B. L.,** Properties of a bactericidal extract from *Brucella abortus, Curr. Microbiol.*, 5, 7, 1981.
258. **Nielsen, K.,** Cleavage of bovine immunoglobulin G1 in whey by an extracellular material from *Brucella abortus, Can. J. Comp. Med.*, 49, 337, 1985.
259. **Cherwonogrodzky, J. W. and Clark, A. G.,** Effect of pH on the production of the Kanagawa hemolysin by *Vibrio parahaemolyticus, Infect. Immun.*, 34, 115, 1981.
260. **Cherwonogrodzky, J. W. and Clark, A. G.,** Production of the Kanagawa hemolysin by *Vibrio parahaemolyticus* in a synthetic medium, *Infect. Immun.*, 37, 60, 1982.
261. **Cherwonogrodzky, J. W. and Bundle, D. R.,** Mouse monoclonal antibodies as substitutes for rabbit monospecific anti-*Brucella* antisera, Abstr. VM3p, *Can. Soc. Microbiol. Conf.*, Toronto, 1986.
262. **Meikle, M. J., Perry, M. B., Cherwonogrodzky, J. W., and Bundle, D. R.,** Fine structure of A and M antigens from *Brucella* biovars, *Infect. Immun.*, 57, 2820, 1986.
263. **Moreno, E. et al.,** unpublished results.
264. **Moreno, E. and Mayer, H.,** unpublished results.

Chapter 3

THE IMMUNE RESPONSE TO *BRUCELLA ABORTUS*: THE HUMORAL RESPONSE

Shelley S. Sutherland and John Searson

TABLE OF CONTENTS

I. INTRODUCTION

Most of the literature on the humoral immune response of animals infected with *Brucella* organisms concerns *Brucella abortus* in cattle. Hence, the immune response discussed in this chapter has most relevance to the disease bovine brucellosis, which for many years has aptly been referred to as "contagious abortion". As the name implies, the disease is very contagious, and frequently, the only clinical sign in cattle is abortion.

Evidence indicates that infected animals act as the main source of the spread of infection.[1] The introduction of infection into a herd previously free from the disease at a time when a large number of animals are pregnant may result in a serious epidemic of abortion.[2,3] The majority of animals are believed to abort once,[1,4] and during subsequent years may conceive and produce normal calves. However, once the organism is established in the sexually mature animal, it may persist indefinitely in the tissues.[5-8]

The greatest concentration of the organism in infected animals is in the uterine contents, the placenta, the fetus, and the fetal membranes; the organisms are shed in the uterine excretions following parturition or abortion.[4] Ingestion of forage and water contaminated by these infected genital discharges is the most common means of transmission.[2,9,10]

II. PATHOGENESIS

The organism was first found in exudate in the intervillous space of the placenta of a pregnant cow in which abortion was imminent.[4] The thick yellow exudate contained the *B. abortus* organism in almost pure culture. Since that time the pathogenesis has been studied or reviewed by many workers.[1,11,12]

The spread of brucellae after experimental infection was studied in 11 pregnant cows infected with *B. abortus* strain 544 (S544) by the conjunctival route, and it was found that the organisms drained from the conjunctiva into the regional lymph nodes.[11] From these foci, a bacteremia occurred with the localization of the organisms in other lymphoid tissue such as the spleen and the iliac, mesenteric, and supramammary lymph nodes. Hyperplasia of cells in these lymph nodes resulted in the formation of granulomata.

Once the bacteria succeeded in breaking out of the reticuloendothelial system, a second bacteremia occurred, resulting in a generalized infection including the predilection sites, i.e., uterus, udder, and associated lymph nodes. The earliest lesion in the uterus appeared at the junction of the placental tissue and cotyledon. It was also reported that localization of *B. abortus* organisms in the cytoplasm of the chorion cells in the placenta of aborting animals occurred.[13] At the same time, it was noted that the chorion was generally edematous, and there were dull, thickened leather-like areas due to the multiplication of this organism. Entry of organisms into the amniotic fluid then resulted in the infection of the fetal stomach and lungs just prior to, or at, abortion.

Finally, it appears that the organism leaves the uterus after abortion and localizes in the udder, and it is in this organ that the disease generally maintains itself from one gestation period to the next. Between gestation periods the organism may also be found in the lymph nodes associated with the udder and uterus, the lymph nodes of the digestive tract, other organs of the body, such as the spleen, and occasionally the bloodstream.[14-16] *B. abortus* has been isolated from hygromas of the knee joint of cattle, from the epididymis, and from testes of bulls.[17,18]

The mechanism for invasion of the body by brucellae in other animal species has been shown to be similar to that in the cow.[19,20] The sequence of events is as follows:

1. Survival at the site of the primary invasion
2. Entrance into the phagocytes and proliferation

3. Invasion of the bloodstream
4. Dissemination into the organs, mainly spleen, liver, bone, marrow, lymph nodes, kidneys, and uterus

The relationship between the distribution of certain nutrients and corresponding distribution of brucellae, when *B. abortus* was isolated from the tissues of pregnant cows 3 to 4 weeks after intravenous injection of the organisms, was studied. It was found that over 90% of the organisms were located on the fetal membrane of the placental barrier.[21] This preferential growth in certain fetal tissues was ascribed to erythritol.[22,23]

Similar concentrations of erythritol in the fetal placentas were found in cattle, sheep, goats, and pigs. Lower concentrations were found in less susceptible species, e.g., man, rabbits, guinea pigs, and rats, where localization in the fetal placentas does not usually occur.[24]

It has been observed that concentrations of erythritol which promote growth of virulent strains do not support growth of attenuated strains such as *B. abortus* strain 19 (S19), *B. abortus* strain 45/20 (S45/20), *B. abortus* strain 205/19, and *B. abortus* strain 58/20.[25,26] In fact, the growth of S19 is inhibited by 2 μg/ml or more of erythritol. Thus, it would seem that the virulent strains are well adapted to utilize the high levels of erythritol localized within the pregnant uterus of the susceptible host.

III. THE HUMORAL IMMUNE RESPONSE — GENERAL

The humoral immune response in cattle is generally similar to that of other species.[27,28] Most of the immunoglobins occurring in the serum and secretions of cattle possess antigenic determinants and physicochemical characteristics in common with immunoglobulin G (IgG), immunoglobulin A (IgA), and immunoglobulin M (IgM) of humans.[29-33] The IgG antibody class in cattle can be divided into two subclasses, IgG_1 and IgG_2, on the basis of differences in electrophoretic migration rates.[31,34]

Studies of the response of cattle to *B. abortus* infection have usually involved an evaluation of the humoral response or the ability to resist challenge after vaccination.[35-38] Serological methods have been widely used in evaluating the humoral response and are used almost exclusively in the detection of the disease for eradication programs.[39,40]

Early in this century the serum agglutination test (SAT) and complement fixation test (CFT) for bovine brucellosis were utilized,[41,42] and since that time an ever-increasing variety of serological techniques has been developed. During this period many of the serological tests have been modified in an effort to devise a method for measuring and distinguishing between the immune response of an animal when vaccinated, latently infected, clinically diseased, or when it has experienced a combination of these events. The improvement in serological methods has meant that the prevalence of false positive reactions caused by nonspecific antibodies and false negative reactions caused by blocking and incomplete antibodies has been reduced.

Until the 1960s the information gained regarding the nature of antibodies produced in response to *B. abortus* infection related to their serological activity. After 1960 new knowledge of immunology was applied to the disease when investigations into the distribution of antibody activity in the sera of cattle vaccinated with S19 or infected with virulent *B. abortus* began.[43-45]

IV. RESPONSE TO STRAIN 19 VACCINATION

S19 vaccine, a live attenuated vaccine, has been shown to be capable of producing protection from infection in cattle in field trials[46,47] as well as in controlled challenge experiments in Australia, the U.S., the U.K., and South Africa.[48-51] A large number of countries used the S19 vaccine as an aid in the control of brucellosis. However, early in the 1950s it was found that vaccination of adult cattle with S19 often led to the production of persistent antibodies which made interpretation of subsequent serological tests difficult.[52,53] In contrast, less persistent antibodies are produced when the vaccine is given to calves under 6 months of age.[53,54]

Antibodies resulting from vaccination are more reactive in some serological tests than in others. It has been shown that IgM measurable by the SAT test is the first antibody produced after S19 vaccination. Indeed, this test is subject to more persistent reactions after S19 vaccination than the CFT, possibly because the SAT is more sensitive to IgM than to other antibody isotypes, whereas the CFT is sensitive to IgG_1.[55] Despite the fact that S19 vaccination continues to cause problems in the serological diagnosis of this disease, the CFT has been generally accepted for its routine diagnosis.[58] A more recent serological test, the indirect hemolysis test (IHLT), is reported to have increased sensitivity and, therefore, the ability to reduce some of the problems associated with S19 vaccination.[56-58] Nevertheless, S19 vaccination continues to cause problems not only because the specificity of certain tests is lower following vaccination,[59,60] but also because of the occurrence of infected but serologically unreactive animals.[61,62]

More recently, attempts have been made to eliminate persistent vaccinal serological reactions by modification of S19 calfhood vaccination.[63] Studies have also been extended into an effective method of vaccination that could be applied to all ages of cattle. This involved studies concentrating on dosage adjustments and methods of vaccination to minimize titers without compromising resistance. Strain 19 has been used conjunctivally,[64] orally,[65] intracaudally,[66] subcutaneously,[66] intravaginally,[67] and via aerosol.[68] Subcutaneous vaccination of mature nonpregnant cattle with 3×10^8 colony forming units (cfu) of S19 produced immunity at least as good as that produced by calfhood vaccination by the same route.[63,69] Subcutaneous vaccination of infected dairy herds with 3×10^9 cfu of S19 organisms has also been shown to reduce effectively the natural infection rate in 6 months.[70] Although these studies, involving vaccination of adult cattle, showed the serological response to be greatly reduced in the majority of vaccinated cattle, other studies showed the occurrence of persistent titers.[71] Some workers, aiming at discriminating serologically between S19 vaccinated cattle and those infected with field strains of *B. abortus*, have utilized the ability of monoclonal antibodies to detect small differences between antigens. Based on the molecular differences between known virulent and avirulent *Brucella* strains,[72] a monoclonal antibody with the ability to bind to field strains of *B. abortus* biotypes 1 and 2, but not S19, *B. suis*, *B. melitensis*, or *B. ovis*, has been produced. It is hoped that when used in a sandwich enzyme-linked immunosorbent assay (ELISA) this system can remove the problem of vaccinated cattle being positive in the indirect ELISA.[73]

V. RESPONSE TO STRAIN 45/20 VACCINATION

A killed dissociated rough strain incorporated with "Falba" liquid paraffin provided a vaccine known as S45/20 which was shown to generate considerable immunity to challenge in guinea pigs.[74,75] In a later experiment where cattle were given two doses of killed S45/20 adjuvant vaccine 6 weeks apart, an effective immunity developed against a subsequent challenge with S544.[76] In addition, there were either no agglutination titers or only low agglutination titers which were of short duration.

A comprehensive account of experiments leading up to the commercial production, the field use, and beneficial effects of this killed S45/20 adjuvant vaccine has been described for both *Brucella*-infected and noninfected herds.[77,78] This and other studies showed that the vaccine was similar to S19 in protective capacity for cattle and produced low levels of serum agglutinins which disappeared within a few weeks of the second vaccination. On the other hand, persistent reactions in the CFT have been more of a problem as a small proportion of vaccinated cattle maintained titers for 12 to 15 months. Other workers have found the positive titers to the CFT to be transient.[50,79,80]

To overcome the problems of differentiating between infected and noninfected cattle previously vaccinated with S45/20, it was proposed that a CFT using an antigen extracted from a rough strain of *B. abortus* be used.[81] Long-lasting titers to this rough antigen complement fixation test (RCFT) were found in all S45/20 vaccinated cattle and effectively identified them as cattle vaccinated with S45/20.[56] However, when challenged, interpretation of the RCFT based on the method described earlier[81] wrongly classified most of the infected and noninfected cattle.[56] Similar results to this test have been recorded under field conditions.[82] It is evident from these studies that the test has little value at present. It should also be noted that the study of Miller and colleagues[81] was not supported by any biological or cultural evidence of infection.

Two commercially produced killed S45/20 adjuvant vaccines have been generally available: "Duphavac Na", manufactured by Philips Duphar Pty. Ltd. and *B. abortus* 45/20 vaccine, manufactured by Arthur Webster Pty. Ltd. They appear to differ in their ability to produce measurable serum antibody titers[59] and in the localized reaction produced in the tissue at the site of inoculation.[83] Which of these vaccines was the more efficient in producing immunity in cattle against the virulent strains of *B. abortus* is unknown.

VI. RESPONSE OF ARTIFICIALLY CHALLENGED CATTLE

In cattle, IgM is the predominant antibody produced immediately after artificial challenge and remains present generally for 1 to 3 weeks.[84-86] In those cattle that become infected, the IgG antibody soon becomes dominant, and there is usually more IgG_1 than IgG_2 antibody.[84,87,88] Significant amounts of IgG_2 antibody have also been reported to exist in some infected cattle.[89] These authors suggest measuring IgG_2 as a means of discriminating between infected and noninfected cattle. The ratio of IgG_2 to IgG_1 increases greatly but temporarily at the onset of lactation when selective secretion of IgG_1 into colostrum begins.[90] This was readily demonstrated when eight experimentally infected cattle showed a decrease in the CFT titer at the time of parturition;[91] the CFT is sensitive to antibody of the IgM and IgG subclasses, not IgG_2.[88]

Artificially challenged cattle that become infected develop a rapid and persistent serological response which may be detected by many serological tests such as the Rose-Bengal test (RBT), SAT, anti-bovine globulin test (ABGT), CFT, indirect hemolysis test (IHLT), and radioimmunoassay (RIA), while most of the cattle that resist challenge show no positive titers to any of these tests.[56,92]

The more sensitive ELISA has also been shown to detect high levels of IgG and IgM from 3 up to 6 weeks after challenge, and these levels persisted until slaughter 50 weeks later.[93] Noninfected but challenged cattle had low positive levels of IgG and IgM for 2 to 7 weeks after challenge.[93,94] These transient but nevertheless false positive reactions in cattle that were challenged and in which no infection developed suggest that cattle experiencing any immunologic stimulation by *B. abortus* antigen through challenge or infection may develop antibodies detectable by the ELISA.

The overall sensitivity and specificity of the ELISA was calculated for a group of cattle which were artificially challenged with the virulent *B. abortus* S544 and the results compared with other serological tests (Table 1).[93]

VII. RESPONSE OF NATURALLY INFECTED CATTLE

Brucellosis in cattle is complicated by many factors, all of which have a marked influence

TABLE 1
Comparison of the Number of Cattle Positive and Negative to 5 Serological Tests and Their Correlation with the Isolation of *B. abortus* S544

Test result[a]	Diagnosis: 27 Infected cattle[b]	Diagnosis: 95 Noninfected cattle[c]	Sensitivity (%)	Specificity (%)
RBT +[d]	21	28	78	71
−[e]	6	67		
SAT +	19	5	70	95
−	8	90		
CFT +	22	2	81	98
−	5	93		
IHLT +	25	3	93	97
−	2	92		
ELISA				
IgG +	26	18[f]	96	81
−	1[f]	77		
IgM +	21	12[f]	77	87
−	6f	83		

[a] RBT = Rose Bengal test; SAT = serum agglutination test; CFT = complement fixation test; IHLT = indirect hemolysis test.
[b] Diseased cattle, cattle from which *B. abortus* was isolated at necropsy.
[c] Nondiseased cattle, cattle from which *B. abortus* was not isolated at necropsy.
[d] +, Positive titer.
[e] −, Negative titer.
[f] All vaccinated cattle.

on its serological detection and on the sensitivity and specificity of the tests currently available. The many tests currently available clearly illustrate that no single test can yet satisfy what is defined as the "ideal" serological test,[95] i.e., one which will

1. Detect infection early during the long and variable incubation period
2. Not be influenced by the presence of nonspecific antibodies
3. Detect latent or chronic carriers
4. Differentiate a response to vaccination from that of field infection

Information has been obtained from eradication programs, field studies based on bacteriologic examination, a variety of studies on the concentration of antibody isotypes present in the serum and milk of cattle following vaccination, studies on the antigenic components against which antibodies are detected (antibodies to different components become detectable at different stages of the immune response), and from the quantitative and qualitative differences in the components of the *B. abortus* isotypes.[72] This is now sufficient to enable some conclusions to be made on the usefulness and limitations of each test.[84,94] There is general agreement that the CFT is the most accurate serological test for the detection of bovine brucellosis as it shows a better correlation with the recovery of organisms at necropsy of artificially and naturally infected cattle.[56,92,95] As such, the CFT is a definitive test in many countries where eradication schemes are in progress.[96,97] Several limitations of this test, however, have been encountered.

1. The CFT often shows anticomplementary reactions as well as prozones where IgG_2 antibodies to *B. abortus* block IgG_1 and IgM antibodies.[98]

TABLE 2
The Number of Cattle with Different Combinations of CFT, RBT, and ELISA Results Giving Culture Positive or Culture Negative Results

Serology			Culture	
CFT	RBT	ELISA[a]	Positive	Negative
		0—9	0	8
≤4	−	10—25	4	4
		0—9		2
≤4	+	10—25	5	3
		26—50	2	
		76—99	1	
		26—50	1	1
4	+	51—75	2	1
		26—50	3	
8	+	51—75	2	1
		76—99		1
		51—75	4	
16	+	76—99		1
		≥100	1	
		26—50	1	
		51—75	1	
32	+	76—99	5	
		≥100	3	
64	+	76—99	2	
		≥100	2	1
		76—99	3	
≥128	+	≥100	7	

[a] Value expressed as a % of positive control.

From Hornitzky, M. and Searson, J., *J. Aust. Vet. J.*, 1986. With permission.

2. Some infected cattle may show aberrant serological reactions and are not detected.[61,62]
3. Naturally infected cattle may be difficult to differentiate from those recently vaccinated with either S19 or S45/20 vaccines.[56,99]
4. There are variations in the test due to differences in technique.[100]

Hence, alternative or supplementary tests have been assessed in recent years for use in eradication schemes to enhance the serological detection of infected cattle. The IHLT and subsequently the ELISA have been used successfully with the CFT in recent years.[101,102] While some workers maintain the ELISA offers no advantages over the CFT in the detection of infected cattle, other workers have shown it offers increased sensitivity and is a valuable tool for detecting infected cattle, particularly those with low CFT titers (Table 2).[91,101,102] These contrasting findings may be the result of the many different methods of performing the test. Different reagents, diagnostic interpretations, and variation in selection of different

TABLE 3
The Relationship between Serological Reactions, Vaccination Status, and the Type of *B. abortus* Recovered at Slaughter from 50 Cattle from New South Wales

Serological reactions				Type of *Brucella abortus* isolated			
CFT	RBT	ELISA 1	ELISA 2	1	2	S19	Total no. of cattle
+	+	+	+	21	13	8	42 (84)
−	+	+	+	1			1 (2)
+	−	+	+	1			1 (2)
−	−	−	+		4		4 (8)
−	−	−	−		2		2 (4)

( ) = percentage of cattle.

populations of cattle can greatly influence the overall findings, some of which were illustrated in a study when two ELISA tests were compared (Table 3).[91]

VIII. PASSIVE IMMUNITY

In animals having an epitheliochorial type of placentation, the young characteristically are born virtually agammaglobulinemic.[103,104] There is little or no transfer of maternal immunoglobulins through the intact placenta, and immunoglobulins from the dams are acquired by the neonate through ingestion and subsequent absorption from the small intestine.[105,106] This absorption of immunoglobulins by the calf occurs mostly during the first 6 h after birth.[107]

It was shown that sera of neonatal calves born from infected cows were devoid of any antibodies at the time of birth, but after colostrum was fed, antibodies were present in high concentration.[1,108] Calves from dams with high and low serum agglutinating antibody levels had high and low levels of agglutinating antibody, respectively.[109] In addition, antibody titers detected in young calves by the CFT and RBT persisted for 5 to 6 months, whereas the SAT titers became negative at about 2 months.[110] In noninfected calves the mean levels of immunoglobulins were one third those in colostrum, whereas in the infected group this ratio was reduced to about one seventh.[111]

The passive transfer of antibody in the calf has been shown to interfere with active antibody production following subsequent immunization with some bacterial antigens.[112-114] In piglets it has been proved that the passive transfer of antibodies through colostrum depressed antibody production against *Brucella*.[115] A similar effect on active production of *B. abortus* agglutinins in young rats has also been observed.[116,117] This work was confirmed when calves, which received no antibodies against *B. abortus* in colostrum, responded to S19 vaccination even on their first day of life.[117] Conversely, in calves receiving antibrucella colostral antibody, response to S19 vaccination may be suppressed.

The protective effects of the colostral antibodies was recently demonstrated. In a group of calves born to infected dams and exposed to large numbers of brucellae via uterine excretions and infected milk, it was observed that the maternal antibody in the colostrum of the infected cows protected 85% of the calves against a heavy challenge. When subjected to an anamnestic test, these animals did not appear to be even sensitized or "primed" by the heavy exposure at birth.[117] This passive transfer of immunity in calves is not consistent with the passive transfer studies in adults which will be discussed in the next section.

IX. LATENT CARRIER CATTLE

In *Brucella* eradication programs, isolated cases of *B. abortus* infection have been reported to occur in pregnant heifers in herds where there is no evidence of infection from external sources, the origin being from infected animals which do not react to the diagnostic tests.[40,118,119] Such animals are usually termed latent carriers and the herds in which they occur "problem herds".[120,121] The terms "chronic"[122] and "symptomless" carriers[123] have also been used to describe such animals.

The concept of the latent carrier animal was confirmed when 167 cows originating from 97 brucellosis problem herds were studied.[15] This study showed that *B. abortus* was isolated from all 167 cows and that none of the seroagglutination tests used identified all infected cattle. The cause of infection in these latent carrier animals was not known.

Calves born to infected dams generally have high serological titers which fall to nil within 4 to 6 weeks after birth. This fall in titer generally has been interpreted as evidence that infection failed to become established; it has been argued that this is because prepubescent animals are innately resistant to infection.[124,125]

More recently, however, *B. abortus* was recovered from tissues of calves born to infected dams even though the serological titers in these calves had fallen to nil in the usual way.[7,122,123] Moreover, under normal circumstances, the presence of infection in these calves was only detected when they matured, became pregnant, and aborted.[7,123] From these observations, it is suggested that infection of the calf may occur at or before birth and may then remain dormant until the mature animal becomes pregnant.

In support of this concept it has been reported that *B. abortus* infection *in utero* may produce a latent carrier state in female calves.[126-128] This was confirmed experimentally;[127] 22 heifers born to artificially infected dams were separated from the dams at birth, deprived of colostrum, and reared in isolation units. All calves remained serologically negative from birth to 1 year. They were artificially inseminated when mature, retained until 6 weeks after calving or abortion, and were then killed and cultured. In addition, the calves or aborted fetuses from these heifers were also cultured. *B. abortus* was isolated from 4 out of 22 heifers and the aborted fetuses were similarly infected. Serological reactions in the heifers remained negative until 8 weeks prior to parturition. The latent carrier condition may also arise from the vaccinated and or nonpregnant animal in which infection becomes established after challenge.[62]

Killed S45/20 vaccine in adjuvant has been used to induce a serologically detectable anamnestic response in the latent carrier animal.[8,129,130] Although this method has proved successful in detecting those animals which have been exposed to field infection, it is not possible to differentiate them from S19-vaccinated animals. To aid the early identification of latent carrier animals and to differentiate between vaccinated and field-infected animals, newer tests such as ELISA have, in Australia, been successful in detecting cattle infected with *B. abortus*, whereas the standard serological tests, RBT, SAT, and CFT, have failed.[131-133]

X. DIAGNOSTIC EFFICIENCY OF SEROLOGICAL TESTS

The immune response as detected by laboratory study of *Brucella* is influenced by many factors. These include the long and variable "incubation period" of the disease during which the serological tests are negative,[134-136] the vaccinal status of the animal, the nature of the challenge,[137-139] variations in serological responses of individual cattle to both vaccination and challenge,[94,138] and stage of gestation at the time of exposure to the disease.[139] Studies on the immune response of cattle to *B. abortus* infection have usually involved an evaluation of the humoral response[48,137,140] or the ability to resist challenge after vaccination.[49,141]

Serological methods have been widely used in evaluating the humoral response and are used almost exclusively in the detection of the disease for eradication programs.

Serological test methods involve the two major biological variables, antigen and antibody, and one is fixed in order to measure variation in the other. In many such tests there is great difficulty in interpreting and evaluating the efficiency of a test because false positive and false negative reactions may occur; the number of false positive and false negative reactions varies according to the minimum diagnostic value set for that particular test. The diagnostic efficiency of a test procedure may be assessed according to the criteria of Cochrane and Holland.[142] Simplicity, acceptability, accuracy, cost, precision and repeatibility, sensitivity, and specificity are qualities listed as important in the efficiency of biological tests; sensitivity and specificity, based on the concepts of Thorner and Remein.[143] The concept of diagnostic efficiency, however, may be misleading if it suggests that sensitivity and specificity are constant characteristics of a test. Both of these parameters can vary between different populations of cattle; they can also vary within a population and, indeed, will change with prevalence as eradication proceeds. In addition, vaccination with S19 or S45/20 can modify the response of an animal and alter the pattern expected in natural disease. The assessment of "the serological tests" for brucellosis should therefore be made with caution and be related to the knowledge of the population being tested.[84]

XI. CROSS-REACTIONS

Antibodies reacting with *B. abortus* antigens are sometimes found in the serum of cattle not known to have exposure to the bacterium. Some groups of bacteria have antigenic similarity to *B. abortus* which in most cases is due to the presence of N-acylated 4-amino-4, 6-dideoxy-D-mannose (N-acylated D perosamine) repeating units in the O chains of the lipopolysaccharide somatic antigens of these organisms. Cross-reactions between *B. abortus* and *Yersinia enterocolitica* serogroup 0:9, *Escherichia coli* 0:116 and 0:157, *Salmonella* serotypes of the Kauffman-White group N, *Francisella tularensis*, *Vibrio cholerae*, and *Pseudomonas maltophilia* are all documented and recently reviewed by Corbel.[144] All smooth strains of the various *Brucella* species cross-react in agglutination tests with unabsorbed antisera to *B. abortus*. However, there is little or no cross-reaction between smooth and rough phase *Brucella* sp.[144,145]

XII. NONSPECIFIC REACTIONS

Occasionally, sera from some cattle contain nonspecific factors which are not caused by exposure to either *B. abortus* or antigenically related organisms but result in positive reactions in serological tests for brucellosis. These nonspecific reactions are more important in the latter stages of brucellosis eradication compaigns. "Natural antibodies" with reactivity for *B. abortus* antigens have been found in IgM and IgG classes of immunoglobulins.[146] Much of the *Brucella* agglutinating activity present in sera of uninfected cattle is labile in the presence of EDTA.[147-149] These immunoglobulins attach to the *Brucella* antigens by the Fc portion of their 7S subunit structures, not by the Fab′ antibody-combining sites.[150]

Approximately 80% of sera from cattle, which have reactions in the serum agglutination test but are negative in the complement fixation test for *B. abortus*, may have EDTA-labile agglutinins.[146]

Antigen material found in silage has been implicated as a cause of cross-reaction in agglutination, complement fixation, and Rose Bengal tests for brucellosis.[145]

It has been found that up to 70% of the sera from uninfected cattle, which gave positive reactions in an indirect ELISA for brucellosis using standard polyclonal conjugates, were negative when retested using a $F(ab')^2$ conjugate (Table 4). Thus, it is suggested that with

TABLE 4
Comparison of Standard and $F(ab')^2$ Horseradish Peroxidase Conjugates in a Screening ELISA Test for *B. abortus* in Herds Known to be Free of Disease

Standard conjugate	$F(ab')^2$ conjugate	No. of samples	Percent
+ve	+ve	421	28.7
+ve	−ve	1042	71.3

some sera nonimmunologic binding due to the hydrophobic Fc portion of the conjugate antibody molecule to the antigen and/or the inert carrier (e.g., microtiter plate) occurs. This highlights the value of retesting using the $F(ab')^2$ conjugate in place of the standard polyclonal conjugate.

XIII. CONCLUSION

After an abortion in the field the isolation of *B. abortus* is required before the disease can be positively diagnosed. The isolation of *B. abortus* is most easily effected from the stomach contents, spleen, meconium, or lungs of the aborted fetus and from the dam's vaginal discharge[16,151] or milk.[152,153] However, while positive diagnosis of *B. abortus* infection is obtained only when the organism is isolated and identified, this ideal is not practicable when large numbers of animals are involved. The only practicable means are serological tests, and a wide range has been developed for the detection of *B. abortus* antibodies in serum and milk,[16,39] many of which are now performed as microtests using automated equipment.

Recent advances in serological tests and the antibody responses to *B. abortus* infection in cattle have permitted quantitative studies of the isotypes involved, antigenic specificity, and antibody affinities to be studied and related to one another.

Accepting that biological variation exists, the serological detection of bovine brucellosis has improved over recent years to the extent that accurate diagnostic levels may be set for eradication schemes according to the stage of eradication.

ACKNOWLEDGMENT

We wish to thank Mr. Leo den Hollander for his assistance with the references during the preparation of this manuscript.

REFERENCES

1. **Huddleson, I. F.,** *Brucellosis in Man and Animals,* rev. ed., Commonwealth Fund, New York, 1943.
2. **Stableforth, A. W.,** Infectious diseases of animals, in *Diseases Due to Bacteria,* Vol. 1, Stableforth, A. W. and Galloway, I. A., Eds., Butterworth, London, 1959, 53.
3. **Wilson, G. S. and Miles, A.,** *Topley and Wilson's Principles of Bacteriology, Virology and Immunity,* Vol. 2, Edward Arnold, London, 1975, 2171.
4. **Bang, B.,** The etiology of epizootic abortion, *J. Comp. Pathol.,* 10, 125, 1897.
5. **Fabyan, M.,** The persistence of *B. abortus*, Bang, in the tissues of inoculated animals, *J. Med. Res.,* 28, 81.

6. **Doyle, T. M.,** The distribution of *B. abortus* in the system of "carrier" cows, *J. Comp. Pathol.*, 48, 192, 1935.
7. **Morgan, W. J. B.,** Some recent advances in the diagnosis of brucellosis, *Ir. Vet. J.*, 25, 214, 1971.
8. **Cunningham, B. and O'Connor, M.,** The use of killed 45/20 adjuvant vaccine as a diagnostic agent in the final stages of the eradication of Brucellosis. The clearance and Brucellosis from problem herds by interpretation of anamnestic serological responses, *Vet. Rec.*, 89, 680, 1971.
9. **Birch, R. R. and Gilman, H. L.,** The pathogenicity for cattle of *Brucella* strains isolated from cases of undulant fever in man, *J. Infect. Dis.*, 56, 78, 1935.
10. **Wray, C.,** Survival and spread of pathogenic bacteria of veterinary importance within the environment, *Vet. Bull.*, 45, 543, 1975.
11. **Payne, J. M.,** The pathogenesis of experimental brucellosis in the pregnant cow, *J. Pathol. Bacteriol.*, 78, 447, 1959.
12. **Olitzki, A. L.,** *Immunological Methods in Brucellosis Research. II. In Vivo Procedures*, S. Karger, Basel, 1970.
13. **Smith, T.,** A characteristic localization of *Bacillus abortus* in the bovine fetal membranes, *J. Exp. Med.*, 29, 451, 1919.
14. **Washko, F. V., Donham, C. R., Hutchings, L. M., and Heimlich, A.,** Recovery of *Brucella* from tissues of cattle exposed to *Brucella abortus*, *J. Am. Vet. Med. Assoc.*, 120, 82, 1952.
15. **Nelson, C. J., Anderson, R. K., Kimberling, C. V., and Pietz, D. E.,** Epizoologic factors of bovine brucellosis: comparative bacteriologic studies of infected herds, *Am. J. Vet. Res.*, 25, 1515, 1966.
16. **Alton, G. G., Jones, L. M., and Pietz, D. E.,** Laboratory techniques in brucellosis, Monogr. Ser. No. 55, World Health Organization, Geneva, 1975.
17. **Anon.,** Pathogenesis, in Brucellosis: a veterinarians guide to the literature, Ministry of Agriculture and Fisheries, New Zealand, 1977, 19.
18. **Fensterbank, R.,** Congenital brucellosis in cattle associated with localization of a hygroma, *Vet. Rec.*, 103, 283, 1979.
19. **Van der Hoeden, J.,** Pathogenesis of brucellosis Bang, *J. Comp. Pathol.*, 46, 232, 1933.
20. **Spink, W. W.,** *The Nature of Brucellosis*, University Minnesota Press, Minneapolis, 1956.
21. **Smith, J., Keppie, J., Pierce, J. H., Fuller, R., and Williams, A. E.,** The chemical basis of the virulence of *Brucella abortus*. I. Isolation of *B. abortus* from bovine foetal tissue, *Br. J. Exp. Pathol.*, 42, 631, 1961.
22. **Pearce, J. H., Williams, A. E., Harris-Smith, P. W., Fitzgeorge, R. B., and Smith, H.,** The chemical basis of virulence of *Brucella abortus*. II. Erythritol, a constituent of bovine foetal fluids which stimulates the growth of *Brucella abortus* in bovine phagocytes, *Br. J. Exp. Pathol.*, 43, 31, 1962.
23. **Keppie, J., Williams, A. E., Witt, K., and Smith, H.,** The role of erythritol in the tissue localization of the brucellae, *Br. J. Exp. Pathol.*, 46, 104, 1965.
24. **Keppie, J.,** Host and specificity in microbial behaviour "*in vivo*" and "*in vitro*", in *Proc. 14th Symp. Soc. Gen. Microbiol.*, London, Cambridge University Press, London, 1964, 44.
25. **Williams, A. E., Keppie, J., and Smith, J.,** The chemical basis of the virulence of *Brucella abortus*. III. Foetal erythritol, a cause of the localization of *Brucella abortus* in pregnant cows, *Br. J. Exp. Pathol.*, 43, 530, 1962.
26. **Keppie, J., Witt, K., and Smith, H.,** The immunization of guinea pigs and mice with a whole-culture extract of a smooth and a rough strain of *Brucella abortus*, *Br. J. Exp. Pathol.*, 53, 518, 1967.
27. **Butler, J. E. and Maxwell, C. F.,** Preparation of bovine immunoglobulins and free secretory components and their specific antisera, *J. Dairy Sci.*, 55, 151, 1971.
28. **Butler, J. E., Winter, A. J., and Wagner, G. G.,** Symposium: bovine immune system, *J. Dairy Sci.*, 55, 151, 1973.
29. W.H.O. Report, Nomenclature for human immunoglobulins, Bull. WHO, 30, 447, 1964.
30. **Aalund, O.,** *Heterogeneity of Ruminant Immunoglobulins*, Munksgaard, Copenhagen, 1968.
31. **Butler, J. E.,** Bovine immunoglobulins: a review, *J. Dairy Sci.*, 52, 1895, 1969.
32. **Mehta, P. D. and Tomasi, J. R.,** Comparative studies of mammalian immunoglobulins, *Fed. Proc.*, 28, 820, 1969.
33. **Mach, J. P. and Pahud, J. J.,** Secretory IgA, a major immunoglobulin in most bovine external secretions, *J. Immunol.*, 106, 552, 1971.
34. **Milstein, C. P. and Feistein, A.,** Comparative studies of two types of bovine immunoglobulin G heavy chains, *Biochem. J.*, 107, 559, 1968.
35. **Gordon, W. S.,** The assessment of brucellosis immunity in cattle, in *Proc. 15th Int. Vet. Congr., Stockholm 1953*, Vol. 2, Gernandts Boktryekeri, Stockholm, 1954, 179.
36. **Jenness, R. and Anderson, R. K.,** Immunological response of cows to *Brucella* antigens, *J. Dairy Sci.*, 54, 21, 1971.
37. **McDiarmid, A.,** The degree and duration of immunity in cattle resulting from vaccination with S19 *B. abortus* vaccine and its implication in the future control and eradication of brucellosis, *Vet. Rec.*, 69, 877, 1957.

38. **Cocks, E. and Davies, G.,** *Brucella abortus* (strain 19) vaccine: potency tests in cattle, *J. Biol. Stand.*, 1, 171, 1973.
39. **Morgan, W. J. B.,** The serological diagnosis of bovine brucellosis, *Vet. Rec.*, 80, 612, 1967.
40. **Christie, T. E., Kerr, W. R., and McCaughey, W. J.,** Brucellosis eradication in Northern Ireland, *Vet. Rec.*, 82, 176, 1968.
41. **McFadyean, J. and Stockman, S.,** Epizootic abortion in cattle. Report of the departmental committee appointed by the Board of Agriculture and Fisheries to inquire into epizootic abortion, Appendix to Part 1, Her Majesty's Stationery Office, London, 1909.
42. **Mohler, J. R. and Traum, J.,** Infectious abortion of cattle and the occurrence of its bacterium in milk. III. Infectious abortion of cattle, Annu. Rep., Bureau Animal Industry, 1911, 147.
43. **Rose, J. E. and Roepke, M. H.,** Physiochemical studies or post-vaccinal *Brucella* agglutinins in bovine serum, *Am. J. Vet. Res.*, 25, 325, 1964.
44. **Rose, J. E., Lambert, G., and Roepke, M. H.,** Ultra-centrifugation and heat. Inactivation studies on serogglutinins of pregnant heifers artificially infected with virulent *Brucella abortus, Am. J. Vet. Res.*, 25, 329, 1964.
45. **Rice, C. E., Tailyor, J., and Cochrane, D.,** Ultracentrifugal studies of sera from cattle vaccinated or naturally infected with *Brucella abortus, Can. J. Comp. Med. Vet. Sci.*, 30, 270, 1966.
46. **Hardenbergh, J. G.,** Calfhood vaccination against Bang's disease. Effect on agglutinating titres and results on first pregnancies, *J. Am. Vet. Med. Assoc.*, 94, 479, 1939.
47. **Buddle, M. B.,** Vaccination against bovine brucellosis, *Aust. Vet. J.*, 24, 171, 1948.
48. **Gordon, W. S.,** The assessment of brucellosis immunity in cattle, *Proc. 15th Int. Vet. Congr., Stockholm, 1953*, Vol. 2, Gernandts Boktryekeri, Stockholm, 1954, 179.
49. **Cocks, E. and Davies, G.,** *Brucella abortus* (strain 19) vaccine: potency tests in cattle, *J. Biol. Stand.*, 1, 171, 1973.
50. **Hendricks, J. B. and Ray, W. E.,** Studies on cattle vaccinated with live strain 19 and/or 45/20 adjuvant *Brucella abortus* vaccines, *Proc. U.S. Animal Health Assoc.*, 1970, 147.
51. **Worthington, R. W., Horwell, F. D., Mulders, M. S. G., McFarlane, I. S., and Schutte, A. P.,** Investigation of the efficiency of three Brucella vaccines in cattle, *J. S. Afr. Vet. Assoc.*, 45, 87, 1974.
52. **Nagy, L. K., Hignett, P. G., and Ironside, C. J. T.,** Bovine brucellosis: a study of an adult vaccinated, *Brucella* infected herd, *Vet. Rec.*, 81, 140, 1967.
53. **Cunningham, B. and O'Reilly, D. J.,** *Brucella abortus* vaccines. Agglutination responses in blood serum and milk following vaccination of cattle of various ages with live S19 and killed 45/20 adjuvant *B. abortus* vaccine, *Vet. Rec.*, 82, 678, 1968.
54. **Morgan, W. J. B.,** Some recent advances in the diagnosis of brucellosis, *Ir. Vet. J.*, 25, 214, 1971.
55. **Beh, K. J.,** Distribution of *Brucella* antibody among immunoglobulin classes and a low molecular weight antibody fraction in serum and whey of cattle, *Res. Vet. Sci.*, 14, 381, 1973.
56. **Sutherland, S. S., Le Cras, D. V., Robertson, A. G., Johnston, J. M., and Evans, R. J.,** Serological response of cattle after vaccination and challenge with *Brucella abortus, Vet. Microbiol.*, 7, 165, 1982.
57. **Alton, G. G., Corner, L. A., and Plackett, P.,** Vaccination of pregnant cows with low doses of *Brucella abortus* strain 19 vaccine, *Aust. Vet. J.*, 56, 369, 1980.
58. **Nicoletti, P. and Carlsen, W. B.,** Indirect haemolysis test in the serodiagnosis of bovine brucellosis, *Am. J. Vet. Res.*, 42, 1494, 1981.
59. **Alton, G.,** Recent developments in vaccination against bovine brucellosis, *Aust. Vet. J.*, 54, 551, 1978.
60. **Plackett, P., Alton, G. G., Carter, P. D., and Corner, L. A.,** The indirect haemolysis test (HLT) for bovine brucellosis — comparison with the complement fixation test (CFT) in vaccinated or experimentally infected cattle, *Aust. Vet. J.*, 56, 405, 1980.
61. **Cordes, D. O. and Cartes, M. E.,** Persistence of *Brucella abortus* infection in six herds of cattle under brucellosis eradication, *N.Z. Vet. J.*, 27, 255, 1979.
62. **Sutherland, S. S., Le Cras, D. V., and Robertson, A. G.,** A study of cattle infected with *Brucella abortus* and which showed aberrant serological reactions, *Aust. Vet. J.*, 59, 132, 1982b.
63. **Plackett, P., Alton, G. G., Carter, P. D., and Corner, L. A.,** Failure of single dose of *Brucella abortus* strain 19 vaccine to protect cattle when given early in calfhood, *Aust. Vet. J.*, 56, 409, 1980.
64. **Plommet, M. and Fensterbank, R.,** Vaccination against bovine brucellosis with a low dose of strain 19 administered by the conjunctival route. III. Serological response and immunity in the pregnant cow, *Ann. Rech. Vet.*, 7, 9, 1976.
65. **Nicoletti, P. and Milward, F. W.,** Protection by oral administration of *Brucella abortus* strain 19 on oral challenge exposure with a pathogenic strain of *Brucella, Am. J. Vet. Res.*, 44, 1641, 1983.
66. **Manthei, C. A.,** Evaluation of vaccinal methods and doses of *Brucella abortus* strain 19, *Proc. Annu. Meet. U.S. Livestock San. Assoc.*, 56, 115, 1952.
67. **Quevedo, J. M., Cedro, V. C. F., Benedetti, L. M. E. de, et al.,** Immunization of cattle against brucellosis with intravaginal pellets of *Brucella abortus* strain 19, *Rev. Invest. Agropecu.*, 11, 1, 1974.

68. **Selivanov, A. V. and Marinkova, V. V.,** Aerosol immunization against brucellosis, *Veterinarya (Moscow),* 12, 27, 1969.
69. **Alton, G. G., Corner, L. A., and Plackett, P.,** Vaccination against bovine brucellosis, *Dev. Biol. Stand.,* 56, 643, 1984.
70. **Nicoletti, P., Jones, L. M., and Berman, D. T.,** Adult vaccination with standard and reduced doses of *Brucella abortus* strain 19 vaccine in a dairy herd infected with brucellosis, *J. Am. Vet. Med. Assoc.,* 173, 1445, 1978.
71. **Beckett, F. W. and MacDiarmid, S. G.,** The effect of reduced-dose *Brucella abortus* strain 19 vaccination in accredited dairy herds, *Br. Vet. J.,* 141, 507, 1985.
72. **Hatten, B. A. and Brodeur, R. D.,** Soluble antigens of virulent and attenuated histotypes of *Brucella abortus, Infect. Immun.,* 22, 956, 1978.
73. **Gorrell, M. D., Milliken, G. L., Anderson, B. J., and Pucci, A.,** An enzyme immunoassay for bovine brucellosis using a monoclonal antibody specific for field strains of *Brucella abortus, Dev. Biol. Stand.,* 56, 491, 1984.
74. **McEwen, A. D. and Samuel, J. McA.,** *Brucella abortus:* heat stable protective antigen revealed by adjuvant and present in a "Rough" variant strain 45/20: immunisation experiments on guinea pigs, *Vet. Rec.,* 67, 546, 1955.
75. **McDiarmid, A. and Sutherland, F. B.,** A comparison of the immunity produced in guinea pigs by the inoculation of living heat-killed and adjuvant vaccines prepared from *B. abortus* strains 45/20 and S19, *Vet. Rec.,* 69, 1067, 1957.
76. **McDiarmid, A.,** Immunisation, contre la brucellose au moyen d'une souche ne déterminant pas la formation d'agglutinines, *Ann. Inst. Pasteur,* 190, 792, 1962.
77. **Roerink, J. H. G.,** Development of a Non-agglutinogenic Killed *Brucella abortus* Adjuvant Vaccine and Its Applicability in the Control of Bovine Brucellosis, Thesis, Rijksuniversiteit, Utrecht, 1966.
78. **Roerink, J. H. G.,** Investigation into the usefulness of the non-agglutinogenic *Brucella abortus* adjuvant vaccine Duphavac N.A. in the control of bovine brucellosis, *Vet. Rec.,* 80, 727, 1967.
79. **Cunningham, B.,** Vaccination of cattle with killed adjuvant vaccine: effects on serological and milk ring tests when used in cattle previously exposed to infection or vaccinated with S_{19}, *Vet. Rec.,* 82, 7, 1970.
80. **Waghela, S. and Philpott, M.,** Serological response of cattle in Kenya vaccinated with a killed *B. abortus* strain 45/20 adjuvant vaccine, *Vet. Rec.,* 99, 505, 1976.
81. **Miller, J. K., Kelly, J. I., and Roerink, J. H. G.,** A complement fixation method for quantitative differentiation of reactions to 45/20 vaccine and *Brucella* infection, *Vet. Rec.,* 98, 210, 1976.
82. Items for the Information of Animal Health Committee, Rep. No. 18, Sub-committee Brucellosis and Tuberculosis, Australian Bureau of Animal Health, Canberra, 1980, 32.
83. **MacIntyre, G. C. and Sizer, D.,** Vaccination against brucellosis, *Vet. Rec.,* 82, 22, 1969.
84. **Chappel, R. J.,** Towards improved serological diagnosis of bovine brucellosis, *W.H.O. Bull.,* 353, 1, 1980.
85. **Rice, C. E., Tailyor, J., and Cochrane, D.,** Ultracentrifugal studies of sera from cattle vaccinated or naturally infected with *Brucella abortus, Can. J. Comp. Med. Vet. Sci.,* 30, 270, 1966.
86. **Plackett, P.,** Enzyme-linked antiglobulin test for bovine brucellosis, *Dept. Agric. Victoria, Agric. Notes Ser.,* 134, 80, 1984.
87. **Butler, J. E.,** The amplified ELISA. Principles and applications for the comparative quantification of class and sub class antibodies and the distribution of antibodies and antigens in biochemical separates, in *Enzymology,* Vol. 73b, Colowick, S., Eds., 1981, 482.
88. **Allen, G. S., Chappel, R. J., Williamson, P., and McNaught, D. J.,** A quantitative comparison of the sensitivity of serological tests for bovine brucellosis to different antibody classes, *J. Hyg. Cambridge,* 76, 287, 1976.
89. **Lawman, M. J. P., Ball, D. R., Hoffman, E. M., Des Jurdin, L. E., and Boyle, M. D. P.,** Production of *Brucella abortus* specific protein A reactive antibodies (IgG2) in infected and vaccinated cattle, *Vet. Microbiol.,* 12, 43, 1985.
90. **Williams, M. R. and Millar, P.,** Changes in serum immunoglobulin levels in Jerseys and Friesians near calving, *Res. Vet. Sci.,* 26, 81, 1979.
91. **Sutherland, S. S., Evans, R. J., and Bathgate, J.,** Application of an enzyme-linked immunosorbent assay in the final stages of a bovine brucellosis eradication programme, *Aust. Vet. J.,* 63, 412, 1986.
92. **Hayes, J. and Chappel, R. J.,** A comparison of the results of the brucellosis radioimmunoassay and other serological tests in experimentally infected cattle, *J. Hyg. Cambridge,* 88, 21, 1982.
93. **Sutherland, S. S.,** Evaluation of the enzyme-linked immunosorbent assay in the detection of cattle infected with *Brucella abortus, Vet. Microbiol.,* 10, 23, 1984.

94. **Morgan, W. J. B.,** The diagnosis of *Brucella abortus* infection in Britain, in *Bovine brucellosis: An International Symposium,* Crawford, R. P. and Hidalgo, R. J., Eds., Texas A & M University Press, College Station, 1977, chap. 2.
95. **Heck, F. C., Deyoe, B. L., and Williams, J. D.,** Antibodies to *Brucella abortus* in sera from strain 19 vaccinated and non-vaccinated cows as determined by enzyme-linked immunosorbent assay and conventional serologic methods, *Vet. Immunol. Immunopathol.,* 3, 629, 1982.
96. **Alton, G. G., Maw, J., Rogerson, B., and McPherson, G. G.,** The serological diagnosis of bovine brucellosis. An evaluation of the complement fixation, serum agglutination and Rose Bengal tests, *Aust. Vet. J.,* 51, 57, 1975.
97. **Meyer, M. E.,** Use of automated complement fixation screening test for serodiagnosis of bovine brucellosis, *Proc. U.S. Anim. Health Assoc.,* 83, 81, 1979.
98. **McNaught, D. J., Chappel, R. J., Allan, G. S., Bourke, J. A., and Rogerson, B. A.,** The effects of IgG2 and of antigen concentrations on prozoning in the complement fixation test for bovine brucellosis, *Res. Vet. Sci.,* 22, 194, 1977.
99. **Cunningham, B.,** The control and eradication of brucellosis. I. Serological response in cattle following vaccination with S_{19} and killed *Brucella* 45/20 adjuvant vaccine, *Vet. Rec.,* 82, 7, 1968.
100. **Alton, G. G., Jones, L. M., and Pietz, D. E.,** Bacteriological methods, *WHO Monogr. Ser.,* 55, 11, 1975.
101. **Sutherland, S. S. and MacKenzie, R. M.,** Applied serology in the latter stages of the eradication of bovine brucellosis, *Aust. Vet. J.,* 60, 240, 1983.
102. **Hornitsky, M. and Searson, J.,** The relationship between the isolation of *Brucella abortus* and serological status of infected, non vaccinated cattle, *Aust. Vet. J.,* 63, 172, 1986.
103. **Fey, J.,** Colibacillosis in calves, Hans Huber, Bern, 73, 1972.
104. **Logan, E. F.,** Colostral immunity to Colibacillus in the neonatal calf, *Br. Vet. J.,* 130, 405, 1974.
105. **McGirr, J. L.,** Colostral transmission of antibody substances from mother to offspring, *Vet. J.,* 103, 345, 1947.
106. **Brambell, F. W. R.,** Transmission of passive immunity from mother of young, in *Frontiers of Biology,* Vol. 18, Newburger, A. and Tatum, E. L., Eds., North-Holland, Amsterdam, 1970, 201.
107. **Husband, A. J., Brandon, M. R., and Lascelles, A. K.,** Absorption and endogenous production of immunoglobulins in calves, *Aust. J. Exp. Biol. Med. Sci.,* 59, 491, 1972.
108. **McDiarmid, A.,** The transference of agglutinins for *Brucella abortus* from cow to calf and their persistence in the calf's blood, *Vet. Rec.,* 58, 146, 1946.
109. **Peiris, G. S.,** Presence and duration of maternal antibodies against *Brucella* infection in calves, *Ceylon Vet. J.,* 20, 18, 1972.
110. **Morgan, W. J. B., MacKinnon, D. J., Lawson, J. R., and Cullen, G. A.,** The Rose Bengal plate agglutination test in the diagnosis of brucellosis, *Vet. Rec.,* 85, 636, 1969.
111. **Cunningham, B.,** The transfer of *Brucella abortus* antibodies from dam to calf, *Vet. Rec.,* 85, 636, 1977.
112. **Smith, A. N. and Ingram, D. G.,** Immunological response of young animals. II. Antibody production in calves, *Can. Vet. J.,* 6, 226, 1965.
113. **Wigzell, Y.,** Antibody synthesis of the cellular level. Antibody-induced suppression of 7S antibody synthesis, *J. Exp. Med.,* 124, 953, 1966.
114. **Richardson, M., Beck, C. C., and Clark, D. T.,** Pre-natal immunization of the lamb to *Brucella*: dissociation of immuno-competence and reactivity, *J. Immunol.,* 101, 1363, 1968.
115. **Hoerlin, A. B.,** The influence of colostrum on antibody response in baby pigs, *J. Immunol.,* 78, 112, 1957.
116. **Halliday, R.,** Effect of passive immunization against *Brucella abortus* on active production of Brucella abortus agglutinins in young rats, *J. Pathol. Bacteriol.,* 96, 137, 1968.
117. **Cunningham, B.,** Protective effects of colostral antibodies by *B. abortus* on strain 19 vaccination and field infection, *Vet. Rec.,* 101, 521, 1977.
118. **Sjollema, P.,** Value of examination of mixed milk samples from a herd for the diagnosis of bovine brucellosis, *Tijdschr. Diergeneeskd.,* 92, 35, 1967.
119. **Christie, T. E.,** Eradication of brucellosis in Northern Ireland: field problems and experience, *Vet. Rec.,* 85, 268, 1969.
120. **Nicoletti, P. and Muraschi, T. F.,** Bacteriologic evaluation of serologic test procedures for the diagnosis of brucellosis in problem cattle herds, *Am. J. Vet. Res.,* 27, 689, 1966.
121. **Luchsinger, D. W., Angus, R. D., Gue, C. S., and Anderson, R. H.,** The utilization of *Brucella abortus* culturing and biotyping results in the epizootiologic investigation of bovine brucellosis, *Proc. U.S. Anim. Health Assoc.,* 1974, 85.

122. **Morgan, W. J. B. and Richards, R. A.,** The diagnosis, control and eradication of bovine brucellosis in Great Britain, *Vet. Rec.*, 94, 510, 1974.
123. **Robertson, F. J.,** Brucellosis: a possible symptomless carrier, *Vet. Rec.*, 88, 313, 1971.
124. **Fitch, C. P., Boyd, W. L., Kelly, M. D., and Bishop, L. M.,** An extended study of female offspring of positive Bang's diseases cattle, *J. Am. Vet. Med. Assoc.*, 99, 413, 1941.
125. **McEwen, A. D.,** The resistance of the young calf to disease, *Vet. Rec.*, 62, 83, 1950.
126. **Hignett, P. G. and Nagy, L. K.,** Effect of exposure on very young calves to virulent *Brucella abortus* on their serological response to re-infection by the same organism at 6 months of age, *Nature (London),* 201, 204, 1964.
127. **Plommet, M., Fensterbank, R., Renoux, G., Gestin, J., and Philippon, A.,** Experimental bovine brucellosis. XII. Persistance to adult age of congenital infection in the heifer, *Ann. Rech. Vet.*, 4, 419, 1973.
128. **Lapraik, R. D., Brown, D. D., Maan, H., and Brand, T.,** Brucellosis: a study of five calves from reactor dams, *Vet. Rec.*, 97, 52, 1975.
129. **Cunningham, B.,** The control and eradication of brucellosis. Serological response in cattle following vaccination with S19 and killed *Brucella* 45/20 adjuvant vaccine, *Vet. Rec.*, 82, 7, 1968.
130. **Reid, M. A. and Harvey, P. R.,** The use of *Brucella abortus* 45/20 adjuvant vaccine as a diagnostic aid in the brucellosis eradication campaign in Papua New Guinea, *Aust. Vet. J.*, 48, 495, 1972.
131. **Nielsen, K., Samagh, B. S., and Stemshorn, B.,** Agglutination of *Brucella abortus* cells by sera from cattle experimentally infected with *Escherchia coli, Vet. Microbiol.*, 5, 123, 1980.
132. **Stuart, F. A. and Corbel, M. J.,** Identification of a serological cross-reaction between *Brucella abortus* and *Escherichia coli* 0:157, *Vet. Rec.*, 110, 202, 1982.
133. **Cioglia, L.,** Agglutinazione della *S. urbana* nelle brucellosi e suo valore diagnostico, *Boll. Soc. Ital. Biol. Sper.*, 24, 1117, 1948.
134. **Nelson, C. J., Anderson, R. K., Kimberling, C. V., and Pietz, D. E.,** Enzoologic factors of bovine brucellosis: comparative bacteriologic studies of infected herds, *Am. J. Vet. Res.*, 25, 1515, 1966.
135. **Robertson, F. J.,** Brucellosis: a possible symptomless carrier, *Vet. Rec.*, 88, 313, 1971.
136. **Morgan, W. J. B. and Richards, R. A.,** The diagnosis, control and eradication of bovine brucellosis in Great Britain, *Vet. Rec.*, 94, 510, 1974.
137. **McEwen, A. D., Priestley, F. W., and Paterson, J. D.,** An estimate of a suitable infective dose of *B. abortus* for immunisation tests of cattle, *J. Comp. Pathol. Ther.*, 52, 116, 1939.
138. **Berman, D. T., Irwin, M. R., and Beach, B. A.,** Statistical considerations of controlled experiments in brucellosis, *Am. J. Res.*, 10, 130, 1949.
139. **Fensterbank, R., Plommet, M., and Pardon, P.,** Traitement de la brucellose bovine par l'oxytetracycline, *Ann. Rech. Vet.*, 6, 43, 1975.
140. **Jenness, R. and Anderson, R. K.,** Immunological response of cows to *Brucella* antigens, *J. Dairy Sci.*, 54, 21, 1971.
141. **McDiarmid, A.,** The degree and duration of immunity in cattle resulting from vaccination with S19 *Br. abortus* vaccine and its implication in the future control and eradication of brucellosis, *Vet. Rec.*, 69, 877, 1957.
142. **Cochrane, A. L. and Holland, W. W.,** Validation of screening procedures, *Br. Med. Bull.*, 27, 3, 1971.
143. **Thorner, R. M. and Remein, Q. R.,** Principles and procedures in the evaluation of screening for diseases, *Public Health Monogr.*, No. 67, 1961.
144. **Corbel, M. J.,** Recent advances in the study of *Brucella* antigens and their serological cross-reactions, *Vet. Bull.*, 55, 927, 1985.
145. **Garbin, B. and Trap, D.,** Serologie de la brucellose bovine. Reactions non specifiques: etat des recherches, *Bull. Inf. Lab. Serv. Vet.*, 15.16, 29, 1984.
146. **Macmillan, A. P. and Cockrem, D. S.,** Reduction of non specific reactions to the *Brucella abortus* serum agglutination test by the addition of EDTA, *Res. Vet. Sci.*, 38, 288, 1985.
147. **Nielsen, K., Samagh, B. S., Speckman, G., and Stemshorn, B.,** The bovine immune response to *Brucella abortus*. II. Elimination of some sporadic serological reactions by chelation of divalent cations, *Can. J. Comp. Med.*, 43, 420, 1979.
148. **Macmillan, A. P. and Bell, R. A.,** Non-specific reactions to the *Brucella abortus* SAT, *Vet. Rec.*, 116, 139, 1985.
149. **Nowlan, P. F. and de Geus, H.,** Use of EDTA modified antigen in the serodiagnosis of bovine brucellosis, *Vet. Rec.*, 116, 13, 1985.
150. **Nielsen, K., Stilwell, K., Stemshorn, B., and Duncan, R.,** Ethylene diaminetetraacetic acid (disodium salt)-labile bovine immunoglobulin M Fc binding to *Brucella abortus:* a cause of nonspecific agglutination, *J. Clin. Micro.*, 14, 32, 1981.

151. **Edwards, S. J., McDiarmid, A., De Ropp, R. S., and McLeod, D. H.,** Immunity in cattle vaccinated with *Brucella abortus* strain 19 and a note comparing this strain with 45/20, *Vet. Rec.*, 58, 141, 1946.
152. **Morgan, W. J. B. and McDiarmid, A.,** Excretion of *Brucella* in the milk of experimentally infected cattle, *Res. Vet. Sci.*, 1, 53, 1960.
153. **Davies, G. and Casey, A.,** The survival of *Brucella abortus* in milk and milk products, *Br. Vet. J.*, 129, 345, 1973.

Chapter 4

THE IMMUNE RESPONSE TO *B. ABORTUS*: THE CELL-MEDIATED RESPONSE TO INFECTIONS

Paul Nicoletti and Alexander J. Winter

TABLE OF CONTENTS

I. INTRODUCTION

Brucellae are facultative intracellular parasites which are capable of multiplication and survival within host phagocytes. There is evidence that protection from infection and elimination of *Brucella abortus* requires cell-mediated immune (CMI) responses,[1,2] defined as those in which T lymphocytes function as effector cells in conferring protective immunity. An example of a CMI response is the activation of macrophages by lymphokines secreted by immune T cells. Early studies demonstrated *in vitro* that acquired resistance to *B. abortus* was expressed by guinea pig macrophages,[3] and macrophages derived from immunized animals have subsequently been found to play an important part in this process. There is also evidence from *in vitro* studies and passive transfer experiments[5-11] that antibodies are involved in protective immunity and that antibodies specific for the O-polysaccharide have a principal role in this process.[5,11]

The most widely used correlates of CMI are lymphocyte stimulation, macrophage migration inhibition, and delayed-type hypersensitivity (DTH skin test) reactions. The types of soluble *Brucella* antigens used to measure CMI reactions greatly affect the results of the measurements. The most commonly used preparations are complex mixtures of antigens which have been extracted from various *Brucella* cells by different methods. Most contain a large number of proteins and some contain lipopolysaccharide (LPS). The presence of LPS complicates the interpretation of DTH reactions both because of its toxicity and its participation in antibody-mediated cutaneous hypersensitivity.[12] Results obtained from studies on lymphocyte-mediated immune assays have usually correlated more closely with infection than with protection. Indeed, the use of these methods has often been proposed as a diagnostic procedure.

The following sections of this chapter will review selected aspects of microbial virulence factors and the host response to *Brucella* infections, followed by a discussion of correlates of CMI which have been used as diagnostic tests. Concluding sections will be devoted to a brief discussion of adjuvants and host factors which may influence the animal's response to *Brucella* vaccines or to infection.

II. HOST-PARASITE INTERACTIONS

A. NEUTROPHILS

The pathogenesis of *Brucella* infections occurs in several phases. Organisms are phagocytized by polymorphonuclear leukocytes (PMNs) in which some survive and multiply. These cells are transported via circulating humors to lymphoid tissues and susceptible tissues such as the fetal placenta. The inability of the PMNs to effectively kill virulent *Brucella* at the primary site of infection is a key factor in the dissemination to regional lymph nodes, other sites of the reticuloendothelial system, and organs such as the uterus and udder.

No complete explanation exists for all phases of the process of permanent infection by *Brucella* organisms. In general, virulent but not avirulent organisms survive and multiply intracellularly. Bacterial cell wall components and ability to utilize nutritive components within phagocytic cells are important. *In vitro* studies with human and guinea pig PMNs using *B. abortus* rough strain 45/20 and smooth strain 45/0 in guinea pigs demonstrated limited killing of the rough (R) strain and little or no killing of the smooth (S) strain. The authors proposed that this was due to failure of degranulation.[13] Bertram et al.[14] found that degranulation and the myeloperoxidase-hydrogen peroxide-halide system of killing were inhibited by low molecular weight nucleotide materials which were identified and characterized by Canning et al.[15,16] as 5′ guanosine monophosphate and adenine.

B. MACROPHAGES

The survival of ingested bacteria within macrophages is similar to that of PMNs. To avoid being killed within the phagocyte, the bacteria must be able to survive the antibacterial activity of the phagolysosome, inhibit phagolysosomal fusion, escape from the phagosome, or prevent the metabolic burst associated with phagocytosis.[17] Such activities may be mediated by surface components of the bacteria or the production and release of intracellular factors. Extracts of virulent *B. abortus* strain 2308, but not attenuated strain 19, contained factors which inhibited phagolysosomal fusion of unelicited mouse macrophages.[18] The factors were not associated with LPS and may have been the nucleotides defined by Canning et al.[15,16]

Most studies on survival and multiplication of *Brucella* spp. within macrophages have been conducted in laboratory animals. Holland and Pickett[3] found that *B. suis, B. abortus*, and *B. melitensis* multiplied extensively within normal rat, mouse, and guinea pig monocytes maintained in cell cultures for 3 d. Intracellular growth of *Brucella* had no observable toxic effects on most phagocytic cells. Nonsmooth brucellae and *B. abortus* strain 19 multiplied slowly. In contrast, ''immune'' monocytes, i.e., those derived from animals previously infected with smooth *Brucella*, greatly restricted the intracellular growth of all *Brucella*. Intracellular growth was unaffected by addition of antiserum to the culture medium. The authors concluded that the *in vitro* studies correlated well with the development of acquired cellular immunity *in vivo*.

In other early studies, divergent results were obtained in *in vitro* studies attempting to correlate virulence of *B. abortus* with growth in monocytes. Braun et al.[19] showed that virulent S cells multiplied in monocytes from normal guinea pigs to a much greater extent than R forms. Smooth cells also caused greater destruction of monocytes. However, Freeman et al.[20,21] found that the cytopathogenic effect of *Brucella* variants was inversely related to smoothness and that virulence was not correlated with the ability to multiply in guinea pig monocytes. Subsequent studies by Smith and Fitzgeorge[22] demonstrated that virulent strains of *B. abortus* resisted the bactericidal power of bovine phagocytes more than did attenuated or avirulent strains, and this antibactericidal activity was related to differences in composition of the cell walls.

Joubert and Valette[23] demonstrated acquired cellular resistance *in vitro* in the macrophages of cattle immunized with *Brucella* vaccines, particularly those of the live attenuated type (strain 19). *Brucella* multiplied more slowly in phagocytes from immunized animals, and the authors suggested that this test could serve as a screening method for the activity of vaccines used in cattle.

C. PHAGOCYTE-SERUM INTERACTION

It is well known that opsonization of bacterial cells enhances phagocytosis. In phagocytosis and intracellular growth studies, opsonization of both virulent and avirulent *B. abortus* increased phagocytosis by normal bovine cells.[24] However, there was no significant effect on their subsequent intracellular behavior. In contrast, Ralston and Elberg[25] demonstrated that immunization of rabbits with Rev. 1 (a *B. melitensis* attenuated S strain) caused production of humoral factors which enhanced ingestion and killing of virulent *Brucella* by macrophages. These factors were associated with the IgM and IgG fractions of serum and were almost certainly antibodies. Recent studies have demonstrated that antibodies opsonizing *B. abortus* stimulate oxidative metabolism in bovine neutrophils[15] and macrophages[26] as well as the myeloperoxidase-hydrogen peroxide-halide system in neutrophils.[16] The latter finding indicates that antibodies facilitate phagolysosomal fusion otherwise inhibited by *B. abortus*. Taken together, these findings provide a basis for the cooperative interaction of antibodies and phagocytes in immunity to *B. abortus*.

D. T CELL AND MACROPHAGE INTERACTION

It has been demonstrated in the mouse model that the interrelationship between specifically sensitized T lymphocytes and macrophages afford one mechanism of protective immunity against *B. abortus*.[1,2] Splitter and Everlith[27] found that *B. abortus*-induced bovine macrophage-T lymphocyte collaboration was a prerequisite for eventual clearance of the organism. *In vivo*-primed T lymphocytes of peripheral blood origin recognized phagocytized bacterial components and contributed to their removal. However, the authors concluded that host mechanisms that contribute to elimination or persistence of sequestered *B. abortus* remain to be fully identified.

III. CORRELATES OF CELL-MEDIATED IMMUNITY

A. LYMPHOCYTE STIMULATION

It is widely recognized that cell-mediated immunity (CMI) is a major defense mechanism in brucellosis. Following contact with Brucella antigens, the *in vitro* transformation of peripheral blood lymphocytes by *Brucella* antigens is one of the methods to measure CMI responses. A linear relationship has been demonstrated between the logarithm of the number of cells tested and the logarithm of tritiated thymidine (^{3}H-thymidine) incorporated into deoxyribonucleic acid. Various terms are used for the assay such as blastogenesis, lymphocyte transformation, and lymphocyte stimulation. The assays have been evaluated for diagnostic purposes since the serologic tests which are easier to perform have not been fully successful in distinguishing infected animals nor in predicting protective immunity.

Since serologic tests which measure humoral responses to exposure to *Brucella* antigens have diagnostic limitations, early studies with the lymphocyte stimulation test examined the temporal development of cell-mediated responses in cattle. Kaneene et al.,[28] using an extract of autoclaved *B. abortus* cells (BASA) as antigen, found that experimentally infected cattle developed cellular responses by 1 to 2 weeks postexposure and several days earlier than seroagglutinins could be detected. These responses persisted in nearly all infected cattle. A study in a naturally infected herd confirmed that lymphocyte stimulation responses preceded development of antibodies.[29]

The effects of various vaccines on cell-mediated responses have been examined by several workers. Kaneene et al.[30] found that cellular responses persisted longer in some calves than humoral responses following *B. abortus* strain 19 vaccination and that there was an absence of correlation between the responses. They suggested that the cellular assay could possibly be used to differentiate persisting serological reactions due to vaccine or infection. In other trials with *Brucella* polypeptide containing some LPS antigen, Chukwu[31] reported that a single dose of strain 19 induced very little of T cell memory whereas repeated vaccination produced prolonged immunological memory as measured by a lymphocyte stimulation test. In contrast to poor responses to a single dose of strain 19, increased and sustained responses were demonstrated in cattle vaccinated with *B. abortus* 45/20 killed adjuvant vaccine.[32] The author concluded this was probably due to stimulation of the reticuloendothelial system by the oil adjuvant.

An original objective of evaluation of the whole blood lymphocyte stimulation test was to determine whether results could be correlated with protection against *Brucella* infection. A lymphocyte blast transformation assay, using heat-killed *B. abortus* as antigen, was performed sequentially after challenge exposure in strain 19-vaccinated and -nonvaccinated heifers.[33] Mean responses were higher in vaccinated heifers, but no association was found between these and clinical responses to challenge exposure. Chukwu[34] evaluated cell-mediated immune responses before and after challenge with a virulent strain of *B. abortus* in strain 19- and 45/20-vaccinated heifers. Infected (culture positive) heifers had very high lymphocyte transformation (LT) responses, whereas those of vaccinated culture-negative heifers were low. He concluded there was no correlation between the lymphocyte stimulation responses and protective immunity, because the strain 19-vaccinated heifers were more resistant to infection, yet had the lowest LT responses. In fact, it is to be expected that the magnitude and duration of detectable immune responses would be decreased in immune animals which eliminate infection sooner and, therefore, remove the antigenic stimulus for a high and protracted response.

A principal problem in the control of bovine brucellosis is the accurate differentiation of cattle infected with virulent *Brucella* from those with postvaccinal antibodies or with infections caused by antigenically related organisms. Some workers have suggested that the lymphocyte stimulation assays could serve this purpose. Kaneene et al.[35] found significant correlation between cell-mediated responses and bacteriological results. Cattle infected with virulent strains had higher responses than those infected with strain 19. Cattle experimentally infected with *Yersinia enterocolitica* serotype 09, having antibodies which reacted to *Brucella* antigens, could be differentiated by a lymphocyte transformation method from *Brucella*-infected cattle.[36]

Baldwin and co-workers[37] examined this question with a lymphocyte stimulation test utilizing as antigen purified *B. abortus* porin which had been denatured and contained less than 1.5% LPS. They used a cell titration assay which provided a more reliable and discriminating method for evaluating lymphocyte stimulation than did single point assays.[38] In comparison with BASA, porin antigen was much more accurate in distinguishing *B. abortus* infected from uninfected, nonvaccinated cattle.[37] However, nonpregnant heifers experimentally infected with *B. abortus* strain 2308 sometimes failed to react in the lymphocyte stimulation assay despite the presence of strong antibody responses. Moreover, positive reactions were detected in some heifers following vaccination with strain 19 or infection with cross-reactive *Escherichia coli* 0116:H31. It was concluded that the lymphocyte stimulation assay with either porin or BASA could serve as a useful adjunct to serological tests but should not be relied upon as a sole method for diagnosis of bovine brucellosis.

Lymphocyte stimulation assays have been studied in animal species other than cattle. Kaneene et al.[39] studied cell-mediated responses in naturally infected swine, suckling pigs, and controls. On a group basis there was a high correlation between serum antibodies and lymphocyte transformation, but on an individual basis the correlation was poor. Jones and Buening[40] found that a *Brucella* antigen prepared from smooth *B. abortus* strain 2308 stimulated an antigen-specific lymphoproliferative response in tests on whole blood from *B. abortus*-infected guinea pigs. The responding cell population was characterized as mostly T lymphocytes. However, a negative response did not exclude infection.

It is not surprising that interpretations of lymphocyte stimulation assays differ, since there has existed no uniformity among investigators in the methods employed for the conduct and analysis of these tests. The choice of antigen and the concentration or concentrations of lymphocytes tested have been major factors contributing to these differences. Results are also affected to an important extent by variations in lymphocyte reactivity independent of their sensitization to the test antigen due to age and health of the animals, stage of gestation, seasonal changes, and the composition of the tissue culture medium including numbers of monocytes in the cultures. Due to all of these variables, results of an individual assay, in particular one relying on a single cell concentration, may be misleading and lead to erroneous conclusions.[37,38]

B. LEUKOCYTE MIGRATION INHIBITION

The search for an *in vitro* method to measure CMI has led to evaluations of assays for migration inhibitory factor (MIF). Azadegan et al.[41] developed a MIF assay under agarose of bovine mononuclear leukocytes from vaccinated (strain 19) and control calves. They found a significant difference between migration of leukocytes from the two groups and concluded the procedure could serve as an alternative or complement to other types of assays for CMI. Subsequent studies[42] in vaccinated and control heifers exposed to experimental infection demonstrated that the degree of leukocyte migration inhibition (LMI) was not a

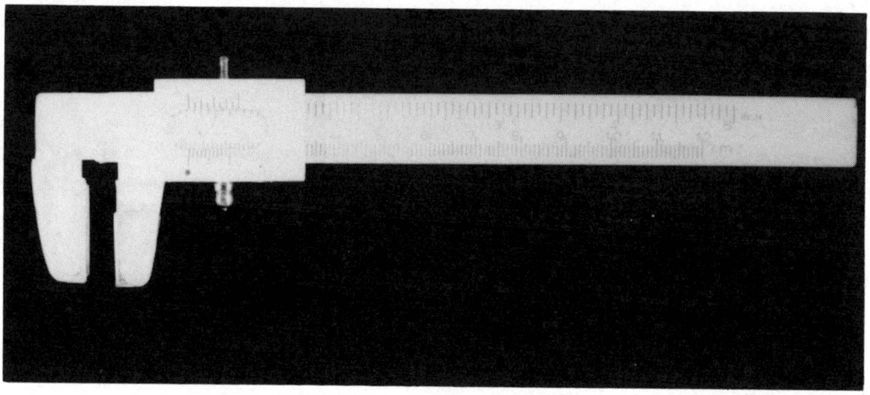

FIGURE 1. A type of caliper used to measure skin thickness for reactions to *Brucella* allergens.

measurement of protection. All heifers had increased LMI after exposure, and controls which aborted had the highest responses. It was concluded that the assay was not practical or useful. A limited study by Patel et al.[43] suggested that a LMI test may be useful in serologically doubtful cases.

Few reports have dealt with the stimulation or inhibition of leukocyte migration by factors isolated from *Brucella* organisms. Birmingham et al.[44] caused no significant leukocyte migration with any of the *Brucella* fractions which they examined in the absence of serum. An intact complement system appeared to be required for the generation of chemotactic factor. They concluded that C5a was necessary for chemotaxis and that a phenol-soluble LPS was the major active component of the *Brucella* cells, which initiates accumulation of mononuclear and polymorphonuclear phagocytes at sites of invasion.

C. DELAYED-TYPE HYPERSENSITIVITY

Delayed-type hypersensitivity is a correlate of CMI produced by interaction of an injected antigen with specifically sensitized T-lymphocytes. DTH reactions may develop as a result of infection with virulent strains or vaccination with strain 19. A cellular infiltration and various degrees of induration are induced, and the size of the lesion is measured by calipers (Figure 1). The cervical area (Figure 2) is considered to be the most sensitive test site in cattle.

Various antigens produced from different *Brucella* species and by various methods have been used to elicit DTH reactions in field and experimental studies. It is generally accepted that antigens which provoke delayed hypersensitivity reactions are proteins. Allergens containing LPS are unsuitable. The O polysaccharide of *B. abortus* elicits Arthrus-type reactions, and lipid A produces a primary toxid reaction which is indistinguishable from DTH without histological examination. Protein antigens prepared from R and S strains of *B. melitensis* were examined in guinea pigs.[45] A smooth LPS antigen used in normal and infected guinea pigs was of lower potency than protein antigens, and the erythematous response occurred earlier and resembled an Arthrus reaction, rather than DTH. The authors concluded that it is preferable to prepare skin test antigens from rough strains of *Brucella* which do not contain O polysaccharide.

Nicoletti[46] studied the above antigen in three cattle herds in an attempt to detect latent brucellosis following reports[47-49] that DTH reactions may precede development of serum antibodies in infected cattle. Responses to the skin test antigen were inferior to serologic tests in the diagnosis of brucellosis. None of the 23 skin test positive/seronegative cows were positive to subsequent blood tests, and all were culture negative.

Chukwu[50] evaluated the brucellin skin test along with serologic and cellular response assays in cattle of various status. The skin test was negative in 9 of 23 infected heifers and

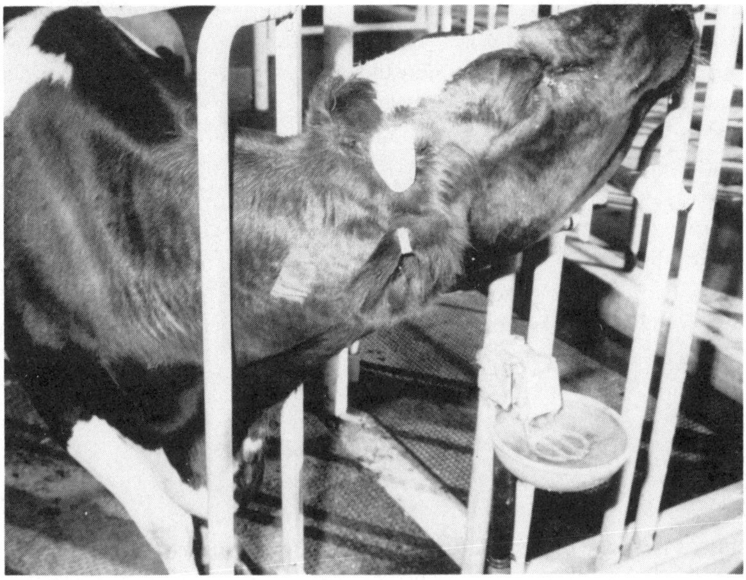

FIGURE 2. The cervical area of the cow is used to determine delayed-type hypersensitivity.

was positive in recently vaccinated (strain 19) heifers. The reactions persisted for many months in heifers immunized with 45/20 adjuvant vaccine. He concluded the skin test was not very reliable for detecting individual infected cattle.

Sutherland[51] examined a protein antigen prepared from vaccine strain 45/20 in 12 controls and in cows which had been sensitized by vaccination with strain 19 or strain 45/20 and were later challenged with a virulent strain. The intradermal test was sensitive but not specific in detecting infected cattle. In an evaluation of an intradermal test in cattle herds, MacDiarmid and Hellstrom[52] found a low sensitivity but high specificity when compared with the complement fixation test.

Use of skin tests in sheep and goats has largely been limited to diagnosis of infected animals. Several studies have compared allergic reactions to serological findings. Jones and Marley[53] titrated a protein antigen in sheep and found the most suitable route of injection was palpebral, with reading of the test at 48 h. Ebadi and Zowghi[54] used an intrapalpebral allergic test in a large flock of infected sheep and compared the results with serologic and bacteriologic findings. There was no constant correlation between serologic and allergic reactions, and both types of diagnostic methods were necessary to detect infected animals. Fensterbank has summarized the use of allergic reactions in diagnosing brucellosis in cattle, sheep, and goats.[55] He stated that the allergic test does not react in noninfected animals except in those previously vaccinated, that neither allergic nor serologic tests detect all infected animals, and that humoral and cellular responses to infection are independent of each other.[48]

The usefulness of skin tests in the diagnosis of brucellosis and as indices of cellular responses is usually based on comparisons with serologic tests or *in vitro* assays such as lymphocyte stimulation. The discrepancies between results of these comparisons are due to a large number of factors. It is well known that some infected animals have long seronegative incubation periods or may have latent infections. Some of these may have DTH reactions. A further problem is anergy, whereby chronically infected animals with serologic titers may be incapable of responding to allergens. The phenomenon of anergy is not well understood, but has been attributed to a lack of antigen-sensitive cells in the peripheral blood and their compartmentalization,[56] to antigen-antibody complexes,[57] to stimulation of suppressor B cells,[58] to production of suppressor T cells or overwhelming infection,[59] and to removal of sensitized lymphocytes from the blood and their suppression in lymphoid tissues.[60]

In addition to the above factors, skin test antigen preparations vary according to the strain of origin, methods of extraction, and concentration of proteins. There are also variations in responses attributable to animal species, site of injection, herd and individual animal status, methods and time of recording results, and criteria for a positive reaction. Serologic and other assays vary in type and in comparisons and evaluation of findings.

In general, strain 19 vaccination has caused only transient and weak DTH responses in cattle, whereas killed adjuvant vaccines such as 45/20 have caused long-lasting responses. Since live vaccines have often produced greater protection against infection, it is questionable whether DTH is a useful indicator of protective immunity, and to date the use of skin tests in cattle for diagnosis has been of limited value.

IV. IMMUNOSTIMULANTS

A. VACCINES

The effects of living and dead vaccines on cellular responses in lymphocyte stimulation and DTH have been cited in previous sections of this chapter. Results of studies have varied widely.

Several workers have examined hematologic changes in cells following strain 19 vaccination or infection in cattle. Lindquist et al.[61] found an increase in leukocyte counts following inoculation of strain 19. Neutrophils were increased, monocytes slightly increased, while lymphocyte numbers decreased. Similar findings were reported by Greatorex.[62] Krishna Murdy and Hajela[63] examined leukocytes in cattle and buffalo heifers and found an initial decrease in total leukocyte counts followed by a gradual increase. They reported a correlation between the appearance and reduction of specific agglutinins and hematologic changes. A decrease in the average number of leukocytes in blood of cattle was found following induced infection.[64]

The numbers of circulating T and B lymphocytes following vaccination or infection has also been studied. Patsula and Kondaurov[65] found a general decrease of T lymphocytes and increase of B lymphocytes in calves immunized and reimmunized with *B. abortus* strains 82 and 104M. The number of B lymphocytes rose sharply with appearance of antibodies. In infected calves, there was an initial decrease of both types of lymphocytes followed by an increase at day 35 postexposure. Iskandarov[66] found that the numbers of B and T lymphocytes were directly related to the dose of strain 19 given to the cattle. As stated above, Chukwu[31] reported that a single dose of strain 19 in cattle was a poor inducer of T cell memory, whereas repeated vaccination provided a secondary antigenic stimulus and prolonged immunologic memory by long-lived T cells.

B. ADJUVANTS

A vaccine composed of killed *B. abortus* 45/20 cells without adjuvant produced lower skin test reactions and lower levels of immunity in guinea pigs in comparison with the same vaccine in adjuvant.[67] Cattle that were given 45/20 adjuvant vaccine had higher skin test and leukocyte stimulation responses than those which were vaccinated with living strain 19.[31] The author concluded that the large responses were due to the effects of the oil adjuvant on cells of the reticuloendothelial system.

Winter et al.[68] measured immune responses by assays for antibodies, DTH, and blastogenesis in adult cattle which had been immunized with various quantities of killed whole cell, outer membrane, or outer membrane protein antigens of *B. abortus* combined with trehalose dimycolate and muramyl dipeptide adjuvants. The immune responses were measured to porin proteins to provide a comparison among groups of cattle and in the belief that the immune response to this antigen was likely to reflect protective immunity. The magnitude of antibody, DTH, and blastogenesis responses exceeded those observed after experimental infections.

Levamisole, an optic isomer of the phenyl-imidothiazole salt of tetramisole, has been used to stimulate host defenses in mice against *B. abortus*.[69] The precise mechanism of action on cells of the immune system is not known but is probably due to action on T lymphocytes. It has been suggested that its immunomodulary action is by induction of maturation of early precursor cells into functional T lymphocytes.[70] It has no direct action on B lymphocytes.[71]

Serological and cell-mediated immune responses were measured in cattle following strain 19 vaccination with and without simultaneous administration of levamisole.[72] Levamisole administered with strain 19 appeared to have no effect on CMI responses as measured by skin tests and lymphocyte transformation. However, antibody titers were moderately elevated in this group of cattle. The author suggested this was possibly mediated through helper T cells and macrophages. Confer et al.[73] did not observe altered antibody responses to simultaneous administration of levamisole and strain 19 in steers. Cellular responses were not studied. *In vitro* enhancement of stimulation of bovine lymphocytes has been found with levamisole[74] and indomethacin.[75]

Antibody titers, blastogenesis, and skin tests were studied in guinea pigs inoculated with strain 19 and killed *Propionibacterium acnes*.[76] No immunopotentiation by *P. acnes* was observed in antibody titers and DTH responses. However, the proliferative response of lymphocytes when measured by blastogenesis was significantly higher. Also, following challenge, there was a significant decrease in splenic bacterial counts in guinea pigs which had received strain 19 with *P. acnes*. The authors concluded that this was due to the effect of macrophages activated by *P. acnes*, to a T cell-mediated cytolytic mechanism, or both.

V. FACTORS AFFECTING THE IMMUNE RESPONSE: HOST STATUS

Few studies have examined the effects of physiological or environmental factors on cellular responses of natural hosts of brucellosis. Soper et al.[77] reported a seasonal fluctuation in blastogenesis of bovine lymphocytes in response to phytohemagglutinin and concanavalin A with mean counts greater in August than in February. They also observed variations due to stage of pregnancy and to age. Winter et al.[68] also noted that lymphocyte responses were much lower in winter months. Winter and co-workers[78] performed another study to determine whether humoral or CMI responses of cattle to a *B. abortus* vaccine were influenced by the stage of gestation. Heifers were vaccinated with cell envelopes of *B. abortus* strain 45/20 in trehalose dimycolate and muramyl dipeptide adjuvants. Blastogenesis and antibody assays to porin and group 3 outer membrane proteins were performed monthly, and dermal hypersensitivity was tested during the last month of gestation. Skin test responses were significantly greater to both proteins in nonpregnant heifers at 24 but not 48 h after injection. Small but significant differences were noted in the relative quantities of IgA antibodies between nonpregnant and pregnant heifers during months 4 to 6 of gestation. Pregnancy had no effect on the response of lymphocytes to phytohemagglutinin or concanavalin A, but there was evidence for a soluble suppressor factor in the plasmas of pregnant heifers which reduced responses to both of these mitogens. The authors concluded that the magnitude, quality, and duration of humoral and CMI responses to a *Brucella* vaccine were, with few exceptions, unaffected by pregnancy.[78]

VI. CONCLUSIONS

Cell-mediated immune (CMI) responses are defined as those in which T lymphocytes function as effector cells in conferring protective immunity. Examples of CMI responses include the activation of macrophages by T cell lymphokines and the direct lysis of target cells by cytotoxic T cells. The results of studies in animal models indicate that protective immunity to *B. abortus* requires both humoral and CMI responses.

Two widely accepted correlates of CMI are DTH reactions and lymphocyte stimulation tests. It should be stressed that lymphocyte stimulation assays are capable of measuring blastogenic transformation of any type of lymphocyte, including T suppressor cells and B cells. When denatured antigens are employed, however, it is much less likely that the latter two cell types will be detected.

The application of DTH and lymphocyte stimulation tests to bovine brucellosis has demonstrated that these assays have more value in demonstrating infection than in predicting a state of acquired resistance to the disease. However, the use of these assays for diagnosis of bovine brucellosis has achieved only partial success due to limitations in specificity, and they are consequently not recommended as definitive means for diagnosing infection.

It remains to be fully established which subsets of host cells are involved in lymphocyte stimulation, in DTH, and in protective immunity in natural host brucellosis. At the present time, a satisfactory *in vivo* or *in vitro* method for predicting the degree of immunity to *Brucella* organisms does not exist.

REFERENCES

1. **Mackaness, G. B.,** The immunological basis of acquired resistance, *J. Exp. Med.,* 120, 105, 1964.
2. **Cheers, C.,** Pathogenesis and cellular immunity in experimental murine brucellosis, *Dev. Biol. Stand.,* 56, 237, 1984.
3. **Holland, J. J. and Pickett, M. J.,** A cellular basis of immunity in experimental Brucella infection, *J. Exp. Med.,* 198, 343, 1958.
4. **Elberg, S. S.,** Immunity to Brucella infection, *Medicine,* 52, 339, 1973.
5. **Montaraz, J. A., Winter, A. J., Hunter, D. M., Sowa, B. A., Wu, A. M., and Adams, L. G.,** Protection against *Brucella abortus* in mice with O-polysaccharide-specific monoclonal antibodies, *Infect. Immun.,* 51, 961, 1986.
6. **Bascoul, S., Cannat, A., Hugnet, F., and Serre, A.,** Studies on the immune protection to murine experimental brucellosis conferred by *Brucella* fractions. I. Positive role of immune serum, *Immunology,* 35, 213, 1978.
7. **Madraso, E. D. and Cheers, C.,** Polyadenylic acid-polyuridylic acid (poly A:U) and experimental murine brucellosis. II. Macrophages as target cells of poly A:U in experimental brucellosis, *Immunology,* 35, 77, 1978.
8. **Pardon, P.,** Resistance against a subcutaneous *Brucella* challenge in mice immunized with living or dead *Brucella* or by transfer immune serum, *Ann. Immunol. (Paris),* 128C, 1025, 1977.
9. **Plommet, M. and Plommet, A. M.,** Immune serum-mediated effects on brucellosis evolution in mice, *Infect. Immun.,* 41, 97, 1983.
10. **Sulitzeanu, D.,** Passive protection experiments with *Brucella* antisera, *J. Hyg.,* 53, 133, 1955.
11. **Limet, J., Plommet, A. M., Dubray, G., and Plommet, M.,** Immunity conferred upon mice by anti-LPS monoclonal antibodies in murine brucellosis, *Ann. Inst. Pasteur Immunol.,* 138, 417, 1987.
12. **Jones, L. M. and Berman, D. T.,** Antibody-mediated and delayed-type hypersensitivity reactions to Brucella skin test antigens in guinea pigs, *Infect. Immun.,* 11, 360, 1975.
13. **Kreutzer, D. L., Dreyfus, L. A., and Robertson, D. C.,** Interaction of polymorphonuclear leukocytes with smooth and rough strains of *Brucella abortus, Infect. Immun.,* 23, 737, 1979.
14. **Bertram, T. A., Canning, P. C., and Roth, J. A.,** Preferential inhibition of primary granule release from bovine neutrophils by a *Brucella abortus* extract, *Infect. Immun.,* 52, 285, 1986.
15. **Canning, P. C., Roth, J. A., Tabatabai, L. B., and Deyoe, B. L.,** Isolation of components of *Brucella abortus* responsible for inhibition of function in bovine neutrophils, *J. Infect. Dis.,* 152, 913, 1985.

16. **Canning, P. C., Roth, J. A., and Deyoe, B. L.,** Release of 5′-guanosine monophosphate and adenine by *Brucella abortus* and their role of the intracellular survival of the bacteria, *J. Infect. Dis.*, 154, 464, 1986.
17. **Quie, P. G.,** Perturbation of the normal mechanisms of intraleukocytic killing of bacteria, *J. Infect. Dis.*, 148, 189, 1983.
18. **Frenchick, P. J., Markam, R. J. F., and Cochrane, A. H.,** Inhibition of phagosome-lysosome fusion in macrophages by soluble extracts of virulent *Brucella abortus, Am. J. Vet. Res.*, 46, 332, 1985.
19. **Braun, W., Pomales-Lebron, A., and Stinebring, W. R.,** Interactions between mononuclear phagocytes and *Brucella abortus* strains of different virulence, *Proc. Soc. Exp. Biol. Med.*, 97, 393, 1958.
20. **Freeman, B. A., Kross, D. J., and Circo, R.,** Host-parasite relationships in brucellosis. II. Destruction of macrophage cultures by Brucella of different virulence, *J. Infect. Dis.*, 108, 333, 1961.
21. **Freeman, B. A., Pearson, G. R., and Hines, W. D.,** Host-parasite relationships in brucellosis. III. Behavior of avirulent Brucella in tissue culture monocytes, *J. Infect. Dis.*, 114, 441, 1964.
22. **Smith, H. and Fitzgeorge, R. B.,** The chemical basis of the virulence of *Brucella abortus*. V. The basis of intracellular survival and growth in bovine phagocytes, *Br. J. Exp. Pathol.*, 45, 174, 1964.
23. **Joubert, L. M. and Valette, L. R.,** Behavior of *Brucella abortus* in phagocytes. Immunophagocytic test for activity of Brucella vaccines and their stimulants, *Bull. Acad. Vet. Fr.*, 40, 111, 1967; Abstr. 454 in *Vet. Bull.*, 38, 1968.
24. **Smith, H. and Fitzgeorge, R. B.,** The effect of opsonization with antiserum on the survival of *Brucella abortus* within bovine phagocytes, *Br. J. Exp. Pathol.*, 45, 672, 1964.
25. **Ralston, D. J. and Elberg, S. S.,** Serum-mediated immune cellular responses to *Brucella melitensis*. VI. The sequence of cellular and serologic changes after immunization of the rabbit with Rev 1 vaccine strain, *J. Infect. Dis.*, 123, 595, 1971.
26. **Harmon, B. G.,** The role of mammary gland macrophages in chronic bovine brucellosis, *Diss. Abstr. Int. B.*, 47, 2695, 1987.
27. **Splitter, G. A. and Everlith, K. M.,** Collaboration of bovine T lymphocytes and macrophages in T lymphocyte responses to *Brucella abortus, Infect. Immun.*, 51, 776, 1986.
28. **Kaneene, J. M. B., Angus, R. D., Johnson, D. W., Muscoplat, C. C., Anderson, R. K., and Pietz, D. E.,** Temporal cell-mediated immune responses of cattle following experimental and natural exposure to living *Brucella abortus, Can. J. Comp. Med.*, 43, 132, 1979.
29. **Kaneene, J. M. B., Johnson, D. W., Anderson, R. K., and Muscoplat, C. C.,** Utilization of a specific in vitro lymphocyte immunostimulation assay as an aid in detection of Brucella-infected cattle not detected by serologic tests, *J. Clin. Microbiol.*, 8, 512, 1978.
30. **Kaneene, J. M. B., Anderson, R. K., Muscoplat, C. C., and Johnson, D. W.,** Cell-mediated immune response in cattle vaccinated with *Brucella abortus* strain 19 vaccine and nonexposed control animals of the same age, *Am. J. Vet. Res.*, 40, 999, 1979.
31. **Chukwu, C.,** Effect of repeated vaccination on cell-mediated immune response to *Brucella abortus* strain 19 vaccine in cattle, *Microbios Lett.*, 34, 35, 1987.
32. **Chukwu, C.,** Comparison of cell-mediated immune responses to *Brucella abortus* strain 19 vaccine and *Brucella abortus* killed 45/20 adjuvant vaccine in cattle, *Microbios Lett.*, 34, 93, 1987.
33. **Confer, A. W., Hall, S. M., Faulkner, C. B., Espe, B. H., Deyoe, B. L., Morton, R. J., and Smith, R. A.,** Effects of challenge dose on the clinical and immune responses of cattle vaccinated with reduced doses of *Brucella abortus* strain 19, *Vet. Microbiol.*, 10, 561, 1985.
34. **Chukwu, C. C.,** Cell-mediated immunity related to challenge exposure of cattle immunized with attenuated or inactivated strains of *Brucella abortus, Microbios Lett.*, 34, 147, 1987.
35. **Kaneene, J. M., Anderson, R. K., Johnson, D. W., Muscoplat, C. C., Nicoletti, P., Angus, R. D., Pietz, D. E., and Klausner, D. J.,** Whole-blood lymphocyte stimulation assay for measurement of cell-mediated immune responses in bovine brucellosis, *J. Clin. Microbiol.*, 7, 550, 1978.
36. **Chukwu, C. C.,** Differentiation of *Brucella abortus* and *Yersinia enterocolitica* serotype 09 infections: use of lymphocyte transformation test, *Int. J. Zoon.*, 12, 126, 1985.
37. **Baldwin, C. L., Verstreate, D. R., and Winter, A. J.,** Immune response of cattle to *Brucella abortus* outer membrane proteins measured by lymphocyte blastogenesis, *Vet. Immunol. Immunopathol.*, 9, 383, 1985.
38. **Baldwin, C. L., Antezak, D. F., and Winter, A. F.,** Cell titration for measuring blastogenesis of bovine lymphocytes, *Vet. Immunol. Immunopathol.*, 9, 319, 1985.
39. **Kaneene, J. M., Anderson, R. K., Johnson, D. W., Angus, R. D., Muscoplat, C. C., Pietz, D. E., Vanderwagen, L. C., and Sloane, E. E.,** Cell-mediated immune responses in swine from a herd infected with *Brucella suis, Am. J. Vet. Res.*, 39, 1742, 1978.
40. **Jones, R. L. and Buening, G. M.,** *Brucella abortus* antigen-induced lymphocyte reactivity in guinea pigs, *Vet. Immunol. Immunopathol.*, 4, 433, 1983.
41. **Azadegan, A., Kaneene, J. M. B., Muscoplat, C. C., and Johnson, D. W.,** Development of a migration inhibitory factor assay under agarose of bovine mononuclear leukocytes using an antigen of *Brucella abortus, Am. J. Vet. Res.*, 42, 122, 1981.

42. **Dorsey, T. A. and Deyoe, B. L.,** Leukocyte migration-inhibition responses of nonvaccinated and vaccinated heifers to experimental infection with *Brucella abortus, Am. J. Vet. Res.*, 43, 548, 1982.
43. **Patel, J. B., Zhala, V. M., and Jaiswal, T. N.,** Leucocyte migration inhibition test as an aid to the diagnosis of brucellosis in cattle, *Indian Vet. J.*, 64, 365, 1987.
44. **Birmingham, J. R., Tabatabai, L. B., Deyoe, B. L., Jeska, E. L., and Nuessen, M. E.,** Generation of chemotactic factor for granulocytes and monocytes from serum by fractions of *Brucella abortus, Immunology,* 46, 17, 1982.
45. **Jones, L. M., Diaz, R., and Taylor, A. G.,** Characterization of allergens prepared from smooth and rough strains of *Brucella melitensis, Br. J. Exp. Pathol.*, 54, 492, 1973.
46. **Nicoletti, P.,** The use of a Brucella protein antigen in dermal hypersensitivity as an adjunct method to diagnose bovine brucellosis, *Vet. Immunol. Immunopathol.*, 5, 27, 1983/1984.
47. **Kasyanov, A. N.,** Significance of allergenic diagnostic methodology in cattle herds acutely infected with brucellosis, *Tr. Vses. Ord. Lenina Int. Exp. Vet.*, 42, 269, 1974.
48. **Fensterbank, R.,** Diagnostic allergique de la brucellose bovine. 2. Utilisation du test allergique dans les troupeaux infectes, *Ann. Rech. Vet.*, 8, 195, 1977.
49. **Cunningham, B., Miler, J. J., Dolan, L., McKeon, F., and O'Meara, M.,** Immunological characteristics in cattle of allergens derived from smooth *Brucella abortus* S99, *Vet. Rec.*, 107, 369, 1980.
50. **Chukwu, C. C.,** Evaluation of brucellin skin test for bovine brucellosis, *Int. J. Zoon.*, 12, 6, 1985.
51. **Sutherland, S. S.,** Assessment of an intradermal test for detection of bovine brucellosis, *Vet. Res. Commun.*, 6, 297, 1983.
52. **MacDiarmid, S. C. and Hellstrom, J. S.,** An intradermal test for the diagnosis of brucellosis in extensively managed cattle herds, *Prev. Vet. Med.*, 4, 361, 1987.
53. **Jones, L. M. and Marly, J.,** Titration of a Brucella protein allergen in sheep sensitized with *Brucella melitensis, Ann. Rech. Vet.*, 6, 173, 1975.
54. **Ebadi, A. and Zowghi, E.,** The use of allergic test in the diagnosis of *Brucella melitensis* infection in sheep, *Br. Vet. J.*, 139, 456, 1983.
55. **Fensterbank, R.,** Allergic diagnosis of brucellosis, in *Brucella Melitensis,* Verger, J. M. and Plommet, M., Eds., Martinus Nijhoff, The Hague, 1985, pp. 167.
56. **Schlossman, S. F., Levin, H. A., Rocklin, R. E., and David, J. R.,** The compartmentalization of antigen-reactive lymphocytes in desensitized guinea pigs, *J. Exp. Med.*, 134, 741, 1971.
57. **Outteridge, P. M. and Lepper, A. W. D.,** Immunosuppressive factors released by transforming lymphocytes in the delayed hypersensitivity skin response to tuberculin, *Immunology,* 25, 981, 1973.
58. **Neta, R. and Salvin, S. B.,** Specific suppression of delayed hypersensitivity: the possible presence of a suppressor B cell in the regulation of delayed hypersensitivity, *J. Immunol.*, 113, 1716, 1974.
59. **Kantor, F. S.,** Infection, anergy and cell-mediated immunity, *N. Engl. J. Med.*, 292, 629, 1975.
60. **Corner, L. A., Outteridge, P. M., Pearson, C. W., and Lepper, A. W. D.,** Immune mechanisms affecting bovine lymphocytes during suppression of tuberculin skin sensitivity in cattle, *Int. Arch. Allergy Appl. Immunol.*, 52, 3, 1976.
61. **Lundquist, N. S., Zehner, C. E., and Phillips, P. H.,** The effect of vaccination with *Brucella abortus* (strain 19) on certain blood constituents in young heifers, *J. Dairy Sci.*, 32, 914, 1949.
62. **Greatorex, J. C.,** Changes in blood picture of calves following vaccination with *Brucella abortus* (strain 19) vaccine, *J. Comp. Pathol.*, 68, 36, 1958.
63. **Krishna Murdy, D. and Hajela, S. K.,** Haematological changes in adult bovine and buffalo heifers following vaccination with *Brucella abortus* (strain 19) vaccine, *Indian J. Vet. Sci.*, 32, 125, 1962.
64. **Ferguson, L. C., Irwin, M. R., and Beach, B. A.,** Interrelationships of the blood cells of cattle in health and following an induced infection with *Brucella abortus, J. Infect. Dis.*, 82, 101, 1948.
65. **Patsula, Yu I. and Kondaurov, B. I.,** Numbers of T and B lymphocytes in the blood of calves following inoculation of *Brucella abortus, Veterinariya, Moscow,* 7, 30, 1981; Abstr. 1468 in *Vet. Bull.*, 1982.
66. **Iskandarov, M. I.,** Status of T and B systems of immunity in cattle immunized with various doses of *Brucella abortus* strain 19 vaccine, *Tr. Vses. Inst. Eksp. Vet.*, 63, 11, 1985; Abstr. 1806 in *Vet. Bull.*, 1987.
67. **Jones, L. M. and Berman, D. T.,** Antibody response, delayed hypersensitivity, and immunity in guinea pigs induced by smooth and rough strains of *Brucella abortus, J. Infect. Dis.*, 124, 47, 1971.
68. **Winter, A. J., Verstreate, D. R., Hall, C. E., Jacobson, R. H., Castleman, W. L., Meredith, M. P., and McLaughlin, C. A.,** Immune response to porin in cattle immunized with whole cell, outer membrane, and outer membrane protein antigens of *Brucella abortus* combined with trehalose dimycolate and muramyl dipeptide adjuvants, *Infect. Immun.*, 42, 1159, 1983.
69. **Renoux, G. and Renoux, M.,** Stimulation of anti-*Brucella* vaccination in mice by tetramisole, a phenylimidothiazole salt, *Infect. Immun.*, 8, 544, 1973.
70. **Amery, W. K.,** A hypothesis: the mechanism of action of levamisole: immune restoration through enhanced cell maturation, *J. Reticuloendothel. Soc.*, 24, 187, 1978.
71. **Brunner, C. J. and Muscoplat, C. C.,** Immunomodulatory effects of levamisole, *J. Am. Vet. Med. Assoc.*, 176, 1158, 1980.

72. **Chukwu, C. C.,** Cell-mediated immune response of cattle following *Brucella abortus* strain 19 vaccination and simultaneous administration of levamisole, *Microbios Lett.*, 34, 7, 1987.
73. **Confer, A. W., Hall, S. M., and Espe, B. H.,** Transient enhancement of the serum antibody response to *Brucella abortus* strain 19 in cattle treated with levamisole, *Am. J. Vet. Res.*, 46, 2440, 1985.
74. **Kaneene, J. M. B., Okino, F. C., Anderson, R. K., Muscoplat, C. C., and Johnson, D. W.,** Levamisole potentiation of antigen specific lymphocyte blastogenic response in *Brucella abortus* exposed but nonresponsive cattle, *Vet. Immunol. Immunopathol.*, 2, 75, 1981.
75. **Kaneene, J. M. B., Anderson, R. K., Johnson, D. W., and Muscoplat, C. C.,** Application of indomethacin as a potentiator of lymphocyte blastogenesis in *Brucella Abortus* exposed cattle, *Vet. Immunol. Immunopathol.*, 4, 375, 1983.
76. **Panangala, V. S., Haynes, T. B., Schultz, R. D., and Mitra, A.,** Immunomodulation with killed *Propionibacterium acnes* in guinea pigs simultaneously vaccinated with *Brucella abortus* strain 19, *Vet. Immunol. Immunopathol.*, 13, 71, 1986.
77. **Soper, F. E., Muscoplat, C. C., and Johnson, D. W.,** *In vitro* stimulation of bovine peripheral blood lymphocytes: analysis of variation of lymphocyte blastogenic response in normal dairy cattle, *Am. J. Vet. Res.*, 39, 1039, 1978.
78. **Winter, A. J., Hall, C. E., Jacobson, R. H., Verstreate, D. R., Meredith, M. P., and Castleman, W. L.,** Effect of pregnancy on the immune response of cattle of *Brucella* vaccine, *J. Reprod. Immunol.*, 9, 313, 1986.

Chapter 5

DETECTION OF *BRUCELLA* CELLS AND CELL COMPONENTS

John E. Mayfield, John A. Bantle, Darla R. Ewalt, Vincent P. Meador, and Louisa B. Tabatabai

TABLE OF CONTENTS

I. INTRODUCTION

Quick and accurate diagnosis is essential for an effective medical or epidemiological response to any disease. Ideally, such identification is accomplished by the observation of unique symptoms and confirmed by the detection of circulating antibodies directed against the disease organism or by the direct detection of the offending organism or its components.

Because brucellosis in animals is generally not characterized by a unique set of symptoms and in ungulates may even be asymptomatic for long periods of time, diagnosis has traditionally relied on the detection of circulating antibodies followed by the bacteriologic isolation of living organisms.[1-4] *Brucella* culture takes several days and requires reasonably fresh samples. Frequently, appropriate samples such as tissue from spleen, lymph node, or mammary gland are not practical to obtain, while blood, milk, or vaginal secretions may only intermittently exhibit living organisms. Antibody detection is faster, but not all infected animals produce significant levels of circulating antibodies,[2,4] and a number of unrelated organisms are known that produce cross-reacting antibodies.[5] Because of these difficulties, there is increasing interest in developing techniques for the direct detection of *Brucella*-derived molecules in tissue, blood, or other samples. To date, these efforts have mostly focused on the detection of *Brucella* antigens, but the detection of *Brucella*-specific nucleic acid sequences also has significant promise. Various techniques for antigen detection have been extensively developed for other diseases, but for brucellosis, such experimentation has been limited and little attention has been paid to protein antigens.

II. BACTERIOLOGIC ISOLATION OF BRUCELLAE

Isolation of living brucellae from tissues or fluid samples such as milk has traditionally provided the most accurate method for the detection of *Brucella* infection. The development of effective culture methods has progressed steadily over the past 75 years, and these techniques have been discussed in detail elsewhere.[4,6,7] In this review we will present only a brief overview of the methods which are most widely used by diagnostic laboratories around the world.

Many media have been developed for the isolation of brucellae, and three are most commonly used for growth of the organism. Each may also serve as the basal medium for various selective media. These are serum-dextrose agar, serum-tryptose agar (Difco Laboratories, Inc., Detroit, MI), and serum-trypticase soy agar (Baltimore Biological Laboratories, Baltimore, MD).[6] Likewise, a variety of selective media have been developed for the isolation of brucellae from potentially contaminated tissue sources or milk. The following seven media are most commonly used today (see Table 1 for details).

1. Tryptose agar plus three antibiotics (TSA) contains cycloheximide, bacitracin, polymyxin B, and 5% bovine serum.[6,8] Brucellae grow very well on this medium, but it is not highly selective, and growth of contaminants can be a major problem.[9]
2. Tryptose agar plus three antibiotics and ethyl violet (TSAEV) contains cycloheximide, bacitracin, polymyxin B, ethyl violet, and 5% bovine serum.[6,10] *Brucella abortus* biovar 2 may be inhibited by this medium. It is more selective than the TSA medium, and contamination is not a major problem.[9]
3. Ewalt's medium contains bacitracin, cycloheximide, lincomycin, nystatin, polymyxin B, and 5% bovine serum.[10] *B. abortus* biovar 2 may be inhibited by this medium, and control of contamination is similar to the TSAEV medium.[9]
4. Farrell's medium contains bacitracin, cycloheximide, nalidixic acid, nystatin, polymyxin B, vancomycin, and 5% horse serum.[11] The size of the *Brucella* colonies is usually small, and occasionally growth of the colonies is inhibited. This medium is highly selective and is recommended for isolation from heavily contaminated sources.[12]

TABLE 1
Composition of Selective Media Used for the Isolation of *Brucella* sp.

Medium:	TSA	TSAEV	Ewalt	Farrell	Ryan	B-S
Basal medium:	Serum tryptose[a]	Serum tryptose[a]	Serum tryptose[a]	Serum dextrose[b]	Oxoid blood agar[c]	Tryptone soya[d]
Amphotericin B (μg/ml)	—	—	—	—	—	4
Bacitracin (U/ml)	7.5	7.5	25	25	—	20
Cetrimide	—	—	—	—	1:40,000	—
Cycloheximide (μg/ml)	30	30	100	100	150	100
Cycloserine (μg/ml)	—	—	—	—	—	312.5
Lincomycin (U/ml)	—	—	6	—	—	—
Nalidixic acid (μg/ml)	—	—	—	5	10	5
Nystatin (U/ml)	—	—	100	100	100	100
Penicillin (U/ml)	—	—	—	—	5	—
Polymyxin B (U/ml)	1.8	1.8	5	5	10	5
Ristocetin (μg/ml)	—	—	—	—	10	—
Vancomycin (μg/ml)	—	—	—	20	—	20
Ethyl violet	—	1:800,000	—	—	—	—

[a] Tryptose agar enriched with 5% bovine serum.
[b] Oxoid nutrient agar plus 5% horse serum and 1% dextrose.
[c] Oxoid No.2 blood agar base enriched with 5% defribrinated horse blood.
[d] Oxoid tryptone soya broth plus 5% horse serum.

5. Ryan's medium contains penicillin, ristocetin, polymyxin B, nalidixic acid, cetrimide, nystatin, cycloheximide, and 5% blood.[13] This medium inhibits *B. abortus* biovar 2. Its effectiveness in controlling contamination is moderate.[12]
6. Brodie-Sinton's medium (B-S) contains bacitracin, polymyxin B, nalidixic acid, vancomycin, nystatin, cyclohexmide, amphotericin B, cycloserine, and 5% horse serum.[14] It is a good enrichment broth and is usually used for the isolation of *Brucella* from milk.
7. Biphasic medium has a solid phase of Farrell's medium and a liquid phase of Brodie-Sinton's medium.[15] It is sometimes used as an additional medium in areas with a low incidence of brucellosis. This medium may require longer incubation times for colonies to grow and is moderately effective in controlling contamination.[16]

In most diagnostic situations, contamination is possible, and it is generally recommended that more than one type of selective medium be used.[7,12,17] Also, *Brucella* may be present in very low numbers. Several laboratories use at least four different selective media per specimen, such as TSA, TSAEV, Ewalt's medium, and Farrell's medium.

Both the choice and the condition of specimens are very important. The supramammary lymph nodes are the best source for isolating brucellae in cattle, sheep, and goats.[6] The mandibular lymph nodes are the best source in swine.[18] A complete set of specimens to culture includes lymph nodes (supramammary, mandibular, suprapharyngeal, internal iliac, parotid, prescapular, bronchial, lumbar, mesenteric, hepatic, and superficial inguinal); and section of each udder quarter, uterus, ovaries, spleen, placental cotyledons, fetal tissue (lungs, stomach contents, liver), testes, accessory nodes, semen, and milk.[6,7,19] Since brucellae may localize in small numbers in tissue or milk, it is important to culture as complete a set of specimens as possible. For humans and dogs, blood is often the only available specimen.[6]

Several different culturing techniques are effective. For all methods, the fatty tissue is removed from the lymph nodes by using sterile scissors and forceps, and the cleaned tissue is dipped in 95% ethanol and flamed to remove surface contaminants. The tissue is then ready for culture by one of the following methods:

Slice and mince: — The cleaned tissue is sliced in half with a sterile scalpel and the exposed inner surface is finely minced. The minced surface is then smeared over the entire surface of each selective medium.[20,21] This is a simple method and requires very little equipment. The isolation rate is approximately 40 to 50% but is decreased by inadequate mincing of the inner surface.[16]

Stomacher: — The cleaned tissue is cut into small pieces and placed in a bag with an equal volume of phosphate-buffered saline. The bag is then placed in a Colworth Stomacher (Tekmar Co., Cincinnati, OH) and macerated for a minimum of 5 min. Using a sterile cotton swab, the tissue suspension is spread over the entire surface of each selective medium.[19] The tissue must be macerated for at least 5 min, and some tissues, particularly uterine tissue, may not macerate well. The isolation rate is approximately 40 to 50%.[16] An additional method is to pipette 1 ml of the tissue suspension onto the solid phase of biphasic medium so that the excess drains into the liquid phase. Usually, a second flask is inoculated with 2 ml of the same tissue suspension. Each flask is observed at 3-, 5-, 7-, and 14- d intervals, and if no growth is observed, each solid phase is reinoculated by tipping the flask.[15] This method is very sensitive, but more prone to contamination than the solid phase methods. The isolation rate is approximately 30 to 40%.[16]

Centrifuge: — Centrifugation is effectively used for the isolation of brucellae from milk. Milk is centrifuged at 6000 to 7000 $\times$ *g* for 15 min. Using a sterile cotton swab, cream and sediment are inoculated onto separate plates of selective media.[6]

Guinea pig inoculation: — This method is rarely used because of the cost and time required. After injecting a suspension of the sample into several guinea pigs, half of the animals are necropsied at 3 weeks and the other half at 6 weeks. Each animal serum is tested for the presence of antibodies, and the spleen, lymph nodes, and any visible lesions are cultured.[7,8]

Culturing from blood: — Heparinized blood can be directly spread onto agar plates or can be inoculated into trypticase soy broth which is frozen, thawed, and incubated for several days before it is subcultured onto solid medium. The second method is slower but more sensitive.[6] Isolation from blood is recommended only if other specimens cannot be obtained.

Brucella ovis and most strains of *B. abortus* require CO_2 for growth. The other species will grow in the presence of CO_2 except *B. canis*, which is somewhat inhibited, and should be cultured aerobically. In standard practice, inoculated plates are incubated in 10% CO_2 at 37°C. Colonies are visible at 3 d, and the final observation can be made at 7 d. Using 10 to 15 $\times$ magnification with oblique light at a 45° angle from the bottom, the colonies are a light greenish-blue with a beige coloration in the central area. The *Brucella* colonies are circular, convex, slightly opaque, and butyrous in nature. The average size after 7 d is 1.5 to 2 mm in diameter. To the experienced observer, the colony morphology is quite distinctive.

The various *Brucella* species and biovars can be identified by their ability to grow on media containing basic fuchsin, thionin, thionin blue, penicillin, and erythritol.[6,7] Other important tests for the complete identification of brucella are lysis by phage, hydrogen sulfide production, urease activity, and determination of the dominant A or M antigen.[6,7,22] Table 2 summarizes these characteristics.

III. ANTIGEN DETECTION

A. ANTIGENS

When either whole smooth killed *B. abortus* or *B. abortus* fractions containing cell

TABLE 2
Differential Growth Characteristics of the Species and Biovars within the Genus *Brucella*[6,23,a]

Species	Biovar	CO_2 for growth	H_2S produced	Growth on dyes[b]: Thiomin A B C	Growth on dyes[b]: Basic fuchsin A B C	Urea hydrolized	Lysis by[c] phage: RTD	Lysis by[c] phage: RTD × 10,000	Agglutination by monospecific antisera: A	Agglutination by monospecific antisera: M	Growth on erythritol: 1 mg/ml	Growth on erythritol: 2 mg/ml[d]	Thionine blue 1:500,000	Penicillin 5 U/ml
abortus	1	(+ −)[e]	+	− − −	+ + +	+ >1 h	+	+	+	−	+	+	+	+
(strain 19)	1	−	+	− − −	+ + +	+ >1 h	+	+	+	−	−	−	−	−
	2	(+ −)[e]	+	− − −	− − −	+ >1 h	+	+	+	−	+ −	+	−	−
	3	(+ −)[e]	+	+ + +	+ + +	+ >1 h	+	+	+	−	+	+	+	+
	4	(+ −)[e]	+	− − −	+ + +	+ >1 h	+	+	−	+	+	+	+	+
	5	−	−	− + +	+ + +	+ >1 h	+	+	−	+	+	+	+	+
	6	−	+ −	− + +	+ + +	+ >1 h	+	+	+	−	+	+	+	+
	7	−	+ −	− + +	+ + +	+ >1 h	+	+	+	+	+	+	+	+
	9	(+ −)[e]	+	− + +	+ + +	+ >1 h	+	+	−	+	+	+		
suis	1	−	+	+ + +	− − −	+ <30 min	—	(+ −)[e]	+	−	+	+	+ −	−
	2	−	+ −	− + +	− − −	+ <30 min	−	(+ −)[e]	+	−	+	+	−	−
	3	−	−	+ + +	− − +	+ <30 min	−	+ −	+	−	+	+	+	+ −
	4	−	−	+ + +	− − +	+ <30 min	−	+ −	+	+	+	+	−	−
	5	−	−	+ + +	− − −	+ <30 min	−	+	−	+				
melitensis	1	−	−	+ + +	+ + +	+ >1 h	−	−	−	+	+	+	+	+
	2	−	−	+ + +	+ + +	+ <30 min	−	−	+	−	+	+	+	+
	3	−	−	+ + +	+ + +	+ >1 h	−	−	+	+	+	+	+	+
neotomae		−	+	− − +	− − +	+ <30 min	−	+ −	+	−	+	+	+	−
canis		−	−	+ + +	− − ±	+ <30 min	−	−	−	−	+	+	+ −	−
ovis[f]		+	−	+ + +	+ + +	− >1 h	−	−	−	−	−	+	−	−

[a] Characteristics which currently serve as the basis for biotyping at the National Veterinary Services Laboratory.
[b] Concentration of dyes: A = 1:25,000, B = 1:50,000, C = 1:100,000.
[c] Tbilisi phage RTD: routine test dilution. See Alton et al.[6] for a description of this procedure.
[d] 5% serum added to medium.
[e] Positive is dominant characteristic, but negative strains occur.
[f] Requires added serum.

envelope components are injected into test animals, a strong humoral antibody response is generated against the smooth lipopolysaccharide (SLPS). These antibodies represent a powerful tool for the identification or detection of *B. abortus* in tissue samples. It should be noted, however, that even though crude LPS is an effective antigen in rabbits, highly purified LPS is an ineffective antigen in this species.[24,25] We have observed that rabbit antisera prepared against partially purified *B. abortus* strain 19 proteins react strongly with a polydisperse material in Western blots of solubilized whole smooth *Brucella* cells, but not solubilized whole rough *Brucella* cells.[26] Superficially, the profile of this material on sodium dodecyl sulfate (SDS) gels looks much like SLPS.[27] These protein preparations are known to contain polysaccharide, but are very low in 2-keto-3-deoxysugars (KDO)[28] (see Table 3). It is not clear at this time whether, in preparations which are low in SLPS and high in protein, the SLPS becomes an extremely effective antigen in rabbits, or whether these protein preparations are contaminated with an undefined polysaccharide. Interestingly, sera from previously infected cattle also react with a similar polydisperse material on Western blots of whole smooth *Brucella* proteins or *Brucella* protein fractions.[40,41] In mice, small amounts of purified SLPS are sufficient to produce high levels of circulating antibody.[42-44] Smooth LPS is known to have a high affinity for proteins;[45,46] so it may be that most test antisera, even those ostensibly against purified protein preparations, also react with LPS antigens.

The "O" chain polysaccharide antigens of *B. abortus* LPS have recently been characterized at the molecular level.[47-49] These structures are not unique to *B. abortus*, but are highly characteristic of the entire *Brucella* genus. Structural similarities also exist with the O chains of some other Gram-negative bacteria which allow antibody cross-reactions.[48] The known cross-reacting species or strains are *Yersinia enterocolitica* serogroup 0:9, *Salmonella* serotypes of Kauffman-White group N, *Escherichia coli* 0:157, *Pseudomonas maltophilia, Francisella tularensis*, and *Vibrio cholerae*.[5] This is not an extensive list, and some are not animal pathogens, but the potential for cross-reaction complicates the use of anti-LPS antiserum as a diagnostic reagent unless the presence of the other known cross-reacting species can be ruled out on other grounds.

The only other known *Brucella* polysaccharide is a poly-D glucose[47,48] which is probably located in the periplasmic space. This polysaccharide has traditionally been called "poly B",[50] and infected animals frequently produce antibodies to it.[51]

Many *Brucella* proteins are also antigenic, and antisera against them represent potentially useful detection reagents. Past work to characterize such antigens, however, has been limited. Several laboratories have made substantial progress in the isolation of outer membranes and the purification of outer membrane proteins.[52-60] Tabatabai and co-workers have worked extensively to characterize and purify a small group of soluble proteins (BCS proteins) which can be extracted from methanol-killed *B. abortus* by 1 *M* NaCl, 0.1 *M* sodium citrate.[35-37] These proteins are probably periplasmic, but may be loosely associated with the outer membrane. The extreme difficulty of separating *Brucella* LPS from *Brucella* proteins[28,45,46] makes it nearly impossible to prepare LPS-free *Brucella* protein or protein fractions from smooth strains. Because of this, it is important to recognize that test antisera produced against proteins or protein fractions isolated from smooth *Brucella* strains may contain anti-LPS antibodies. This point is very important, and investigators working in this field must provide proof that their antigen is SLPS or carbohydrate free, or that their antiserum does not react with SLPS, if the presence of such antibodies would affect the experimental interpretation. Unfortunately, such proof may not be trivial, since undetectable levels of KDO may be insufficient.[28] The limulus lysate gelation assay[27,34] may provide greater sensitivity (Table 3). Another approach is to preabsorb all sera with whole smooth *Brucella* cells prior to use.[54,61] To be assured that a *Brucella* protein is free of *Brucella* SLPS it is best either to isolate the protein from rough *Brucella* strains which do not contain O chain antigen or to clone the gene and isolate the protein from recombinant *E. coli*. This latter approach intro-

TABLE 3
Chemical Composition of Some Selected LPS and Protein Fractions[27-29]

Fraction name	Total carbohydrate[a] (mg/100 mg)	Total protein[b] (mg/100 mg)	Total KDO[c] (mg/100 mg)	KDO/ carbohydrate (μg/mg)	KDO/ protein (μg/ mg)	Carbohydrate/ protein (mg/mg)	LLGA[d] (ng/ml)
f5(crude SLPS)	10.7	24	0.32	29.9	13.3	0.44	1
f5p(pure SLPS)	11.6	6.3	0.62	53.4	98.4	1.84	0.1
R3(crude RLPS)	13.6	1.3	0.25	18.3	192	10.5	10
R3a(pure RLPS)	6.5	1.5	0.74	114	493	4.3	1
E. coli LPS (0128:B12)	18.4	4.4	1.7	92.3	386	4.18	0.2
	(mg/ml)[e]	(mg/ml)	(μg/ml)	(μg/mg)	(μg/ mg)	(mg/ml)	
BCSP extract[f]	0.1	1.69	7.0	70	4.1	0.06	ND[g]
BCSP20[f]	0.1	0.5	0.0	—	—	0.20	ND
Recombinant BCSP20[f]	0.0	1.0	0.0	—	—	0.0	ND
BCSP31[f]	0.01	1.0	<1.0	<100	<1.0	0.01	ND
Recombinant BCSP31[f]	0.0	1.0	0.0	—	—	0.0	ND

[a] Based on the phenol-sulfuric acid method of Dubois,[30] using equal amounts of D-glucose, D-mannose, D-galactose, and D-rhamnose as standard.

[b] Total protein determined by the method of Lowry et al.,[31] using bovine serum albumin as a standard.

[c] Based on the measurement of malonaldehyde formation with thiobarbituric acid after a preliminary hydrolysis step with 0.25 *N* sulfuric acid.[32] 2-keto-3-deoxy octulosonic acid was used as a standard, corrected for 2-deoxy aldoses.[33]

[d] LLGA: *Limulus* lysate gellation assay. Number shown is the amount required for gellation.[34] *E. coli* 0127:B12 LPS served as standard.

[e] Based on the phenol-sulfuric acid method of Dubois,[30] using alpha-D-glucose as a standard.

[f] BCSP extract: *Brucella abortus* salt-extractable proteins;[35-37] BCSP20: 20-kDa protein purified from the BCSP extract; recombinant BCSP20: same protein purified from recombinant *E. coli* cells; BCSP31: 31-kDa protein purified from the BCSP extract; recombinant BCSP31: same protein purified from recombinant *E. coli* cells.

[g] Not determined.

duces the possibility that the protein may be contaminated with *E. coli* antigens. Fortunately, *E. coli* LPS seems to have much less affinity for *Brucella* proteins than *Brucella* LPS (Table 3). Yet another alternative is to use monoclonal antibody instead of antiserum as the test reagent.

At the present time, the number of cloned *B. abortus* genes is quite small. Mayfield and co-workers have cloned the genes for proteins of 20, 31, 34, 45, 58, and 65 kDa.[38,62] These genes code for proteins present in the high salt extract characterized by Tabatabai et al.[35-37] Ficht and co-workers have reported the cloning of the class 2 porin genes,[63] and Sanborn has reported the cloning of a 43-kDa outer membrane protein.[64] Several laboratories have verbally reported the cloning of less well-characterized genes or gene fragments.[65-68] Three of the genes cloned by Mayfield and co-workers, BCSP20, BCSP31, and BCSP45 (20-, 31-, and 45-kDa proteins, respectively), have been expressed at high levels and purified from *E. coli* cells.[38,62] Rabbit antisera against these recombinant proteins do not react with *Brucella* SLPS. Thus far, these sera have only been used to detect *Brucella* antigens in Western blot experiments, where the protein antigens are present in very high concentrations. During the next few years, we can expect reports of the cloning and expression of increasing numbers of outer membrane, periplasmic, and cytoplasmic proteins from *B. abortus*. Antisera to these proteins should represent highly specific probes for the detection of *Brucella* cells or cell fragments.

Antisera to individual *B. abortus* proteins, or monoclonal antibodies to individual proteins, may also prove to be specific to certain *B. abortus* biovars or to certain *Brucella* species. The conservation of individual proteins across the various species and strains of the genus *Brucella* has been little studied. Diaz et al.[69] showed that the majority of bands seen by immunoelectrophoresis were common to all species within the genus *Brucella*. The identity of the bands, however, was unknown. Verstreate and Winter[46] compared the one-dimensional SDS gel electrophoresis profiles of outer membrane proteins from 49 *B. abortus* strains. Generally, the profiles were similar, but some differences were noted. Bricker and co-workers[39,70] have examined the conservation of BCSP20, BCSP31, and BCSP45 in all *Brucella* species and available biovars by SDS-PAGE and Western blot, using antisera to the recombinant proteins. Conservation of these proteins is surprisingly varied. BCSP20 is present at fairly constant levels in all strains and species investigated except for *B. ovis*, in which it is missing; BCSP31 is present in highly variable amounts and is missing from *B. neotomae*; and BCSP45 is missing from *B. suis*. Thus, none of the three antigens is specific to any *Brucella* species or strain, and none is found in all the strains tested either. Also, none of the three proteins is found in *Y. enterocolitica* 0:9 or *E. coli* K12. Clearly, the use of antisera to individual proteins or monoclonal antibodies will require careful investigation of the reactivity of these reagents to the various strains and species of *Brucella* as well as other potentially cross-reacting genera.

A potential complication is presented by the recently discovered IgG binding protein present in *B. abortus* cells.[71] The isotype and species specificity of this protein is not yet known, but its presence suggests that test antisera to any antigen will bind to *B. abortus* if the test is sufficiently sensitive.

Monoclonal antibodies represent an alternative approach to the laborious process of cloning genes and purifying proteins from heterologous cells, but their extreme specificity may create problems. Of particular concern are antibodies specific to epitopes which are obscured in some situations or strains, but not others, even though the protein is always present. Epitopes which are present in native proteins but not in denatured ones are very common[72] and fairly easy to screen for, but epitopes which vary with the physiological state of the cell could be very troublesome.

Earlier efforts to detect *Brucella* antigens in tissues and fluids have employed antisera to either whole *Brucella* cells or cell fractions which contain mixtures of components. Generally, in such mixtures LPS must be considered to be the dominant antigenic component.

B. HISTOLOGICAL TECHNIQUES

Brucellae can be detected histologically in tissues by a variety of specific and nonspecific staining techniques. Modified Koster, modified Ziehl-Neelson, and Gram's staining methods[73] will nonspecifically stain organisms, but cannot differentiate brucellae from other bacteria of similar size, shape, and staining affinity. Antibody labeling techniques using anti-*Brucella* antibody to attach fluorescein, enzyme, or colloidal gold to brucellar antigen can specifically label or stain brucellae. Thus far, these procedures have primarily been used in research. Immunofluorescence and immunoenzymatic techniques are suitable, however, for diagnostic purposes. Immune staining techniques are particularly applicable to brucellosis pathogenesis studies and for identifying organisms in samples from which culture is impossible, such as tissue overgrown with contaminating bacteria or tissue stored in fixatives. Antigen-specific staining techniques also have the advantage of detecting *Brucella* antigen in tissues where viable brucellae are no longer present due to host degradation of organisms, but in which brucellar antigen is still present.

The specificity of immunologic staining for *Brucella* is dependent on the specificity of the primary anti-*Brucella* antibody. Polyclonal antibody can be prepared against either whole cells or *Brucella* cell fractions. As was previously discussed, such sera usually react strongly with LPS, but may react with *Brucella* proteins as well. Since cross-reacting antigens have been reported between *Brucella* and other bacteria,[5] primary antibody attachment to non-*Brucella* organisms is a concern. Monoclonal antibodies should help insure the specificity of primary antigen labeling, but the reduced number of sites for antigen-antibody reaction may be expected to decrease the intensity of specific staining, and therefore, limit their use. At the present time there has been only one preliminary report of monoclonal antibodies use to label brucellae in histological samples.[74]

1. Immunofluorescence

Immunofluorescent labeling of brucellae has been performed on tissues, cytological smears, and cultured cells.[75-78] Histological examination of tissues to be evaluated by this technique requires use of fresh or fresh-frozen tissue. Fluorescent antibody labeling can be very specific and is rapidly and easily performed. Meyer reported that smears of *B. abortus, B. suis, B. melitensis*, and *B. neotomae* strains could be stained using anti-*B. abortus* or anti-*B. melitensis* serum as the primary antibody. A few *B. suis* biovar 3 isolates were refractory to staining, however.[78] Brucellae were not labeled with *Bordatella bronchiseptica* or *pertussis* antisera. Meyer also tested the specificity of her anti-*Brucella* antisera and found no labeling of *Bordatella bronchiseptica, B. pertussis*, or *Pasteurella multocida*.

Corbel[75] demonstrated that fluorescein-conjugated antibodies directed against whole *B. abortus* provided an effective means for labeling brucellae in tissue smears of bovine abortion material. Antisera were prepared by inoculating rabbits with live *B. abortus* strain 544 and conjugating the IgG fraction with fluorescein isothiocyanate. The method detected all species of *Brucella* except *B. ovis*, but it also detected *Y. enterocolitica* serotype 0:9 and *Staphylococcus aureus*. The detection of *S. aureus* was presumably through the action of protein A. *Francisella tularensis* was among the various species tested which were not detected.

Comparison of the sensitivity of bacterial culture to fluorescent antibody staining of tissues or tissue impression smears indicate fluorescent antibody detection is not as sensitive as culture,[76,78] but is applicable for tissues or cells from sites containing numerous brucellae (e.g., placenta, uterine exudate). Deyoe[76] reported that of 100 lymph nodes collected from infected swine, 14 were identified by fluorescent antibody examination as containing brucellae, while 84 were determined to contain brucellae based on bacterial isolation.

2. Immunoenzymatic Staining

This technique involves direct or indirect attachment of enzyme to *Brucella* antigenic sites with anti-*Brucella* antibody and then reaction with an appropriate substrate-chromogen solution to precipitate a colored end product at the antibody attachment site.[79,80] It is a very useful technique for staining brucellae in fixed tissue sections and has numerous advantages over fluorescent antibody staining. Antigen can be stained in the formalin-fixed, paraffin-, or plastic-embedded tissues routinely processed for histological examination. These procedures result in superior morphological detail of cells and antigen localization sites as compared to frozen or fresh processed tissues. The precipitated colored end product can provide a permanent record on the slide. The variety of chromogens allows variation in the staining color and choice of counterstain. Evaluation of the staining requires no special microscopic equipment. Use of preservative-fixed tissues also limits the exposure of laboratory personnel to live brucellae. Fixatives do, however, have the potential to alter antigenic structure and change antibody binding sites. Therefore, brucellar antigens must be evaluated to find the proper fixative that adequately preserves both antibody binding and tissue morphology.

Horseradish peroxidase (HRP) is commonly used as the enzyme for this procedure, but others including alkaline phosphatase and glucose oxidase are also available. While enzyme molecules can be directly conjugated to the primary antibody, it is preferable to increase the number of enzyme molecules bound to each antibody attachment site. This can be accomplished by either the peroxidase-antiperoxidase method[81] or the avidin-biotin-peroxidase complex method.[82,83]

In the first method, three types of antibody are required. The sample is first incubated with, for example, rabbit anti-*Brucella* antisera, followed sequentially by goat anti-rabbit antisera, and HRP-(rabbit)-anti-HRP complex. In the second method, the sample is incubated successively with anti-*Brucella* antibody, biotinylated secondary antibody which is directed against the primary antibody, avidin, and HRP-biotin conjugate.[82] Avidin has a very high affinity for biotin and provides a bridge between the biotinylated antibody and the biotinylated enzyme. Reagents for these procedures, other than the anti-*Brucella* primary antibody, are commercially available. In experiments to date, primary antibody prepared against whole *B. abortus* strain 2308, *B. abortus* LPS fractions, *B. abortus* strain 19 BCSP extract, and whole *B. suis* biovar 4 have all been found suitable for labeling *B. abortus* in formalin-fixed tissues using an avidin-biotin-peroxidase technique.[84]

The relative sensitivity of the avidin-biotin procedure using HRP has been evaluated in comparison with bacterial culture for paired mouse spleen specimens.[83] Primary antibody was prepared against BCSP extract.[35-37] Approximately 1.0×10^6 colony forming units (cfu) of *B. abortus* per gram of spleen were required before specific staining could be positively identified in histologic sections. In general, staining sensitivity in these and other tissues varies according to the site of brucellae localization within the tissue, the number of nonviable brucellae, preservation of the brucellar antigen, and the primary antibody preparation and dilution. As with immunofluorescence, tissues containing few organisms may be poor candidates for diagnosing brucellosis with immunoenzymatic staining; however, the procedure has been used to successfully localize brucellar antigen in tissues in pathogenesis studies[85-88] (see Figure 1). In tissues with sparse brucellar antigen, small chromogen precipitates are difficult to detect and, especially, with the HRP chromogen 3,3-diaminobenzidine which produces a brown-colored end product, may be confused with other similar colored substances in fixed tissues such as hemosiderin and acid hematin.[83]

Brucellae have been stained by the avidin-biotin-HRP technique in formalin-fixed, paraffin-embedded tissues from a variety of hosts including bovine, caprine, murine, and avian species.[83-87] Most tissues are suitable for staining; however, there has been some difficulty with staining conjunctival tissue for brucellar antigen.[84]

3. Immunogold Labeling

Colloidal gold has been used to label specifically brucellar antigen in tissues prepared for transmission electron microscopic examination. Gold particles can be directly or indirectly attached to anti-*Brucella* antibody and appear in tissue specimens as electron-dense spherical particles which vary in size according to the diameter of gold particles used. Gold labeling can also be visualized by light microscopy and appears red in color.[89]

These specific labels are a tremendous asset for electron microscopic localization of brucellae and brucellar antigen in the cell cytoplasm. Lysosomal degradation frequently eliminates identification of brucellae by morphological criteria, but specific antigen labeling is still capable of locating the site of infection (see Figure 2). A notable accomplishment by this technique has been the intracellular localization of *B. abortus* in the rough endoplasmic reticulum of placental trophoblastic epithelial cells.[90] The identification of this unique site for intracellular bacteria was made possible by the electron microscopic examination of immunogold-labeled brucellae using gold-conjugated immunoglobulin from milk whey.[90]

C. OTHER TECHNIQUES BASED ON ANTIGEN-ANTIBODY SPECIFICITY

There are several reports of nonhistological techniques for detecting *Brucella*-derived antigens. In 1976, Sadykov[91] successfully employed a neutralization reaction with a *Brucella* polysaccharide antigen. The test could detect 75 to 250×10^3 brucellae per milliliter in solution. There was no discussion of cross-reacting organisms.

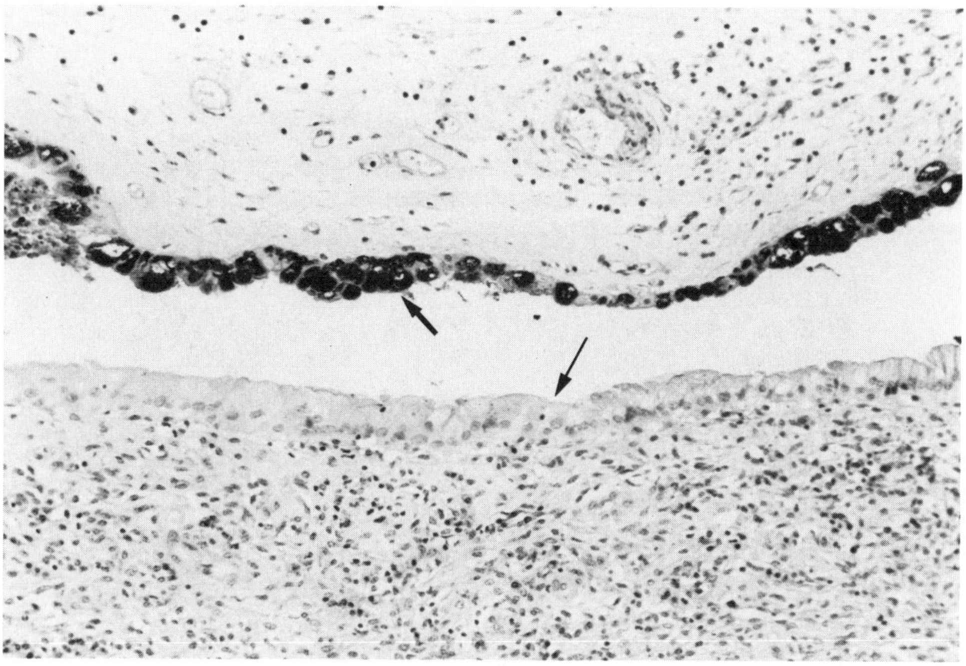

FIGURE 1. Light microscope photograph of placenta (top) and endometrium (bottom) of a *Brucella abortus*-infected cow. Avidin-biotin-peroxidase staining of intracytoplasmic *Brucella abortus* in placental trophoblastic epithelial cells (dark staining cells — wide arrow) using 3,3-diaminobenzidine as the chromogen. The adjacent uterine epithelium does not stain (thin arrow). (Hematoxylin counterstain; magnification = × 75.)

Chen et al.[92] described an ELISA assay for the presence of *Brucella abortus* antigens in vaginal swabs taken from postparturient cows. In this assay, diluted, heat-killed samples were adhered to plastic cuvettes and then incubated with guinea pig anti-*Brucella* serum. The assay was completed by the addition of HRP conjugated goat anti-guinea pig IgG and reaction with a chromogenic substrate. The guinea pig serum was collected from guinea pigs given subcutaneous injections of *B. abortus* 2308. Test results indicated a lower limit of detection of about 250 cfu per 1 ml cuvette. This method is faster than bacteriologic culture and more quantitative than the method described by Corbel.[75]

Mahajan and Kulshreshtha[93] described the use of counterimmunoelectrophoresis to detect soluble *B. melitensis* antigens in the stomachs of aborted sheep fetuses. The method was reported to be somewhat more reliable than the culture of living *B. melitensis* strain from this source. The test antisera were produced in rabbits injected with sonicated *B. melitensis* strain 16M cells. The method is rapid, but may be difficult to scale up for a large number of samples.

The most sensitive method of *Brucella* antigen detection yet described employed an indirect sandwich ELISA using antibody covalently coupled to nylon beads.[94] With a fluorogenic substrate, this technique was capable of detecting 100 fg of solubilized LPS in buffer or 10 brucellae added to a suspension of leukocytes previously isolated from 100 ml of bovine blood. The detection of low picogram concentrations of crude LPS added to undiluted plasma or serum was not possible in this experiment. However, application of the procedure of Reinhold and Fine[95] provided substantially increased sensitivity.[96] In preliminary experiments this bead-ELISA system was successfully used to detect *Brucella* antigens in circulating leukocytes from pregnant heifers during the first 2 weeks after infection.[97]

Recently, Limet[98] evaluated the use of an assay based on the inhibitory activity of LPS

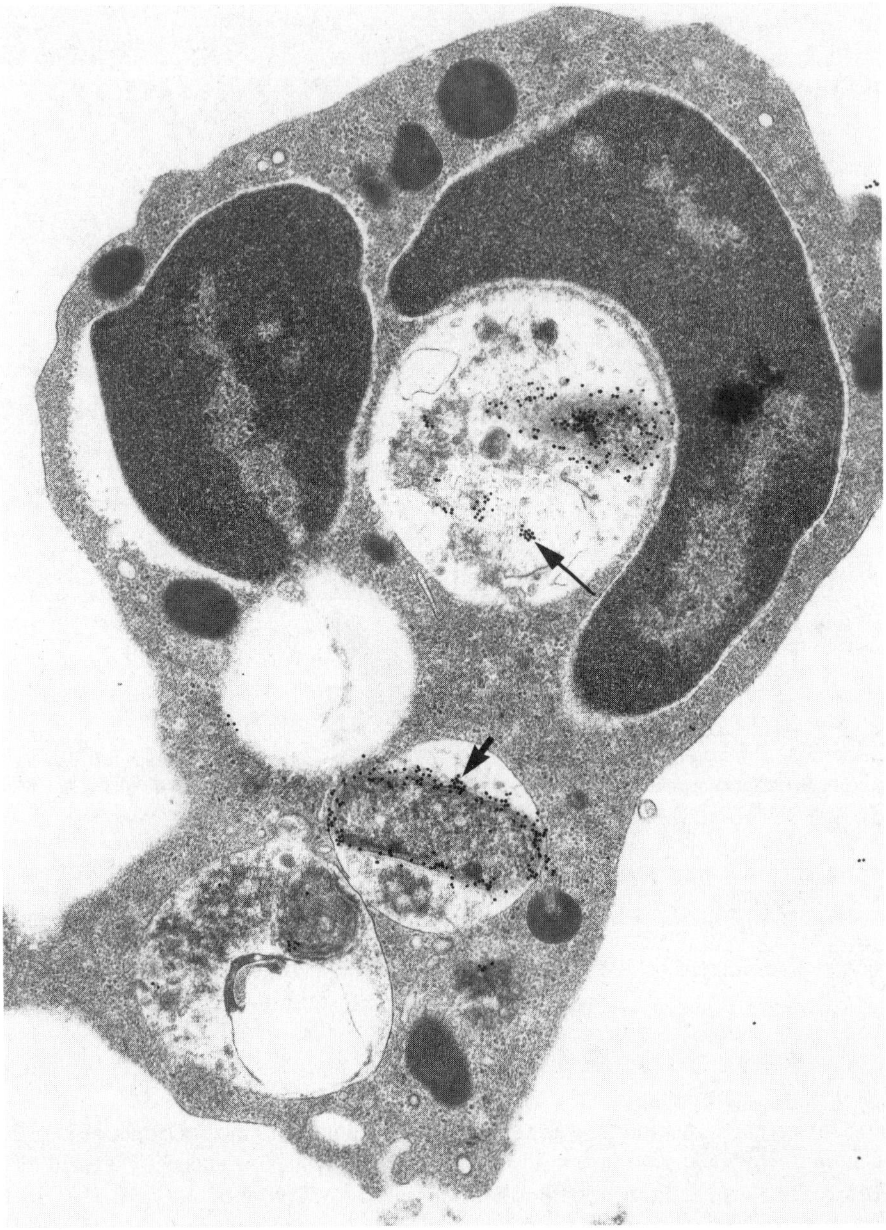

FIGURE 2. Electron micrograph of a neutrophil from the milk of a *Brucella abortus*-infected goat. *Brucella* antigen is labeled with 20-nm colloidal gold-conjugated antibody. One membrane-bound vacuole contains an intact, gold-labeled brucella (wide arrow), while another vacuole contains gold-labeled brucellar fragments (thin arrow).

toward the agglutination of LPS-coated latex beads by anti-LPS antibodies for the detection of *Brucella* LPS in mouse blood and urine. He found that antigens were present from day one after injection of 7.5×10^5 *B. abortus* biovar 3. He also showed that detectability of SLPS antigens in blood or serum was rapidly lost on storage, but that a simple heat or alkaline treatment effectively stabilized or exposed the antigen. A monoclonal antibody was also described which distinguished between *B. abortus* biovar 3 and *Y. enterocolitica* and *B. melitensis*. If confirmed, this result may imply the existence of an SLPS epitope which is unique to *B. abortus*. Monoclonals to such epitopes might represent the ideal reagent for detection of smooth *Brucella* strains, since during infection, SLPS most probably represents the dominant *Brucella* antigen present.

IV. NUCLEIC ACID HYBRIDIZATION

Nucleic acid hybridization has a completely different physical basis than antigen-antibody binding and has the potential for equally good specificity. The technique has been used to good advantage with a number of diseases,[99] but, as yet, there have been only two preliminary reports for *B. abortus*.[100,101] Diagnostic nucleic acid probes offer several potential advantages. These include the choice of high, moderate, or low specificity with the same sample; no need for prior cell culture; direct detection of the pathogen; a high tolerance for contaminating organisms; and compatibility with tissue fixation. Because of the specialized equipment required, the technology is only likely to be used in the research or sophisticated diagnostic laboratory in the near future. Of the different methods available, only the blot and *in situ* methods will be discussed here because these techniques are most applicable to diagnosis.

A. GENERAL ASSAY REQUIREMENTS

Any diagnostic test must be specific, sensitive, reliable, and able to be replicated in a large number of laboratories. Indeed, by these criteria, current diagnostic DNA techniques are comparable to antibody and culture-based tests. However, the key to the adoption of any diagnostic DNA test will be either superior performance or some other clear advantage over established test methods. An illustration of this may be seen in the case of the DNA dot blot test for human malaria employing a ^{32}P-labeled *Plasmodium* repetitive DNA probe developed by Franzen et al.[102] Despite their claiming a sensitivity equivalent to light microscopy and a theoretical sensitivity equal to the current solid phase radioimmunoassay, Bruce-Chwatt[103] commented that the method held no particular advantage over current tests and might even prove to be more costly and time consuming than present assays.

In the final analysis, diagnostic DNA tests will be adopted when they prove to be faster and more cost effective than current tests without sacrificing sensitivity or specificity. These latter qualities of diagnostic DNA tests are only now being realized. Any requirement for a large investment in laboratory equipment and retaining can only be justified if more than one disease can be diagnosed. Consequently, any diagnostic DNA test for brucellosis must be coupled with tests for other diseases. In fact, diagnostic DNA-based tests have been developed for disease organisms such as *Plasmodium*,[102] *Babesia*,[104] pseudorabies virus,[105] bovine herpesvirus,[106] *Leishmania*,[107] *Echinococcus*,[108] *Trichinella*,[109] *Brugia*,[110] *Crithidia*,[111] enterotoxigenic *Escherichia coli*,[112] *Bacteroides* sp.,[113] and *Schistosoma*,[114] to name just a few. For recent reviews, see Tenover[99] and Lowe.[115] As diagnostic DNA techniques are developed for increasing numbers of human and animal diseases, techniques will be standardized, and utilization in the clinical laboratory should increase as a result.

B. PROBE REQUIREMENTS

1. Selection of DNA Sequences for Use as a Probe

The simplest possible probe is one made of whole cell *Brucella* DNA. For most bacteria, the potential for between *Brucella* and sequence overlap with other bacterial species leading to false positives would eliminate this type of probe from consideration. However, studies by Hoyer and McCullough[116,117] showed essentially no sequence homology between *Brucella* and other organisms. Another measure of this evolutionary divergence from other genera is the sequence dissimilarity of *Brucella* ribosomal RNA cistrons from a number of other bacteria.[118] For many dot blot applications, whole cell DNA probes should prove acceptable. However, for *in situ* diagnostic tests, where the sensitivity of the reaction is extremely high, even a cross-reaction of 2 to 3% would be unacceptable. In order to obtain sufficient specificity, cloned *Brucella* sequences must be employed and the cross-reactivity of each cloned sequence with the DNA of other organisms examined. While it would be desirable to focus on those genes responsible for pathogenesis, already cloned genes[38,62-68] may prove suitable.

The complexity of the *Brucella* genome has been estimated by De Ley et al.[118] to be as high as 4.3×10^6 base pairs (bp) using initial hybridization kinetics. Hopper and Bantle[119] used solution hybridization kinetics in comparison to a known *E. coli* standard in order to arrive at a complexity of 4.6×10^6 bp for *B. abortus* strain 19. Since numerous unpublished studies, including those in our laboratory,[120] have failed to discover any plasmids in *Brucella*, it follows that each cell contains about 5 fg of DNA. This determines the limits of sensitivity on a per cell basis for most diagnostic DNA probes.

Reducing the complexity of the probe lowers the sensitivity of the assay because any labeled probe only hybridizes with complementary sequences. Thus, a 1-kilobase probe (about 1 ag) has only 1/4600 the sensitivity of a probe made of labeled whole cell DNA. Some of this lost sensitivity can be regained by utilizing "sandwich" techniques where labeled concatemers are built up on the initial hybrid.[121] Another possibility is to use a probe complementary to ribosomal RNA sequences that are not evolutionarily conserved. Since each cell contains several thousand rRNA molecules, this would allow many more hybridization targets, and sensitivity would be greatly increased. The work of De Ley et al.[118] suggests that *Brucella* ribosomal cistrons contain a number of sequences that will not cross-react with other species. Polymerase chain reaction (PCR) also has great potential for increasing sensitivity. Preliminary results in our laboratory (J.A.B.) indicate that amplification of a small target sequence (several hundred base pairs) allows the detection of 50 brucellae in 0.1 gm of tissue.

If it should prove desirable to use diagnostic DNA to detect different species or strains within the genus, it is likely that oligonucleotide probes or short 100- to 300-bp probes will be required. This will be necessary because the degree of polynucleotide homology is so great among the species[116-118,122] that Verget et al.[123] have recommended the establishment of only one species, *B. melitensis*. The least amount of complexity overlap is reported for *B. ovis*, which in one study showed 94% homology to *B. abortus*; but much of this difference might be due to deletion[117] or to the normal variance of the experimental method used. Long sequences are, therefore, likely to share too much sequence homology between species to be easily discriminated by hybridization techniques. The choice of a probe then depends on whether the method of detection is by blot or *in situ* hybridization and whether it is any brucellae or a particular species of *Brucella* that is to be detected.

2. Radioactive vs. Nonradioactive Probes

Radioisotopes such as ^{32}P, ^{35}S, and ^{131}I have long been used to label probes, offering sensitivities of 0.1 to 5 pg depending on specific activity. For the research laboratory, radioactive probes are a good choice because of the high sensitivity and the existence of established protocols. Unfortunately, the typical animal disease diagnostic laboratory is poorly equipped to work with radioisotopes. Licensing, radioactive waste removal, and worker safety considerations would keep many of these laboratories from ever adopting diagnostic DNA tests requiring radioisotopes.

Nonradioactive probes such as biotinylated DNA solve many of the above-mentioned problems. In several comparison studies, the sensitivity of biotinylated DNA probes approaches that of ^{32}P-labeled DNA. Terpstra et al.[124] found that a ^{32}P-labeled leptospiral DNA detected 1.5 pg of target DNA, while a biotin-labeled probe detected 5 pg in a dot blot assay. Neumann et al.[125] found almost the same level of sensitivity using ^{32}P and biotin-labeled adenovirus type 2 DNA probes. Biotinylated DNA probes can present high background problems because contaminants in the test sample (usually endogenous biotin) can interact with the color development components (see Kuritza et al.[113] for a discussion of this problem). Aside from blocking and washing protocols designed to reduce this background, the only solution is to prepare very pure DNA from the test sample.

C. DNA DOT BLOT TESTS

1. Test Description

In this test procedure, the test sample, consisting of DNA, RNA, DNA and RNA, or crude homogenate, is denatured and applied to a solid support, usually nitrocellulose. Nylon membranes may also be used.[121] Using a dot blot apparatus that divides the surface of the nitrocellulose into a number of 0.1-cm^2 wells, each well area can hold approximately 10 μg of single-stranded nucleic acid. After a baking step to fix the target nucleic acid to the paper and a blocking step to stop the binding of any additional material, the hybridization is carried out at 42 to 65°C in the presence of varying amounts of formamide. The hybridization time can be varied from 2 to 48 h, depending on probe concentration and complexity. Following hybridization and washing to remove nonhybridized material, detection is carried out. If the label is radioactive, detection is by autoradiography. If the label is biotin, then streptavidin-alkaline phosphatase is usually bound to the biotinylated DNA. Nitro-blue tetrazolium/5-bromo-4-chloro-3-indolyl phosphate is added, and a blue-colored precipitate forms. Streptavidin has an extremely high specificity and affinity for biotin. The dissociation constant (K_d) of biotin and streptavidin is 10^{-15} *M*, and because streptavidin is polyvalent, it can form large complexes with additional biotin groups. We consider alkaline phosphatase superior as an enzyme label to HRP because it produces a stronger signal.[115,125,126] The entire detection sequence can easily be accomplished in only 4 h.[126]

2. Potential Applications to the Diagnosis of Brucellosis

Dot blot tests can be used in any situation where the number of bacteria is reasonably large. Tissue or fluid samples from abortion material are an excellent choice, but it is also possible to use infected vaginal discharges, placentas, milk, or lymph node samples. It should also be possible to detect the organism in the spleens and livers of dead animals. Only in cases where large numbers of organisms are present would detection be likely. Two important points need to be made. First, while detection in milk is entirely feasible, it is unlikely that the current milk ring test would be replaced because it is rapid, cheap, and has a >98% predictive accuracy rate.[127] Second, an animal suffering from chronic brucellosis would probably not test positive because of the low numbers of organisms usually found in the tissues. We believe that the best use of the dot blot test for brucellosis is in the diagnosis of the cause of fetal abortion in conjunction with tests for other bovine pathogens that cause abortion. Currently, *Brucella* must be cultured from the abortus or other specimen, and it takes several days for a test result. A complicating factor can be the growth of other microorganisms making identification difficult. Dot blot analysis would be faster and potentially more accurate than current methods. Because ruminants are very prone to fetal invasions by infectious agents, the test would be performed frequently enough to justify the required effort and expense.

3. Specificity and Sensitivity

We have examined the sensitivity and specificity of dot blot analysis in the detection of *B. abortus* in fetal tissue.[119] A biotinylated DNA probe from whole genome *B. abortus* strain 19 DNA was prepared by photobiotinylation.[128] This process is inexpensive, and large amounts of probe can be quickly prepared. High probe concentrations (4 μg/ml) allow reaction times to be as short as 4 h. Calf thymus and rat liver DNA applied in amounts as high as 80 μg to the nitrocellulose (10 μg bound) failed to react significantly with the probe (see Figure 3). In separate experiments, *E. coli* and *Pasteurella haemolytica* DNA in 10-

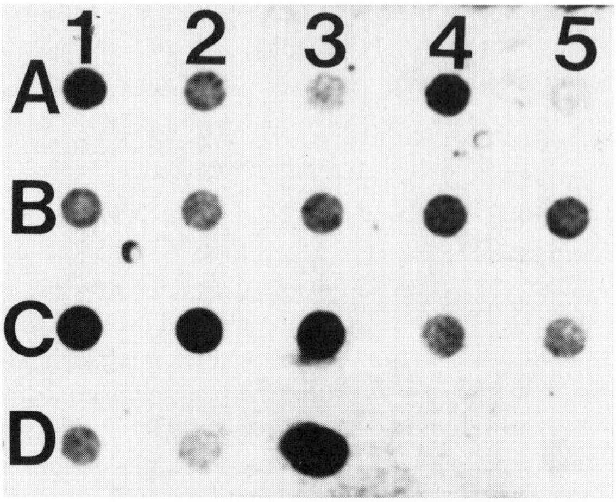

FIGURE 3. Detection of *Brucella abortus* using biotinylated DNA dot blot analysis. Row A: *Brucella* strain 19 cells were diluted serially, mixed with 1 g of rat liver, homogenized, filtered, and extracted with phenol-chloroform. Al, 10^9 *Brucella* cells per gram of liver; A2, 10^8 *Brucella* cells per gram of liver; A3, 10^7 *Brucella* cells per gram of liver; A4, 10^6 *Brucella* cells per gram of liver; A5, 10^5 *Brucella* cells per gram of liver. Row B: the same as row A except that the DNA was purified using guanidinium isothiocyanate. Row C: the same as row A except that no rat liver was used. Row D: comparison of different target DNAs. D1, 0.1 μg of *B. abortus* DNA; D2, 40 μg of yeast tRNA; D3, 1 μg of *B. abortus* DNA; D4, 80 μg of rat liver DNA; D5, 80 μg of purified calf thymus DNA.[119]

μg amounts also failed to react with these DNA probes. Should future specificity tests show minor cross-reaction with other pathogens, hybridization subtraction chromatography[129] could be performed to eliminate cross-reactive sequences from the probe. A low background required extensive purification of the DNA by phenol:chloroform extraction and ethanol precipitation to remove contaminating materials. The specificity appears to be very high for this test.[119]

Potential sensitivity was measured with the BRL (Bethesda Research Laboratories, Bethesda, MD) DNA detection kit. Using this kit, only 5 pg of biotinylated lambda DNA spotted onto nitrocellulose produced an obvious blue dot. However, using the hybridization probe, 1 ng of DNA, but not 0.1 ng, could be detected. Taking into account normal losses during the procedure, DNA from 1×10^5 *Brucella* was required for detection (see Figure 3). Severe problems were encountered when *Brucella* cells were mixed with host tissue such as liver because the host DNA competed with *Brucella* DNA for nitrocellulose binding sites, causing the sensitivity of the assay to decline rapidly. To solve this problem, a filtration procedure has been developed which separates much of the host DNA from the *Brucella* DNA.[119] When this filtration step is included, a sensitivity of at least 1×10^6 cells per gram of tissue (1-g tissue samples were spiked with known amounts of *Brucella* cells) is obtained (see Figure 3). Although this is a significant reduction in sensitivity compared to purified *Brucella* DNA, it should be more than adequate, since very high fetal tissue counts have been reported.[130]

4. Relative Cost, Time, and Ease of Assay

The current test can be performed in less than 48 h by a technician. Further time savings can be realized by simplifying the DNA purification steps. The cost for DNA probe technology should be no higher than for ELISA. Required equipment includes a microcentrifuge for the DNA extraction, an oven for the baking of the filters, and a constant temperature bath for the hybridization.

D. *IN SITU* HYBRIDIZATION TESTS

1. Test Description

The test sample consists of tissue or cell samples from the host that have been fixed (Carnoy's) for 5 min and mounted on a microscope slide. The probe can be added in concentrations as high as 3 μg/ml in 50% formamide-2X SSC (1X SSC is 0.15 *M* NaCl plus 0.015 *M* sodium citrate)-10% dextran sulfate-0.4 μg/ml sonicated salmon sperm DNA. Both probe and target DNA are heat denatured at 80°C for 10 min and then hybridized at 37°C for 1 h. Following hybridization the sample is washed in TP (0.05% Triton X-100 in phosphate buffered saline) and immersed in 50% formamide-2X SSC for 10 min at 37°C. The sample is then rinsed three times in TP (2 min each time).[131] Detection is carried out by conventional autoradiography[132] or by using a biotin detection protocol as previously discussed for dot blot analysis.[131] This technique can also be used for electron microscopy.[133] The technique takes about the same time as the dot blot analysis.

2. Potential Applications to the Diagnosis of Brucellosis

Because *in situ* hybridization can potentially detect a single *Brucella* cell, it is ideally suited to the diagnosis of brucellosis in adult animals and may even be able to detect *Brucella* in peripheral blood samples. Any sample (including milk or vaginal exudate) that can be affixed to a microscope slide can be examined. While the technique is unlikely to be faster or less expensive than the Rivanol or complement fixation tests,[127] it would directly detect the pathogenic agent. Vaccinated animals would not cause false positives so long as the vaccine strain was cleared by the vaccinated animal prior to testing.

The potential of *in situ* hybridization in research should not be overlooked. Extremely sensitive investigations can be conducted on the localization of *Brucella* in host tissues and intracellular reservoirs of the disease uncovered. The technique may be particularly advantageous for tissues or samples which are difficult to investigate by immunohistochemical methods.

3. Specificity and Sensitivity

The sensitivity of this technique can be extremely high, as demonstrated by Lawrence et al.[134] Using a biotinylated probe against Epstein-Barr virus integrated into the genome of a lymphoma cell, they were able to detect as little as 5 kb (5 ag) of viral sequence using a strepavidin-fluorescein-biotin detection method.

There are no reports of attempts to use this technique to detect *Brucella*. Because of the extreme sensitivity, we feel it would be best to choose a single sequence as probe. One potential difficulty is the requirement that the probe and other components of the biotin detection system penetrate the *Brucella* cell wall. It is encouraging that both *Chlamydia* and *Neisseria* infections have been diagnosed using *in situ* hybridization techniques.[131,135] Horn et al.[135] used a solvent treatment to permeabilize *Chlamydia* and *Neisseria* in their diagnostic test. There is no reason to believe the cell wall of *Brucella* is any more difficult to penetrate than *Neisseria*. If this problem can be solved, *in situ* hybridization offers significant possibilities as a diagnostic test.

4. Relative Cost, Time, and Ease of Assay

Animal disease diagnostic laboratories are already well equipped for standard microscopy and should not have trouble using a streptavidin-biotin based detection system that generates a colored precipitate. Fluorescence microscopy is often used in the animal diagnostic lab-

TABLE 4
Comparison of *Brucella* Antigen Detection Methods

Method	Tissue	Detection limit[a] (brucellae)	Time required to complete assay (h)	Ref.
Bacterial culture	All	1	72—168	6
BRT[b]	Milk	3×10^4/ml	1	6
Coagulation	Milk (whey)	3×10^4/ml	3	33
Counterimmunoelectrophoresis	Fetal stomach contents	—[c]	6—7	93
DNA probes				
Blot	Various	1×10^6/g tissue	48	This chapter
In situ hybridization[d]	Any	1 in tissue section	24—48	134
PCR	Any	50/0.1 gm tissue	24—48	This chapter
ELISA	Blood	8—10	6—8	94
	Milk	2.5×10^3	6—8	136
	Vaginal swabs	2.5×10^2	6—8	92
FA[e]	Infected cells	1 in infected leukocytes	1	77
	Tissue smears	Less sensitive than culture	1	78
Immunohistochemistry	Various	1×10^6/g tissue	48—72	83
		1 in tissue section	48—72	This chapter
Neutralization of agglutination	All	$7—25 \times 10^3$/ml	—	91
	Urine, blood	7.5×10^5/mouse	6—8	98

[a] Estimated or calculated from data presented in reference given.
[b] BRT; brucella ring test.
[c] Reported to be more sensitive than culture from this source, probably because of detection of cell fragments.
[d] Not yet demonstrated for *Brucella*, by analogy with results from other organisms.
[e] FA, fluorescent antibody.

oratory. This means that the highly sensitive streptavidin-biotin technique could also be used. In summary, *in situ* hybridization should prove neither difficult nor time consuming for the average histotechnician.

V. SUMMARY

Bacteriologic culture remains the most sensitive method for the detection of *Brucella* in infected tissues, but other detection methods offer advantages in particular situations. Antigen-antibody based detection methods are generally much faster than culture techniques, but are one to five orders of magnitude less sensitive (see Table 4). The exception to this is immunogold-based electron microscopy which can clearly identify a single brucella in a tissue section. Unfortunately, the difficulty of finding an infected region of the tissue limits this and other histological techniques to at least moderately infected tissues. As a general rule, if the number of brucellae is less than about 1×10^6 organisms per gram, histological techniques will require extensive time in screening tissue sections to find a region of infection. The use of DNA as a diagnostic tool is in its infancy, but the technique promises highly specific and sensitive tests for a variety of diseases. Blot tests for tissue infection suffer from interference by tissue DNA, and so are not as sensitive as tests for purified DNA. Extra purification steps can often solve this problem. *In situ* hybridization is capable of detecting a single bacterial cell in a tissue section, although the usefulness of this technique for the study of brucellosis has yet to be explored. One clear advantage of all of the nonculture techniques is the potential to detect nonliving *Brucella* cells or cell fragments. This is advantageous and invaluable in pathogenesis studies where fixed tissue can be studied at a later time and may also improve the apparent sensitivities of the various immunoenzymatic, fluorescent antibody, and histological techniques. Table 4 summarizes the various detection techniques discussed in this paper.

One subject that should receive greater attention is the characterization of various antigens for the immunologic detection of *Brucella* cells or cell fragments. The "O" antigen of the smooth LPS is clearly the dominant antigen in the *Brucella* cell, and most test antisera probably react with this cellular component. Unfortunately, these epitopes are neither species nor genus specific. Various proteins are also good antigens, but little has been done to characterize the suitability of individual proteins as diagnostic reagents. With the growing availability of cloned *Brucella* genes and SLPS free *Brucella* proteins isolated from *E. coli* or other bacterial cells, information about the protein antigens should increase rapidly in the next few years.

ACKNOWLEDGMENTS

The authors wish to acknowledge support by the USDA-ARS (Midwest Area) cooperative research agreements No. 58-6125-4-17 (to J.A.B.) and No. 55-6125-4-2 (to J.E.M.); and USDA research grant No. 85-CRCR-1-1842 (to L.B.T. and J.E.M.).

REFERENCES

1. **McCullough, N. B.,** Microbial and host factors in the pathogenesis of brucellosis, in *Infectious Agents and Host Reactions,* Mudd, S., Ed., W. B. Saunders, Philadelphia, 1970, 324.
2. **McCullough, N. B.,** Immune Response to *Brucella,* in *Manual of Clinical Immunology,* Rose, N. R. and Friedman, H., Eds., American Society for Microbiology, Washington, D.C., 1976, 304.
3. **Nicoletti, P.,** The epidemiology of bovine brucellosis, *Adv. Vet. Sci. Comp. Med.,* 24, 69, 1980.
4. **Meyer, M. E.,** The genus *Brucella,* in *The Prokaryotes, A Handbook on Habitats, Isolation, and Identification of Bacteria,* Vol. I, Starr, M. P., Stolp, H., Truper, H. G., Balows, A., and Schlegel, H. G., Eds., Springer-Verlag, Berlin, 1981, 1063.
5. **Corbel, M. J., Stuart, F. A., and Brewer, R. A.,** Observations on serological cross-reactions between smooth *Brucella* species and organisms of other genera, *Dev. Biol. Stand.,* 56, 341, 1984.
6. **Alton, G. G., Jones, L. M., and Pietz, D. E.,** *Laboratory Techniques in Brucellosis,* 2nd ed., World Health Organization, Geneva, 1975.
7. **Corbel, M. J. and Hendry, D. M. F. D.,** Brucellas, in *Isolation and Identification of Micro-Organisms of Medical and Veterinary Importance,* Collins, C. H. and Grange, J. M., Eds., Academic Press, New York, 1985.
8. **Nelson, C. J., Anderson, R. K., Kimberling, C. V., and Pietz, D. E.,** Epizootiologic factors of bovine brucellosis: comparative bacteriologic studies of infected herds, *Am. J. Vet. Res.,* 27, 1515, 1966.
9. **Ewalt, D. R.,** A New Selective Medium for Isolating *Brucella* sp. from Bovine Milk, M.S. thesis, Iowa State University, Ames, 1983.
10. **Ewalt, D. R., Packer, R. A., and Harris, S. K.,** An improved selective medium for isolating *Brucella* sp. from bovine milk, *Proc. 3rd Int. Symp. Vet. Lab. Diag.,* 1983, 577.
11. **Farrell, I. D.,** The development of a new selective medium for the isolation of *Brucella abortus* from contaminated sources, *Res. Vet. Sci.,* 16, 280, 1974.
12. **Farrell, I. D. and Robertson, L.,** A comparison of various selective media, including a new selective medium for the isolation of brucellae from milk, *J. Appl. Bacteriol.,* 35, 625, 1972.
13. **Ryan, W. J.,** A selective medium for the isolation of *Brucella abortus* from milk, *Mon. Bull. Minist. Health,* 26, 33, 1967.
14. **Brodie, J. and Sinton, G. P.,** Fluid and solid media for isolation of *Brucella abortus, J. Hyg. Cambridge,* 74, 359, 1975.
15. **Corner, L. A., Alton, G. G., and Iyer, H.,** An evaluation of a biphasic medium for the isolation of *Brucella abortus* from bovine tissue, *Aust. Vet. J.,* 62, 187, 1985.

16. **Ewalt, D. R.,** unpublished data.
17. **Painter, G. M., Deyoe, B. L., and Lambert, G.,** Comparison of several media for the isolation of brucella, *Can. J. Comp. Med. Vet. Sci.*, 30, 218, 1966.
18. **Deyoe, B. L. and Manthei, C. A.,** Sites of localization of *Brucella suis* in swine, *71st Annu. Proc. U.S. Livestock Sant. Assoc.*, 1967, 102.
19. **Forbes, L. B.,** The eradication of bovine brucellosis: an effective sample acquisition protocol for a low incidence area, in *23rd Annu. Proc. Amer. Assoc. Veterinary Laboratory Diagnosticians*, 1980, 195.
20. **McCullough, N. B., Eisele, C. W., and Byrne, A. F.,** Incidence and distribution of *Brucella abortus* in slaughtered Bang's reactor cattle, *Public Health Rep.*, 66, 341, 1951.
21. **Washko, F. V., Donham, C. R., Hutchings, L. M. and Heimlick, A.,** Recovery of brucella from tissues of cattle exposed to *Brucella abortus, J. Am. Vet. Med. Assoc.*, 120, 82, 1952.
22. **Hennager, S. G., Harris, S. K., Ewalt, D. R., and Jarnagin, J. L.,** Rapid identification of dominant *Brucella* antigens, using a microagglutination test, *Am. J. Vet. Res.*, 44, 2418, 1983.
23. **Jones, L. M., Montgomery, V., and Wilson, J. B.,** Characteristics of carbon dioxide independent cultures of *Brucella abortus* isolated from cattle vaccinated with strain 19, *J. Infect. Dis.*, 15, 312, 1965.
24. **Anderson, P., Insel, R. A., and Smith, D. H.,** Immunogenicity of a protein-carbohydrate complex prepared from membrane released in cultures of *Haemophilus influenzae* type b, in Weinstein, L. and Fields, B. N., *Seminars in Infectious Disease*, Vol. 4, *Bacterial Vaccines*, Robbins, J. B., Hill, J. C. and Sadoff, J. C. Eds., Thieme-Stratton, New York, 1982, 327.
25. **Berman, D. T.,** personal communication.
26. **Bricker, B. J. and Mayfield, J. E.,** unpublished data.
27. **Moreno, E., Pitt, M. W., Jones, L. M., Schurig, G. G., and Berman, D. T.,** Purification and characterization of smooth and rough lipopolysaccharides from *Brucella abortus, J. Bacteriol.*, 138, 361, 1979.
28. **Tabatabai, L. B.,** unpublished data.
29. **Tabatabai, L. B., Deyoe, B. L., and Patterson, J. M.,** Immunogenicity of *B. abortus* salt extractable proteins and fractions, *Vet. Microbiol.*, 20, 49, 1989.
30. **Dubois, M., Gilles, K. A., Hamilton, J. K., Rebers, P. A., and Smith, F.,** Colorometric method for determination of sugars and related substances, *Anal. Chem.*, 28, 350, 1956.
31. **Lowry, O. H., Rosenbush, N. J., Farr, A. L., and Randall, R. J.,** Protein measurement with the Folin phenol reagent, *J. Biol. Chem.*, 193, 265, 1951.
32. **Ashwell, G.,** New colorometric methods of sugar analysis, *Methods Enzymol.*, 8, 85, 1966.
33. **Garrett, L. K.,** A Coagulation Test for the Detection of *Brucella abortus* Antigen and Antibody in Cows Milk, M.S. thesis, Iowa State University, Ames, 1981.
34. **Sullivan, J. D. and Watson, S. W.,** Factors affecting the sensitivity of *Limulus* lysate, *Appl. Microbiol.*, 28, 1023, 1974.
35. **Tabatabai, L. B., Deyoe, B. L., and Ritchie, A. E.,** Isolation and characterization of toxic fractions from *Brucella abortus, Infect. Immun.*, 26, 668, 1979.
36. **Tabatabai, L. B. and Deyoe, B. L.,** Biochemical and biological properties of soluble protein preparations from *Brucella abortus, Dev. Biol. Stand.*, 56, 199, 1984.
37. **Tabatabai, L. B. and Deyoe, B. L.,** Characterization of salt extractable protein antigens from *Brucella abortus* by crossed immunoelectrophoresis and isoelectricfocusing, *Vet. Microbiol.*, 9, 549, 560, 1984.
38. **Mayfield, J. E., Bricker, B. J., Godfrey, H., Crosby, R. M., Knight, D. J., Halling, S. M., Balinsky, D., and Tabtabai, L. B.,** The cloning, expression, and nucleotide sequence of a gene coding for an immunogenic *Brucella abortus* protein, *Gene*, 63, 1, 1988.
39. **Bricker, B. J., Tabatabai, L. B., Deyoe, B. L., and Mayfield, J. E.,** Conservation of antigenicity in a 31 kDa *Brucella* protein, *Vet. Microbiol.*, 18, 313, 1988.
40. **Belzer, C. A.,** Bovine Humoral Immune Response to Salt-Extractable Protein Antigens after Vaccination and Infection with *Brucella abortus*, M.S. thesis, Iowa State University, Ames, 1986.
41. **Belzer, C. A., Tabatabai, L. B., and Deyoe, B. L.,** Differentiation of the immune response of cattle vaccinated or infected with *Brucella abortus* by Western blot, *Vet. Micro.*, submitted.
42. **Michaud, R. A.,** Characterization and Immunogenicity of *Brucella abortus* Ribosomes and 'LPS' in CF1 and C3H/HeJ Mice, M.S. thesis, Iowa State University, Ames, 1982.
43. **Spellman, J. M. and Reed, N. D.,** Immune and mitogenic responses by BALB/c, C3H/HeJ, and nude mice to *Brucella abortus* bacterin and lipopolysaccharide, *Infect. Immun.*, 24, 371, 1979.
44. **Pugh, G. W., Tabatabai, L. B., Bricker, B. J., Mayfield, J. E., Phillips, M., McDonald, T. J., and Zehr, E. S.,** Identification of a possible virulence factor for *Brucella abortus* infection in BALB/c mice, *Am. J. Vet. Res.*, 50, 887,1989.
45. **Moriyon, I. and Montanes, M. A.,** *In vitro* interactions between lipopolysaccharides and heterologous outer membrane proteins, *Curr. Microbiol.*, 12, 229, 1985.
46. **Verstreate, D. R. and Winter, A. J.,** Comparison of sodium dodecyl sulfate-polyacrylamide gel electrophoresis profiles and antigenic relatedness among outer membrane proteins of 49 *Brucella abortus* strains, *Infect. Immun.*, 46, 182, 1984.

47. **Perry, M. B., Bundle, D. R., and Cherwonogrodzky, J. W.,** The structures and serology of the A and M antigens of *Brucella abortus* and *Brucella menitensis* and the structure of polysaccharide B of *Brucella* species, presented at 14th Int. Congr. Microbiol., Manchester, U.K., September 1986.
48. **Bundle, D. R., Cherwonogrodzky, J. W., Caroff, M., and Perry, M. B.,** The lipopolysaccharides of *Brucella abortus* and *B. melitensis, Ann. Inst. Pasteur Microbiol.*, 138, 92, 1987.
49. **Moreno, E., Borowiak, D., and Mayer, H.,** *Brucella* lipopolysaccharides and polysaccharides, *Ann. Inst. Pasteur Microbiol.*, 138, 102, 1987.
50. **Diaz, R., Jones, L. M., Leong, D., and Wilson, J. B.,** Surface antigens of smooth brucellae, *J. Bacteriol.*, 96, 893, 1968.
51. **Moreno, E., Speth, S. L., Jones, L. M., and Berman, D. T.,** Immunochemical characterization of *Brucella* lipopolysaccharides and polysaccharides, *Infect. Immun.*, 31, 214, 1981.
52. **Winter, A. J.,** Outer membrane proteins of *Brucella, Ann. Inst. Pasteur Microbiol.*, 138, 87, 1987.
53. **Verstreate, D. R., Creasy, M. T., Caveney, N. T., Baldwin, C. L., Blab, M. W., and Winter, A. J.,** Outer membrane proteins of *Brucella abortus*: isolation and characterization, *Infect. Immun.*, 35, 979, 1982.
54. **Santos, J. M., Verstreate, D. R., Perera, V. Y., and Winter, A. J.,** Outer membrane proteins from rough strains of four *Brucella* species, *Infect. Immun.*, 46, 188, 1984.
55. **Dubray, G. and Bezard, G.,** Isolation of three *Brucella abortus* cell-wall antigens protective in murine experimental brucellosis, *Ann. Rech. Vet.*, 11, 367, 1980.
56. **Dubray, G. and Charriaut, C.,** Evidence of three major polypeptide species in the *Brucella* outer membrane, *Ann. Rech. Vet.*, 14, 311, 1983.
57. **Douglass, J. T., Rosenberg, E. Y., Nikaido, H., Verstreate, D. R., and Winter, A. J.,** Porins of *Brucella* species, *Infect. Immun.*, 44, 16, 1984.
58. **Gomez-Miguel, M. J. and Moriyon, I.,** Demonstration of a peptidoglycan-linked lipoprotein and characterization of its trypsin fragment in the outer membrane of *Brucella* spp., *Infect. Immun.*, 53, 678, 1986.
59. **Moriyon, I., Gamazo, C., and Diaz, R.,** Properties of the outer membrane of *Brucella, Ann. Inst. Pasteur Microbiol.*, 138, 89, 1987.
60. **Moriyon, I. and Berman, D. T.,** Isolation, purification and partial characterization of *Brucella abortus* matrix protein, *Infect. Immun.*, 39, 394, 1983.
61. **Hindsdill, R. D. and Berman, D. T.,** Antigens of *Brucella abortus, J. Bacteriol.*, 93, 544, 1967.
62. **Mayfield, J. E.,** unpublished data.
63. **Ficht, T. A., Beardon, S. W., Sowa, B. A., and Adams, G. L.,** A major outer membrane protein of *Brucella abortus* is encoded by a multiple copy gene, presented at the 40th Annu. Brucellosis Research Conf., Chicago, November 1987.
64. **Sanborn, M. R.,** Cloning and expression of a 43K *Brucella abortus* outer membrane protein gene in *E. coli,* presented at the 39th Annu. Brucellosis Research Conf., Chicago, November 1986.
65. **Essenberg, R. C.,** Cloning of Genes from *Brucella abortus* for glucose transport and utilization, presented at the 40th Annu. Brucellosis Research Conf., Chicago, November 1987.
66. **Sanborn, M. R., Sherwood, J. L., Phillips, M., Ayola, B., Keyser, G., Bantle, J., Weldon, B., and Essenberg, R. C.,** Cloning and expression of a putative carbohydrate epitope of *Brucella abortus* LPS in *E. coli*, presented at the 40th Annu. Brucellosis Research Conf., Chicago, November 1987.
67. **Sowa, B. A., Ficht, T. A., and Adams, G. L.,** Preparation and characteristic of selected Brucella abortus membrane protein immunogens, presented at the 40th Annu. Brucellosis Research Conf., Chicago, November 1987.
68. **Rossetti, O.,** Cloning of *B. abortus* genes in gt-11, presented at the 39th Annu. Brucellosis Research Conf. Chicago, November 1986.
69. **Diaz, R., Jones, L. M., and Wilson, J. B.,** Antigenic relationship of the Gram-negative organism causing canine abortion to smooth and rough Brucellae, *J. Bacteriol.*, 95, 618, 1968.
70. **Bricker, B. J., Tabatabai, L. B., and Mayfield, J. E.,** unpublished data.
71. **Bricker, B. J., Tabatabai, L. B., and Mayfield, J. E.,** An IgG binding protein on the surface of *Brucella abortus,* submitted.
72. **Benjamin, D. C., Berzotsky, J. A., East, I. J., Gard, F. R. N., Hannum, C., Leach, S. J., Margoliash, E., Michael, J. G., Miller, A., Prager, E. M., Reichlin, M., Sercarz, E. E., Smith-Gill, S. J., Todd, P. E., and Wilson, A. C.,** The antigenic structure of proteins: a reappraisal, *Annu. Rev. Immunol.*, 2, 67, 1984.
73. **Sheehan, D. C. and Hrapchak, B. B.,** *Theory and Practice of Histotechnology,* C. V. Mosby, St. Louis, 1980.
74. **Ritchie, A. E., Patterson, J. M., and Tabatabai, L. B.,** Direct immunoelectron microscopy of Brucellas sp. particulate antigen(s), Abstr. 22nd Annu. Electron Microscopy Colloq., Ames, Iowa, *J. Electron Microsc., Tech.*, 2, 656, 1985.
75. **Corbel, M. J.,** The direct fluorescent antibody test for detection of *Brucella abortus* in bovine abortion material, *J. Hyg. Cambridge,* 71, 123, 1973.

76. **Deyoe, B. L.,** Research findings applicable to eradication of swine brucellosis, *Proc. U.S. Anim. Health Assoc.*, 76, 108, 1972.
77. **Janney, G. C. and Berman, D. T.,** Staining intracellular *Brucella* organisms by means of fluorescent antibodies, *Am. J. Vet. Res.*, 23, 596, 1962.
78. **Meyer, M. E.,** Identification of *Brucella* organisms by immunofluorescence, *Am. J. Vet. Res.*, 27, 424, 1966.
79. **Bullock, G. R. and Petrusz, P., Eds.,** *Techniques in Immunocytochemistry,* Vol. 2, Academic Press, New York, 1983.
80. **Bullock, G. R. and Petrusz, P., Eds.,** *Techniques in Immunocytochemistry,* Vol. 3, Academic Press, New York, 1985.
81. **DeLellis, R. A., Sternberger, L. A., Mann, R. B., Banks, P. M., and Nakane, P. K.,** Immunoperoxidase techniques in diagnostic pathology, *Am. J. Clin. Pathol.*, 71, 483, 1979.
82. **Hsu, S. M., Raine, L., and Fanger, H.,** Use of avidin-biotin-peroxidase complex (ABC) in immunoperoxidase techniques: a comparison between ABC and unlabeled antibody (PAP) procedures, *J. Histochem. Cytochem.*, 29, 577, 1981.
83. **Meador, V. P., Tabatabai, L. B., Hagemoser, W. A., and Deyoe, B. L.,** Identification of *Brucella abortus* in formalin-fixed, paraffin-embedded tissues of cows, goats, and mice with an avidin-biotin-peroxidase complex immunoenzymatic staining technique, *Am. J. Vet. Res.*, 47, 2147, 1986.
84. **Meador, V. P.,** unpublished data.
85. **Ackermann, M. R., Cheville, N. F., and Deyoe, B. L.,** Ileal dome lymphoepithelial cell endocytosis and transport of *B. abortus* strain 19 in calves, *Vet. Pathol.*, 25, 28, 1988.
86. **Anderson, T. D., Meador, T. D., and Cheville, N. F.,** Pathogenesis of placentitis in the goat inoculated with *Brucella abortus*. I. Gross and histologic lesions, *Vet. Pathol.*, 23, 219, 1986.
87. **Detilleux, P. G., Cheville, N. F., and Deyoe, B. L.,** Pathogenesis of *Brucella abortus* in chicken embryos, *Vet. Pathol.*, 25, 138, 1988.
88. **Meador, V. P., Hagemoser, W. A., and Deyoe, B. L.,** Histopathologic findings in *Brucella abortus* infected pregnant goats, *Am. J. Vet. Res.*, 49, 274, 1988.
89. **Polak, J. M. and Van Noordan, S.,** *Immunocytochemistry: Modern Methods and Applications,* 2nd ed., Wright, Bristol, 1986.
90. **Anderson, T. D., Cheville, N. F., and Meador, T. D.,** Pathogenesis of placentitis in the goat innoculated with *Brucella abortus*. II. Ultrastructural studies, *Vet. Pathol.*, 23, 227, 1986.
91. **Sadykov, S. Z.,** Detecting *Brucella* antigens in pathological material with the neutralization reaction, *Vet. Bull.*, 47 (Abstr. No. 5404), 741, 1976.
92. **Chen, I.-M., Thoen, C. O., Pietz, D. E., and Harrington, R.,** Application of an enzyme-linked immunosorbent assay for detection of *Brucella* antigens in vaginal discharge of cows, *Am. J. Vet. Res.*, 45, 32, 1984.
93. **Mahajan, N. K. and Kulshreshtha, R. C.,** Couunterimmunoelectrophoresis detection of brucella antigens in fetal stomachs of aborted sheep, *Vet. Rec.*, 119, 455, 1986.
94. **Perera, V. Y., Creasy, M. T., and Winter, A. J.,** Nylon bead enzyme-linked immunosorbant assay for detection of sub-picogram quantities of *Brucella* antigens, *J. Clin. Microbiol.*, 18, 601, 1983.
95. **Reinhold, R. B. and Fine, J.,** A technique for quantitative measurement of endotoxin in human plasma, *Proc. Soc. Exp. Biol. Med.*, 137, 334, 1971.
96. **Winter, A. J.,** unpublished data.
97. **Perera, V. Y. and Winter, A. J.,** Detection of *Brucella* antigens in leukocytes of infected heifers by an enzyme-linked immunosorbent assay sensitive to sub-picogram levels, *Abstr. Annu. Meet. Am. Soc. Microbiol.*, 83, C160, 1983.
98. **Limet, J. N.,** Immunoassay of *Brucella* lipopolysaccharide-derived antigen in plasma: a new method of diagnosis, *Ann. Inst. Pasteur Microbiol.*, 138, 126, 1987.
99. **Tenover, F. C.,** Diagnostic deoxyribonucleic acid probes for infectious diseases, *Clin. Microbiol. Rev.*, 1, 82, 1988.
100. **Hopper, B. R., Essenberg, R. C., Morton, R. J., Craven, R. C., and Bantle, J. A.,** Detection of *Brucella abortus* in tissue homogenates using biotinylated DNA as a molecular probe, *Abstr. Annu. Meet. Am. Soc. Microbiol.*, 87, 356, 1987.
101. **Douglas, J.,** Development of DNA probes for *Brucella* , presented at the 37th Annu. Brucellosis Research Conf., Chicago, November 1984.
102. **Fransen, L., Shabo, R., Perlman, H., Wigzell, H., Westin, G., Aslund, L., Persson, T., and Pettersson, U.,** Analysis of clinical specimens by hybridisation with probes containing repetitive DNA from *Plasmodium falciparum, Lancet,* 2, 1234, 1984.
103. **Bruce-Chwatt, L. J.,** DNA probes for malaria diagnosis, *Lancet,* 2, 1046, 1984.
104. **McLaughlin, G. L., Edlind, T. D., and Ihler, G. M.,** Detection of *Bebesia ovis* using DNA hybridization, *J. Protozool.*, 33, 125, 1986.
105. **McFarlane, R. G., Thawley, D. G., and Soloranzo, R. F.,** Detection of latent pseudorabies virus in porcine tissue, using a DNA hybridization dot-blot assay, *Am. J. Vet. Res.*, 47, 740, 1986.

106. **Dunn, D. C., Blair, C. D., Ward, D. C., and Beaty, B. J.,** Detection of bovine herpesvirus-specific nucleic acids by *in situ* hybridization with biotinylated DNA probes, *Am. J. Vet. Res.,* 47, 470, 1986.
107. **Lawrie, J. M., Jackson, P. R., Stitteler, J. M., and Hockmeyer, W. T.,** Identification of pathogenic *Leishmania* promastigotes by DNA: DNA hybridization with kintoplast DNA cloned into *Escherichia coli* plasmids, *Am. J. Trop. Med. Hyg.,* 34, 257, 1985.
108. **McManus, D. P. and Simpson, A. J. G.,** Identification of the *Echinococcus (hydatid disease)* organisms using cloned DNA markers, *Mol. Biochem. Parasitol.,* 17, 171, 1985.
109. **Chambers, A. E., Almond, N. M., Knight, M., Simpson, A. J. G., and Parkhouse, R. M. E.,** Repetitive DNA as a tool for the identification and comparison of nematode variants: application to *Trichinella* isolates, *Mol. Biochem. Parasitol.,* 21, 113, 1986.
110. **Sim, B. K. L., Mak, J. W., Cheong, W. H., Sutanto, I., Kurniawan, L., Marwoto, H. A., Franke, E., Campbell, J. R., Wirth, D. F., and Piessens, W. F.,** Identification of *Brugia malayi* in vectors with species-specific DNA probe, *Am. J. Trop. Med. Hyg.,* 35, 559, 1986.
111. **Chiang, M., Chia, D., and Barnett, E. V.,** Evaluation of fluorescent antinuclear antibody assays, *Crithidia luciliae* substrate, and single stranded DNA-binding capacity in diagnosis of four rheumatic diseases, *J. Clin. Microbiol.,* 15, 684, 1982.
112. **Maddox, C. W. and Wilson, R. A.,** High technology diagnostics: detection of enterotoxigenic *Escherichia coli,* using DNA probes, *J. Am. Vet. Med. Assoc.,* 188, 47, 1986.
113. **Kuritza, A. P., Getty, C. E., Shaughnessy, P., Hesse, R., and Salyers, A. A.,** DNA probes for identification of clinically important *Bacteroides* species, *J. Clin. Microbiol.,* 23, 343, 1986.
114. **Walker, T. K., Rollinson, D., and Simpson, A. J. G.,** Differentiation of *Schistosoma haematogium* from related species using cloned ribosomal RNA gene probes, *Mol. Biochem. Parasitol.,* 20, 123, 1986.
115. **Lowe, J. B.,** Clinical applications of gene probes in human genetic disease, malignancy, and infectious disease, *Clin. Chim. Acta,* 157, 1, 1986.
116. **Hoyer, B. H. and McCullough, N. B.,** Polynucleotide homologies of *Brucella* deoxyribonucleic acids, *J. Bacteriol.,* 95, 444, 1968.
117. **Hoyer, B. H. and McCullough, N. B.,** Homologies of deoxyribonucleic acids from *Brucella ovis,* canine abortion organisms, and other *Brucella* species, *J. Bacteriol.,* 96, 1783, 1968.
118. **De Ley, J., Mannheim, W., Segers, P., Lievens, A., Denijn, M., VanHoucke, M., and Gillis, M.,** Ribosomal ribonucleic acid cistron similarities and taxonomic neighborhood of *Brucella* and CDC group Vd, *Int. J. Syst. Bacteriol.,* 37, 35, 1987.
119. **Hopper, B. R. and Bantle, J. A.,** unpublished data.
120. **Bantle, J. A. and Hopper, B. R.,** unpublished data.
121. **Meinkoth, J. and Wahl, G.,** Hybridization of nucleic acids immobilized on solid supports, *Anal. Biochem.,* 138, 267, 1984.
122. **O'Hara, M. J., Collins, D. M., and Lisle, G. W.,** Restriction endonuclease analysis of *Brucella ovis* and other *Brucella* species, *Vet. Microbiol.,* 10, 425, 1985.
123. **Verger, J. M., Grimont, G., Grimont, P. A. D., and Grayon, M.,** Taxonomy of the genus *Brucella, Ann. Inst. Pasteur Microbiol.,* 9, 549, 1987.
124. **Terpstra, W. J., Schoone, G. J., and Ter Schegget, J.,** Detection of leptospiral DNA by nucleic acid hybridization with ^{32}P- and biotin-labelled probes, *J. Med. Microbiol.,* 22, 23, 1986.
125. **Neumann, R., Rudloff, P., and Eggers, H. J.,** Biotinylated DNA probes: sensitivity and applications, *Naturwissenschaften,* 73, 553, 1986.
126. **Buckland, R. M.,** Strong signals from streptavidin-biotin, *Nature (London),* 320, 557, 1986.
127. **Huber, J. D. and Nicoletti, P.,** Comparison of the results of card, Rivanol, complement-fixation, and milk ring tests with the isolation rate of *Brucella abortus* from cattle, *Am. J. Vet. Res.,* 47, 1529, 1986.
128. **Forster, A. C., McInnes, J. L., Skingle, D. C., and Symons, R. H.,** Non-radioactive hybridization probes prepared by the chemical labelling of DNA and RNA with a novel reagent, photobiotin, *Nucleic Acids Res.,* 13, 745, 1985.
129. **Vitek, M. P., Kreissman, S. G., and Gross, R. H.,** The isolation of ecdysterone inducible genes by hybridization subtraction chromatography, *Nucleic Acids Res.,* 8, 1191, 1981.
130. **Alexander, B., Schnurrenberger, P. R., and Brown, R. R.,** Numbers of *Brucella abortus* in the placenta umbilicus and fetal fluid of two naturally infected cows, *Vet. Rec.,* 108, 500, 1981.
131. **Pollice, M. and Yang, H. L.,** Use of nonradioactive DNA probes for the detection of infectious bacteria, *Clin. Lab. Med.,* 463, 1985.
132. **Gall, J. G. and Pardue, M. L.,** Nucleic acid hybridization in cytological preparations, *Methods in Enzymology,* Vol. 21(D), Grossman, L. and Moldave, K., Eds., Academic Press, 1971, 470.
133. **Hutchinson, N. J., Langer-Safer, P. R., Ward, D. C., and Hamkalo, B. A.,** *In situ* hybridization at the electron microscope level: hybrid detection by autoradiography and colloidal gold, *J. Cell. Biol.,* 95, 609, 1982.
134. **Lawrence, J., Villnave, C. A., and Singer, R. H.,** Sensitive, high-resolution chromatin and chromosome mapping *in situ*: presence and orientation of two closely integrated copies of EBV in a lymphoma line, *Cell,* 52, 51, 1988.

135. **Horn, J. E., Quinn, T., Hammer, M., Palmer, L., and Falkow, S.,** Use of nucleic acid probes for detection of sexually transmitted infections agents, *Diag. Microbiol. Infect. Dis.*, 4, 101S, 1986.
136. **Pritchard, W. D.,** Detection of *Brucella abortus* in Tissue by the Fluorescent Antibody Method, M.S. thesis, University of Wisconsin, Madison, 1965.

Chapter 6

THE BRUCELLAPHAGES

Charlotte E. Rigby

TABLE OF CONTENTS

I. INTRODUCTION

The isolation of several new brucellaphages during recent years has made phage typing a convenient technique for rapid identification of *Brucella* species. With the brucellaphages now available, the diagnostic bacteriologist can rapidly confirm the identity of an isolate as *Brucella* and in most cases can easily determine the species to which it belongs. This is a welcome advance, since only a limited number of the standard biochemical and physiological tests are useful for brucellae. The oxidative metabolic profiles on which definitive speciation depends are too difficult and hazardous to be used outside of reference laboratories.

The relationships between brucellae and their phages are not well understood. Research in this area is needed; better knowledge of the genetics of the brucellaphages and of the phage-host cell relationship will be important in understanding *Brucella* genetics and applying the new techniques of molecular biology and genetic engineering to the control and prevention of brucellosis.

This chapter summarizes the characteristics of the currently recognized brucellaphages, presents selected procedures for their preparation and use, and summarizes recent advances in our knowledge about their properties and genetics. Several excellent reviews and monographs dealing with the brucellaphages have been published,[1-5] and attempts have been made to avoid excessive duplication of the information they present.

II. THE BRUCELLAPHAGES AND THEIR ORIGINS

A. HISTORY

After the first reports from Russia in the 1930s of the discovery of phages active against *Brucella abortus*, several brucellaphages were reported by workers in Europe, England, and North America.[1-5] However, none of these phages were stable enough to be well characterized or widely used, and the story of the brucellaphages of today really begins with the isolation by the Russian workers Pophadze and Abashidze in the late 1950s of a group of phages which were active against smooth strains of *B. abortus*.[1,2] These phages were obtained from feces, blood, fetuses, and milk of cattle infected with *B. abortus* and were named the Tbilisi (Tb) phages after the city in southern Russia where they were isolated. One of the Tb phages was made available to workers in other countries and quickly became accepted as the reference brucellaphage.[3] Its properties have been well studied, notably by Morgan and Corbel and co-workers in England,[4,5] and by Drozevkina, Ostrowskaja, and other workers in Russia.[1,3,4]

Since the 1960s, several more brucellaphages have been obtained, either by isolation from field strains of brucellae or by obtaining derivatives of Tb phage under laboratory conditions. There are now brucellaphages available which lyse all six *Brucella* species. Smooth-rough (S-R) variation is important in brucellae, and the brucellaphages are highly specific for either smooth or rough strains.

B. BRUCELLAPHAGE GROUPS

The brucellaphages, most of which are named for the places where they were first isolated, have been allocated into groups based on their host specificities.[4] Table 1 lists the brucellaphages currently available and shows their propagating host strains and their host ranges.

Group 1 consists of Tb and some similar isolates, such as A422.[6] They replicate only in smooth strains of *B. abortus*, but at high concentrations they lyse smooth strains of *B. suis* and *B. neotomae*. This phenomenon, termed "lysis from without", occurs when the bacterial cell is lysed by enzymes elaborated by large numbers of phages, but there is no replication and no production of infective phage.[7]

TABLE 1
The Groups, Propagating Host Strains, and Host Range of Some Currently Available Brucellaphages

Group	Phage	Propagating host strain(s)	Host range Lysis at: RTD[a]	Host range Lysis at: $10^4 \times$ RTD Only
1	Tb	*B. abortus* S19 *B. abortus* 544	*B. abortus*	*B. suis*[b] *B. neotomae*[b]
2	Fi	*B. abortus* S19	*B. abortus* *B. neotomae*	*B. suis*[b]
3	Wb	*B. abortus* S19 *B. suis* 1330	*B. abortus* *B. neotomae* *B. suis*	*B. melitensis*
4	Bk2	*B. melitensis* 16M	*B. abortus* *B. neotomae* *B. suis* *B. melitensis*	
5	R/C	*B. canis* RM 6/66	*B. canis* *B. ovis*	*B. abortus*[c] *B. melitensis*[c]
6	Iz	*B. abortus* S19 *B. melitensis* 16M *B. melitensis* B115	*B. abortus* *B. neotomae* *B. suis*[d] *B. melitensis*[d,e]	
7 (proposed)	Np	*B. abortus* S19	*B. abortus*	*B. neotomae*

[a] RTD = routine test dilution (highest dilution producing lysis on the propagating strain).
[b] "Lysis-from-without"; there is no phage replication.
[c] Lysis of some rough strains.
[d] Lysis of both smooth and rough strains.
[e] RTD determined on *B. melitensis* = 10^2 RTD on *B. abortus*.

Group 2 contains a single phage, Firenze (Fi). It is active against *B. abortus* and, to a limited extent, *B. suis* biovar 4.[8] Fi was associated with *B. abortus* cultures isolated in Italy from humans and sheep. When first isolated, the cultures were typical smooth *B. abortus* isolates, but some were resistant to lysis by Tb. Phage plaques appeared in lawns of growth of some cultures, which apparently were persistently infected phage carriers; their smooth colonial morphology, coupled with their resistance to lysis by Tb, suggested that the Fi phage infection was lysogenic.

Group 3 phages are typified by Weybridge (Wb) phage, isolated (as were the similar phages M51 and S708) from laboratory strains of *Brucella* spp.[9,10] The group also includes MC/75 and D, both derived from Tb.[11,12] Group 3 phages are active against smooth *B. abortus, B. suis, B. neotomae*, and some strains of *B. melitensis*.[13,14]

The Berkeley (Bk) phages of group 4 are active against smooth strains of *B. melitensis* as well as the other smooth brucellae. Bk0 was developed by Douglas and Elberg from stocks of Wb plated on *B. melitensis* Isfahan.[15] Bk2, selected after repeated passages of BK0 on *B. melitensis* 16M, is the most stable variant.

Group 5 phages are active against rough strains and species of brucellae.[16] R phage, which lyses rough *B. abortus*, was selected from a mixture of Wb, MC/75, and D phages.[17] Unfortunately, both R and its derivative R/O, obtained after repeated passage of R on *B. ovis*, are rather unstable. However, R/C, a derivative obtained by passaging R/O on *B. canis* Mex 51, has proved to be very stable and is active against most rough brucellae.[18] A new phage, R/732, which is also active against rough brucellae, has recently been described.[14]

Izatnagar (Iz) phage (group 6) originated in India where it was associated with caprine and ovine abortion material.[5,19] It appears to be active against all smooth strains of brucellae, as well as rough strains of *B. melitensis* and *B. suis*. However, variable results reported from different laboratories suggest that its host range may vary with the propagating host.

We propose the addition of group 7 for the Nepean (Np) phage, which was associated with atypical strains of *B. abortus* isolated from cattle in Ontario.[20] These strains, which resemble *B. abortus* biovar 4 but are sensitive to dyes and penicillin, have been described.[21,22] They were initially smooth and were lysed by Tb phage, but after continued subculture, growth of some strains became nonsmooth. Np phage was isolated from lytic plaques which appeared in lawns of growth of these strains. At routine test dilution (RTD: see below) Np is active only against *B. abortus*.

All these phages are highly specific for brucellae. They do not adsorb to, replicate in, or lyse any other bacteria that have been tested so far, even those which share common antigenic determinants, such as *Yersinia enterocolitica* 0:9, *Salmonella* serogroup N (0:30) *Escherichia coli* 0:157, *Pseudomonas maltophilia, Francisella tularensis,* or *Vibrio cholerae* serovars Anaba and Ogawa.[5] Similarly, brucellae are not lysed by any other bacteriophages. Lysis by brucellaphage is, thus, a useful rapid test to confirm the identity of *Brucella* spp. and for speciation within the genus.

III. PROPAGATION AND PHAGE TYPING

A. TITRATION OF PHAGE

Brucellaphages are titrated on a lawn of the propagating strain inoculated on serum dextrose agar (SDA, tryptose agar with 5% bovine serum and 1% dextrose). Tenfold dilutions of the phage suspension to be tested are prepared in 1.0-ml volumes of 0.15 *M* NaCl (it is important to use a fresh pipette or pipette tip for each dilution) and 5 μl of each dilution placed on the surface of the freshly inoculated plate. The cultures are incubated 24 to 48 h at 37°C, then examined for phage plaques. The routine test dilution (RTD) is the highest dilution of phage which gives an area of confluent lysis. A satisfactory phage preparation should have a titer of at least $10^4 \times$ RTD. The titer of the phage suspension (plaque-forming units [pfu]/ml) is calculated by multiplying the number of plaques in 5 μl $\times$ 200 $\times$ dilution.

B. PROPAGATION METHODS

It is important to use propagating host strains that are as smooth as possible (except, of course, for R/C phage). Small volumes (10 to 30 ml) of brucellaphage may be propagated conveniently on agar plates. Overlay methods have been recommended,[4,23,24] but spread plate cultures are more convenient and usually give equally good results. In this method, 5 to 10 SDA plates are each inoculated with 0.1 ml of a suspension of the propagating strain (Table 1) containing about 1×10^9 organisms and spread to produce a smooth lawn of growth. The plates are allowed to dry for 10 to 15 min; then 0.1-ml phage suspension at RTD or about 10^5 pfu is spread evenly over the surface. The cultures are incubated for 24 to 48 h at 37°C (in 10% CO_2 if the strain requires it) and then examined. Plates in which lysis is confluent or almost so are harvested. Each plate is flooded with 3 to 5 ml of trypticase soy broth (TSB) and allowed to stand for 10 to 15 min. The fluid is collected and centrifuged at 6000 $\times$ *g* to sediment cellular debris. Then the supernatant is filtered through a 0.22-μm filter. The addition of toluene (about 0.2%) is recommended to control bacterial contamination.[4]

For larger volumes of phage, propagation in aerated liquid culture may be appropriate. *Brucella* host strains grow considerably better in TSB that has been sterilized by filtering, rather than by autoclaving, and addition of glutamine (0.005 *M*) or 5% bovine serum may improve the yield of some phages. Glucose should not be included because it may stimulate

TABLE 2
A Simplified Phage Typing Scheme to Aid in Rapid Speciation of Brucellae

		Lysis by phage		
		Tb		
	***Brucella* sp.**	**RTD[a]**	**10^4 RTD**	**R/C RTD**
B. abortus	Smooth	L[b]	L	—
	Rough	—	—	(L)
B. suis	Smooth	—	L	—
	Rough	—	—	—
B. melitensis	Smooth	—	—	—
	Rough	—	—	(L)
B. neotomae	Smooth	—	L	—
B. ovis	Rough	—	—	L
B. canis	Rough	—	—	L

[a] RTD = routine test dilution, determined on *B. abortus* S19 for Tb and *B. canis* RM6/66 for R/C.
[b] L = lysis; (L) = lysis of some strains; — = no lysis.

the growth of rough variants. We use 100 ml of TSB in a 250-ml screw-cap flask inoculated with 1 ml of host strain (10^{10} colony-forming units [cfu]) and 1 ml of phage suspension (10^9 pfu). For host strains that require CO_2 we preincubate the medium in the flask in 10% CO_2 overnight before inoculation and screw the caps on tightly during incubation. Cultures are incubated at 37°C at 140 rpm for 24 to 48 h, then placed at 4°C for 24 to 48 h. The cultures are often turbid after incubation, indicating incomplete cell lysis, but such cultures usually become clear after 24 to 48 h at 4°C. It is convenient to titrate the phage in the culture fluid (lysate) at this point to ensure the titer is acceptable before proceeding. Viscosity of the lysate may be reduced by incubation with DNase and RNase (1 μg/ml) at room temperature for 2 to 18 h. The lysate is then centrifuged at 6000 × *g* to sediment cellular debris, and the supernatant is filtered through a 0.22-μm filter and stored at 4°C with added toluene (0.2%).

Brucellaphages are stable at refrigerator temperatures, and we have stored suspensions of Tb for several years with no loss of titer.

C. HOST RANGE DETERMINATION

The host range (Table 1) of each brucellaphage preparation must be checked before the phage is used for identification or characterization of brucellae. This is done by titrating each phage on as many *Brucella* strains as possible, using the procedures described above. For use in phage typing, satisfactory suspensions must lyse only the appropriate strains at RTD.

D. PHAGE TYPING PROCEDURES

Since the brucellaphages are highly specific for either smooth or rough brucellae, it is important to determine the S-R status of a culture before attempting to determine its sensitivity to brucellaphages. This is easily done by microscopic examination or agglutination tests.[4,23,24] For phage typing, the culture to be tested is inoculated onto an SDA plate so as to produce an even lawn of growth. Phage suspensions diluted to RTD (and 10^4 RTD in the case of Tb) are spotted onto the lawn in 5-μl drops, and the plates incubated 24 to 48 h. In our laboratory the two phages Tb and R/C enable the rapid identification of most *Brucella* isolates. The results are interpreted as shown in Table 2. As discussed above, the host range

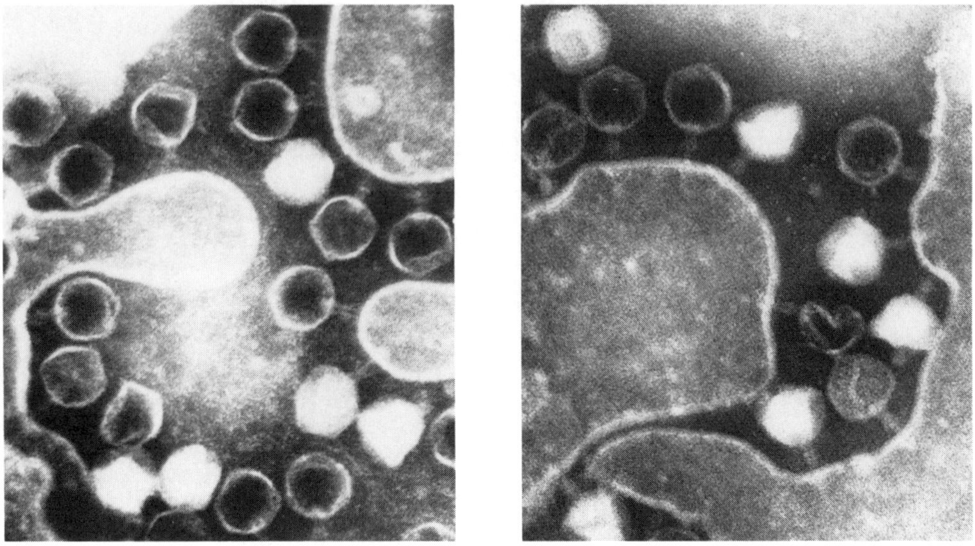

A B

FIGURE 1. Brucellaphages Tb (A) and Np (B) negatively stained with 2% phosphotungstic acid. Full and empty phage heads are absorbed to fragments of host *Brucella abortus* cell membranes. (Magnification × 150,000.)

of Iz phage needs to be better established before it can be recommended for definitive speciation of brucellae.

IV. CHARACTERIZATION OF BRUCELLAPHAGES

A. ULTRASTRUCTURAL MORPHOLOGY

For rapid examination of phages by electron microscopy, impression grids are conveniently made by simply inverting a carbon-Formvar-coated 400-mesh copper grid on a phage plaque, then negatively staining with 2% phosphotungstic acid. All brucellaphages that have been studied are morphologically identical[4,25,26] and consist of an icosahedral head (50 to 65 nm in diameter) with a short (10 to 30 nm) tail (Figure 1). These features place them in the morphological family Podoviridae (footed viruses),[26,27] whose other members include the well-studied phages T7 of *Escherichia coli* and P22 of *Salmonella typhimurium*. Tail fibers seen in some preparations may mediate attachment to the host cell.[26] Disk-like structures, sometimes seen attached to the tail fibers, probably consist of fragments of host cell membrane.[4] Brucellaphages are often seen in arrays attached to fragments of host cell membranes, and full and empty phage heads are common. Large tubular polyheads have been described in some preparations.[26]

B. CHEMICAL AND PHYSICAL PROPERTIES

Susceptibility of brucellaphages to chemical and physical agents varies.[4,20] They are generally resistant to freezing, ether, toluene (which may increase the infective titer by dispersing clumps of phage), EDTA, and nonionic detergents such as Triton X-100, but sensitive to periodate, phenol, formaldehyde, and cationic detergents. Sensitivity varies to chloroform, anionic detergents (sodium deoxycholate), reducing agents (dithiothreitol) and heat (60°C), and to the enzymes proteinase K, trypsin, and chymotrypsin.[4,20] R/C is unique in its susceptibility to DNase, possibly because this mutant phage has a "leaky" capsid which permits entry of the enzyme to the DNA core.[20]

C. ANTIGENIC RELATIONSHIPS

These have been studied by comparing the neutralization rate constants (K values) of rabbit antisera raised against each of the phages and absorbed to remove antibodies against host *Brucella* antigens.[4,26] The results demonstrate that all brucellaphages are antigenically related to some degree, but clear-cut differences or relationships are difficult to establish. However, the results indicate some serological relationship between Tb and Fi, but little relationship between these and Wb; these results may reflect differences in the host ranges of these phages.

D. REPLICATION

Studies of the kinetics of adsorption (K values from 3 to 4.9×10^{-11} ml/min) and replication (for which estimates of the latent period vary from 30 to 150 min and the burst size varies from 10 to 26) have shown that brucellaphages cannot easily be differentiated by these criteria.[5]

Phage receptors are associated with the lipopolysaccharide-protein complex of the *Brucella* cell membrane. However, the receptors are not identical in all smooth brucellae, and adsorption of the same phage to different brucellae varies considerably. For example, Tb adsorbs to smooth strains of all biovars of *B. abortus* and *B. suis*, but not to any biovars of *B. melitensis*, even though identical A and/or M antigens occur in different biovars of all three species.[4,5,25] Receptors for the R phages of group 5, unlike those for the smooth-specific phages of groups[1-4], are heat labile and detectable only in live cultures.[4]

The relatively low efficiency of phage replication may reflect the slow-growing nature of brucellae. However, the frequent observation by electron microscopy of many empty phage heads, even in preparations of phages grown on the appropriate host under apparently ideal conditions (Figure 1), suggests that brucellaphage replication is not particularly efficient.

E. NUCLEIC ACID STUDIES

In common with other members of the Podoviridae, brucellaphages have linear double-stranded DNA, with a molecular weight of approximately 25×10^6 (about half the weight of the infective phage).[20,27] Brucellaphage DNA composition is about 45.3 to 46.7% G + C (compared to 57.3 to 58.5% G + C for *Brucella* DNA).[25]

We have reported restriction endonuclease analyses of the DNA of some brucellaphages.[20] The DNAs of Fi, Wb, Iz, and R/C could not be differentiated from Tb DNA on the basis of digestion profiles with AvaI, BglI, EcoRI, HindIII, and PvuII restriction endonucleases. Digestion profiles of Np DNA were similar to those of the other brucellaphages, but in each profile there were small but consistent differences in the numbers and/or sizes of fragments, frequently with extra fragments in either Tb or Np DNA appearing as doublet bands close to common fragments (Figure 2).

In Southern hybridizations, Tb and Np DNA probes hybridized equally well to all HindIII fragments of both Tb and Np DNA, but neither hybridized to HindIII-digested chromosomal DNA of propagating strains of *B. abortus*, or to *B. suis* or *B. melitensis*.[20] There is, thus, no evidence for incorporation of host cell DNA in the brucellaphage genome or of phage DNA in the chromosome of these brucellae. When HindIII-digested DNA extracted from *B. abortus* strains that excreted Np phage was electrophoresed, the entire characteristic digestion profile of Np DNA was superimposed on a background of chromosomal DNA fragments. These results indicate that in these strains, at least, lysogeny is not the mechanism of phage carriage.

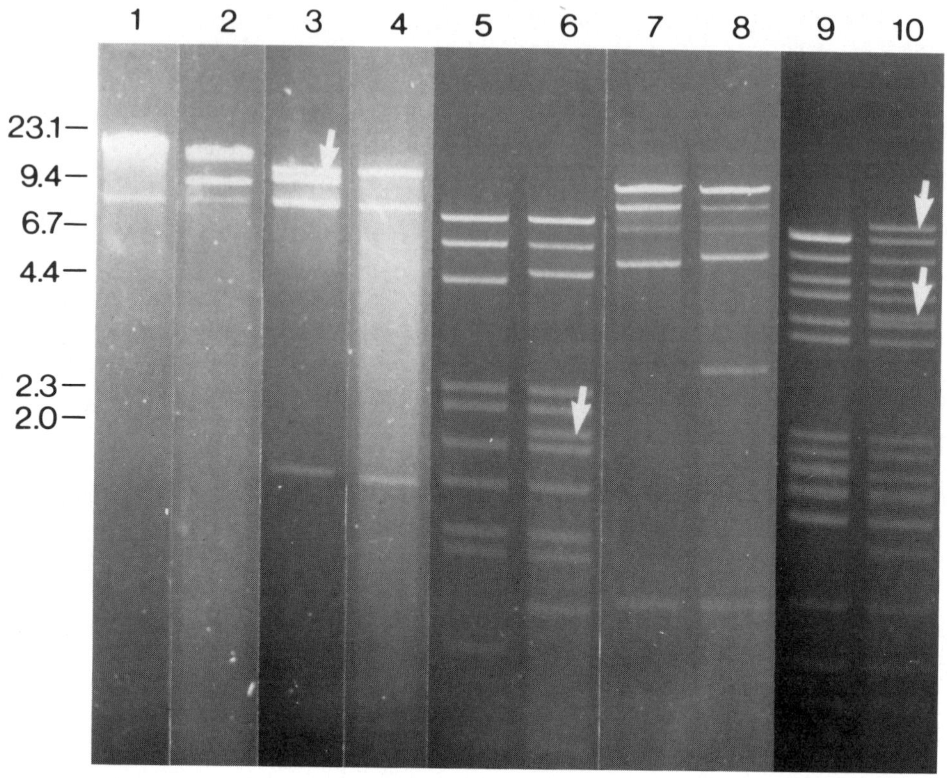

FIGURE 2. Agarose gel electrophoresis of DNA of Tb (lanes 1, 3, 5, 7, and 9) and Np (lanes 2, 4, 6, 8, and 10) phages. DNA was digested with AvaI (lanes 1 and 2); EcoRI (lanes 3 and 4); BglI (lanes 5 and 6); PvuII (lanes 7 and 8); and HindIII (lanes 9 and 10). Arrows indicate additional "doublet" bands close to common fragments.

V. DISCUSSION

The ease and simplicity of methods for the propagation and use of brucellaphages make them valuable reagents for the bacteriologist involved in the isolation and biotyping of brucellae, and they have been valuable in strain characterization and epidemiological studies.[14] Biovar-specific brucellaphages would clearly be useful for taxonomy and epidemiology, and it can be expected that new members of the group will continue to be isolated and characterized in the future.

The remarkable genetic similarity of the brucellaphages, in spite of their different sources and geographic origins, is consistent with the suggestion that they are best considered as host range mutants within a single species.[4,5] In this respect they reflect the remarkable genetic homogeneity of the genus *Brucella* itself.[28]

Very little is known about the molecular mechanisms involved in replication of the brucellaphages, but the replication of some other members of the Podoviridae such as T7 and P22, which produce both lytic and lysogenic infections, has been intensively studied.[29] For example, P22 DNA is circularly permuted with terminal redundancies accounting for 2.5% of its genome. In lytic infections, P22 mediates generalized transduction, in which random host DNA segments are incorporated into progeny phage. In lysogenic P22 infections, a portion of the genome (prophage) integrates into the host cell chromosome, and when it excises during induction of lytic infection it carries with it neighboring host genes which are transferred by specialized transduction.

If brucellaphage replication is similar to that of P22, generalized transduction may be possible during lytic infections; and if lysogenic infections occur in brucellae by the integration of prophage, then specialized transduction may occur as well. However, transduction in brucellae has not yet been reported, nor has inserted brucellaphage DNA has been detected in *Brucella* chromosomal DNA.

Another mechanism of lysogeny is that of phage P1 of *Bacillus subtilis*, in which the phage DNA circularizes and behaves as an extrachromosomal plasmid. However, repeated attempts in many laboratories, including our own, to demonstrate plasmid DNA in brucellae have all been negative.

Phage-controlled synthesis and expression of toxins and other virulence factors is well known in other bacteria such as clostridia and salmonellae. It is tempting to speculate that brucellaphages may play a similar role, but supportive evidence is lacking. Better knowledge of the relationships of brucellaphages and their host bacteria is required, especially of the possible existence and importance of lysogenic mechanisms. Such studies will open new possibilities for genetic manipulations and better techniques for diagnosis, control, and prevention of brucellosis.

REFERENCES

1. **Parnas, J.,** Brucella phages, properties and application, *Bibliotheca Microbiologica,* Fasc. 3, S. Karger, Basel, 1963.
2. **Stableforth, A. W. and Jones, L. M.,** Report of the subcommittee on taxonomy of the genus "Brucella". Speciation in the genus "Brucella", *Int. Bull. Bacteriol. Nomencl. Taxon.,* 13, 145, 1963.
3. **Jones, L. M.,** Report to the international committee on nomenclature of bacteria by the subcommittee on taxonomy of 'brucellae'. Minutes of meeting, 22—23 July 1966, *Int. J. Syst. Bacteriol.,* 17, 371, 1967.
4. **Corbel, M. J. and Thomas, E. L.,** The brucella-phages; their properties characterisation and applications, Booklet 2266, Ministry of Agriculture, Fisheries and Food, London, 1980.
5. **Corbel, M. J.,** Recent advances in brucella-phage research, *Vet. Bull.,* 54, 65, 1984.
6. **Moreira-Jacob, M.,** New group of virulent bacteriophages showing differential affinity for *Brucella* species, *Nature (London),* 219, 752, 1968.
7. **Jones, L. M., Merz, G. S., and Wilson, J. B.,** A lytic factor associated with brucellaphage causing 'lysis-from-without', *Experientia,* 24, 20, 1968.
8. **Corbel, M. J. and Thomas, E. L.,** Properties of some new brucella phage isolates; evidence for lysogeny within the genus, *Dev. Biol. Stand.,* 31, 38, 1976.
9. **Morris, J. A. and Corbel, M. J.,** Properties of a new phage lytic for *Brucella suis, J. Gen. Virol.,* 21, 539, 1973.
10. **Morris, J. A., Corbel, M. J. and Phillip, J. I. H.,** Characterization of three phages lytic for *Brucella* species, *J. Gen. Virol.,* 20, 63, 1973.
11. **Corbel, M. J. and Thomas, E. L.,** Description of a new phage lytic for several *Brucella* species, *J. Biol. Stand.,* 4, 195, 1976.
12. **Thomas, E. L. and Corbel, M. J.,** Isolation of a phage lytic for several *Brucella* species following propagation of Tbilisi phage in the presence of mitomycin C, *Arch. Virol.,* 54, 259, 1977.
13. **Corbel, M. J.,** Brucella phages: advances in the development of a reliable phage typing system for smooth and non-smooth *Brucella* isolates, *Ann. Inst. Pasteur Microbiol.,* 138, 70, 1987.
14. **Gargani, G. and Pacetti, A.,** Attempts to use phage-typing in the epidemiology of brucellosis, *Curr. Top. Vet. Med. Anim. Sci.,* 32, 7, 1985.
15. **Douglas, J. T. and Elberg, S. S.,** Properties of the Berkeley phage lytic for "Brucella melitensis" and other species, *Ann. Sclavo,* 20, 681, 1978.
16. **Corbel, M. J.,** Properties of brucella-phages lytic for non-smooth *Brucella* strains, *Dev. Biol. Stand.,* 56, 55, 1984.
17. **Corbel, M. J.,** Production of a phage variant lytic for non-smooth *Brucella* strains, *Ann. Sclavo,* 19, 99, 1977.
18. **Corbel, M. J. and Thomas, E. L.,** Use of phage for the identification of *Brucella canis* and *Brucella ovis* cultures, *Res. Vet. Sci.,* 38, 35, 1985.

19. **Corbel, M. J., Tolari, F., and Yadava, V. K.,** Characterization of a new phage lytic for both smooth and non-smooth *Brucella* species, *Res. Vet. Sci.,* 44, 45, 1988.
20. **Rigby, C. E., Cerqueira-Campos, M. L., Kelly, H. A., and Surujballi, O. P.,** Properties and partial genetic characterization of Nepean (Np) and other lytic phages of *Brucella* species, *Can. J. Vet. Res.,* submitted.
21. **Ewalt, D. R. and Forbes, L. B.,** Atypical isolates of *Brucella abortus* from Canada and the United States characterized as dye sensitive with M antigen dominant, *J. Clin. Microbiol.,* 25, 698, 1987.
22. **Garcia, M. M., Brooks, B. W., Ruckerbauer, G. M., Rigby, C. E., and Forbes, L. B.,** Characterization of an atypical biotype of *Brucella abortus, Can. J. Vet. Res.,* in press.
23. **Alton, G. G., Jones, L. M., and Pietz, D. E.,** *Laboratory Techniques in Brucellosis,* 2nd ed., World Health Organization, Geneva, 1975.
24. **Verger, J. M.,** Bacteriological methods, in *Techniques for the Brucellosis Laboratory,* Alton, G. G., Jones, L. M., Angus, R. D. and Verger, J.-M., Eds., Institut National de la Recherche Agronomique, Paris, 1988.
25. **Calderone, J. G. and Pickett, M. J.,** Characterization of brucellaphages, *J. Gen. Microbiol.,* 39, 1, 1965.
26. **Ackermann, H.-W., Simon, F., and Verger, J.-M.,** A survey of Brucella phages and morphology of new isolates, *Intervirology,* 16, 1, 1981.
27. **Matthews, R. E. F.,** Classification and nomenclature of viruses. Fourth report of the International Committee on Taxonomy of Viruses, *Intervirology,* 17, 23, 1982.
28. **Verger, J. M., Grimont, F., Grimont, P. A. D., and Grayon, M.,** *Brucella,* a monospecific genus as shown by deoxyribonucleic acid hybridization, *Int. J. Syst. Bacteriol.,* 35, 292, 1985.
29. **Birge, E. A.,** *Bacterial and Bacteriophage Genetics,* Springer-Verlag, New York, 1981, chap. 7.

Chapter 7

EPIDEMIOLOGY AND SURVEILLANCE

Richard P. Crawford, Jan D. Huber, and Bruce S. Adams

TABLE OF CONTENTS

I. INTRODUCTION

Since the published information on epidemiology and surveillance of *Brucella abortus* is voluminous, the authors elected to cite four standard references, to select pertinent epidemiological factors for review, and to supplement published information with the most recent data available. The standard references are as follows: *Bovine Brucellosis — An International Symposium,* published in 1977;[1] Brucellosis Research: An Evaluation, published in 1977;[2] Report of the National Brucellosis Technical Commission, published in 1978;[3] and a review article, "The Epidemiology of Bovine Brucellosis", by Nicoletti and published in 1980.[4] The epidemiological factors selected were demographics, reservoir, transmission, host susceptibility, surveillance and case finding, and program components and their scientific basis.

II. DEMOGRAPHICS

Bovine brucellosis was recorded in 120 of 175 countries in 1986 (Table 1).[5] Thirty-three countries did not record the disease, and data were not available from the remaining 22. Of the 33 countries that did not record brucellosis, 16 reported the year of last occurrence.

Those 16 are listed chronologically as follows:

1.	Cyprus	1932
2.	U.K./Channel Islands	1935
3.	Norway	1952
4.	Sweden	1957
5.	Bulgaria	1958
6.	Korea D.P.R.	1959
7.	Finland	1960
8.	Denmark	1962
9.	Switzerland	1963
10.	Czechoslovakia	1964

TABLE 1
Distribution of Brucellosis in the World in 1986

	Number of countries			
Continent	Recorded	Not recorded	Data not available	Total
Africa	42	4	8	54
Americas	26	7	4	37
Asia	27	6	9	42
Europe	18	12	0	30
Oceania	6	4	1	11
U.S.S.R.	1	—	—	1
Total	120	33	22	175

11.	Romania	1969
12.	Papua New Guinea	1974
13.	Belize	1980
14.	French Polynesia	1984
15.	Israel	1984
16.	Hungary	1985

The other 17 countries that did not record brucellosis in 1986 and also did not give the year of last occurrence are listed by continent as follows:

Africa

1. Chad
2. Ghana
3. Madagascar
4. Somalia

Americas

5. Bahamas
6. Bermuda
7. Canada*
8. Falkland (Malvinas)
9. Guyana
10. Trinidad and Tobago

Asia

11. Malaysia (Sabah)
12. Philippines
13. Singapore

Europe

14. Iceland
15. U.K./Isle of Man

Oceania

16. New Caledonia
17. Solomon Islands

This information is not intended as a documentation of freedom from bovine brucellosis. Recognition of brucellosis-free status is probably best left up to each individual country based on criteria it wishes to use. The reporting of brucellosis varies with the thoroughness of surveillance in a country.

III. RESERVOIR

Brucella abortus has been isolated from at least nine species of domestic livestock.[7-21] These are listed in Table 2 as the cow, horse, pig, sheep, goat, bactrian camel, dromedary camel, water buffalo, and yak. The isolations were made in 37 countries; however, the geographical location of the dromedary camel infection was not reported.[11] Cattle are the natural or primary host for *B. abortus*. Of 266 isolations identified to a source, 241 (90.6%) were from cattle.[14] Of 1722 isolations in France, 1683 (97.7%) were from cattle.[13] Also, 395 (84.8%) of 466 *B. abortus* cultures in a third survey were from cattle.[22] These data support the belief that cattle (especially female cattle) are primarily responsible for maintaining *B. abortus* infection. Bison and a limited population of elk are considered an active biological reservoir in the U.S.[7] Brucellosis in wildlife will be discussed in much more detail in a Chapter 13.

Nine biovars of *B. abortus* have been identified.[17] Biovar 8 is not currently known to exist.[23] *B. abortus* strain 19, utilized as a vaccine, originated from a biovar 1. Rough strains are rarely isolated and cannot be characterized as to biovar.[13] Other atypical isolates are occasionally reported. Table 2 shows the geographical distribution of the biovars, although they were not reported in every case. Biovar 1 appears to be the most common and widely distributed. Only biovars 1, 2, and 4 have been reported from the Americas. The U.S.S.R., England, and France have the greatest reported variety of biovars. Biovars 2 and 8 have not been reported in the U.S.S.R. Biovars 3 and 7 have not been reported in England. Biovars 5, 7, and 8 have not been reported in France.

The biovars of *B. abortus* are differentiated primarily by CO_2 requirement, H_2S production, growth on media containing dyes, agglutination with monospecific antiserum, phage lysis, and substrate oxidation.[23] Biovar identification may provide helpful epidemiological information in some outbreaks of brucellosis.[24]

Differences between biovars as to pathogenicity and other such characteristics have not been documented.[25] However, some epidemiologists from the western region of the U.S. believe that biovar 2-affected herds are characterized by longer incubation periods compared to biovars 1 and 4.

IV. TRANSMISSION

The exposure of cattle to *B. abortus* is the genesis of the transmission process. The outcome of this exposure is dependent on the exposed animal's susceptibility to infection and the amount of exposure. A diverse range of factors is capable of influencing both host susceptibility and exposure level, making it inappropriate to attempt to present a dogmatic description of transmission. The intent of this section on transmission is to provide information on the cattle husbandry and environmental factors which may influence the level of exposure to brucellosis. It is these factors which ultimately determine how and why brucellosis is maintained and spread in cattle populations. While factors associated with host susceptibility are an integral part of the transmission process, they will be dealt with in more detail in a later section of this chapter.

A. SOURCE AND LEVEL OF EXPOSURE

In most circumstances the primary escape route of *B. abortus* is the uterine fluid and placenta expelled by infected cows when they abort or have a full-term parturition. The risk posed to susceptible animals following parturition of infected cattle depends on three factors: the number of *Brucella* excreted, the survival of these organisms under the prevailing environmental conditions, and the probability of susceptible animals being exposed to enough organisms to establish infection. These three factors are the determinants of the level of exposure.

TABLE 2
Brucella abortus Isolations from Domestic Livestock

Species	Country	Biovar
Cattle	U.S.	1,2,4
	Argentina	1,4
	Brazil	1,2
	Colombia	1,2
	Chile	1,4
	Cuba	1,4
	Equador	4
	Peru	1
	Uruguay	1
	Venezuela	1,2
	India	1,3
	Iran	3
	Israel	1,2,6
	Japan	2
	England	1,2,4,5,8,9
	France	1,2,3,4,6,9
	Germany	2,5
	Italy	6,7
	Malta	3
	Poland	6
	Cameroon	2,3,6
	Chad	2,3,5,6
	Cote d'Ivoire	1,4
	Egypt	3,7
	Rwanda	Not reported
	Senegal	1,3
	Somalia	6
	Sudan	Not reported
	Tanzania	Not reported
	Togo	3
	Uganda	3,8,9
	U.S.S.R.	1,3,4,5,6,7,9
Horse	U.S.	1,2,4
	Brazil	1
	Columbia	1
	Australia	1
Swine	U.S.	1
	France	1
Sheep	Canada	1
	U.S.	Not reported
	England	6
	France	3
	Tunisia	6
Goat	France	1
	Switzerland	6
Camel (*Camelus dromedarius*)	Not reported	Not reported
(*C. bactrianus*)	Mongolia	Not reported
Buffalo (*Bubalus bubalus*)	Eastern Mediterranean area	3
	Egypt	3
	Italy	1
	U.S.	1
	Brazil	Not reported
	India	Not reported
Yak	Mongolia	Not reported

1. The Number of *B. abortus* Excreted

While infected cows shed large numbers of *Brucella* at parturition, there are a number of factors which may influence the relative numbers of organisms excreted. It has been asserted that strain 19-vaccinated infected cows excrete fewer organisms at parturition than nonvaccinated infected cows.[26] A field-based study of California dairy herds reported that the presence of one or more nonvaccinated reactors was associated with an eightfold increase in the percentage of the herd that eventually became reactors when compared to herds in which all reactors were calfhood strain 19 vaccinated.[27] The authors of this latter study contend that calfhood strain 19 vaccination essentially eliminates the infected cow as a source of the disease. In contrast, no statistically significant difference occurred in the distribution of *B. abortus* in tissues from culture positive cows that were calfhood, adult, or nonvaccinated.[28] Unpublished data from the Texas A&M Brucellosis Research Project using placental tissue and uterine swabs collected from experimentally infected cows at their first parturition postchallenge also contradict this effect of strain 19 vaccination.[29] Using the relative number of brucella colonies on bacteriologic culture media as an indicator of the number of organisms excreted, the Wilcoxon rank sum test did not demonstrate a statistically significant difference between the colony counts from nine infected strain 19 vaccinates and 21 nonvaccinates. This observation of no effect is being pursued with additional strain 19 and nonvaccinated cattle exposed to pathogenic *B. abortus*.

The duration of shedding may be another factor influencing the number of organisms excreted. Unpublished observations of experimentally infected cows indicate that decreasing numbers of organisms are detected when uterine discharges are cultured at sequential parturitions, and a substantial proportion of the uterine samples from infected cows are culture negative at the second and third parturition following challenge.[29] The relevance of this apparent "self cure" has not been explored under field conditions.

2. Survival in the Environment

It is generally accepted that the growth of *B. abortus* outside the mammalian host plays no part in the epidemiology of the disease, i.e., it does not multiply but merely persists.[7] The viability of *B. abortus* outside the host is influenced by the prevailing environmental conditions. Numerous studies have assessed the persistence of *B. abortus* under a variety of environmental conditions. These have been comprehensively reviewed by Wray.[30] In general, viability is enhanced by cool temperatures and moisture and limited by high temperatures, direct sunlight, and dryness. Table 3 summarizes some of the recorded survival times.

Reports focusing on the survival of *Brucella* in manure, which is used extensively as a pasture fertilizer in many production systems, have been reviewed by Verger.[31] It is of interest to note the conflicting survival times for *B. abortus* in liquid manure. This appears to have its basis in the temperature encountered in the manure storage tanks: at high temperatures (45 to 50°C) survival time is short, about 4 h, but at lower temperatures (15°C), survival times of at least 8 months have been recorded.[32,33] It has been reported that xylene at a concentration of 1000 ppm is an effective disinfectant for liquid manure, killing *Brucella* in 1 month.[32] Acid fermentation of feed mixtures containing bovine manure to a pH of 4.5 or less will ensure the destruction of *B. abortus*.[34] Although manure represents a possible vehicle of exposure to *Brucella*, there is scant field evidence to substantiate its occurrence.

3. The Probability of Exposure

The risk that a susceptible animal will be exposed to an infective dose of *B. abortus* depends largely on the husbandry practices under which the cattle are managed. A full discussion of factors contributing to this risk is given in the section on inter- and intraherd transmission later in this chapter.

TABLE 3
Survival Time of *Brucella abortus* under Various Environmental Conditions

Environment	Survival time
Direct sunlight	4.5 h
Soil	
Dry	4 d
Moist	66 d
Cold	151—185 d
Manure	
Slurry	8—240 d
High temperature	2 d
Water	
Drinking	5—114 d
Polluted	30—150 d
Fetus in shade	180 d

4. Other Sources of Infection

This section is not a comprehensive review of all reported sources of infection. For example, the role of wildlife is discussed elsewhere in this text (Chapter 13). The intent of this section is to review briefly selected potential sources of bovine brucellosis.

a. Bulls

The role of the infected bull in the transmission of *Brucella* has been the subject of a number of investigations.[35-37] *Brucella* infection in bulls can produce orchitis, epididymitis, and seminal vesiculitis, and localization in these organs generally results in *Brucella* being shed in the semen. Infected bulls often remain functionally fertile.

The potential for bulls to transmit infection to cows depends on the method of insemination employed. When used for natural mating, the risk of infected bulls transmitting the disease to susceptible cows is considered to be negligible.[35,36] However, when used as semen donors for artificial insemination, there is a considerable risk of inseminated cows becoming infected. Manthei et al. reported that 8 of 12 cows became infected after insemination with semen from an infected bull.[37] Bendixon and Blom recorded the spread of *Brucella* attributable to infected donor bulls — in one instance 41 herds became infected over a 2-month period following the use of an infected bull, while another infected bull was responsible for transmitting the disease to 20 herds over a 9-d period after his semen was used.[38]

b. Horses

While domestic equidae are susceptible to infection with *B. abortus*, there is little evidence to suggest that they have a significant role in the epidemiology of the disease. In infected horses *Brucella* tend to localize in the bursae, tendons, muscles, and joints rather than the tissues of the reproductive tract.[7] While the classical clinical features of equine brucellosis are the chronically discharging abscesses associated with fistulous withers and poll evil, other clinical syndromes, including abortion, have been described.[15,39,40] One case report suggested that a mare with recurring abscesses acquired *Brucella* infection 6 years previously on a neighbor's pasture and subsequently transmitted it to two cows and a dog on the owner's premises.[41] It is likely that abortions in mares due to *B. abortus* pose a substantial risk for transmission to cattle, but documentation of this is lacking.

c. Embryo Transfer

The increasing use of embryo transfer technology in cattle herds in recent years has stimulated interest in the risk of transmitting *Brucella* from infected donor cows to recipients. In one study, workers failed in 36 attempts to isolate *B. abortus* from the uterine flushings of culture-positive and serologic-reactor cows despite the fact that at 17 of these attempts, the donor cows were excreting *Brucella* in their udder secretions.[42] Other studies involving the contamination of embryos with *B. abortus in vitro* have shown that:

1. *B. abortus* did not adhere to zona pellucida (ZP)-intact or ZP-free embryos after ten sequential antibiotic-free washings, although isolations were made from all groups of ZP-defective embryos subjected to the same washing regime.[43,44]
2. Standard embryo freezing and thawing techniques in the absence of antibiotics caused a 64% reduction in the number of *Brucella* after either 2 weeks or 6 months of storage. Addition of antibiotics produced a 99.9% reduction in viability. Glycerol protected the organism during freezing and thawing in the absence of antibiotics, but viability was greatly reduced with antibiotics present. Dimethylsulfoxide (DMSO) protected the organism both with and without antibiotics present.[45]

The general consensus of opinion is that the transfer of embryos from *Brucella*-infected cows poses little risk to recipients.

d. Dogs

Infection of dogs with *B. abortus* has been recorded. A study in which 9 dogs orally infected with *B. abortus* were housed with 11 pregnant heifers demonstrated that infected dogs can be a source of *Brucella* for cattle.[46] In this study, 2 of the dogs aborted with *B. abortus* being isolated from abortion substrates, and 7 of the 11 heifers developed brucellosis titers within 5 months. Six of these heifers subsequently aborted, and *B. abortus* was isolated. Bacteriologic monitoring of these dogs for a period of 2 years indicated that *B. abortus* was only excreted at the time of abortion.

A serologic survey of 216 dogs on 18 dairy farms in Chile revealed that 44% of dogs on 16 farms with bovine brucellosis had positive brucellosis titers; 1 of 18 (6%) dogs had a positive reaction on a farm from which the disease had been eliminated 1 year previously; and on a farm with no history of bovine brucellosis, all 12 dogs were negative.[47]

There may be a need to consider the role of dogs in situations where they are kept in close proximity to cattle.

B. EXPOSURE ROUTES

The oral route is considered the most common means by which *B. abortus* reaches a susceptible host, and entry point is believed to be the nasopharynx area.[48] For a more complete description of pathogenesis see Chapter 12. It has been suggested that the conjunctival route is important in intensively housed cattle and the oral route is important under range conditions.[49] More pathogenic *B. abortus* organisms administered orally are required to infect cattle experimentally, and, therefore, the most commonly used experimental challenge route is conjunctival.[48] The data suggest that the minimum infective dose may depend on the exposure route, e.g., fewer organisms required to establish infection via the conjunctival route.

Congenital infection is of major epidemiological significance. As high as 20% of heifer calves born to infected cows are persistently infected with *Brucella*.[50-53]

Although feeding *Brucella*-contaminated milk to young calves from *Brucella*-free dams has not been shown to result commonly in persistent infection, the authors have observed pregnant heifers nursing other heifers in the same pen, and this could result in transmission

* The last case of brucellosis was detected by positive serotests in late December of 1986 but not recorded until confirmed by bacteriologic culture in January 1987.[6]

of *B. abortus*. The possibility of transmission of *Brucella* during machine milking in a similar fashion to the transmission of common mastitis pathogens has been raised, and there is experimental evidence to show that systemic infection will follow udder infusion of *Brucella*.[54] However, there is no field evidence available to support these milk-borne routes of exposure as major risk factors under field conditions.

C. INTER- AND INTRAHERD TRANSMISSION

The literature abounds with reports attributing a wide range of cattle husbandry factors with promoting or restricting the spread of brucellosis. A comprehensive review of these factors has been published by Salmon and Meyer.[55] The relative importance of a particular factor is dependent on the prevailing cattle husbandry and management practices. This point is illustrated by studies conducted in two separate regions of northwest Mexico with different industry infrastructures.[56-58] While both studies identified six factors contributing significantly to brucellosis infection, only two were common to both regions.

From a practical viewpoint, factors influencing the transmission of brucellosis in any given geographic region can be classified into two fundamental categories: factors associated with the transmission of the disease between herds and factors influencing the maintenance and spread of infection within herds. Both factors must be addressed to exert effective control over disease transmission.

1. Factors Influencing Interherd Transmission

a. Replacement Animals

The purchase of infected replacement animals is widely reported as the major factor responsible for introducing brucellosis into previously free herds.[24,59-63] A study in Ontario, Canada, established that owners of infected herds purchased more cattle than owners of noninfected herds.[64] Another study of west Scotland dairy herds revealed that 72.2% of herds that purchased replacements were infected compared to 34.6% of those that reared their own replacements.[61] A survey of *Brucella*-infected herds in the U.S. attributed 55.6% of herd infections to the purchase of infected cattle.[65]

''Purchasing practices'' is an all-encompassing phrase embracing at least three components which may operate independently or interact to influence the risk of purchasing infected cattle. These are the frequency, source, and brucellosis test history of purchased cattle.

i. Frequency of Purchase

In general terms, the more often cattle are introduced into the herd, the greater the risk of introducing infected cattle. The Ontario study, noted previously, recorded a positive association between the frequency of purchase and *B. abortus* infection.[64] Observations that dairy herds were more likely to be infected than beef herds have been attributed to the greater interchange of cattle between dairy herds.[60,64,66]

ii. Source of Purchase

Few studies on the relative risks involved in purchasing from different sources have been made. The Ontario researchers recorded differences between infected and noninfected herds with respect to the source of replacement cattle.[64] They concluded that, in Ontario, purchases from cattle dealers and directly from other herds posed greater risks than purchases from auction markets or from purebred sales. The risks associated with different sources of purchase in a given region will depend on the infrastructure of the cattle industry and the effectiveness of regulatory measures exerted to control brucellosis.

iii. Brucellosis Test History

Testing cattle for brucellosis prior to purchase has been widely promoted as a means of reducing the risk of introducing infected cattle into a herd. While the rationale for this is soundly based, it does not completely eliminate the risk of purchasing infected animals. Schilf reported that 10% of herds receiving negative test cattle that had been exposed to infection developed brucellosis infection because of these cattle.[67] Meyer and Vanderwagen stated that "despite meeting the vaccination, status of herd of origin and pre- and postentry testing requirements, it is obvious that California is buying brucellosis via the serologically negative exposed animal in the incubation stage of the disease."[68] To decrease the risk associated with purchasing these infected cattle, most brucellosis control authorities encourage herd owners to keep purchased breeders isolated from the main herd and to conduct a postcalving test.

b. Proximity to Infected Herds

The tendency of *Brucella* infection to spread from infected herds to neighboring clean herds has been widely recorded. Cattle contacts at fence lines, confinement failures (strays), and sharing of common pasture are the main methods by which transmission to adjacent herds occurs.[59]

Geographic clustering of infected herds is often encountered and forms the basis of using adjacent herd testing as an effective method of case finding in brucellosis eradication programs. The Ontario study demonstrated a cluster effect and showed that proximity to infected herds was the best statistical discriminator between infected and noninfected herds.[64]

Christie's observations on brucellosis in Northern Ireland led to the comment that "brucellosis does not have much respect for fences no matter how good they may be".[60] He further stated that if one herd in a clean area became reinfected, it was the exception rather than the rule that the infection would be contained within that one herd. Luchsinger et al. cited a case in which brucellosis was introduced into a herd in a certified free county that eventually resulted in 23 new herds being infected over a 2-year period, an attack rate of 50%.[20]

c. Waterways

Evidence for the role of common waterways in the interherd transmission of *B. abortus* is equivocal. Transmission via streams was proposed as a likely explanation for some brucellosis outbreaks in France: the disease spread from herds upstream to those downstream but did not spread upstream from infected herds.[69] In the U.S., although spread of infection to herds downstream of infected herds has been documented, it was considered that cattle contacts rather than waterborne infection was the most probable means of transmission.[70] While water provides a favorable environment for the survival of *Brucella*, the dilution factor associated with running waterways make it an unlikely means of interherd transmission.

d. Scavengers

Scavenging animals and birds have been cited as a means by which *Brucella*-contaminated material may be carried between herds, although substantial evidence for this means of spread actually occurring is lacking.

2. Factors Influencing Intraherd Transmission

A variety of cattle husbandry practices have been shown to be associated with the spread of *B. abortus* infection within herds. Most of these practices influence the probability of exposure to infection. Identification and modification of practices contributing to disease transmission is fundamental to successful control of brucellosis in infected herds. Some of the important factors are reviewed here.

a. Vaccination Level

While discussed in greater detail in Chapter 11, strain 19 vaccination has been shown to reduce substantially the transmission of brucellosis in herds. The consensus of opinion is that strain 19 vaccination confers a relative immunity, with 65 to 75% of animals protected against exposure under field (range) conditions resulting in a marked reduction in the number of reactor animals detected in infected vaccinated herds compared to infected nonvaccinated herds. The benefits of vaccination are twofold: it decreases an animal's susceptibility to infection and the ensuing reduction in within-herd prevalence decreases the level of exposure by reducing the number of animals excreting *Brucella*.

b. Herd Size

A number of studies have shown a positive association between herd size and the presence of brucellosis.[8,59,61,64] While herd size is a relative concept with distinction between large and small herds being dependent on location and agricultural practices, this association has been demonstrated over a diverse range of situations. Although animals within large herds are not intrinsically more susceptible to infection, there are aspects associated with large herd size which may facilitate the transmission of brucellosis.

A fundamental aspect is the possibility of an association between herd size and the purchase of replacement cattle. In many cases (particularly dairy herds), large herd size is often maintained by the introduction of replacement cattle from outside the herd.[59] In these circumstances it is possible that the introduction of either infected or susceptible animals contributes more to brucellosis transmission than herd size per se.

A second aspect relates to logistic and managerial problems associated with controlling the disease as herd size increases. Control of brucellosis in larger herds often demands concerted managerial input which may be difficult to implement. Mistakes are more likely to occur in larger herds than smaller herds.

A third aspect is the possible influence of herd size on disease dynamics. As the population base in which the disease resides increases in size, it is likely that disease transmission is enhanced as a result of greater opportunity for more complex interactions between the infected, susceptible, and resistant sections of the population. It may also increase the overall possibility of aberrant manifestations of the disease (e.g., latent infection) being encountered.

c. Population Density

The stocking density (cattle-to-land area) at which a herd is maintained has been cited as an important factor influencing the transmission and persistence of brucellosis.[59,60] There appears to be a positive association between stocking density and disease prevalence which can be attributed to the fact that stocking density is an important determinant of the potential for contact between susceptible and infected animals.

d. Method of Housing

Associations between housing and stabling methods and brucellosis infection have been reported by several workers. Studies in Ontario and Scotland demonstrated that brucellosis was more prevalent in herds in which loose housing was practiced as opposed to those in which cattle were tethered.[61,64] It was postulated that the basis for this difference was the greater opportunity for exposure to contaminated material and a generally lower standard of cleanliness associated with loose housing.

e. Use of Maternity Pens

Under intensive systems of cattle husbandry, the use of maternity stalls has been shown to reduce the level of infection markedly.[27] The benefits of individual calving stalls can be attributed to decreasing the exposure of susceptible animals to infective materials.

TABLE 4
Age-Specific Prevalence of Brucellosis Reactors in a Dairy Herd Containing Cattle Infected with *Brucella abortus*

Age (years)	No. cattle	No. reactors	Percent
<2	36	3	8
3—4	220	19	9
5—6	243	26	11
>7	28	2	7

TABLE 5
Calves Weaned from Reactor Cows in Three Cattle Herds in Texas and Cultured for *Brucella abortus*

Sex	No. cultured*	No. infected
Females	37	2
Males	27	0

* Females cultured at parturition and males cultured at breeding soundness examination.

V. HOST SUSCEPTIBILITY

A. AGE

Most data suggest that young cattle are less susceptible to *B. abortus* than older sexually mature cattle. Susceptibility appears to be more strongly associated with sexual maturity than age, and, therefore, young sexually immature cattle generally do not become infected following exposure or recover quite rapidly. After sexual maturity, age does not appear to be a factor. The data in Table 4 are from a dairy herd that was infected with *B. abortus* and support the concept that age does not affect susceptibility in mature female cattle.

B. SEX

Controlled studies on the effect of sex have not been extensive, probably because males are not as important as females in the epidemiology of *B. abortus* in cattle. Males do become infected, and a relative susceptibility could be immature calves, mature bulls, and mature cows. Latency or persistent infection in calves is also discussed in the section of this chapter on transmission, and although inconclusive, the data in Table 5 support females as more likely than males to be persistently infected following exposure as calves. On the other hand, Rankin has concluded that bulls often become infected in calfhood, and infected males are detected at earlier ages than females in herds in Ireland.[71] Males in infected herds do not appear to be as likely to become reactors as females, as evidenced by the data in Table 6.

When herd bulls remain with the cows continuously, the bulls appear to be less susceptible, as evidenced by the data in Table 7. All 438 card test-negative females were strain 19 vaccinated, and in the following year 70 new reactors were detected in the breeding cows, but no new reactors occurred in the herd bulls.[72] In Australia, it is common practice to slaughter any male with detectable *Brucella* antibodies because infected bulls have been found that were nonreactors or had low antibody titers, and, therefore, serologic data may underestimate *B. abortus* infection in males.

TABLE 6
Sex-Specific Cumulative Incidence (CI) in Three Texas Beef Herds Containing Cattle Infected with *Brucella abortus*

Sex	No. cattle	No. reactors	CI
Males	123	4	3%
Females	649	126	19%

TABLE 7
Sex-Specific Prevalence of Card Test Positives in Six Texas Beef Herds Containing Both Bulls and Cows

Sex	Card positive	Card negative	% positive
Herd bulls	1	16	6
Breeding cows	65	438	13

C. BREED

No conclusive data to support breed susceptibility are available. It has been reported that small breeds such as the Jersey are more susceptible, but this apparent breed susceptibility may be confounded by earlier sexual maturity.

D. GESTATION

Susceptibility increases with pregnancy. To date, this observation is unexplained. Hormones such as estrogen or progesterone do not affect susceptibility.[73] Susceptibility increases as stage of gestation increases in cattle. The probability of isolation of *B. abortus* at parturition increased from 0.22 to 0.90 as fetal age at exposure of nonvaccinated heifers increased from 60 to 150 gestation days.[74] Elapsed time between *B. abortus* exposure and culture attempt did not confound the effect of fetal age at exposure on risk of infection in strain 19-vaccinated heifers.[75]

E. OTHERS

Induced resistance to infection via vaccination has been reported by a number of research groups. Vaccination will be discussed in a separate chapter. Inate or nonspecific resistance may have a genetic basis, but conclusive data are not available at this time.

VI. SURVEILLANCE AND CASE FINDING

Planning, management, and evaluation of brucellosis control and eradication programs require accurate information on the prevalence, incidence, and trends of infection within the designated cattle population. The effectiveness and nature of methods developed to provide this information are determined by many factors. These include:

1. The infrastructure of the cattle industry (husbandry methods, marketing systems and patterns, type of production, herd size distribution)
2. The resources available to allocate to surveillance, encompassing finances, facilities, manpower, and regulatory resources
3. The scope and accuracy demanded of the surveillance system
4. The objectives of the program: control vs. eradication
5. The prevalence and distribution of the disease in the population

No single surveillance system is appropriate to all situations, and the demands placed on a system change as the program shifts from one phase to another. Excellent discussions of the principles and philosophies underlying the development of animal disease surveillance systems have been published.[76-85]

The remainder of this section is devoted to a discussion of the methods currently used to monitor and detect brucellosis in cattle populations. Since these methods rely on serologic tests to indicate infection, it is opportune to review briefly the factors influencing the validity of these tests.

A. TEST EVALUATION

The basic characteristics of tests (sensitivity, specificity, and predictive values) are important considerations in test selection and in evaluation of the information generated by a particular surveillance method. The sensitivity of a test can be defined as the conditional probability that the test will identify all individuals with brucellosis in a given population. A test with a high sensitivity will identify most diseased individuals resulting in only a few false negatives. Specificity is defined as the conditional probability that the test will identify all nondiseased individuals. A test with a high specificity will correctly identify most of the nondiseased individuals resulting in only a few false positives. In general terms, the sensitivity and specificity of a test are inversely related: as the sensitivity increases the specificity decreases, resulting in an increased proportion of false positive reactions. Conversely, a decrease in sensitivity increases the proportion of false negative reactions.

The use of a highly sensitive but lowly specific test will inflate the apparent disease prevalence because of the high number of false positives. For example, agglutination tests (e.g., Rose Bengal and card tests) are highly sensitive tests that are widely used as rapid and inexpensive screening tests. However, their relative lack of specificity, particularly when used in populations in which a large proportion of cattle are strain 19 vaccinated, can lead to inflated estimates of the true prevalence of brucellosis. In many circumstances, supplementary tests with high specificity are employed in an attempt to obtain a definitive determination of infection status. These tests (e.g., Rivanol, complement fixation tests) are based on detecting antibody in serial dilutions of serum — the greater the dilution at which antibody is detected, the more likely that the animal is actually infected. Establishing the titer levels for positive and negative reactions directly influences the sensitivity and specificity of these tests. Higher titers for the positive level lower the sensitivity (more false negatives), and lower titers reduce the specificity which leads to more false positives. Considerations of the sensitivity vs. specificity of diagnostic tests become important as control programs progress. The final eradication stages demand highly sensitive tests when the small number of false positives can be tolerated.

Allied to sensitivity and specificity is the predictive value of a test. The predictive value of a positive test is defined as the proportion of test positive animals that are actually infected, while the predictive value of a negative test is the proportion of test negative animals that are truly nondiseased. Although the predictive values of a test are tied to its sensitivity and specificity, they are also influenced by the prevalence of disease in the population. For a test with a given sensitivity and specificity, as the disease prevalence decreases the predictive value of a positive test also declines, meaning that a larger proportion of the test positive animals are false positives. At low prevalence levels the large majority of the cattle tested are actually nondiseased, and the number of false positives may exceed the number of true positives (diseased). For further information on the effect of disease prevalence, consult References 86 to 88.

B. CASE FINDING METHODS

1. Testing Animals at Slaughter or Markets

This is the primary surveillance method employed in many countries. Its effectiveness is dependent on having a system that allows tested cattle to be traced back to their herd of origin. With market testing (e.g., First Point of Concentration testing as employed in the U.S.) it is relatively easy to determine the herd from which the tested cattle originated. With slaughter cattle, the task becomes more difficult and a number of countries have devised methods to address the problem.

Australia has implemented a system based on attaching a tail tag to all cattle consigned to sale or slaughter. In that system all cattle owners are issued an individual herd identification number which is printed on plastic ratchet-type self-locking tags or plastic adhesive tags. These tags are attached to each animal when it leaves the herd, thus allowing rapid identification of the herd of origin at both market and slaughter.

The U.S. has adopted a system that utilizes numbered back tags that are placed on animals at markets. This permits traceback to the market from which the cattle were consigned, and further examination of the market records allows identification of the original consignor of the cattle.

Apart from the logistics of achieving adequate traceback capability, the effectiveness of market and slaughter testing programs are influenced by a number of other factors. Amosson and Dietrich studied the strengths and weaknesses of the U.S. Market Slaughter Testing program.[89] Their major conclusions were

1. There is a distinct herd size bias in its effectiveness in detecting infected herds, with large herds having a higher probability of being detected more promptly following infection than small herds. They suggest that secondary testing (on-farm whole herd testing) may be efficiently used by concentrating testing in smaller herds when funds for secondary testing are limited.
2. Detection probabilities can vary 30% between the buildup and liquidation phases of the cattle cycle, i.e., the secular upswings and downswings in the cattle inventory. This effect can be partially offset through the use of first point of concentration (marketplace) testing.
3. A 95% herd strain 19 vaccination level can reduce the probability of detection by as much as 40% when compared to a similar herd that is not vaccinated. This masking effect of strain 19 vaccination is one of the two main reasons why vaccination has been discouraged or prohibited in some areas and countries that have achieved a very low herd prevalence. The other reason is the confusion created by vaccine antibodies on the interpretation of serological tests.

While slaughter and market testing is a useful system for monitoring and detecting brucellosis infection when the herd prevalence is relatively high, its value diminishes as the herd prevalence declines. When the herd prevalence is low, there are great difficulties in sampling at a sufficient level to ensure reliable detection of infected herds.[66]

2. Bulk Milk Ring Tests

The bulk milk ring test (BMRT) is widely used to monitor the brucellosis status of dairy herds. Its usefulness as a surveillance method for dairy herds derives from the ease of regularly and readily obtaining samples and its simplicity and low cost to perform.

An Australian study by Rolfe and Sykes made comparisons between the BMRT, whole herd testing, and abattoir monitoring in 2698 dairy farms to evaluate their effectiveness at detecting *Brucella* infection when the herd prevalence is low.[66] At the time of the study the herd prevalence was 0.67% (18 herds). They concluded that:

1. The BMRT was effective in identifying herds infected with *B. abortus* at low herd and animal prevalence. In no herds were culture-positive cows from the milking herd

detected without the BMRT being positive, although in 6 of the 18 infected herds, the infected animals were in the nonmilking portion of the herd. The predictive value of a positive BRMT was 8%, with much of the high false positive rate being attributed to testing seasonally calving herds in very early or late lactation.

2. Whole herd testing using the Rose Bengal test (RBT) and complement fixation test (CFT) had a positive predictive value of 18%, with many of the false positive reactions attributed to the persistence of strain 19 vaccination titers. In herds in which the culture-positive infected animals were in the milking herd, whole herd testing and the BMRT detected infection with similar efficiency.
3. Abattoir monitoring is not useful for dairy herds when brucellosis prevalence is low. None of the infected herds were identified by this method, and nearly 2% of the herds were placed under suspicion because of positive abattoir monitoring (false positives).

Herd size and frequency of testing need consideration when evaluating a BMRT program. With increasing herd size the ability of the test to detect one infected animal in a herd decreases due to the dilution effect.[90] Increasing the volume of the test sample as herd size increases has been utilized as a method of maintaining test sensitivity.[91] Sampling frequency should be at least three, and preferably four, times per year to ensure that all cows are in the milking herd on at least two or three sampling dates.[91] Positive and suspicious BMRT results need to be promptly investigated using whole herd serologic testing, epidemiologic investigation, and individual milk ring tests where indicated.

3. Survey Testing

When herd and animal prevalence or geographic distribution are unknown, survey testing of randomly selected herds can be a useful adjunct to other monitoring procedures in the early stages of a control and eradication program. This type of information is essential for determining the logistical requirements of a program and for setting realistic priorities and objectives. The sampling method employed and the sample size tested is dictated to a large extent by such considerations as the expected herd prevalence, herd size distribution, geographic distribution of herds, and the accuracy required of estimates derived from the survey. In general terms, and depending on what is already known about the cattle population, an appropriately planned multistage sampling procedure will yield more useful information than a simple random sample.

4. Testing Herds Adjacent to Infected Herds

The propensity for *Brucella* infection to be present in adjacent herds has been well documented. This results in a geographic clustering of infected herds. Thus, the testing of herds adjacent to known infected herds is an efficient and epidemiologically essential method of case finding. In areas of high farm density it is often appropriate to extend the test area beyond those herds that are immediately adjacent to the initial case herd to include herds at least two herds removed.

5. Testing Indicated by Epidemiologic Evidence

Effective control of brucellosis demands a thorough epidemiological investigation of newly detected infected herds. These investigations should attempt to identify the origin of infection and the herds to which it may have spread. Prompt testing of all suspect or at-risk herds is essential.

6. Area Testing

This entails the systematic testing of all herds within a given geographic area. In an eradication program it is essential that the brucellosis status of every herd be determined, and the available monitoring systems are not capable of achieving this goal with a sufficient degree of confidence. Area testing is usually instituted after effective control measures have been implemented in herds known to be infected, and the area prevalence has been reduced to a low level.

7. Abortion Investigation

Due to the lack of sensitivity of the traditional surveillance methods when the prevalence of brucellosis is reduced to very low levels, some eradication programs have adopted an abortion notification and investigation scheme in an attempt to detect infection. A comprehensive literature search did not identify any reports on the efficacy of such schemes. It could be presumed that the reliability of detection would be heavily dependent on the diligence of herd owners in reporting abortions. It could also be expected that the efficiency (number of confirmed cases/number investigated) from such schemes would be very low, thus raising the issue of their cost effectiveness. However, abortions and stillbirths are the major clinical manifestations of brucellosis, and every effort should be made to encourage herd owners to seek a prompt investigation of any such occurrence in their herd.

VII. PROGRAM COMPONENTS AND THEIR SCIENTIFIC BASIS

Most countries with brucellosis problems have programs designed to control and/or eliminate the infection in cattle and protect the public from exposure and disease. These programs generally have several components. In order to insure that the program is effective and successful, each component needs to be scientifically sound and accepted by all concerned. The following is an attempt by the authors to list the major components of a program to eliminate the disease in the cattle population and briefly describe the scientific basis for the component. We recognize that there is danger in attempting to simplify brucellosis program components, but in view of the widespread distribution of *B. abortus* and the absence of a formal program in some countries, we have outlined the following menu of components for consideration if the reader finds infection and wants some help with what to do. The following components are not ranked in importance because importance depends on many factors.

Quarantine — This is a period of time during which cattle movement is restricted and the cattle are tested and observed for evidence of brucellosis. Quarantine is used to prevent the interherd transmission of *B. abortus* via cattle that are infected, especially test-negative cattle that are incubating brucellosis. The quarantine period is based on the maximum period of time that elapses between exposure and detectable disease or diagnosis. This time may be as short as 2 to 3 weeks, or in the case of latent heifers 2 to 3 years. Quarantine periods have generally been compromises between what regulatory officials request and what producers will tolerate. The time should be sufficient so that all exposed cattle have had sufficient time to develop brucellosis and assurance can be given that the remaining cattle will not be a source for interherd transmission of *B. abortus*. The period of time usually ranges from 120 d to 1 year or until all breeding cattle have completed a gestation without laboratory evidence of infection.

Test and reservoir reduction — This is commonly referred to as test and slaughter. All breeding cattle in the herd are tested with positive cattle being sent to slaughter. This component is based on the thesis that test-positive cattle are infected and if they are removed from the herd, then exposure and transmission are prevented or at least reduced. A major factor to reduce exposure and transmission is detection and removal of infected cows prior to parturition.

Depopulation — This is slaughter of all cattle. Depopulation is based on the same premise as test and slaughter except that once infection in a herd is established, all cattle are assumed to have been exposed, capable of becoming infected, and acting as a source of new infection. All cattle are slaughtered, and none are retained for breeding.

Vaccination — This is administration of immunizing agents such as strain 19. Properly vaccinated cattle have an increased resistance against field strain infection following exposure. If vaccinated cattle are more resistant, then they are not as likely to be infected and, therefore, cannot be a source of pathogenic *B. abortus*. As the number of infected cattle in the herd are reduced, then exposure potential should be reduced, and if exposure is reduced, then new cases of infection are reduced. The value of vaccination is also important when exposure is always a possibility and could be a good insurance policy to prevent unwanted infection following occasional unknown exposures.

Herd management — This is management of the herd to reduce contact between infected and susceptible cattle. All infected cattle in a herd are difficult, if not impossible, to detect with current tests, and all vaccinated cattle do not develop sufficient resistance to prevent infection if exposed. Therefore, transmission is less likely if highly susceptible cattle are not allowed to come in contact with cattle during their most contagious time, i.e., parturition.

Scientifically sound flexibility — All infected herds should not be treated the same. Each infected herd may be different, and individual herd plans are advisable. The herd plan developed to eliminate brucellosis and prevent spread to other cattle, either intra- or interherd, should be composed of scientifically sound components that are within the capabilities of the producer.

Education — This involves complete understanding of the herd plan. All participants, including producers, veterinarians, and regulatory officials, need to understand and adhere to all components of the plan. Many programs are unsuccessful because the participants didn't understand the problem or the role of others. Wholehearted cooperation and a desire to succeed will overcome many obstacles.

Guidelines — Any program needs general rules. An example is the Uniform Methods and Rules (UMR) for brucellosis eradication developed by Veterinary Services of the U.S. Department of Agriculture in cooperation with the U.S. Animal Health Association. The UMR is continually updated as new scientific information is made available, and the document is used as a guide for state programs in the U.S. Many states adopt the UMR, and others modify it to meet the needs of that particular state.

REFERENCES

1. *Bovine Brucellosis — An International Symposium,* Crawford, R. P. and Hidalgo, R. J., Eds., Texas A&M University Press, College Station, 1977.
2. Brucellosis Research: An Evaluation, a Rep. Subcommittee on Brucellosis Research, National Academy of Sciences, Washington, D.C., 1977.
3. Rep. National Brucellosis Technical Commission, U.S. Animal and Plant Health Inspection Service and U.S. Animal Health Association, Hyattsville, MD, 1978.
4. **Nicoletti, P.,** The epidemiology of bovine brucellosis, *Adv. Vet. Sci. Comp. Med.,* 24, 69, 1980.
5. **Bellver-Gallent, M., Ed.,** Animal Health Yearbook, FAO-WHO-OIE, Geneva, 1986.
6. **Kellar, J. A.,** Agriculture Canada, personal communication, 1988.
7. **Ray, W. C.,** The epidemiology of *Brucella abortus*, in *Bovine Brucellosis — An International Symposium,* Crawford, R. P. and Hidalgo, R. J., Eds., Texas A&M University Press, College Station, 1977, 103.
8. Subcommittee on Brucellosis Research, Epidemiology of bovine brucellosis, in Brucellosis Research: An Evaluation, Washington, D.C., National Academy of Sciences, 1977, chapter 8.
9. **Garcia, C., Szyfres, B., and Tome, J. G.,** Types of *Brucella* isolated from man and animals in Latin America, *Rev. Latinoam. Microbiol.,* 14(3), 117, 1972.
10. **Feinhaken, D. and Dafni, I.,** Identification of *Brucella* isolates in Israel, 1970—1979, *Refu. Vet.,* 37, 117, 1980.

11. Epidemiology, in *Joint FAO/WHO Expert Committee on Brucellosis,* 6th Rep., World Health Organization, Geneva, 1986, 48.
12. **Akakpo, A. J. and Bornarel, P.,** Epidemiology of animal brucellosis in tropical Africa: clinical, serological, and bacteriological surveys, *Rev. Sci. Tech.,* 6(4), 981, 1987.
13. **Verger, J. M., Duee, J. P., and Grayon, M.,** *Brucella* isolations in France: during 10 years of typing, *Ann. Microbiol. Inst. Pasteur,* 133B, 433, 1982.
14. **Meyer, M. E.,** The epizootiology of brucellosis and its relationship to the identification of *Brucella* organisms, *Am. J. Vet. Res.,* 25(105), 553, 1964.
15. **Carrigan, M. J. and Cockram, F. A.,** *Brucella abortus* biotype 1 arthritis in a horse, *Aust. Vet. J.,* 64, 190, 1987.
16. USDA/APHIS/VS, National Veterinary Services Laboratory, Brucellosis: an isolation and typing laboratory summary, Fiscal Years 1980, 1984.
17. **Meyer, M. E. and Morgan, W. J. B.,** Designation of neotype strains and of biotype reference strains for species of the genus *Brucella, Int. J. Syst. Bacteriol.,* 23(2), 135, 1973.
18. **Meyer, M. E.,** Advances in research on brucellosis, 1957—1972, *Adv. Vet. Sci. Comp. Med.,* 18, 231, 1974.
19. **Andreani, E., Prosperi, S., Salim, A. H., and Arush, A. M.,** Serological and bacteriological investigation on brucellosis in domestic ruminants of the Somali Democratic Republic, *Rev. Elev. Med. Vet. Pays Trop.,* 35(4), 329, 1982.
20. **Ray, W. C.,** Section A: bacterial, rickettsial, and mycotic diseases, in *CRC Handbook Series in Zoonoses,* Vol. 1, CRC Press, Boca Raton, FL, 1979, 99.
21. **Salem, T. F.,** Biological typing of local *Brucella* species isolated from farm animals, Thesis, Cairo University, Faculty of Veterinary Medicine, Department of Microbiology, 1987, 84.
22. **Meyer, M. E.,** Host-parasite relationships in brucellosis. I. Reservoirs of infection and interhost transmissibility of the parasite, in *Proc. U.S. Livestock Sanit. Assoc.,* 70, 1966, 129.
23. **Corbel, M. J., Gill, K. P. W., Thomas, E. L., and Hendry, D.,** *Methods for the Identification of Brucella,* Publication 2085, Ministry of Agriculture, Fisheries, and Food, Weybridge, England, 1983, 8.
24. **Luchsinger, D. W., Angus, R. D., Gue, C. S., and Anderson, R. K.,** The utilization of *Brucella abortus* culturing and biotyping results in the epizootiologic investigation of bovine brucellosis, in *Proc. 77th Annu. Meet. U.S. Anim. Health Assoc.,* Virginia, 1973, 85.
25. **Espe, B.,** *B. abortus* (biotype 2) abortion in a cow, *Brucellosis Vet. Newsl.,* 5, 4, 1988.
26. **Nelson, C.,** Immunity to *Brucella abortus,* in *Bovine Brucellosis — An International Symposium,* Crawford, R. P. and Hildago, R. J., Eds., Texas A&M University Press, College Station, 1977, 177.
27. **Vanderwagen, L. C., Sharp, J., and Meyer, M. E.,** A retrospective study of the relationships of vaccination status of reactor animals, management practices at calving and herd size to eradicating brucellosis in 79 dairy herds, *Proc. Annu. Meet. U.S. Anim. Health Assoc.,* 1977, 83.
28. **Crawford, R. P., Williams, J. D., Childers, A. B., Hidalgo, R. J., Huber, J. D., and Boyd, C. L.,** The effects of *Brucella abortus* on serology, bacteriology, and production in three Texas cattle herds, in *Proc. Annu. Meet. U.S. Anim. Health Assoc.,* 89, 1978.
29. **Crawford, R. P.,** unpublished data, 1988.
30. **Wray, C.,** Survival and spread of pathogenic bacteria of veterinary importance within the environment, *Vet. Bull.,* 45, 543, 1975.
31. **Verger, J. M.,** Prevalence and survival of *Brucella* species in manure, in *Proc. CEC Workshop; Communicable Diseases Resulting from Storage, Handling, Transport and Landspreading of Manures,* Walton, J. R. and White, E. G., Eds., Commission of European Communities, Luxembourg, 1982, 157.
32. **Plommett, M.,** Survival of *Brucella abortus* in liquid manure from cattle. Disinfection with xylene, *Ann. Rech. Vet.,* 3, 621, 1972.
33. **King, N. B.,** The survival of *Brucella abortus* (USDA strain 2308) in manure, *J. Am. Vet. Med. Assoc.,* 131, 349, 1957.
34. **McCaskey, T. A.,** Evaluation of an ensiling process to achieve safety from *Brucella abortus* in cattle manure-feed mixtures, in *Proc. 5th Int. Symp. on Agricultural Wastes: Agricultural Waste Utilization and Management,* American Society of Agricultural Engineers, St. Joseph, MI, 1985, 167.
35. **Rankin, J. E. E.,** *Brucella abortus* in bulls, *Vet. Rec.,* 77, 132, 1965.
36. **King, R. O. C.,** *Brucella* infection in the bull: a progress report of mating experiments with naturally infected bulls, *Aust. Vet. J.,* 16, 117, 1940.
37. **Manthei, C. A., Detray, D. E., and Goode, E. R.,** *Brucella* infection in bulls and the spread of brucellosis in cattle by artificial insemination. I. Intrauterine injection, *Sci. Proc. Annu. Meet. Am. Vet. Med. Assoc.,* 87, 177, 1950.
38. **Bendixon, H. C. and Blom, E.,** Investigations on brucellosis in the bovine male, with special regard to the spread of the disease by artificial insemination, *Vet. J.,* 103, 337, 1947.
39. **Denny, H. R.,** A review of brucellosis in the horse, *Equine Vet. J.,* 5, 121, 1973.

40. **McCaughey, W. J. and Kerr, W. R.,** Abortion due to brucellosis in a thoroughbred mare, *Vet. Rec.*, 80, 186, 1967.
41. **Wisniewski, E.,** Brucellosis in a horse, *Med. Wet.*, 40, 419, 1984.
42. **Stringfellow, D. A., Howell, V. L., and Schnurrenberger, P. R.,** Investigations into the potential for embryo transfer from *Brucella abortus* infected cows without transmission of infection, *Theriogenology*, 18, 733, 1982.
43. **Mallek, Z., Guerin, B., Nibart, M., et al.,** Effects of *Brucella abortus* contamination in vitro of mouse and cow embryos, *Bull. Acad. Vet. France*, 57, 479, 1984.
44. **Stringfellow, D. A., Wolfe, D. F., Lauerman, L. H.,** et al., Resistance of preimplantation bovine animals to infection with *Brucella abortus*, *Am. J. Vet. Res.*, 47, 1924, 1986.
45. **Stringfellow, D. A., Howell, V. L., and Schnurrenberger, P. R.,** Investigations into the potential for embryo transfer from *Brucella abortus* infected cows without transmission of infection, *Theriogeniology*, 18, 733, 1982.
46. **Kick, P., Grunbaum, E. G., Letz, et al.,** The dog as a source of re-infection in brucellosis-free dairy herds, *Monatsh. Veterinaermed.*, 33, 700, 1978.
47. **Martin, R. and Kuschel, S. F.,** Brucellosis in dogs in a rural area of Llanquique Province, Chile, *Gac. Vet.*, 40, 121, 1978.
48. **Nicoletti, P. and Milward, F. W.,** Protection by oral administration of *Brucella abortus* strain 19 against an oral challenge exposure with a pathogenic strain of *Brucella*, *Am. J. Vet. Res.*, 44, 1641, 1983.
49. **Russell, L.,** Epidemiology discussion, in *Bovine Brucellosis — An International Symposium*, Crawford, R. P. and Hildago, R. J., Eds., Texas A&M University Press, College Station, 1977, 168.
50. **Plommett, M., Fensterbank, R., Renoux, G.,** et al., Experimental bovine brucellosis. XII. Persistence to adult age of congenital infection in the heifer, *Ann. Rech. Vet.*, 4, 419, 1973.
51. **Crawford, R. P., Huber, J. D., and Sanders, R. B.,** Brucellosis in heifers weaned from seropositive dams, *J. Am. Vet. Med. Assoc.*, 189, 547, 1986.
52. **Catlin, J. E. and Sheehan, E. J.,** Transmission of bovine brucellosis from dam to offspring, *J. Am. Vet. Med. Assoc.*, 188, 867, 1986.
53. **Wilesmith, J. W.,** The persistence of *Brucella abortus* infection in calves: retrospective study of heavily infected herds, *Vet. Rec.*, 103, 149, 1978.
54. Brucellosis Research: An Evaluation, Rep. Subcommittee on Brucellosis Research, National Academy of Sciences, Washington, D.C., 1977, chap. 7.
55. **Salmon, M. D. and Meyer, M. E.,** Epidemiology of bovine brucellosis in the Mexicali Valley, Mexico: literature review of disease associated factors, *Am. J. Vet. Res.*, 45, 1557, 1984.
56. **Salmon, M. D., Meyer, M. E., and Hird, D. W.,** Epidemiology of bovine brucellosis in the Mexicali Valley, Mexico: data gathering and survey results, *Am. J. Vet. Res.*, 45, 1561, 1984.
57. **Salmon, M. D., Meyer, M. E., and Cramer, J. C.,** Epidemiology of bovine brucellosis in the Mexicali Valley, Mexico: results of path analysis, *Am. J. Vet. Res.*, 45, 1567, 1984.
58. **Salmon, M. D. and Meyer, M. E.,** Epidemiology of brucellosis in the coastal region of Baja California Norte, Mexico: results of path analysis in an area of high prevalence, *Prev. Vet. Med.*, 4, 485, 1987.
59. **Nicoletti, P.,** The epidemiology of bovine brucellosis, *Adv. Vet. Sci. Comp. Med.*, 24, 69, 1980.
60. **Christie, T. E.,** Eradication of brucellosis in Northern Ireland: field problems and experience, *Vet. Rec.*, 85, 268, 1969.
61. **Sargent, E. D., Wilson, A. L., and Munro, R. F.,** Brucellosis, an investigation on dairy farms in the west of Scotland, Rep. No. 141, The West of Scotland Agricultural College, Auchincruive, 1973.
62. **Harrison, M. A., Morgan, I. R., and Rogerson, B. A.,** Problem herds in brucellosis eradication in north-eastern Victoria, *Victorian Vet. Proc.*, 37, 33, 1979.
63. **Van Wavern, G. M.,** The control of brucellosis in the Netherlands, *Vet. Rec.*, 72, 928, 1960.
64. **Kellar, J., Marra, R., and Martin, W.,** Brucellosis in Ontario: a case control study, *Can. J. Comp. Med.*, 40, 119, 1976.
65. **Dietrich, R. A., Amosson, S. H., and Crawford, R. P.,** *Operational Characteristics, Herd Health Management Practices and Associated Epidemiological Data, Brucellosis Quarantined Herds, United States, 1980—82*, Texas Agricultural Research Station, College Station, 1987.
66. **Rolfe, D. C. and Sykes, W. E.,** Monitoring of dairy herds for *Brucella abortus* infection when prevalence is low, *Aust. Vet. J.*, 64, 97, 1987.
67. **Schilf, E. A.,** Completing the eradication of brucellosis by 1975, *J. Am. Vet. Med. Assoc.*, 161, 1525, 1972.
68. **Meyer, M. E. and Vanderwagen, L. C.,** Comparison of factors influencing the control and eradication of bovine brucellosis in two ten year periods, 1960 to 1970 and 1970 to 1980, *Proc. Annu. Meet. U.S. Anim. Health Assoc.*, 1982, 152.
69. **Plommett, M.,** Epidemiology discussion, in *Bovine Brucellosis — An International Symposium*, Crawford, R. P. and Hildago, R. J., Eds., Texas A&M University Press, College Station, 1977, 150.

70. **Ray, W. C.,** Epidemiology discussion, in *Bovine Brucellosis — An International Symposium,* Crawford, R. P. and Hildago, R. J., Eds., Texas A&M University Press, College Station, 1977, 151.
71. **Rankin, J. E. F.,** Brucella abortus in bulls: a study of twelve naturally-infected cases, *Vet. Rec.,* 77, 132, 1965.
72. **Crawford, R. P., Sanders, R. B., and Adams, L. G.,** Early removal of cows with brucellosis and the effect of strain 19 vaccinated cattle herds, *Prev. Vet. Med.,* 5, 191, 1988.
73. Brucellosis Research: an Evaluation, Rep. Subcommittee on Brucellosis Research, National Academy of Sciences, Washington, D.C., 1977, chap. 7.
74. **Crawford, R. P., Adams, L. G., and Williams, J. D.,** Relationship of fetal age at conjunctival exposure of pregnant heifers and Brucella abortus isolation, *Am. J. Vet. Res.,* 48, 755, 1987.
75. **Crawford, R. P., Adams, L. G., Williams, J. D.,** Relationship of days in gestation at exposure and development of brucellosis in strain 19-vaccinated heifers, *Am. J. Vet. Res.,* 49, 1037, 1988.
76. **Beal, V. C.,** Perspectives on animal disease surveillance, in *Proc. Annu. Meet. U.S. Anim. Health Assoc.,* 1983, 359.
77. **Beal, V. C.,** The use of mathematical models in animal disease program evaluation, in *Proc. Annu. Meet. U.S. Anim. Health Assoc.,* 1983, 386.
78. **Anderson, R. K.,** Surveillance: criteria for evaluation and design of epidemiologic surveillance systems for animal health and productivity, in *Proc. Annu. Meet. U.S. Anim. Health Assoc.,* 1982, 321.
79. **McCallon, W. R. and Beal, V. C.,** The fallacy of drawing inferences from biased data: some case examples, in *Proc. Annu. Meet. U.S. Anim. Health Assoc.,* 1982, 350.
80. **Beal, V. C.,** *Regulatory Statistics,* Vol. 1, 6th ed., VS, APHIS, USDA, Hyattsville, MD, 1983.
81. **Beal, V. C.,** *Regulatory Statistics,* Vols. 2A and B, 2nd ed., VS, APHIS, USDA, Hyattsville, MD, 1986.
82. **Leech, F. B.,** A critique of methods and results of the British national surveys of diseases in farm animals, I. Discussion of the surveys, *Br. Vet. J.,* 127, 511, 1971.
83. **Leech, F. B.,** A critique of methods and results of the British national surveys of diseases in farm animals. II. Some general remarks on population surveys of farm animal disease, *Br. Vet. J.,* 127, 587, 1971.
84. **Ray, W. C.,** Problems in developing valid data for performing mathematical studies of bovine brucellosis programs, in *Proc. Vet. Prevent. Med. and Epidem. Work Conf.,* Feb. 4 to 6, 1980, VS, APHIS, USDA, Hyattsville, MD, 1983, 3.
85. **Davies, G.,** Animal disease surveillance, in *Proc. 2nd Int. Symp. Vet. Epidem. and Econ.,* Canberra, Australia, 1979.
86. **Martin, S. W.,** The evaluation of tests, *Can. J. Comp. Med.,* 41, 19, 1977.
87. **Feinstein, A. R.,** Clinical biostatistics. XXI. On the sensitivity, specificity, and discrimination of diagnostic tests, *Clin. Pharm. Ther.,* 17, 109, 1975.
88. **Gerstman, B. B. and Capucci, D. T.,** Evaluating the reliability of diagnostic test results, *J. Am. Vet. Med. Assoc.,* 188, 248, 1986.
89. **Amosson, S. H. and Dietrich, R. A.,** Theoretical basis and empirical implications of the bovine market slaughter testing program in the United States, *Prev. Vet. Med.,* 3, 53, 1984.
90. **Roepke, M. H. and Stiles, F. C.,** Potential efficiency of the milk ring test for detection of brucellosis, *Am. J. Vet. Res.,* 31, 2145, 1970.
91. **Pietz, D.,** Brucella antigens and serologic test results, in *Bovine Brucellosis — An International Symposium,* Crawford, R. P. and Hidalgo, R. J., Eds., Texas A&M University Press, College Station, 1977, 49.

Chapter 8

CONVENTIONAL SEROLOGICAL TESTS

Alastair MacMillan

TABLE OF CONTENTS

I. TESTING STRATEGY

The strategy selected for the control of brucellosis in an individual country or region is dependent on many factors. Areas with extensive range conditions of agriculture, possibly with groups of people accompanied by their herds and flocks moving freely and with little control across national boundaries, will pose different problems than areas in which farms are well fenced, stock movement well controlled, and animal identity well recorded. International agreements, such as those in force within the European Community, may also dictate methods of brucellosis control.

All of these factors will ultimately decide the most appropriate strategy for control. Large-scale, expensive centralized testing regimes, using sophisticated serological tests, as have been used in Europe, North America, and Australasia, may not be feasible elsewhere. Better there, perhaps, should be a greater decentralization and, perhaps, greater reliance on simple screening tests such as the milk ring test (MRT), the Rose Bengal plate test (RBPT), or for the future, "dipstick"-type immunoassays appropriate for decentralized usage by semiskilled personnel.

Whatever strategy for eradication is chosen, it must be remembered that the only obvious clinical sign of brucellosis in domestic animals is abortion, and this is not pathognomic. Diagnosis can only be made on the basis of laboratory testing, the demonstration of the causal organism, and the detection of significant levels of specific antibody being complementary methods for diagnosis. Standardized procedures must be adhered to so that meaningful results are obtained, and there should not be reliance on a single test. Complete eradication will involve repeated testing over many years if it is possible at all, and then the national herds and flocks will require monitoring to ensure that they remain free of the disease should it be reintroduced.

II. COLLECTION OF BLOOD AND MILK SAMPLES

Good blood sampling technique is essential in order to minimize hemolysis and bacterial contamination which may lead to the serum rapidly becoming untestable by many tests. It is recommended that silicone-coated evacuated glass tubes without additives be used to collect the blood sample, as they help ensure good clotting and clot retraction, to provide an easy source of serum without the need for centrifugation. Clotting is also aided by maintaining the sample at 25° to 37°C for 1 to 2 hours. Experience at this laboratory has been that factors affecting the silicone coating process can seriously reduce the efficiency of clot retraction, leading to a severe increase in the effort needed to prepare the samples for testing. It is essential that a fresh needle be used for each animal to prevent the spread by this route of bloodborne pathogens. Plasma can be used for serological testing (although it should be avoided for the complement fixation test), but a blood sample put into a tube with anticoagulants will almost certainly need centrifugation.

Serum need not be removed from the clot before testing. Samples taken from all over Great Britain are tested centrally in one laboratory. Because of the temperate climate and reasonably efficient postal communications, samples of whole blood are dispatched and the serum is tested at the laboratory without separation. Following receipt, the blood samples are maintained at 4°C and remain suitable for testing for a period of about 2 weeks. After this time, significant hemolysis and bacterial growth in some tubes occurs, making them unsuitable for testing in many conventional serological tests. For longer periods of storage, it is best to remove serum from the clot as soon as possible and freeze at −20°C or below.

Milk samples are best collected in plastic containers. If milk is to be collected directly from the udder, the teats must be thoroughly cleansed and the first milk stripped out and examined for signs of mastitis before the sample is taken. It is important to ensure that milk from all four quarters is included, unless that from each quarter is to be sampled separately. If the milk is to be collected from the bulk tank or churn, it is essential that the contents are well mixed before the sample is taken.

Excessive shaking, the presence of colostrum or milk from mastitic cows, or the testing of milk following pasteurization may produce false positive or negative reactions to the MRT. Samples should be allowed to stand overnight at 4°C before testing, but must then be tested as soon as possible.

Great care and attention must be taken in recording the identity of the animal and the corresponding sample which should be uniquely labeled. When test results are entered on the form, there should be a clear association between animal, sample, and test result to avoid confusion of identity.

III. ACTIVITY OF IMMUNOGLOBULIN ISOTYPES

Many reports have been published on the rate of appearance and decline of antibody isotypes and their reactions in the various serological tests for brucellosis.[1-27]

After vaccination of cattle with S19, IgM antibodies appear after about 5 d, reaching peak values after 13 d after vaccination. IgG_1 antibodies appear a little later or simultaneously with IgM, and peak values are reached at 28 to 42 d, after which they decline. The same general pattern follows experimental infection with virulent strains and also in chronic field cases except that IgM antibody activity declines to low levels, and residual activity resides in IgG_1 and IgG_2 as well as in IgA which remain at higher levels.[14,17,18,20,23,26,28]

Although it is reported that IgM is able to fix complement, more recent work has revealed that in order to fix guinea pig complement, the presence of bovine complement is required.[29] The agglutinating activity of the IgM antibodies is reduced in a low pH (3.65).[25,26,28]

The IgG_1 isotype does not appear to be a good agglutinating antibody at a neutral pH, but fixes guinea pig complement unassisted. Its activity is unimpaired at low pH.

IgG_2 agglutinates efficiently, and evidence has been presented that IgG_2 present in sufficient quantities can induce prozoning and, at higher concentrations, can abrogate reactions to the complement fixation test (CFT).[30,31] It was claimed that IgG_2 could react at low pH.[14,30] It was also shown that, at certain concentrations, IgG_1 could inhibit the activity of IgG_2 in the serum agglutination test (SAT) and suggested that the prozone phenomenon seen with certain sera could be due to an excess of IgG_1 over IgG_2. There are also differences in the proportion of activity due to IgG_1 and IgG_2 in infected and vaccinated animals, the level of IgG_1 being higher in infected than in vaccinated cattle.[15,19,26,32]

IV. FALSE POSITIVE REACTIONS

The terms "specificity" and "sensitivity" are widely used, and are useful terms for comparing the performance of serological tests.[33] Ideally, results of serological tests should be related to the results of cultural examination, but this information is often not available, and these tests are often compared with one another, particularly with the complement fixation test. The following definitions are intended to be used:

$$\text{Sensitivity} = \frac{\text{true positive}}{\text{true positive} + \text{false negative}} \times 100$$

$$\text{Specificity} = \frac{\text{true negative}}{\text{true negative} + \text{false positive}} \times 100$$

Sera from a proportion of animals that are not infected or have not been vaccinated give positive reactions with serological tests. In the Islands of Jersey and in Guernsey where no vaccination or infection has occurred over many years, 0.59% of 4072 animals over 12 months of age showed antibody levels greater than 30 IU in the SAT.[34] Kovacic, in a study of 53,100 cattle apparently free of brucellosis,[35] noted a seasonal trend in nonspecific SAT reactors, with the rate being highest in spring and summer. In the author's experiences in Great Britain, however, there is a seasonal trend in nonspecific reactors to the RBPT, SAT, and CFT, with the rate being highest in the winter months. In addition, there is a highly significant geographical distribution in the proportion of animals positive to these tests, with, for instance, over 12% of samples from animals in northwest England being positive to the automated RBPT, compared to less than 5% in some other areas. This distribution is also reflected by samples falsely positive to the CFT. These reactions assume significance when the prevalence of brucellosis has fallen to a very low level where they may outnumber the cases of brucellosis.[36]

A. NONSPECIFIC REACTIONS

In many cases, positive reactions cannot be attributed to antibodies arising from infection with *Brucella* or organisms sharing structurally similar antigens. Some may be associated with so-called natural antibodies, which may be antibodies or their catabolic products, although the nature of the inciting antigen is unknown.[37] They have been detected in the serum of a wide range of species, especially with cattle, pigs, and horses.

In bovine sera, natural antibody activity toward *B. abortus* has been detected in IgM and IgG_1, but not IgG_2 or IgA isotypes.[37] A high proportion of the *Brucella* agglutinating activity present in the serum of uninfected cattle has been shown to be abrogated by the chelating agent ethylenediamine tetraacetic acid disodium salt (EDTA).[38] These EDTA-labile agglutinins are mainly IgM molecules, although in a small proportion of animals IgG may also be involved. These immunoglobulins bind to *Brucella* cells via the Fc region of their monomeric subunits and not via the antibody combining site of their Fab portions.[37] This interaction can be blocked by EDTA and by some, but not all, structurally related chelating agents.[39]

Although an excess of calcium, barium, strontium, or magnesium ions can neutralize the blocking action of EDTA or its analogues, the agglutinating activity is not dependent on the presence of divalent cations or complement. Furthermore, unlike specific or cross-reacting IgM antibodies, the EDTA-labile agglutinins are relatively resistant to reduction with 2-mercaptoethanol. In addition, the binding site does not appear to be associated with *Brucella* LPS.[37,40]

Of the agglutination reactions given by sera from cattle with negative complement fixation reactions to *Brucella*, 70 to 80% are completely or partially EDTA labile.[37,41] Cross-reacting antibodies probably account for most of the remainder of these reactions as well as the majority of those given by brucellosis-free animals showing positive CFT titers. Differentiation of EDTA-labile agglutinating antibodies from EDTA-resistant antibodies can be achieved by performing the standard SAT in a diluent composed of 10 m*M* EDTA or a mixture of EDTA and EGTA in phosphate-buffered saline, pH 7.2, instead of phenol saline. The incubation conditions and reading of the test are otherwise the same as for the standard procedure. The titers of the SAT and modified SAT are compared. In sera in which the agglutinating activity is entirely attributable to EDTA-labile agglutinins, a complete, or almost complete, loss in titer will occur in the presence of the chelating agent. Other sera containing a mixture of EDTA-labile and nonlabile agglutinins will show only a partial loss in titer in the presence of EDTA. Such reactions must be interpreted in the light of supporting evidence provided by the herd history and other serological tests, particularly the CFT.

B. CROSS-REACTIONS

Infection of animals with a number of bacterial species unrelated to the genus *Brucella* has been shown to provide an antibody response causing false positive reactions in serological tests using *Brucella* antigens. These reactions may be confused with brucellosis resulting in overcondemnation which will prove costly, whereas if the reactions given by samples from infected animals are incorrectly attributed to cross-reaction, then this may result in the spread of brucellosis.

The organisms implicated in such cross reactions and the nature of the similarity of their surface antigens is extensively reviewed by Corbel.[36] There are likely to be many other species of organism, most likely nonpathogenic commensals capable of cross-reacting, and the species already implicated are probably of partial significance, at least in the bovine host.

Animals experimentally infected with cross-reacting organisms generally show a much more rapid and prolonged rise in SAT titers than in the CFT, although even these are affected in some animals.[42] This is due to the much greater IgM component of the response, the rise in specific IgG_1 being much more transient.

At present, there is no single test that can differentiate between antibodies evoked by *Brucella* from those stimulated by cross-reacting species on a single sampling. The antibody isotype response is not a sufficient means of discrimination, but studies of the differences and similarities in antigens could lead to a means of discrimination using a more specific antigen. Where the prevalence of brucellosis is at a low level, it may be wise to isolate and retest reactor animals rather than slaughtering immediately, as the titers of animals infected with cross-reacting organisms will usually have fallen dramatically if resampled after 1 or 2 months, whereas those from *Brucella*-infected animals will have remained high.[36]

1. *Salmonella urbana*

Cross-agglutination between *Brucella* species and *Salmonella urbana* was first reported by Cioglia.[43] Antisera to the A and M determinants of smooth *Brucella* strains cross-react weakly with the *Salmonella* 0:30 antigen.[44] In each case, the cross-reacting titer is usually much lower than the titer toward the homologous antigen. Cross-absorption tests will usually identify the homologous antigen quite readily.[44] Apart from the agglutination reactions, the *Salmonella* 0:30 antigen may also evoke low antibodies reacting with *Brucella* antigens in the CFT, although the antiglobulin test (AGT) and the mercaptoethanol test (MET) are usually negative.[42,45] Natural infection of cattle by *S. urbana* has been observed to produce a relatively low-grade serological response to *B. abortus*. In cattle previously vaccinated with strain 19, low to moderate SAT titers were produced,[46] but few animals developed detectable CFT titers. The cross-reacting antibody response was transient, and most animals reverted to negative reactions to *B. abortus* within a month. About 7 d after the experimental oral inoculation of 10^{13} viable organisms, the animal developed *B. abortus* titers detectable in the SAT and RBT. The MRT with *B. abortus* antigen also became positive in most animals. CFT titers for *B. abortus* also rose in most animals but did not reach high titers. In contrast, SAT titers of up to 1/1280 (1641 IU) were attained. The AGT and MET remained negative. *Brucella* antibody titers declined to diagnostically insignificant levels within 4 to 6 weeks, although all animals developed high titers of agglutinating antibodies for *S. urbana* O and H antigens.[45]

In contrast, pigs inoculated orally with *S. urbana* failed to develop cross-reacting antibodies to *B. abortus*. Even repeated subcutaneous injection of the organism only induced low agglutination titers for *B. abortus*. This was not attributable to failure to establish infection as *S. urbana* was subsequently recovered from lymph nodes of the animals at autopsy, even though no overt disease had been produced.[45]

2. *Escherichia coli*

The development of antibodies to *B. abortus* in cattle experimentally infected with the strain of *Escherichia coli* 0:116 was reported by Nielsen et al.[47] The pattern of response to repeated injections differed from the other cross-reactions so far described in that antibodies to *B. abortus* were only produced by cattle with preexisting, "natural" agglutinins to the organism and not by seronegative calves or guinea pigs.

A serological cross-reaction between *E. coli* serogroup 0:157 and *Brucella* species has been reported. Antisera to *E. coli* 0:157 agglutinated smooth *Brucella* species, but not nonsmooth strains. Conversely, antiserum to smooth *Brucella* species agglutinated *E. coli* 0:157, but to lower titer. Antiserum to *E. coli* 0:157 also reacted in the RBPT and CFT.[48]

The oral inoculation of unvaccinated cattle with *E. coli* 0:157 evoked substantial titers of antibodies cross-reacting with *B. abortus*. SAT titers of up to 1/640 (820 IU) developed 10 to 14 d after infection, and the RBPT and MRT became positive, although only low transient CFT titers developed (up to 1/20 or 83.3 IU). Much higher SAT titers to the homologous *E. coli* antigen developed from the 5th day following inoculation. Most of the cross-reacting antibody was of the IgM class. Differentiation could be achieved by cross-absorption sera with *B. abortus* and *E. coli* 0:157 antigens. Oral inoculation of pigs with *E. coli* 0:157 did not induce detectable agglutination to *B. abortus* although agglutination titers of up to 1/640 toward a homologous antigen were developed. Reinoculation of these animals by the subcutaneous route gave substantial titers of cross-reacting antibodies. SAT titers of up to 1/320 (410 IU) and CFT titers of up to 4/40 (266 IU) developed by the 12th day. The RBPT also became positive. Detectable CFT activity toward *B. abortus* persisted for at least 2 months, but agglutination activity usually declined to insignificant levels after 1 month. Antibody titers toward *E. coli* 0:157 antigen were always very much higher than those to *B. abortus*.[45]

3. *Pseudomonas maltophilia*

A strain of *Pseudomonas maltophilia* (555) isolated from bovine milk by Diaz in Pampona, Spain, was found to cross-react with smooth *Brucella* species. Following oral inoculation with *P. maltophilia* 555, unvaccinated cattle developed very low levels of agglutinating antibodies to *B. abortus*. Following reinoculation by the same route agglutinin titers of up to 1/80 (102.5 IU) developed, and in some cases the MRT became positive. It is not known how significant the role of *P. maltophilia* is as the cause of serological cross-reactions to *Brucella* in the field, but the organism is widespread in the environment.[45]

4. *Yersinia enterocolitica*

A serological cross-reaction between *Yersinia enterocolitica* 0:9 and *Brucella* species was first reported by Ahvonen and colleagues.[49,50] This cross-reaction is the most complete so far described for any organism sharing antigenic determinants with *Brucella*. Potentially, it presents the most serious source of confusion in the serological diagnosis of *Brucella*. Both natural and experimental infections with *Y. enterocolitica* 0:9 in a variety of species give rise to *Brucella* SAT titers which may equal those produced with the homologous antigen. Other serological tests, including the AGT, the CFT, the RBPT, and others may also be positive.[51-54] Monospecific antisera for the *Brucella* A antigen, but not the M antigen, cross-agglutinate *Y. enterocolitica* 0:9.[44,55] Oral inoculation of *Y. enterocolitica* 0:9 did not produce antibodies cross-reacting with *B. abortus* in one reported study.[56] However, in an experiment conducted by Corbel,[57] *Brucella*-negative milking cows inoculated with a strain of *Y. enterocolitica* 0:9 developed significant serological responses to *B. abortus*. Following reexposure to the organism by the same route, high antibody levels developed in the SAT with rather lower titers developing the CFT. In one animal diagnostically significant titers of cross-reacting antibodies persisted for at least 6 months.[57] Cross-reacting antibodies evoked by *Y. enterocolitica* 0:9 are usually mainly of the IgM class, but substantial titers of IgG antibodies can also be produced. Pigs injected parenterally or inoculated by the oral route with *Y. enterocolitica* 0:9 developed diagnostically significant titers of antibodies cross-reacting with *Brucella*.[45] Antibodies active in the SAT and RBPT became detectable 7 to 10 d after inoculation with the CFT following.

V. RESPONSE TO VACCINATION

A. CALFHOOD STRAIN 19 VACCINE

Subcutaneous inoculation of a standard dose of strain 19 vaccine results in the appearance of an SAT response within 5 to 7 d, reaching a peak of around 2000 IU at 2 to 3 weeks before declining at a rate largely dependent on the age of the animal. Of animals vaccinated at 6 to 9 months of age, 80% had diagnostically insignificant levels within 12 months, whereas if vaccinated at more than 9 months of age, only 50% will pass 12 months later.[58,59] Reactions to the CFT develop at the same time as or a little after the SAT, but decline sooner.[60-64]

B. STRAIN 45/20 ADJUVANT VACCINE

The pattern following the use of the 45/20 adjuvant vaccine differs in many important respects from that of S19. In animals not previously sensitized to *Brucella* (neither infection nor vaccination), vaccination is followed by a low SAT and CFT response which soon becomes negative. There is a more active and prolonged response to the AGT. However, by using rough cells as antigen, significant titers in both the SAT and AGT can be found.[65,66]

The second vaccination, usually given 6 to 10 weeks later, elicits a response to both SAT and CFT as well as reactions to the RBPT. Reactions to the SAT rarely exceed 100 IU and fall to less than 30 IU after 3 to 4 months. However, reactions to the RBPT and CFT persist much longer. The CFT reaction often does not become negative until 6 to 9 months later.[66-74] Reactions to the AGT persist for the longest period of time.

After the second dose of vaccine, antibodies to rough antigens rapidly increase in titer, especially in the AGT, persisting for many months.[66]

Antibodies detected by the AGT appear about the same time as reactions to the CFT, reaching a minimum of 4 weeks after vaccination. They fall much more slowly, however, being significantly raised after a period of up to 6 months.[75-78] Kiss reports that they may persist for 4 to 5 years,[79] making the test difficult to use for diagnosis in vaccinated animals.

The use of the 45/20 vaccine in animals that have already received calfhood S19 gives a similar pattern except that the levels are higher, and, in particular, CFT and AGT reactions persist very much longer.[66,67,72,75,80-83] The nature of the immunoglobulin response to 45/20 vaccine has been studied by Corbel in animals not previously sensitized to *Brucella*.[75] The initial response consisted of IgM followed by IgG. A similar pattern was produced when 45/20 vaccine was administered to animals previously sensitized as calves with S19, although IgG appeared somewhat sooner. Repeated vaccinations with 45/20 led to only a slight IgM response but to an enhanced IgG response, especially IgG_2, which contrasts with the position after S19 vaccination.

The importance of the adjuvant on the antibody response has been noted by numerous workers.[68,84-86] Some adjuvants not only cause a more persistent local reaction, but also a more prolonged antibody response. This effect is even more pronounced in animals previously vaccinated as calves with S19. Although considerable variations are reported by the various workers, reactions to the CFT, RBPT, and AGT can persist in some animals for periods of up to 18 months.

C. REV 1 VACCINE

There is now a large amount of information available on the use of this vaccine. In one series of observations, after vaccination of young goats, agglutination titers rose rapidly and peaked at about 2 weeks, with titers up to 1280 IU. Subsequently, they fell rapidly, but at 6 months after vaccination, 14% were still positive (80 IU), and at 1 year 2%. In the CFT, there was a similar rise, the subsequent fall was more rapid, and all the goats became negative at 1 year following vaccination, with the exception of one which had a titer of 1/10. In subsequent years, all the goats were negative to the CFT.[87] Jones et al. found the serological response in sheep, using the same tests, to be almost identical.[88] Fensterbank (quoted by Alton) found more persistent reactions in sheep, reporting that 30% of ewes were still positive 12 months after vaccination and 10% still positive after 20 months.[89]

VI. ANTIBODIES FOLLOWING INFECTION

There are numerous reports on the rate of appearance of antibodies detected with various diagnostic tests in both nonvaccinated animals following experimental infection, as well as under field conditions. In the latter cases, there is no control over the size, the route, or the timing of the challenge dose, but, on the whole, the results are in general agreement with those observed following experimental infection.

Following infection via the conjunctival sac, there is a longer period before diagnostically significant titers are produced as compared with subcutaneous inoculation. Antibodies detected by the CFT and RBPT rise at the same time as or sometimes sooner than those detected by the SAT. Often, reactions do not reach significant levels until after abortion has occurred. In long-standing chronic infections, reactions to the SAT tend to wane to or below diagnostically significant levels, whereas reactions to the CFT and RBPT persist at diagnostic levels almost indefinitely. The AGT reactions also often reach significant levels before the SAT and persist for much longer.

Under field conditions, the rapidity with which an antibody response rises to any of the tests will depend on the size and virulence of the challenge strain, route of infection, and the susceptibility of the host. It has been the experience of most workers in the field that the response detected by the SAT takes somewhat longer to reach diagnostically significant levels than the CFT or RBPT. However, high levels to all the tests are eventually reached, except in some cases and in chronic long-standing infections where the SAT results tend to wane. In the study of a brucellosis outbreak in Great Britain, of 58 infected animals identified, 4 did not become serologically positive until they aborted at between 225 to 438 d following exposure to infection. It was presumed that these had received a suboptimal dose of infecting organisms.

No test or combination of tests currently available will differentiate an infected from a recently vaccinated animal. The basis for differential diagnosis is the observation that after vaccination of young animals less than 6 months of age with S19, reactions, especially to the CFT, disappear in the large majority of cases by the time the animal reaches 18 months of age. After infection, the reactions, especially to the CFT, MET, Rivanol, and AGT persist. During a herd investigation involving animals in varying stages of infection, this pattern will be reflected by somewhat varying responses to the battery of tests involved, coupled with the fact that, as the disease progresses, antibodies to subsurface antigens will be produced which, by the very nature of the tests themselves, will be detected by some tests and not by others.

VII. DEMONSTRATION OF ANTIBODIES IN BODY FLUIDS

Many serological tests have been described that are designed to demonstrate the presence of *Brucella* antibodies in body fluids, including serum, milk, vaginal mucus, semen, and muscle extract.[90] The perfect serological test would be one which would detect infection early in the incubation stage of the disease before abortion has occurred and one which would discriminate antibodies following vaccination and field infection. It would not be subject to false positive reactions due to nonspecific antibody. However, at present, such an assay is not available, although multiple tests increase the confidence in the diagnosis, and sequential tests over time also provide a better insight than single tests.

A. BUFFERED *BRUCELLA* ANTIGEN TESTS

The group of tests that can be considered to comprise buffered *Brucella* antigen tests relies on the principle that the ability of IgM antibodies to bind is markedly reduced in a low pH. The buffered *Brucella* antigen (BBA) tests now play a major role in the serological diagnosis of brucellosis. The name BBA includes the card test and RBPT in which the same Rose Bengal-stained antigen is used and the buffered plate agglutination (BPAT) test, performed with a different antigen. There may be some difference in the sensitivity between different procedures, even when the same antigen is used. The RBPT is a simple spot agglutination test that is effective in the diagnosis of bovine brucellosis in many species when used as a screening test. The test is oversensitive for diagnosis in individual animals, especially those vaccinated. However, in the U.S. the card test is often used as a definitive test. The RBPT may be used for testing sheep and goats for *B. melitensis* infection, but it is less proved in these species, and those animals vaccinated with Rev. 1 vaccine remain positive to the test for many months. In humans, the RBPT and card tests are effective as rapid presumptive tests, and in pigs they are at least as efficient as any other established serological procedure, but even then need to be used as a herd test rather than for diagnosis in individual animals. The BPAT recently introduced in the U.S. appears to be efficient as a rapid screening test in bovine and porcine brucellosis.[91] It is emphasized here that the sensitivity of serological reactions is greatly influenced by the temperature at which the reaction takes place. It is particularly important in the case of procedures such as the RBPT, BPAT, and card test that the reagents be at ambient temperature. Sera and antigen, if used straight from the refrigerator, will react poorly.

Specificity and sensitivity are also markedly affected by variations in the proportion of antigens and serum used in the test. The relationship between sensitivity and specificity is not linear.

The acid plate agglutination test was first introduced by Rose and Roepke in 1957 following observations that the antibody activity of "nonspecific" agglutinins was destroyed at a low pH (3.6), whereas that of the specific antibody remained unaffected.[92] Since then the test has been widely evaluated in many countries, under many different conditions of herd management and laboratory testing.[51,93-97] A card test developed by Pietz and evaluated by Nicoletti was shown to be efficient in the detection of culturally positive animals.[93] Since then, this basic principle has served as the precursor for a variety of tests currently in use. The tests are cheap and simple to perform and lend themselves both to the laboratory and for field use, although greater consistency can be expected in the former.

1. Rose Bengal Plate Test

The antigen consists of *Brucella* cells stained with Rose Bengal and suspended in buffer at pH 3.65 ± 0.05. The antigen remains stable for long periods when kept at 4°C, but may lose sensitivity slowly at higher temperatures, and for this reason particularly, the test is more satisfactory when carried out in the laboratory. The antigen used in the countries of the European Community (EC) is standardized without reference to the cell concentration, but so that a positive reaction is achieved at a dilution of 1:47.5 of the international anti-*B. abortus* serum (ISaBS), but a negative reaction at a dilution of 1:55. Elsewhere, the packed cell volume of the antigen is adjusted to 8% and the sensitivity compared to that of previous batches with a panel of sera.[98] Detailed methods for the preparation and standardization of the antigens are described by Hendry et al. and Alton et al.[98,99]

The test is usually conducted on ruled enamel strips, on a glass or ceramic tile, or in a WHO hemagglutination plate. Thirty microliters of the serum and antigen are applied, each having been allowed sufficient time to attain room temperature. The reaction mixture is rocked or oscillated for 4 min, and then the tests are read by examining for agglutination in good light.

The test is usually interpreted as positive if any agglutination is apparent, although the reaction may be graded according to its rapidity or appearance. The experience and attentiveness of the operator will have a bearing on the performance, an important point to remember about this test which appears to be a simple procedure.

Morgan mentions circumstances in which the RBPT may be negative, but the CFT positive:[100]

1. Early stages following infection
2. Following the ingestion of colostrum from reactor dams, the CFT is often the last test to become negative in the newborn.
3. Following S19 vaccination in a small proportion of calfhood vaccinated animals, the so-called "problem heifer" reactor. In a small number of cases, perhaps amounting to 0.5% of vaccinates, reactions to the RBPT and CFT persist much longer than normal. Reactions to the CFT are often the last to become negative, in those cases where the reactions become resolved.
4. Similarly, following 45/20 vaccination, CFT reactions usually extend for longer than those to the RBPT.

The situation where the CFT is negative but the RBPT positive, a rather more common occurrence, may arise following vaccination or infection with other cross-reacting genera.[101] It may also arise because it has been shown that bovine IgG_2 may inhibit the ability of IgG_1 to fix complement, although the RBPT reaction is unaffected.[30]

The manual RBPT was introduced in Britain as the official screening test in 1970. Negative sera are not normally tested further, but positive sera are then subjected to the SAT and CFT. In a trial published by Morgan using S19-vaccinated animals, a specificity of 98.82% was achieved in 11,000 sera from cattle presumed free of brucellosis.[102] Of 15,000 samples from animals in herds with brucellosis, 11% of samples were positive; of these, 61.6% were positive and 26.6% negative to the CFT. Of the 13,000 RBPT negative sera, 0.9% were positive to the CFT. This excellent measure of apparent agreement is due to the similarity in isotype sensitivity of the two tests.

Of 300 sera from cases of abortion confirmed bacteriologically, 91.4% were positive to the RBPT, compared to 92.7% positive to the CFT and 66.9% positive to the SAT (>100 IU).

Hormitzky and Searson report similar sensitivity with the RBPT detecting 45/49 (92%) of culture-positive animals, including 8 that were CFT negative.[103] Stryszak compared the performance of the standard RBPT and a variant using 15 μl and not 30 μl of serum.[104] The conventional test agreed with the SAT/CFT interpretation in 87.9% of cases compared with 99.4% for the variant. Not surprisingly, the converse was true for the specificity with figures of 97.1% and 83.1%, respectively. No indication was given on the method of antigen standardization.

Since the mid-1970s an automated system for conducting the RBPT has been in use in Britain.[97] The ADAM system used 35-mm transparent plastic tape on which to conduct the tests during the reaction period. The tape travels at a constant speed with the sample tubes moving in a chain alongside. Five microliters each of antigen and sera are added and mixed with fine jets of humidified air in an enclosed chamber. After 2 min of incubation at room temperature, a narrow beam of light is passed through the center of each test, and the emergent light is collected by a photomultiplier. Agglutinates larger than 10 μm are counted, and if the number exceeds a predetermined threshold count, the sample is recorded as positive. A single machine is capable of testing up to 8000 samples per day. Under the Brucellosis Eradication Scheme, over 2 million samples are tested each year by the automated RBPT at the Central Veterinary Laboratory, Weybridge, the rate of throughput exceeding 20,000 samples daily in the winter months. Studies have revealed a sensitivity in detecting culture positive animals of 97% and an overall specificity of 92.7%.

A number of studies have revealed the effectiveness of the RBPT in the diagnosis of *B. melitensis* infection in sheep and goats,[105-107] although Falade found a reduced efficiency.[108,109] Several found that the RBPT detects infected animals earlier in the immune response than the CFT and that the sensitivity was good.[110-112] There is no doubt, however, that more work is required to standardize the RBPT preparations internationally.

Corbel investigated the relative performance of RBPT antigens prepared from *B. abortus* and *B. melitensis* for the serological diagnosis of cattle infected with M antigen dominant *B. abortus* biovar 5.[101] Both antigens were successful in detecting similar numbers of sera taken from herds infected with biovar 5 in the automated procedure,[97] although substantially fewer samples were reactive with the *B. abortus* antigen in the manual test. Unfortunately, the bacteriological status of these individual animals was not provided. This evidence was taken to suggest that a *B. melitensis* antigen should be used in addition to a *B. abortus* antigen for diagnosis of animals in which a *B. abortus* biovar 5 infection is a possibility.

The joint FAO/WHO Expert Committee on Brucellosis recommends the use of the RBPT as a useful screening test for the diagnosis of *B. melitensis* infection in sheep and goats,[113] but concludes that the use of *B. melitensis* antigen should be further investigated. The Joint FAO/WHO Expert Committee on Brucellosis highlights the problems with the serological diagnosis of *B. suis* infection in pigs.[113] The RBPT has, however, proved useful as a herd screening test.

2. The Buffered Plate Agglutination Test

This test, described by Angus and Barton, is used as a screening test.[91] The antigen consists of an 11% suspension of strain 1119-3 stained with crystal violet and brilliant green and buffered to pH 3.63. Thirty microliters of antigen are mixed with 80 μl of serum on 4-cm squares of a glass plate ensuring that the reagents are at room temperature. The pH of this resulting mixture is about 4.1. The reagents are well mixed with a wooden applicator or spreader, and the plate is incubated in a humid chamber at 37°C. The plate is rocked with a circular motion on three occasions: after mixing, after 4 min, and after 8 min prior to reading. Reactions are read by eye and scored as positive if complete or partial agglutination has occurred.

Dohoo et al. extensively analyzed the sensitivity and specificity of the test and its agreement with other tests in various categories of animals.[114] The specificity of the test varied from a minimum of about 93% among S19 vaccinates to 98.8% among nonvaccinates. The sensitivity among 174 culture-positive animals was 95.4%, exceeding the sensitivity of the CFT (92.9%). The agreement between the tests was measured in two ways. First, the percentage of samples that yielded the same result on each test was determined, and second, agreement was determined. There was good agreement between the BPAT and all other tests except perhaps the strictest interpretations of the SAT, agreement being particularly good (93%) with the CFT. However, this may have been due to most of the samples being negative on both tests. The authors concluded that the BPAT was a good screening test both in terms of performance and economy.

3. The Card Test

This test is essentially the same as the RBPT. All materials used in the card test are disposable and provided as field or laboratory kits. These kits provide plasma collection equipment that allows rapid separation of plasma using phytohemagglutinin and an anticoagulant. Thirty microliters each of plasma and Rose Bengal-stained antigen are placed on a teardrop area of a card, mixed, and spread. The test is read immediately after rocking for 4 min.[93]

Huber and Nicoletti examined 4553 cows using the card test[115] and bacteriological culture. The card test was positive for 1661 of the 1666 culture-positive animals, a sensitivity of 99.7%. The test was positive for 1308 of the 1311 (99.8%) adult vaccinates and 103 of the 104 (99%) nonadult vaccinated cows shedding field strains. The specificity of the card test was 8.4% in 2432 culture-negative, adult vaccinated cows and 30.8% in 351 nonadult vaccinated cows. These parameters may have been influenced by the method of sample selection for the trial. Nicoletti stated that in experimentally infected animals it took an average of 62 d after exposure for the tube test to detect infection, while it took only 43 d for the card test to detect the same animals.[116] The evidence indicates that the card test should be an effective diagnostic test for detecting early infection, and this achieves eradication more rapidly.

B. SERUM AGGLUTINATION TEST

This test, introduced by Wright and Smith in 1897,[117] has been very widely used and evaluated for diagnosis.[24,90,93,102,118,119,121] There is no complete agreement between the results of the SAT and other tests, since these may detect antibodies of other isotypes; and because the antigen used in the SAT is a whole cell, the antibodies detected are those directed against the surface molecules. The SAT is normally conducted by making doubling dilutions of the serum in phenol saline in round-bottomed tubes and adding an equal volume of the standard antigen. After mixing, the tubes are incubated overnight at 37°C, and the degree of agglutination is then read by comparing the opacity against standards representing various degrees of agglutination. The resulting titer should then be converted to international units.[122,123]

Antigens may be prepared from suspensions grown in a liquid medium by batch or continuous culture and are equally satisfactory, provided the satisfactory precautions are taken to prevent dissociation.[125,126] Tube test antigens are normally suspended in phenol saline without buffering, but if buffered at pH 5, titers may be reduced, especially nonspecific reactions such as those observed in clean herds.[127] Patterson et al. also described a tube agglutination test using antigen buffered at pH 3.6.[26]

Standard strains of *B. abortus* are normally used for antigen production, e.g., strain 99 or strain 1119-3. Original work by Huddleson[128] and Feinberg and Wright[129] indicated that antigen prepared from *B. abortus* could be used for the satisfactory diagnosis of other smooth strains. However, Alton found that sera from animals infected with *B. melitensis* gave different results when tested against antigen prepared from *B. abortus* or *B. suis*, even though all the antigens had been standardized against the ISaBS.[130] His conclusion was that a homologous antigen should be used. Farrell and Robertson recommended the use of a *melitensis* antigen as well as the *abortus* antigen for the serological examination of guinea pigs inoculated with milk samples for diagnosis.[131] The use of the *abortus* antigen alone missed some guinea pigs infected with the M dominant *abortus* biotypes 5 and 9. The first ISaBS was established in 1952, and its use was recommended by a joint FAO/WHO Expert Committee on Brucellosis in 1953.[132,133] The adoption of a unitage system to express antibody levels rather than expressing them in titers represents significant progress toward international harmonization. The second ISaBS was established in 1969 to replace the diminished stocks of the first ISaBS.[134] This standard was established after an extensive international collaborative assay. Antigens are normally standardized visually in comparison with a turbidity reference, and usually a 50% end-point reading is chosen. There is a linear relationship between the ranges of 5 and 90% agglutination between antigen density and antibody titer,[135] so that turbidity references covering 25, 50, and 75% agglutination are appropriate for standardizing antigens. Under the EC directive on brucellosis, *brucella* antigens for the SAT have to be standardized against the ISaBS to give end-point readings at final dilutions of 1/600 to 1/1000 of that serum if the reading is made at 50% and between dilutions of 1/500 and 1/750 if the reading is made at 75%. In Britain, the antigen is standardized to give 50% agglutination at a final dilution of 1/650 of the ISaBS. It is essential to include with every days's tests a working standard serum standardized againsg the ISaBS to ensure daily consistency. Because of the less than perfect specificity of the SAT and the use of vaccines, there may be some difficulty in deciding upon the pass-fail threshold. In the EC, in order for a herd to be classified officially as brucellosis free, all animals must show an antibody level of less than 30 IU. In brucellosis-free herds, however, if the animals have been vaccinated with strain 19, they must show an antibody level of less than 80 IU. In the U.S., antibody levels of 200 IU or more are considered positive for official S19 vaccinates and less than 50 IU negative. For nonvaccinates the corresponding values are 100 IU or more as positive and less than 25 IU as negative. Intermediate values in both cases are classed as suspect. The SAT has been the most widely used serological test for brucellosis diagnosis, but much evidence has been gathered to demonstrate that it is not the most effective test for individual animals. Alton et al. and Berman found that 11 and 12%, respectively, of culture-positive cattle had less than 100 IU of antibody, and 4% had less than 30 IU.[124,136] These observations have been made by other workers.[96] Conversely, the SAT can lead to a severe degree of overcondemnation. In an analysis of 11,000 sera from herds apparently free of brucellosis, Morgan and Richards showed that 588 would be classed as reactors on the basis of the SAT alone (greater than 30 IU).[96] The SAT is still largely used on both sheep and goats, although it is necessary to use 5% sodium chloride in the buffer to avoid the prozone phenomenon. This phenomenon is due to a high concentration of the nonagglutinating immunoglobulin IgG_1 which competes for binding by the agglutinating antibody IgG_2. It has been well recognized that in these species, too, the SAT titer may wane in chronic infections to negative levels.[108,114,137] Dohoo et al. found the specificity of the SAT to be 94 and 99.5% when interpreted at the 30- and 60-IU levels, respectively.[114] This specificity dropped to 85 and 97.5% among strain 19 vaccinates. The sensitivity in detecting culture-positive animals was 98.3 and 93.1% for each of these levels. Stemshorn et al. found a specificity of 99.2% but a sensitivity of only 68.9%.[138] The experimental infection of five pigs did not result in the appearance of diagnostically significant SAT titers.[139]

C. THE EDTA-MODIFIED SAT

A phenomenon giving rise to nonspecific agglutination has been discovered which involves the nonimmune binding of bovine IgM to cells of *B. abortus*. This binding is sensitive to the action of certain divalent cation chelators, as shown by Scheibner[141] and Nielsen et al.[142] As the level of bovine brucellosis in a national herd is reduced, the relative importance of nonspecific reactions to serological tests for brucellosis increase, and such reactions caused by the presence in the serum of so-called natural (nonspecific) antibodies to *Brucella* continue to pose a significant diagnostic problem.[37] Nielsen et al. found that 0.6% of 2000 unexposed cattle in Canada had SAT titers considered to be of diagnostic significance.[37] In the studies by MacMillan and Cockrem[41] and Nielsen et al.[38] it has been shown that a large proportion of positive SAT reactions to *Brucella* given by sera taken from uninfected animals are sensitive to the action of EDTA. Such sera are thought to contain IgM molecules which cause agglutination of the *B. abortus* test antigen in a nonimmune manner, namely, by binding by their Fc region. This binding is inhibited by EDTA and by some structurally related compounds, while specific immune binding by the IgM, Fab region (i.e., induced by infection with *B. abortus*) is unaffected. Some other calcium ion chelators do not inhibit nonspecific agglutination, which suggests that the phenomenon does not involve the chelation of calcium but is dependent on the actual structure of EDTA.[39] It is unlikely that it is the antigenic determinant that has been altered because Nielsen and Duncan showed that prior washing of the cells in EDTA did not inhibit binding, although this still leaves the possibility that the alteration is reversible on removal of the EDTA.[39] A proportion of the false positive SAT titers sensitive to the action of EDTA varies from study to study. MacMillan and Cockrem,[41] working in Great Britain, showed that of 199 samples with SAT titers greater than 100 IU but CFT titers of 8.3 international CFT units or less, 127 (63.8%) proved to be sufficiently sensitive to the action of EDTA for the titer to fall below 100 IU. Indeed, in 103 of these samples the titer dropped by at least one doubling dilution. However, animals with previous SAT titers greater than 100 IU and CFT titers of at least 20 ICFTU, the titer fell to pass levels (less than 100 IU) in only three samples (5.3%). None of the sera taken from infected animals and giving an SAT greater than 100 IU dropped to below 100 IU in the modified SAT. Significant EDTA sensitivity using the modified SAT was revealed in 3 of 37 of the sera taken from known infected animals. In these cases, however, there was a significant CFT titer.

Nielsen et al. and MacMillan and Bell also reported observations that the SAT titers of some sera from cattle infected with *B. abortus* are reduced considerably by the addition of 10 m*M* EDTA.[141,142] These workers agreed that it was essential that the EDTA-modified SAT be used in conjunction with the CFT, lest some infected animals be missed. In other cases, the titers of animals that were SAT negative became positive with the addition of EDTA, and prozones were abrogated by the addition of EDTA.

D. PLATE AGGLUTINATION TEST

This test has been widely adopted in North America as a rapid and effective test used instead of the tube test. The antigen is stained to facilitate the reading usually with crystal violet and brilliant green. The test is conducted by placing decreasing volumes of sera on a clean glass plate and adding antigen. The reagents are thoroughly mixed, spreading over an area of 2-cm diameter and the plate rotated. The results are read in an illuminated box after 8 min.[24] The dilutions generally used are 1/25, 1/50, 1/100, and 1/200 of serum, and the results are interpreted according to the vaccination status of the animals. Reactions at 1/200 or over for vaccinates and 1/100 or more for nonvaccinates are considered as reactors.[24] Although the test has the advantage of speed and in not being affected by the prozone phenomenon, it is slightly less sensitive than the tube test.[143] In other respects, the comments made on the tube agglutination test also apply to the plate agglutination test. The antigen supplied by the U.S. Department of Agriculture gives a titer of 1/1000 with the ISaBS, so that a titer of 1/100 with the test represents 100 IU.[120]

George and Carmichael have described a rapid plate test for the diagnosis of *B. canis* infection using a Rose Bengal-stained antigen and doubling serum dilutions.[144] Badakhsh et al. reported improved specificity following the incorporation of mercaptoethanol.[145] The technique has recently been modified to include an antigen prepared from the mucoid variant (M-) of *B. canis*.[146] The test is both rapid and reasonably specific, although it has been discovered in field use that noninfected dogs may react occasionally to the test.

E. COMPLEMENT FIXATION TEST

1. *B. abortus*

M'Fadyean and Stockman described the use of this test for diagnosis,[147] but because of the difficult technical procedures and the occurrence of anticomplementary sera, it was not used routinely, and the simpler tube agglutination test with standardized antigens became adopted as the routine test. Early reports showed that the CFT could classify animals more accurately than the SAT.[61]

Observations that the test was a better indicator of infection than the SAT and that reactions to the CFT receded sooner than those to the SAT test after S19 calfhood vaccination, together with technical improvements in reagents and apparatus, increased interest in the test.

The complexity of the test protocol has meant that considerable variations have occurred in doing the tests. For example, in the first phase of the test, the fixation of complement may be made either in the cold, 4°C overnight, or at 37°C for 30 min. Consequently, there is some diversity of opinion between various workers, especially regarding the levels at which diagnostic titers ought to be set. Burki,[62] Hill,[60] and Bertschinger[148] recommended the use of the ISaBS for both standardizing the antigen as well as interpreting the result in relation to its activity and suggested that sera with 1/24th of the CF activity of the ISaBS should be considered positive. MacKinnon recommended that animals officially vaccinated with S19 showing 1/24th and 1/12th of the CF activity of the ISaBS should be considered "suspicious" and "positive", respectively.[149] For nonvaccinated animals, the values should be 1/48th and 1/24th, respectively. In the U.S. titers of 2+ at 1/40, representing around 1/30th the activity of the ISaBS, are taken as positive for nonvaccinates and titers of 1/80 for official vaccinates.[150]

Morgan et al. reported an international collaborative assay intended to determine the suitability of the 2nd ISaBS as a standard for the CFT. Each of the participating laboratories conducted a series of tests using its own methods on a panel of sera including the 2nd ISaBS.[151] Of 25 laboratories in 18 countries, no two laboratories used exactly the same technique. Differences were noted in the species of *Brucella* used as antigen and whether whole cells or extracts were used. There were wide differences in the method of estimating antigen units. The two common factors to all were the use of sheep red blood cells and of guinea pig serum as a source of complement.

Nielsen and Duncan, using purified bovine isotypes, found that only IgG_1 fixed guinea pig complement in a CFT.[29] The addition of bovine complement resulted in the activation of guinea pig complement by IgM, IgG_2, and IgA, although a greater quantity of these isotypes was required. Not unexpectedly, there were wide variations in the titers of the sera, varying up to 400-fold in some cases. Although there are reports that cold fixation results in slightly higher titers,[150] the results of the collaborative exercise, using mean relative potencies, showed that there were no significant differences between those using cold fixation and those using warm fixation, and the use of the ISaBS greatly reduced any variability. Krolak et al. compared the techniques of cold and warm fixation.[152] A similar degree of complement fixation occurred in 2 h at 37°C, 8 h at 20°C, and 18 h at 4°C. The titer of a standard serum using cold fixation was double that using the warm fixation technique, although the titer of the latter could be raised by increasing the antigen dilution. There were wide differences in the interpretation of the test in terms of diagnostic criteria, but it was recommended that the ISaBS should be used as a basis for comparing results. The WHO Expert Committee on Biological Standardization concluded that the ISaBS was suitable for this purpose[153] and in 1976 assigned it a unitage of 1000 IU of CF activity per ampule (as well as 1000 IU agglutinating activity per milliliter).[154]

The EC Commission has adopted the German Standard for anti-*B. abortus* serum as the EC Standard for the CFT, it having been calibrated against the ISaBS. Its CF activity was defined as 1000 EEC CF units per ampule, and an international collaborative assay was made to compare the CF activity of the First, Second, and the EC Standard.[155] The results showed that the EC CF unit was equivalent to the IU of CFT activity and that there was no significant difference between the overall means for three laboratories using cold fixation and the five laboratories using warm fixation.

Most laboratories use whole cells as antigen, mostly prepared from *B. abortus*. In Italy a soluble antigen is used.[156] This is prepared by alternately freezing and thawing suspensions of either *B. abortus* or *B. melitensis*, whereby it is claimed that it is possible (certainly, in experimental animals) to distinguish between *B. abortus* and *B. melitensis* infections. Other authors have experimented with chloroform extracts,[157] trichloroacetic acid extracts,[158] or supernatant following centrifugation of whole cell suspensions.[159]

In sheep and goats, the CFT is considered to be the most specific and sensitive method for the serological diagnosis of *B. melitensis* infection.[160]

Sera that show anticomplementary activity are sometimes a problem, especially if sera become contaminated (either chemically or bacteriologically). Prozoning sometimes occurs and can give rise to false negative reactions if a single dilution test is performed. It occurs more commonly if a dilute antigen is used, and excess IgG_2 may play an important part in this phenomenon.[31,161] However, Morgan[102] and Nicoletti[116] state that prozones are not a significant problem for diagnosis, especially where the CFT is used in parallel with other tests such as the SAT.

Much information has accumulated over the last 25 years on the value of the CFT as a definitive test for diagnosing infected animals and for differentiating calfhood vaccinated animals from infected animals and from nonspecific reactions. The test is widely used in eradication programs and has reduced the time required for eradication and the number of overcondemnations.[24,61,63,150,162-175]

To cite a few reports in the literature, Alton et al. found that all culturally positive animals reacted to the CFT.[24] Nicoletti reported that the CFT correctly identified 98% of culturally positive animals,[116] and Berman, using the cold-fixation technique, isolated *Brucella* from 134 of 303 animals with CF titers of 2+ at 1/40 or more, but from only 1 of 82 cows with CF titers between 2+ at 1/20 and 1+ at 1/40 and from none with titers below 1+ at 1/20.[136]

However, some culturally positive animals fail to show a positive CF reaction, especially at the time of abortion. This phenomenon can be found for most of the serological tests. Nonspecific reactions are rare, although recently a phenomenon was recognized in Great Britain in which 0.5% of all samples tested in the county of Lancashire failed the CFT. All these properties make the test an extremely valuable one, and, next to the isolation of the causal organism, it is still regarded as the best indicator of infection.

In the tests so far considered, either smooth cells or extracts of smooth cells have been used as antigen. In order to differentiate CF reactions following 45/20 vaccination from those following infection, Miller et al. evaluated the use of an antigen derived from the rough 45/20 strain and claimed that by comparing titers to the CFT using a smooth antigen and the rough antigen, many reactions could be resolved.[176] However, these data were not supported by cultural evidence. Automation of the CFT has found it in increasing favor as a sensitive and specific screening test for bovine brucellosis.[177-179]

2. *B. ovis*

Worthington et al. found that the CFT for the diagnosis of *B. ovis* infection in rams had a sensitivity of 96.3% and a specificity of 99.3%.[180] They concluded that there was no immediate need to replace the CFT for diagnosis. Worthington et al. described cultural and serological studies on 686 rams infected with *B. ovis*.[181] They found 11% to be negative to the CFT, although previously the CFT had been found to have high sensitivity.[182] This result was apparently due to the biased selection of the rams, and no evidence was found that the CFT reactivity changed during the course of the disease.

A hot saline extract of *B. ovis* is used for the CFT to diagnose *B. ovis* infection in sheep.[183] It is reported to contain mostly protein, but also significant amounts of rough lipopolysaccharide. The antigen, once made, must be thoroughly checked for anti-complementary activity, as this occurs more commonly than with *B. abortus* antigen.

Either the warm (37°C for 30 min) or cold (4°C for 14 to 18 h) fixation technique can be used. Burgess and Norris, however, report that higher titers and the detection of positive reactions earlier in the course of the disease occur more with the former protocol, which, however, is more prone to anti-complementary reactions.[184]

An international anti-*B. ovis* serum has been established containing 1000 IU/ml, and it is recommended that in future, all antigens should be titrated against this serum to facilitate international trade.

3. *B. suis*

The experimental infection of five pigs using *B. suis* biotype 2 by a single subcutaneous injection produced a low and very transient CFT titer in two animals. Three injections induced a more significant titer in four, but the fifth remained negative.

4. *B. melitensis*

The joint FAO/WHO Expert Committee on Brucellosis recommended the CFT as the definitive test for the diagnosis of *B. melitensis* infection in sheep and goats, although inactivation of the serum at 62°C may be necessary to reduce the occurrence of anticomplementary reactions.[113]

F. ANTIGLOBULIN TEST

The antiglobulin test (AGT) was developed to detect antibodies which, although they combine with the antigen, do not give rise to agglutination. The presence of these so-called "incomplete agglutinins" can be detected by using an antibody directed against the IgG fraction of the species of animal being tested. These reagents can be obtained commercially. The test is performed in two phases. First, the conventional SAT is carried out, and after reading, those tubes that do not show any agglutination are centrifuged and the deposited cells washed thoroughly.

The presence, if any, of the nonagglutinating *Brucella* antibody on the *Brucella* antigen is detected by adding the antiglobulin reagent at its working dilution and either reincubating the tubes and reading after 24-h of incubation, or conducting the test on a slide (rapid or plate method).[185] The latter tends to give reactions at least a dilution tube lower than the former.[175,186]

Depending on the specificity of the antiglobulin reagent, all the major immunoglobulins can take part in the test and it is undoubtedly more sensitive than the ordinary tube agglutination test. It is reported to be of value in the early stages of the disease as well as in clearing nonspecific agglutination reactions, but its use is not recommended in vaccinated herds,[187-189] since reactions to the test persist for many years after S19 and, especially, 45/20 vaccinations.

The AGT is most useful in sera with inconclusive reactions to routine SAT and/or those giving consistently anticomplementary reactions to the CFT. Those giving at least twofold higher titers to the AGT, compared with the SAT, may be considered as positive. However, some workers consider a titer of 1/40 or more to AGT as being positive. When used in chronically infected herds alongside the SAT and CFT, it can increase the number of infected animals detected.[79,189-191]

Anczykowski, however, found the AGT to perform poorly.[192] Although 97% of infected animals showed a 2 or more dilution increase over the SAT titer and 88% showed 3 or more dilution increase, the specificity levels were 70 and 95%, respectively, for nonvaccinated and 50 and 70%, respectively, for S19 vaccinated cattle.

G. DISULFIDE BOND REDUCTION TESTS

Mercaptoethanol (ME) breaks down the disulfide bonds holding together the pentameric structure of IgM, leading to a loss of agglutinating activity. Dithiothreitol (DTT) has a similar disulfide-bond reducing activity and, because of the potential carcinogenicity of ME, DTT is now preferred. Sensitivity to these agents implies the presence of IgM, and the test can be of value to resolve animals following vaccination with S19 and in the chronic stage of the disease where even low titers,[2,193] consisting of DTT or ME resistant antibodies, are significant. For example, Nicoletti reported that the ME test correctly classified 97% of culturally positive cattle.[171]

The test is carried out by mixing equal volumes of serum and 0.2 *M* DTT, incubated 1 h at 37°C, after which serial dilutions are made in saline and antigen is added. The DSR test is read in the usual way after overnight incubation.[123] A reduction in the titer after treatment with DTT would indicate that the major Ig present is IgM and not, therefore, indicative of infection. If there is no reduction in titer, this would indicate that the major immunoglobulin would be IgG and, therefore, more indicative of infection. Often, there is only a slight reduction in titer, in which case a retest should be done.

The mercaptoethanol tube agglutination test has been used to resolve suspect SAT reactions in the diagnosis of *B. canis* infection in dogs.[145]

H. RIVANOL TEST

For the test, 3 volumes of a 0.4% solution of Rivanol (2-ethoxy-6,9-diamino-acridine lactate) is added to 1 volume of serum to be tested, and the precipitate is removed by centrifugation. A tube agglutination test or a rapid plate test can be carried out on the resulting supernatant. The interpretation of the test can be by the comparison of the titer before and after treatment of the serum or by taking any reaction after treatment of the serum at the 1/25 dilution or more as a positive reaction. Anczykowski and Murat-Skwarek found the test much less effective in detecting infected animals than the CFT and described it as time consuming to carry out and difficult to read.[194] Other authors report that, in the main, the test results compare with those of the CFT.[94,171,195] The correlation of the test with bacteriologically proved infection has been recorded only by Nicoletti, who found that it correctly identified 96 to 98% of culturally positive cattle.[171] However, a number of cattle from which *Brucella* was not isolated also reacted to the test, and it appears to offer few advantages over the now well-established and standardized CFT.

I. HEAT INACTIVATION TEST

The incubation of serum-antigen mixture at 56°C reduces titers, especially the so-called "nonspecific" titers. This reduction rarely happens with high-titered sera.[206] The effect on titer is more pronounced with sera from vaccinated than sera from nonvaccinated cattle. Shibata et al. also found that heating sera to 56 or 65°C for 15 min before conducting the SAT often decreased the titers of both some high-reacting and many low-reacting sera.[197]

A heat inactivation test (HIT) for differentiating specific from nonspecific antibodies was first described by Amerault et al.[198,199] and Lambert et al. by following the results in calves vaccinated with S19 and in experimentally and naturally infected cattle.[121,200] After experimental exposure, the test became positive (reaction at 1/25 dilution) sooner than the SAT or CFT, and it correlated well with the ability to recover the organism.

The HIT has been evaluated by a number of other authors.[201-205] The results confirm those reported earlier, although in most cases, bacteriologic tests were not done. The HIT seemed to be of value in sera giving "suspect" reactions. However, because of technical difficulties in actually doing the test, it has now been largely superseded by other tests.

J. MILK RING TEST

Sutra et al. examined the antibody isotype involved in the milk ring formation in 17 experimentally infected cows. IgA was always involved: IgM in 15 cases and IgG in 11 cases.[206]

The MRT can be used effectively as a screening test in bovine dairy herds. It has not yet proved effective with the milk of sheep and goats, but it is the most practical method for locating infected dairy herds in many countries. On many farms milk is now collected in bulk tanks, and dilution factors are greatly increased as compared with milk collected in cans. However, the MRT is still a valuable test when used on bulk samples, provided that the test is carried out with sufficient frequency.

1. Collection of Milk Samples

Milk samples may be collected in 14- × 100-mm tubes, or in small screw-capped bottles, with or without a preservative, such as formalin or mercuric chloride. The milk contained in cans or bulk tanks is thoroughly mixed so that the cream is evenly dispersed. A quantity of milk is obtained from each can of milk, but milk from no more than three cans should be pooled in each collection tube. When collecting milk from bulk tanks, milk from no more than one tank should be put in each collection tube. A small metal dipper is commonly used for collecting milk from cans or a bulk tank. The samples should be refrigerated for 24 h and then tested as soon as possible. The test is not effective if the milk has been pasteurized.

2. Procedure for the Milk Ring Test

Briefly, the milk is mixed thoroughly to disperse the cream evenly before the transfer of 1 ml to a narrow test tube. One drop (0.03 ml) of ring test antigen is added and while closing the top of the tubes with a polyethylene-coated foam-rubber pad, the tubes are inverted several times. The mixture is allowed to stand for about 1 min and is then examined to make sure that the antigen is thoroughly mixed with the milk. Incubation is at 37°C for 1 h.

The test is read using a uniform light source. If the intensity of blue color in the cream layer is equal to or deeper than in the skim portion, the test is considered positive. If the intensity of the color in the cream layer is less than in the skim portion, the test is considered negative. False positives may occur at periods where a high proportion of the animals in the herd are drying off, or where mastitis is prevalent.

3. Adjusting the Milk Ring Test for Herd Size

When milk is collected in bulk tanks, some workers think it necessary to adjust the MRT to compensate for the dilution factor. However, it has been the experience in Great Britain that in the role as a screening test, this extra complication is not necessary.

K. SERIAL DILUTION MILK RING TEST

It is possible to detect infected animals in a herd by using the serial dilution MRT, but the test is generally used to supplement the serological tests when questionable results are obtained. Serial doubling dilutions of the milk under tests are made using MRT-negative milk, and the antigen is added to each. A titer of the 1:16 or more can often be correlated with the presence of *Brucella* organisms in the milk. Ring tests carried out on undiluted milk samples from individual cows may give false positive results shortly after parturition, near the end of lactation, and when mastitis is present.

APPENDIX

THE MILK RING TEST (MRT)

Procedure for Test

Milk Sample Collections

1. Bulk tank milk on the farm. A 10-ml sample of milk is collected after the bulk tank milk has been agitated for 3 min. The sample is identified to include date of collection and is transported to the receiving station in the refrigerated compartment of the tanker.
2. Can milk at creamery/receiving plant. A representative 10-ml sample of milk from all the cans is collected from the weighing bowl.
3. Can milk on the farm. The milk in the can should be stirred vigorously with the sampling dipper at several levels for at least 10 s. The dipper should then be plunged deeply up and down at least six times and then emptied. The sample should be taken finally from well below the surface of the well-mixed milk and transferred into the sample tube. There is no need to rinse the dipper between cans; the mixing process will prevent carryover of agglutination reaction from one sample to another.

Antigen

The MRT antigen must be stored at 4°C when not in use and remains stable under these conditions for many months.

The volume of antigen delivered for each test must be 0.03 ml, using special "dropping" bottles, delivering 0.03 ml per drop. The antigen must be well mixed within 1 min of its addition to the milk.

Technique of Testing

Milk samples should be stored at 4°C until at least 12 h have elapsed since the time of collection.

The milk samples and antigen are removed from the refrigerator and left at room temperature for 1 h before setting up the tests.

1. Gently mix the milk in the sample tube to ensure even distribution of the cream.
2. Place 1.0 ml of the milk in a plastic "agglutination" tube to give a column of milk about 2 cm high.

3. Add 1 drop (0.03 ml) of MRT antigen (the dropping pipette must be held vertically).
4. Gently mix the contents within 1 min of adding the antigen.
5. Place in a 37°C incubator for 1 h.
6. Readings are recorded as follows:

Color of cream ring	Color of milk column	MRT reading
Definitely blue	White	+ + +
Definitely blue	Slightly blue	+ +
Definitely blue	Definitely blue	+
Slightly bluer than or the same color as the milk column	Definitely blue	±
Incomplete blue ring	Definitely blue	Trace
White	Definitely blue	–

Note should be made of any loss of blue color from both cream and milk columns. If there is insufficient cream, as shown by the presence of a thin blue line at the top of the milk column, the sample should be retested by resubmitting the sample to test as described above but add 4 to 5 drops of cream from a known MRT-negative milk sample just before adding the antigen. (''Negative'' cream should be obtained by pooling MRT negative milk samples and collecting the gravity cream.)

MRT DILUTION TEST

The test may be indicated where reactions to the herd MRT are not confirmed by herd blood tests; milk from individual animals in the herd is subjected to the following:

1. Routine MRT.
2. MRT, after diluting 1:5 and 1:10 in MRT negative milk.

Procedure for Test

Tube	1	2
Volume of milk to be tested (ml)	1	1
Volume of known MRT negative milk (ml)	4	9

Both the ''test'' and ''MRT-negative'' milk must be well mixed by inverting the tubes gently several times both before and after the dilutions have been made.

The milk after dilution is tested in the usual way, using 1 ml of the diluted milk, and the results are recorded as for the routine test.

SERUM TESTS

Rose Bengal Plate Test (RBPT)

Procedure for Test

Antigen

The antigen consists of *Brucella* cells stained with Rose Bengal and suspended in buffer at pH 3.6. While the antigen remains stable for long periods when stored at 4°C, it deteriorates when constantly removed from the refrigerator and left standing on the bench for prolonged periods.

Antigen Dispenser

A 10-ml plastic dispenser fitted with a special delivery tip which delivers 30 ml per drop is used. This should be filled and used for each day's tests, the stock antigen being stored at 4°C. If this is not available, a microtiter pipette can be used.

Plates

These consist of ruled, enamel strips marked 1 to 10, which fit into the platform of the rocking machine. Alternatively, the test is done in WHO hemagglutination (HA) plates or on a flat glass plate or ceramic tile divided into 15-mm squares. A piece of toweling should be placed on the bench underneath the HA plates to prevent scratching of the plastic.

The rocking machine is fitted with a platform and timer which allows for the automatic mixing of serum-antigen mixture for 4 min, after which the tests are immediately read. A rotary agitator is used for the HA plates.

Technique of Testing

Using a 30-ml pipette fitted with a disposable plastic tip, 0.03-ml volumes of sera are dispensed onto the strip. Alternatively, an automatic pipette delivering 25 or 30 μl can be used.

Remove the sera from the refrigerator. Fill the antigen dispenser with antigen from the stock bottle (the latter must then be returned to the refrigerator). Leave the antigen for use and sera at room temperature for 0.5 to 1 h before the test.

The sensitivity of the test is affected by the temperature at which the reaction takes place. If antigen and serum are tested immediately after removal from the refrigerator, the test will be less sensitive than if the reactions were at room temperature. The antigen may deteriorate if not kept under refrigeration.

1. Holding the antigen dispenser upright, place one drop (30 ml) of antigen on each square of the enamel strip.
2. With the 30-μl pipette, place one drop of the serum alongside (but not into) the antigen.
3. With the wooden applicator stick, mix the antigen and serum thoroughly. The serum and antigen must be properly mixed.
4. When all 10 sera have been added and mixed, place the strip on the rocking machine and mix for 4 min.
5. Read the tests with great care by examining for agglutination in a good light. Reading is often facilitated when the mixture is observed as it flows away from the operator. With some sera, agglutination takes place almost immediately after the serum and antigen have been mixed, whereas in other cases, agglutination is delayed until the end of the 4-min period.
6. Record as either
   − = no agglutination or
   + = any degree of agglutination.

A more detailed evaluation may be used, as recommended for the test done in HA plates, i.e., 0 = no agglutination, no rim formation, uniform pink color; 1 = barely perceptible agglutination and/or some rim formation; 2 = fine agglutination with a definite rim, some clearing; 3 = course clumping, definite clearing.

A known positive and negative serum must be included in each day's tests.

Unless disposable strips are available, the enamel strips should be well washed with running water, dried, and reused.

Plastic tips, for reasons of economy, can be reused by washing in running tap water and drying.

The Buffered Plate Agglutination Test
Procedure for Test

1. Bring the serum samples and antigen to room temperature.
2. Place 80 μl of each serum under test on one of the squares of a marked glass plate.
3. Shake the antigen bottle until the contents are thoroughly mixed; then drop 30 μl of antigen on each drop of serum.
4. Mix the serum and antigen in each drop with a toothpick or spreader. Spreaders can be designed to mix a whole row of drops at the same time.
5. Rotate the plate in the hand for three rotations on each of three occasions: (1) after mixing, (2) after 4 min, and (3) after 8 min. Then read.

Any agglutination is recorded as positive.

Serum Agglutination Test (SAT)
Procedure for Test

A working standard strain calibrated against the Second International Standard anti-*B. abortus* serum must be included in each day's tests. Also essential is a negative control serum and opacity standards.

Phenol Saline

To 85 g NaCl and 50 g phenol add 10 l of distilled water and mix thoroughly.

Plastic agglutination tubes holding 2 ml are used.

Syringes are used for delivering saline and antigen. A repeating pipette can be used.

For serial dilutions, 1-ml all glass tuberculin syringes can be used; these can be fitted with fine polyethylene tubing into the nozzle of the syringe and cemented in place. The "dead space" with this type is negligible.

Antigen

This is normally issued as a concentrate antigen. It must be diluted in phenol saline for use at the appropriate dilution.

Technique

1. With the automatic syringe, add 0.5 ml of phenol saline to the last four tubes of a five-tube test.
2. Flush out the syringe in phenol saline so that the dead space tubing only is filled with saline.
3. Draw up to 0.8-ml mark with phenol saline and then up to the 1-ml mark with serum.
4. Transfer to the first tube.
5. Mix without frothing.
6. Transfer 0.5 ml to the next tube and carry on to the last tube and discard 0.5 ml.
7. Mix contents of antigen bottle.
8. Using the automatic syringe add 0.5 ml antigen to each tube.
9. Mix contents of the agglutination tubes.
10. Incubate in 37°C incubator for 18 h (±1 h).

For testing the sera of sheep and goats it is recommended that a 5% sodium chloride solution containing 0.5% of phenol be used, both for making the serum dilution and for diluting the antigen concentrate.

Setting up of Opacity Standards

These should be set up each day and the tubes incubated with the tests.

The antigen as used must be diluted $^1/_2$ in phenol saline (to bring it to the same opacity as in the test where it is diluted $^1/_2$ with the serum dilutions).

Mix 2 ml antigen + 2 ml phenol saline (antigen $^1/_2$).

Tube	**1**	**2**	**3**	**4**	**5**
Phenol saline (ml)	1	0.75	0.5	0.25	0-.0
Antigen $^1/_2$ (ml)	0	0.25	0.5	0.75	1-.0
Representing	100	75	50	25	0
Agglutination score	4+	3+	2+	1+	—

Reading the Tests

The degree of agglutination is read by comparing with the standard against a black background with a source of light (fluorescent strip light) coming from above and behind the tubes.

End point reading — The titer of the serum is the highest dilution giving a definite 1+ (25% agglutination) reading or more. Traces or less than 1+ readings should be ignored. If two successive serum dilutions give a 1+ reading, e.g., 4 + 4 + 1 + 1±, the test on this serum should be repeated on the following day. If the same pattern is again found, the end point will be taken as the lower dilution giving 1+, in this case, 1/40 (the third tube).

Recording of results — The results must be recorded to show the degree of agglutination (numerator) and the dilution (denominator). Thus, 3/40 is 3+ agglutination at a dilution of 1/40.

EDTA-Modified SAT

Procedures for Test

The technique is essentially the same as that used for the conventional SAT except that EDTA is incorporated with the antigen.

Preparation of EDTA Antigen

This should be prepared by mixing 1 part SAT Antigen plus 9 parts 10 m*M* EDTA (disodium salt) in PBS.

Ingredients for 10 m*M* EDTA in PBS are 900 ml phosphate-buffered saline at pH 7.2 and 3.024 g EDTA (disodium salt). The pH of the resulting suspension should exceed 6.0.

Stir the ingredients, heating slightly until dissolved. Decant into bottles and store at +4°C until ready to use. When required, reconstitute the standardized SAT antigen concentrate with the EDTA solution. Once antigen has been added, the EDTA antigen has a shelf life of 6 weeks. The EDTA antigen should be used exactly as the conventional SAT antigen.

The USDA Tube Agglutination Test

This version of the agglutination test is often called the standard tube agglutination test (STAT).

Procedure for Test

Two methods are given for carrying out the test.

The Decimal Dilution Method

This is used for dilutions of not greater than 1:400.

1. The required number of clear glass test tubes (13 or 14 mm × 100 mm) are arranged in rows of four in suitable racks.
2. With a 0.2-ml pipette withdraw serum from the blood tube. Deliver 0.08 ml serum into the first tube, deliver 0.04 ml of serum into the second tube, 0.02 ml into the third tube, 0.01 ml into the fourth tube, and 0.005 ml may be added to a fifth tube.
3. Successive samples are pipetted into each row of tubes.
4. With an automatic pipette, 2.0 ml of properly diluted antigen is delivered into each tube. This results in dilutions of 1:25, 1:50, 1:100, 1:200, and 1:400.
5. The racks of tubes are gently shaken and placed in an incubator at 37°C for 48 h (±3 h).

The Multiple (Doubling) Dilution Method

This is used for determining the end titer.

1. Place tubes in racks in rows of five or more tubes, one row for each sample.
2. Using a 0.2-ml pipette, withdraw serum, deliver 0.16 ml into the first tube, and withdraw the pipette as described previously.
3. After delivering 0.16 ml of serum from each sample into the first tube of the appropriate row with a clean dry pipette, add 0.4 ml of properly diluted tube antigen into each tube containing serum (the first tube of each row) and 2.0 ml into each succeeding tube.
4. With a 2.0-ml pipette, thoroughly mix the serum-antigen mixture in the first tube by filling and emptying the syringe at least three times. Transfer 2.0 ml into the second tube. The contents of the second tube are mixed in the same manner and 2.0 ml transferred to the third tube, and so on. Discard 2.0 ml from the last tube.
5. Perform this procedure using a clean pipette with each serum or taking care to rinse the rheometer syringe thoroughly between samples. This results in dilutions of 1:25, 1:50, 1:100, and so on.
6. The tubes are kept in the incubator at 37°C for 48 h.

Reading the Results

The tubes are observed against a dull black background with the light coming from behind the tubes. Extraneous light should be reduced to a minimum.

A *positive* reaction is one in which the serum-antigen mixture is clear and gentle shaking does not disrupt the flocculi.

An *incomplete* reaction is one in which the serum-antigen mixture is partially clear and gentle shaking does not disrupt the flocculi.

A *negative* reaction is one in which the serum-antigen mixture shows no sign of clearing and gentle shaking reveals no flocculi.

Recording USDA Agglutination Reactions

This section applies also to recording and interpreting the plate agglutination test described below.

A positive reaction is recorded as +, an incomplete reaction as "I", and a negative reaction as −.

The Plate Agglutination Test

Procedure for Test

Equipment Required

1. A testing box about 48 cm long, 33 cm wide, and 12 cm deep is employed. The glass testing plate, etched to make 60 4-cm squares, is supported near the top of the box and can be easily lifted from it. A light source, either a fluorescent tube or two electric bulbs, is partially encased below the plate near the front of the box, so that oblique light may be thrown from beneath toward the serum-antigen mixture. The inside of the box should be painted a dull black. Also, the box should be provided with a hinged cover to prevent too rapid evaporation of the mixtures.
2. Good quality pipettes delivering 0.2 ml and graduated in divisions of 0.01 ml are used.
3. An antigen dropper that delivers exactly 0.03 ml of antigen per drop is needed.
4. Toothpicks may be used for spreading the serum-antigen mixtures, but a strong wire bent to obtain a straight 1-cm bearing surface is more satisfactory. Multiple spreaders capable of spreading series of up to five serum-antigen mixtures at a time are more satisfactory, since drying is minimized by reducing the time required for spreading. These spreaders may be constructed from various metal or plastic substances. They should be rinsed in water and wiped dry between samples.

Technique of the Test

If the test is carried out carefully, the results obtained with the four-dilutions in the plate test will correspond with those obtained with the 1:25, 1:200, and 1:200 dilutions of the USDA tube test, although occasionally sera are encountered that give a somewhat different titer in the two tests.

Both serum and antigen should be brought to approximate room temperature. With the pipette held at an angle of 45° to the plate and touching it, amounts of 80, 40, 20, and 10 ml of the serum sample are placed in a column of four squares on the test plate. The antigen bottle is gently shaken to ensure a homogeneous suspension. Holding the dropper in a vertical position, one drop of plate antigen (0.03 ml) is dropped on each quantity of serum. Beginning with the drop containing 0.01 ml of serum, the serum and antigen are mixed thoroughly with the spreader, using a circular motion and spreading over an area 2 cm in diameter; the same procedure is followed for the 20, 40, and 80 μl drops, increasing the diameter of the spread progressively up to about 3 cm for the 80 μl serum drop. With this method it is not necessary to wipe the spreader between different drops of the same sample. The test plate is now lifted from the box and gently rocked in a rotating movement to ensure thorough mixing. The plate is replaced and the lid closed to inhibit evaporation.

The test plate should be allowed to stand for 8 min before readings are made; most samples reach the peak of agglutination in that time. About 4 min before reading, the test plate should be taken up, gently rotated four times, and replaced on the box. This is important. After 8 min the test plate is again rotated four times and replaced on the box. The light should be turned on and the box tilted slowly to allow the mixture to flow from side to side while readings are being made. Readings are best made against the black background of the box. Only three classifications of agglutination reaction should be made: complete agglutination (+), where clumps of agglutinated antigen are separated by water-clear liquid; incomplete agglutination (I), where there is definite agglutination of varying degrees without complete clearing in the background fluid; and negative (−), where there is no sign of any agglutination having taken place.

The Rivanol Test

This is a supplementary serological test. The rivanol solution selectively precipitates

IgM in the serum; this is then removed by centrifugation, and the supernatant tested in a plate agglutination test.

Procedure for Test

Rivanol solution — Rivanol powder (2-ethoxy-6,9-diaminoacridine lactate) is dissolved in sterile distilled water to make a 1% w/v solution by stirring on a magnetic stirrer for at least 1 h. It should be bottled under aseptic conditions in an amber bottle and stored in the dark.

1. Allow the serum and reagents to warm to room temperature for at least 1 h.
2. With a 1-ml pipette, measure 0.5 ml of Rivanol solution into the required number of suitably sized (say 13 or 14 × 100 mm) tubes.
3. Using a fresh 1.0-ml pipette for each sample, measure 400 μl of serum under test into each tube and mix immediately by shaking the tube.
4. Leave the tubes at room temperature for at least 5 min but not more than 1 h. Then centrifuge them at approximately 1000 *g* for 5 min to pack the precipitate and leave a clear yellow supernatant.
5. Using a 200-μl pipette, measure 80, 40, 20, and 10 μl of supernatant fluid onto separate squares or a clear glass plate such as that described for use in the USDA plate agglutination test.
6. Add 1 drop (30 μl) of Rivanol antigen to each quantity of supernatant fluid, and mix with a stirrer, starting with the lowest dilution. Each quantity should be spread to approximately the sizes described for the plate test.
7. Rinse and dry the stirrer between samples.
8. Tilt the plate in a circular motion for four rotations.
9. Replace the plate in the box and close the lid to prevent evaporation.
10. After 6 min, rotate the plate as before for four rotations.
11. After another 6 min, again rotate the plate for four rotations and read the reaction against the illuminated background of the box.
12. The test is read and the results recorded in the same way as described for the USDA plate agglutination test. (Although they are not true dilutions, for convenience the mixtures of supernatant and antigen are taken as 1:25, 1:50, 1:100, and 1:200.)

Complement Fixation Test (CFT)

Procedure for Test

Cleaning and Preparation of Glassware and Plastic Plates

After use, glassware and reusable plastic plates are thoroughly rinsed to remove all reagents and placed in detergent solution for at least 2 h, preferably overnight. All traces of reagent should then be removed with warm running tap water, with scrubbing where appropriate. Then the materials are rinsed at least twice in distilled water and dried in the hot-air oven at 56°C.

Reagents for the Test

Diluent

The diluent is Veronal-buffered saline with the addition of calcium, magnesium, and gelatin. It is made by adding one part of buffer concentrate to four parts of 0.04% gelatin and storing at 4°C.

CF Buffer Concentrate

1.	Sodium chloride	425 g
2.	Barbituric acid	28.75 g

3.	Sodium barbiturate	18.75 g
4.	Magnesium sulfate	10.18 g
5.	Calcium chloride	1.47 g
6.	Distilled water	10,000 ml

Fill a 10-l container with 6 l of distilled water. Dissolve the barbituric acid in 1 l of boiling water. Dissolve the sodium chloride, sodium barbiturate, magnesium sulfate, and calcium chloride in 3 l of water and add each to the container. Add the dissolved barbituric acid to the container. The solution should be water clear and should not need filtering. Store at 4°C.

0.04% w/v Gelation Solution

1.	Bacto gelatin	1600 mg
2.	Distilled water	4000 ml

Add gelatin to 1 l of distilled water in a measuring cylinder and dissolve in a 58°C water bath. When dissolved, pour into a 5-l glass flask and add 3 l distilled water. Add 1 l of buffer concentrate and mix thoroughly before use. This diluent is now ready for use in the test and should be stored at 4°C.

Satisfactory results are also obtained with a 0.85% solution of sodium chloride containing calcium and magnesium ions, provided the pH lies between 7.3 and 7.4. This solution is prepared by adding 1 ml of the stock solution of calcium and magnesium salts to each liter of saline solution. If 0.25 ml of a 5% solution of sodium azide is added per liter to control bacterial contamination, the solution can be stored at room temperature.

Antigen

The antigen should be standardized against the Second International Standard anti-*B. abortus* serum so as to give 50% fixation at a dilution of 1/200 of this serum. The antigen is usually issued as a concentrate and is diluted in CF diluent before use in each day's test.

Complement

The source of complement is freeze-dried guinea pig serum (reconstitution of the freeze-dried product should be carried out according to the manufacturer's instructions using sterile distilled water). This reagent is preserved by Richardson's method such that 8 parts of serum are mixed with 2 parts of preservative. A 1/10 dilution of complement is made by adding 1 ml of the preserved serum to 7 ml of distilled water.

The following method is for those laboratories that wish to prepare their own complement.

Adult guinea pigs receiving adequate green food produce good quality complement. Withhold all food, but not water, during the 12 h preceding bleeding. Do not use pregnant females or those that have recently given birth.

1. Bleed at least four guinea pigs, preferably many more.
2. Collect the serum as soon as it has separated from the clot, separation should not take more than 1 h.
3. Pool the sera and centrifuge to remove any erythrocytes. The clear serum is used as complement.

Preservation and Storage of Complement

Untreated complement may be stored at −40°C or below. It does not store satisfactorily at higher temperatures. Freeze-drying is recommended. When freeze-drying complement, it is advisable to add Richardson's preservative, as described below, before drying; this gives the product better keeping qualities after reconstitution.

Richardson's method is useful for preserving complement in the liquid state where adequate facilities for freezing or drying complement are not available. If kept in the refrigerator, complement so preserved should maintain a reasonable titer for at least 6 months, and even at room temperature loss of titer is not rapid.

Two stock solutions that keep indefinitely are used:

Solution A

Boric acid	0.93 g
Borax	2.29 g
Sorbitol	11.74 g
Saturated sodium chloride solution to	100 ml

Solution B

Borax	0.57 g
Sodium axide	0.81 g
Saturated sodium chloride solution to	100 ml

To preserve complement, mix 8 parts of guinea pig serum with 1 part of solution B, followed by 1 part of solution A. Before use, each 1 ml of preserved complement is added to 7 ml of distilled water to restore tonicity; this produces a 1:10 dilution of complement. Further dilutions are made in CF diluent.

Hemolytic System

Sheep cells — One or more sheep, known to produce erythrocytes of a consistently satisfactory level of sensitivity, should be chosen and subsequently used exclusively. For routine tests 130 ml of blood is taken into 270 ml of Alsever's solution every week. It is stored at 4°C for 7 d before use. A 3% suspension of cells is used in the test.

Alsever's solution — This is made up as follows:

Glucose	18.66 g
Sodium chloride	4.18 g
Sodium citrate	8.00 g
Citric acid	0.55 g
Distilled water to	1000 ml

The solution may be autoclaved or passed through a sterilizing (0.45 μm) filter and contained in a collecting bottle which has been previously sterilized in the autoclave.

Take approximately 20 ml of sheep red blood cells from the Alsever's solution and wash them three times centrifugally in CF diluent (2500 rpm for 5 min). Decant the supernatant fluid each time by a water vacuum pump. Make a 1 in 3 dilution of cells in diluent, i.e., 15 ml washed cells plus 30 ml diluent. If possible the hemoglobin content of the cells should be measured using a hemoglobinometer. Any reading obtained should be converted to a percentage; the percentage of hemoglobin required is 50. If it is not, the cell concentration must be adjusted.

Dilute the standardized cells to 1 in 11. The dilution of cells 1 in 3 before standardization and 1 in 11 after gives a final dilution of 1/33 representing 3%.

Hemolysis

This is a serum containing a high level of antibody against sheep erythrocytes. When this antibody is combined with erythrocytes in suspension, the erythrocytes are said to be sensitized, i.e., in the presence of complement they will lyse.

For the CF test hemolysin is best prepared in rabbits. Satisfactory commercial preparations are widely available. For those wishing to prepare their own hemolysin, the following method is recommended:

1. Prepare a 10% suspension of washed sheep erythrocytes, then after step 5 has been completed, suspend the deposit in 9 times its volume of CF diluent.
2. Inoculate a number of well-grown rabbits, intravenously, with this suspension at the rate of 1 ml/kg body weight on days 1, 2, 3, 4, 5, 8, and 11.
3. Remove a sample of blood from each rabbit on day 14 and titrate the serum individually.
4. Exsanguinate those rabbits giving satisfactory titers.
5. Separate the sera, pool, and filter.
6. Freeze-dry or preserve in an equal volume of glycerol for storage at refrigerator temperature.

Five MHD (units) of hemolysis are used. A volume at the appropriate dilution, equal to the volume of sheep cell suspension, e.g., 110 ml, is prepared. Mix together the hemolysis cells and allow them to sensitize at room temperature for 30 min (while the complement is fixing in the routine tests). Two volumes of the sensitized cells are added to each serum dilution.

Technique
Reagents:

1. Veronal buffered saline (CF diluent)
2. Working strength antigen
3. Complement at $1^1/_4$ units
4. Rabbit anti-sheep hemolytic serum (HIB) at 5 U
5. 3% Sheep red blood cells

Apparatus:

1. Inactivation tubes
2. Waterbath racks
3. 0—100-μl Adjustable single-channel pipette
4. 25-μl Multichannel pipette
5. 200-μl Single-channel pipette
6. Round-bottomed microtiter plates
7. Plate sealers (tape)
8. Shaker/incubator
9. Tips for pipette
10. Hemocue hemometer or equivalent
11. Plastic troughs for reagents
12. Measuring cylinders
13. Glass and plastic beakers
14. 10-ml Graduated glass pipettes
15. Reading box or mirror

Inactivation of Serum

Fresh serum contains natural complement and usually also has some anticomplementary activity, which may be increased by bacterial contamination. In serum samples to be tested, complement is destroyed and the anticomplementary activity reduced or eliminated by heating the serum. The process is known as inactivation. Hemolysis in serum has little, if any, anticomplementary effect but may interfere with the reading of the test in the lower serum dilutions.

Cattle serum may be inactivated at 58°C. In the case of sera from sheep and goats, especially the former, anticomplementary reactions may occur frequently. In this case the sera may be inactivated at 60°C, though it should be borne in mind that there will be considerable inactivation of IgM at this temperature.

Sera should be inactivated on the day they are to be tested.

Method

The standard time for inactivation is 30 min. The sera may be inactivated in the microtiter plates, in an incubator provided that (1) the plates are not stacked in the incubator, (2) the door of the incubator is not opened during the period of inactivation, and (3) the period of incubation is taken to begin when the temperature of the incubator regains the correct operating temperature after being opened to receive the plates. The incubator should be equipped with a fan. If a suitable incubator is not available, the plates may be floated in a water bath, or the sera may be inactivated in tubes in a water bath before being dispensed into the microtiter plates.

Method for Screening

1. Make $^1/_2$ dilutions of each serum sample in CF diluent. Dilutions are made by mixing 200 μl of test serum with 200 μl diluent in inactivation tubes or direct on the plate.
2. The diluted samples for then inactivated by heating at 58°C for 50 min for bovine sera and at 60°C for 30 min for ovine sera.
3. Most test samples are screened initially at $^1/_2$. Any that give a positive reaction are then titrated out to four dilutions, or more.
4. For the initial screening test add 25 μl of each test sample to wells on the plate so that they contain 96 samples.
5. Prepare antigen ready for use and add 25 μl to all wells on the test plate using multichannel pipette.
6. Add 25 μl of $1^1/_4$-Unit complement to all wells using a multichannel pipette.
7. Seal the plates with tape, shake, and incubate for 30 min at 37°C. (The plates need to be shaken for the first 10 min only.)
8. Mix together equal volumes of 3% sheep RBCs and 5 U hemolysis for 30 min before they are required and leave at room temperature.
9. Add 50 μl sensitized cells to each well with a multichannel pipette promptly after the first incubation.
10. Seal, shake, and incubate the plates for a further 30 min. Read the plates over a mirror or reading box, assessing the fixation in every well.

Method for Titration

1. Holding the microtiter plate sideways and using a multichannel pipette, deliver 25 μl diluent into all wells except for those in rows 4 and 8. It is advisable to mark the plate with lines separating individual tests.
2. Using a single-channel pipette deliver 25 μl inactivated sample at $^1/_2$ into the first and second wells of every test and the AC control row.
3. Make doubling dilutions changing tips in between each sample. Pick up 25 μl from

second well and transfer to the third. Mix well and pick up 25 μl from this well and transfer to the fourth. Mix well and discard 25 μl. Repeat for the AC control well using a clean tip.

4. There should now be 25 μl in all wells. Add an extra 25 μl diluent to all wells in the AC control row; this is in place of antigen.
5. Prepare antigen for use, and using a multichannel pipette add 25 μl to all wells except those in the AC control rows.
6. Add 25 μl $1^{1}/_{4}$-U complement to all wells using a multichannel pipette.
7. Seal the plates with tape, shake, and incubate as before.
8. Add 50 μl sensitized sheep red blood cells to each well using a multichannel pipette. Incubate as before.
9. Plates are read over a mirror or reading box assessing fixation in each well. Record the end point and any anticomplementary reaction. Ignore traces of fixation in interpretation.

Alternative Method

1. Dispense sequentially with a hand-held pipetting device 50 μl of each serum under test into the first well of each row of the microtiter plate, using 12 rows per plate.
2. Similarly, dispense 50 μl of the positive control serum into an appropriate number of first wells (it is convenient to repeat the control serum in all the rows remaining in the plate after the last serum under test has been dispensed). Every 20th plate should have the control serum in every row to provide a check on the correct functioning of the diluting and dispensing equipment.
3. Cover the wells containing serum with adhesive tape and inactivate.
4. Remove the tape, add 25 μl of diluent to each well, except the first of each row which already contains serum.
5. Make doubling dilutions of the sera, using the 25 μl microdilutors, starting with the undiluted serum in the first wells and continuing right across the plate. This produces dilutions up to 1:128.
6. Add 25 μl of freshly diluted antigen, starting at the third wells and continuing to the end of the rows.
7. The second well in each row is an anticomplementary control for each serum. Add 25 μl of diluent ot these wells to compensate for lack of antigen.
   The test is read from the 1:4 dilution, i.e., from the third well onward. The fact that the serum in the anticomplementary control wells is more concentrated (1:2) than the first dilution of the test proper is not considered to have practical significance.
8. Add 25 μl of complement containing 5 $C'H_{50}$s to each well except the first of each row, which, being a reservoir for serum, plays no further role in the test. Antigen and complement may be mixed and added together.
9. Tap each plate gently to mix the reagents, cover the plates with the special microtiter lid or with another plate, then place them in the incubator at 37°C for 30 min. Alternatively, if the cold fixation method is preferred the plates may be placed in the refrigerator (about 4°C) for 18 h.
10. Complement controls need to be set up at this point.
11. Prepare the sensitized erythrocyte suspension by mixing equal volumes of standardized erythrocyte suspension and hemolysin dilution and keeping the mixture at room temperature for 15 min with agitation, preferably mechanical.
12. Remove the plates from the incubator, and add 25 μl of sensitized erythrocyte suspension to each well except the first of each row.
13. Cover the plates again and replace them in the incubator at 37°C for 30 min, during which time it is essential that the erythrocytes are prevented from settling in the wells.

They may be kept in suspension by fixing the plates on a mechanical agitator installed in the incubator.

14. Remove the plates from the incubator and either centrifuge them in the special microtiter carrier plates or place them in the refrigerator for at least 2 h to deposit any unlysed erythrocytes.
15. Read the results from below, using a magnifying mirror, or from above with the plates over a diffused white light. With complete fixation (a 4+ reaction), there is no lysis, a button or erythrocytes is seen at the bottom of each well and the supernatant is clear and colorless. A negative reaction (recorded as 0) is indicated by complete lysis; there are no erythrocytes and the liquid in the well is clear and colored with hemoglobin. Various degrees of incomplete lysis are recorded as 1+, 2+, and 3+ reactions. The anticomplementary control well should show complete hemolysis, otherwise the result for that particular serum is not valid.

Interpretation of the Results

The end point of the reaction is taken as the highest dilution showing a 1+ reaction or greater, i.e., 25% or more fixation. This dilution is the titer of the serum.

Standardization of the Complement Fixation Test

The European Economic Community (EEC) has established a unitage system based on a standard serum which is required to conform to the ISABS. This serum is taken to contain 1000 international complement-fixation test units (ICFTU) and if it is tested by a given method and gives a titer of, say, 500, then the factor for an unknown serum tested by that method can be found from the formula:

$$1000/500 \times \text{titer if test serum} = \text{no. of ICFTU of antibody in the test serum}$$

The EEC considers any serum containing 20 ICFTU, or more, as positive.

Difficulties in standardization arise because different techniques selectively favor complement fixation by different immunoglobulin isotypes. It is recommended that any country using the complement fixation test on a national scale obtain agreement on a standard method by the different laboratories performing the test. This allows the same level of sensitivity to be obtained. The periodic distribution of a collection of sera can then be used to monitor the level of agreement between laboratories.

Problems Encountered with the Complement Fixation Test

Anticomplementary Sera

An anticomplementary result from any serum sample may invalidate that result, but the following notes are provided as a guide to the action to be followed.

1. If the AC control shows only a trace, then this can be ignored.
2. If the result is at least three dilutions greater than the AC control, then the AC effect may be ignored, e.g., 4/16 AC 4/2.
3. If the AC control result is the same as the test result, then a fresh sample should be collected and retested.

The absence of complete hemolysis in the serum anticomplementary control well or tube indicates that the serum under test possesses some nonspecific anticomplement activity. Thus, the test result is not valid. There are a number of ways of dealing with this situation.

The following method is often successful for retesting anticomplementary sera:

1. Make a 5% solution of bovine serum albumin (BSA) fraction V, or 5% guinea pig serum, in CF diluent.
2. Prepare the first serum dilution to be used in the test in one of these solutions (say 0.2 ml of serum and 0.6 ml of solution).
3. Incubate in the water bath at 37°C for 30 to 60 min.
4. Inactivate at the temperature appropriate to the type of serum; then proceed with the test as usual.

Alternatively, the serum may be retested in duplicate:

1. With antigen as in the ordinary test
2. The same procedure with an equal volume of diluent in lieu of antigen

Any significant increase in titer in Step 1 over Step 2 may be taken to indicate specific positivity.

Prozones

When a prozone occurs, i.e., when some of the wells or tubes with more concentrated serum show hemolysis and higher dilutions greater fixation, the highest dilution showing fixation is taken as the titer of the serum; lesser or negative reactions in the earlier wells or tubes are ignored. It is possible, though of extremely rare occurrence, for the prozone to blot out the positive reaction altogether. Such sera are often highly positive on agglutination.

Hemagglutination

When CF tests in microtiter plates are being read, hemagglutination of unlysed erythrocytes may be seen in the early serum dilutions, preventing the formation of a button of cells in the center of the well. More common with partial fixation, the phenomenon shows as a thin layer of erythrocytes spread over the bottom of the well. The presence of unlysed erythrocytes indicates a positive reaction.

Control Sera

Controls are set up in duplicate with each batch of tests. These consist of negative serum controls diluted $^1/_2$, positive serum controls at an appropriate dilution. A short complement titration is also prepared with each batch of tests.

1. The negative control is a known *Brucella* negative serum treated in an identical way to the test sera except that there is only a single dilution serum control well at $^1/_2$. It is normally diluted $^1/_2$ with diluent and frozen at −20°C in 800-μl aliquots. When required, an aliquot is thawed out and then inactivated as usual before use; afterward treat as a normal test serum and put 25 μl of $^1/_2$ dilution into the AC control.
2. The positive controls should give the correct titers each day in the test, and the negative control should give no reaction at any dilution.
3. The complement titration consists of dilutions of complement from working strength to a quarter of working strength plus a hemolytic system control.

Use empty wells on the control plate to mix the dilutions of complement before adding them to complement controls.

Washing-Up Routine

Microtiter Plates and Glassware

1. Thoroughly rinse in running hot water to remove all traces of reagents.

2. Soak overnight in Decon 90: Decon 90, 120 ml; distilled water, 5000 ml.
3. Rinse well to remove Decon 90.
4. Examine plate wells for cleanliness and remove any soil with a mop-headed test-tube brush.
5. Rinse in distilled water three times.
6. Dry at 37°C.

Tuberculin Syringes and Automatic Pipettes

Wash out thoroughly with distilled water after use.

All Other Glassware

1. Soak in Decon 90 overnight.
2. Rinse well.
3. Rinse in distilled water.
4. Dry in hot air.

Titration of Hemolysis

Hemolysis is the most stable of the test reagents. It is always used in excess such that circumstances cannot arise where there could be insufficient hemolysis to cause complete lysis of a known number of sheep cells in the presence of adequate complement. Hemolysis is prepared in rabbits or horses by injecting them with sheep red blood cells; it is heat treated to destroy its native complement and preserved by the addition of an equal volume of glycerine.

Reagents

1. Veronal buffered saline
2. Complement (1/10 dilution)
3. Rabbit anti-sheep red blood cells of unknown titer
4. 3% Sheep red blood cells

Apparatus

1. Plastic or glass tubes for HIB dilutions
2. 0- to 100-μl adjustable single-channel pipette
3. 25-μl multichannel pipette
4. Round-bottomed microtiter plates
5. Plate sealer (tape)
6. Shaker/incubator
7. Tips for pipette
8. Hemometer
9. Plastic troughs for reagents
10. 10-ml Glass measuring cylinder
11. 50-ml Glass beakers
12. 10-ml Graduated glass pipettes
13. Reading box or mirror

Method

1. Make a 3% suspension of sheep red blood cells by adding 1 ml standardized cells (at 33%) to 10 ml diluent.

2. Prepare 1/10 and 1/15 dilutions of hemolysin in tubes.

	Hemolysin	Diluent
1/10	100 μl	0.9 ml
1/15	100 μl	1.4 ml

From these prepare 1/1000 and 1/1500 dilutions as follows:
10 μl of HIB (1/10) + 0.9 ml diluent to give 1/1000, and 10 μl of HIB (1/15) + 0.9 ml diluent to give 1/1500.

3. Add 25 μl diluent to 2 to 7.
4. Add 25 μl of 1/1000 hemolysin dilution to column 1 A—D and column 2 A—D.
5. Add 25 μl of 1/1500 hemolysin dilution to column 1 E—H and column 2 E—H.
6. From these, prepare doubling dilutions across the plate by mixing the contents of column 2 and transfer to column 3 and repeating the procedure up to column 7. Discard 25 μl from column 7.
7. Add 25 μl sheep cells at 3% to each well.
8. Incubate for 30 min at room temperature ensuring that the cells are kept in suspension by shaking at intervals.
9. Mix the contents of the wells thoroughly before adding 50 μl diluent to each well (in lieu of serum and antigen).
10. Add 25 μl of a 1/10 dilution of complement to each well.
11. Incubate the plate at 37°C in an incubator, shaking gently for the first 10 min.
12. Read the results against a white background or light source. One unit of MHD is the highest dilution to show complete hemolysis. In the test 5 U are used.

Titration of Complement

Reagents

1. Veronal buffered saline
2. Working strength antigen
3. Complement (freeze-dried guinea pig serum)
4. Hemolytic system

Apparatus

1. 0- to 100-μl adjustable single-channel pipette
2. 25-μl multichannel pipette
3. Round-bottomed microtiter plate
4. Plate sealers (tape)
5. Shaker/incubator
6. Tips for pipettes
7. Hemometer
8. Plastic troughs for reagents
9. 10-ml Graduated glass pipettes
10. 50-ml Glass beakers
11. Reading box or mirror

Method

1. Reconstitute the complement according to manufacturers instructions and make a 1/10 dilution in distilled water.

2. From the 1/10 dilution make a series of complement dilutions in tubes as shown in Table 1. Store these at 4°C until ready to use.
3. Add 25 μl diluent to rows B to D and F to H.
4. Add 25 μl of:

1/20	Dilution to column	1 A & B and E & F
1/25		2 A & B and E & F
1/30		3 A & B and E & F
1/35		4 A & B and E & F
1/40		5 A & B and E & F

5. Prepare doubling dilutions; i.e., mix contents of row B and transfer 25 μl to row C; mix contents of row C and transfer 25 μl to row D; mix contents of row D and discard 25 μl.
6. Repeat of rows F to H.
7. Add 25 μl diluent to all wells in lieu of serum.
8. Add 25 μl antigen at working dilution to rows A to D.
9. Add 25 μl diluent to rows E to H in lieu of antigen.
10. Incubate the plate at 37°C for 30 min, shaking gently for the first 10 min.
11. Add 50 μl hemolytic system, sensitized for 30 min at room temperature, to all wells.
12. Incubate as in Step 10.
13. Read the results. 1 U or 1 MHD is the highest dilution to show complete hemolysis.

$$1 \text{ MHD} = 1/60$$

$$11/4 \text{ MHD} = 1/60 \times \frac{5}{4} = \frac{1}{48} \text{ (ca. 1/50)}$$

Coombs Antiglobulin Test

Technique

Take the four tubes of an SAT succeeding a 4+ agglutination and carry out the following instructions.

1. Label tubes, balance, and spin for 30 min/2000 rpm.
2. Tip off the supernatant, add 1 ml of phenol saline, and vibrate to mix.
3. Balance the tubes and spin again from 30 min/2000 rpm.
4. Tip off supernatant, add 1 ml of phenol saline, and vibrate to mix.
5. Balance and spin again for 30 min/2000 rpm.
6. Tip off supernatant, add 0.5 ml anti-species globulin, and vibrate to mix.
7. Incubate overnight at 37°C and read as a normal SAT.
8. A negative and a positive control should be included.

Interpretation

Coombs tests must be 3 dilutions higher than SAT to be positive, e.g.

SAT	1/20	Coombs	1/40	Pass
SAT	1/20	Coombs	1/80	Inconclusive
SAT	1/20	Coombs	1/320	Fail

REFERENCES

1. **Rose, J. E. and Amerault, T. E.,** Electrophoretic and ultracentrifugation studies on serum from pregnant heifers after exposure to virulent *Brucella abortus, Am. J. Vet. Res.*, 25, 998, 1964.
2. **Rose, J. E. and Roepke, M. H.,** Physicochemical studies on postvaccinal *Brucella* agglutinins in bovine serum, *Am. J. Vet. Res.*, 25, 325, 1964.
3. **Rose, J. E., Roepke, M. H., and Briggs, D. R.,** Physicochemical properties of nonspecific bovine seroagglutinins for *Brucella abortus, Am. J. Vet. Res.*, 25, 118, 1964.
4. **Rose, J. E., Lambert, G., and Roepke, M. H.,** Ultracentrifugation and heat inactivation studies on seroagglutinins of pregnant heifers artificially infected with virulent *Brucella abortus, Am. J. Vet. Res.*, 25, 329, 1964.
5. **Hajdu, S. and Beseda, I.,** A study of the heterogeneity of *Brucella* antibodies. I. Determination of heterogeneity of *Brucella* post vaccination antibodies by means of gel filtration on Sephadex G200, *Vet. Cas.*, 12, 153, 1969.
6. **Hajdu, S.,** A study of the heterogeneity of *Brucella* antibodies. II. The proving of brucella post-infection antibodies by means of gel filtration on Sephadex G200, *Vet. Cas.*, 13, 121, 1970.
7. **Hajdu, S.,** A study of the heterogeneity of *Brucella* antibodies. III. Contribution to the formation of "heavy" and "light" antibodies of IgG type in cattle, *Vet. Cas.*, 13, 159, 1970.
8. **Najdu, S.,** Studium heterogenity porcinnych brucelovych postvakcinacnach protilatok gelovou filtracion na sephadexe G-200, *Vet. Cas.*, 14, 65, 1972.
9. **Hajdu, S., Beseda, I., and Polck, B.,** Some physico-chemical and serological properties of *Brucella* IgG_1 and IgG_2 antibodies of cattle, *Vet. Med. (Praha)*, 19, 141, 1974.
10. **Corbel, M. J.,** Characterization of antibodies active in the Rose Bengal plate test, *Vet. Rec.*, 90, 484, 1972.
11. **Corbel, M. J.,** Identification of the immunoglobulin class active in the Rose Bengal plate test for bovine brucellosis, *J. Hyg.*, 70, 779, 1972.
12. **Corbel, M. J.,** The nature of the antibody response to *Yersinia enterocolitica* serotype IX in cattle, *J. Hyg.*, 71, 309, 1973.
13. **Diaz, R. and Levieux, D.,** Role respectif en serologie de la brucellose bovine des antigenes et des immunoglobulins G1 et G2 dans les tests d'agglutination, de Coombs et au Rose Bengale ainsi que dans le phenomene de zone, *C. R. Acad. Sci.*, 274D, 1593, 1972.
14. **Beh, K. J.,** Distribution of *Brucella* antibody among immunoglobulin classes and a low molecular weight antibody fraction in serum and whey of cattle, *Res. Vet. Sci.*, 14, 381, 1973.
15. **Beh, K. J.,** Quantitative distribution of *Brucella* antibody amongst immunoglobulin classes in vaccinated and infected cattle, *Res. Vet. Sci.*, 17, 1, 1974.
16. **Rice, C. E., Tailyour, J., and Cochrane, D.,** Ultracentrifugal studies of sera from cattle vaccinated or naturally infected with *Brucella abortus, Can. J. Comp. Med. Vet. Sci.*, 30, 270, 1966.
17. **Rice, C. E., Cochrane, D., and Tailyour, J.,** Electrophoretic studies of sera from cattle vaccinated or naturally infected with *Brucella abortus, Can. J. Comp. Med. Vet. Sci.*, 30, 161, 1966.
18. **Rice, C. E., Alexander, D. C., and Barrett, B. B.,** Chromatographic studies of sera from calves vaccinated with *Brucella abortus* Strain 19, *Can. J. Comp. Med. Vet. Sci.*, 31, 114, 1967.
19. **Rice, C. E. and Boyes, B.,** Serum immunoglobulins in bovine brucellosis, *N.Z. Vet. J.*, 19, 146, 1971.
20. **Renoux, G., Plommet, M., and Philippon, A.,** Brucellose bovine experimentale. VIII. Evolution des anticorps agglutinants ou fixant le complement apres inoculation par *Brucella abortus* au 5^e-6^e mois et demi de la gestation, *Ann. Rech. Vet.*, 2, 159, 1971.
21. **Renoux, G., Plommet, M., and Phillippon, A.,** Microreactons d'agglutination et de fixation du complement pour le diagnostic des brucelloses, *Ann. Rech. Vet.*, 2, 263, 1971.
22. **Levieux, D.,** Immunoglobulines bovines et brucellose. I. Purification des immunoglobulines et preparation de leur antiserums specifiques, *Ann. Rech. Vet.*, 5, 329, 1974.
23. **Levieux, D.,** Immunoglobulines bovines et brucellose. II. Activite des IgG_1, IG_2 et IgM du serum dans les reactions d'agglutination, de fixation du complement et dans le test au Rose Bengale, *Ann. Rech. Vet.*, 5, 343, 1974.
24. **Alton, G. G., Maw, J., Rogerson, B. A., and McPherson, G. G.,** The serological diagnosis of bovine brucellosis. An evaluation of the complement fixation, serum agglutination and Rose Bengal tests, *Aust. Vet. J.*, 51, 57, 1975.
25. **Allan, G. S., Chappel, R. J., Williamson, P., and McNaught, D. J.,** A quantitative comparison of the sensitivity of serological tests for bovine brucellosis to different antibody classes, *J. Hyg.*, 76, 287, 1976.
26. **Patterson, J. M., Deyoe, B. L., and Stone, S. S.,** Identification of immunoglobulins associated with complement fixation, agglutination and low pH buffered antigen tests for brucellosis, *Am. J. Vet. Res.*, 37, 319, 1976.
27. **Erler, W. and Schimmel, D.,** Untersuchungen an Rindersera, die in der Brucellose-Langsamagglutination verdachtig reagieren. IV, *Mh. Vet. Med.*, 22, 212, 1967.

28. **Chappel, R. J., Allan, G. S., Williamson, P., and McNaught, D. J.,** A quantitative comparison of the sensitivity of serological tests for bovine brucellosis to different antibody classes, *Victorian Vet. Proc.*, 34, 48, 1976.
29. **Nielsen, K. and Duncan, J. R.,** Bovine IgM: does it fix guinea pig complement in the absence of bovine complement components?, *Vet. Immunol. Immunopathol.*, 14, 335, 1987.
30. **Plackett, P. and Alton, G. G.,** A mechanism for prozone formation in the complement fixation test for brucellosis, *Aust. Vet. J.*, 51, 374, 1975.
31. **McNaught, D. J., Chappel, R. J., Allan, G. S., Bourke, J. A., and Rogerson, B. A.,** The effects of IgG_2 and of antigen concentration on prozoning in the complement fixation test for brucellosis, *Res. Vet. Sci.*, 22, 194, 1977.
32. **Lawman, M. J. P., Ball, D. R., Hoffman, E. M., Desjardin, L. E., and Boyle, M. D. P.,** Production of *Brucella abortus* — specific protein A-reactive antibodies (IgG_2) in infected and vaccinated cattle, *Vet. Microbiol.*, 12, 43, 1986.
33. **Crawford, R. P. and Hidalgo, R. J., Eds.,** *Bovine Brucellosis: an International Symposium,* Texas A&M University Press, College Station, 1977.
34. Report, Animal Health Services in Great Britain 1968, Her Majesty's Stationery Office, London, 1968, 47.
35. **Kovacic, H.,** Ucestalost pojave nespecifinih aglutinina za Brucele u serumu goveda., *Vet. Arh.*, 40, 50, 1970.
36. **Corbel, M. J.,** Recent advances in the study of *Brucella* antigens and their serological cross reactions, *Vet. Bull.*, 55, 927, 1985.
37. **Nielsen, K., Duncan, J. R., Stemshorn, B., and Ruckerbauer, G.,** Relationship of humoral factors (antibody and complement) to immune responsiveness, resistance and diagnostic serology, in *The Ruminant Immune System,* Butler, J. E., Ed., Plenum Press, New York, 1981, 367.
38. **Nielsen, K. H., Samagh, B. S., Speckmann, G., and Stemshorn, B.,** The bovine immune response to *Brucella abortus*. Elimination of some sporadic serological reactions by the chelation of divalent cations, *Can. J. Comp. Med.*, 43, 420, 1979.
39. **Nielsen, K. and Duncan, J. R.,** Demonstration that non-specific bovine *Brucella abortus* agglutinin is EDTA-labile and not calcium-dependent, *J. Immunol.*, 129, 366, 1982.
40. **Nielsen, K., Stilwell, K., Stemshorn, B., and Duncan, R.,** Ethylenediaminetetracetic acid (disodium salt)-labile bovine immunoglobulin M Fc binding to *Brucella abortus*: a cause of nonspecific agglutination, *Clin. Microbiol.*, 14, 32, 1981.
41. **MacMillan, A. and Cockrem, D. S.,** Reduction of non-specific reactions to the *Brucella abortus* serum agglutination test by the addition of EDTA, *Res. Vet. Sci.*, 38, 288, 1985.
42. **Corbel, M. J.,** Serological relationship between *Brucella* spp. *Yersinia enterocolitica* serotype IX and *Salmonella* serotypes of Kauffmann-White group N., *J. Hyg.*, 75, 151, 1975.
43. **Cioglia, L.,** Agglutinazione della *S. urbana* nell brucellosi e suo valcre diagnostico, *Boll. Soc. Ital. Biol. Sper.*, 24, 1117, 1948.
44. **Corbel, M. J.,** Comparison of *Brucella abortus* and *B. melitensis* antigens for the Rose Bengal plate test on sera from cattle infected with *B. abortus* biovar-5, *Vet. Rec.*, 117, 385, 1985.
45. **Corbel, M. J., Stuart, F. A., and Brewer, R. A.,** Observations on serological cross-reactions between smooth *Brucella* species and organisms of other genera, *Dev. Biol. Stand.*, 56, 341, 1984.
46. **Corbel, M. J. and Wray, C.,** Effect of natural infection with *Salmonella urbana* on the serological status of cattle in relation to tests for brucellosis, *Br. Vet. J.*, 131, 324, 1975.
47. **Nielson, K., Samagh, B. S., and Stemshorn, B.,** Agglutination of *Brucella abortus* cells by sera from cattle experimentally infected with *Escherichia coli, Vet. Microbiol.*, 5, 123, 1980.
48. **Stuart, F. A. and Corbel, M. J.,** Identification of a serological crossreaction between *Brucella abortus* and *Escherichia coli* 0:157, *Vet. Rec.*, 110, 202, 1982.
49. **Ahvonen, P. and Sievers, K.,** *Yersinia enterocolitica* infection associated with *Brucella* agglutinins. Clinical features of 24 patients, *Acta Med. Scand.*, 185, 121, 1969.
50. **Ahvonen, P., Jansson, E., and Aho, K.,** Marked cross-agglutination between brucellae and a subtype of *Yersinia enterocolitica, Acta Pathol. Microbiol. Scand.*, 75, 291, 1969.
51. **Corbel, M. J. and Cullen, G. A.,** Differentiation of the serological response to *Yersinia enterocolitica* and *Brucella abortus* in cattle, *J. Hyg.*, 68, 519, 1970.
52. **Hurvell, B., Ahvonen, P., and Thal, E.,** Serological cross-reactions between different *Brucella* species and *Yersinia enterocolitica:* agglutination and complement fixation, *Acta Vet. Scand.*, 12, 86, 1971.
53. **Akkermans, J. P. W. M. and Hill, W. K. W.,** *Yersinia enterocolitica* (serotype 9) infection as a disturbing factor in serological demonstration of *Brucella* infection in swine, *Tijdschr. Diergeneeskd.*, 96, 1654, 1971.
54. **Mittal, K. R. and Tizard, I. R.,** A simple technique to differentiate between animals infected with *Yersinia enterocolitica* IX and those infected with *Brucella abortus, Res. Vet. Sci.*, 26, 248, 1979.
55. **Corbel, M. J. and Philip, J. I. H.,** Relationship of *Brucella abortus* agglutinogenic antigens to the receptor sites for Tbilisi phage, *Res. Vet. Sci.*, 13, 91, 1972.

56. **Mittal, K. R. and Tizard, I. R.,** Serological cross-reactions between *Brucella abortus* and *Yersinia enterocolitica* serotype 09, *Vet. Bull. Weybridge,* 51, 501, 1981.
57. **Corbel, M. J.,** in International Symposium on Human and Animal Brucellosis, Taipei, Taiwan, 1984.
58. **King, B. N. and Frank, N. A.,** Effect of age on resistance and retention of titre in cattle vaccinated with Strain 19 *Brucella abortus* vaccine, *J. Am. Vet. Med. Assoc.,* 139, 100, 1961.
59. **Redman, D. R., Deyoe, B. L., and King, N. B.,** Resistance of cattle to *Brucella abortus* following vaccination at two and three months of age, *J. Am. Vet. Med. Assoc.,* 150, 403, 1967.
60. **Hill, W. K. W.,** Een Vergelijking tussen de Widal-reactie en een gemodificeerde complement-binding bij brucellose van runderen, *Tijdschr. Diergeneeskd.,* 85, 3, 1960.
61. **Hill, W. K. W.,** Standardization of the complement fixation test for brucellosis, *Bull. Off. Int. Epiz.,* 60, 401, 1963.
62. **Burki, F.,** Standardisierung der Komplementbindungsreaktion auf Brucellose, *Zentralbl. Bakteriol. Parasitenkd. Infektionskr. Hyg. Abt. 1,* 183, 225, 1961.
63. **Burki, F.,** Emploi de l'epreuve de deviation du complement pour distinguer les animaux infectes de Brucellose de ceux qui ont ete vaccines contre la maladie, *Bull. Off. Int. Epizoot.,* 60, 419, 1963.
64. **Cunningham, B.,** The national brucellosis programme of the Republic of Ireland, in *Bovine Brucellosis: an International Symposium,* Crawford, R. P. and Hidalgo, R. J., Eds., Texas A&M University Press, College Station, 1977, 356.
65. **Shimizu, T. and Shibata, S.,** A technique of agglutination test with R-type *Brucella* as antigen, *Natl. Inst. Anim. Health Q.,* 2, 15, 1962.
66. **Corbel, M. J. and Bracewell, C. D.,** The serological response to rough and smooth Brucella antigens in cattle vaccinated with *Brucella abortus* strain 45/20 adjuvant vaccine, in Regamey, R. H., Hulse, E. C., and Valett, L., Eds., *International Symposium on Brucellosis (II),* Rabat, 1975, Karger, Basel, *Dev. Biol. Stand.,* 31, 351, 1976.
67. **Roerink, J. H. G.,** Investigation into the usefulness of the nonagglutinogenic *Brucella abortus* adjuvant vaccine Duphavac NA in the control of bovine brucellosis, *Vet. Rec.,* 80, 727, 1967.
68. **Hendricks, J. B. and Ray, W. C.,** Studies on cattle vaccinated with live strain 19 and/or 45/20 adjuvant *Brucella abortus* vaccines, *Proc. Annu. Meet. U.S. Anim. Health Assoc.,* 74, 147, 1971.
69. **Joseph, P. G., Cheah, P. F., Arunsaalam, V., and Hant, T. C.,** Studies on bovine brucellosis. I. Serum antibody response following calfhood vaccination with live S19 and killed 45/20 adjuvant vaccines, *Kajian Vet.,* 8, 63, 1976.
70. **Kiss, Z.,** Experimental study of the Duphavac N.A. (BR 45/20) brucellosis vaccine. I. Serological responses of animals in brucellosis free herds, *Acta Vet. (Budapest),* 21, 171, 1971.
71. **Kiss, Z.,** Vizsgalatok a bengal-voros targylemezprola ertekelesere, *Mag. Allatorv. Lapja,* 30, 182, 1975.
72. **Hall, W. T. K., Ludford, C. G., and Ward, W. H.,** Infection and serological responses in cattle given 45/20 vaccine and later challenged with *Brucella abortus, Aust. Vet. J.,* 52, 409, 1976.
73. **McKeon, F. W.,** A recent trial comparing two 45/20 adjuvant *Brucella* vaccines, in Regamey, R. H., Hulse, E. C., and Valett, L., Eds., *International Symposium on Brucellosis (II),* Rabat, 1975, Karger, Basel, *Dev. Biol. Stand.,* 31, 343, 1976.
74. **Wisniowski, J., Krolak, M., and Drozdzynzka, M.,** Standardization trial of the antiglobulin (Coombs) test applied in the serological diagnosis of brucellosis in cattle, *Bull. Off. Int. Epizoot.,* 75, 217, 1971.
75. **Corbel, M. J.,** The immune response to *Brucella abortus* 45/20 adjuvant vaccine in terms of immunoglobulin class, in Regamey, R. H., Hulse, E. C., and Valette, L., Eds., *International Symposium on Brucellosis (II),* Rabat, 1975, Karger, Basel; *Dev. Biol. Stand.,* 31, 141, 1976.
76. **Cunningham, B.,** The Coombs test for brucellosis in cattle, *Vet. Rec.,* 80, 527, 1967.
77. **Cunningham, B.,** Serological response in cattle following vaccination with S19 and killed *Brucella* 45/20 adjuvant vaccine, *Vet. Rec.,* 82, 7, 1968.
78. **Beh, K. J. and Lascelles, A. K.,** The use of the antiglobulin test in the diagnosis of bovine brucellosis, *Res. Vet. Sci.,* 14, 239, 1973.
79. **Kiss, Z.,** Value of antiglobulin test in brucellosis eradication in a one-year field study, *Acta Vet. (Budapest),* 20, 365, 1970.
80. **Roerink, J. H. G.,** Development of a non-agglutinogenic killed *Brucella abortus* adjuvant vaccine and its applicability in the control of bovine brucellosis, *Thesis, Rijksuniversiteit, Utrecht,* 1966.
81. **Cunningham, B.,** Vaccination of cattle with killed 45/20 adjuvant vaccine. Effects of serological and milk ring tests when used in cattle previously exposed to infection or vaccinated with S19, *Vet. Rec.,* 86, 2, 1970.
82. **Waghela, S. and Philpott, M.,** Serological response of cattle in Kenya vaccinated with a killed *B. abortus* 45/20 adjuvant vaccine, *Vet. Rec.,* 99, 505, 1976.
83. **Joseph, P. C., Arumsaalam, V., and Kim, L. G.,** Studies on bovine brucellosis. II. Serum antibody response following re-vaccination with killed 45/20 adjuvant vaccine and adult vaccination with live S19 vaccine, *Kaijian Vet.,* 9, 8, 1977.

84. **de Keyser, P. J. and Florent, A.,** Vergelijkende studie van de serologische reakties en de immuniteit bij kalveren ingeent tegen brucellose met stam 19 en niet agglutinogene entstoffen, *Vlaams Diergeneeskd. Tijdschr.*, 36, 175, 1967.

85. **McDiarmid, A.,** Vaccination against brucellosis with particular reference to the use of killed adjuvant vaccines prepared from the non-agglutinogenic Strain 45/20 *Brucella abortus*, in *Some Diseases of Animals Communicable to Man in Britain: Proceedings of a Symposium Organized by the British Veterinary Association and the British Small Animal Veterinary Association,* London, Graham-Jones, O., Ed., 1968, 281.

86. **Morgan, W. J. B. and McDiarmid, A.,** Adjuvant vaccines prepared from killed *Brucella abortus* Strain 45/20, *Vet. Rec.*, 83, 184, 1968.

87. **Alton, G. G., Elberg, S. S., and Crouch, D.,** Rev. 1 *Brucella melitensis* vaccine. The stability of the degree of attenuation, *J. Comp. Pathol.*, 77, 293, 1967.

88. **Jones, L. M., Entessar, F., and Ardalan, A.,** Comparison of living vaccines in producing immunity against natural *Brucella melitensis* infection in Iran, *J. Comp. Pathol.*, 74, 17, 1964.

89. **Alton, G. G., Fensterbank, R., Plommet, M., and Verger, J. M.,** La brucellose de la chevre, in *Les Maladies de la Chevre Les Colloques de l'INRA No. 28,* INRA, Paris, 1984, 69.

90. **Morgan, W. J. B.,** The serological diagnosis of bovine brucellosis, *Vet. Rec.*, 80, 612, 1967.

91. **Angus, R. D. and Barton, C. E.,** The production and evaluation of a buffered plate antigen for use in a presumptive test for brucellosis, *Dev. Biol. Stand.*, 56, 349, 1984.

92. **Rose, J. E. and Roepke, M. H.,** An acidified antigen for detection of nonspecific reactions in the plate-agglutination test for bovine brucellosis, *Am. J. Vet. Res.*, 18, 550, 1957.

93. **Nicoletti, P.,** Utilization of the card test in brucellosis eradication, *J. Am. Vet. Med. Assoc.*, 151, 1778, 1967.

94. **Nicoletti, P. and Muraschi, T. F.,** Bacteriologic evaluation of serologic test procedures for the diagnosis of brucellosis in problem cattle herds, *Am. J. Vet. Res.*, 27, 689, 1966.

95. **Pietz, D.,** *Brucella* antigens and serologic test results, in *Bovine Brucellosis: an International Symposium,* Crawford, R. P. and Hidalgo, R. J., Eds., Texas A&M University Press, College Station, 1977, 40.

96. **Morgan, W. J. B. and Richards, R. A.,** The diagnosis, control and eradication of bovine brucellosis in Great Britain, *Vet. Rec.*, 94, 510, 1974.

97. **Gower, S. G. M., Wright, E. C., Davies, G., Morgan, W. J. B., Hopkinson, W. I., Gibbs, D. F., and Bennet, E. J.,** An automated Rose Bengal agglutination test using the ADAM system, *Vet. Rec.*, 95, 544, 1974.

98. **Alton, G. G., Jones, L. M., Angus, R. D., and Verger, J. M.,** *Techniques for the Brucellosis Laboratory,* l'Institut National de la Recherche Agronomique, Paris, in press.

99. **Hendry, D. M. F. D., Corbel, M. J., Bell, R. A., and Stack, J. A.,** Brucella antigen production and standardization, Ministry of Agriculture, Fisheries and Food, Alnwick, U.K., 1985.

100. **Morgan, W. J. B.,** *Brucella abortus* (review), in *Handbuch der bacteriellen Infektionen bei Tieren,* Vol. 4, Blobel, H. and Schliesser, T., Eds., Fischer, Frankfurt am Main, 1982, 53.

101. **Corbel, M. J.,** Comparison of *Brucella abortus* and *B. melitensis* antigens for the Rose Bengal plate test on sera from cattle infected with *B. abortus* biovar 5, *Vet. Rec.*, 117, 385, 1985.

102. **Morgan, W. J. B.,** The diagnosis of *Brucella abortus* infection in Britain, in *Bovine Brucellosis: an International Symposium,* Crawford, R. P. and Hildago, R. J., Eds., Texas A&M University Press, College Station, 1977, 21.

103. **Hornitzky, M. and Searson, J.,** The relationship between the isolation of *Brucella abortus* and serological status of infected, non-vaccinated cattle, *Aust. Vet. J.*, 63, 172, 1986.

104. **Stryszak, A.,** Serological activity of cattle anti-*Brucella abortus* immunoglobulin in Rose Bengal Plate Test (RBPT), *Pol. Arch. Weter.*, 26, 7, 1986.

105. **Agrimi, P., Andreani, E., Legrottaglie, R., and Redini, S.,** Il test a Rosa Bengala nella diagnosi sierologica di brucellosi bovina, *Ann. Fac. Med. Vet. Univ. Studi Pisa,* 30, 109, 1977.

106. **Contini, A., Coni, V., and Casu, A.,** Diagnosi sierologica di brucellosi negli ovini e caprini con l'antigene tamponato, *Atti. Soc. Ital. Sci. Vet.*, 27, 640, 1973.

107. **Fensterbank, R. and Maquere, M.,** Assainissement d'un troupeau ovin atteint de brucellose par les moyens de la prophylaxie sanitaire en utilisant l'epreuve au Rose Bengale, *Recl. Med. Vet.*, 154, 657, 1978.

108. **Falade, S.,** A comparison of three serological tests in the diagnosis of caprine brucelloses, *Res. Vet. Sci.*, 24, 376, 1978.

109. **Falade, S.,** Some observations on the use of the Rose Bengal plate, tube agglutination, heat inactivated and rivanol tests in caprine brucellosis, *Trop. Vet.*, 1, 49, 1983.

110. **Shimi, A. and Tabatabayi, A. H.,** Pathological, bacteriological and serological responses of ewes experimentally infected with *Brucella melitensis, Bull. Off. Int. Epizoot.*, 93, 1411, 1981.

111. **Trap, D. and Gaumont, R.,** Comparaison entre electrosynerese et epreuves serologiques classiques dans le diagnostic de la brucellose ovine, *Ann. Rech. Vet.*, 13, 33, 1982.

112. **Waghela, S., Wandera, J. G., and Wagner, G. G.,** Comparison of four serological tests in the diagnosis of caprine brucellosis, *Res. Vet. Sci.*, 28, 168, 1980.
113. World Health Organisation, Joint FAO/WHO Expert Committee on Brucellosis, 6th Rep., Tech. Rep. Ser. No. 740, 1986.
114. **Dohoo, I. R., Wright, P. F., Ruckerbauer, G. M., Samagh, B. S., Robertson, R. J., and Forbes, L. B.,** A comparison of five serological tests for bovine brucellosis, *Can. J. Vet. Res.*, 50, 485, 1986.
115. **Huber, J. D. and Nicoletti, P.,** Comparison of ten results of card, Rivanol, complement-fixation, and milk ring tests with the isolation rate of *Brucella abortus* from cattle, *Am. J. Vet. Res.*, 47, 1529, 1986.
116. **Nicoletti, P.,** Use of 45/20 bacterin to detect latent infection in brucellosis, in *Bovine Brucellosis: an International Symposium,* Crawford, R. P. and Hidalgo, R. J., Eds., Texas A&M University Press, College Station, 1977, 72.
117. **Wright, A. E. and Smith, F.,** Application of serum test to the differential diagnosis of typhoid and Malta fever, *Lancet,* 1, 656, 1897.
118. **Stableforth, A. W.,** Brucellosis, in *Infectious Disease of Animals: Diseases Due to Bacteria, Vol. 1,* Stableforth, A. W. and Galloway, I. A., Eds., Butterworths, London, 1959, 53.
119. **Alton, G. G.,** Development and evaluation of serological tests, in *Bovine Brucellosis: an International Symposium,* Crawford, R. P. and Hidalgo, R. J., Eds., Texas A&M University Press, College Station, 1977, 61.
120. **Jones, L. M.,** *Brucella* antigens and serologic test results, in *Brucellosis: an International Symposium,* Crawford, R. P. and Hidalgo, R. J., Eds., Texas A&M University Press, College Station, 1977, 40.
121. **Lambert, G. and Amerault, T. E.,** Comparative study of three serologic tests for detecting the response in cattle to virulent *Brucella abortus, Am. J. Vet. Res.*, 23, 529, 1962.
122. **Stableforth, A. W.,** Laboratory methods for detection of brucellosis, in *Advances in Control of Zoonoses, WHO Monogr. Ser.*, 19, 89, 1953.
123. **Morgan, W. J. B., Mackinnon, D. J., Gill, K. P. W., Gower, S. G. M., and Norris, P. I. W.,** *Brucellosis Diagnosis — Standard Laboratory Techniques,* 2nd ed., Central Veterinary Laboratory, Weybridge, U.K., 1978.
124. **Alton, G. G., Jones, L. M., and Pietz, D. E.,** Laboratory techniques in brucellosis, 2nd ed., *WHO Monogr. Ser.*, 55, 1975.
125. **Collins, J. M., Boyce, K. J., and Edgar, A. W.,** Preparation of the living vaccine from continuous culture produced cells, in Regamey, R. H., Hennessen, H., Ikic, D., and Ungar, J., Eds., *Int. Symp. on Biotechnical Developments in Bacterial Vaccine Production,* Stockholm, Karger, Basel; *Symp. Ser. Immunobiol. Stand.*, 3, 171, 1967.
126. **Boyce, K. J., Edgar, A. W., and Maley, A. D.,** the continuous culture of *Brucella abortus* (Strain 19), *Mitt. Versuchsstn, Gaerungsgewerbe Wien,* 21(11—12), 122, 1967.
127. **Kotsche, W.,** Der Einfluss des pH — Wertes auf den Ausfall der Brucellose Serumlangsam-agglutinations-Reaktion, *Mh. Vet. Med.*, 22, 14, 1967.
128. **Huddleson, I. F.,** *Brucellosis in Man and Animals,* New York Commonwealth Fund, New York, 1939, 204.
129. **Feinberg, R. J. and Wright, G. G.,** Factors influencing the agglutination titration in human brucellosis, *J. Immunol.*, 67, 115, 1951.
130. **Alton, G. G.,** Standardization of agglutinating antigens for the diagnosis of brucellosis, *Res. Vet. Sci.*, 12, 330, 1971.
131. **Farrell, I. D. and Robertson, L.,** The isolation of *Brucella abortus* from milk, I. Biological examination (guineapig inoculation), *Mon. Bull. Min. Health Pub. Health Lab. Serv.*, 26, 52, 1967.
132. World Health Organization, Joint FAO/WHO Expert Committee on Brucellosis, 6th Rep., *WHO Tech. Rep. Ser.*, 68, 9, 1953.
133. World Health Organization, Joint FAO/WHO Export Committee on Brucellosis, 2nd Rep., *WHO Tech. Rep. Ser.*, 67, 7, 14, 16, 1953.
134. **Davidson, I., Hebert, C. N., and Morgan, W. J. B.,** The Second International Standard for anti-*Brucella abortus* serum, *Bull. W.H.O.*, 40, 129, 1969.
135. **Shimizu, T.,** Relationship between concentration and antigen titer in *Brucella abortus* agglutination suspensions, *Natl. Inst. Anim. Health Q.*, 1, 212, 1961.
136. **Berman, D. T.,** Discussion, in *Bovine Brucellosis: an International Symposium,* Crawford, R. P. and Hidalgo, R. J., Eds., Texas A&M University Press, College Station, 1977, 82.
137. **Renoux, G. and Alton, G.,** Etude sur la brucellose ovine et caprine. IV. Reactions serologiques dans le sang et le lait de chevres recemment infectees par *Br. melitensis, Arch. Inst. Pasteur Tunis,* 32, 523, 1955.
138. **Stemshorn, B. W., Forbes, L. B., Eaglesome, M. D., Nielsen, K. H., Robertson, F. J., and Samagh, B. S.,** A comparison of standard serological tests for the diagnosis of bovine brucellosis in Canada, *Can. J. Comp. Med.*, 49, 391, 1985.
139. **Tewes, S.,** Untersuchungen uber Ursachen und Erkennung unspezifischer Reaktionen bei der serologischen Diagnose der Schweine brucellose, *Dissertation, Hannover,* University of Hannover, 1986.

140. **Scheibner, E.,** Titriplexplasma in der Brucellosediagnostik, Berlin, *Muench. Tieraerztl. Wochenschr.,* 89, 281, 1976.
141. **Nielsen, K. H., Duncan, J. R., and Stemshorn, B. W.,** Non-specific reactions to the *Brucella abortus* SAT (letter), *Vet. Rec.,* 116, 550, 1985.
142. **MacMillan, A. P. and Bell, R. A.,** Non-specific reactions to the *Brucella abortus* SAT (letter), *Vet. Rec.,* 116, 139, 1985.
143. **Sacco, T. F.,** Indagine sieroepidemiologica sulla brucellosi con il test al Rosa Bengala (card test), *Atti Soc. Ital. Buiatria,* 8, 141, 1976.
144. **George, L. W. and Carmichael, L. E.,** Development of a Rose Bengal stained plate test antigen for the rapid diagnosis of *Brucella canis* infection, *Cornell Vet.,* 68, 530, 1978.
145. **Badakhsh, F. F., Carmichael, L. E., and Douglass, J. A.,** Improved rapid slide agglutination test for presumptive diagnosis of canine brucellosis, *J. Clin. Microbiol.,* 15, 286, 1982.
146. **Carmichael, L. E. and Joubert, J. C.,** A rapid slide agglutination test for the serodiagnosis of *Brucella canis* infection that employs a variant (M-) organism as antigen, *Cornell Vet.,* 77, 3, 1987.
147. **M'Fadyean, J. and Stockman, S.,** in Report of the Departmental Committee to Inquire into Epizootic Abortion. Part I. Epizootic abortion in cattle, (Cd.4742)., Appendix to Part I (Cd 4863), Board of Agriculture and Fisheries, 1909.
148. **Bertschinger, H. U.,** Erfahrungen mit der hollandischen Technik der Komplement-bindungsreaktionsdiagnose der Rinderbrucellose, *Schweiz. Arch. Tierheilkd.,* 103, 260, 1961.
149. **MacKinnon, D. J.,** The complement fixation test in brucellosis, *Bull. Off. Int. Epizoot.,* 60, 383, 1963.
150. **Jones, L. M., Hendricks, J. B., and Berman, D. T.,** The standardization and use of the complement-fixation test for the diagnosis of bovine brucellosis, with a review of the literature, *Am. J. Vet. Res.,* 24, 1143, 1963.
151. **Morgan, W. J. B., Davidson, I., and Hebert, C. N.,** The use of the second International Standard for anti-*Brucella abortus* serum in the complement fixation test, *J. Biol. Stand.,* 1, 43, 1973.
152. **Krolak, M., Truszczynski, M., and Blaszczyk, B.,** Distinct fixation of defined amounts of complement by the *Brucella abortus* antigen and bovine antibodies under different incubation conditions in the CFT, *Comp. Immunol. Microbiol. Infect. Dis.,* 9(4), 325, 1986.
153. World Health Organization, WHO Expert Committee on Biological Standardization. 23rd Rep., *Tech. Rep. Ser.,* 463, 22, 1971.
154. World Health Organization, WHO Expert Committee on Standardization, *Tech. Rep. Ser.,* 594, 18, 1976.
155. **Davidson, I. and Hebert, C. N.,** International Standard for anti*Brucella abortus* serum: comparison of the complement-fixing activity of the first and second International Standard and the EEC Standard, *Bull. W.H.O.,* 56, 123, 1978.
156. **Gargani, G.,** Preparation and titration of *melitensis* and *abortus* antigen for the complement fixation test at the Italian Brucellosis Centre, *Proc. 8th Int. Congr. Permanent Sect. Microbiol. Stand. Ass. Microbiol. Soc. Berne.,* Karger, Basel; Progr. Immunobiol. Stand. 1, 189, 1964.
157. **Trbic, B.,** Specificity of the CF reaction using chloroform extracted *Brucella* antigen, *Acta Vet. (Belgrade),* 10, 103, 1960.
158. **Elberg, S. S.,** Immunization against Brucella infection. VII. Immunological and epidemiological studies in Cordoba, Spain, *Bull. W.H.O.,* 20, 1033, 1959.
159. **Pilet, C., Toma, B., and Bonneau, M.,** Sur le choix d'une technique de fixation de complement adaptee au diagnostic de la brucellose, *Ann. Inst. Pasteur (Paris),* 113, 947, 1967.
160. **Farina, R.,** Current serological methods in *B. melitensis* diagnosis, in *Brucella melitensis,* Verger, J. M. and Plommet, M.,Eds., Nijhoff, for Comm. European Commun., Luxembourg, 1985, 139.
161. **Plackett, P. and Alton, G. G.,** A mechanism for prozone formation in the complement fixation test for brucellosis, *Aust. Vet. J.,* 51, 374, 1975.
162. **Burki, F.,** Die differentialdiagnostische Beurteilung der Seren von 1000 Problemkuhen mittels der Komplementbindungsreaktion auf Brucellose, Berlin, Muench., *Tieraerztl. Wochenschr.,* 73, 444, 1960.
163. **Schaaf, A. van der, and Jaartsveld, F. H. J.,** The relative immunizing value of mucoid L and S19 strains of *Brucella abortus* with comparative observations on the agglutination and complement fixation tests, *Vet. Rec.,* 74, 1070, 1962.
164. **Waveren, G. M. van,** The value of the complement fixation test for the eradication of bovine brucellosis in the Netherlands, *Bull. Off. Int. Epizoot.,* 63, 1015, 1965.
165. **Wisniowski, J.,** The role of the complement fixation test in the serological diagnosis of brucellosis in cattle, *Br. Vet. J.,* 120, 15, 1964.
166. **Wisniowski, J. and Krolak, M.,** Ocena metody odezynu wiazania dopetniacza stosowanej w polsce w rozpoznawaniu brucelozy bydta w swietle propozyeji standardu miedzynarodowego, *Pol. Arch. Weter.,* 9, 637, 1966.
167. **Redaelli, G.,** Isolamento delle brucelle da bovine con reazione positiva alla fissazione del complemento, *Atti Soc. Ital. Sci. Vet.,* 18, 620, 1964.

168. **Redaelli, G.,** Sull'isolamento di microorganismi del genere *Brucella* dal latte e dagli organi di bovine reagenti positivamente alla fissazione del complemento, *Arch. Vet. Ital.*, 16, 287, 1965.
169. **Preuss, S.,** Kritische Bewertung der Ergebnisse von Blutuntersuchungen im Rahmen der Bruzellosebekampfung 1956—1963 mit besonderer Berucksichtigung der Komplementbindung, Inaug.-Diss., Tierarztl. Fak. Munchen, 1965.
170. **Goyon, M.,** L'epreuve de fixation du complement dans le diagnostic de la brucellose bovine: application simultanee dans la Sarthe de l'epreuve de fixation et de l'epreuve d'agglutination lente, *Bull. Mens. Soc. Vet. Prat. Fr.*, 51, 218, 1967.
171. **Nicoletti, P.,** Further evaluation of serologic test procedures used to diagnose brucellosis, *Am. J. Vet. Res.*, 30, 1811, 1969.
172. **Nicolas, J. A. and Pestre-Alexandre, M.,** Resultats compares des deux reactions de sero-agglutination de Wright et de fixation du complement dans le diagnostic serologique de la brucellose, *Rev. Med. Vet.*, 122, 17, 1971.
173. **Mylrea, P. J.,** The diagnosis of brucellosis in dairy herds, *Aust. Vet. J.*, 48, 369, 1972.
174. **McPherson, G. G.,** Rapid screening tests for the diagnosis of bovine brucellosis by the complement fixation test, *Aust. Vet. J.*, 50, 108, 1974.
175. **Mylrea, P. J. and Fraser, G. C.,** The use of supplementary tests in the serological diagnosis of bovine brucellosis, *Aust. Vet. J.*, 52, 261, 1976.
176. **Miller, J. K., Kelly, J. I., and Roerink, J. H. G.,** A complement fixation method for quantitative differentiation of reactions to 45/20 vaccine and brucella infection, *Vet. Rec.*, 98, 210, 1976.
177. **Timbs, D. V., Moxham, J. W., and Liberona, H. H.,** The use of automated complement fixation techniques in the brucellosis eradication scheme, *N. Z. Vet. J.*, 26, 52, 1978.
178. **Meyer, M. E.,** Use of automated complement fixation screening test for serodiagnosis of bovine brucellosis, in *Proc. 83rd Annu. Meet. U.S. Animal Health Assoc.*, San Diego, California, 1979, 81.
179. **Meyer, M. E. and Kasten, R. W.,** Comparison of results of the automated complement fixation screening tests with conventional serologic tests in two dairy herds — one with recent and one with chronic brucellosis, in *Proc. 84th Annu. Meet. U.S. Animal Health Assoc.*, Louisville, Kentucky, 1980, 128.
180. **Worthington, R. W., Weddell, W., and Penrose, M. E.,** A comparison of three serological tests for the diagnosis of *B. ovis* infection in rams, *N. Z. Vet. J.*, 32, 58, 1984.
181. **Worthington, R. W., Stevenson, B. J., and de Lisle, G. W.,** Serology and semen culture for the diagnosis of *Brucella ovis* infection in chronically infected rams, *N. Z. Vet. J.*, 33, 84, 1985.
182. Animal Health Division (New Zealand), The complement fixation test for the diagnosis of *Brucella ovis* infection in rams, *N. Z. Vet. J.*, 31, 157, 1983.
183. **Riezu-Boj, J. I., Moriyon, I., Blasco, J. M., Marin, C. M., and Diaz, R.,** Comparison of lipopolysaccharide and outer membrane protein-lipopolysaccharide extracts in an enzyme-linked immunosorbent assay for the diagnosis of *B. ovis* infection, *J. Clin. Microbiol.*, 23, 938, 1986.
184. **Burgess, G. W. and Norris, M. J.,** Evaluation of the cold complement fixation test for the diagnosis of ovine brucellosis, *Aust. Vet. J.*, 59, 23, 1982.
185. **Hajdu, S.,** La specificite et la valeur des epreuves a l'antiglobuline, modifiees par l'auteur, dans le diagnostic de la brucellose des bovins, *Bull. Off. Int. Epizoot.*, 60, 433, 1963.
186. **Krolak, M.,** Bodonia nod wartoschia odezynu antyglobinowego (Coombsa) w rozpoznawaniu brucelozy bydla, *Pol. Arch. Weter.*, 17, 95, 1974.
187. **Buchmeiser, R.,** Beitrag zur serologischen Diagnostik der Rinderbrucellose mittels der Serum-Langsamagglutination, der Komplementbindungsreaktion und des Coombs-Schnell-testes, *Rindertuberk., Brucell.*, 10, 143, 1961.
188. **Fenske, G.,** Empfindlichkeit und Spezifitat serologischer Untersuchungsverfahren auf Rinderbrucellose, *Mh. Vet. Med.*, 32, 18, 1977.
189. **Fenske, G.,** Der Wert von Antiglobulintest und indirektem Fluoreszenztest für die serologische Diagnose der Rinderbrucellose, *Arch. Exp. Vet. Med.*, 31, 211, 1977.
190. **Lewkowicz, H.,** Odezyn antyglobulinowy w rozpoznaniu brucelozy bydla, *Med. Weter.*, 28, 536, 1972.
191. **le Pennec, J. and Goyon, M.,** Contribution a l'etude des anticorps antibrucelliques bovins par le Coombs au serum chauffe et al deviation du complement, *Bull. Mens. Soc. Vet.*, 49, 299, 1965.
192. **Anczykowski, F.,** Studies on the usefulness of the antiglobulin test (AG) in cattle brucellosis, *Pol. Arch. Weter.*, 26, 47, 1986.
193. **Anderson, R. K., Jenness, R., Brumfield, H. P., and Gough, P.,** *Brucella*-agglutinating antibodies. Relation of mercaptoethanol stability to complement fixation, *Science*, 143, 1334, 1964.
194. **Anczykowski, F. and Murat-Skwarek, P.,** On the value of the Rivanol test in routine diagnosis of bovine brucellosis, *Pol. Arch. Weter.*, 15, 205, 1972.
195. **Badnjevic, B.,** Znacaj rivanol testa u pronalazenju bruceloznih reaktora kod svinja, *Vet. Glas.*, 19, 315, 1965.
196. **Morse, E. V., Schneider, D. W., and McNutt, S. H.,** The effect of incubation at 56°C on the tube agglutination test for bovine brucellosis, *Am. J. Vet. Res.*, 16, 269, 1955.

197. **Shibata, S., Suzuki, K., and Isayama, Y.,** The stability of *Brucella* agglutinin serum, *Natl. Inst. Anim. Health Q.*, 1, 74, 1961.
198. **Amerault, T. E., Manthei, C. A., Goode, E. R., and Lambert, G.,** A heat inactivation test for differentiating specific and non-specific agglutination reactions for bovine brucellosis, *Am. J. Vet. Res.*, 22, 564.
199. **Amerault, T. E., Lambert, G., and Manthei, C. A.,** The heat stability of *Brucella* agglutins in bovine serum, *Am. J. Vet. Res.*, 23, 1023, 1962.
200. **Lambert, G., Amerault, T. E., and Rose, J. E.,** Studies on *Brucella* agglutinins in bovine serum, *Proc. Annu. Meet. U.S. Livestock Sanit. Assoc.*, 66, 104, 1963.
201. **Hellmann, E. and Hein, W.,** Untersuchungen an Rinderseren aus brucelloseverseuchten Bestanden und Impfbestanden mittels der Serumlangsamagglutination und des Hitzetestes, *Dtsch. Tieraerztl. Wochenschr.*, 67, 459, 1960.
202. **Beinhauer, W.,** Beitrag zur Differenzierung fraglicher Reaktionen in der Agglutination auf Abortus Bang mittels sog. Hitzetest, *Dtsch. Tieraerztl. Wochenschr.*, 72, 369, 1965.
203. **Klein, H. and Langkamp, J.,** Der Hitze-Test in der Brucellose Diagnostik, *Dtsch. Tieraerztl. Wochenschr.*, 71, 384, 1964.
204. **Pathak, R. C. and Singh, C.,** Comparative study of standard and supplemental serological tests for the diagnosis of brucellosis in herds of varying infection status, *Veterinarian*, 6, 35, 1969.
205. **Kunter, E.,** Die diagnostische Aussagekraft des Hitzetestes in der Serodiagnostik der Rinderbrucellose, *Mh. Vet. Med.*, 29, 135, 1974.
206. **Sutra, L., Caffin, J. P., and Dubray, G.,** Role of milk immunoglobulins in the *Brucella* milk ring test, *Vet. Microbiol.*, 12, 359, 1986.

Chapter 9

PRIMARY BINDING TECHNIQUES FOR THE SERODIAGNOSIS OF BOVINE BRUCELLOSIS: ENZYME IMMUNOASSAY

Peter F. Wright, Klaus H. Nielsen, and Walter A. Kelly

TABLE OF CONTENTS

I. INTRODUCTION

Enzyme immunoassay (EIA) techniques have been extensively applied in veterinary medicine to the detection of antibody for the indirect or presumptive diagnosis of infection. To a somewhat lesser extent, EIA techniques have been applied to the detection of antigens for direct demonstration of pathogens. This chapter will focus primarily on aspects of EIA development, standardization, and application as related to the serological diagnosis of bovine brucellosis.

Detection of antibody has been the mainstay of bovine brucellosis eradication, control, and surveillance programs throughout the world. Based on serological test results, decisions have been made to slaughter, quarantine, depopulate, or vaccinate large numbers of cattle. A number of "conventional" serological techniques (described in Chapter 5) have been successfully applied in these specific programs. The choice of conventional techniques and their combination in testing is essentially driven by the impact of diagnostic sensitivity and specificity on program objectives. Individually, these assays, although useful, are limited with respect to their diagnostic capabilities, thus, the need for serial and/or parallel testing. In addition, no combination of conventional tests has yet been successfully applied in diagnostic programs for the discrimination of vaccinated from infected cattle, an age-old problem in brucellosis serology.

EIA techniques offer a multitude of well-known advantages over conventional and even other primary binding techniques.[90] In particular reference to bovine brucellosis, perhaps the most important considerations are related to the inherent diagnostic flexibility of EIA techniques and their adaptability to meet exacting diagnostic requirements of sensitivity, specificity, and discrimination.

II. ASSAY TECHNIQUES FOR ANTIBODY DETECTION

A. INDIRECT ENZYME IMMUNOASSAY

The indirect EIA, initially described as the enzyme-linked immunosorbent assay (ELISA) nearly two decades ago,[30] is still, perhaps, the most commonly used EIA technique for the detection of antibody. Its principle is well known and has been discussed elsewhere.[57,59] The extent to which the indirect EIA has been applied to brucellosis serology is demonstrated in Table 1. This table also summarizes two important factors that are pertinent to the diagnostic performance of the indirect EIA: antigen and antiglobulin.

A wide variety of antigen preparations have been used in the indirect EIA, ranging from whole cells to crude and semipurified smooth lipopolysaccharide (SLPS) preparations to polysaccharide and protein extracts. The antigens of the SLPS are likely the most immunodominant in all of these preparations with the possible exception of the polysaccharide and protein extracts. Coupled with the fact that SLPS is very strongly absorbed by most plastics,[61] the major reaction detected in the indirect EIA is most likely that of the SLPS antigens and bovine antibody.

The diagnostically relevant antigenic epitopes appear to reside on the O chain component rather than the lipid A or core components of the SLPS. The structure of the O chains of both *B. abortus* 1119-3 and *B. melitensis* 16M have recently been characterized and compared.[9] Both O chains consist of a linear homopolymer of 4,6-dideoxy-4-formamido-α-D-mannose. Whereas the O chain of *B. abortus* 1119-3 is linked throughout by 1,2-glycosidic linkages, every fifth linkage in the O chain of *B. melitensis* 16M is substituted by a 1,3-glycosidic linkage. However, it is important to realize that the frequency of 1,3-glycosidic linkages is variable between the O chains of other *Brucella* species and strains, but does not appear to occur more frequently than in the O chain of *B. melitensis* 16M.[94] Information derived from monoclonal antibodies, raised against the cross-reacting organism, *Yersinia*

TABLE 1
Compendium of Indirect EIA Techniques Used for Antibody Detection in Brucellosis Diagnosis

Antigen preparation	Antiglobulin specificity	Ref.
Whole cell	Gamma globulin	86, 87
	IgG	14
	IgG,G_1,A	85
	NS[a]	74
Sonicated	Gamma globulin	76
	IgG,M	62
	NS[a]	75
Autoclaved	Gamma globulin	38, 77, 86, 87
	IgG	13, 36, 37
	IgG,G_2,M	5
	IgG,G_2,M,A	11
Filtered	IgG	6
SLPS	Gamma globulin	76
	IgG	28, 59, 93
	IgG_1,G_2	81
	IgG_1,G_2,M	45, 61
	IgG_1,G_2,M,A	60
Poly B	IgG_1,G_2,A	7
Protein	IgG	83
NS[a]	Gamma globulin	88

[a] NS = not specified.

enterocolitica 0:9 (to be discussed in further detail), and used in precipitation studies, would suggest that two basic types of epitopes exist on the O chain. Some monoclonal antibodies are capable of precipitating purified preparations of both O chain and SLPS, while others will precipitate only SLPS. Those capable of precipitating O chain are likely directed to multivalent epitopes along the length of the chain. Given the tendency of purified SLPS to form micelles in aqueous solution, those capable of precipitating SLPS only are likely directed toward epitopes that are univalent and may represent terminal portions of the O chain.[10]

Monoclonal antibodies raised independently against *B. abortus* 1119-3 and against *B. melitensis* 16M further suggest that there are both shared and unique determinants for both the multivalent and univalent epitopes on the O chains of these two *Brucella* strains.[94] It could be hypothesized that the shared, multivalent epitope is related to those segments of their respective O chains having in common a series of 1,2-glycosidic linkages. The unique multivalent and univalent epitopes of both strains are likely related to the 1,3-glycosidic linkages. Experiments are currently underway to determine the position of these linkages on the O chain and their effect on these epitopes.[94]

The antigenic cross-reactivity of other Gram-negative bacteria with *Brucella* is well known and is discussed in Chapter 2 and 8 and will only be discussed here from a diagnostic point of view. The cross-reactivity between *Y. enterocolitica* 0:9 and *B. abortus* 1119-3 is understandable in that their O chains are identical in all respects.[15] This would suggest that the SLPS from *Yersinia* could be substituted in the indirect EIA for the serological diagnosis of *Brucella* infection. In a comparison of SLPS preparations from both bacteria in the indirect EIA, we have observed that some sera from apparently healthy cattle will preferentially react with the SLPS of *Yersinia* and not with that of *Brucella*. This is supported by monoclonal antibody studies[10] where it is suggested that this may be related to epitopes residing in the core region of the *Yersinia*. Because of the risk of false positive reactions, the SLPS of *Y. enterocolitica* 0:9 is not a suitable substitute for use in the indirect EIA. Other cross-reactions are related to N-acyl derivatives of 4-amino-4,6-dideoxy-D-mannose units present in the O chains of some bacteria.[9] Antibody to these antigens will cause false positive reactions not only in conventional assays, but in the indirect EIA using *Brucella* SLPS as well. Therefore, the use of SLPS in the indirect EIA does not circumvent diagnostic problems related to antibody elicited by cross-reacting antigens.

It is highly likely that both the multi- and univalent epitopes of the SLPS of *B. abortus* are functional in the indirect EIA. If an epitope which was unique to the *Brucella* species O chain could be "functionally isolated" and used in the EIA, then possibly the diagnostic specificity could be markedly improved. Given that the O chains of *Brucella* (and *Y. enterocolitica* 0:9) are unbranched homopolymers, then a synthetic epitope defined by a linear sequence of 1,2-linked, 4,6-dideoxy-4-formamido-α-D-mannose might serve well as a highly specific diagnostic antigen in the indirect EIA. The N-acyl derivatives found in the O chains of other bacteria do not appear to elicit antibodies which cross-react with this shared, multivalent epitope of *Brucella*.[94] Another diagnostic approach lies in the use of monoclonal antibodies in competitive EIA techniques and will be discussed in the next section. Sera from cattle vaccinated against *B. abortus* are also reactive with SLPS in the indirect EIA, and this, too, will be discussed in the next section.

Nevertheless, the many epitopes of SLPS make it a highly sensitive antigen for use in brucellosis serology. As diagnostic antigen, it has been used over a broad range of purities in the indirect EIA, as indicated in Table 1. We have used both highly purified SLPS and semipurified preparations and have not found them to differ significantly in terms of diagnostic performance. However, it is important to point out that high diagnostic specificity becomes more critical as the prevalence of disease decreases to very low levels and that the purity of the SLPS preparation used may have a bearing on the number of nonspecific reactions observed. We prefer to use a semipurified preparation based on the hot water/phenol extraction method of Redfearn.[1,59] We suspect that highly purified SLPS preparations, having a tendency to form spontaneous micelles in aqueous solution, adsorb irregularly to the surface of polystyrene and result in unacceptable assay variability and high negative backgrounds. Less-purified preparations may contain LPS-associated proteins[53,64] that may inhibit micelle formation yet do not interfere with diagnostic performance. In our hands, this semipurified SLPS preparation is extremely stable and well suited for use in the indirect EIA.

Antibody isotypes produced in response to *B. abortus* infection in cattle include IgM, IgG_1, IgG_2, and IgA.[2,3,4,12,43,49,50,68,69,72,73] Although IgM may be the major antibody isotype produced very early in the immune response, the switch to the IgG isotypes occurs very rapidly in our experience. In addition, we have observed that in serum the IgG_1 response is highly prevalent, whereas, the IgG_2 response is far more variable. Again, all four isotypes are present in milk,[4,23] although IgG_1 and/or IgA predominate. All isotypes are theoretically detectable in the indirect EIA depending on the immunological specificity of the antiglobulin used. As may be seen in Table 1, antiglobulins ranging from broad to single isotype specificities have been used. For a global estimate of antibody activity, a Coomb's reagent with specificity for all heavy (H) and both light (L) chains could be used. Although not true Coomb's-type reagents, anti-γ globulin reagents have been used (Table 1). However, this type of reagent tends to introduce a bias in favor of detection of IgM which is notorious for its nonspecific reactivity in many assays. This bias could be reduced by using an anti-IgG (H + L) reagent which would still be capable of detecting IgM (and IgA) based on anti-L chain activity but would favor the detection of the IgG isotypes which are likely more specific. Broad immunological specificity may also be attained by using an anti-L chain reagent, but care must be taken to ensure that it is capable of detecting both kappa and lambda L chains. Monospecific (anti-H chain) reagents have also been used in the indirect EIA. Given the high prevalence of IgG_1 in both serum and milk, a monospecific reagent of this immunological specificity may be the best choice favoring high diagnostic specificity.

Given the diversity and multiplicity of epitopes on SLPS, it is likely that the immune response to this antigen will be very heterogeneous with respect to the antibody isotypes elicited and their relative specificities and affinities. In the indirect EIA, two important factors must be considered. First, there may be competition among the various isotypes for binding to a limited amount of SLPS adsorbed to the microplate well. It has been reported that IgM antibody activity could not be assessed due to competition by other isotypes of higher concentration.[45] Second, the extent of competition may be affected not so much by the relative concentration of these isotypes but by their relative affinities[11] for specific multi- and univalent epitopes. Therefore, it should not be assumed that the specificity(ies) of an antiglobulin reagent will guarantee detection of all or any one isotype(s). In all likelihood, the indirect EIA favors the detection of high affinity antibody to the immunodominant epitopes of the SLPS. This would suggest that the indirect EIA using SLPS is probably best suited for the detection of high-affinity IgG_1, the most prevalent isotype in the immune response to *Brucella* infection. Unfortunately, the immunodominant epitopes of SLPS do not confer to the indirect EIA the unique ability to discriminate between cattle which are infected and those which have been vaccinated or exposed to cross-reacting bacterial antigens. However, the choice of antiglobulin does have an effect in reducing the number of false positive reactions detected in the indirect EIA. We have extensively field-tested three antiglobulins: a polyclonal anti-IgG (H + L), a mouse monoclonal anti-L chain (kappa and lambda), and a mouse monoclonal anti-IgG_1. Our data (which will be presented in a later section) indicate that the latter reagent is optimal when used in combination with SLPS in the diagnostic indirect EIA. Our protocol for the indirect EIA using these reagents is presented in the appendix of this chapter.

B. COMPETITIVE ENZYME IMMUNOASSAY

Competitive EIAs are generally based on a competition between two antibody populations for a limited number of antigenic determinants. Thus, the most common type of assay involves the immobilization of antigen on a matrix and co-incubation of test material (serum) and an antibody of known specificity and quantity which is conjugated with an enzyme. Displacement of the known antibody would reflect the presence of antibody in the test sample and would ultimately result in less substrate utilization. The degree of competition is directly proportional to the amount of antibody in the unknown sample except when conjugate displacement is total.

There are two approaches to enhancing the specificity of the competitive EIA. One is to select an exquisitely specific antibody, such as a monoclonal, to a single epitope on the antigen and the second is to select a highly purified and specific antigen. Although both approaches have been tried in the detection of antibody to *B. abortus*, particularly in attempts to discriminate the antibody response of field-infected from strain 19-vaccinated cattle, no specific antigenic epitope has yet been isolated which will allow discrimination. The use of monospecific antibody has been successful, particularly but not exclusively, in assays employing monoclonal antibodies. Schurig[79] and Schurig et al.[78] have reported a monoclonal antibody capable of discriminating antibody responses induced by *B. abortus* and a cross-reactive *E. coli* using *B. abortus* SLPS as the antigen. The possibility of using a monoclonal antibody for a competitive test for antibody to *B. abortus* was also suggested by Quinn et al.[66]

In a limited study, Rylatt et al.[76] compared the complement fixation test (CFT), the indirect EIA, and the competitive EIA, using different monoclonal antibodies. Using alkali-treated SLPS as antigen, monoclonal antibodies of four different specificities (as described

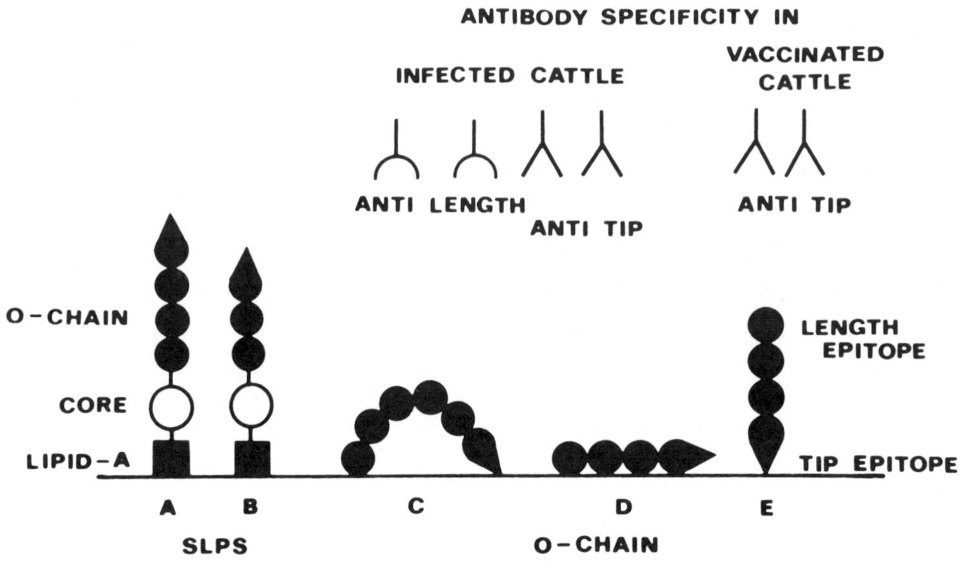

FIGURE 1. Proposed antigen attachment to a "medium binding" polystyrene microplate. Some of the various configurations of antigen binding for SLPS and its O chain (PS antigen). Antibody to the O chain "tip" epitope found in sera from both vaccinated and infected cattle can interact with the SLPS antigen. Antibody with specificity for the "length" epitope, presumably found only in sera from infected cattle, can also interact with the SLPS antigen. In contrast, the "tip" epitope may not be exposed in the attachment of the purified O chain to "medium binding" polystyrene (although it appears to be exposed when "high-binding" polystyrene is used). Therefore, immune complex formation can only occur with sera from infected cattle since only they contain antibody to the "length" epitope. Intact SLPS, presumably attached to the matrix by the Lipid A region, is depicted in A and B. In C, the tip and the core end of the O chain are attached to the polystyrene. In D, the O chain is lying flat, and in E, the O chain is attached only by the tip. In all instances, the epitope associated with the tip of the molecule is occluded, resulting in no reactivity with sera from vaccinated cattle.

by Bundesen et al.)[8] were tested against a panel of 30 sera from vaccinated and control cattle, some of which had been challenged with virulent bacteria and field-infected cattle. In summary, the indirect EIA and three of the four monoclonals used in the competitive EIA demonstrated much higher diagnostic sensitivities than the CFT. However, this was offset by diagnostic specificities which were considerably lower than the CFT. The fourth monoclonal matched the CFT in terms of both diagnostic sensitivity and specificity.

Our studies[54,56] of a competitive EIA for discrimination of the antibody response of field infected from strain 19-vaccinated animals makes use of a polysaccharide (PS) derivative of the SLPS as antigen rather than intact SLPS itself. The PS is predominantly O chain; however, a small portion of the core of the SLPS still remains attached after the isolation and purification process. The PS appears to interact sufficiently well with polystyrene to allow its hydrophobic binding in the absence of linkage reagents as previously reported.[7] This finding may be related to the method and purity of the PS preparation (outlined in Chapter 2). Differences between the SLPS and the PS antigens in their interactions with polystyrene appear to play an important role in the actual discrimination of the antibody response of field-infected from vaccinated cattle. In our hands, discrimination can only be obtained if "medium binding" polystyrene microplates are used for antigen attachment. If "high binding" polystyrene microplates are used as the antigen matrix, the PS behaves similarly to SLPS, and discrimination is not possible. A proposed explanation for this phenomenon is depicted in Figure 1. The use of "low binding" capacity polystyrene results in an unacceptable increase in the coefficient of variation between duplicate samples. The monoclonal antibody used in this competitive EIA was derived from a fusion of murine plasmacytoma cells with spleen cells from a mouse immunized with *Yersinia enterocolitica* 0:9 SLPS[10] and conjugated with horseradish peroxidase.[39] The details of the assay procedure are outlined in the appendix following this chapter.

In a preliminary study,[54] groups of 25 sera of various titers (determined by the indirect EIA and by conventional serological tests) were selected from sera of field cattle either vaccinated or infected and from sequential serum samples from cattle experimentally vaccinated with *B. abortus* strain 19 or infected with strain 2308. All were compared by indirect EIAs using SLPS and PS antigens and a competitive EIA using PS antigen only. Antibody was not detected by the indirect or competitive EIAs using PS antigen in the 25 sera from the field-vaccinated cattle. Of the 25 sera from field-infected cattle, all were considered positive by the competitive EIA, while two gave doubtful reactions by the indirect PS EIA. Sequential samples from an experimentally vaccinated cow did not at any time react with the PS antigen in either competitive or indirect EIA. Serum samples taken sequentially from a cow experimentally infected with strain 2308 became positive 1 to 2 weeks earlier in the competitive EIA and the indirect SLPS EIA than in the indirect PS EIA. On the basis of these and other data (not reported) the competitive EIA using PS antigen is presently being evaluated for use as a confirmatory test.

While the exact mechanism of discrimination is unknown, it has been known for some time that the "poly-B" antigen reacted only with sera from infected cattle in the gel diffusion test.[27,42] The "poly-B" antigen, while consisting primarily of polyglucose, contains a small amount of the PS, which is presumably the active component in gel diffusion (see Chapter 1). The large amount of polyglucose likely interferes with the binding of the PS to polystyrene for EIA purposes, and a satisfactory sensitization of polystyrene had not been obtained until now that the PS is available in a purified form. Our hypothesis[54] takes into account the antigen presentation both to the animal and as an antigen on the surface of the microplate well. Thus, when cattle are vaccinated with strain 19, they have the ability to overcome the temporary infection and are therefore exposed to the antigens of *B. abortus* only for a relatively short period of time. Given that this particular monoclonal will precipitate both the SLPS and O chains (PS) of *B. abortus* 1119-3, *B. melitensis* 16M, and *Y. enterocolitica* 0:9, it is probably specific for the multivalent, linear epitope defined by the 1,2-glycosidic linkages along the "length" of the O chain. We therefore suggest that this epitope is not highly immunodominant and that only after prolonged exposure, such as that offered by infection, will a humoral response be elicited. On the other hand, vaccinated as well as infected cattle likely respond to (an) immunodominant epitope(s) which we would suggest may represent the univalent epitope(s) at the "tip" of the O chain, because sera from vaccinated cattle do not react in the gel diffusion test, which requires antibody to multivalent epitopes for precipitation to occur. As previously mentioned, antigen presentation on the surface of the microplate well is important. The indirect PS EIA would suggest that the "tip" epitopes are no longer available for binding when the PS is adsorbed to "medium binding" polystyrene (see Figure 1), whereas the "length" epitope is functionally exposed. Both epitope types are likely functional when PS is adsorbed to "high binding" polystyrene, and sera from both vaccinated and infected cattle react alike in the indirect PS EIA (analogous to SLPS). The role of antibody affinity may be important as well and remains to be determined.

In our ongoing assessment of the competitive EIA, we have observed an inability to differentiate approximately 15% of sera from strain 19-vaccinated cattle. These sera gave results comparable to those obtained with sera from field infected animals. There could be a number of reasons for this. First, the high degree of competition may be a result of actual infection with strain 19. Of 18 sera obtained from cattle from which strain 19 was isolated, all behaved as sera from infected cattle (i.e., competed strongly). Second, these reactive sera from vaccinates may have been obtained at the peak of the antibody response to strain

19, in which case transient competition may occur. Third, a small percentage of individual cattle may produce antibody of very high affinity which could compete better than the population average. Fourth, some cattle may have a mixed infection. A second problem may arise with the sensitivity of our present competitive EIA in that it may not detect antibody as early in the response to *B. abortus* infection as the indirect EIA using SLPS antigen. This is possibly a result of low affinity antibody, synthesized very early in the response, not being able to compete efficiently with the monoclonal antibody. This reaction pattern could be mistaken for that of a strain 19-vaccinated animal.

The competitive EIA does offer several advantages in that it requires fewer steps than indirect EIA, it negates the need for antispecies reagents, it lends itself to automation, and it is capable of differentiating at least 85% of vaccinated cattle from field infected cattle.

C. OTHER PRIMARY BINDING ASSAYS

There are four principal variables which differ among primary binding assays for antibody detection. First, the formats may differ: indirect, competitive, or capture types. Second, the detection systems may vary: enzyme-substrate interaction, radioisotope enumeration, fluorescence quantitation, or a combination of enzyme conversion and fluorogenic substrate quantitation. Of the above, most have been applied to *Brucella abortus* serology with varying degrees of success. The indirect and competitive EIAs have been discussed in detail and will therefore be omitted. Third, antigen preparations are highly variable. Since the antigens of *B. abortus* have been described in previous sections and in detail in Chapter 2, they will not be dealt with further. It suffices to say that as a rule, the antigens applied in a given assay should be as well defined as possible. A very important variable is the source of the antibody used as the detection reagent. Needless to say, monoclonal antibodies with their inherent qualities lend themselves to assay standardization, interlaboratory comparison of data, and intralaboratory quality control. Rather than attempting to evaluate all the various primary binding assays described in the literature, we have presented in brief a cross-section of the types, the detection systems, the antigens which have been applied in brucellosis serology (Table 2).

Generally, primary binding assays that involve radioisotopes are less suited for large-scale diagnostic use due to the inherent problems with radioactive materials. Individually, fluorescent immunoassays and enzyme immunoassays are very useful with the combination of the two, enzyme immunoassay utilizing a fluorogenic substrate, being the most sensitive. All of the various primary binding assays described in the literature involve immobilization of reagents on a solid matrix (i.e., microplate) and prolonged washing procedures are necessary between steps. This technology lends itself moderately well to automation and is best suited for use in a laboratory setting. To increase cost efficiency and to reduce turn-around time in field testing, it will be necessary to develop more robust, less labor intensive, and more rapid assays that are at least as specific and sensitive as those developed for the laboratory. Ideal candidates for rapid field serology are homogeneous or nonseparation-based assays; however, none have presently been developed for brucellosis.

III. ASSAY DEVELOPMENT AND STANDARDIZATION

A. STANDARDIZATION

As stated in the introduction, EIA techniques demonstrate a high degree of inherent diagnostic flexibility and adaptability. Although we would propose that this is an all-important feature, it may also be responsible for the lack of standardization which is apparent in Table 1. However, these two factors, antigen and antiglobulin, only represent the tip of the iceberg with respect to standardization.

The majority of EIA techniques utilize 96-well polystyrene microplates. Literally dozens

TABLE 2
Compendium of Other Primary Binding Assays Used in Brucellosis Diagnosis

Assay type	Detection system	Antigen	Refs.
Indirect RIA[a]	^{125}I[b] mouse anti-bovine isotype	DMSO[c] extract on SRBC[d]	84
Indirect RIA	Rabbit anti-bovine + ^{125}I sheep anti-rabbit	Whole cells dried	48
Indirect RIA	^{125}I anti-isotype	Whole cell suspension	63
Indirect RIA	^{125}I protein A[e]	Whole cells dried to ps.	29
Indirect RIA	^{125}I protein A	BASA[f]	46, 47
Competitive RIA	^{125}I IgG_1 and IgG_2 or globulin prep.	Whole cells	16—21, 35
Capture RIA for antigen	^{125}I anti-*Brucella*	Antigen assay	89
Indirect FIA[g]	FITC anti-bovine[h]	Whole cells	24
Indirect FIA	FITC anti-bovine	SDS[i] extract on agarose	67
Inhibition FIA	FITC anti-*Brucella* or -*Yersinia*	Whole cells	24
Quantitative FIA (FIAX)	FITC anti-bovine	BASA	34
Quantitative FIA (TRACK XI)	FITC anti-bovine	Cell sonicate	33
Indirect ELISA amplified	HRPO[j] anti-guinea pig C3[k]	SLPS[l]	40, 41
Indirect ELISA amplified	Rabbit anti-bovine + βGL[m] goat anti-rabbit	Cell supernatant	6
Indirect ELISA "dipstick"	HRPO anti-bovine	SLPS	58
Indirect ELISA dot assay	HRPO anti-bovine	SLPS	71
Indirect ELISA fluorogenic	HRPO anti-bovine	Commercial	88
Sandwich ELISA	? Anti-bovine	MCAB[n] to *B. abortus* + antigen	31
Sandwich ELISA for antigen (fluorogenic)	Rabbit. anti-*Brucella* + HRPO goat anti-rabbit	Antibody on nylon beads	65

[a] RIA, radioimmunoassay.
[b] ^{125}I, iodine-125.
[c] DMSO, dimethyl sulfoxide.
[d] SRBC, sheep red blood cells.
[e] Protein A, staphylococcal protein.
[f] BASA, *Brucella abortus* soluble antigen.
[g] FIA, fluorescent immunoassay.
[h] FITC, fluorescein isothiocyanate.
[i] SDS, sodium dodecyl sulfate.
[j] HRPO, horseradish peroxidase.
[k] C3, third component of complement.
[l] SLPS, smooth lipopolysaccharide.
[m] βGL, beta galactosidase.
[n] MCAB, monoclonal antibody.

of these plates, from many manufacturers, are available throughout the world. These plates vary tremendously with respect to their antigen and antibody adsorption characteristics. As mentioned, plates are now available in broadly classified "high, medium, or low (protein) binding" capacities. Modification of the polystyrene polymer with the introduction of reactive groups, by proprietary means, is usually the basis for changing the adsorption characteristics. No plate is universally applicable to all EIA techniques, and to generalize would be misleading. However, the choice of plate will have an affect on the diagnostic performance of any assay. The interaction of the PS antigen with different polystyrene plates is a case in point. In our experience, "medium binding" plates are best suited for use with both the SLPS and PS antigens. A simple method for identifying this standard type of plate has been described.[91]

All separation-type EIA techniques utilize an assortment of buffers for coating, washing, dilution, and substrate addition. The common element in all of these is water. We have found that water quality is an important variable in assay standardization. So much so that we have established "downstream" water purification systems in all of our diagnostic EIA laboratories and have adopted a standard of Type 2, pyrogen-free water for all of our buffers. The degree of water treatment required will depend greatly on the ionic and organic content of the source water. For laboratory-based assays, a minimum requirement of distilled and heat- or filter-sterilized water may be acceptable. As mentioned, various buffers are used in EIA techniques, even for any given step in the assay procedure. We do not propose that any one buffer system is better than another as the choice of buffer composition and concentration may be, in part, related to water quality, antigen purity, plate type, test sample and dilution, and enzyme/substrate system. However, the most generally accepted buffer systems should be incorporated into a standard reference protocol by which all others could be compared for acceptability in terms of diagnostic performance. There are a number of other physicochemical factors which may affect assay performance and many of these have been discussed elsewhere.[59,90,91] Of these, temperature is perhaps one of the most important. All reactions which take place in the indirect EIA are temperature and, therefore, time dependent. Again, we would not propose that any given combination of temperature and incubation time is better than another. This will depend very much on the sample load, expected turnaround times, and equipment used. As with the buffer system, generally accepted temperatures and times should be incorporated into a standard reference protocol which is defined in terms of diagnostic performance.

As suggested in Table 1, there is a need to standardize the method of antigen preparation which is used in both the indirect and competitive EIA. A minimum requirement for SLPS and PS antigen purity should be established, and reference standards should be available for comparison. It is likely that more than one strain of *B. abortus* may be used for antigen preparation, but strains, such as S1119-3 for which the PS antigen has been characterized, should be chosen as a reference. The amount of antigen available for binding in the assay will be influenced by the concentration and volume of antigen placed in the well, as well as other factors previously discussed. At best, a generally accepted concentration and volume should be incorporated into a standard reference protocol.

Because all of the reactions involved in separation-based EIA techniques occur at a solid/fluid interface, concentration and volume become interdependent. The reactive surface area at the bottom of the microplate well remains constant, but the reactive surface area on the sides of wells will change as volumes are increased or decreased. Therefore, the surface area-to-volume ratio is not a simple relationship, and concentrations of not only antigen but test samples, conjugate, and substrate need to be adjusted accordingly if a standard of diagnostic performance is to be maintained.

In addition to the above considerations, the dilution of the test sample will have a dramatic effect on diagnostic sensitivity and specificity. In general, diagnostic specificity will decrease and sensitivity will increase as test sample dilution decreases and vice versa. This is also true of the competitive assay, although it differs in principle from the indirect EIA. It may be appropriate to consider different dilutions of test sample for use in a standard reference protocol, and these will be dependent on the proposed diagnostic applications.

Many enzyme labels have been used in separation-type EIA techniques, including horseradish peroxidase, alkaline phosphatase, beta galactosidase, urease, and others. We do not propose that any one label is universally superior to another, and the pros and cons of each may be found in the literature. One enzyme in particular, horseradish peroxidase, is probably, by far, the most widely used label. By virtue of this fact alone, it may be the logical choice for use in a standard reference protocol. However, it should not preclude the use of other enzyme labels. One complication associated with this particular enzyme is the broad choice hydrogen donors (the chromogen) which may be used with its substrate, hydrogen peroxide.[59] There are many considerations in the choice of chromogen which include availability, ease of preparation, stability, type of reading, environmental safety, and others. The choice of chromogen will also have a bearing on diagnostic sensitivity and specificity based on inherent reaction characteristics. In weighing all of these factors, we have chosen 2,2′-azino-di-(3 ethylbenzothiazoline-6-sulfonic acid), or ABTS, for our use, while some prefer *ortho*-phenylene diamine (OPD) or 5-aminosalicylic acid (5-AS) or others to a lesser extent. The choice is not simple, and again, the most generally accepted enzyme (and conjugation method) and substrate system should be incorporated into a standard reference protocol.

Given that an enzyme label has been chosen, one must then consider the antiglobulin reagent (or the particular antibody used in a competitive assay) and the conjugation procedures used. These problems can be resolved by the use of monoclonal antibodies because of the inherent advantages of such antibodies in terms of uniformity of specificity, reproducibility of production, and uniformity of labelling using a given conjugation procedure. Thus, identical antibody-enzyme conjugates can be produced indefinitely. It should be realized that monoclonal antibodies have certain disadvantages. In particular, the specificity to be used must be ascertained very carefully to ensure reactivity with immunoglobulin from all breeds of cattle, and storage of conjugates must be determined for each conjugate preparation as many are susceptible to freezing and can best be stored sterilized at 4°C. In spite of these shortcomings, monoclonal antibodies appear to be the most reasonable way to ensure uniformity of diagnostic performance between laboratories, given that agreement can be reached with respect to their specificities and that they are made available for use by all laboratories.

Most laboratories use at least positive and negative serum samples, run at frequent intervals, to ensure assay uniformity. Often more than one of each sample is included on each 96-well plate, and in our particular case, a target positive, a low positive, a negative, and a buffer control sample is included four times on each plate. The use of serum samples is associated with the same problems in assay standardization as the use of nonstandardized polyclonal antibody reagents for detection. Eventually all such reagents run out, and it is impossible to produce an exact replacement. To solve this problem, bovine monoclonal antibodies may be used. We have presently produced a stable mouse × bovine hybrid cell line which produces bovine monoclonal IgG_1 antibody with specificity for *B. abortus* SLPS. The IgG_1 produced by this cell line has no measurable reactivity with antibody to mouse immunoglobulins (heavy or light chains), and it is capable of fixing guinea pig complement (it may also be used as a standard reagent in the CFT for antibody to *B. abortus*). It is hoped that this particular reagent can be used to standardize indirect EIA procedures and data expression in the near future, and we would welcome an opportunity to collaborate with other laboratories in the thorough testing of this antibody as well as the monoclonal antibody-enzyme conjugates presently employed in our diagnostic laboratory.

B. DATA EXPRESSION AND HANDLING

The end result of any EIA technique is the enzymatic conversion of a substrate and the elaboration of a quantifiable product. Chromogenic substrate systems are the most commonly used, and although reactions may be observed by eye, they are most often quantitated using one of the many microplate photometers on the market today. Numerical raw data in the form of absorbance or optical density units must then be processed and/or expressed in terms that can be diagnostically interpreted.

For diagnostic application, single dilution assays are preferred for economic and logistic reasons. As single dilution assays, both the indirect and competitive EIA techniques are quantitative over a relatively limited range. However, this range is more than adequate for diagnostic purposes. Beyond this, a dilution series must be used if an accurate estimate of antibody activity is to be achieved. Many methods of data expression have been proposed and compared elsewhere.[25,26,32,36,44,51,70]

In its raw form, data is expressed in OD units. The OD unit is useful, from a seroepidemiological standpoint, in defining and monitoring healthy or vaccinated populations and establishing threshold criteria for determining seropositive and negative status. However, this form of data expression is not suitable for reporting to the field as interpretation requires an in-depth knowledge of individual assay performance. If initial interpretation is to be made by laboratory staff, then in their simplest form the data may be reported as negative, positive, or suspicious. We are currently using this form of data expression for our diagnostic indirect EIA. It is perhaps most suitable for screening assays given that confirmatory testing will be conducted on the same or a second test sample if required. This reduces the onus on field staff in the initial phase of diagnosis since this form of data expression may be applied to all EIA screening techniques for any disease and is easily interpreted without the need for knowledge of individual assay diagnostic criteria. However, alone it would impose certain limitations in the final decision-making process if serological data are to be combined with clinical signs and herd history by field staff.

Quantitative, or at least semiquantitative, expression of antibody activity from a single dilution assay is an important attribute of EIA techniques. In competitive EIA techniques, results are most often calculated as percent inhibition of the binding of an antibody-positive reference reagent (e.g., monoclonal). A mathematical baseline of 0% inhibition is usually defined by the reactivity of this reagent in the presence of an antibody-negative reference serum which should be run on every plate. The calculated results are then expressed on a relative scale of 0 to 100. In general, competitive assays are extremely sensitive in the lower range of detection of high-affinity antibody, and the difference between negative and positive may be far removed on the scale. This serves to expand the scale of low titered reactions and allows a greater flexibility in their interpretation.

A similar type of approach has been taken in the expression of data from the indirect EIA.[82,95] In this method, results are calculated as percent positivity relative to an antibody positive reference serum which should be run on every plate. Using the OD value of the positive reference as the denominator, all results are expressed on a relative scale of 0 to 100 plus. Again, this serves to expand the scale of low titered reactions and allows a greater flexibility in their interpretation.

The two methods just described are probably the simplest and easiest to calculate and standardize between laboratories. Each requires only one antibody-positive reference reagent. However, care should be taken in the choice of these reference reagents and the dilution used. It is important that these reagents do not saturate all available antigen binding sites as this will compromise the diagnostic sensitivities of these assays. It is also important that their reactions result in values which fall in the midrange of OD detection to ensure that the calculations are based on the linear portion of the detection curve. For example, in our current indirect EIA when antibody-positive sera with end-point titers greater than 1:800 are tested at a single dilution of 1:100, the antigen-antibody system begins to saturate, and OD values of greater than approximately 1.4 units become very erratic and are not predictive of antibody activity. Therefore, we have chosen a positive reference serum and dilution which repeatedly results in an OD of approximately 1.0. This is a very important consideration if this method of data expression is to be used since all results are expressed in terms which are relative to the performance of the antibody-positive reference.

The relative expression of antibody activity in both the indirect and competitive EIA techniques does serve to simplify data from a field perspective. It is semiquantitative on a unified scale of 0 to 100 in the diagnostically critical range of antibody activity. Diagnostic threshold ranges can be defined numerically for each assay type and disease. Changes in the levels of antibody activity are readily apparent. This form of data expression is also well suited for seroepidemiological studies.

The methods thus described are simple but not highly quantitative. If quantitative results are required, then it is necessary to use some form of dilution series to obtain a more accurate estimate of antibody activity, especially if a single point OD value is used as the measure. The simplest method is the classic end-point titration, and this may be applied to both the indirect and competitive techniques. End-point criteria should be established in the linear portion of the response curve. We tend to favor an end point which corresponds to the high OD value range, in the indirect EIA, expected from antibody-negative sera at the initial dilution of the titration series and the lower OD range in the competitive assay. This compensates for nonspecific factors which may be present at low but not high dilutions if negative serum is not used in the diluent buffer. The data are then simply expressed as the reciprocal or log of the reciprocal of the dilution which represents the end point.

End-point titrations are somewhat limited in that they do not quantitatively compensate for the effects of antibody concentration and antibody affinity. Two sera may give identical endpoints, but their titration curves may be markedly different. Whether or not this is important will depend on what level of sensitivity to change in antibody activity is required from the data. We have recently described for the indirect EIA a method which demonstrates a high degree of correlation with end-point titrations ($R = 0.975$).[91] The first two dilutions in the linear portion of titration series are used to define antibody activity as a function of slope and concentration. Although this is not a novel concept, we have developed a system whereby, in a series of only 5 fourfold dilutions, antibody activities with end points as great as 1:204,800 may be predicted on a continuous and linear scale. This eliminates the need for a lengthy dilution series and increases the sensitivity of measurement in that both the titration slope and the concentration of antibody are considered in the calculation of activity. This approach could also be adapted for the quantitative expression of antibody activity in the competitive EIA. It must be remembered that the measurement of antibody activity is influenced by a great number of factors. In the indirect EIA technique using SLPS antigen, antibody activity measurement is a function of a variety of epitopes and their exposure to binding, the affinities and concentrations of the immunoglobulin isotypes which compete for binding, and the isotype specificity(ies) of the conjugate in the detection of binding. In the competitive EIA technique using PS antigen, antibody activity measurement is a function of the availability of a particular epitope of a specificity defined by the monoclonal antibody reference reagent and the affinities and concentrations of the antibody isotypes which compete for binding. Therefore, if the quantitative expression of antibody activity is to be meaningful between laboratories, there is a need to standardize the many factors which will influence measurement.

C. QUALITY CONTROL AND AUTOMATION

The EIA is a labor-intensive test requiring steps that involve liquid handling (serum dilution and addition to microplate wells), reagent dispensing (antigen, conjugate, substrate), plate washing, incubation, shaking, and reading. In addition, 96 OD values are generated per microplate, and these data must be manipulated and reduced to a form which is usable by the researcher or diagnostician.

Hand-operated apparatus, such as single and multichannel pipettors, dispensers, and washers may be used to perform manually many of the steps in the EIA procedure. However, under conditions where the technician must process many test plates per day or where many other duties are also expected to be performed, the use of these apparatuses may prove too slow and too prone to human error. Interruptions, loss of concentration, unfamiliarity with the apparatus, and differences in technique between technicians are only a few of the factors that can contribute to error, variability, and difficulty in maintaining quality control.

Fortunately, with the advent of microcomputers and electronically controlled instrumentation designed to handle microtiter plates, much of the tedium and error associated with labor-intensive testing can be alleviated. Automated and semiautomated instrumentation can perform most of the steps in a microtiter EIA process more quickly, and with greater precision and accuracy, than can a human. Automation also allows much of the work involved in sample tracking, data reduction, and quality control to be handled by the microcomputer which can monitor the EIA process and product and perform any calculations, charting, and much of the decision-making required to determine both the status and acceptability of a test result.

The instrumentation which can be used in automating a microplate-based EIA falls into three broad categories:

1. Dedicated instruments designed to handle only one (or perhaps two) steps in the assay: this would include microprocessor-controlled plate washers, plate reagent dispensers, plate readers, and liquid-handling systems. These instruments can usually be operated in a stand-alone mode with selection of programs or options accomplished via buttoms located on the instrument itself or on an attached control unit. Some of the instruments may also be interfaced to or controlled by an external microcomputer, usually via a serial port. This feature is often found in plate readers and liquid-handling systems where a high degree of versatility in reading protocols and data reduction or in dilution and placement of samples may be required. This laboratory currently employs a number of dedicated instruments to automate the EIA techniques (Figure 2). A computer-controlled Gilson 212B Liquid Handling System is used to pick up, dilute, and deliver serum samples from vacutainer tubes into microplate wells. An eight-channel piston-type Autodrop from Flow Labs is used to dispense reagents into microplates, while a 96-well washer from the same company is used to wash the plates after each incubation period. The plates are read with a Multiskan MCC plate photometer from Flow Labs which is interfaced to an IBM microcomputer running in-house EIA software designed to control the action of the photometer, store data, calculate test results and statistics, and produce test reports (described in more detail later in this chapter). When automating an EIA protocol using dedicated instruments, the user is not obliged to purchase all the instruments from the same manufacturer but is free to choose the products which will best suit the particular assay and testing load. This is generally the most flexible and least expensive way of building a system, but it also offers the lowest degree of automation, since these instruments, even when purchased from the same manufacturer, are not designed to work together as a coordinated system. The user must still move the test plate from one instrument to the next and usually must initiate the action of each instrument by pressing a start button or pressing a key on the computer. Complete automation of an EIA procedure with this type of instrumentation is virtually impossible.
2. Integrated (all-in-one) instruments designed to take one microplate at a time through several steps of an EIA procedure: this type of instrument is illustrated by the Boehringer EIA Processor II and the Beckman Biomek 1000. The Biomek, for example, features an XYZ arm and a stable of modular tools each designed to perform one particular function such as dispensing, pipetting, washing, or photometry. Bench space is saved since many different functions are handled by one machine rather than spread out over a number of dedicated instruments. Operation of this type of instrument is often under external microcomputer control via proprietary software. Versatility and ease of use are as much a function of the software supplied by the manufacturer as they are of the design of the instrument itself. This sort of instrument is very useful if the entire EIA process can be completed on one plate before the next plate is started. However, if plates must be batched, unattended operation is not possible for longer than one function per plate. This is because the instrument is not capable of automatically moving one plate to a holding or incubation area while it performs the same

AUTOMATION OF INDIRECT *BRUCELLA* EIA

Instrument	**Function**	
Dispenser	Dispense antigen into plate	
	↓	Incubate overnight
Plate washer	Wash plate with buffer	
	↓	
Liquid handler	Dilute and dispense Control and Test samples into plate	
	↓	Incubate 3 h
Plate washer	Wash plate with buffer	
	↓	
Dispenser	Dispense conjugate into plate	
	↓	Incubate 1 h
Plate washer	Wash plate with buffer	
	↓	
Dispenser	Dispense substrate into plate	
	↓	Shake plate
Photometer + PC	Read plate at 4-min development; calculate "Target" time	
	↓	Shake plate
Photometer + PC	Read plate at "Target" time automatically	
	↓	
PC	Calculate and output test results	

Note: Dispenser = Titertek AutoDrop Dispenser;[a] Plate washer = Titertek Microplate Washer;[a] Photometer = Titertek Multiskan MCC Microplate Reader;[a] Liquid handler = Gilson 212B;[b] PC = IBM XT Microcomputer;[c] Plate shaker = Titertek Microplate Shaker;[a] Plates = NUNC Microplate 96F (2-69620).[d]

[a] Flow Laboratories Inc., McLean, VA.
[b] Gilson Medical Electronics Inc., Middleton, WI.
[c] IBM Corp., Boca Raton, FL.
[d] NUNC Inc., Naperville, IL.

FIGURE 2. Flowchart of the microplate-based indirect *Brucella* EIA showing the instrumentation currently used and the assay steps which it automates. The model name and manufacturer for each instrument, as well as for the microplates, are listed at the bottom of the flowchart.

function on the next plate. Also, because it is integrated, a breakdown in a key element such as the power supply or the XYZ arm could cripple the operation of the entire instrument. Generally, a high degree of unattended automation is possible, depending on the requirements of the particular EIA protocol and subject to the limitations described above for batch processing.

3. Robotic arm systems designed to perform many EIA operations or functions on many plates in any order: while these systems have in the past been designed for use in the analytical chemistry lab, manufacturers are now adding capabilities for microplate EIA work. Examples of robotic systems are the Zymate system from Zymark and the HP Genenchem system from Hewlett Packard. These particular systems feature a robot arm which is capable of XY and rotational-Z movement and a set of function modules or work stations positioned at various locations within the cylindrical area accessible by the arm. A module can be a multichannel dispenser/washer, a plate "hotel", a

plate shaker, etc. The arm can pick up and use various tools, such as single and multichannel pipettors, but it can also move microtiter plates between work stations and incubation areas. This is an important feature because it allows unattended batch processing of plates. Most robotic systems operate completely under the control of a microcomputer and/or proprietary controller and come with software that contains a built-in programming language and subroutines which are used to build an EIA protocol. Often a teaching pendant or similar device is included to help the user teach the robot arm the required movements and locations. The software is designed such that any change made to the physical system, such as the introduction of an additional tool or a change in location of a workstation may be accommodated with minimal reprogramming and little impact on existing protocols. Some systems will also allow interfacing to dedicated instruments such as bar code readers, liquid-handling systems, and plate photometers from other manufacturers; this may require some customization and programming on the part of the robot arm manufacturer. Robotic arm systems are potentially capable of automating an entire microplate-based EIA assay from sample dilution to test report output, thus reducing human intervention to replenishing sample racks and biological reagents and performing various system maintenance tasks.

Regardless of the type of instrumentation used to automate an EIA procedure, one must realize that all instruments have certain inherent limitations and maintenance requirements which must be taken into consideration when designing and implementing the system. It is important to look at the function that the instrument is performing to determine where potential problems may occur.

Systematic variability can be a problem with instruments that perform operations (such as dispensing, washing, and reading) on multiple (usually 8, 12, or 96) wells at a time.[91] Reagent dispensers can be periodically checked by dispensing, for example, 200 μl of $NiSO_4$ into each well and checking the volume visually or measuring the optical density (OD) at 414 nm with a plate reader. The function of plate washers can be checked by stopping them at various points in the wash cycle to check visually that all wells are being properly aspirated and filled with wash fluid and that no overflow is occurring that could cause cross-contamination of wells. Plate photometers generally employ fiber optics to blank and read on multiple wells. While this technology is highly reliable, some multichannel plate readers may exhibit small interchannel differences when reading wells containing fluid exhibiting a meniscus.[92] This effect can be minimized by ensuring that the photometer is blanked on fluid exhibiting the same type and degree of meniscus formation as the test samples and that the number of blanking wells matches the number of optic channels.

Plate shakers should exhibit uniform shaking across all wells of each microtiter plate to allow uniformity of substrate development and meniscus formation. Multiplate shakers may be more susceptible to this problem than single or dual plate shakers.

Carryover can be a problem with liquid-handling systems that use a nondisposable tip. This problem can be somewhat alleviated by using a liquid-sensing-type probe which can be programmed to draw a sample from just below the meniscus, thus minimizing sample carryover on the outside of the tip and by using high sample dilutions or by programming probe rinse cycles between sample pickups. In order to maintain accuracy and consistency, liquid-handling systems should not be relied on to aspirate sample volumes which are smaller than the recommended minimum for the given sample syringe size (often 5% of the total syringe volume) or for the particular probe (to avoid capillary action).

When using automated reagent dispensers, separate feed lines and dispensing heads and manifolds should be maintained, if possible, for each reagent (antigen, conjugate, substrate, buffer) and should be clearly marked. This would reduce, for example, the chance of spontaneous development occurring in a substrate which was being dispensed through a manifold that had previously been used for conjugate.

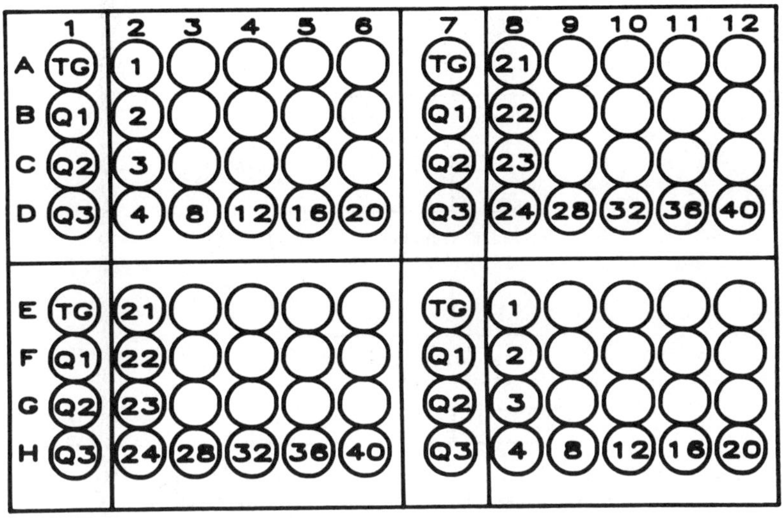

FIGURE 3. Sample placement configuration used in the indirect *Brucella* EIA: TG, target serum (high positive quality control serum); Q1, positive quality control serum; Q2, negative quality control serum; Q3, buffer control (no serum).

All instruments which dispense or aspirate buffers should be thoroughly rinsed with distilled water after use to reduce ''salt'' buildup which can block delivery and aspiration tubes and can be extremely difficult to remove once dried, especially from metallic surfaces.

Given that an EIA protocol has been standardized with respect to buffers, reagents, physicochemical parameters, and instrumentation, there is still a certain degree of inherent variability which must be monitored. A set of criteria must be established to maintain quality control of both intra- and interassay variability.

In order to increase confidence in EIA test results, our laboratory runs all test samples in duplicate and quality control sera in quadruplicate. A microplate sample placement pattern (Figure 3), first described by Stemshorn et al.,[80] has been chosen which minimizes intraplate variability. The plate is divided horizontally and vertically into four equal quadrants using a felt pen, and sample replicates are placed at two corresponding locations in diagonally opposed quadrants. This helps to average any variability which may be caused by thermal gradients by maximizing the distance between all duplicates. It also ensures that sample duplicates do not share the same X or Y coordinates, thus minimizing any systematic variability associated with multichannel instruments. The mean OD and percent coefficient of variation of each set of replicates is determined and must fall within preset specification limits in order to be accepted. The specification limits are based on repeated testing of many sera having a wide range of antibody activities and incorporate a sliding scale which takes into account the fact that the expected variability of replicate ODs will increase as mean OD decreases.

Interplate variability is minimized by using a timing protocol to read plates.[93] Under this protocol, each plate is read at its own calculated final development time, at which point the mean OD of a high titer seropositive ''target'' control serum located in quadruplicate on the plate should equal 1.0. Within limits, this reading method compensates for the compounded variability caused by such factors as weight and volume measurement errors, temperature fluctuations, pH errors, and errors in timing incubation periods and allows direct comparison of ODs between plates without the need for standard curves or data normalization. The final development time for each plate is calculated by the following equation:

$$Y = A + (B/X)$$

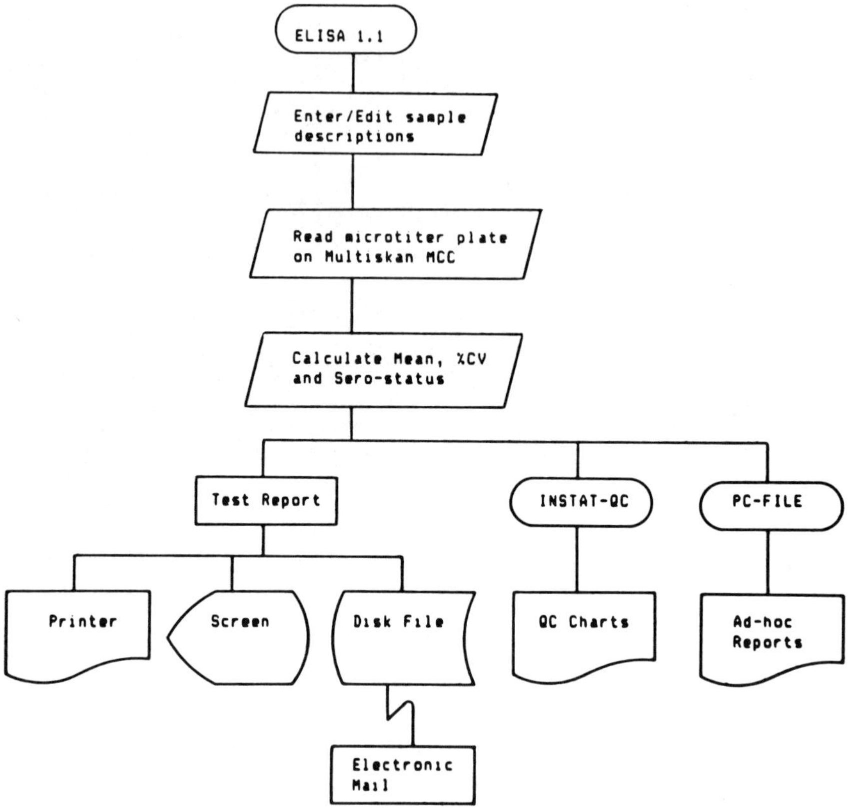

FIGURE 4. Flow diagram of a microcomputer software system (ELISA 1.1) developed to automate the reading of microplates using a timing protocol; the calculation, storage, and presentation of test results; and the performance of intra- and interassay quality control for the indirect *Brucella* EIA.

where Y = final development time in minutes; X = mean OD of "target" control at 4 min development; A = 0.1394; and B = 4.5939.

In addition to the "target" control serum, each quadrant contains a low titer positive control, a negative control, and a buffer control (no serum). The positive and negative controls are pooled sera meant to be representative of field samples, while the buffer control is included to monitor potential nonspecific binding of the conjugate and spontaneous development of the substrate. The mean ODs of the control reagents for each plate are monitored on a continuing basis, over the last 40 plates, and must fall within established specification limits for interplate variability. Since this variability has been found to be normally distributed about the mean, specification limits have been set at two or three standard deviations, depending on the particular control reagent.

As mentioned previously, an EIA software program has been developed by our laboratory to automate the reading of microtiter plates, the calculation and presentation of results, and the performance of intra- and interassay quality control. The program runs any IBM compatible microcomputer interfaced via the RS-232 serial port to a Flow Labs Multiskan Plus, MCC, or Twinreader computer-controllable plate photometer. The essential functions of the program are outlined in Figure 4. The program allows the user to enter and edit test sample descriptions for each plate prior to testing. After automatically selecting the correct filter and blanking the photometer, the program instructs the photometer to read the plate at 4 min of substrate development, calculates the final development time (using previously described equation), and automatically reads the plate again at the calculated final development time. Raw OD values from this final reading are stored in a disk file. The program automatically performs the required calculations and statistics and puts out a test report (Figure 5) to either the screen, the printer, or an ASCII file. Along with information regarding plate and test identification, the report also shows the test results for each control reagent and test sample, including ID and description, replicate ODs, mean OD, replicate % CV, acceptability, and serological status. Samples with unacceptably high replicate % CV, according to the previously mentioned sliding scale, are flagged with an asterisk in the STATUS column. The serological status is determined from the mean OD, shown in the MEAN column, and is flagged in the STATUS column as either positive (P) if greater than 0.350, suspicious (S) if between 0.250 and 0.350, or negative (blank) if less than 0.250. Samples which show unacceptably high variability or which are flagged as suspicious are normally retested.

The program has been integrated with two commercially available software packages. Each time a plate is read, all raw data for the target and quality control reagents are automatically appended to a file which is directly accessed by a commercial quality control program (INSTAT-QC, V1.A, Mikel Aiken, Statistical Consulting Services, 517 E Lodge Dr., Tempe, AZ 86283). This program is used on a daily basis to produce a QC chart (Figure 6) for each control reagent in order to monitor interplate variability. The QC chart shows the plate mean OD (denoted by an asterisk) for the control reagent over the last 40 plates and identifies those which lie outside the specification limits (denoted by colons). In addition, all sample identification and test results are automatically appended to another file which is imported on a regular basis into a commercial database management package (PC-FILE III, V4.0, Buttonware Inc., P.O. Box 5786, Bellevue, WA 98006) for the preparation of ad hoc and monthly reports.

An executable version (not source code) of the *Brucella* EIA software and documentation is available by sending a 5.25 or 3.5-in. disk directly to the authors of this chapter. In order to demonstrate the programming methods used to interface a computer-controllable plate reader to an IBM compatible microcomputer, the source code for a very simple BASICA program is included in Appendix 3. The program listing is heavily documented to allow it to be self-explanatory.

IV. DIAGNOSTIC VALIDATION AND APPLICATION

Once a serological assay has been developed and standardized, it must be thoroughly validated before it can be effectively put to diagnostic use. The two principal parameters which must be defined are diagnostic specificity and diagnostic sensitivity.[52] Both are estimates of the probability that the test results will truly reflect the infection status of the animal. Diagnostic specificity represents the probability that an animal will test seronegative given that it is not infected. Diagnostic sensitivity represents the opposite situation. The key point here is that the infection states must be known in order to derive these estimates. In theory, the selection criteria by which samples are chosen for these estimates should be based on parameters which are not serologically based as they may introduce a bias into the sample selection. However, in practice this is not always possible or desirable when dealing with infectious diseases for which serologically based eradication, control, or surveillance programs have been in existence for many years. In the case of brucellosis in developed countries, literally generations of cattle have been under a selection pressure which has been based on their serological activity in conventional assays. This is true of both vaccinated and nonvaccinated populations. Therefore, with the introduction of the indirect or competitive EIA technique, the serological biases already in existence must be considered.

Brucella ELISA

Filename: 86101501 Target Time: 10.30 min.

	1	2	3	4	5	6	7	8	9	10	11	12
A	1.067	0.181	0.126	0.175	0.134	0.217	1.047	0.448	0.178	0.139	0.161	0.124
B	0.725	0.175	0.140	0.192	0.149	0.356	0.703	0.261	0.132	0.103	0.153	0.208
C	0.101	0.200	0.201	0.155	0.125	0.194	0.085	0.141	0.143	0.186	0.138	0.136
D	0.014	0.175	0.209	0.105	0.120	0.386	0.011	0.255	0.146	0.101	0.252	0.075
E	0.990	0.409	0.181	0.126	0.150	0.101	0.997	0.145	0.085	0.128	0.098	0.193
F	0.677	0.326	0.126	0.088	0.136	0.165	0.604	0.114	0.119	0.155	0.101	0.258
G	0.098	0.161	0.143	0.215	0.143	0.120	0.068	0.177	0.196	0.123	0.100	0.157
H	0.009	0.236	0.186	0.123	0.277	0.073	0.012	0.154	0.186	0.101	0.101	0.337

ID	DESCRIPTION	MEAN	%CV	STATUS
TG	____________	1.025	3.7	P
Q1	____________	0.677	7.8	P
Q2	____________	0.088	17.1	
Q3	____________	0.012	18.1	
1	____________	0.163	15.6	
2	____________	0.145	29.9	
3	____________	0.189	8.6	
4	____________	0.165	9.0	
5	____________	0.106	27.5	
6	____________	0.130	11.5	
7	____________	0.199	1.8	
8	____________	0.198	8.2	
9	____________	0.152	21.9	
10	____________	0.174	15.1	
11	____________	0.139	16.3	
12	____________	0.103	2.7	
13	____________	0.116	21.9	
14	____________	0.125	27.2	
15	____________	0.113	15.7	
16	____________	0.111	12.2	
17	____________	0.205	8.3	
18	____________	0.307	22.6	* S
19	____________	0.176	14.9	
20	____________	0.362	9.6	P
21	____________	0.429	6.4	P
22	____________	0.294	15.7	S
23	____________	0.151	9.4	
24	____________	0.246	5.5	
25	____________	0.180	1.2	
26	____________	0.129	3.3	
27	____________	0.143	0.0	
28	____________	0.166	17.0	
29	____________	0.132	6.9	
30	____________	0.096	11.1	
31	____________	0.201	10.2	
32	____________	0.112	13.9	
33	____________	0.156	5.0	
34	____________	0.145	8.3	
35	____________	0.141	2.5	
36	____________	0.265	6.7	S
37	____________	0.113	14.5	
38	____________	0.187	16.3	
39	____________	0.128	8.8	
40	____________	0.074	1.9	

FIGURE 5. Test report produced by ELISA 1.1 software showing results for one *Brucella* EIA test plate, including plate and test sample identification, and statistics (including intraassay variability) and serological status for each quality control and test sample.

```
NEGATIVE CONTROL - PLATE MEAN

Data file:SYSDATA  Chart file:QCQ2  Date:06-29-1988

# Observations =   160     # Batches =   40
Min =   .09025    Max =   .14525    Mean =   .1149812
SDW =   1.627875E-02  SDB =   .0123467   SDU =   1.540745E-02
dfw =   120  dfb =   39   dfu =   159
SSW =   3.179975E-02    SSB =   5.945195E-03   F =   .5752533

                 Chart for Mean
           -------------------------------------------------------------------------
88032201  I                :            *  I        :                             I   0.108
88032202  I                :               *        :                             I   0.114
88032203  I                :               I  *     :                             I   0.128
88032301  I                :            *  I        :                             I   0.105
88032302  I                :              *I        :                             I   0.109
88032303  I                :               *        :                             I   0.113
88032304  I                :               I     *  :                             I   0.137
88032401  I                :              *I        :                             I   0.110
88032901  I                :        *      I        :                             I   0.090
88032902  I                :          *    I        :                             I   0.097
88032903  I                :               *        :                             I   0.113
88032904  I                :               I*       :                             I   0.120
88033001  I                :              *I        :                             I   0.109
88033002  I                :               *        :                             I   0.114
88033003  I                :               I    *   :                             I   0.130
88033101  I                :             * I        :                             I   0.103
88033102  I                :              *I        :                             I   0.109
88040601  I                :               *        :                             I   0.115
88040602  I                :               *        :                             I   0.116
88040603  I                :               I       *:                             I   0.145
88040701  I                :            *  I        :                             I   0.102
88041201  I                :             * I        :                             I   0.105
88041401  I                :               I*       :                             I   0.120
88041501  I                :               I   *    :                             I   0.127
88042101  I                :            *  I        :                             I   0.102
88042102  I                :          *    I        :                             I   0.099
88042103  I                :          *    I        :                             I   0.095
88042104  I                :               I  *     :                             I   0.124
88042601  I                :            *  I        :                             I   0.101
88042602  I                :            *  I        :                             I   0.101
88042603  I                :              *I        :                             I   0.109
88042604  I                :               I  *     :                             I   0.121
88042701  I                :             * I        :                             I   0.104
88042801  I                :               I  *     :                             I   0.122
88042802  I                :               I   *    :                             I   0.126
88042901  I                :               I     *  :                             I   0.134
88042902  I                :               I     *  :                             I   0.137
88050301  I                :               I  *     :                             I   0.123
88050501  I                :               I     *  :                             I   0.137
88050601  I                :               I  *     :                             I   0.125
           -------------------------------------------------------------------------

Minimum =  0    Maximum =   .25
Specified St.Dev. =  0
: Specification limits ( .059    .149 )
     0  Above
     0  Below
I Target ( .1149812 )
     17  Above
     23  Below
```

FIGURE 6. QC chart produced by INSTAT-QC program showing performance of the negative quality control serum (Q2) over the last 40 plates. Each asterisk represents the mean OD of the four Q2 replicates in one test plate.

Serological tests are developed for two basic applications: screening and confirmation. Screening tests are "first line" assays which are generally rapid and cost effective and demonstrate a high diagnostic sensitivity, usually at the expense of diagnostic specificity. Only the reactors, true and false positive alike, from the screening process are subjected to confirmatory testing. The latter assays demonstrate a higher diagnostic specificity and a relatively lower diagnostic sensitivity. This is referred to as series testing and reactors which

TABLE 3
Comparison of Three Indirect Enzyme Immunoassays (EIA) to the Complement Fixation Test (CFT)

Diagnostic criteria	Test group	EIA conjugate specificity		
		IgG (H + L)	L chain (κ + λ)	H chain (G_1)
Specificity	BPAT (−)[a] NV[b]	↓[c]	↓	↓
	BPAT (−)V	↓ ↓ ↓	↓ ↓	↓ ↓
	BPAT (+)NV	↓ ↓	↓ ↓ ↓	↓
	BPAT (+)V	↓ ↓ ↓	↓ ↓	↓ ↓
Sensitivity	BPAT (+)C[d]	↑	↑	↑ ↑

[a] BPAT (−/+); reactor status on buffered plate antigen test.
[b] NV/V; nonvaccinated, vaccinated.
[c] Comparison against CFT performance; lower than (↓, ↓ ↓, ↓ ↓ ↓) or greater than (↑, ↑ ↑, ↑ ↑ ↑) by 0—1%, 1—2% or >2%, respectively.
[d] C; positive on bacterial culture.

are seropositive in the screening and then in the confirmatory assay are usually culled. As the prevalence of disease falls to low levels, the majority of reactions observed in screening will certainly fall into the false positive category. This places a greater emphasis on the need for high diagnostic specificity. To increase diagnostic specificity, two (or more) conventional confirmatory tests may be applied. Only those animals which are seropositive in both confirmatory assays are then culled. However, we now have the potential to develop both new screening and new confirmatory assays to replace the conventional tests and increase diagnostic specificity without compromising diagnostic sensitivity. Therefore, in considering development it is important to establish the diagnostic objective of the new assay as related to screening or confirmation. Only then can the appropriate samples be selected for meaningful estimates of diagnostic specificity and sensitivity.

Prior to the development of the competitive EIA, we had initially developed our indirect EIA as a confirmatory test to replace the complement fixation test (CFT) which we had been using in combination with the buffered plate antigen test (BPAT) as a laboratory screening test. In the process of validation, we tested enzyme conjugates of three specificities: a polyclonal anti-bovine IgG (H + L), a monoclonal anti-bovine L chain (kappa and lambda), monoclonal anti-bovine IgG_1 (H chain specific). The comparative results are presented in Table 3. It should be pointed out that our observations on diagnostic specificity are representative of the population of dairy cattle in eastern Canada after several years of national brucellosis-free status. Observations on diagnostic sensitivity are based on historical serum banks.

Four groups of sera were selected for the evaluation of diagnostic specificity. As a confirmatory test, only BPAT reactors would be subjected to the indirect EIA and CFT. However, export certification often requires that all animals be tested by the CFT and, hence, the need to test BPAT negative sera as well. Although calfhood vaccination with S19 is for the most part not a routine practice in Canada, a small number of cattle are tested which have been vaccinated, primarily for export purposes. Therefore, sera from vaccinated cattle, irrespective of BPAT reactor status, would also require testing. It is important that not only the correct sample groups be chosen, but that the sampling strategy and sample size be appropriate to ensure the required confidence in the estimate.[90]

The comparison in Table 3 has been stylized to emphasize the differences in performance of the conjugates. All three conjugates demonstrated a higher diagnostic sensitivity than the CFT, and the greatest increase was observed with anti-IgG_1. In the nonvaccinated groups, all three conjugates demonstrated comparable diagnostic specificities in the testing of BPAT-negative cattle (>99%), but none could match the diagnostic specificity of the CFT which approached 100%. In the testing of BPAT-positive cattle (false positives) a more pronounced difference was observed. Again, none could match the diagnostic specificity of the CFT which again approached 100%, and only the anti-IgG_1 was observed to come close to the performance of the CFT. In the vaccinated groups, the differences between the three conjugates and the CFT were even greater. Although the CFT demonstrated a decrease in diagnostic specificity in the BPAT-positive group (99.3%) and still approached 100% in the BPAT-negative group, all three conjugates, especially the anti-IgG (H + L), demonstrated even greater decreases.

The conjugate specificity which most closely approximated the performance of the CFT was the monospecific anti-IgG_1. This is perhaps not surprising since it has been shown that the CFT preferentially detects IgG_1 as well.[55] The diagnostic sensitivity of this conjugate was not compromised by its restricted immunological specificity when compared to the broader specificity conjugates. Its lower diagnostic specificity in the testing of vaccinated cattle is most likely a function of its increased immunological sensitivity in the detection of low levels of residual antibody in calfhood vaccinates. It approached but did not match the specificity of the CFT in the testing of nonvaccinated cattle. This may not be a fair comparison as the CFT has been used as a confirmatory assay for many years in this population and as such may have introduced a bias in favor of the CFT. Therefore, if an indirect EIA were to be used to replace the CFT, then an anti-IgG_1 conjugate would be the best choice. However, in our particular situation where brucellosis has been eradicated, the indirect EIA does not offer any distinct diagnostic advantages. Although the data are not presented here, we have shown that vaccinated cattle represent a distinct subpopulation at the high end of the OD range of seronegative status. Therefore, in areas where vaccination is commonplace or mandatory, it may be expected that the indirect EIA will detect more vaccinates than the CFT. This may be partially circumvented by establishing a higher diagnostic threshold for this group provided that vaccination practices are reliable and records are accurate.

We are currently modifying our indirect EIA using the anti-IgG_1 conjugate for use as a laboratory-based screening assay. The modifications are not intended to change its diagnostic performance but rather to increase volume capabilities and cost effectiveness through automation. Based on our observations using the current protocol, it is expected that the number of BPAT reactors, reactors which would require further testing, may be significantly reduced without compromising diagnostic sensitivity. It is intended that confirmation of indirect EIA screening reactors will be done with the competitive EIA. From an immunological standpoint, this would appear to be a reversal of the current testing strategy in that a monospecific (IgG_1) assay is used prior to confirmation on a broad specificity assay such as the competitive assay. However, the key to these assays lies in the specificities of the immunoglobulins detected, not their isotypes. Therefore, the immunological specificity of the indirect EIA using SLPS is much broader than that of the competitive EIA using PS antigen. Although the CFT could still be used as a confirmatory test, it would not offer the advantage of discrimination of infected cattle from those which have been vaccinated or exposed to organisms which would otherwise cause cross-reactions. At the time of writing, our field studies have not been completed, and conclusive diagnostic data may not be presented as yet.

V. CONCLUSIONS

The tremendous potential of enzyme immunoassay techniques is perhaps no more apparent than in the application to the serodiagnosis of bovine brucellosis. Described in this chapter are two EIA techniques which in combination provide the most comprehensive detection system yet developed for eradication, control, and surveillance programs.

The indirect EIA using a SLPS antigen and a monoclonal anti-bovine IgG_1 conjugate demonstrates a higher diagnostic specificity than does the BPAT without compromising diagnostic sensitivity. Thus, the indirect EIA offers a distinct diagnostic advantage as a laboratory-based screening assay.

The competitive EIA using a PS antigen and a unique monoclonal antibody is capable of discriminating between infected cattle and those which have been vaccinated or exposed to a cross-reacting organism. Thus, the competitive EIA demonstrates a greater diagnostic specificity than the currently used confirmatory assays.

Much of the groundwork and understanding which has gone into the development of these assays has come from the chemical characterization of the O chain component of the PS antigens of *Brucella* and antigenically related organisms. Both assays have also benefited from monoclonal antibody development. The enhanced diagnostic specificity of the indirect EIA is primarily due to the refinements in antiisotype specificity. The ability to discriminate infected cattle in the competitive EIA is attributable to a monoclonal antibody specific for a defined linear epitope on the O chain.

The range of equipment and instrumentation which is now available for microplate-based EIA techniques runs the entire gamut from manual manipulation to full automation. This serves to make EIA techniques available at all levels of laboratory sophistication. Computer-interfaced plate readers have greatly enhanced data analysis and expression. This expands the potential of EIA techniques in terms of diagnostic interpretation and reporting.

Despite the advances that have been made in brucellosis serology, there are still areas which warrant further investigation. Although the O chain has been chemically characterized, the whole story with respect to epitope mapping has not yet unfolded. The diagnostic implications of the univalent and multivalent epitopes may allow for the serotyping of *Brucella* where we have relied on biotyping in the past. Modification of these epitopes through genetic engineering may provide a distinct marker for the identification of vaccinated cattle. The potential availability of highly specific subunit antigens may allow for the development of nonseparation-based EIA techniques which would greatly enhance the versatility of testing with respect to automation. In many ways, the information presented in this chapter may represent only the beginning of many advances yet to be realized in *Brucella* serology.

ACKNOWLEDGMENTS

The authors wish to acknowledge the expert technical capabilities of David Gall and Sofija Balsevicius. The antigens used in our tests were prepared by Dr. John Cherwonogrodzky. The enzyme conjugates were prepared by Diane Henning.

The authors also wish to acknowledge the patience and the expertise of Joan Graham in the production of this manuscript.

APPENDICES

APPENDIX 1: INDIRECT EIA FOR DETECTION OF BOVINE SERUM ANTIBODY TO *B. ABORTUS*

Method

1. Prepare a 1.0 μg/ml solution of *B. abortus* SLPS antigen in coating buffer (see Reagents and Buffers sections). It has been our experience that in the titration of many batches of this antigen, the effective working concentration has always been 1.0 μg/ml. However, you may wish to titrate this antigen to prove this point. Stock solutions of SLPS

at 1 mg/ml in low molarity buffer may be stored in small volumes at $-20°C$. It is very important that upon thawing of stock SLPS (or PS) the antigen be vigorously mixed because these preparations tend to form micelles at high concentration. Immediately upon dilution, dispense 200-μl volumes into all wells, seal microplates, and incubate overnight (~18 h) at room temperature (20 to 25°C).

2. Prepare dilutions of test samples and controls in wash/diluent buffer. Test sera dilutions of 1:100 in a minimal final volume of 1.0 ml are recommended. Control sera, stored frozen as neat aliquots, should be appropriately diluted to a final volume sufficient for the day's assay. Controls should consist of four types (see Figure 3): sera from A, positive reactor ("target", see Step 13); B, weakly positive reactor; C, healthy animal, and D, buffer (no serum) control. All dilutions should be prepared and maintained at room temperature.
3. Immediately prior to sample application, wash with equal efficiency all wells of the antigen-coated plate with wash/diluent buffer. After five consecutive washes, discharge residual buffer by inverting plate and rapping onto a lint-free absorbent surface.
4. Dispense 200-μl volumes of test and control samples into microplate wells in the configuration described in Figure 3. Seal the plate and incubate for 3 h at room temperature.
5. Immediately prior to the end of the serum incubation, prepare a solution of conjugate in a volume of wash/diluent buffer sufficient for the day's assay. Maintain the working dilution of conjugate at room temperature. Stock solutions of conjugate should be stored under appropriate conditions. We have found that filter-sterilized aliquots stored at $+4°C$ are reliable for our HRPO-conjugated monoclonal anti-bovine IgG_1 antibody. The addition of 50% glycerol and storage at colder temperatures is optional. *Note:* The working dilution of conjugate must first be determined by titration. Optimally, that dilution should be chosen which results in either (1) an OD of approximately 1.0 after 10 min of substrate development after the conjugate has been reacted for 1 h at room temperature with 100 ng/ml of ammonium sulfate-precipitated bovine gamma-globulin that has been passively adsorbed to the microplate as described for SLPS in Step 1; or (2) an OD of approximately 2.0 after 10 min of substrate development after the conjugate has been reacted, as above, with a very strong seropositive reactor diluted and applied as in Step 4.
6. Immediately prior to conjugate addition, wash the microplates as described in Step 3.
7. Dispense 200-μl volumes of conjugate to all wells of microplate, seal the plate, and incubate for 1 h at room temperature.
8. Immediately prior to the end of the conjugate incubation, prepare a working solution of enzyme substrate in a volume of fresh substrate buffer sufficient for the day's assay. Maintain the working solution at room temperature.
9. Immediately prior to substrate addition, wash the microplates as described in Step 3.
10. Dispense 200-μl volumes of substrate into all wells of the microplate, and immediately begin timing the development with a stopwatch. As well, the microplate should be transferred to an orbital-action microplate shaker for the duration of the assay to ensure uniform enzyme-substrate contact and uniform meniscus formation.
11. Immediately prior to use, the microplate photometer should be "blanked" on wells containing undeveloped substrate solution. For this purpose we use a separate, uncoated plate which has been rinsed with a wash/diluent buffer. Prior to blanking, the plate is shaken on a plate shaker to ensure uniform meniscus formation. *Note:* Photometric microplate scanners should be allowed a warmup time of 15 to 20 min before use.
12. At precisely 4 min, the average amount of product development (OD_{414} nm) in the wells containing the "target" control should be calculated. The target development time for each plate may then be determined either mathematically or from a standard curve based on the following equation:

$$Y = A + (B/X)$$

where Y = final development time in minutes, X = mean OD of target control at 4 min, A = 0.1394, and B = 4.5939. Each individual plate should be read at the final development time determined by its target control. *Note*: The target control is selected such that, at a dilution of 1:100 and after repeated testing, it results in an average OD of approximately 1.0 under standard conditions of testing and a substrate development time of 10 min. For best results, choose a strongly positive serum which may be prediluted in negative serum prior to a 1:100 dilution in wash/diluent buffer.

13. The mean ODs of the replicate controls should be used to define and monitor assay function in accordance with standard quality control procedures.
14. The mean ODs of duplicate test samples are used to determine seropositive and seronegative status based on a threshold OD (s) determined by seroepidemiological methods.

Reagents

1. Antigen — Semi-purified SLPS antigen prepared from *B. abortus* strain 413[59]
2. Conjugate — Mouse monoclonal anti-bovine IgG_1 conjugated with horseradish peroxidase[39]
3. Substrate — Hydrogen peroxide (H_2O_2), 3% solution (Catalogue No. H-324, Fisher Scientific Ltd., Mississauga, Ontario) — ABTS, hydrogen donor (Catalogue No. A1888, Sigma Chemical Co., St. Louis)

Buffers

1. Coating buffer — 0.06 *M* carbonate buffer, pH 9.6: $NaHCO_3$*, 3.80 g/l; Na_2CO_3, 1.93 g/l.
2. Wash/diluent buffer — 0.01 *M* phosphate buffer saline, pH 7.2 + 0.05% Tween 20: NaCl, 8.50 g/l; NaH_2PO_4, 0.315 g/l; Na_2HPO_4, 1.10 g/l; Tween 20, 0.5 ml/l.
3. Substrate buffer — 0.05 *M* citrate buffer, pH 5.0: $H_3C_6H_5O_7$, 9.6 g/l; adjust pH to 5.0 with NaOH pellets.
4. Substrate solution — required per microplate: citrate buffer, 20 ml; H_2O_2 (3% stock), 0.10 ml; ABTS (40 m*M* in H_2O), 0.50 ml.

Microtiter Equipment**

1. Plate washer — Titertek*** Microplate Washer
2. Dispenser — Titertek Autodrop Dispenser
3. Plate shaker — Titertek Plate Shaker
4. Photometer — Titertek Multiskan MCC Microplate Reader and Interface
5. Microtiter plates — NUNC† Microplate 96F (2-69620)

* All chemicals obtained from Fisher Scientific Ltd. except Tween 20 (Catalogue No. P1379, Sigma Chemical Co.).

** The equipment named below was used in the development of this protocol, and its listing does not constitute an endorsement of these products.

***™ Flow Laboratories Inc., McLean, VA.

† Gibco/BRL, Burlington, Ontario.

APPENDIX 2: COMPETITIVE EIA FOR THE DIFFERENTIATION OF *B. ABORTUS*-INFECTED FROM VACCINATED (S19) CATTLE

Method

1. Prepare a solution of *B. abortus* PS antigen in coating buffer (see Reagents and Buffers sections, Appendix 1). The working concentration of PS antigen will vary from batch to batch and, therefore, must be determined against the monoclonal anti-PS conjugate in a checkerboard titration. Optimally, that concentration of antigen should be chosen which is in the initial phase of the plateau portion of the titration curve. It should be expected that the working dilution of this antigen should be less than 2.5 μg/ml. Stock solutions of PS antigen at 1 mg/ml may be stored in small volumes at −20°C. Dispense 200μl volumes into all wells, seal microplates, and incubate overnight (~18 h) at room temperature (21 to 22°C). *Note:* The only plates which we recommend for use in this assay are the NUNC microplates described in Appendix 1. Use of other plates may result in antigen performance characteristics similar to SLPS.
2. Prepare dilutions of test samples and controls in wash/diluent buffer. Test sera dilutions of 1:50 in a minimal final volume of 0.5 ml are recommended. Control sera, stored frozen as neat aliquots, should be appropriately diluted to a final volume sufficient for the day's assay. Controls should consist of four types (see Figure 3): A, buffer (no serum) control and sera from B, nonvaccinated; C, vaccinated; and D, infected cattle. All dilutions should be prepared and maintained at room temperature.
3. Prepare a solution of conjugate in wash/diluent buffer. The working dilution of conjugate must be determined by titration as indicated above. Optimally, that dilution should be chosen which results in an OD of approximately 1.0 after 10 min of substrate development after the conjugate has been coincubated with the buffer control (described below). Maintain working dilution of conjugate at room temperature. Stock solutions of conjugate should be stored under optimal conditions. We have found that filter-sterilized aliquots stored at +4°C are adequate for our HRPO-conjugated monoclonal anti-PS antibody.
4. Immediately prior to sample and conjugate application, wash plates as described in Appendix 1.
5. Dispense 100-μl volumes of test sera and controls into microplate wells as described in Figure 3. Immediately dispense 100-μl volumes of conjugate into all wells of the plate. The plate should be transferred to an orbital-action plate shaker for a period of 2 min to ensure thorough but gentle mixing. Seal the plate and incubate for 2 h at room temperature.
6. Immediately prior to the end of the sample/conjugate incubation, prepare a working solution of enzyme substrate, as described in Appendix 1. Maintain the substrate
7. Immediately prior to substrate addition, wash microplates as described solution at room temperature in Appendix 1.
8. Dispense 200-μl volumes of substrate into all wells of the plate and proceed with the development protocol as described in Appendix 1. We prefer to use the buffer control as the "target" because the reaction of the conjugate is not influenced by the presence of negative sera.
9. The mean ODs of the replicate control sera should be used to define and monitor assay function in accordance with standard quality control procedures.
10. The mean ODs of duplicate test samples are used to determine seropositive and seronegative status based on percent inhibition of the conjugated anti-PS antibody. A baseline of 0% inhibition may be set using either the buffer or negative serum control. It should be expected that sera from healthy, vaccinated, and nonvaccinated cattle should result in inhibitions of less than 20 to 25% (with the exceptions noted in the text).

Reagents

1. Antigen — Purified PS antigen prepared from *B. abortus* strain 1119-3.[22]
2. Conjugate — Mouse monoclonal YsT9-2 anti-PS conjugated with horseradish peroxidase (prepared in the laboratory of Nielsen).
3. Substrate — As per Appendix 1.
4. Buffers — As per Appendix 1.
5. Equipment — As per Appendix 1.

APPENDIX 3A

```
100 '=========================================================================
102 '
104 ' Program:  BRUC-EIA  - by W.A.Kelly, Agriculture Canada  1988
106 '
108 ' Descrip:  Simple program to automate microtiter plate reading and data
110 '           reduction for the EIA test for Brucella antibody. Plate is read
112 '           using a timing protocol to minimize inter-plate variability.
114 '
116 '           - TEST plate: must be set up according to following layout
118 '
120 '                      1  2  3  4  5  6   7  8  9 10 11 12
122 '
124 '                   A  TG  1  5  9 13 17| TG 21 25 29 33 37
126 '                   B  Q1  2  6 10 14 18| Q1 22 26 30 34 38
128 '                   C  Q2  3  7 11 15 19| Q2 23 27 31 35 39
130 '                   D  Q3  4  8 12 16 20| Q3 24 28 32 36 40
132 '                   ----------------------------------------
134 '                   E  TG 21 25 29 33 37| TG  1  5  9 13 17
136 '                   F  Q1 22 26 30 34 38| Q1  2  6 10 14 18
138 '                   G  Q2 23 27 31 35 39| Q2  3  7 11 15 19
140 '                   H  Q3 24 28 32 36 40| Q3  4  8 12 16 20
142 '
144 '           - BLANKING plate: must contain undeveloped substrate in column1
146 '
148 '           - SEROLOGICAL STATUS: determined by mean OD as follows
150 '
152 '                   STATUS          Mean OD
154 '                   --------------  ----------------
156 '                   Positive (P)    >.350
158 '                   Suspicious (S)  >=.250 and <=.350
160 '                   Negative        <.250
162 '
164 '           - samples flagged with an asterisk * show high %CV and should
166 '             be retested especially if they are borderline positive
168 '
170 '           - User may specify name of disk file into which raw OD values
172 '             are stored after plate is read. Format of file is ASCII with
174 '             each 6-character OD value stored on a separate line.
176 '             96 OD values are stored for each plate.
178 '
180 '
182 ' Language: IBM BASICA or GWBASIC Ver 2.0 or later
184 '
186 ' Hardware: IBM-PC or compatible with RS-232 serial port COM1 connected via
188 '           straight-thru serial modem cable to Flow Labs Multiskan MCC set
190 '           for 1200 baud communication speed.
192 '           Dot matrix or daisy wheel printer.
194 '
196 ' *Caution: The program presented here is a demonstration version only!
198 '           Many validation and error-trapping routines which MUST be
200 '           included in a production version have been intentionally
202 '           omitted for the sake of brevity and readability.
204 '           In order to obtain a full-featured production version,
206 '           please contact the author.
208 '
1000 '========================================================================
1010 '<LOGIC>
1020 ' Main Program Logic - this shows overall program flow
1030 '------------------------------------------------------------------------
1040 ON ERROR GOTO 20100
1100 GOSUB 2100     '<INIT>      - INITIALIZE VARIABLES
1110 GOSUB 3100     '<MCCBLNK>   - SETUP AND BLANK PLATE READER
1120 GOSUB 4100     '<MCC4MIN>   - READ PLATE AT 4 MIN DEVELOPMENT
1130 GOSUB 5100     '<CALCTGTM>  - CALCULATE TARGET TIME; START COUNTDOWN
1140 GOSUB 6100     '<MCCTGTM>   - READ PLATE AT TARGET TIME
1150 GOSUB 7100     '<SAVODR>    - SAVE RAW ODS IN A DISK FILE IF DESIRED
1160 GOSUB 8100     '<CALCREPS>  - EXTRACT REPLICATES AND CALCULATE RESULTS
1170 GOSUB 10100    '<PRNRPT>    - PRINT TEST RESULTS
1180 GOSUB 15100    '<RERUN>     - DOES OPERATOR WANT TO READ ANOTHER PLATE?
1190 IF RERUN THEN 1120
1200 CLOSE
1210 END
```

APPENDIX 3A. Source code for simple BASIC program (BRUC-EIA) to automate reading and data-reduction of the *Brucella* EIA using an IBM compatible microcomputer interfaced to a Flow Lab Multiskan MCC plate reader.

```
2000 '----------------------------------------------------------------------
2010 '<INIT>
2020 ' Display opening screen, initialize and dimension variables
2030 '----------------------------------------------------------------------
2100 DEFINT I-L:KEY OFF
2110 DIM OD(96), MN(44), SD(44), CV(44), REP(4)
2120 DIM COL$(12), OD$(96), SAMPLES$(4), STATUSCV$(44), STATUSBR$(44)
2130 FALSE=0:TRUE=NOT FALSE
2140 LF$=CHR$(10):FF$=CHR$(12)           'linefeed and formfeed
2150 SUSP=.25                            'OD cutoff - Suspicious
2160 POSV=.35                            'OD cutoff - Positive
2170 CVLIMIT=1000:CVOVER$=">999.9"       'upper %CV for reporting
2180 FOR I=1 TO 4:READ SAMPLE$(I):NEXT
2190 DATA TG, Q1, Q2, Q3,                'ID strings for Target and controls
2200 CLS:BEEP:PRINT TAB(30);"** BRUC-EIA **"
2210 PRINT:PRINT:PRINT "Please ensure that Multiskan MCC and printer are turned
'ON'":PRINT:INPUT "Press <ENTER> when ready to continue ...",Z$
2230 RETURN
3000 '----------------------------------------------------------------------
3010 '<MCCBLNK>
3020 ' Open RS-232 communications via COM1 to Multiskan MCC; setup and blank
3030 '----------------------------------------------------------------------
3100 OPEN "COM1:1200,N,8,1,CS,DS" AS #1 'comm parameters, don't check CTS,DSR
3110 CMD$="R":GOSUB 11100                'enter computer-control (CC) mode
3120 CMD$="O":GOSUB 11100                'move plate carrier to OUT position
3130 CMD$="F2":GOSUB 11100               'choose Filter #2 (414 NM)
3140 PRINT:INPUT "* BLANKING *          Put BLANKING plate onto carrier and press
 <ret>...",Z$
3150 CMD$="S1":GOSUB 11100               'move plate to column 1
3160 CMD$="B":GOSUB 11100                'BLANK the reader on plate column 1
3170 CMD$="O":GOSUB 11100                'move plate carrier to OUT position
3180 RETURN
4000 '----------------------------------------------------------------------
4010 '<MCC4MIN>
4020 ' Read test plate at 4 min development
4030 '----------------------------------------------------------------------
4100 PRINT:INPUT "* 4 MIN READING *    Put TEST plate onto carrier; at 4 min. pr
ess <ret>...",Z$:PRINT
4110 GOSUB 13100    '<MCCREAD> - read TEST plate at 4 min.
4120 RETURN
5000 '----------------------------------------------------------------------
5010 '<CALCTGTM>
5020 ' Calculate Target-Time from mean Target serum OD at 4 min.
5030 '----------------------------------------------------------------------
5100 TARGETMEAN=(OD(1)+OD(5)+OD(49)+OD(53))/4
5110 TARGETIME=.1394+(4.5939/TARGETMEAN):DEVTIME=TARGETIME
5120 TARGETIME=INT(TARGETIME*60-240+STARTREAD)
5130 PROMPTIME=TARGETIME-20   'prompt operator 20 seconds before Targetime
5140 PRINT:PRINT "Calculated Target Time = ";:PRINT USING "##.##";DEVTIME;:PRINT
 " min."
5150 LOCATE 25,20:PRINT "Seconds to Target Time:";
5160 IF INT(TIMER)<PROMPTIME THEN GOSUB 12100:GOTO 5160
5170 RETURN
6000 '----------------------------------------------------------------------
6010 '<MCCTGTM>
6020 ' Prompt the Operator; automatically read plate at Target-Time
6030 '----------------------------------------------------------------------
6100 BEEP:PRINT:PRINT "* TARGET READING *  Make sure TEST plate is on carrier"
6110 IF INT(TIMER)<>TARGETIME THEN GOSUB 12100:GOTO 6110
6120 LOCATE 25,1:PRINT SPACE$(79);
6130 GOSUB 13100    '<MCCREAD> - read TEST plate at Target Time
6140 RETURN
7000 '----------------------------------------------------------------------
7010 '<SAVODR>
7020 ' Store the raw ODs in an ASCII disk file; or bypass storage if desired
7030 '----------------------------------------------------------------------
7100 PRINT:BEEP:INPUT "Enter a FILENAME .. (press <ret> for no file): ",ODFIL$
7110 IF ODFIL$<>"" THEN OPEN ODFIL$ FOR OUTPUT AS #2:FOR I=1 TO 96:PRINT #2,OD$(
I):NEXT:CLOSE #2
7120 RETURN
```

APPENDIX 3A (continued)

```
8000 '------------------------------------------------------------------------
8010 '<CALCREPS>
8020 ' Get replicates for the 44 sera (TG, Q1, Q2, Q3 and 40 Test-sera)
8030 '------------------------------------------------------------------------
8100 K=0                          ' k = sample #
8110 REPS=4                       ' Target and QC sera done in quadruplicate
8120 FOR I=1 TO 4: K=K+1
8130   REP(1)=OD(I):REP(2)=OD(I+4):REP(3)=OD(I+48):REP(4)=OD(I+52)
8140   GOSUB 9100                                     '<CALCSTAT>
8150 NEXT I
8160 REPS=2                       ' Test-sera done in duplicate
8170 FOR I=8 TO 40 STEP 8:FOR J=1 TO 4:K=K+1        ' Test-sera 1 to 20
8180   REP(1)=OD(I+J):REP(2)=OD(I+J+52):GOSUB 9100 '<CALCSTAT>
8190 NEXT J:NEXT I
8200 FOR I=12 TO 44 STEP 8:FOR J=1 TO 4:K=K+1       ' Test-sera 21 to 40
8210   REP(1)=OD(I+J):REP(2)=OD(I+J+44):GOSUB 9100 '<CALCSTAT>
8220 NEXT J:NEXT I
8230 RETURN
9000 '------------------------------------------------------------------------
9010 '<CALCSTAT>
9020 ' Calculate Mean, SD, %CV, CVstatus and Serostatus of QC and Sample reps
9030 '------------------------------------------------------------------------
9100 SUM=0:SUMSQ=0
9110 FOR L=1 TO REPS
9120   SUM=SUM+REP(L): SUMSQ=SUMSQ+(REP(L)*REP(L))
9130 NEXT
9140 MN(K)=SUM/REPS:M=MN(K)                       'Mean OD
9150 SD(K)=SQR((SUMSQ-REPS*M*M)/(REPS-1))         'Std. Dev
9160 STATUSCV$(K)=NUL$:STATUSBR$(K)=NUL$
9170 IF M=0 THEN CV(K)=CVLIMIT:RETURN             'bypass possible div by zero
9180 CV(K)=ABS(SD(K)/M)*100                       '% Coeff of Variation
9190 IF M<=.6 THEN A=2.27974: B=4.52842: ELSE A=2.63708: B=6.08114
9200 IF CV(K)>(A+B/M) THEN STATUSCV$(K)="*"      'flag high %CVs
9210 IF M>=.25 AND M<=.35 THEN STATUSBR$(K)="S" ELSE IF M>.35 THEN STATUSBR$(K)=
"P"                                               'flag Positive and Suspicious
9220 RETURN
10000 '-----------------------------------------------------------------------
10010 '<PRNRPT>
10020 ' Print test results: ID, Description, Mean-OD, %CV, Sero-status
10030 '-----------------------------------------------------------------------
10100 LPRINT TAB(30);"Brucella ELISA":LPRINT TAB(30);STRING$(14,45):LPRINT
10110 LPRINT "Filename: ";:IF ODFIL$=NUL$ THEN LPRINT "no file"; ELSE LPRINT ODF
IL$;
10120 LPRINT TAB(50);"Target Time: ";:LPRINT USING "##.##";DEVTIME;:LPRINT " min
.":LPRINT
10125 GOSUB 14100                                 'print raw OD matrix
10130 LPRINT:LPRINT "ID";TAB(6);"DESCRIPTION";TAB(24);"MEAN";TAB(33);"%CV";TAB(4
0);"STATUS":LPRINT STRING$(45,45)
10140 FOR K=1 TO 44
10150   IF K>4 THEN LPRINT USING "##";K-4; ELSE LPRINT SAMPLE$(K);
10160   LPRINT TAB(6);"_______________";
10170   LPRINT TAB(23);:LPRINT USING "##.###";MN(K);
10180   LPRINT TAB(31);:IF CV(K)<CVLIMIT THEN LPRINT USING " ###.#";CV(K); ELSE
LPRINT CVOVER$;
10190   LPRINT TAB(40);STATUSCV$(K);TAB(42);STATUSBR$(K)
10200   IF K=4 THEN LPRINT
10210 NEXT K:LPRINT FF$;
10220 RETURN
11000 '-----------------------------------------------------------------------
11010 '<MCCCMD>
11020 ' Send command to plate reader and get response
11030 '-----------------------------------------------------------------------
11100 PRINT #1,CMD$;LF$;:INPUT #1,ERRSTATUS$
11110 RETURN
12000 '-----------------------------------------------------------------------
12010 '<COUNTDN>
12020 ' Display countdown of seconds to Target-Time at bottom of screen
12030 '-----------------------------------------------------------------------
12100 LOCATE 25,45:PRINT TARGETIME-INT(TIMER);
12110 RETURN
```

APPENDIX 3A (continued)

```
13000 '-----------------------------------------------------------------------
13010 '<MCCREAD>
13020 ' Send command to MCC to read plate; extract OD values from MCC output
13030 '-----------------------------------------------------------------------
13100 STARTREAD=TIMER
13110 PRINT #1,"M12";LF$;                                   ' issue "READ" command
13120 FOR I=1 TO 12:INPUT #1,COL$(I):PRINT COL$(I):NEXT ' get 12 columns of ODs
13130 WELL=0
13140 FOR I=1 TO 12:COL$(I)=RIGHT$(COL$(I),55)
13150   FOR K=1 TO 50 STEP 7:WELL=WELL+1                    ' get 8 ODs per column
13160     OD$(WELL)=MID$(COL$(I),K,6):OD(WELL)=VAL(OD$(WELL))
13170   NEXT K:NEXT I
13180 RETURN
14000 '-----------------------------------------------------------------------
14010 '<PRN8X12>
14020 ' Print 8x12 plate matrix of raw OD values
14030 '-----------------------------------------------------------------------
14100 FOR I=1 TO 12:LPRINT TAB(I*6-1);I;:NEXT:LPRINT:LPRINT
14110 FOR I=1 TO 8:LPRINT CHR$(64+I);:FOR J=0 TO 11:LPRINT TAB(J*6+3);OD$(J*8+I)
;:NEXT:LPRINT:NEXT:LPRINT
14120 RETURN
15000 '-----------------------------------------------------------------------
15010 '<RERUN>
15020 ' Determine if operator wants to read another plate
15030 '-----------------------------------------------------------------------
15100 BEEP:PRINT:INPUT "Run another test plate (Y/N)";Z$
15110 IF (Z$="Y" OR Z$="y") THEN RERUN=TRUE ELSE IF (Z$<>"N" OR Z$<>"n") THEN RE
RUN=FALSE ELSE 15100
15120 RETURN
20000 '-----------------------------------------------------------------------
20010 '<ERRTRAP>
20020 ' Trap printer and filename errors
20030 '-----------------------------------------------------------------------
20100 BEEP
20110 IF ERL=7110 THEN PRINT "*** Invalid filename or path":RESUME 7100
20120 IF ERL=10100 THEN PRINT "*** Turn printer 'ON' and 'ON-LINE'":RESUME
20130 PRINT "*** Error #";ERR;" on Line ";ERL
```

APPENDIX 3A (continued)

APPENDIX 3B

ELISA B-abort

Filename: 86101501	Test Date: 10-15-1986
Plate#: 1	Test Time: 14:33:49
Dev.Time: 10.30 min.	Technician: WP

CONTROLS:

ID	Quad1	Quad2	Quad3	Quad4	MEAN	%CV	STATUS
TG	1.067	0.997	1.047	0.990	1.025	3.7	P
Q1	0.725	0.604	0.703	0.677	0.677	7.8	P
Q2	0.101	0.068	0.085	0.098	0.088	17.1	
BC	0.014	0.012	0.011	0.009	0.012	18.1	

TEST SAMPLES:

ID	DESCRIPTION	Quad1	Quad2	MEAN	%CV	STATUS
1	G/K30#18	0.181	0.145	0.163	15.6	
2	G/K30#26	0.175	0.114	0.145	29.9	
3	G/K30#43	0.200	0.177	0.189	8.6	
4	G/K32#15	0.175	0.154	0.165	9.0	
5	G/K32#39	0.126	0.085	0.106	27.5	
6	G/K34#16	0.140	0.119	0.130	11.5	
7	G/K34#43	0.201	0.196	0.199	1.8	
8	G/K34#45	0.209	0.186	0.198	8.2	
9	G/K34#73	0.175	0.128	0.152	21.9	
10	G/K34#115	0.192	0.155	0.174	15.1	
11	H/K34#130	0.155	0.123	0.139	16.3	
12	G/K34#165	0.105	0.101	0.103	2.7	
13	G/K36#53	0.134	0.098	0.116	21.9	
14	G/K37#23	0.149	0.101	0.125	27.2	
15	G/K37#51	0.125	0.100	0.113	15.7	
16	G/K37#74	0.120	0.101	0.111	12.2	
17	G/K40#1	0.217	0.193	0.205	8.3	
18	G/K40#2	0.356	0.258	0.307	22.6	*S
19	G/K40#3	0.194	0.157	0.176	14.9	
20	G/K40#4	0.386	0.337	0.362	9.6	P

ID	DESCRIPTION	Quad3	Quad4	MEAN	%CV	STATUS
21	G/K40#5	0.448	0.409	0.429	6.4	P
22	G/K44#46	0.261	0.326	0.294	15.7	S
23	G/K44#95	0.141	0.161	0.151	9.4	
24	G/K44#116	0.255	0.236	0.246	5.5	
25	G/K44#131	0.178	0.181	0.180	1.2	
26	G/K44#142	0.132	0.126	0.129	3.3	
27	G/K45#49	0.143	0.143	0.143	0.0	
28	G/K46#2	0.146	0.186	0.166	17.0	
29	G/K48#66	0.139	0.126	0.132	6.9	
30	G/K50#7	0.103	0.088	0.096	11.1	
31	G/K50#15	0.186	0.215	0.201	10.2	
32	G/K50#66	0.101	0.123	0.112	13.9	
33	G/K51#34	0.161	0.150	0.156	5.0	
34	G/K52#32	0.153	0.136	0.145	8.3	
35	G/K52#33	0.138	0.143	0.141	2.5	
36	G/K53#26	0.252	0.277	0.265	6.7	S
37	G/K57#57	0.124	0.101	0.113	14.5	
38	G/K60#14	0.208	0.165	0.187	16.3	
39	G/K60#55	0.136	0.120	0.128	8.8	
40	G/K64#6	0.075	0.073	0.074	1.9	

APPENDIX 3B. Sample test report produced by BRUC-EIA program showing results for one *Brucella* EIA test plate, including plate identification, raw OD for each well, and statistics (including intraassay variability) and serological status for each quality control and test sample.

REFERENCES

1. **Baker, P. J. and Wilson, J. B.,** Hypoferremia in mice and its application to the bioassay of endotoxin, *J. Bacteriol.*, 90, 903, 1965.
2. **Beh, K. J.,** Immunoglobulin class specificity of non-agglutinating antibody produced in cattle following *Brucella abortus* 45/20 vaccination, *Aust. Vet. J.*, 51, 481, 1975.
3. **Beh, K. J.,** Quantitative distribution of *Brucella* antibody amongst immunoglobulin classes in vaccinated and infected cattle, *Res. Vet. Sci.*, 17, 1, 1974.
4. **Beh, K. J.,** Distribution of *Brucella* antibody among immunoglobulin classes and a low molecular weight antibody fraction in serum and whey of cattle, *Res. Vet. Sci.*, 14, 381, 1973.
5. **Berman, D. T., Wilson, B. L., Moreno, E., Angus, R. D., and Jones, L. M.,** Characterization of *Brucella abortus* soluble antigen employed in immunoassay, *J. Clin. Microbiol.*, 11, 355, 1980.
6. **Boraker, D. K., Stinebring, W. R., and Kunkel, J. R.,** BrucELISA: an enzyme-antibody immunoassay for detection of *Brucella abortus* antibodies in milk: correlation with the *Brucella* ring test and with shedding of viable organisms, *J. Clin. Microbiol.*, 14, 396, 1981.
7. **Bruner, J. A., Thoen, C. O., and Hall, M. R.,** Evaluation of poly-B in an ELISA for detection of brucella antibodies in cow's milk, in ASM Annu. Meet., New Orleans, 80, 83 (Abstr. E21), 1983.
8. **Bundesen, P. G., Wyatt, D. M., Cottis, L. E., Blake, A. S., Massingham, D. A., Fletcher, W. A., Street, G., Welch, J. S., and Rylatt, D. B.,** Monoclonal antibodies directed against *Brucella abortus* cell surface antigens, *Vet. Immunol. Immunopathol.*, 8, 245, 1985.
9. **Bundle, D. R., Cherwonogrodzky, J. W., Caroff, M., and Perry, M. B.,** The lipopolysaccharides of *Brucella abortus* and *B. melitensis, Ann. Inst. Pasteur Microbiol.*, 138, 92, 1987.
10. **Bundle, D. R., Gidney, M. A., Perry, M. B., Duncan, J. R., and Cherwonogrodzky, J.,** Serological confirmation of *Brucella abortus* and *Yersinia enterocolitica* 0:9 O-antigens by monoclonal antibodies, *Infect. Immun.*, 46, 389, 1984.
11. **Butler, J. E.,** The amplified ELISA: principles of and applications for the comparative quantitation of antibodies and antigens in biochemical separates, in *Methods in Enzymology,* Vol. 73, Colowick, S. P. and Kaplan, N. O., Eds., Academic Press, New York, 1981, 482.
12. **Butler, J. E., Seawright, G. I., McGivern, P. L., and Gilsdorf, M.,** Class and subclass antibody response of *B. abortus* strain 19 and field strain challenged cattle: evidence for predominant IgG_1 response in infected animals, in *The Ruminant Immune System,* Butler, J. E., Ed., Plenum Press, New York, 790 (Abstr.), 1981.
13. **Byrd, J. W., Heck, F. C., and Hidalgo, R. J.,** Evaluation of the enzyme-linked immunosorbent assay for detecting *Brucella abortus* antibodies, *J. Vet. Res.*, 40, 896, 1979.
14. **Cargill, C. M., Lee, K., and Clarke, I.,** Use of an enzyme-linked immunosorbent assay in a bovine brucellosis eradication program, *Aust. Vet. J.*, 62, 49, 1985.
15. **Caroff, M., Bundle, D. R., and Perry, M. B.,** Structure of the O-chain of the phenol-phase soluble lipopolysaccharide of *Yersinia enterocolitica* serotype 0:9, *Eur. J. Biochem.*, 139, 195, 1984.
16. **Chappel, R. J. and Hayes, J.,** Comparison of radioimmunoassay with the complement fixation test and the indirect haemolysis test in the field diagnosis of bovine brucellosis, *J. Hyg.*, 90, 67, 1983.
17. **Chappel, R. J., Hayes, J., Brain, G. J., and McNaught, D. J.,** A modified radioimmunoassay for antibodies against *Brucella abortus, J. Hyg.*, 88, 1, 1982.
18. **Chappel, R. J., Hayes, J., Rogerson, B. A., and Shenfield, L. J.,** The serological response of cattle to vaccine against brucellosis as measured by the brucellosis radioimmunoassay and other tests, *J. Hyg.*, 88, 11, 1982.
19. **Chappel, R. J., McNaught, D. J., Bourke, J. A., and Allan, G. S.,** The diagnostic efficiency of some serological tests for bovine brucellosis, *J. Hyg.*, 80, 373, 1978.
20. **Chappel, R. J., McNaught, D. J., Bourke, J. A., and Allan, G. S.,** Comparison of the results of some serological tests for bovine brucellosis, *J. Hyg.*, 80, 365, 1978.
21. **Chappel, R. J., Williamson, P., McNaught, D. J., Dulling, M. J., and Allan, G. S.,** Radioimmunoassay for antibodies against *Brucella abortus*: a new serological test for bovine brucellosis, *J. Hyg.*, 77, 369, 1976.
22. **Cherwonogrodzky, J. W. and Nielsen, K. H.,** *Brucella abortus* 1119-3 O-chain polysaccharide to differentiate sera from *B. abortus* S-19-vaccinated and field-strain-infected cattle by agar gel immunodiffusion, *J. Clin. Microbiol.*, 26, 1120, 1988.
23. **Collin, J. C.,** Role des differentes immunoglobulines dans l'epreuve de l'anneau utilisee pour le depistage de la brucellose, *Ann. Microbiol.*, 127B, 177, 1976.
24. **Corbel, M. J. and Day, C. A.,** Assessment of fluorescent antibody absorption procedures for differentiation of the serological response to *Yersinia enterocolitica* serotype IX and *Brucella abortus* in cattle, *Br. Vet. J.*, 129, 17, 1973.

25. **Cremer, N. E., Cossen, C. K., Hansen, C. V., and Shell, G. R.,** Evaluation and reporting of enzyme immunoassay determinations of antibody to Herpes simplex virus in sera and cerebrospinal fluid, *J. Clin. Microbiol.,* 15, 815, 1982.
26. **de Savigny, D. and Voller, A.,** The communication of ELISA data from laboratory to clinician, *J. Immunoassay,* 1, 105, 1980.
27. **Diaz, R., Garatea, P., Jones, L. M., and Moriyou, I.,** Radial immunodiffusion test with a *Brucella* polysaccharide antigen for differentiating infected from vaccinated cattle, *J. Clin. Microbiol.,* 10, 37, 1979.
28. **Dohoo, I. R., Wright, P. F., Ruckerbauer, G. M., Samagh, B. S., Robertson, F. J., and Forbes, L. B.,** A comparison of five serological tests for bovine brucellosis, *Can. J. Vet. Res.,* 50, 485, 1986.
29. **Douglas, J. T. and Kennedy, R. C.,** Comparison of enzyme linked immunosorbent assay and solid phase radioimmunoassay in the detection of antibodies against *Brucella abortus, Abstr. Annu. Meet. Am. Soc. Microbiol.,* 81, 285, 1981.
30. **Engvall, E. and Perlmann, P.,** Enzyme-linked immunosorbent assay (ELISA): quantitative assay of immunoglobulin G, *Immunochemistry,* 8, 871, 1971.
31. **Gorrell, M. D., Milliken, G. L., Anderson, B. J., and Pucci, A.,** An enzyme immunoassay for bovine brucellosis using monoclonal antibody specific for field strains of *Brucella abortus, Dev. Biol. Stand.,* 56, 491, 1984.
32. **Gripenberg, M. and Gripenberg, G.,** Expression of antibody activity characterized by the shape of the dose-response curve, *J. Immunol. Methods,* 62, 315, 1983.
33. **Hall, S. M. and Confer, A. W.,** Comparison of Track XI fluorometric immunoassay system with other serologic tests for detection of serum antibody to *Brucella abortus* in cattle, *J. Clin. Microbiol.,* 25, 350, 1987.
34. **Hall, S. M., Confer, A. W., Tabatabai, L. B., and Deyoe, B. L.,** Detection of serum antibody to *Brucella abortus* in cattle by use of a quantitative fluorometric immunoassay, *J. Clin. Microbiol.,* 20, 1023, 1984.
35. **Hayes, J. and Chappel, R. J.,** A comparison of the results of the brucellosis radioimmunoassay and other serologic tests in experimentally infected cattle, *J. Hyg.,* 88, 21, 1982.
36. **Heck, F. C., Williams, J. D., and Pruett, J.,** Interpretation of spectrophotometric absorbance values to define results of enzyme-linked immunosorbent assays, *J. Clin. Microbiol.,* 11, 398, 1980.
37. **Heck, F. C., Williams, J. D., Pruett, J., Sanders, R., and Zink, D. L.,** Enzyme-linked immunosorbent assay for detecting antibodies to *Brucella abortus* in bovine milk and serum, *Am. J. Vet. Res.,* 41, 2082, 1980.
38. **Heck, F. C., Williams, J. D., Crawford, R. P., and Flowers, A. I.,** Comparison of serological methods for the detection of *B. abortus* antibody in sera from vaccinated and non-vaccinated cattle, *J. Hyg.,* 83, 491, 1979.
39. **Henning, D. M. and Nielsen, K.,** Peroxidase labelled monoclonal antibodies for use in enzyme immunoassay, *J. Immunoassay,* 8, 297, 1987.
40. **Hinchliffe, D. M. and Robertson, L.,** Detection of complement fixation by enzyme linked immunosorbent assay (COMPELISA), *J. Clin. Pathol.,* 36, 100, 1983.
41. **Hinchliffe, P.,** The detection of complement fixing antibodies by ELISA (COMPELISA) in brucella serology, *Dev. Biol. Stand.,* 56, 465, 1984.
42. **Jones, L. M., Berman, D. T., Moreno, E., Deyoe, B. L., Gilsdorf, M. J., Haber, J. D., and Nicoletti, P.,** Evaluation of a radial immunodiffusion test with polysaccharide B antigen for diagnosis of bovine brucellosis, *J. Clin. Microbiol.,* 12, 753, 1980.
43. **Jones, L. M.,** Brucella antigens and serological test results, in *Bovine Brucellosis,* Crawford, R. P. and Hidalgo, R. J., Eds., Texas A&M University Press, College Station, 1977, 40.
44. **Kramer, S. M., Jewell, N. P., and Cremer, N. E.,** Discriminant analysis of data in enzyme immunoassay, *J. Immunol. Methods,* 60, 243, 1983.
45. **Lamb, V. L., Jones, L. M., Schurig, G. G., and Berman, D. T.,** Enzyme-linked immunosorbent assay for bovine immunoglobulin subclass-specific response to *Brucella abortus* lipopolysaccharide, *Infect. Immun.,* 26, 240, 1979.
46. **Lawman, M. J. P., Ball, D. R., Hoffmann, E. M., Desjardin, L. E., and Boyle, M. D. P.,** Production of *Brucella abortus*-specific Protein A-reactive antibodies (IgG_2) in infected and vaccinated cattle, *Vet. Microbiol.,* 12, 43, 1986.
47. **Lawman, M. J. P., Thurmond, M. C., Reis, K. J., Gauntlett, D. R., and Boyle, M. D. P.,** Solid phase radioimmunoassay for detection of immunoglobulins against bovine *Brucella abortus, Vet. Immunol. Immunopathol.,* 6, 291, 1984.
48. **Levieux, D.,** A solid-phase radioimmunoassay for the determination of bacterial-specific antibodies within different immunoglobulin classes: application to bovine *Brucella abortus* antibodies, *Ann. Rech. Vet.,* 9, 523, 1978.
49. **Levieux, D.,** Immunglobulines bovines et brucellose. I. Purification des immunglobulines et preparation de leurs antiserums specifiques, *Ann. Rech. Vet.,* 5, 329, 1974.

50. **Levieux, D.,** Immunoglobulines bovines et brucellose. II. Activité des IgG_1, IgG_2 et IgM du serum dans les réactions d'agglutination, de Coomb's, de fixation du complement et dans le test au Rose Bengale, *Ann. Rech. Vet.*, 5, 343, 1974.
51. **Malvano, R., Boniolo, A., Dovis, M., and Zannino, M.,** ELISA for antibody measurement: aspects related to data expression, *J. Immunol. Methods,* 48, 51, 1982.
52. **Martin, S. W.,** The evaluation of tests, *Can. J. Comp. Med.*, 41, 19, 1977.
53. **Moreno, E., Borowiak, D., and Mayer, H.,** Brucella lipopolysaccharides and polysaccharides, *Ann. Inst. Pasteur Microbiol.*, 138, 102, 1987.
54. **Nielsen, K., Cherwonogrodzky, J., Duncan, J. R., and Bundle, D. R.,** Enzyme immunoassay for differentiation of the antibody response of *Brucella abortus* infected or strain 19 vaccinated cattle, *Am. J. Vet. Res.*, 50, 5, 1989.
55. **Nielsen, K. and Duncan, J. R.,** Bovine IgM: does it fix guinea pig complement in the absence of bovine complement components?, *Vet. Immunol. Immunopathol.*, 14, 335, 1987.
56. **Nielsen, K., Wright, P. F., Cherwonogrodzky, J., Duncan, J. R., and Stemshorn, B.,** Enzyme immunoassay for diagnosis of bovine brucellosis, *Ann. Inst. Pasteur,* 138, 75, 1987.
57. **Nielsen, K. H., Wright, P., Kelly, W. A., and Cherwonogrodzky, J. H.,** *Brucella* sp. Antibodies, in *Methods of Enzymatic Analysis,* 3rd Ed., Bergmeyer, H. U., Ed., VCH Publishers, Weinheim, West Germany, 1986, 152.
58. **Nielsen, K., Ballinger, R., Stiller, J., and Rosenbaum, B.,** A 'dipstick' enzyme immunoassay for detection of antibody to *Brucella abortus* in cattle sera, *Can. J. Comp. Med.*, 49, 298, 1985.
59. **Nielsen, K. and Wright, P. F.,** *Enzyme Immunoassay and Its Application to the Detection of Bovine Antibody to Brucella abortus,* Agriculture Canada/Animal Diseases Research Institute, Nepean, Ontario, 1984, 1.
60. **Nielsen, K., Heck, F. C., Wagner, G. G., Stiller, J., Rosenbaum, B., Pugh, R., and Flores, E.,** Comparative assessment of antibody isotypes to *Brucella abortus* by primary and secondary binding assays, *Prev. Vet. Med.*, 2, 197, 1984.
61. **Nielsen, K., Stiller, J., Adams, G., and Williams, R.,** Binding of *Brucella abortus* whole cells and lipopolysaccharide antigens to plastics, *Res. Vet. Sci.*, 34, 131, 1983.
62. **Oliver, D. G. and Cooper, R. S.,** Comparative study of enzyme-linked immunoassay (IgG and IgM) with standard serological tests for diagnosis of brucellosis in cattle, *Proc. Annu. Meet. U.S. Anim. Health Assoc.*, 85, 90, 1981.
63. **Parratt, D., Nielsen, K., White, R. G., and Payne, D. J. H.,** Radioimmunoassay of IgM, IgG and IgA brucella antibodies, *Lancet,* 1, 1075, 1977.
64. **Perera, V. Y., Winter, A. J., and Ganem, B.,** Evidence for the covalent bonding of native hapten-protein complexes to smooth lipopolysaccharide of *Brucella abortus, FEMS Microbiol. Lett.*, 21, 263, 1984.
65. **Perera, V. Y., Creasy, M. T., and Winter, A. J.,** Nylon bead enzyme linked immunosorbent assay for the detection of sub-picogram quantities of *Brucella* antigens, *J. Clin. Microbiol.*, 18, 601, 1983.
66. **Quinn, R., Campbell, A. M., and Phillips, A. P.,** A monoclonal antibody specific for the A antigen of *Brucella* spp., *J. Gen. Microbiol.*, 130, 2285, 1984.
67. **Raybould, T. J. G. and Chantler, S.,** Serological differentiation between infected and vaccinated cattle by visual and quantitative immunofluorescence using *Brucella abortus* antigen coupled to Sepharose beads, *J. Immunol. Methods,* 30, 37, 1979.
68. **Rice, C. E. and Boyes, B.,** Serum immunoglobulins in bovine brucellosis, *N.Z. Vet. J.*, 19, 146, 1971.
69. **Rice, C. E., Tailyour, J., and Cochrane, D.,** Ultracentrifugal studies of sera from cattle vaccinated or naturally infected with *Brucella abortus, Can. J. Comp. Med.*, 30, 270, 1966.
70. **Richardson, M. D., Llewellyn, P. A., and Warnock, D. W.,** Antibody to *Aspergillus fumigatus* antigens in normal sera: influence on positive-negative discrimination, *J. Immunoassay,* 5, 205, 1984.
71. **Rojas, X., Alonso, O., Quinones, A., and Soto, L.,** A new enzyme linked immunosorbent assay on nitrocellulose membranes for brucellosis diagnosis, *Arch. Met. Vet. Chile,* 18, 149, 1986.
72. **Rose, J. E. and Amerault, T. E.,** Electrophoretic and ultracentrifugation studies on serum from pregnant heifers after exposure to virulent *Brucella abortus, Am. J. Vet. Res.*, 25, 998, 1984.
73. **Rose, J. E. and Roepke, M. H.,** Physiochemical studies on post-vaccinal *Brucella* agglutinins in bovine serum, *Am. J. Vet. Res.*, 25, 325, 1964.
74. **Ruppanner, R., Meyer, M. E., Willeberg, F., and Behymer, D.,** Comparison of ELISA with other tests using sera from heifers experimentally infected with *Brucella abortus, Proc. 2nd Int. Symp. Vet. Lab. Diag.*, 7, 232, 1980.
75. **Ruppanner, R., Meyer, M. E., Willeberg, F., and Behymer, D.,** Comparison of the enzyme-linked immunosorbent assay with other tests for brucellosis in experimentally infected heifers, *Am. J. Vet. Res.*, 41, 1329, 1980.

76. **Rylatt, D. B., Wyatt, D. M., and Bundesen, P. G.,** A competitive enzyme immunoassay for the detection of bovine antibodies to *Brucella abortus* using monoclonal antibodies, *Vet. Immunol. Immunopathol.,* 8, 261, 1985.
77. **Saunders, G. C., Clinard, E. H., Bartlett, M. L., and Sanders, W. M.,** Application of the enzyme-labelled antibody microtest to detection and surveillance of animal diseases, *J. Infect. Dis.,* 136, s258, 1977.
78. **Schurig, G. G., Hammerberg, C., and Finkler, B. R.,** Monoclonal antibodies to *Brucella* surface antigens associated with the smooth lipopolysaccharide complex, *Am. J. Vet. Res.,* 45, 967, 1984.
79. **Schurig, G. G.,** Use of competitive-ELISA with monoclonal antibodies to *Brucella* for diagnostic purposes: a model with cross-reactive *E. coli* antigens, Conf. Res. Workers Animal Dis., 64th Annu. Meet., Chicago, (Abstr. 193), 1983, 37.
80. **Stemshorn, B. W., Buckley, D. J., St. Amour, G., Lin, C. S., and Duncan, J. R.,** A computer-interfaced photometer and systematic spacing of duplicates to control with in plate enzyme immunoassay variation, *J. Immunol. Methods,* 61, 367, 1983.
81. **Stemshorn, B. W., Nielsen, K., Samagh, B. S., Forbes, L. B., and Ingram, D. G.,** Evaluation of an enzyme-labelled antiglobulin test for anti-Brucella immunoglobulin G among 3 cattle populations, *Am. J. Vet. Res.,* 41, 1779, 1980.
82. **Suther, D. E., Cooper, R. S., and Vanderwagen, L. C.,** Description and evaluation of an enzyme-linked immunosorbent assay test used for brucellosis screening of dairy cattle in Southern California, *Proc. Annu. Meet. U.S. Anim. Health Assoc.,* 90, 167, 1986.
83. **Tabatabai, L. B. and Deyoe, B. L.,** Specific enzyme-linked immunosorbent assay for detection of bovine antibody to *Brucella abortus, J. Clin. Microbiol.,* 20, 209, 1984.
84. **Tedder, T. F. and Hoffmann, E. M.,** Immunoradiometric assay for examination and quantitation of *Brucella abortus* specific antibodies reactive with the antigen(s) used in the indirect hemolysis test, *J. Clin. Microbiol.,* 14, 415, 1981.
85. **Thoen, C. O., Bruner, J. A., Luchsinger, D. W., and Pietz, D. E.,** Detection of *Brucella* antibodies to different immunoglobulin classes in cow's milk by enzyme-linked immunosorbent assay, *Am. J. Vet. Res.,* 44, 306, 1983.
86. **Thoen, C. O., Hopkins, M. P., and Pietz, D. E.,** Detection of Brucella antibodies in cow's milk using an automated enzyme-linked immunosorbent assay, *Proc. 2nd Int. Symp. Vet. Lab. Diag.,* 7, 228, 1980.
87. **Thoen, C. O., Pietz, D. E., Armbrust, A. L., and Harrington, R., Jr.,** Enzyme immunoassay for detecting *Brucella* antibodies in cow's milk, *J. Clin. Microbiol.,* 10, 222, 1979.
88. **Van Aert, A., Brioen, P., Dekeyser, P., Uytterhaegen, L., Sijens, R. J., and Boeye, A.,** A comparative study of ELISA and other methods for the detection of *Brucella* antibodies in bovine sera, *Vet. Microbiol.,* 10, 13, 1984/85.
89. **Wilson, D. V., Thornley, M. J., and Coombs, R. R. A.,** A solid phase assay with radioactively labelled antibody for the detection of *Brucella abortus, J. Med. Microbiol.,* 10, 281, 1977.
90. **Wright, P. F. and Nielsen, K. H.,** Application of enzyme immunoassay in veterinary medicine: serodiagnosis of bovine brucellosis, in *Nonisotopic Immunoassay,* Ngo, T. T., Ed., Plenum Press, New York, 1988, 129.
91. **Wright, P. F.,** Enzyme immunoassay: observations on aspects of quality control, *Vet. Immunol. Immunopathol.,* 17, 441, 1987.
92. **Wright, P. F., Gall, D. E. J., and Kelly, W. A.,** Effect of meniscus formation and duplicate sample placement configurations on the variability of measurement by three microtiter plate photometers, *J. Immunol. Methods,* 81, 83, 1985.
93. **Wright, P. F., Kelly, W. A., and Gall, D. E. J.,** Application of a timing protocol to the reduction of inter-plate variability in the indirect enzyme immunoassay for detection of anti-'*Brucella*' antibody, *J. Immunoassay,* 6, 189, 1985.
94. **Bundle, D.,** personal communication.
95. **Moreno, E., et al.,** 1979.
96. **Searson, J.,** Department of Agriculture, New South Wales, Australia, personal communication.

Chapter 10

AN AUTOMATED FLUORESCENCE-BASED BRUCELLOSIS TEST SYSTEM WITH A PROVEN TRACK RECORD

Michael L. Snyder, Philip L. McMahon, and Erwin F. Workman, Jr.

TABLE OF CONTENTS

I. INTRODUCTION

Laboratory diagnosticians have attempted to employ automation in the performance of various assay procedures for more than 20 years. With the advent of radioimmunoassays, a need was recognized to minimize worker exposure to some of the isotopes of choice: thus evolved some of the pioneering automation. Methods for numerous clinicial chemistries were eventually developed in the form of continuous flow instrumentation. These developments in automation were then modified to perform such laboratory methodologies as complement-fixation, hemagglutination-inhibition, hematology, histology, enzyme immunoassay, and immunofluorescent assays.

In recent years, there has been an explosion of technological improvements to test systems development as a result of new solid phases in addition to improved microprocessor control systems and solid-state circuitry in automated hardware. This trend is a result of increasing demands in both the human and veterinary medical professions for shorter assay times, improved reliability, increased sample testing load, and, in many applications, improved sensitivity and/or specificity.

Commercially developed and manufactured diagnostic test kits have added yet another benefit to the diagnostician, namely, standardization. Performance control of commercial test kits varies among manufacturers, however, companies with the appropriate developmental and manufacturing experience can and do maintain high levels of standardization from serial to serial.

The purpose of this article is to discuss an innovative immunofluorescent test system for the serodiagnosis of *Brucella abortus* and *Brucella suis*.

II. GENERAL DESCRIPTION OF TECHNOLOGY

Particle concentration fluorescence immunoassay, or PCFIA, is a technology that utilizes submicron polystyrene particles as the solid phase to which antigens or antibodies are attached. The label is comprised usually of antibodies to which fluorescent markers such as fluorescene, rhodamine, or Texas Red have been attached. The coated particles, fluorescent label, and wash solution are added to appropriate compartments of a five-section reagent tray (Figure 1). These reagents are dispensed by a series of eight pipettors into reaction plates containing the test samples, as shown in Figure 2. Following the reaction period of the assay components in 96-well reaction plates (Figure 3) unreacted reagents and wash solution are separated from the particles by filtration through membranes at the bottom of each well. Figure 4 shows a cross-section of the special well structure with the incorporated filter membrane. The particles, being too large to penetrate the pores, layer on top of the membranes. After washing, the well is then moved into position below a front-surface fluorimetric read system (Figure 5), and the amount of particle-bound fluorescence is measured in each well. Concentration of the particles on the surface of the membranes gives rise to a 1000-fold increase in the concentration of the bound label and, therefore, a sensitive immunoassay technique. The PCFIA instrument has the optical capacity to read fluorescent labels at four different wavelengths and, therefore, the potential to measure four different analytes. PCFIA is more sensitive than other fluorescent immunoassay techniques and is equivalent to the sensitivity of RIA and ELISA. The PCFIA test for antibody to cattle and/or swine brucellosis described below is a totally automated, sample-in answer-out immunoassay when performed on the IDEXX "Screen Machine" instrument shown in Figure 6. The PCFIA technology is summarized in the four-panel illustration of Figure 7.

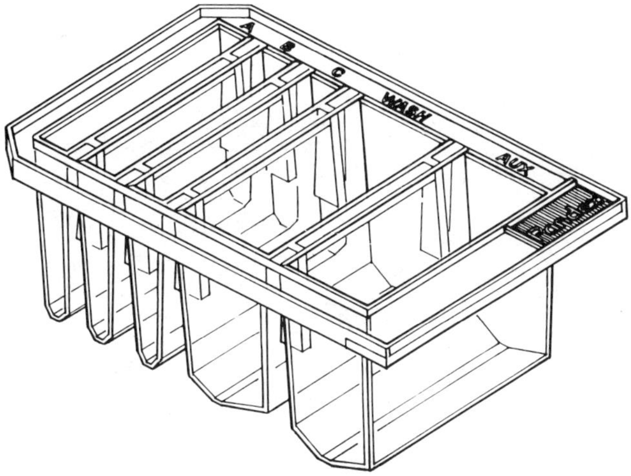

FIGURE 1. Five-section tray for use in the PCFIA screen machine contains antigen or antibody-coated particles, fluorescent label, and wash solution.

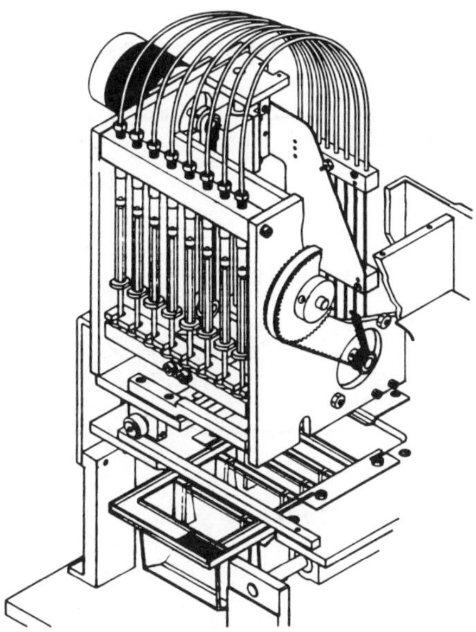

FIGURE 2. Microprocessor-controlled, syringe-metered pipetting unit for the PCFIA Screen Machine.

III. ASSAY CONFIGURATION FOR DETECTION OF ANTIBODY TO *BRUCELLA ABORTUS* AND/OR *B. SUIS*

Brucellosis in swine is caused almost exclusively by *Brucella suis* and is currently diagnosed using the same test procedures employed for *Brucella abortus*. As directed by the Uniform Methods and Rules (UMR), most states screen-test swine samples for brucellosis using the *B. abortus* plate agglutination (BAPA) test. Positive samples are then retested using the buffered brucella antigen (BBA) card test. Occasionally, card-positive samples are subsequently tested using the rivanol precipitation (RIV) test, complement fixation (CF) test,

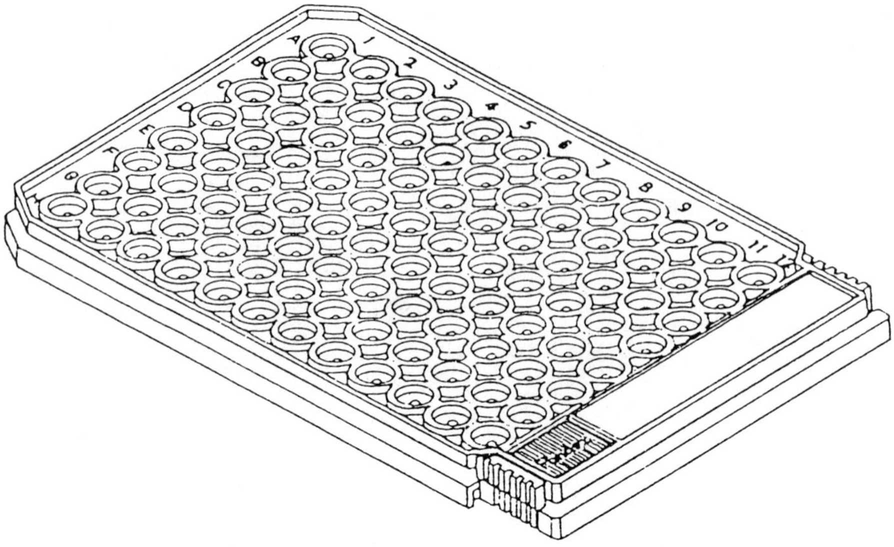

FIGURE 3. Overview of the special 96-well microplate used for PCFIA in the Screen Machine.

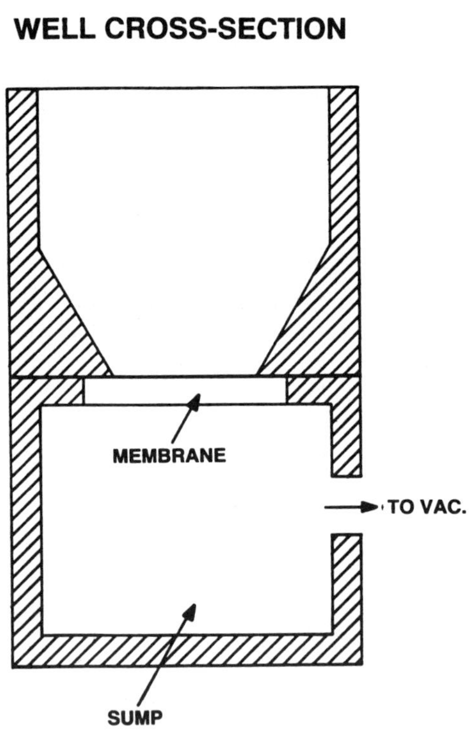

FIGURE 4. Cross-section of PCFIA microplate structure illustrating the unique filter at the base of the microwell and vacuum port located on the end of the plate.

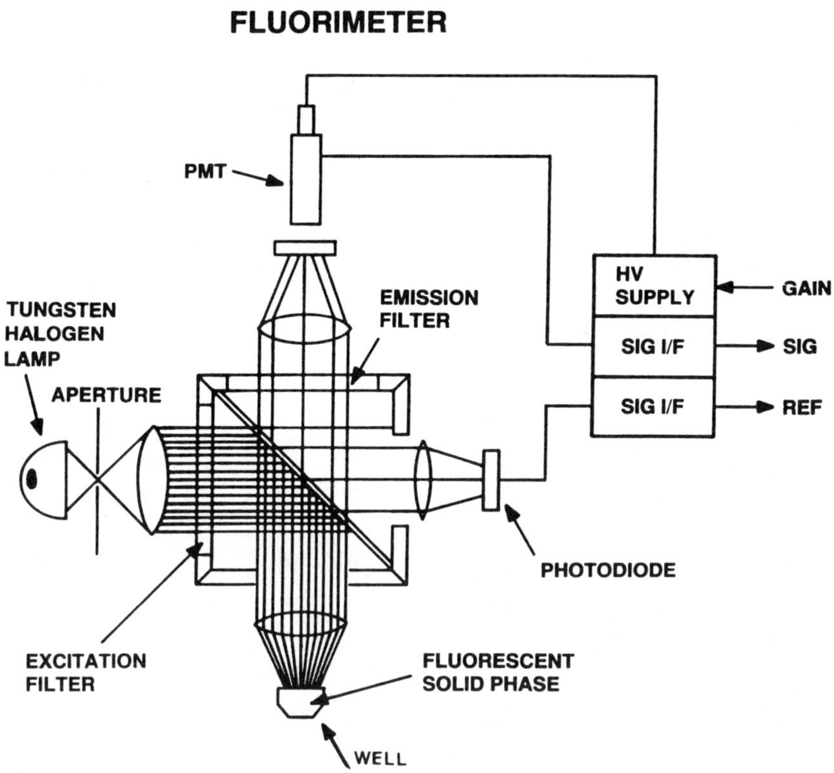

FIGURE 5. Schematic diagram of the enhanced PCFIA fluorimeter based on epiluminescent detection of bound fluorescent label.

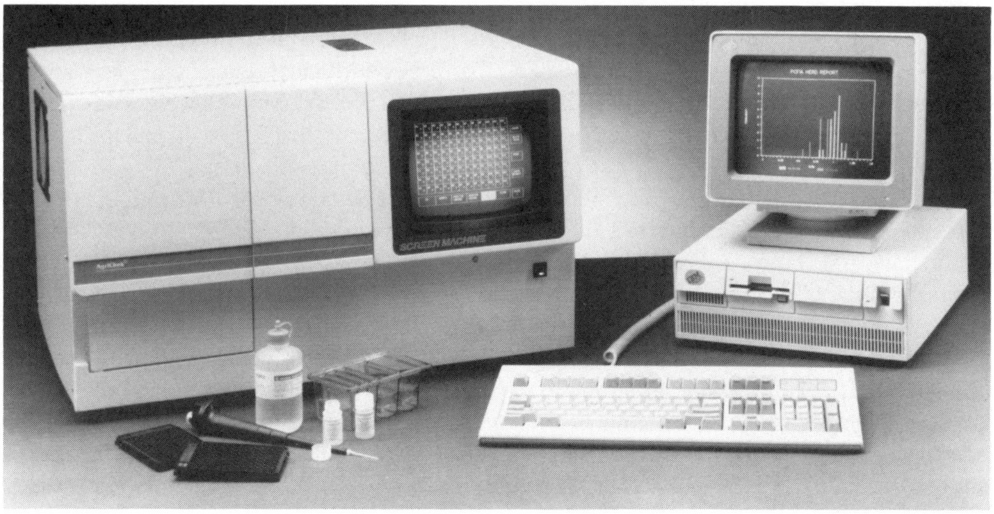

FIGURE 6. The Particle Concentration Fluorescence Immunoassay (PCFIA) Screen Machine and supporting computer system.

and/or standard tube agglutination test (STT) to aid in determining the status of swine for *B. suis* infection.

The diagnostic test procedures outlined above all utilize *Brucella abortus* organisms as the source of test antigens. The application of these test procedures utilizing *B. abortus*, the organism responsible for infections in cattle, is appropriate, due to the homology of antigens

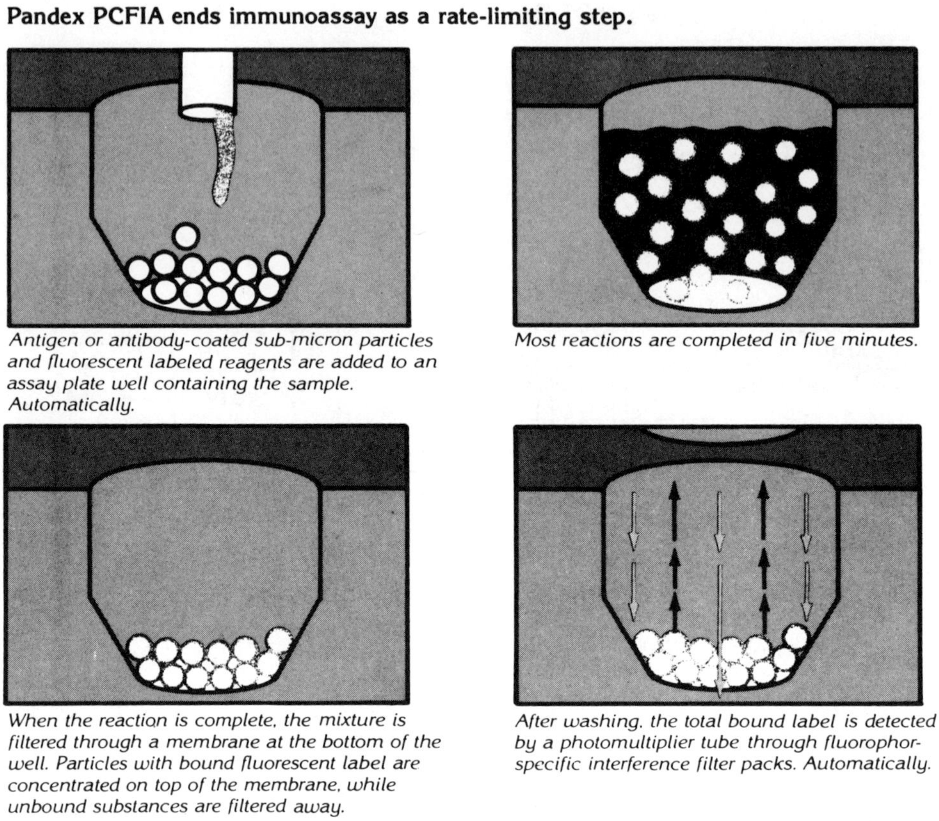

FIGURE 7. A four-step summary of PCFIA technology.

between *B. abortus* and *B. suis*. Several reports have appeared in the literature describing extensive DNA homology between the species of *Brucella*,[1-3] including *B. abortus* and *B. suis*. Therefore, due to the homology of expressed antigens between most species of *Brucella*, we can apply the IDEXX® PCFIA *Brucella abortus* antibody test kit to diagnose *Brucella suis* infection in swine.

There are two basic immunochemical approaches for detecting specific antibodies to *B. abortus* and/or *B. suis*. The first is a "two-step" or indirect double-antibody sandwich assay as illustrated in Figure 8. The solid phase coated with a phenol-extracted and purified *B. abortus* antigen is mixed with the serum sample. Antibodies specific for *B. abortus* and/or *B. suis* in the sample will bind to the antigen due to major antigens common to both species of *Brucella*. The presence of these bound antibodies is determined by the binding of a second antibody made against bovine immunoglobulins, which is labeled with a fluorescent probe. Notice also in Figure 8 the binding of nonspecific antibodies to the solid phase or to the antigen per se and their subsequent detection with the second antibody. This nonspecific binding inevitably occurs and varies from sample to sample and assay to assay. Another disadvantage of this assay is that the label, or anti-bovine IgG-conjugated second antibody, is species specific; therefore, *B. suis* antibodies present in swine sera would not be detected in the indirect assay.

The second immunochemical approach and the one utilized in this system, is a competitive assay, as illustrated in Figure 9. The solid phase, again coated with *B. abortus* antigen, is now mixed with both the serum sample and fluorescene-labeled *B. abortus* specific antibodies. In this method, the specific antibodies in the sample compete with labeled specific antibodies for antigen binding sites on the solid phase. The amount of labeled antibody

—Indirect Sandwich Assay

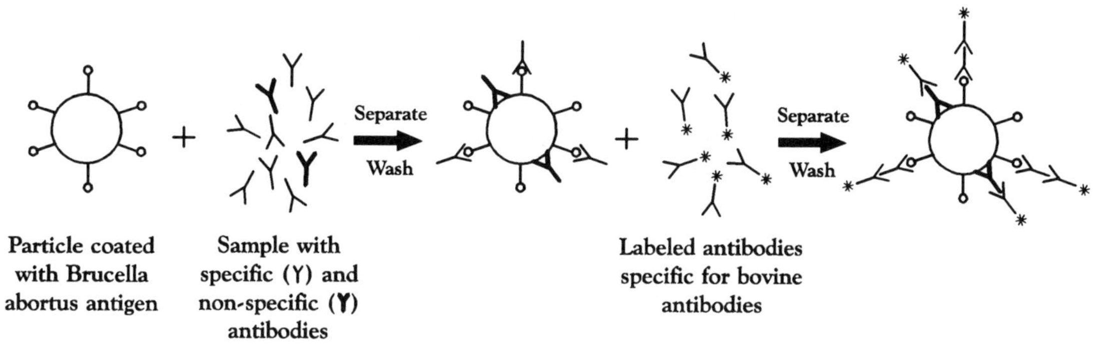

FIGURE 8. Flow diagram of an indirect two-step PCFIA.

—Competitive Assay

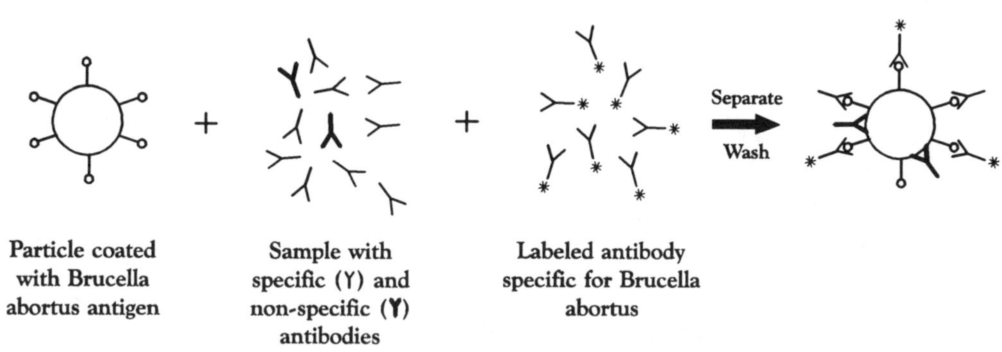

FIGURE 9. Flow diagram of a competitive PCFIA.

bound to the solid phase will decrease if specific antibody is present in the sample, thus providing an inverse measurement of specific antibody. Binding of nonspecific antibody to the solid phase does not affect the competition between specific antibody in the sample and labeled antibody, and thus, does not affect the resulting signal. Nonspecific binding to the antigen on the solid phase is instantly displaced by the higher affinity-labeled antibody and does not affect the resulting signal. For this reason, the competitive assay is more specific than the indirect double-antibody assay and is the method of choice for an immunoassay that measures antigen-specific antibodies. Due to the fact that in the competitive assay format the fluorescein-labeled anti-*B. abortus* conjugate is competing for specific antigenic sites with the sample antibodies and is not required to recognize a specific species class of antibody, one may assay for both *B. abortus* and *B. suis*.

Figure 10 shows a standard curve for the PCFIA competitive assay for antibodies to brucellosis. Note again, photon counts from the fluorescing label decrease with increasing concentration of *B. abortus*-specific antibody. These results may be expressed as a ratio by

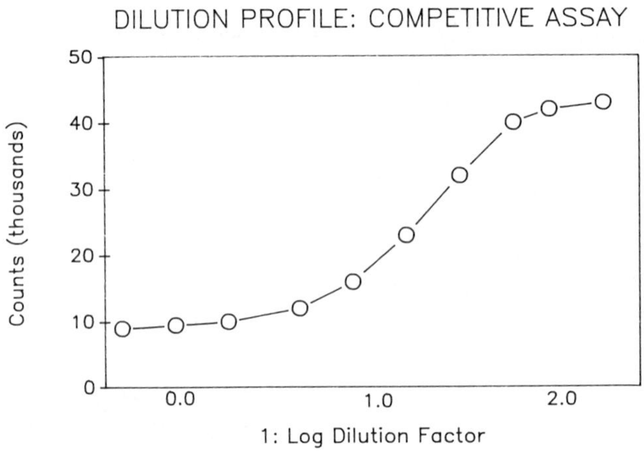

FIGURE 10. Standard curve for PCFIA channel counts using a 10-member serum sensitivity panel.

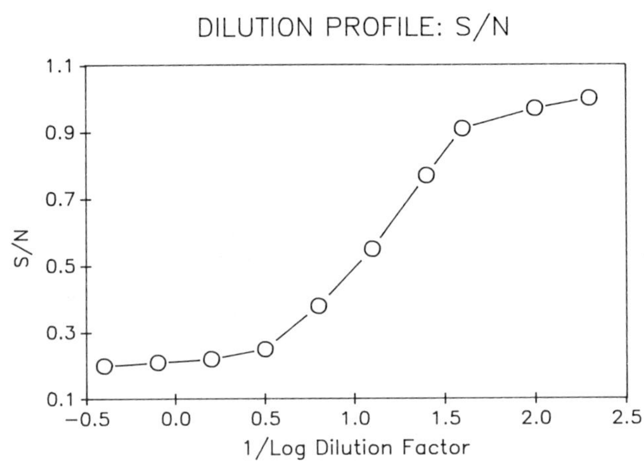

FIGURE 11. Standard curve for PCFIA sample/negative (S/N) values using a 10-member serum sensitivity panel.

dividing the counts of the sample by the average counts of the negative control; this ratio is expressed as the S/N value, which ranges from approximately 1.0 for negative samples to approximately 0.15 for strong positive samples:

N = Average of counts in negative control wells; S = counts in sample well; therefore,

$$S/N = \frac{\text{Counts in sample well}}{\text{Average counts of negative control}}$$

A standard curve plotting S/N vs. sample dilution is shown in Figure 11.

As previously noted, the PCFIA instrument has the potential to detect emission spectra for four different conjugated fluorescent dyes through the use of a four-channel filter pack. We routinely use two fluorescent dyes in our *B. abortus* Antibody Test Kit. One dye is conjugated to purified *B. abortus* antibodies and is measured by the Channel A filter. The second fluorescent dye is bound to particles which provide a constant reference count and is measured by the Channel B filter. The reference particles validate assay performance. The test kit is manufactured to provide reference particles in a concentration that produces a certain count range.

The S/N formula can then be further defined as:

$$S/N = \frac{\text{Sample (channel A counts)/sample (channel B counts)}}{\text{Neg. contr. (channel A counts)/neg. contr. (channel B counts)}}$$

An example of the utility of the Channel B reference particles is seen when a sample of very poor quality is tested. Such a sample may "clog" the membrane pores, thus preventing the vacuuming of reagents from that test well. Fluids remaining in the well during frontal surface reading produce varying degrees of light scatter, thus reducing the counts. Without the reference particles that show a similar reduction in counts, the sample could be incorrectly classified as a reactor.

IV. THE AUTOMATED PCFIA SYSTEM

The PCFIA assay for *B. abortus*- and/or *B. suis*-specific antibodies is completely automated using the IDEXX Screen Machine and interfaced computer. The screen machine is capable of storing a total of 50 protocols by entering the appropriate parameters into the microprocessor via an active touch-screen. Therefore, while the *B. abortus* test protocol is standardized for all instruments, protocols for other assay applications can be stored and are readily available to the operator.

The technician first fills the reagent tray with volumes sufficient for the batch. The instrument automatically checks reagent volumes against the number of plates to be run to assure adequate volumes. Sample diluent is then added to all assay plate wells except the control wells of plate 1. Fifty μl of prediluted controls and 2 μl of samples are then pipetted into respective wells by the technician. After the assay plates are inserted into the instrument's plate magazine and the operator initiates the assay protocol via the active touch screen (Figure 12), the instrument performs automatically all of the steps necessary for carrying out the assay (see Table 1). Data are then sent to the computer, where results are calculated and printed out upon request. The printout lists all identification numbers (current software allows each sample to be identified with three separate ID numbers), S/N values, result interpretation (negative, positive, or suspicious) for each sample, and a positive and invalid report (examples in Tables 2 and 3). Samples with the same herd ID number (or Vet Code number) may then be grouped and profiled. Such profiles have proven very useful in evaluating the overall status of the herd. Retests of the same herd show an overall trend toward positive or negative values.

Since the assay is performed by the instrument, it is inherently more reliable because it is not subject to human error. In addition, each step performed by the instrument is independently verified, thus assuring that each sample was tested correctly. This is accomplished by adding reference particles to the antigen-coated particles during the manufacturing process. The fluorescence of the label and reference particles are both read and analyzed. The relative intensity of each fluorophor can be used to verify that each step carried out by the instrument was performed correctly.

The throughput capability of the system is outlined in Table 4. Note that one technician using one instrument can assay approximately 750,000 samples per year. Throughput for other combinations of manpower and number of instruments are also shown. These estimates are based on information obtained at the Veterinary Diagnostic Laboratory in Auburn, Alabama.

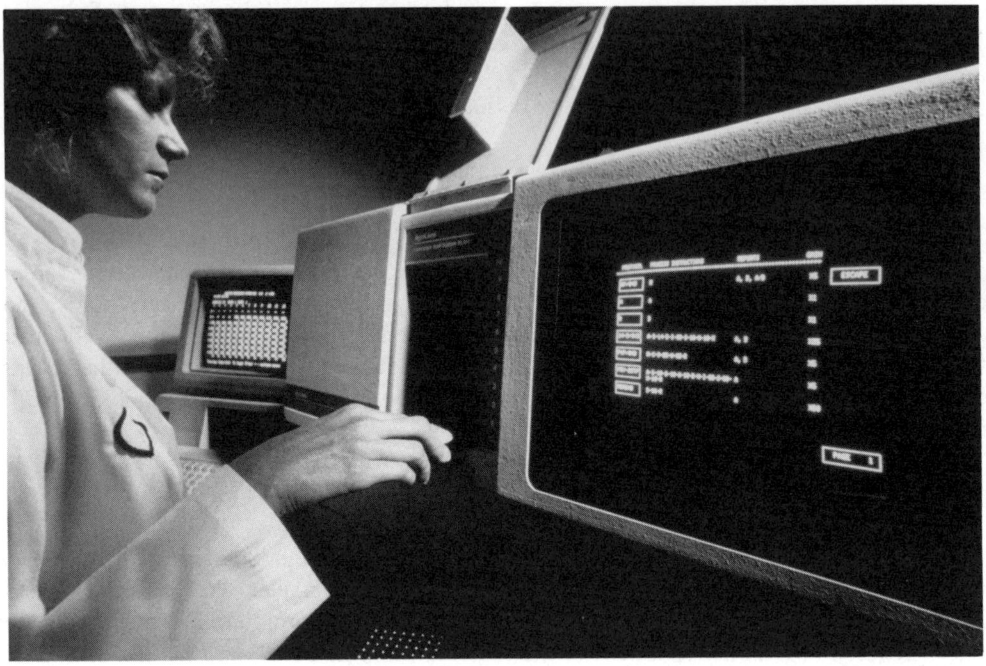

FIGURE 12. Technician inserting the PCFIA reaction plate into the Screen Machine for automated processing and data analysis. Notice the touch-screen to the right of the plate elevator displaying stored protocols.

V. REQUIRED FEATURES OF AN AUTOMATED SYSTEM

Many commercial equipment and diagnostic companies claim high performance for their products, but it is the responsibility of the potential user to thoroughly investigate "real-world" performance data before considering capital expenditure and implementation in his/her laboratory (i.e., "Buyer beware!"). Areas of critical importance are:

1. **Specificity**: Is the false positive rate of the candidate diagnostic system equivalent to or lower than current tests?
2. **Sensitivity**: Does the candidate diagnostic system have equivalent or improved sensitivity compared to currently used standard tests?
3. **Reliability**: Can the laboratory diagnostician depend upon the system to provide not only a precise answer, but also an accurate result day to day and month to month?
4. **Standardization**: Can a laboratory expect to obtain the same basic result as an affiliated laboratory for the same sample(s)?

The laboratory diagnostician not only should expect these qualities in a diagnostic test system, but should demand them to meet the quality requirements of today's needs in rapid and accurate diagnosis.

The following discussions will present actual assay validation and field laboratory performance data obtained using IDEXX Corporation's *B. abortus* PCFIA Antibody Test System. While this discussion focuses on *B. abortus* infection in cattle, we have presented similar data validating this assay for the diagnosis of *B. suis* infection in swine.[4]

TABLE 1
Automated Immunoassay for Antibody to *B. abortus* and/or *B. suis*

Instrument Protocol

Instrument Performs the Following Automatically

Adds antigen-coated particles
Adds fluorescent-labeled antibody
Filters reaction mixture, separating bound from unbound labeled antibody
Washes particles
Measures bound fluorescent-labeled antibody by front-surface fluorimetry
Sends data to IBM computer, which calculates and prints-out results

VI. DETERMINATION OF THE CUTOFF FOR THE PRESUMPTIVE TEST

In a population of animals exposed to *B. abortus* via vaccination or natural field exposure, a range of antibody concentrations will be found in the peripheral blood from very high to very low. Antibody levels will vary depending on many factors. A presumptive test should be designed to maximize sensitivity (i.e., number of positives/number of confirmed positives) while keeping specificity (number of negatives/number of confirmed negatives) at an acceptable level.

A distribution of antibody levels in a hypothetical positive population are presented in Figure 13. Levels range from very high to approaching zero. The ideal assay should have a cutoff infinitely close to the zero antibody point, thus having perfect sensitivity. The ideal assay should also have perfect specificity which requires that the population of true negatives have an infinitely tight distribution around the zero antibody line. The assay presented herein does not meet the criteria for an ideal assay in that all true negatives do not fall exactly at the zero antibody point. Consequently, the cutoff is not set infinitely close to that point.

A more realistic distribution of true negatives is shown in Figure 14, which represents a hypothetical population of true positive and negative animals. The shaded area indicates overlap between negative and positive populations. If the cutoff is set at the negative control in order to have ideal sensitivity, slightly less than half of the true negatives will be called positive. The farther the cutoff is moved from the negative control, the more true positives will be called negative, and the fewer true negatives will be called positive.

The presumptive cutoff of the PCFIA test has been established to provide equivalent sensitivity and improved specificity as compared to the traditional presumptive tests.

Negative population distribution — Negative (BAPA negative) samples from multiple herds in Alabama and Missouri were tested by PCFIA. The results were combined for each state and plotted in the form of a frequency distribution.

The negative population (BAPA negative) distribution from Alabama is shown in Figure 15. A total of 3710 results have been grouped in 0.05 S/N increments and plotted as a function of frequency. The mean S/N for this population is 1.02, with a standard deviation of 0.07. A negative population (BAPA negative) of 3756 samples from Missouri is shown in Figure 16. Again, these results have been grouped in 0.05 S/N increments and plotted as a function of frequency. The mean S/N for this population is 0.98, with a standard deviation of 0.079. All samples have S/N values of 0.75 or above; thus, 0.7 has been set as the presumptive test cutoff. Further validation of this cutoff in terms of sensitivity and specificity is presented below. Thus, samples with a S/N less than or equal to 0.70 are considered presumptive positives.

Positive population distribution — A positive population (culture positive) distribution

TABLE 2
IDEXX Brucellosis Test System, Screening Test Report

Negative controls

	A1	A2	A3	A4	A5	A6	A7	A8	A9	A10	A11	A12	MEAN
Ch A	55882	54542	52582	49468	48252	49650	45594	50120	49770	47896	53988	55730	51122
Ch B	27086	27393	27462	25542	25256	25708	25198	26132	27162	27042	25042	25550	26215

Weak positive controls

	B1	B2	MEAN
Ch A	29782	31164	30473
Ch B	23988	23736	23862
S/N	0.636	0.673	0.655

Strong positive controls

	B3	B4	MEAN
Ch A	5580	5824	5702
Ch B	27436	24044	25740
S/N	0.104	0.124	0.114

Plate # 1

A FORM #	1	2	3	4	5	6	7	8	9	10	11	12
VET #	NEG	NEG	NEG	NEG	NEG	NEG	NEG	NEG	NEG	NEG	NEG	NEG
TUBE #												
Ch A	55882	54542	52582	49468	48252	49650	45594	50120	49770	47896	53988	55730
Ch B	27086	27393	27462	25542	25266	25708	25198	26132	27162	27042	25042	25550
S/N	1.057	1.020	0.981	0.993	0.979	0.990	0.927	0.983	0.939	0.908	1.105	1.118
RESULT												

B FORM #	POS	POS	POS	POS	vet	vet	vet	vet	vet	vet	vet	vet
VET #					herd	herd	herd	herd	herd	herd	herd	herd
TUBE #					1	2	3	4	5	6	7	8
Ch A	29782	31164	5580	5824	29782	31164	29986	37770	42330	8376	9350	20586
Ch B	23988	23736	27436	24044	25934	26400	26042	26380	26444	24954	22204	23160
S/N	0.636	0.673	0.104	0.124	0.589	0.605	0.590	0.734	0.820	0.172	0.216	0.456
RESULT					POS.	POS.	POS.	NEG.	NEG.	POS.	POS.	POS.

C FORM #	vet	vet	vet	vet	vet	vet	vet	vet	vet	vet	vet	vet
VET #	herd	herd	herd	herd	herd	herd	herd	herd	herd	herd	herd	herd
TUBE #	9	10	11	12	13	14	15	16	17	18	19	20
Ch A	50972	48456	44598	44334	52446	35142	43094	47896	9532	22410	31164	35900
Ch B	28346	28086	5890	24866	23410	24206	26262	26294	26100	25754	24720	23742
S/N	0.922	0.884		0.914	1.148	0.744	0.840	0.934	0.187	0.446	0.646	0.775
RESULT	NEG.	NEG.	INVALID	NEG.	NEG.	NEG.	NEG.	NEG.	POS.	POS.	POS.	NEG.

D FORM #	vet	vet	vet	vet	vet	vet	vet	vet	vet	vet	vet	vet
VET #	herd	herd	herd	herd	herd	herd	herd	herd	herd	herd	herd	herd
TUBE #	21	22	23	24	25	26	27	28	29	30	31	32
Ch A	40824	47734	47026	42330	37700	29986	21596	9422	8698	19312	26450	37966
Ch B	27844	26934	27654	23456	24130	26556	25314	25968	25838	24640	26334	27655
S/N	0.751	0.908	0.872	0.925	0.801	0.579	0.437	0.186	0.173	0.402	0.515	0.704
RESULT	NEG.	NEG.	NEG.	NEG.	NEG.	POS.	POS.	POS.	POS.	POS.	POS.	NEG.

E FORM #	vet	vet	vet	vet	vet	vet	vet	vet	vet	vet	vet	vet
VET #	herd	herd	herd	herd	herd	herd	herd	herd	herd	herd	herd	herd
TUBE #	33	34	35	36	37	38	39	40	41	42	43	44
Ch A	38256	40042	8376	18166	25698	33156	38538	47452	9350	19250	28202	34692
Ch B	27124	28000	26543	27345	26876	23134	25345	26345	28567	24234	26278	25234
S/N	0.723	0.733	0.162	0.340	0.490	0.735	0.779	0.923	0.168	0.407	0.550	0.705
RESULT	NEG.	NEG.	POS.	POS.	POS.	NEG.	NEG.	NEG.	POS.	POS.	POS.	NEG.

TABLE 2 (continued)
IDEXX Brucellosis Test System, Screening Test Report

F FORM #	**vet**	**vet**	**vet**	**vet**	**vet**	**vet**	**vet**	**vet**	**vet**	**vet**	**vet**	**vet**
VET #	**herd**	**herd**	**herd**	**herd**	**herd**	**herd**	**herd**	**herd**	**herd**	**herd**	**herd**	**herd**
TUBE #	**45**	**46**	**47**	**48**	**49**	**50**	**51**	**52**	**53**	**54**	**55**	**56**
Ch A	40042	47654	45984	9350	19250	28202	34692	40042	47654	45984	38956	36432
Ch B	26234	26235	27456	26456	27234	28000	25345	26576	26465	27354	25660	26007
S/N	0.782	0.931	0.858	0.181	0.362	0.516	0.702	0.772	0.923	0.862	0.778	0.718
RESULT	NEG.	NEG.	NEG.	POS.	POS.	POS.	NEG.	NEG.	NEG.	NEG.	NEG.	NEG.
G FORM #	**vet**	**vet**	**vet**	**vet**	**vet**	**vet**	**vet**	**vet**	**vet**	**vet**	**vet**	**vet**
VET #	**herd**	**herd**	**herd**	**herd**	**herd**	**herd**	**herd**	**herd**	**herd**	**herd**	**herd**	**herd**
TUBE #	**57**	**58**	**59**	**60**	**61**	**62**	**63**	**64**	**65**	**66**	**67**	**68**
Ch A	29180	20586	6636	5578	93845	9290	20586	20758	29180	36432	38956	45984
Ch B	27665	27445	27663	27665	27334	27445	25334	25667	23456	27654	10567	27653
S/N	0.541	0.384	0.123	0.103		0.173	0.416	0.414	0.638	0.675		0.852
RESULT	POS.	POS.	POS.	POS.	INVALID	POS.	POS.	POS.	POS.	POS.	INVALID	NEG.
H FORM #	**vet**	**vet**	**vet**	**vet**	**vet**	**vet**	**vet**	**vet**	**vet**	**vet**	**vet**	**vet**
VET #	**herd**	**herd**	**herd**	**herd**	**herd**	**herd**	**herd**	**herd**	**herd**	**herd**	**herd**	**herd**
TUBE #	**69**	**70**	**71**	**72**	**73**	**74**	**75**	**76**	**77**	**78**	**79**	**80**
Ch A	50972	42376	35782	48456	43256	36258	27944	44598	44330	44337	52446	53425
Ch B	27323	9678	26445	28111	17995	25663	27446	27552	25664	18000	26778	27000
S/N	0.956		0.693	0.883	1.232	0.724	0.522	0.830	0.885	1.262	1.004	1.014
RESULT	NEG.	INVALID	POS.	NEG.	NEG.	NEG.	POS.	NEG.	NEG.	NEG.	NEG.	NEG.

TABLE 3
IDEXX Brucellosis Test System, Positive and Invalid Reports

Positive Report

Form #	Vet #	Tube #	S/N
vet	herd	1	0.589
vet	herd	2	0.605
vet	herd	3	0.590
vet	herd	6	0.172
vet	herd	7	0.216
vet	herd	8	0.456
vet	herd	17	0.187
vet	herd	18	0.446
vet	herd	19	0.646
vet	herd	26	0.579
vet	herd	27	0.437
vet	herd	28	0.186
vet	herd	29	0.173
vet	herd	30	0.402
vet	herd	31	0.515
vet	herd	35	0.152
vet	herd	36	0.340
vet	herd	37	0.490
vet	herd	41	0.168
vet	herd	42	0.407
vet	herd	43	0.550
vet	herd	48	0.181
vet	herd	49	0.362
vet	herd	50	0.515
vet	herd	57	0.541
vet	herd	58	0.384
vet	herd	59	0.123
vet	herd	60	0.103
vet	herd	62	0.173
vet	herd	63	0.416
vet	herd	64	0.414
vet	herd	65	0.638
vet	herd	66	0.675
vet	herd	71	0.693
vet	herd	75	0.522

Invalid report

Form #	Vet #	Tube #
vet	herd	11
vet	herd	61
vet	herd	67
vet	herd	70

is presented in Figure 17. Data were generated on samples obtained from field trial sites in Alabama, Florida, Texas, and Missouri. Note that the peak of the positive population is positioned at an S/N of approximately 0.15, indicating that the majority of infected animals have extremely high titers. The distribution skews from this peak toward negativity, with 4 of the 505 samples falling into the negative range.

TABLE 4
Instrument Through-put and Technician Time for Automated PCFIA Test for Antibody to *Brucella abortus*

Instrument(s)	Technician(s)	Batch runs per day[a]	Samples per day	Tech time per sample (sec)[b]	Samples per year
1	1	3	2880	10	748,800
1	1 to 1/2	4	3840	11	998,400
1	2	5	4800	12	1,248,000
2	2	7	6720	9	1,747,200
2	3	10	9600	9	2,496,000

[a] Instrument completes batch run (960 samples) in 1 h, 45 min. Initial results (96 samples) available in 30 min.

[b] Based on 8-h day and includes sample preparation (centrifugation/pipetting/tracking).

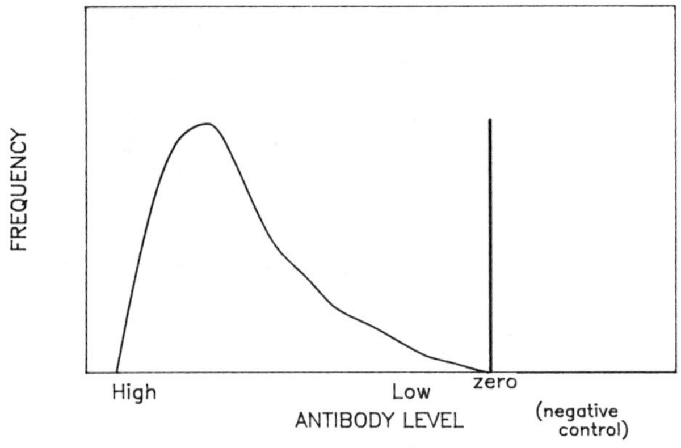

FIGURE 13. Distribution of antibody levels in a hypothetical positive population. The vertical bar represents an ideal assay cutoff infinitely close to the zero antibody point.

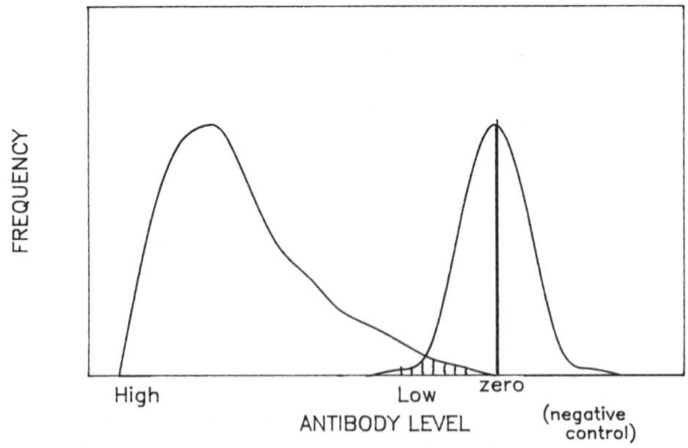

FIGURE 14. Realistic distribution of a hypothetical population of negative sera around the zero antibody point. The shaded area represents overlap between the negative and positive populations.

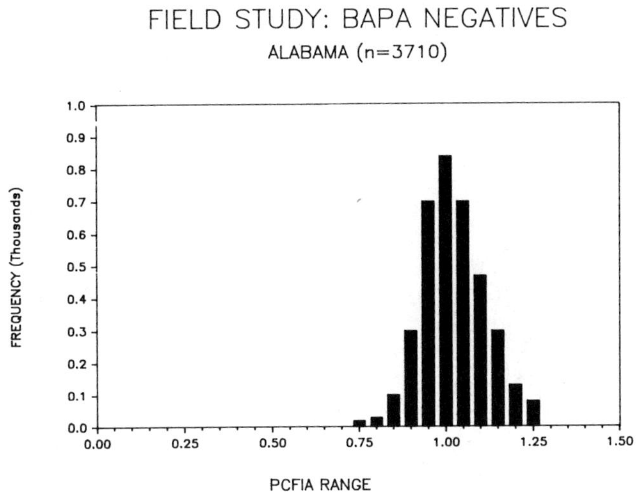

FIGURE 15. PCFIA S/N frequency distribution for Alabama animals negative for *B. abortus* antibodies by the BAPA test (n = 3710).

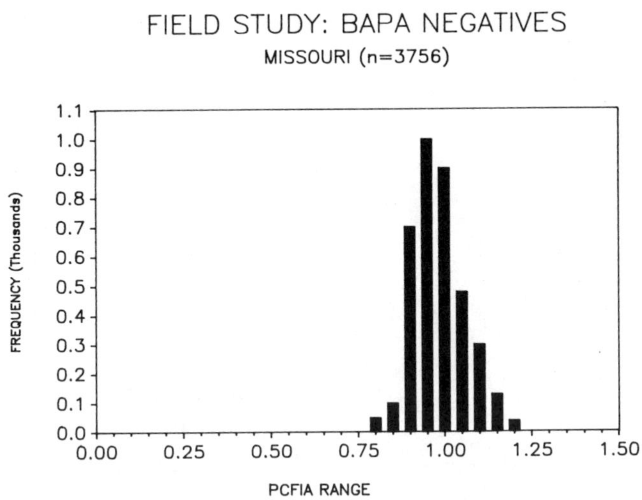

FIGURE 16. PCFIA S/N frequency distribution for Missouri animals negative for *B. abortus* antibodies by the BAPA test (n = 3756).

Combined population distribution — Figure 18 presents the above-described negative and positive population distributions in one histogram. Based on the substantial separation between positive and negative population peaks, the rather tight distribution of the negative population, and the apparently small overlap between the two populations, it would appear that the 0.70 S/N cutoff should provide a high level of sensitivity and specificity. Data that follow support this cutoff and the validity of the assay.

VII. ASSAY VALIDATION

A. SPECIFICITY

1. Brucellosis-Free Population

The specificity of a test may be defined as the number of true test negatives divided by the number of true or confirmed negatives. In the absence of a gold standard test for true

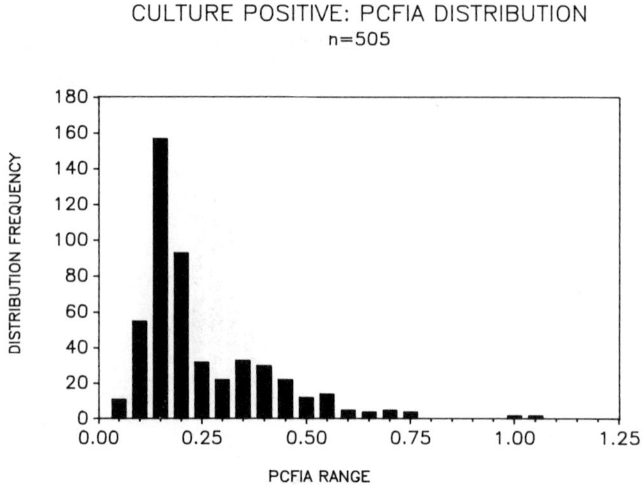

FIGURE 17. PCFIA S/N frequency distribution of *B. abortus*-positive animals confirmed by means of culture.

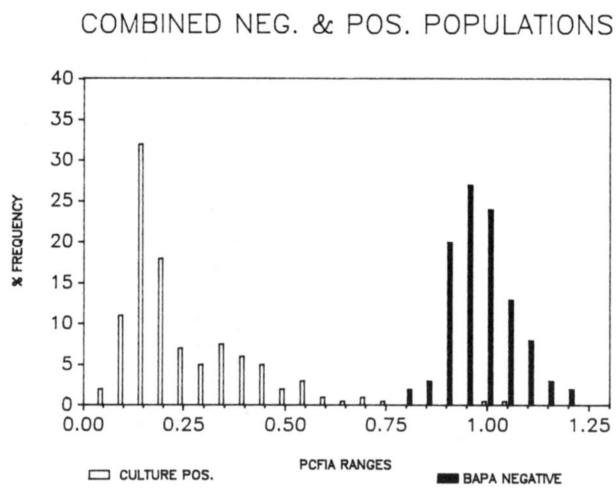

FIGURE 18. PCFIA S/N frequency distribution of combined negative and positive populations previously shown in Figures 15, 16, and 17.

negativity, an effort was made to test populations in areas classified as free of brucellosis. Assuming that all cattle tested are true negatives, one may obtain a "worst case" estimate of the specificity of the *B. abortus* PCFIA. Also, an estimate of the specificity of current tests may be made for comparison.

Pennsylvania was chosen as the brucellosis-free state in which program herds would be tested using the *B. abortus* PCFIA. To this end, a total of 52,856 samples were tested at the Pennsylvania laboratory. All samples were drawn from program herd animals, a population consisting of approximately 70 to 80% calfhood vaccinates. The PCFIA frequency distribution is presented in Figure 19. From this population, BAPA test results were available for 45,851 animals. The agreement of BAPA and PCFIA is shown in Table 5. There were a total of 563 positive BAPA tests and 98 positive PCFIA tests. Thirty-five (35) samples were both BAPA and PCFIA positive and 45,225 samples were negative by BAPA and

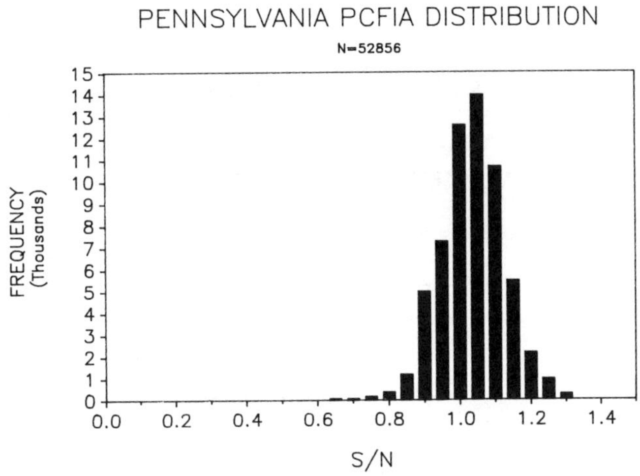

FIGURE 19. PCFIA S/N frequency distribution for Pennsylvania *B. abortus*-free animals (n = 52,856).

PCFIA. These represent a 1.23% positive rate for the BAPA test, compared to a 0.21% positive rate for PCFIA. A McNemar chi-squared test for statistical significance of the difference of the two tests reporting samples as negative was performed. This value is

$$\chi^2 = \frac{(528 - 63)^2}{591} = 365.9$$

For comparison, the $\chi^2_{0.95} = 3.84$ value at the 95% level of confidence; a calculated value greater than 3.84 indicates a significant difference in the specificity of the two tests if the calculated χ^2 is greater than this reference number.

In Pennsylvania, BAPA positives are subsequently tested by the "card" test. Card results were available on 514 of the 563 BAPA positives. Of this subset of samples, 150 were card positive and 32 were PCFIA positive as shown in Table 6. Twenty-three samples were both card and PCFIA positive; 355 were negative by both card and PCFIA. The McNemar chi-square statistic is 103.9, indicating a significant difference in the specificity of these two tests.

All card-positive samples are then assayed using Rivanol. Ninety-two samples fell into this category, resulting in 6 Rivanol positives (5 at 25, 1 at I50). These data for Rivanol are presented in Table 7. The PCFIA mean S/N for the 25 group is 0.638 ± 0.139; the one Rivanol I50 had a PCFIA S/N = 0.441.

The comparison of the card test with PCFIA is from a biased data set, that is, on samples previously screened by BAPA. In order to remove this bias, 1000 sera were randomly selected from Pennsylvania testing and sent to the Texas and Missouri state laboratories for card and PCFIA testing. Following receipt by the laboratory, 924 samples had sufficient volume for testing. Assaying those samples resulted in 10 card positives and 3 PCFIA positive samples. One sample was both card and PCFIA positive. The data are presented in Table 8. The McNemar statistic is equal to 4.45, indicating a significant difference in the specificity of those two tests in this randomly selected population.

From the tests performed on Pennsylvania's free population, one can conclude that PCFIA has greater specificity than both the BAPA and card tests. The application of PCFIA as the screening test would have resulted in an 82.6% decrease in the number of presumptive positive tests.

TABLE 5

		PCFIA −	PCFIA +
BAPA	−	45,225	63
BAPA	+	528	35

TABLE 6

		PCFIA −	PCFIA +
CARD	−	355	9
CARD	+	127	23

TABLE 7
PCFIA and Rivanol Comparison Listing (Pennsylvania)

RIV	PCFIA	BAPA	Card	Date	ID	Tube
25	0.513	+	+	9/03/87	23DAR4903	2
25	0.524	+	+	8/11/87	23VNPO632	3
25	0.539	+	+	8/20/87	23VHK5186	1
25	0.800	+	+	8/19/87	23VLZ6542	3
25	0.816	+	+	8/19/87	23DIE3658	1
I50	0.441	+	+	8/11/87	23VNP0G47	5

2. Certified Negative Herds

Herds in Missouri and Texas that were certified to be free of brucellosis were tested by PCFIA and BAPA; BAPA positives were tested by the card test. A combined population distribution (n = 661) is presented in Figure 20. In addition, a comparison of results among the three testing methodologies is presented in Table 9. No samples were positive by PCFIA, while 37 or 5.59% were positive by BAPA. None of the 37 BAPA positives were positive by the card test.

3. Individual Negative Herds

A number of herds have been followed as a function of time to establish the classification of the herd. Data from two of these herds are presented below.

Herd profiles from a Texas herd, TX-K, are presented in Figure 21. This herd contained

TABLE 8

CARD	PCFIA –	PCFIA +
–	912	2
+	9	1

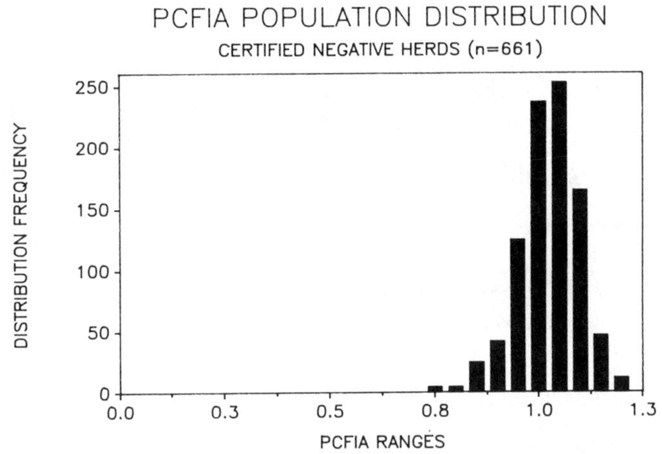

FIGURE 20. PCFIA S/N frequency distribution of Missouri and Texas certified negative animals (n = 661).

both calfhood- and adult-vaccinated animals. Note that portions of this herd were bled in October, November, December, and January. All samples at all bleed dates were negative by PCFIA. Comparisons with the card test are shown in Table 10. In October, 2 of 173 were positive by the card test, but negative by PCFIA as well as Rivanol (RIV), CF, and ELISA. On successive bleed dates, either 2 or 3 samples were card positive. As with the October sampling, all were negative by PCFIA, Rivanol, CF, and ELISA. It should be noted that the TX-K herd was certified as negative in the summer of 1987.

These data suggest that PCFIA is more specific than the card test. Furthermore, PCFIA appears to provide consistent results on individual herds from test date to test date.

4. Negative Adult Vaccinated Herds

Adult vaccinated herds were followed as a function of time at our Auburn field trial site. Herd profiles for four herds are presented in Figures 22A through 22D. Different colors represent the different bleed dates as noted in the legends. Raw data are shown in Tables 11 through 14.

Three of four herds (Figures 22 A, B, and C) were vaccinated in the 1st quarter of 1986 and first bled and tested late 1st quarter or 2nd quarter. The second bleeding took place 2 to 3 months later. The fourth herd (Figure 22D) was vaccinated in the 4th quarter of 1985 and tested in this study on June 13, 1986 and August 6, 1986.

TABLE 9
Field Study, Certified Negative Herds (n = 661; Texas and Missouri)

Test	No. Positive	% Positive
BAPA	37	5.59
PCFIA	0	0
Card	0	0

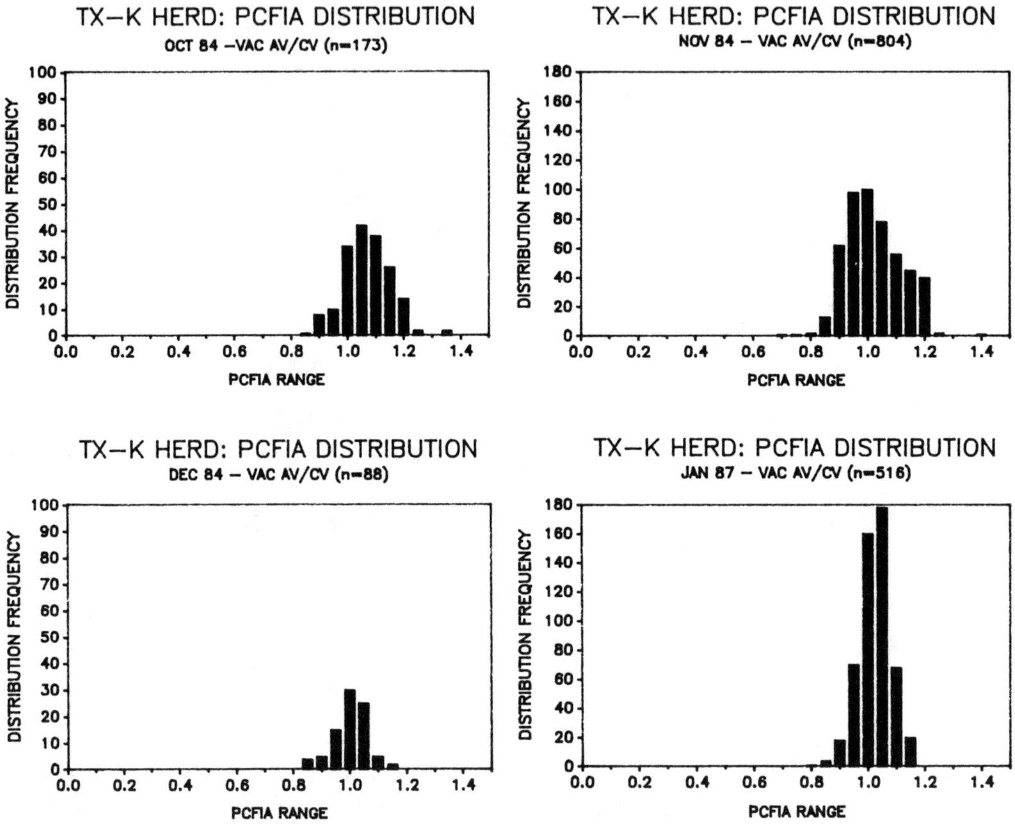

FIGURE 21. PCFIA S/N frequency distributions for four monthly tests of a *B. abortus* negative Texas herd (TX-K).

As demonstrated in Figures 22A and 22B, two of the three herds vaccinated in the 1st quarter of 1986 had become negative by the second test point. Both retests resulted in profiles distributed normally around the negative control with no positive samples by PCFIA. It should be noted, however, that the card test was positive on the second test date for one animal in the herd shown in Figure 22A and 10 animals in the herd shown in Figure 22B. The herd profile depicted in Figure 22C indicates that the vaccine titers are still elevated after the 2nd test date. On a per animal basis, however, all animals showed a reduction in titer by PCFIA. On July 10, four animals were still positive by PCFIA, while 9 remained positive by the card test. The herd represented in Figure 22D had become negative by the first test date and remained negative through the second testing. There was, however, one card positive on June 13, and four on August 6.

TABLE 10

TX-K HERD
10/20/86

PCFIA	CARD −	CARD +
−	171	2
+	0	0

TX-K HERD
11/17/86

PCFIA	CARD −	CARD +
−	502	3
+	0	0

TX-K HERD
12/15/86

PCFIA	CARD −	CARD +
−	86	2
+	0	0

TX-K HERD
1/21/86

PCFIA	CARD −	CARD +
−	515	3
+	0	0

With respect to adult-vaccinated herds, it is apparent that PCFIA provides advantages over the current methodologies. First, it facilitates the generation of herd profiles, which, if used with serial bleeding, allows better assessment of whether antibody levels are rising or falling on a herd basis. Second, PCFIA appears somewhat less sensitive to vaccine strain antibodies compared to the card test, thus more specific. This has eliminated a percentage of borderline samples.

5. Market Cattle

To further assess the specificity of PCFIA relative to that of BAPA, market cattle samples were tested in Missouri and Iowa. In both states, such samples were first tested by BAPA. Positive samples were subsequently tested by the card test and Rivanol test, if appropriate.

The population distribution for Missouri market cattle is presented in Figure 23. Of the 5473 samples tested, 72 or 1.3% were at or below the presumptive cutoff. With regard to BAPA, 457 or 8.4% of the total were positive. Of the 457 positives, 51 were positive by the card test; of those, 8 were positive by Rivanol. Seven of the 8 Rivanol positives were positive by PCFIA; the eighth sample was a +50 and had a S/N of 0.819. Of the 72 PCFIA positives, 39 were tested by the card test. Of these 39, 12 were card positive; seven of the latter were Rivanol positive.

The Iowa market cattle population distribution is presented in Figure 24. Of the 3384 samples tested, 42 or 1.2% were at or below the presumptive cutoff for PCFIA, while 95 or 2.8% were BAPA positive. Of the 95 positives, 10 were positive by the card test; Rivanols were run on 8 of the 10, of which 5 were positive. Four of the 5 were PCFIA positive; the 5th was a Rivanol I50 with an S/N of 0.97. Of the 42 PCFIA positives, 14 were tested by the card test. Four of the 14 were positive, all of which were Rivanol positive.

Of the 8857 MCI samples tested, there were 6 serological reactors having Rivanol titers of I200 or greater. PCFIA had S/N values of 0.25 or less for all 6 animals.

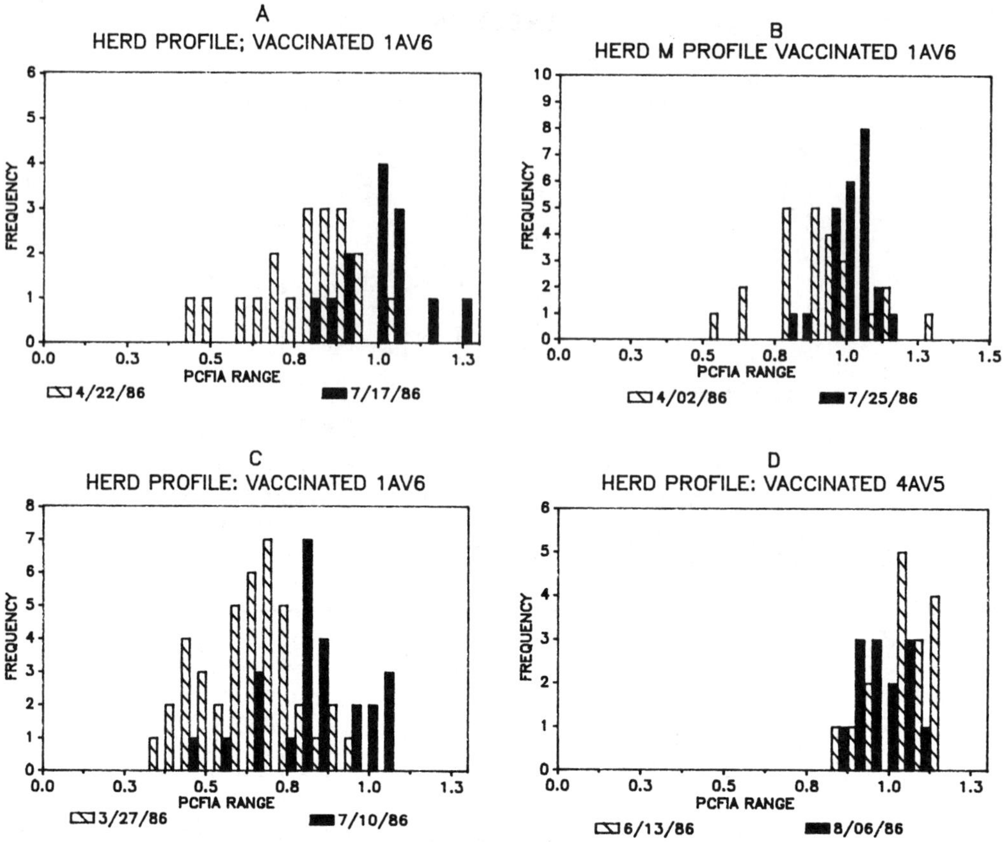

FIGURE 22. PCFIA S/N frequency distribution for the Auburn study.

With regard to market cattle testing, it appears that PCFIA is two to six times more specific than BAPA. Furthermore, no reactor animals were missed.

6. Heterologous Antisera

A 26-member panel of sera from calves experimentally infected with common viral and bacterial bovine pathogens was assayed using PCFIA. Results are shown in Figure 25 along with identification of the challenge organisms. None of these samples were positive in the *B. abortus* PCFIA.

In a separate study, two animals were challenged with either IBR or BVD and sequentially bled over a 2- to 3-month time course. These samples were assayed by PCFIA, BAPA, SPT, STT, card, CF, and Rivanol. PCFIA results from these animals are shown in Figure 26 as a temporal series and as a frequency distribution. Results are summarized in Tables 15 and 16 for all methodologies. None of these samples were positive by PCFIA, although BAPA, STT, and SPT showed some reactivity early in the time course, most likely reflecting nonspecific reactivity as a function of high IgM levels.

PCFIA appears more specific than SPT, STT, and BAPA in this study.

B. SENSITIVITY

The sensitivity of PCFIA relative to other methodologies has been evaluated in four different ways as follows: (1) performance on a panel of serially diluted samples; (2) ability to detect seroconversion in experimentally infected animals; (3) performance on sera from culture-positive animals; and (4) performance on infected herds. Results are presented below.

TABLE 11
Adult VAC 1AV6

Test date 5/22/87				Test date 7/17/86			
Card	Rivanol	CF	PCFIA	Card	Rivanol	CF	PCFIA
Pos	I200	3 + 40	0.446				
Neg	N	2 + 10	0.468	Neg			1.204
Pos	25	−10	0.579	Neg	N	−10	0.830
Neg	N	−10	0.635				
Neg			0.669	Neg			0.864
Neg	N	−10	0.691	Neg	N	−10	0.784
Neg	I25	−10	0.707	Neg	N	−10	1.001
Neg	N	−10	0.760	Pos	N	−10	0.890
Neg	N	−10	0.773	Neg	N	−10	0.990
Neg			0.796				
Neg			0.815				
Neg	N	−10	0.821	Neg	N	−10	0.958
Neg			0.846	Neg			0.974
Neg			0.869	Neg			1.050
Neg			0.873	Neg			1.146
Neg			0.889	Neg			0.964
Neg			0.932	Neg			1.127
Neg			0.936	Neg			1.010
Neg			1.014				

TABLE 12
Adult VAC 1AV6

Test date 4/2/86				Test date 7/25/86			
Card	Rivanol	CF	PCFIA	Card	Rivanol	CF	PCFIA
Pos	I200	1 + 640	0.388				
Pos	N	3 + 40	0.546	Neg			1.052
Pos	I25	3 + 10	0.645	Pos	N	−10	0.959
Pos	N	−10	0.646	Pos	N	−10	0.848
Pos	I25	3 + 20	0.752	Pos	N	−10	0.941
Pos	50	4 + 20	0.754	Pos	N	−10	0.788
Pos	N	−10	0.762	Neg			0.918
Pos	N	4—10	0.768				
Pos	I25	3 + 10	0.776	Neg			0.977
Pos	N	2 + 10	0.799	Neg			0.965
Pos	N	−10	0.864	Neg			0.936
Pos	25	4—20	0.870	Neg			1.152
Pos	N	−10	0.872	Pos	N	−10	1.033
Pos	N	2 + 10	0.885	Neg			1.012
Pos	N	2 + 10	0.886				
Pos	N	−10	0.897	Pos	N	−10	0.941
Pos	N	3 + 10	0.908	Pos	N	−10	1.012
Pos	N	−10	0.914	Neg			1.030
Pos	N	−10	0.937	Neg			1.056
Pos	N	3 + 10	0.947	Neg			0.994
Pos	N	−10	0.954	Neg			1.043
Pos	N	−10	0.974	Pos	N	−10	1.015
Pos	N	−10	0.991	Neg			1.042
Pos	N	−10	1.059	Pos	N	−10	0.950
Pos	N	−10	1.116	Neg			0.961
Pos	N	−10	1.122	Neg			1.089
Pos	N	−10	1.273	Pos	N	−10	0.987

TABLE 13
Adult VAC 1AV6

Test date 3/27/86				Test date 7/10/86			
Card	**Rivanol**	**CF**	**PCFIA**	**Card**	**Rivanol**	**CF**	**PCFIA**
Pos	100	1 + 80	0.316				
Pos	I100	2 + 40	0.374	Neg			0.435
Pos	N	2 + 20	0.383				
Pos	I100	4 + 40	0.406				
Pos	N	4 + 20	0.426	Neg			0.830
Pos	I200	4 + 40	0.439				
Pos	I50	4 + 40	0.448	Neg			1.045
Pos	I100	4 + 40	0.462	Pos	25	−10	0.607
Pos	25	1 + 40	0.479	Neg			0.784
Pos	N	2 + 40	0.500	Neg			0.638
Pos	N	4 + 40	0.544	Pos	N	−10	0.531
Pos	N	4 + 20	0.545				
Pos	I50	1 + 40	0.552				
Pos	N	2 + 20	0.555	Pos	N	−10	0.648
Pos	N	3 + 20	0.557	Neg			0.732
Pos	N	2 + 20	0.558				
Pos	I25	3 + 20	0.587				
Pos	N	2 + 20	0.606				
Pos	25	3 + 20	0.609	Neg			0.915
Pos	N	4 + 40	0.610	Pos	N	−10	0.775
Pos	I25	2 + 40	0.617				
Pos	N	2 + 20	0.624				
Pos	N	2 + 40	0.624	Neg	N	−10	0.763
Pos	I50	3 + 20	0.667				
Pos	N	3 + 20	0.668	Pos	N	−10	0.817
Pos	N	3 + 20	0.670	Neg			0.829
Pos	N	4 + 20	0.683				
Pos	N	2 + 20	0.684	Pos	N	−10	0.769
Pos	N	3 + 20	0.692				
Pos	25	1 + 40	0.700	Pos	25	−10	0.767
Pos	25	4 + 40	0.701				
Pos	25	2 + 10	0.722	Neg			0.901
Pos	N	−10	0.743	Neg			1.020
Pos	N	3 + 20	0.750	Neg			0.996
Pos	I25	1 + 40	0.750	Neg	N	−10	0.786
Pos	N	1 + 10	0.766				
Pos	N	1 + 10	0.789	Pos	N	−10	0.779
Pos	N	2 + 10	0.845	Neg			1.028
Pos	N	−10	0.854	Neg			0.981
No Test	N	1 + 20	0.881				
Pos	N	2 + 20	0.916	Pos	N	−10	0.818

TABLE 14
Adult VAC 1AV6

Test date 6/13/86				Test date 8/06/86			
Card	Rivanol	CF	PCFIA	Card	Rivanol	CF	PCFIA
Neg	Neg	−10	0.839	Neg			0.934
Pos	Neg	−10	0.895				
Neg	Neg	−10	0.915	Pos	Neg	−10	0.813
Neg	Neg	−10	0.935	Pos	Neg	−10	0.885
Neg	Neg	−10	1.011	Neg			0.987
Neg	Neg	2 + 10	1.022	Neg			0.919
Neg	Neg	−10	1.027	Neg			0.861
Neg	Neg	−10	1.035	Pos	Neg	1 + 10	0.922
Neg	Neg	−10	1.035	Neg			1.016
Neg	Neg	−10	1.077				
Neg	Neg	1 + 10	1.080	Neg			0.948
Neg	Neg	−10	1.100	Neg			1.028
Neg	Neg	−10	1.102	Neg			1.054
Neg	Neg	−10	1.104	Neg			1.012
Neg	25	−10	1.131	Neg			0.899
Neg	Neg	3 + 10	1.135	Pos	Neg	−10	0.967

1. Dilution Panel

Two identical 44-member panels were prepared by serially diluting serum with high levels of antibodies to strain 2308 into negative serum. The samples were then randomly numbered and both panels (A and B) were sent to our field trial sites in Missouri, Texas, Pennsylvania, and Iowa. Results from this study are shown in Table 17. The last detectable panel member is listed by site for each methodology. PCFIA appears essentially equivalent to the plate test, slightly more sensitive than the card test, and more sensitive than Rivanol and CF.

2. Experimental Infection

Serum samples were obtained from Drs. Stringfellow and Sparling at Auburn University, who had experimentally infected 16 Holstein heifers with Strain 2308 and followed their response serologically (standard tube test, card test, Rivanol, CF) as a function of time. Details of the study are presented in Table 18. Mean value results are presented for 12 animals in Figure 27. The reciprocal of the S/N Values, 1/(S/N), and titer are plotted on the Y axis while time in days is plotted on the X axis. Results are plotted for PCFIA (1/[S/N]), RIV, CF, and STT. In general, PCFIA and STT became positive in parallel whereas RIV and CF lagged behind, typically by 2 weeks. It should be noted that all animals were culture positive at some point during the study. Furthermore, the card test exhibited a 98% correlation with PCFIA on these samples, although the results are not plotted since titers are not obtained.

These results are consistent with the sensitivity data presented previously. PCFIA will provide sensitivity equivalent to, or better than, the card test and other screening tests based on agglutination. Rivanol and CF methodologies, again, are less sensitive than PCFIA.

3. Performance on Culture-Positive Animals

Serum samples from 505 culture-positive animals were obtained from our field trial sites in Alabama, Florida, and Texas, and tested by PCFIA. Additionally, a panel of sera from culture-positive cattle was provided by the USDA.

PCFIA results for sera obtained from field sites were compared, when possible, to those obtained using STT, card, RIV, and CF. The comparisons are shown in Table 19. Included

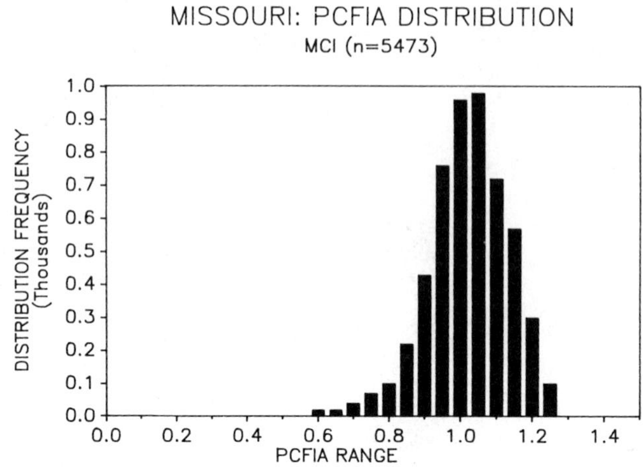

FIGURE 23. PCFIA S/N frequency distribution for Missouri market cattle (n = 5473).

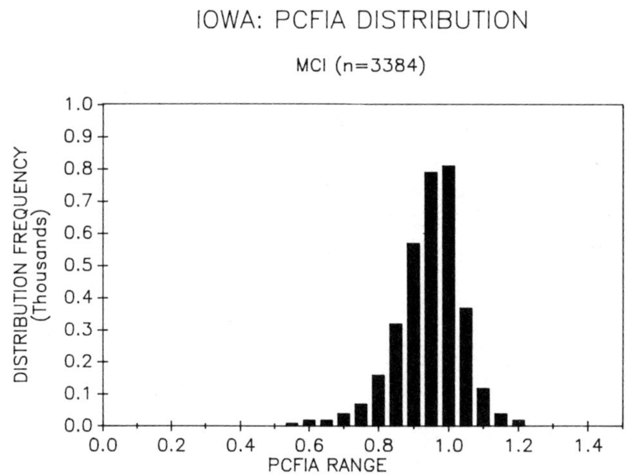

FIGURE 24. PCFIA S/N frequency distribution for Iowa market cattle (n = 3384).

in the comparison are results of the McNemar (chi)^2 test for equivalency. PCFIA, card, and STT demonstrated essentially equivalent sensitivity on culture-positive animals. CF appears less sensitive but not significantly so, whereas RIV is substantially less sensitive than PCFIA. These results are consistent with data generated on diluted positives and on sera from experimentally infected animals. Infection with biotypes 1, 2, and 4 were detected equally by all methodologies. Table 20 lists only discrepant samples from the various methodologies, i.e., any sample which tested negative by any or all methods. Three of the four samples missed by PCFIA were missed by all other tests. The fourth was positive by Card, RIV (+200), and CF (4 + 640); sample volume for this sample was not sufficient for retesting. Based on our sensitivity data for PCFIA, the only plausible explanation is a sample labeling error.

Results on the USDA panel are presented in Table 21. A total of 214 samples was tested as a blind panel. PCFIA and the card test performed equivalently. Unlike the previous panel, STT was 6% more sensitive than PCFIA and Card, which is statistically significant. PCFIA was significantly more sensitive than Rivanol and CF.

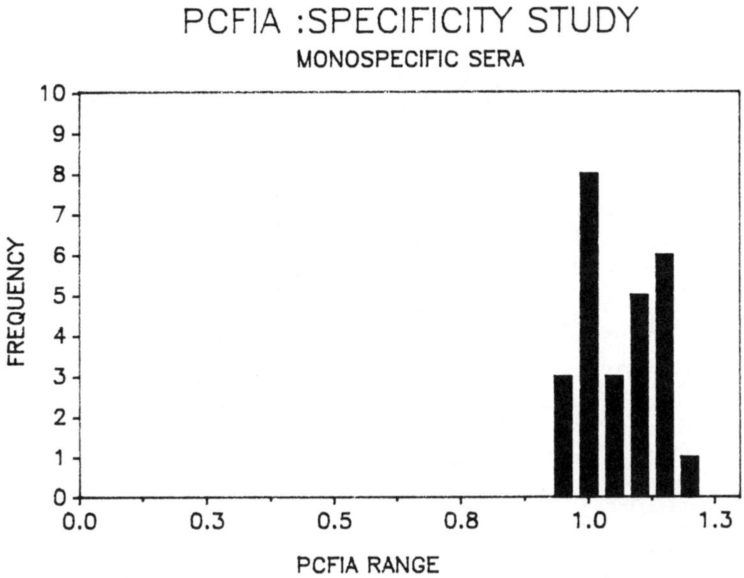

Samples consisted of hyperimmune bovine sera to:

Bovine Enterovirus (Type 1)
Vesicular Stomatitis Virus (Type NJ and IN)
Herpes Mammalitis Virus
Bovine Herpes Virus (Types I and III)
Malignant Catarrhal Fever (OK strain and African strain)
Pseudocowpox
Bovine Parvovirus
Bovine Respiratory Syncytial Virus
Bovine Rotavirus
Bovine Viral Diarrhea Virus
M. paratuberculosis
M. bovis
Bovine polyvalent (hyperimmune to 22 viruses)
Normal Calf Serum

FIGURE 25. PCFIA S/N frequency distribution of non-*B. abortus* hyperimmune serums.

Data from the two panels are combined in Table 22. A total of 719 samples from culture-positive animals were tested by PCFIA, 28 of which were called negative. This translates to a sensitivity of 96.1%, which is slightly better (not statistically significant) than card at 95.8% and slightly worse (statistically significant) than STT at 97.7%. Sensitivity of CF and Rivanol were 92.4 and 91.8%, respectively, significantly worse than PCFIA. These combined results are consistent with results obtained from the testing of a sensitivity panel and the testing of experimentally infected animals as a function of time; PCFIA as a presumptive test is essentially equivalent in sensitivity to the agglutination-based screening tests and substantially more sensitive than Rivanol and CF.

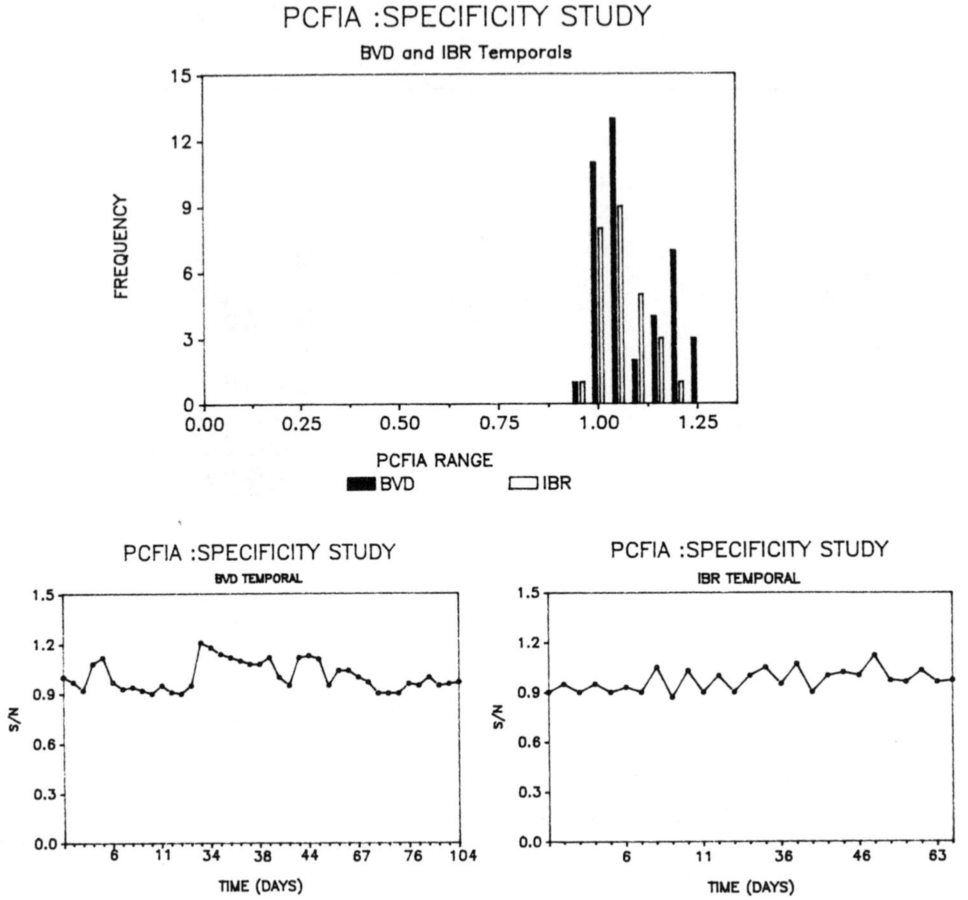

FIGURE 26. PCFIA S/N frequency distribution of bovine viral diarrhea (BVD) virus and infectious bovine rhinotracheitis (IBR) virus temporal studies (top). PCFIA S/N temporal studies for BVD (bottom left) and IBR (bottom right).

TABLE 15
IBR Temporal Series
Comparative Assay Specificity

Infectious agent	Result	PCFIA[a]	BAPA	Card	SPT[b]	STT[c]	RIV[d]	CF[e]
IBR	Neg.	27	18	27	5	11	27	21
	Pos.	0	7	0	22	16	0	1
	NA[f]	0	2	0	0	0	0	5

[a] PCFIA, S/N mean = 0.986; CV = 6.1%.

[b] SPT

Test dil.	−25	I25	25	I50	≥50
Frequency	5	14	2	6	0

[c] STT

Test dil.	−25	I25	25	I50	≥50
Frequency	11	8	8	0	0

[d] RIV (Rivanol), all negative at 1:25 dilution.

[e] CF, 21 negative at 1:10 dilution; 1 positive (1+) at 1:10 dilution; 5 anticomplementary.

[f] NA, Result not available.

TABLE 16
BVD Temporal Series
Comparative Assay Specificity

Infectious agent	Result	PCFIA[a]	BAPA	Card	SPT[b]	STT[c]	RIV[d]	CF[e]
BVD	Neg.	41	38	41	18	35	40	38
	Pos.	0	1	0	22	6	0	0
	NA[f]	0	1	0	0	0	1	3

[a] PCFIA, S/N mean = 1.01; CV = 8.8%.

[b] SPT

Test dil.	−25	I25	25	I50	⩾50
Frequency	18	18	3	1	0

[c] STT

Test dil.	−25	I25	25	I50	⩾50
Frequency	35	5	1	0	0

[d] RIV (Rivanol), all negative at 1:25 dilution.

[e] CF, 38 negative at 1:10 dilution; 3 anticomplementary.

[f] NA, Result not available.

TABLE 17
Field Study: Dilution Panel Relative Sensitivities

Site	PCFIA	BAPA	Card	Rivanol	CF	SPT
1A	6	ND	5	4.75	5	ND
1B	6	ND	4.75	4.75	5	ND
2A	6	6	4.75	ND	ND	ND
2B	6	7	4.50	ND	ND	ND
3A	6.25	6	ND	ND	ND	ND
3B	5.75	6	N1	ND	ND	ND
4A	6.25	5.75	4	5	ND	3.5
4B	6	5.75	3.5	ND	ND	3
5A	7	6	5	4	ND	3
6B	7	5.75	5	4	ND	3

Note: Two identical panels were prepared by serially diluting serum from a high titered infected animal (Strain 2308) into negative serum. The consecutive two-fold dilutions were identified 1 through 11 corresponding to serum dilutions 1:2 through 1:2048. The numeric values for the table indicate the average result for 4 replicates per dilution panel number representing the last detectable panel member for each methodology.

TABLE 18
Experimental Parameters in Strain 2308 *Brucella abortus* Temporal Study[a]

1. Animals: 16 Holstein heifers 12 to 24 months of age, good condition. Two heifers were pregnant (#15 and #18).
2. Exposure: Animals were administered 110 million cells.[b]
3. Sampling: Animals were bled weekly via the jugular vein.
4. Infection: Infection was confirmed by positive culture results from all 16 heifers.

[a] Conducted by Drs. David Stringfellow and Phyllis Sparling, School of Veterinary Medicine, Auburn University, Auburn, Alabama.

[b] Conducted by Dr. Billy Deyoe, NADC, U.S. Department of Agriculture, Ames, Iowa.

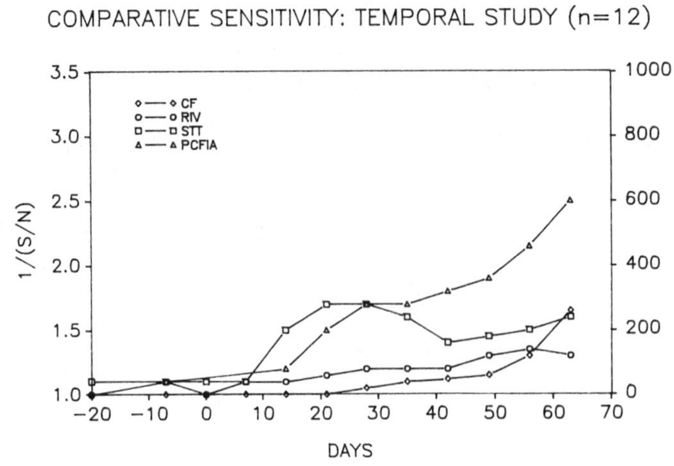

FIGURE 27. Arithmetic mean for 12 *B. abortus* experimentally infected calves assayed by PCFIA, RIV, CR, and STT.

TABLE 19
Methodology Comparison of Culture Positives

Methodology	No. tested	No. missed	Sensitivity (%)	$\chi^2_{0.95}$[a]
PCFIA	505	4	99.2	—
Card	505	6	98.8	(0.2) accept
STT	435	3	99.3	(0) accept
Rivanol	504	21	95.8	(17) reject
CF	484	8	98.3	(3) accept

[a] McNemar χ^2 statistic to test the hypothesis that PCFIA has equivalent sensitivity to the other methodologies. Results expressed as "reject" or "accept" hypothesis. Cutoff for $\chi^2_{0.95}$ at 95% confidence is 3.84.

4. Performance on Infected Herds

Infected herds were tested by PCFIA in parallel with other methodologies to further validate the presumptive cutoff with respect to sensitivity and to validate a less sensitive but highly specific diagnostic cutoff. Recall that PCFIA responds to a continuum of antibody levels such that each S/N value represents a discrete measurement of antibody in the sample. The presumptive cutoff was set at a level to maximize sensitivity while retaining acceptable specificity. The intent of setting a diagnostic cutoff is to maximize specificity so that animals having antibody levels at or below this cutoff have a very high probability of being true reactors. Animals falling between 0.25 and 0.7 should be regarded as suspicious and should be further evaluated with additional tests. Data are presented below which validate the diagnostic cutoff as well as further validate the sensitivity of PCFIA at the presumptive cutoff.

Data on program herds from Missouri and Texas are presented and analyzed based on the presumptive cutoff of 0.7 for the correlation with card and the diagnostic cutoff of 0.25 for correlation with Rivanol and CF. The herds were chosen at the discretion of the epidemiologist of the state, and samples were assayed in their laboratories.

TABLE 20
Culture Positive: Discrepant Samples (All Methodologies)

ID	Card	STT	Rivanol	CF	PCFIA
10/30/78	Pos	100	I25	4 + 20	0.237
10/17/78	Pos	I1600	200	AC	0.02
10/19/78	Pos	800	200	AC	0.114
11/16/78—9	Pos	50	Neg	2 + 40	0.368
1/22/79—110	Neg	25	25	3 + 10	0.577
1/19/79—13	Pos	50	Neg	1 + 20	0.606
2/15/79—493	Pos	50	Neg	4 + 20	0.196
3/13/79—1	Pos	I1600	200	AC	0.108
3/17/79—13	Pos	50	Neg	3 + 20	0.211
3/13/79—109	Neg	50	Neg	Neg	0.566
3/13/79—101	Pos	200	50	AC	0.132
3/27/79—1	Pos	100	Neg	3 + 40	0.378
5/7/79—23	Pos	I1600	200	AC	0.127
5/28/79—10	Pos	50	Neg	Neg	0.729
4/23/79—53	Pos	I200	I200	AC	0.114
5/16/79—22	Pos	100	25	AC	0.286
5/16/79—24	Pos	I100	Neg	3 + 20	0.224
7/16/79—13	Pos	800	200	Neg	0.098
8/14/79—43	Pos	1600	200	AC	0.083
11/27/79—5	Neg	I25	Neg	Neg	0.685
12/29/79—1	Pos	I200	Neg	1 + 80	0.146
1/21/80—7	Pos	6400	200	AC	0.107
1/21/80—15	Pos	I400	50	AC	0.104
1/21/80—1	Pos	1600	200	AC	0.120
1/23/80—21	Pos	1600	200	AC	0.115
1/24/80—1	Pos	800	100	AC	0.129
4/10/79—27	Pos	50	Neg	2 + 10	0.336
4/23/79—41	Pos	400	200	AC	0.155
8/17/79—7	Pos	1600	200	AC	0.177
4/24/76—6	Pos	200	Neg	AC	0.436
4/21/78—7	Pos	100	Neg	3 + 20	0.488
10/25/78—55	Neg	50	Neg	Neg	1.009
8/30/77—41	Neg	Neg	Neg	Neg	1.302
5/31/78—9	Pos	I200	I25	4 + 80	0.346
4/12/78—1	Pos	100	25	AC	0.531
10/24/78—39	Pos	100	I50	AC	0.585
4/27/78—4	Pos	200	I200	AC	0.305
4/5/78—54	Pos	200	200	AC	0.404
5/7/79—37	Pos	I400	200	AC	0.464
393	Pos	NT	200	4 + 640	0.951
8936	Neg	NT	NT	Neg	1.020
8796	Pos	NT	Neg	5	0.290
8711	Pos	NT	Neg	Neg	0.730
8815	Pos	NT	Neg	1 + 10	0.260
9811	Pos	NT	Neg	1 + 10	0.230
8264	Pos	NT	Neg	1 + 80	0.090
8804	Pos	NT	Neg	1 + 20	0.210

Note: AC, anticomplementary; NT, no test.

The MO-J herd was bled May 23, July 31, and September 12, 1986. Herd profiles are presented in Figure 28; correlation with card is presented in Table 23. On May 23, 1986, 26 animals were BAPA positive; 5 were PCFIA positive. Six animals were card positive and were culled. Animal AY02382 had an S/N = 0.698 and was card negative. On the July 31, 1986 bleeding, 33 animals were BAPA positive, 5 of which were card positive and

TABLE 21
Methodology Comparison of Culture Positives (USDA Panel)

Methodology	No. tested	No. missed	Sensitivity (%)	$\chi^2_{0.95}$
PCFIA	214	24	88.8	—
Card	214	24	88.8	(0) accept
STT	214	12	94.4	(8) reject
CF	214	45	78.9	(18) reject
Rivanol	197	38	80.7	(12) reject

TABLE 22
Methodology Comparison of Culture Positives (Combined)

Methodology	No. tested	No. missed	Sensitivity (%)	$\chi^2_{0.95}$
PCFIA	719	28	96.1	—
Card	719	30	95.8	(0.07) accept
STT	649	15	97.7	(6.5) reject
Rivanol	719	59	91.8	(27.9) reject
CF	698	53	92.4	(20.6) reject

PCFIA positive (0.259, 0.165, 0.095, 0.391, and 0.084) and were culled. One animal was card negative, PCFIA positive (S/N = 0.663), and CF (23) positive. Animal AY02382 became a reactor by all methodologies. On September 12, 1986, 19 animals were BAPA positive; none of the animals were card or PCFIA positive. Table 24 shows other test data for animals below S/N 0.25.

Another MO-J (J2) herd was tested on June 12 and July 28, 1986. Profiles are shown in Figure 29; correlation with the card test is shown in Table 25. On June 12, 12 animals were BAPA positive, 9 of which were card positive. All 9 were PCFIA positive (0.176, 0.347, 0.243, 0.162, 0.433, 0.573, 0.282, 0.258, 0.587) and were culled. On July 28, 5 animals were BAPA positive, card positive, and PCFIA positive (0.291, 0.574, 0.669, 0.110, 0.094). Two animals were card negative, PCFIA positive (0.594, 0.696) with a CF of 22 for both animals. Animals below the diagnostic cutoff of 0.25 are shown in Table 26.

Figure 30 depicts a herd profile from an infected herd (MO-H) with two calfhood vaccinates tested on June 17, 1986; comparison with the card test is shown in Table 27. Of the 14 PCFIA-positive samples, 9 were card positive at this bleeding. The 9 Card positives were removed from the herd. Three additional animals became reactive by current methodology on the July 29, 1986 bleeding. These three new reactors are identified numerically in both herd profiles as 1, 2, and 3. Animals 1 and 3 were PCFIA positive and card negative on June 17, while animal 2 was negative by both tests on June 17. Thus, PCFIA detected as positive on June 17 two of the three animals that became card positive on July 29. One calfhood vaccinate was negative at both bleeds; the second was considered positive by all methods on July 29 (numbered 1 and 3 in Figure 30). Animals below the diagnostic cutoff are shown in Table 28.

5. Texas Program Herd

Working in conjunction with Dr. Scott Reynolds, a Texas herd (TX-W) was extensively studied over a 6-month period. The testing regimen included PCFIA, card, Rivanol, automated CF, and ELISA. Reactors were culled and cultured, while suspects were initially separated from the main herd. However, a number of suspect and negative animals were

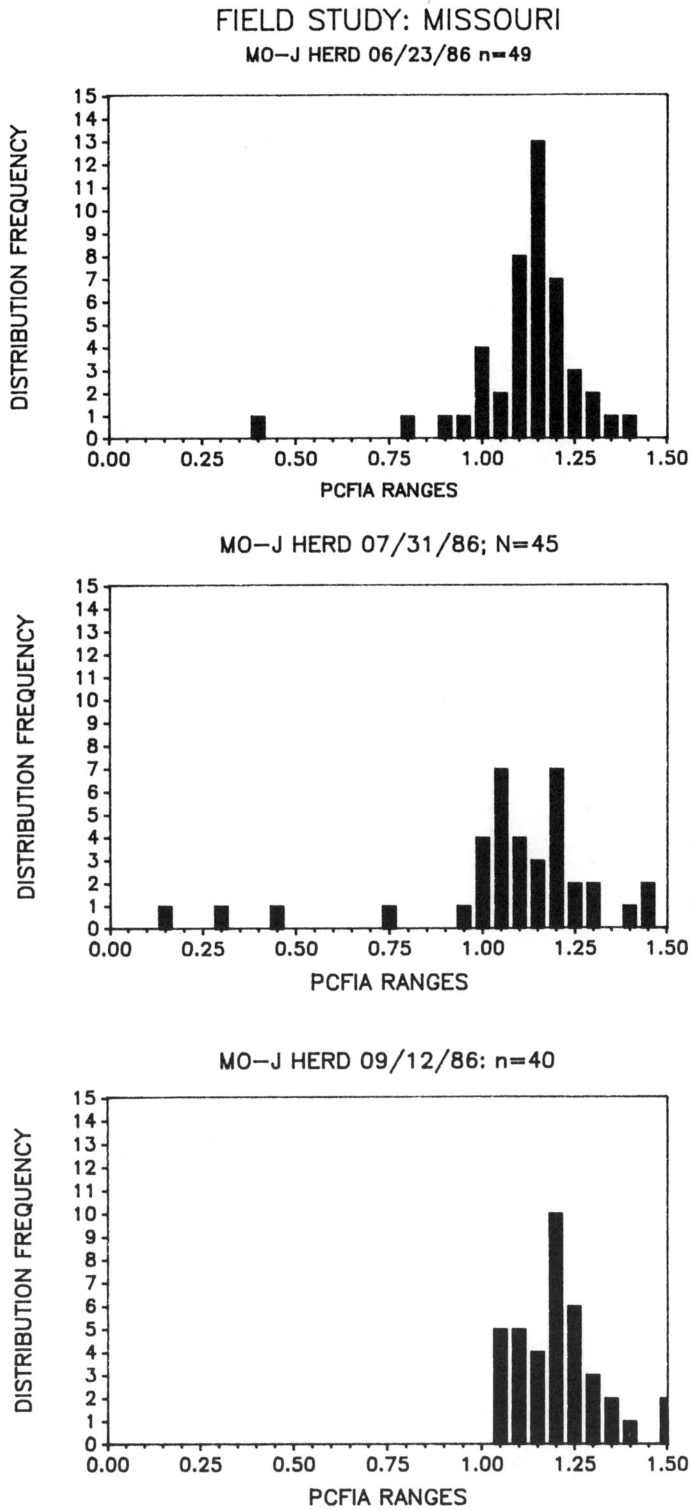

FIGURE 28. PCFIA S/N frequency distribution for a Missouri herd (MO-J) tested over the course of 3 months.

TABLE 23
MO-J Herd

5/23/86

PCFIA	CARD −	CARD +
−	42	2
+	1	4

7/31/86

PCFIA	CARD −	CARD +
−	39	0
+	1	5

9/12/86

PCFIA	CARD −	CARD +
−	40	0
+	0	0

TABLE 24
MO-J Herd, Animals below the Diagnostic Cutoff of 0.25 S/N

Animal ID	Card	SPT	Rivanol	CF	PCFIA(S/N)
AX07902	Pos	+200	+200	1:160	0.165
AX02382	Pos	I200	+200	1:160	0.095
AX02396	Pos	+200	+200	1:160	0.084

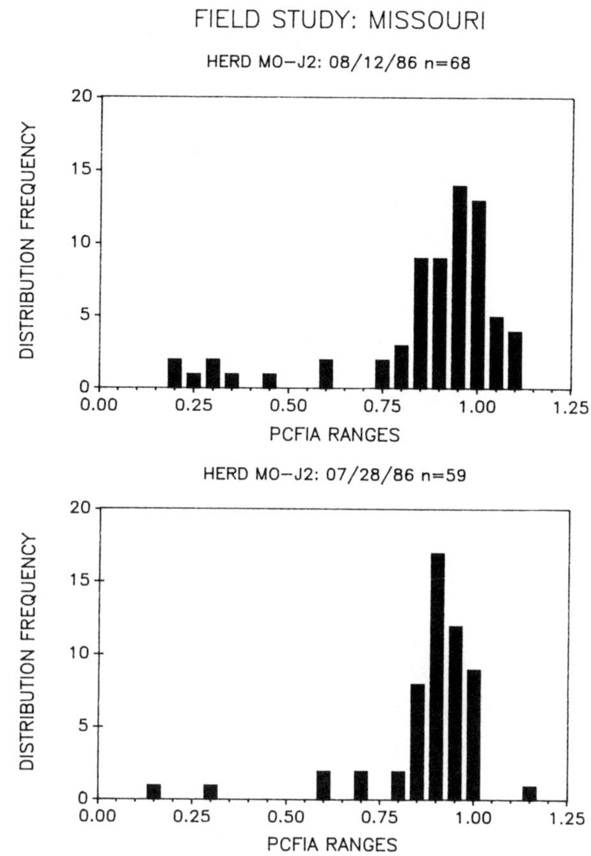

FIGURE 29. PCFIA S/N frequency distribution for a Missouri herd (MO-J2) retested after a 1-month interval, following removal of *B. abortus*-positive animals, based on Card test results.

also cultured. The herd was first tested in August with subsequent tests occurring in September, October, November, December, and January. Herd profiles are presented in Figure 31. The reader will note that the herd size presented in the frequency distributions for August through December changed only as a function of reactor animal removal. The January frequency distribution, however, represents a quarantined portion of the TX-W herd (n = 240) that had consistently tested negative for antibodies to *B. abortus*. The herd cleanup plan was to continue monthly testing and eventually designate this subpopulation of animals as *B. abortus* "certified free".

Due to the nature of this epidemiological study and the extensive amount of culturing, the following discussion will be directed primarily toward the culture results and the time course of the animals cultured.

Lymph nodes were taken from 61 animals at time of slaughter. Eleven were serological negatives; no isolation was made (Table 29). Twenty-seven were serological suspects; biotype 1 was isolated from five. Twenty-three were serological reactors; biotype 1 was isolated from 21.

Serological test results are presented as a function of time for each culture-positive animal. Eighteen animals were slaughtered immediately after the second test date. Biotype 1 was isolated from each of these animals. Serological results are presented for each bleed date in Table 30.

TABLE 25
J2 Herd

6/12/87

PCFIA \ CARD	−	+
−	50	0
+	0	9

7/28/86

PCFIA \ CARD	−	+
−	52	0
+	2	5

TABLE 26
MO-J2 Herd, Animals below the Diagnostic Cutoff of 0.25 S/N

Animal ID	Card	SPT	Rivanol	CF	PCFIA (S/N)
43BDF4615	Pos	+200	+200	1:160	0.110
43VVM2427	Pos	QNS	+200	1:160	0.094
43BDF4590	Pos	+100	I200	ND	0.176
43BDF4600	Pos	I100	I100	ND	0.242
43BDF4610	Pos	I100	I200	ND	0.162

Note: QNS, quantity not sufficient for testing; ND, no determination available.

These results may be summarized as follows:

Test	No. Positive on 8/13/86	No. positive on 9/17/86
PCFIA	16	17
Card	3	18
CF	10	17
Rivanol	1	14

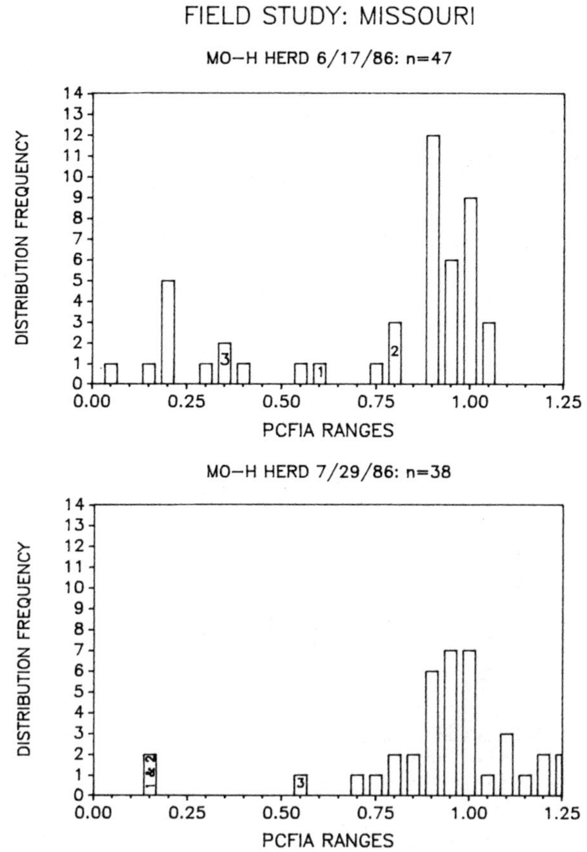

FIGURE 30. PCFIA S/N frequency distributions for a *B. abortus* infected herd (MO-H) retested after removal of card positive reactors. The numbers in the bars represent the location of the PCFIA value of three animals (1, 2, and 3) on two successive tests.

TABLE 27
MO-H Herd

6/17/87 CARD

PCFIA	CARD –	CARD +
–	33	0
+	5	9

7/29/86 CARD

PCFIA	CARD –	CARD +
–	34	0
+	1	3

TABLE 28
MO-H Herd, Animals below the Diagnostic Cutoff of 0.25 S/N

Animal ID	Card	SPT	Rivanol	CF	PCFIA (S/N)
43AYL6509	Pos	+200	+200	ND	0.175
43AYL6639	Pos	I200	I200	ND	0.180
43AYL6651	Pos	+200	I200	ND	0.227
43BDC5525	Pos	+200	+200	ND	0.182
43BFL4188	Pos	I50	ND	ND	0.162
43BES8200	Pos	+200	+200	ND	0.151
43BDG5699	Pos	+200	+200	1:160	0.101
43BFL4193	Pos	+200	+200	1:160	0.118

Note: ND, no determination available.

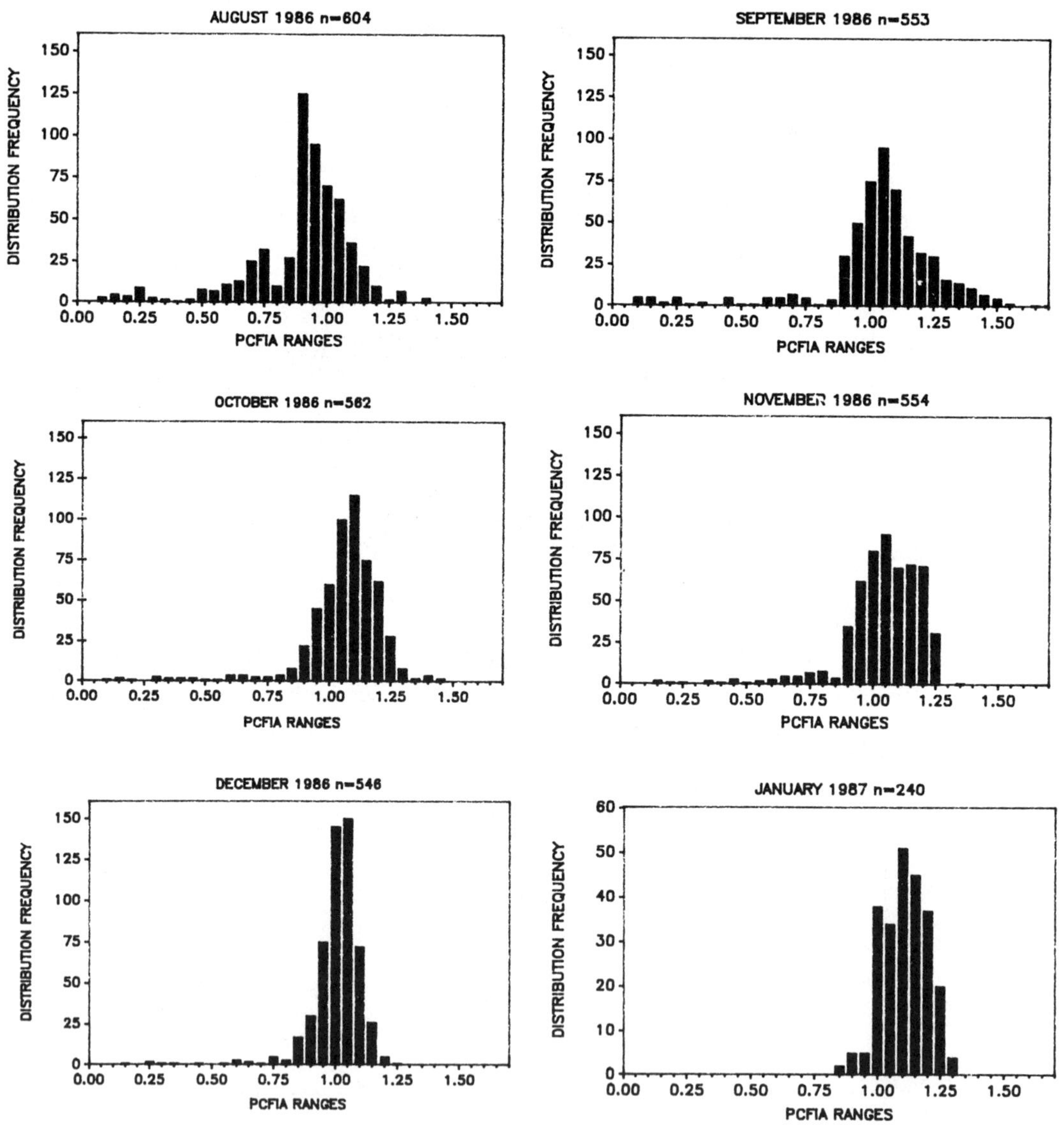

FIGURE 31. PCFIA S/N frequency distributions for tests over a 6-month period on a *B. abortus* infected Texas herd (TX-W).

TABLE 29
Bacteriological Results of Culled Animals in Suspect Herd Negative on All Tests at Time of Slaughter

ID	Card	Rivanol	CF	PCFIA	Culture
8393	N	NT	N	1.042	N.I.
8745	N	NT	N	1.103	N.I.
8779	N	NT	N	1.153	N.I.
8381	N	NT	N	1.046	N.I.
8472	N	NT	N	1.146	N.I.
8496	N	NT	N	1.061	N.I.
8788	N	NT	N	1.021	N.I.
8891	N	NT	N	1.04	N.I.
8329	N	NT	N	0.929	N.I.
8962	N	NT	N	1.046	N.I.
8390	N	NT	N	1.116	N.I.
8315	N	NT	N	1.099	N.I.
8345	N	NT	N	1.082	N.I.

Note: N.I., no *Brucella abortus* isolated.

PCFIA identified as positive on the first test date 16 of 18, whereas Card, CF, and Rivanol detected only 3, 10, and 1 as positive, respectively. In addition, seven animals had PCFIA S/N ≤0.25 on the first bleed date; however, none of these animals were reactor level by CF; furthermore, four were card negative. In general, all animals showed an increase in reactivity for each methodology from the August 13 bleed to the September 17 bleed. The exceptions were 8804, which remained Rivanol negative, and 8711, which remained CF negative and became PCFIA negative (0.65 to 0.73). Animals with reactor level CF at time of slaughter were all below the 0.25 PCFIA diagnostic cutoff.

Serological results for animals followed for 60 to 180 days prior to slaughter are shown in Table 31. Eight of these animals were culture positive for biotype 1. PCFIA detected as positive 6 or 8 at the August 13, 1986 bleed date. Two of the six have S/N values of ≤0.25. A seventh animal was detected as positive at the November 19, 1986 bleeding. The eighth animal was not detected by PCFIA, card, Rivanol, or CF, but was a suspect by ELISA. The card became positive after PCFIA on all seven positive samples. Rivanol was negative at the time of culturing of four of the eight animal samples and became positive after PCFIA on the four it detected. CF also became positive later than PCFIA in four of eight cases.

From the Texas data base, 142 samples with CF titers of 5 or greater were extracted. A graph of the mean PCFIA S/N ± one standard deviation is shown in Figure 32. For animals with reactor level titers, all but three had PCFIA SN values ≤0.250; these three had S/Ns of 0.270, 0.302, and 0.352 and had CF titers of 21.

On the TX-W herd, PCFIA was more sensitive than the card test as a presumptive test. In addition, at the reactor level titer of the automated CF, 94.3% of the PCFIA S/Ns were below the 0.25 diagnostic cutoff (50/53).

C. REPRODUCIBILITY

The reproducibility of assay results was assessed by sharing panels of sera between laboratories to then compare and analyze the reported results. Two panels were used for this study. The first panel consisted of 88 coded members sent to four laboratories. The panel consisted of 11 members of varying antibody level assayed in 8 replicates. Analysis of linear regression using IDEXX's results as the baseline site yielded correlations shown in Table 32. All correlation coefficients are greater than 0.99. The mean slope for all sites

TABLE 30
Comparison of Serodiagnostic Test Methods to Detect *B. abortus* 30 D Prior to Slaughter

Cow ID	Test date	Card	Rivanol	ACF[a]	PCFIA	Culture
8204	8-13-86	N	NT	N	0.70	
	9-17-86	P	50	81	0.14	Bio 1
8218	8-13-86	N	NT	N	0.55	
	9-17-86	P	50	41	0.18	Bio 1
8251	8-13-86	N	NT	N	0.49	
	9-17-86	P	200	81	0.10	Bio 1
8264	8-13-86	N	NT	5	0.36	
	9-17-86	P	N	81	0.10	Bio 1
8276	8-13-86	N	NT	5	0.46	
	9-17-86	P	100	161	0.10	Bio 1
8238	8-13-86	N	NT	N	0.65	
	9-17-86	P	50	81	0.16	Bio 1
8804	8-13-86	P	N	11	0.18	
	9-17-86	P	N	21	0.21	Bio 1
8811	8-13-86	N	NT	11	0.24	
	9-17-86	P	100	81	0.10	Bio 1
8815	8-13-86	N	NT	5	0.26	
	9-17-86	P	N	11	0.26	Bio 1
8817	8-13-86	N	NT	11	0.22	
	9-17-86	P	100	161	0.09	Bio 1
8819	8-13-86	P	N	11	0.25	
	9-17-86	P	50	81	0.10	Bio 1
8820	8-13-86	P	50	11	0.15	
	9-17-86	P	100	41	0.11	Bio 1
8297	8-13-86	N	NT	11	0.11	
	9-17-86	P	50	161	0.09	Bio 1
8467	8-13-86	N	NT	N	N	
	9-17-86	P	50	41	0.15	Bio 1
8484	8-13-86	N	NT	N	N	
	9-17-86	P	200	161	0.06	Bio 1
8711	8-13-86	N	NT	N	0.65	
	9-17-86	P	N	N	0.73	Bio 1
8787	8-13-86	N	NT	N	0.25	
	9-17-86	P	200	161	0.09	Bio 1
8797	8-13-86	N	NT	N	0.52	
	9-17-86	P	200	161	0.08	Bio 1

Note: N, negative; P, positive, NT, no test.

[a] Automated CF.

is 1; the mean intercept is 0.0025. The expected results would be a slope of 1 and an intercept of zero (0).

A second panel consisting of the *B. abortus* serial release test reagents was coded and sent to participating laboratories. Results of this panel are shown in Table 33. All laboratories agreed with an average CV of 6.7%.

In general, instrument and reagent performance resulted in excellent agreement between laboratories. The serial release panel, for example, represents four manufacturing serials containing different lots of particles and label. All of these components were manufactured and quality controlled separately over a time period of 1 to 2 months. Therefore, we conclude that manufacturing and instrument performance is excellent for generating consistent performance in the field.

TABLE 31
Comparison of Serodiagnostic Test Methods to Detect *B. abortus* 60 to 180 D Prior to Slaughter

ID	Test date	Card	Rivanol	ACF[a]	PCFIA	Culture
8814	8-13-86	N	NT	5	0.32	
	9-17-86	P	+25	11	0.212	
	10-29-86	P	N	11	0.368	
	11-19-86	P	N	11	0.135	
	12-18-86	P	200	161	0.139	Bio 1
8911	8-13-86	N	NT	5	0.25	
	9-17-86	P	N	11	0.208	
	10-29-86	P	N	11	0.263	
	11-19-86	P	N	11	0.188	
	12-13-86	P	N	11	0.231	Bio 1
8796	8-13-86	N	NT	5	0.24	
	9-17-86	P	N	5	0.485	
	10-29-86	P	N	N	0.567	
	11-19-86	P	+25	N	0.42	
	12-18-86	P	N	N	0.176	
	1-20-87	P	N	N	0.49	
	2-03-87	P	N	5	0.29	Bio 1
8802	8-13-86	N	NT	5	0.3	
	9-17-86	P	N	5	0.343	
	10-29-86	P	N	N	0.614	
	11-19-86	P	N	N	0.634	
	12-13-86	P	N	N	0.744	
	1-20-87	P	N	N	0.68	
	2-03-87	P	N	81	NT	Bio 1
8936	8-14-86	N	NT	N	0.95	
	9-17-86	N	NT	N	0.912	
	10-29-86	N	NT	N	1.169	
	11-19-86	N	NT	N	0.911	
	12-18-86	N	NT	N	1.016	
	1-20-87	N	NT	N	N	
	2-10-87	N	NT	N	N	Bio 1
8721	8-13-86	N	NT	N	0.67	
	9-17-86	P	N	11	0.18	
	10-29-86	P	50	41	0.08	Bio 1
8258	8-13-86	N	NT	N	0.87	
	9-17-86	N	NT	N	0.75	
	10-29-86	P	100	81	0.116	Bio 1
8262	8-13-86	N	NT	N	0.68	
	9-17-86	N	NT	N	0.876	
	10-29-86	N	NT	N	0.949	
	11-19-86	P	50	41	0.12	Bio 1

Note: N, negative; P, positive, NT no test.

[a] Automated CF.

VIII. FUTURE APPLICATIONS OF PCFIA

With the development of a semi-automated version of the PCFIA Screen Machine, the FCA, it is now feasible for laboratories with fewer numbers of samples to utilize this technology. Further, the FCA is more portable and is currently being evaluated for use in brucellosis testing by at least one state in a mobile laboratory testing application.

Eight states with a total of 24 Screen Machines are using the PCFIA for *B. abortus*

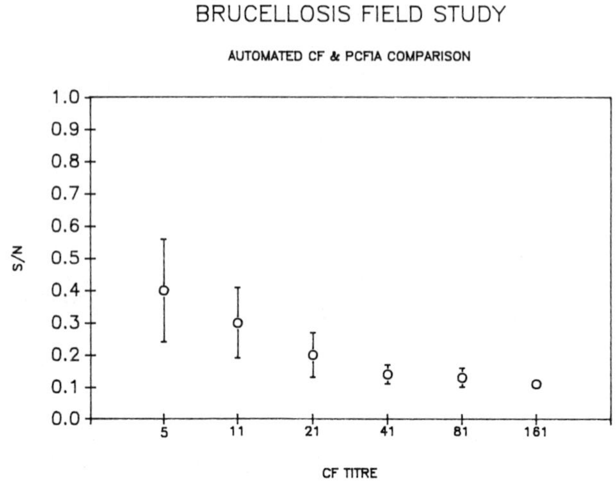

FIGURE 32. Correlation plot of PCFIA S/N values to automated CF titers for animals in the TX-W herd with reactor level CF titers. Error bars indicate ± 1 S.D.

TABLE 32
Field Study: Dilution Panel
Site-to-Site Reproducibility

Site-to-site comparison	R2	Slope	Intercept
SITE 1 (IDEXX)			
Vs. site 2	0.997	1.06	0.007
Vs. site 3	0.993	0.97	−0.018
Vs. site 4	0.999	0.98	−0.017
Vs. site 5	0.997	0.99	0.038

TABLE 33
Coded Panel Results Assayed by Six (6) Brucellosis PCFIA Test Sites, November 20, 1987

AgriTech ID	USDA ID	IA	PA	TN	MO	TX	ASI	Mean	% CV
BaSR-01	1	0.86	0.93	0.83	0.85	0.85	0.89	0.87	4.2
BaSR-02	2	0.22	0.27	0.22	0.25	0.27	0.25	0.25	9.1
BaSR-03	3	0.80	0.79	0.75	0.76	0.79	0.76	0.78	2.7
BaSR-04	4	0.18	0.20	0.16	0.18	0.18	0.18	0.18	7.0
BaSR-05	5	0.89	0.89	0.83	0.80	0.87	0.86	0.86	4.2
BaSR-06	6	0.16	0.21	0.18	0.20	0.21	0.17	0.19	11.4
BaSR-07	7	0.32	0.37	0.28	0.28	0.30	0.31	0.31	10.8
BaSR-08	9	0.18	0.23	0.18	0.19	0.22	0.18	0.20	11.5
BaSR-09	8302A	1.11	1.01	1.02	1.00	1.05	1.00	1.03	4.1
BaSR-10	8302B	1.04	1.05	1.06	1.02	1.05	1.02	1.04	1.6

antibody testing at this writing, with more states preparing to implement the system in the near future. Since we demonstrated the feasibility of a dual analyte PCFIA using a *B. abortus* and infectious bovine rhinotracheitis (IBR) virus prototype (Figure 33), prospective as well as current users of PCFIA have seen the economic benefits for this approach. Current research and development is directed towards a combination swine brucellosis and pseudorabies virus

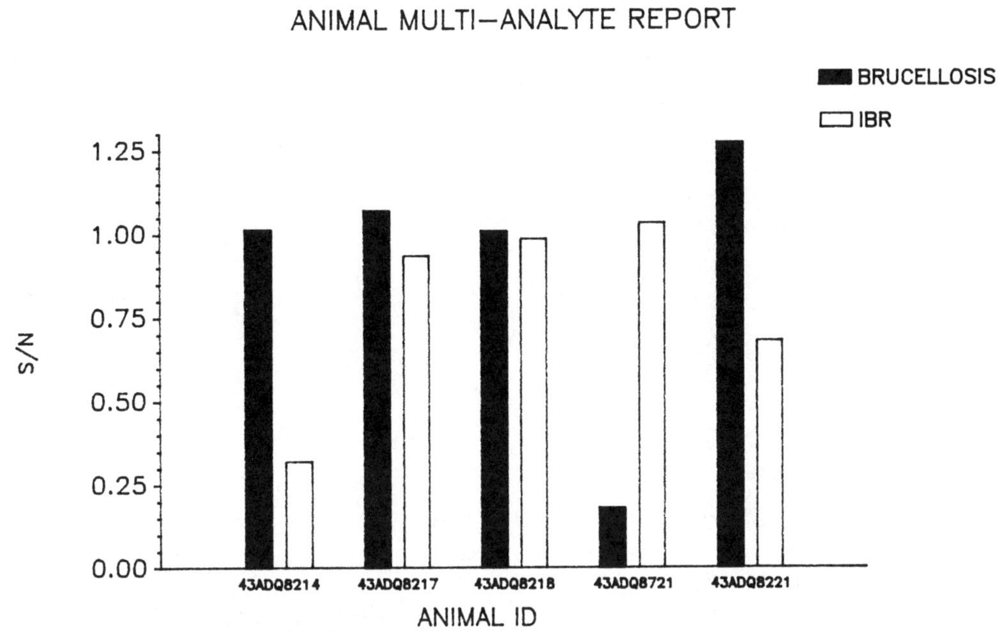

Bleed Date: 9/17/86
Vet Code: 224
Herd I.D. Anderson

ANIMAL ID	Brucellosis Result		IBR Result	
43ADQ8214	1.015	Negative	0.320	Positive
43ADQ8217	1.071	Negative	0.934	Negative
43ADQ8218	1.009	Negative	0.985	Negative
43ADQ8721	0.180	Positive	1.032	Negative
43ADQ8221	1.273	Negative	0.680	Positive

FIGURE 33. PCFIA S/N results for a prototype dual analyte antibody assay for IBR and *B. abortus*.

(PRV) antibody test kit which would decrease the unit cost of sample collection by providing additional diagnostic/epidemiologic information.

IX. SUMMARY

IDEXX Corporation has developed a *B. abortus* assay system with the following characteristics:

1. Automated system with sample-in/answer-out capability
2. High reproducibility, approximately 99% in field evaluations
3. High throughput; 3000 samples per day per instrument with one technician performing all sample handling and routine administrative functions
4. Reliability; all steps of assay verified by instrument
5. Automatic data reduction; generates tabulated results and herd profiles, facilitating management of infected and vaccinated herds
6. Flexibility — easily programmed to add additional assays resulting in multianalyte test results from a single sample dilution
7. Provides a presumptive cutoff as well as a diagnostic cutoff

The data used to validate this assay represent cross-sections of the cattle population in the U.S. This report contains data for more than 80,000 sera from brucellosis free, A, B, and C states that have been categorized and tested as follows:

Classification	No. of sera tested
Culture positive	719
Known challenge/temporals	
B. abortus	192
Not *B. abortus*	94
Free population (PA)	52,856
Negative herds — infected States	>2500
(MO, TX)	8857
Market cattle (MO, IA)	
Infected/program herds	>15000
(TX, MO, AL)	

In terms of specificity, the *B. abortus* PCFIA, when used as a presumptive test, is clearly more specific than BAPA. This improved specificity has been demonstrated on an assortment of negative herds.

With respect to sensitivity, the presumptive cutoff provides equivalent or better sensitivity than the current presumptive methodologies. Its relative sensitivity has been demonstrated by direct comparison to other methodologies in detecting culture-positive animals.

Using the diagnostic cutoff, the assay has a high correlation with reactor level titers of Rivanol and CF. Samples falling between the 0.25 diagnostic cutoff and the 0.7 presumptive cutoff should be regarded as suspicious. Such samples should be retested in 30 days or tested with another assay methodology if a 30-d retest is not practical.

REFERENCES

1. **Hoyer, B. H. and McCullough, N. B.,** Polynucleotide homologies of *Brucella* deoxyribonucleic acids, *J. Bacteriol.,* 95, 444, 1968.
2. **Hoyer, B. H. and McCullough, N. B.,** Homologies of deoxyribonucleic acids from *Brucella suis,* canine abortion organisms, and other *Brucella* species, *J. Bacteriol.,* 96, 1783, 1968.
3. **Verger, J. M., Grimont, F., Grimont, P., and Grayon, M.,** *Brucella,* a monospecific genus as shown by deoxyribonucleic acid hybridization, *Int. J. Syst. Bacteriol.,* 35, 292, 1985.
4. **Snyder, M. L., Workman, E. F., and McMahon, P. L.,** Particle concentration fluorescence immunoassay (PCFIA) for swine brucellosis antibodies, presented to USAHA Swine Brucellosis Subcommittee, Little Rock, AR, October 17, 1988.

Chapter 11

VACCINATION

Paul Nicoletti

TABLE OF CONTENTS

I. INTRODUCTION

Concurrent with the discovery of the cause of infectious abortion in cattle by Bang, it was known that infected animals often developed resistance to subsequent clinical signs. Many workers later conducted experiments using both killed and live cultures. The killed ones conferred poor immunity. The live vaccines were usually of virulent strains administered before the cattle became pregnant. Some provided good immunity against abortion but frequently led to excretion of the pathogenic organisms and infection of susceptible cattle.

The development of a satisfactory product for vaccination was changed forever with the description in the early 1930s of a strain of *Brucella abortus* of low virulence but good immunizing properties. Although many strains have subsequently been described and studied, none has found such wide application as strain 19.

Because of the recognized importance of vaccination in the control of brucellosis, investigations have been carried out for many decades to find a vaccine which confers good resistance without disadvantages. A vaccine has not been developed which has the qualities attributed to an ideal biologic. These include (1) immunogenesis of long duration, (2) minimum interference with diagnostic tests, (3) easy production with long stability and easy storage, and (4) a minimum of adverse effects in vaccinated animals and harmlessness to humans.

Vaccines are often classified as either living or dead. In general, with *Brucella* antigens the living vaccines give a more complete and lasting immunity. The bacteria multiply in the host, usually for limited periods, and the resistance conferred is largely of the cell-mediated type. Antibodies are considered to play a minor role in protection, and numerous studies in cattle have shown a poor relationship of postvaccinal circulating antibody to subsequent immunity.

The effectiveness of vaccines can be evaluated by several methods. These are usually by three criteria: (1) use in laboratory animals and data extrapolation to natural hosts, (2) use in natural hosts in controlled conditions and results compared with unvaccinated controls, and (3) use in field conditions where prevaccination and postvaccination prevalences are compared and/or those of subpopulations of vaccinates and controls.

There are no standardized criteria for evaluation of vaccine effectiveness since number and age of animals, challenge strain and dose, clinical signs, and other indices of infection have varied. Therefore, results of several studies have been inconsistent and difficult to compare with those of others.

This chapter will be largely limited to studies of Bovidae, since vaccination of other animal species will be a part of other chapters.

II. STRAIN 19

A. GENERAL CHARACTERISTICS

In 1930, Buck[1] reported encouraging results of a vaccine of medium virulence called *Bacterium abortus* strain 19 in a few heifers. The isolate was recovered from the milk of a Jersey cow in June 1923 and was virulent. The culture was maintained at room temperature for 1 year or more and, when tested in guinea pigs, had lost its virulence. It was the 19th of the stock culture series isolated by Buck.

The outstanding characteristics of *B. abortus* strain 19 are (1) low stable pathogenicity, (2) relatively high immunogenicity, and (3) antigenicity. The attenuation, cultural, and other biological characteristics of strain 19 are stable and were not altered by several passages through guinea pigs or by intravenous passages through pregnant cattle.[2,3]

Strain 19 resembles other strains of biovar 1 in sensitivity to thionine, growth in presence of basic fuchsin, production of H_2S, agglutination by monospecific antiserum, susceptibility to lysis by phage at routine test dilution, and oxidative metabolism. It differs in requirement for CO_2, sensitivity to thionine blue, penicillin, safranin O, and erythritol, and is less virulent.[4]

The antigenic qualities of strain 19 are directly associated with degree of dissociation. The S-I index should be at least 95% when used as a vaccine. The maximum viability is maintained at a pH of approximately 6.3. Several studies have examined the loss of viability during lyophilization and at various storage temperatures. Losses of viable cells have been approximately 40 to 70%, and the best storage temperature is 0 to 4°C.

Strain 19 was first used to produce a standardized liquid vaccine in the U.S. in 1939. It was introduced for field usage in 1941, and this usage increased greatly during that decade. The improvements in lyophilization methods during the next several years resulted in almost all strain 19 vaccine being produced in that form. It is the only strain of *Brucella* which has gained approval for field usage in bovine brucellosis in the U.S.

In spite of considerable research efforts to find a superior alternative, strain 19 remains the most acceptable vaccine against brucellosis in cattle because of its safety, immunogenecity, practicality of production, and convenience of use in cattle.

B. ANTIBODY FORMATION

The occurrence and persistence of serum antibodies following strain 19 vaccination have been described by many workers. This problem is the major disadvantage since antibodies may interfere with detection of infected cattle.

The magnitude and duration of serum antibodies following vaccination are directly related to many factors. These include age at vaccination, dosage, method of administration, and status of pregnancy. The diagnostic problem is also related to types of tests used and interpretation of results.

The initial antibodies formed following vaccination are of the IgM class, followed within a few days by IgG_1 and IgG_2 subclasses. IgM antibodies persist longer and are most active in agglutination tests at neutral pH. Antibodies may also be present in the milk for a few months and, unless strain 19 has localized in the udder, are of serum origin.

Until recent years, the tube agglutination test or a more rapid modification, the plate test, was the principal diagnostic method. Due to the persistence of vaccinal antibodies, especially of the IgM class, many procedures have been developed with the hope of differentiating the classes of the antibodies (see Chapters 3, 8 and 9). Since antibodies of chronically infected cattle are largely IgG, methods which detect these are most useful. The superiority of these so-called supplemental tests in correctly classifying vaccinated cattle has been shown by many workers.

The rapidity of seroconversion to negative status following strain 19 vaccination is inversely related to age, dose, and stage of pregnancy at time of vaccination.[5,6] There is a greater tendency to retain titers if cattle are pregnant when vaccinated, regardless of vaccinal methods.[5] The percentage of nonpregnant cows negative 2 years following vaccination was higher than that of those which were pregnant.[7]

Vaccination in early calfhood has usually been recommended so that serologic tests will be negative by breeding age. Over 90% of calves vaccinated at 3 to 8 months of age can be expected to be negative within 9 months. In one study,[8] the effect of age on retention of agglutinins in calves vaccinated at different ages was determined. The mean titers in calves vaccinated at 3 months returned to negative by 2 months postvaccination, while those vaccinated at 6 months took 6 months. In calves vaccinated at 9 months, there were titers that were suspicious after 15 months.

The effect of immaturity of calves on immunological responses to strain 19 was studied.[9] Calves vaccinated at 45 d of age had much reduced titers compared with calves vaccinated at 75 d of age. The author concluded that calves under 75 d of age are relatively immunologically immature in response to strain 19. Alton et al.[10] found that vaccination at 1 month of age reduced serologic responses, but effective immunity could only be achieved by a booster dose of vaccine at 12 months of age.

An additional effect on antibody formation is the simultaneous injection of oxytetracycline and strain 19.[11] Therapeutic doses of oxytetracycline reduced humoral antibodies, and the authors suggested this could reduce immunogenesis. No effects on antibody formation were observed when chlorotetracycline was given orally at subtherapeutic levels.[12]

C. IMMUNITY

A large number of experiments have been conducted to examine the resistance conferred by strain 19. These have been with controlled and field conditions. Some of the aspects which are of interest are doses and methods, effects of age, challenge dose, effects of revaccination, and length of immunity. Table 1 summarizes some selected studies conducted under controlled conditions. The effects of strain 19 in field studies are given in Section II.E and II.F. Further data on the effects of doses and of methods of vaccination are given in section II.E.

The results of experiments on the effects of age at time vaccination on immunity have varied. In general, when strain 19 was administered to calves above 2 months of age, few differences except in serologic responses were found.[20,23] Recent studies[10,21] have found that protection given to sexually mature cattle is greater than that of cattle when vaccinated as calves. These findings are consistent with an earlier report that the percentage of cattle that developed brucellosis when vaccinated over 16 months of age was less than that of those vaccinated when sexually immature.[24]

The necessity for revaccination and duration of immunity has been considered. Berman et al.[16] did not demonstrate an increment of resistance by revaccination. Manthei et al.[25] studied the duration of resistance. A total of 93 vaccinated cattle were challenged, of which 80 were vaccinated between 6 and $8^1/_2$ months and the others as yearlings or 2-year-olds. These were compared with 59 controls. All cattle were divided into five groups and challenged in one of five pregnancies. Under the conditions of the experiment, the authors concluded that resistance did not decrease with time.

McDiarmid[26] reported from 7 years of experience with about 500 cows that one dose of vaccine will confer adequate immunity for at least five pregnancies. He concluded that revaccination gave better immunity but that the disadvantages of serology outweighed the benefits.

It is well known that the resistance engendered by strain 19 may not be complete and that the degree of exposure influences the host-parasite interaction. In a summary of controlled research with strain 19, Manthei[27] reported that the relationship of infection to conjunctival exposure with *B. abortus* strain 2308 was directly related to dose. A dose of 15×10^6 which infected 100% of 210 controls also infected 61% of 254 vaccinates with abortion rates of 100 and 47%, respectively. However, reducing the dose to 7×10^5 resulted in 87% infection and 74% abortion in nonvaccinated cattle and 26% infection and 20% abortions in vaccinated cattle. More recently, Confer et al.[17] found that an increase of less than a logarithm of challenge dose resulted in large increases in abortions in both vaccinated and control cattle (see Table 1, Reference 17).

Suppression of antibody formation in calves by maternal antibodies up to approximately 150 d of life was found and could affect resistance from vaccination.[28] Another study[18] showed that hyperimmune serum administered simultaneously with strain 19 reduced protection and suppressed a humoral response.

The results of studies on the immunity produced by strain 19 under controlled conditions have often varied and are influenced by many factors including the virulence of the challenge strain. There are no universally accepted criteria for evaluations. Researchers have generally

TABLE 1
Results of Selected Studies on Immunity in Cattle Vaccinated with Strain 19

Age at vaccination	Vaccine dose	Challenge dose	Live calves	Number infected	Index of infection	Ref. ( )[a]
Adult	5.6×10^4	2×10^7		1/9	6.0	13
	2.8×10^8			2/9	2.3	(a)
	Controls			9/9	30.6	
14—23 mo.	2.25×10^8	1.3×10^7	6/9	8/10	10.8	14
	Controls		2/10	10/10	28.3	(a,b)
1 mo.	110×10^9	2×10^7	2/9	8/9	0—54	15
1 and 12 mo.	+1/20th		5/10	8/10	0—29	(a,b,c)
5 mo.			4/10	9/10	0—25	
	Controls		2/10	10/10	10—55	
8 mo.	Standard	12×10^6	10/13	3/14		16
8 and 14 mo.	dose		15/17	5/18		(c)
8 and 20 mo.			13/16	4/16		
	Controls		3/18	19/22		
10—12 mo.	10^9	9.4×10^6	8/9	2/11		17
	10^{10}		10/11	7/10		(a,d)
	Controls		1/9	7/9		
10—12 mo.	10^9	5.2×10^7	2/10	8/10		17
	10^{10}		0/8	7/8		(a,d)
	Controls		0/9	8/9		
3—6 mo.	9×10^{10}	10^7 (10 mo.)		2/10		18
	4.5×10^9	postvacc.)		1/9		(a)
	9×10^7			8/10		
	Controls			9/10		
Yearling	1.03×10^7	6×10^6	15/15	1/17	0.25	19
	1.00×10^8		15/15	4/17	0.35	(a)
	1.14×10^9		12/13	4/17	1.24	
	1.14×10^{10}		15/16	1/17	0.65	
	9.02×10^{10}		16/16	1/17	0.47	
	Controls		11/16	11/17	4.15	
4 mo.	Standard	7×10^5	19/24	8/24	2.9	20
6 mo.	dose		17/22	8/22	3.0	(b)
8 mo.			17/23	7/23	2.4	
	Controls		7/22	20/22	8.0	
3—6 mo.	4×10^{10}	1×10^7	6/7	4/12		21
14—16 mo.			9/11	1/12		(b)
	Controls		2/8	9/12		
Heifers	100×10^9	1.5×10^8		3/12		22
	Controls			10/12		(e)

[a] ( ) Reason for experiment: a = effects of dosage of vaccine; b = effects of age; c = effects of booster vaccination; d = effects of challenge dose; e = vaccine types.

suggested that for a vaccine to be effective, it should offer approximately 90% protection against infection in vaccination cattle, and the remainder may have fewer clinical signs or a partial immunity.

It has been the author's experience that the duration of immunity under some field conditions may not be as lengthy as under more controlled environments. Revaccination has proved to be necessary to control brucellosis in many large dairy herds.

D. OTHER EFFECTS ON HOST

In addition to the well-known problem of causing postvaccinal antibodies, there are other disadvantages to strain 19. These include abortion, orchitis, pyrexia, anorexia, and occasional persistent udder infection.

1. Pathology

Gross temporary or permanent tissue changes caused by strain 19 are rare. A syndrome of strain 19-induced arthritis of one or both femorotibial joints has been described by several workers.[29-32] Pathological changes include erosion of articular surfaces. Antibodies persist long after viable organisms have apparently been eliminated.

Postvaccinal persistence of strain 19 causing orchitis was described as early as 1943.[33] Lambert et al.[34] reported on strain 19 infections in two bulls. While the frequency of permanent infections is apparently low, recommendations or restrictions against vaccinating bulls are common. Higher postvaccinal antibodies in bulls compared with heifers is an additional reason to limit use of strain 19 to female cattle.

Further studies on the effects of vaccination with strain 19 on bulls were made in 34 bulls.[35] Histopathological examinations found lesions in testicles and epididymides which affected spermatogenesis. Strain 19 was isolated from three bulls from approximately 2 to 23 months postvaccination.

2. Persistence

Strain 19 multiplies only a short time after inoculation and rarely persists for an extended period of time, and this may be decreased by vaccinating cattle at an early age.[2] Strain 19 is markedly less virulent than other biovars of *B. abortus*. However, strain 19 may occasionally persist in the udder. The author summarized several publications on frequency of strain 19 infections and included data from those following different methods of adult cattle vaccination.[36] Shedding of strain 19 was highest among conjunctival vaccinates (1.4%) and was related to the dose in cattle vaccinated subcutaneously. Based on serologic and bacteriologic tests, the persistence of infection for 1 year or longer was low.

Meyer and Nelson[37] studied 11 shedders of strain 19. At least 9 were culture positive in milk for up to 3 years, but no organisms were isolated from fluids or tissues following parturition in 8 cows which calved.

The presence of strain 19 results in persistence of antibodies, and serologic tests are used to detect possible infected cows. Using the complement fixation test to identify seropositive cows, strain 19 was isolated from approximately 2.5% of 2 cattle herds slaughtered 9 and 12 months, respectively, following vaccination.[38]

It seems obvious that the frequency of persistence of strain 19 infections is related to many factors. The most important is age at time of vaccination since calves vaccinated prior to sexual maturity rarely are permanently infected with strain 19. Additional influences may be dosage, method of vaccination, and pregnancy status.

3. Physiology

Vaccination of cattle may sometimes have effects on pregnancy, milk production, feed consumption, or cause other physiological changes. Illnesses in cattle, especially calves, such as anorexia, dullness, and muscular stiffness have been observed a few days following vaccination. More severe signs of paresis and anaphylaxis have also occurred.

Dale and English[39] observed the effects on rectal temperature and feed consumption in dairy and beef calves raised under controlled environmental conditions. Rectal temperatures were elevated 3 to 4°F on days 2 to 3 postvaccination and did not return to normal until the 5th day. Feed consumption was depressed about 30%.

In adult cows, an average reduction in milk yield of 17% and 15% in feed consumption were recorded in a large dairy herd which received a full dose of strain 19 subcutaneously.[40] The effects were observed beginning on the 2nd day postvaccination and were most severe on days 2 to 3 when the milk production loss was 29%. The effects of strain 19 were less when the vaccine was given intradermally in another herd. Other workers[41] did not observe a reduction in milk yield. The physiological effects are lessened if the vaccine dose is reduced.

The published studies on the abortifacient effects of strain 19 were summarized by Manthei.[6] There is general agreement that abortions occur following vaccination, but they are rare. In a field study in which strain 19 was administered to more than 10,000 cows which where 7 to 8 months pregnant, fewer than 1% aborted for any reason.[40]

Beckett and McDiarmid[42] reported that 15 of 307 (4.9%) of cows aborted following vaccination, and strain 19 was isolated from 8. Four of the isolates were able to grow in the presence of erythritol. Strain 19 was recovered at necropsy from four of nine cows vaccinated with 5.8×10^9 cells during pregnancy. In two of these, vaccination had led to colonization of the uterus with erythritol-utilizing mutants of strain 19 with consequent abortion. No adverse effects were observed in the course of pregnancy in more than 700 cattle vaccinated with 3×10^8 cells.

E. METHODS OF ADMINISTRATION AND DOSES

The objectives of a large number of studies have been to determine the best method of administration and dose of strain 19 that would provide the optimum immunity and cause the least interference with diagnostic procedures. Strain 19 has been given by the following routes: subcutaneous (SC), intradermal (ID), intracaudal (IC), conjunctival (C), oral, intravaginal, and intrauterine. Manthei[6] summarized data on immunity resulting from SC, ID, and IC experiments and concluded the recedence of vaccinal titers appears to be more closely related to age at time of vaccination than to the dose of vaccine or method of vaccination.

The standard and a reduced dose of strain 19 given SC, ID, and conjunctivally were compared in adult cows in large Florida dairy herds.[40] There were large differences in serologic responses among methods and doses, but all methods resulted in reductions of brucellosis greater than 90% when prevaccination and postvaccination prevalences were compared. The use of the C route[44] followed reports by French workers[45,46] that under controlled conditions this method resulted in good protection against exposure with a minimum of serologic responses. These and other studies were summarized by Plommet and Fensterbank.[47] The best immunity was produced by an SC dose followed by a C booster. Two doses of strain 19 given by the C route were superior to one SC dose.

Twenty heifers vaccinated orally with a large dose (500×10^9) of strain 19 and 21 unvaccinated controls were challenged orally with a pathogenic strain.[48] Ten of 19 pregnant control heifers aborted, and 14 were culture positive. The pathogenic strain was recovered at necropsy from four vaccinated heifers and none aborted. All serologic tests were negative in heifers by 82 d after vaccination.

The minimum protective dose of strain 19 given by the SC route has not been fully established. Early studies[6] in a limited number of cattle had shown that a 0.2-ml dose may be as effective as a 5-ml dose. The author cautioned that since strain 19 was viable and the minimum dose unknown, the larger dose should be used. For many years the 5-ml dose of vaccine was restricted to calfhood usage because of the problems in differentiating serologic titers caused by vaccine and infection.

In 1976,[50] 5-ml and 0.25-ml doses of strain 19 were compared in a large infected dairy herd. There were no differences in prevalence rates in the two groups of cattle, but the reduced dose caused less interference with all serologic tests. Subsequent studies[10,14,18,19] confirmed that the dose of strain 19 could be greatly reduced without compromise of immunogenesis. A reduced dose of strain 19 is at present the only vaccine available for use in cattle in the U.S.

F. EFFECTIVENESS

This section will be limited to effectiveness under field conditions. In general, this is evaluated by two procedures: a comparison of the prevalence of diseases in vaccinated and unvaccinated populations and a comparison of the incidences prior to vaccination and at time intervals following vaccine administration.

Mingle[51] compared the numbers of reactors (the criterion used for infection) among a large number of cattle in several states. There were approximately 82% fewer reactors among vaccinated than unvaccinated cattle. Vanderwagen et al.[52] examined factors contributing to brucellosis in 79 large dairy herds in California. The presence of one or more unvaccinated reactors increased eightfold the percent of the population becoming reactors (0.69 to 5.95%) and nearly doubled the quarantine period for the herd. Koh and Morley[53] compiled data from over 42,000 cattle in 694 herds to examine the effects of strain 19 vaccination. The prevalence of infection in herds was 1.8% compared with 9.1% in unvaccinated herds.

The author[54] compared the incidences of brucellosis in 153 dairy herds before and following adult cattle vaccination. There was a reduction of greater than 85% in infected cattle removed from the herds by the third postvaccinal herd test. In Mexico, Flores-Castro et al.[55] reported on the effectiveness of vaccination in over 18,000 cows in 110 herds. Prior to vaccination, there were 1034 cows which were seropositive, and 18 to 24 months later only 87 cows had to be slaughtered.

G. PUBLIC HEALTH

There are numerous reports of adverse reactions to accidental or intentional inoculation of strain 19 in humans. Strain 19 does not appear to be hazardous if ingested or present on the dermis.

Two general syndromes in humans have been described.[56] The immediate type is an allergic reaction in humans with previous exposure to *Brucella* antigen. There may be massive swelling and local necrosis of affected tissue. The infectious type simulates reactions which are observed with infections caused by more virulent field strains. These consist of headache, fever, chills, and malaise. Accidental inoculation of strain 19 should be immediately followed with medical care.

III. STRAIN 45/20

A. GENERAL CHARACTERISTICS

Strain 45 was first isolated in 1922 in Great Britain from a cow. It is a rough strain and was developed by 20 passages in guinea pigs.[57] The strain was shown to have good immunizing properties in guinea pigs and cattle.[58] It was used for some years in cattle as a live strain, but studies[3,59] showed it could revert to virulent smooth mutants.

Since killed cells in saline produced little or no immunity, many types of adjuvants were tried. These have been shown to play a major role in effectiveness of strain 45/20 and in local reactions following injection.

B. ANTIBODY FORMATION

The rough strain 45/20 was developed on the basis that it was nonagglutinogenic. It is generally agreed that most 45/20 vaccines have produced little agglutinating antibody response to smooth *Brucella* antigens but that antibodies are produced which react to some serologic tests. These are, particularly, the Coombs antiglobulin and complement fixation tests. Animals previously exposed to smooth organisms such as strain 19 or field strains develop an anamnestic serologic response when injected with strain 45/20. This evidence suggests that strain 45/20 shares a surface antigenic component with smooth strains.

Cattle vaccinated with 45/20 adjuvant vaccine develop agglutinins for rough *Brucella* antigens which are largely of the IgG class.[60] Only small amounts of IgM rough agglutinins are produced at the earliest stage. Nonagglutinating antibodies detectable in the antiglobulin test are mainly of the IgG_1 and IgG_2 subclasses. In cattle previously vaccinated with strain 19, the majority of nonagglutinating antibodies are IgG_1, but significant amounts of IgG_2 are detected after the first and second injections of 45/20.[61]

Roerink[62] vaccinated 935 calves, and 9 months later less than 1% had positive or doubtful reactions to the tube agglutination test. The differences were considerable if cattle had been previously vaccinated with strain 19. Of 105 cattle examined at 6 to 8 months after the second injection of strain 45/20, approximately 4% were positive and 24% suspicious. Cunningham[63] also confirmed that previous sensitization affects serological responses. Approximately 50% of the cattle previously exposed to infection or vaccinated with strain 19 retained positive titers to the antiglobulin or complement fixation tests for at least 12 months. The milk ring test was unaffected.

The serum antibody reactions were monitored for 18 months in 18 heifers vaccinated with 45/20 adjuvant vaccine.[64] Four serological procedures were used: Rose Bengal, serum agglutination, complement fixation, and Coombs antiglobulin tests. The Rose Bengal and serum agglutination tests were negative throughout the experiment. Titers to the complement fixation test rose sharply 4 weeks following the second dose of vaccine and fluctuated in an unpredictable fashion. The Coombs test was positive after the second dose for up to 34 weeks.

Hall et al.[65] conducted the serum agglutination and complement fixation tests on serums of 15 cows given two doses of 45/20 vaccine. Following the first inoculation, none of the serum samples were positive or suspicious in either test. After the second inoculation, the highest agglutinin titers were found 2 weeks postvaccination when one serum was positive (100 IU/ml).

A summary of serological and immunological data on studies of 45/20 vaccines was reported by Ray.[66] Significant increases in agglutinating and complement-fixing antibodies were stimulated by one or two doses of bacterins. Variations were found among cattle and types of previous exposures such as strain 19 or herd infection.

Differences among experiments which have found variations in serological responses to 45/20 vaccines are due to the types of procedures and interpretations of results and the antigenic properties of different vaccines.

C. IMMUNITY

Early investigators showed that immunizing potency of killed *B. abortus* was strongly increased by incorporating the bacteria into a water-in-oil emulsion called incomplete Freund adjuvant. Guinea pigs inoculated with this product developed an immunity equal to that obtained by vaccination with live cultures. Subsequent experiments in cattle found comparable resistance of strains 45/20 and strain 19. However, the 45/20 bacterins often caused marked local reaction at the site of injection.

Table 2 summarizes results of selected studies on the immunity produced by strain 45/20. Many of these experiments compared 45/20 bacterins with strain 19 and unvaccinated

TABLE 2
Results of Selected Experiments on the Immunity Induced in Cattle by *B. abortus* Strain 45/20 Bacterins

Vaccination (2 doses)	Challenge	Normal calf	Infected	Ref.
Duphavac (1)[a]	155 × 10^7	10/10	3/10	67
Abortox[b]		7/8	2/8	
Controls		4/11	7/11	
Duphavac (1)[a]				
15 and 18 mo.	1.5 × 10^7		10/29	68
6 and 9 mo.			2/9	
Controls			16/18	
Duphavac (2)[a]				
Mature cows	1.5 × 10^7	8/10	7/13	65
Controls		2/9	10/12	
Duphavac[a]	7.5 × 10^5		4/10	69
Bortin 45[c]			1/10	
Controls			10/12	
Duphavac (2)[a]				
3—6 mos.	10^7—10^8	6/10	5/11	70
14—16 mos.		6/10	8/11	
Controls		2/8	9/12	
"A4"[c]				
3—6 mos.	1.5 × 10^7	6/11	8/11	71
6—9 mos.		2/12	2/12	
Controls		6/22	20/22	
Locally prepared				
Heifers	1.5 × 10^8		7/10	22
Controls			10/12	
Duphavac (3)[a]				
2 doses	1.3 × 10^7	4/5	4/5	72
1 dose		9/10	5/10	
Controls		2/5	5/5	

[a] Philips Duphar, Amsterdam (1), Sydney (2), Lane Cove, NSW Australia (3).
[b] Institut IFFA Merieux, Lyon.
[c] Glaxo Laboratories, Greenford, Middlesex, England.

controls. As with controlled experiments, immunity conferred by strain 19 (Section II.C) depends on many factors, and evaluation criteria may be different.

The nature of the adjuvant plays a vital role in effectiveness and in local reactions. It also affects the duration of antibodies. Vaccines based on an incomplete Freund (water-in-mineral oil emulsion) appear to have the best immunizing properties. Intramuscular injections are preferred.

Most studies have shown that a minimum of two injections spaced 6 weeks to 3 months are necessary for maximum immunogenicity. A yearly booster is considered necessary.[65] Following studies on a limited number of cattle, Alton et al.[71] questioned if two doses of 45/20 were superior to one dose. The latter method resulted in similar protection and caused fewer serologic responses.

It has been suggested that cattle may not reach full immunologic maturity to strain 45/20 bacterin until they reach sexual maturity.[68] Based on antiglobulin tests this should be after 280 d.[9] Calves less than 6 months of age are apparently not well protected.[66]

The results of studies which have compared the efficacy of strains 19 and 45/20 vaccines have varied. Many have found similar protection of two doses of 45/20 with one dose of strain 19. Others[21,22] have found protection from 45/20 vaccines to be inferior to strain 19.

D. OTHER EFFECTS ON HOST

The effects of *B. abortus* strains 19 and 45/20 adjuvant vaccine on cell-mediated immune response have been studied (see Chapter 4). The 45/20 bacterin elicited higher responses than strain 19.[72] These were considered to be due to the stimulation of the reticuloendothelial system by the oil adjuvant. Strain 45/20 cells without adjuvant do not elicit cell-mediated responses and produce a very poor humoral response.[9,72] There is agreement that postvaccinal serologic responses are not related to degree of protection with 45/20 vaccines.

Roerink[62] summarized the results of observations on several physiological aspects of strain 45/20 usage. In over 10,000 cattle vaccinated intramuscularly, only slight local reactions were observed which did not appear to be painful nor contain abscesses. There was no reported loss of appetite. The milk yield was unaffected after the first injection, but following the second injection, the mean production was reduced 1 to 2 l for a few days. Elevated changes in rectal temperatures of approximately 1°C and of short duration (1 to 3 d) have been reported.[62,69] There are no adverse effects on pregnancy[62] or the milk ring test[63] following strain 45/20 administration.

E. EFFECTIVENESS

The development and use of strain 45/20 vaccines were almost exclusively to overcome the disadvantage of the agglutinogenic property of strain 19. For many years, use of strain 19 was limited to calfhood, and 45/20 vaccines offered an alternative so that adult cattle could be vaccinated and, thus, rapidly develop herd resistance. In recent years, the reduction of the dosage of strain 19, together with improved serologic tests, have reduced the need for 45/20.

A strain 45/20 vaccine was widely used in the Netherlands.[62] It was considered to provide herd resistance equal to strain 19 without serious diagnostic problems and was very beneficial to the control program. Plenderleith[73] used 45/20 vaccines in a Jersey herd. He concluded that it was useful in an infected herd, but when infection is severe, its usefulness is limited. In another herd maintained under natural exposure conditions,[66] the rate of spread of brucellosis among cattle vaccinated with 45/20 bacterin was rapid. There was a higher incidence of brucellosis among those cattle vaccinated with 45/20 than among those with strain 19 or unvaccinated controls.

F. DIAGNOSIS OF LATENCY

A method which has been proposed to identify seronegative but infected cattle (incubative, latency) is to elicit an anamnestic serologic response. This is based upon the principle that exposure to a smooth strain of Brucella (field exposure, strain 19) results in sensitization. Subsequent inoculation with strain 45/20 causes a secondary response which can be measured by the complement fixation test (CFT) or antiglobulin test (AGT). This anamnestic procedure was first proposed by Tacken[74] for use in heifers on infected farms. He estimated that 519 of 554 heifers were salvaged by use of one injection of 45/20 followed by the CFT.

The anamnestic test was used in 53 problem herds to detect possible latent carriers.[75] One injection was followed in 6 weeks by the CFT and AGT. The procedure was considered highly successful. Further studies in Australia on three herds found the procedure useful in three infected herds.[76] Subsequent studies in the U.S. and Australia found that the AGT was more sensitive than the CFT and that the role of the anamnestic test would be limited, especially where strain 19 has been used.[77,78]

IV. OTHER VACCINES FOR CATTLE

A. LIVE

1. Rev. 1

B. melitensis Rev. 1 is an attenuated strain that has been widely used in the control of brucellosis in sheep and goats. It is agglutinogenic and shares many of the advantages and disadvantages of *B. abortus* strain 19.

The evaluation of Rev. 1 for possible use in cattle has been for two major reasons. In countries where *B. melitensis* infection in sheep and goats is widespread, cattle also become infected with this species. Therefore, vaccination of cattle with a homologous species of *Brucella* might be advantageous. Second, since Rev. 1 is a potent immunogen, it was considered possible that doses of this vaccine lower than those of strain 19 could provide equal or superior immunity and postvaccinal antibodies could be reduced.

Van Drimmelen and Horwell[79] were apparently the first workers to use Rev. 1 in cattle. All 11 vaccinated heifers resisted challenge with a pathogenic strain; 3 controls aborted. No vaccine organisms were found in milk or placenta following calving. No lesions were found in the testes of vaccinated bulls, and blood titers in all vaccinates disappeared within 6 months. Subsequent studies by the same workers evaluated different doses of Rev. 1 in 61 heifers.[80] There appeared to be no relationship between postvaccinal titers and resistance to infection. Although the numbers in each group were small, calfhood vaccination with Rev. 1 in doses of one tenth and one hundreth of that recommended for strain 19 provided over 80% protection against abortion.

Other studies compared Rev. 1 and strain 19 in cattle.[81] Both vaccinated groups along with controls were challenged with two different doses of a virulent strain. Rev. 1 was superior to strain 19 in both groups, and numbers of abortions were related to the challenge dose.

In spite of results of these experiments which found Rev. 1 to be a superior immunogen to strain 19, its use in cattle has been very limited.

2. M Vaccine

In the late 1940s a mucoid growth phase of *B. suis* was used in guinea pigs and later in cattle under experimental and field conditions. Some studies found no protection, while others concluded it was not entirely devoid of value as an immunizing agent.[82] While the postvaccinal agglutinins were less than with strain 19, the inferior immunogenicity and other factors were sufficient to prevent licensure of M vaccine.

3. 104-M Vaccine

This strain was isolated from a cow which aborted in 1929 and developed as a vaccine in Russia in 1950. It is reported to be stable in antigenic structure, degree of virulence, and immunogenicity.[83] The postvaccinal antibodies are less than strain 19. Al'bertyan[84] studied vaccine 104-M in 43 calves which were later challenged. It was concluded that strain 104-M caused more profound morphological changes in organs than strain 19, and more calves resisted challenge.

4. *B. suis* strain 2

This attenuated strain of biovar 1 was developed in China. Its virulence is approximately the same as strain 19 and is stable.[85] It is usually given orally via drinking water. A dose of 25 to 50 $\times$ 10^9 live cells in cattle gave a protection rate of 71.4% in 21 cattle, and 20 of 20 controls became infected. Excellent results were reported when used on dairy farms, and the vaccine did not induce persisting antibodies.

B. KILLED

1. Whole Cells

A number of experiments in laboratory animals and in cattle have been conducted using whole cells which were usually combined with various adjuvants. Live and Danks[86] compared strain 19 with an ether-killed field isolate of *B. abortus* in adjuvant (falba — a lanolin-like substance) in an infected herd. The titers persisted in the experimental vaccine group much longer than with strain 19, and results on immunogenicity were inconclusive. Subsequent studies by Gwatkin et al.[87] found protection of an ether-killed saline-oil suspension was comparable to that of strain 19. However, titers following vaccination receded more quickly with strain 19.

Brucella melitensis H38 is a formol-killed smooth strain in an adjuvant and was first developed for immunizing sheep and goats. Renoux et al.[88] compared strains Rev. 1, 19, 45/20, and H38 in calves which were later challenged. The killed H38 in an excipient gave the best results. Later studies by Plommet et al.[89] compared strain 19, H38 adjuvant vaccine, and controls. The percentages of abortion were lowest in the H38-vaccinated heifers. Meyer and Gibbons[90] assessed the effectiveness of H38 vaccine using one or two doses in cattle. One dose protected 50% of cattle against abortion, but only 19% against infection. Two doses protected 70 and 18%, respectively. The vaccine induced high and persistent titers and caused persistent and unacceptable local reactions at the site of injection.

A new type of *Brucella* vaccine was prepared from strain 19 that was heat killed and treated with formol. The antigenic sites were saturated with hyperimmune bovine serum to deprive them of the ability to produce agglutinins. The vaccine (called P.B.) was used in a limited number of cattle, and the authors concluded the immunizing properties were equal to strain 19 or H38, but the level of postvaccinal antibodies was low.[91] Subsequent studies in controlled[22] and field herds[93] found poor protection and that while P.B. vaccine had advantages in low agglutinogenicity, it was poorly immunogenic.

2. Extracts

Extracts of fractions of *Brucella* organisms can induce a substantial immunity to challenge in laboratory animals. The assessment of protection varies but is usually based on the ratio of the numbers of organisms found in spleens of nonimmunized and immunized animals. Very few preparations have been tested in natural host species. A properly organized trial takes many months and is expensive. It has proved to be very difficult to separate the immunogenic fractions of *Brucella* cells from those which cause production of antibodies.

A trial involving small numbers of cattle was conducted by Sterne et al.[93] An extract prepared by treatment of living cells of strain 45/20 with sodium dodecyl sulfate followed by ethanol precipitation induced immunity in three of five cows without causing postvaccinal agglutinins or complement fixing antibodies. Four of five controls aborted. The high protein and carbohydrate content suggested that the immunogen could be a glycoprotein.

Cattle were vaccinated twice with either a salt-extractable protein preparation (CSP) or chemically modified CSP (dCSP) from strain 19 in Freund's complete adjuvant.[94] Both preparations stimulated marked antibody responses which persisted for many months. Following challenge, 60 to 83% of the vaccinated heifers aborted and 70 to 83% were infected. The authors concluded that neither preparation was an effective immunogen and that preparation of a nonliving vaccine which has the desirable property of immunogenesis without adverse effects will be difficult.

V. SUMMARY

The control of brucellosis depends largely on two of the main principles of disease management: prevention of exposure to susceptible animals and increasing resistance of the population through vaccination. The best results are achieved through a combination of these, but vaccination, especially in large cattle populations, is far more effective.

After the discovery of the cause of brucellosis in cattle, there were many efforts to find practical and useful vaccines. These generally consisted of living or killed isolates of field strains, and they had many disadvantages.

The role of vaccination was greatly advanced with the discovery of *B. abortus* strain 19. Despite more than a half century of research, it remains the most widely accepted immunizing agent against bovine brucellosis. There has been no proved change in virulence or immunogenicity. The main disadvantage of postvaccinal antibodies can be minimized by reducing the dosage and the use of serologic tests which help discriminate antibody classes.

Many research studies have been conducted on alternate products for immunization including live and dead cells and extracts. Research is active to produce subunit vaccines. To date, results have not been encouraging.

It is certain that vaccination will remain a major aspect of the control and eradication of bovine brucellosis in many countries and that live strain 19 organisms will be the principal immunogen used in the efforts.

REFERENCES

1. **Buck, J. M.,** Studies of vaccination during calfhood to prevent bovine infectious abortion, *J. Agric. Res. (Washington, D.C),* 41, 667, 1930.
2. **Mingle, C. K., Manthei, C. A., and Jasmin, A. M.,** The stability of reduced virulence exhibited by Brucella abortus strain 19, *J. Am. Vet. Med. Assoc.,* 99, 203, 1941.
3. **Wilson Taylor, A. and McDiarmid, A.,** The stability of the avirulent characters of *Brucella abortus*, strain 19 and strain 45/20 in lactating and pregnant cows, *Vet. Rec.,* 61, 317, 1949.
4. **Jones, L. M., Montgomery, V., and Wilson, J. B.,** Characteristics of carbon dioxide-independent cultures of *Brucella abortus* isolated from cattle vaccinated with strain 19, *J. Infect. Dis.,* 115, 312, 1965.
5. **Berman, D. T., Beach, B. A., and Irwin, M. R.,** A comparison of the effects of subcutaneous and intracaudal vaccination of sexually mature cattle with *Brucella abortus* strain 19, *Am. J. Vet. Res.,* 15, 406, 1954.
6. **Manthei, C. A.,** Evaluation of vaccinal methods and doses of *Brucella abortus* strain 19, in *Proc. 56th Annu. Meet. Livestock Sanit. Assoc.,* 1952, 115.
7. **Moore, T. and Mitchell, C. A.,** Vaccination of sexually mature cows with *Brucella abortus* strain 19 vaccine, *Can. J. Comp. Med.,* 14, 209, 1950.
8. **King, N. B. and Frank, N. A.,** Effect of age on resistance and retention of titer in cattle vaccinated with strain 19 *Brucella abortus* vaccine, *J. Am. Vet. Med. Assoc.,* 91, 100, 1961.
9. **Cunningham, B.,** The effect of immaturity of the calf on immunological responses to strain 19 and killed 45/20 adjuvant vaccines, *Vet. Rec.,* 101, 283, 1977.
10. **Alton, G. G., Corner, L. A., and Plackett, P.,** Vaccination against bovine brucellosis, *Dev. Biol. Stand.,* 56, 643, 1984.
11. **Smith, R. A., Thedford, T. R., Espe, B. H., Woodson, P. D., and Burrows, G. E.,** Effect of oxytetracycline administration on antibody response to *Brucella abortus* vaccination in calves, *J. Am. Vet. Med. Assoc.,* 183, 70, 1983.
12. **Nicoletti, P., Mason, R. M., Jr., and Tehrani, J.,** Antibodies in calves on feed supplemented with chlortetracycline after vaccination with *Brucella abortus* strain 19, *J. Am. Vet. Med. Assoc.,* 90, 1002, 1987.
13. **Alton, G. G., Corner, L. A., and Plackett, P.,** Vaccination of pregnant cows with low doses of *Brucella abortus* strain 19 vaccine, *Aust. Vet. J.,* 56, 369, 1980.
14. **Alton, G. G. and Corner, L. A.,** Vaccination of heifers with a reduced dose of *Brucella abortus* strain 19 vaccine before first mating, *Aust. Vet. J.,* 57, 548, 1981.

15. **Plackett, P., Alton, G. G., Carter, P. D., and Corner, L. A.,** Failure of a single dose of Brucella abortus strain 19 vaccine to protect cattle when given early in calfhood, *Aust. Vet. J.*, 56, 409, 1980.
16. **Berman, D. T., Beach, B. A., and Irwin, M. R.,** Studies on repeated vaccination of cattle with Brucella abortus strain 19. III. the response of vaccinated and revaccinated cattle to conjunctival exposure with a virulent strain of *Brucella abortus* during the third gestation period, *Am. J. Vet. Res.*, 13, 351, 1952.
17. **Confer, A. W., Hall, S. M., Faulkner, C. B., Espe, B. H., Deyoe, B. L., Morton, R. J., and Smith, R. A.,** Effects of challenge dose on the clinical and immune responses of cattle vaccinated with reduced doses of *Brucella abortus* strain 19, *Vet. Microbiol.*, 10, 561, 1985.
18. **Davies, G., Cocks, E., and Hebert, N.,** *Brucella abortus* (strain 19) vaccine: protective dose in cattle, *J. Biol. Stand.*, 8, 165, 1980.
19. **Deyoe, B. L., Dorsey, T. A., Meredith, K. B., and Garrett, L.,** Effect of reduced dosages of *Brucella abortus* strain 19 in cattle vaccinated as yearlings, in *Proc. 83rd Annu. Meet. U.S. Anim. Health Assoc.*, 1979, 92.
20. **Lambert, G., Amerault, T. E., Manthei, C. A., and Goode, E. R., Jr.,** Immunogenic response of calves vaccinated at different ages with Brucella abortus strain 19, in *Proc. 65th Annu. Meet. U.S. Livestock Sanit. Assoc.*, 1961, 93.
21. **Sutherland, S. S., Robertson, A. G., LeCras, D. V., Robertson, G. M., Johnston, J. M., and Evans, R. J.,** The effect of challenge with virulent *Brucella abortus* on beef cattle vaccinated as calves or adults with either *Brucella abortus* strain 19 or 45/20, *Aust. Vet. J.*, 57, 470, 1981.
22. **Worthington, R. W., Horwell, F. D., Mulders, M. S. G., McFarlane, I. S., and Schutte, A. P.,** An investigation of the efficacy of three *Brucella* vaccines in cattle, *J. S. Afr. Vet. Assoc.*, 45, 87, 1974.
23. **Redman, D. R., Deyoe, B. L., and King, N. B.,** Resistance of cattle to *Brucella abortus* following vaccination at two and three months of age, *J. Am. Vet. Med. Assoc.*, 150, 403, 1967.
24. **Haring, C. M. and Traum, J.,** The effect of *Brucella abortus* strain 19 on cattle of various ages and its bearing on adult cattle vaccination, in *Proc. 47th Annu. Meet. U.S. Livestock Sanit. Assoc.*, 1943, 42.
25. **Manthei, C. A., Mingle, C. K., and Carter, R. W.,** Duration of immunity to brucellosis induced in cattle with strain 19 vaccine, in *Proc. 88th Annu. Meet. Am. Vet. Med. Assoc.*, 1951, 128.
26. **McDiarmid, A. A.,** The degree of duration of immunity in cattle resulting from vaccination with S19 *Br. abortus* vaccine and its implication in the future control and eventual eradication of brucellosis, *Vet. Rec.*, 69, 877, 1957.
27. **Manthei, C. A.,** Summary of controlled research with strain 19, in *Proc. 63rd Annu. Meet., U.S. Livestock Sanit. Assoc.*, 1959, 91.
28. **Cunningham, B.,** Protective effects of colostral antibodies to *Br. abortus* on strain 19 vaccination and field infection, *Vet. Rec.*, 101, 521, 1977.
29. **Bracewell, C. D. and Corbel, M. J.,** An association between arthritis and persistent serological reactions to *Brucella abortus* in cattle from apparently brucellosis-free herds, *Vet. Rec.*, 106, 99, 1980.
30. **Wyn-Jones, G., Baker, J. R., and Johnson, P. M.,** A clinical and immunological study of Brucella abortus strain 19-induced arthritis in cattle, *Vet. Rec.*, 107, 5, 1980.
31. **Nicoletti, P., Crowly, A. M., Richardson, J. A., and Farrar, J. A.,** Suspected *Brucella abortus* strain 19-induced arthritis in a dairy cow, *Agri-Practice*, 7, 5, 1986.
32. **Rogerson, B. A. and Morgan, I. R.,** Investigation of aberrant positive reactions to serological tests for bovine brucellosis, *Aust. Vet. J.*, 63, 227, 1986.
33. **Danks, A. G.,** *Brucella abortus* strain 19 and orchitis in a bull, *Cornell Vet.*, 33, 381, 1943.
34. **Lambert, G., Deyoe, B. L., and Painter, G. M.,** Postvaccinal persistence of *Brucella abortus* strain 19 in two bulls, *J. Am. Vet. Med. Assoc.*, 145, 909, 1964.
35. **Yantorno, D. M., Mazza, L. A., Balatti, A. P., Ciprian, F., and Aquirre, W.,** Efectos de la vacunacion con *Brucella abortus* cepa 19 en bovinos machos. Estudio serologico, bacteriologico y anatomo e histo-partologico, *Rev. Med. Vet. (Buenos Aires)*, 60, 59, 1979.
36. **Nicoletti, P.,** Prevalence and persistance of *Brucella abortus* strain 19 infections and prevalence of other biotypes in vaccinated adult dairy cattle, *Am. J. Vet. Res.*, 178, 143, 1981.
37. **Meyer, M. E. and Nelson, C. J.,** Persistence of Brucella abortus, strain 19 infection in immunized cattle, in *Proc. 73rd Annu. Meet. U.S. Anim. Health Assoc.*, 1969, 159.
38. **Duffield, B. J., Streeten, T. A., and Spinks, G. A.,** Isolation of *Brucella abortus* from supramammary lymph nodes of cattle from infected herds vaccinated with a low dose strain 19, *Aust. Vet. J.*, 61, 411, 1984.
39. **Dale, H. E. and English, J. E.,** Effect of brucellosis vaccination on rectal temperature and feed consumption of beef and dairy calves, *Vet. Med.*, 52, 477, 1957.
40. **Nicoletti, P.,** A preliminary report on the efficacy of adult cattle vaccination using strain 19 in selected dairy herds in Florida, in *Proc. 80th Annu. Meet. U.S. Anim. Health Assoc.*, 1977, 91.
41. **Worthington, R. W., Mulders, S. G., McFarlane, I. S., and Becker, D.,** A serological investigation on adult cattle vaccinated with *Brucella abortus* strain 19, *Onderstepoort, J. Vet. Res.*, 40, 7, 1973.

42. **Beckett, F. W. and McDiarmid, S. C.,** The effect of reduced-dose *Brucella abortus* strain 19 vaccination in accredited dairy herds, *Br. Vet. J.*, 141, 507, 1985.
43. **Corner, L. A. and Alton, G. G.,** Persistence of *Brucella abortus* strain 19 infection in adult cattle vaccinated with reduced doses, *Res. Vet. Sci.*, 31, 342, 1981.
44. **Nicoletti, P., Jones, L. M., and Berman, D. T.,** Comparison of the subcutaneous and conjunctival route of vaccination with *Brucella abortus* strain 19 vaccine in adult cattle, *J. Am. Vet. Med. Assoc.*, 173, 1450, 1978.
45. **Plommet, M. and Fensterbank, R.,** Vaccination against bovine brucellosis with a low dose of strain 19 administered by the conjunctival route. III. Serological response and immunity in the pregnant cow, *Ann. Rech. Vet.*, 7, 9, 1976.
46. **Fensterbank, R. and Plommet, M.,** Vaccination against bovine brucellosis with a low dose of strain 19 administered by the conjunctival route. IV. Comparison between two methods of vaccination, *Ann. Rech. Vet.*, 10, 131, 1979.
47. **Plommet, M. and Fensterbank, R.,** La vaccination antibrucellique administree par voie conjonctivale, *Dev. Biol. Stand.*, 56, 681, 1983.
48. **Nicoletti, P. and Milward, F. W.,** Protection by oral administration of *Brucella abortus* strain 19 against an oral challenge exposure with a pathogenic strain of *Brucella, Am. J. Vet. Res.*, 44, 1641, 1983.
49. **McDiarmid, A.,** A comparison of the intradermal and subcutaneous routes in producing immunity to brucellosis in cattle, *J. Comp. Pathol.*, 64, 384, 1954.
50. **Nicoletti, P., Jones, L. M., and Berman, D. T.,** Adult vaccination with standard and reduced doses of *Brucella abortus* strain 19 vaccine in a dairy herd infected with brucellosis, *J. Am. Vet. Med. Assoc.*, 173, 1445, 1978.
51. **Mingle, C. K.,** Status of the state-federal brucellosis eradication program, in *Proc. 67th Annu. Meet. U.S. Livestock Sanit. Assoc.*, 1963, 87.
52. **Vanderwagen, L. C., Sharp, J., and Meyer, M. E.,** A retrospective study on the relationships of vaccination status of reactor animals, management practices at calving and herd size to eradicating brucellosis in 79 dairy herds, in *Proc. 82nd Annu. Meet. U.S. Anim. Health Assoc.*, 1977, 83.
53. **Koh, S. H. and Morley, F. H. W.,** The effect of calfhood vaccination with strain 19 on the serological diagnosis and eradication of bovine brucellosis, *Aust. Vet. J.*, 57, 551, 1981.
54. **Nicoletti, P.,** The effects of adult cattle vaccination with strain 19 on the incidence of brucellosis in dairy herds in Florida and Puerto Rico, in *Proc. 83rd Annu. Meet. U.S. Anim. Health Assoc.*, 1979, 75.
55. **Flores-Castro, R., de Cordoba, L. F., Trejo-Salomon, J., and del Rio-Vargas, J.,** Adult cattle vaccination and revaccination with strain 19 reduced doses for the control of brucellosis: a field experience in Mexico, *Int. J. Zoon.*, 12, 299, 1985.
56. **Revich, S. J., Walder, A. W., and Pivnick, H.,** Human infection by *Brucella abortus* strain 19, *Can. J. Public Health,* 52, 285, 1961.
57. **McEwen, A. D. and Priestley, F. W.,** Experiments on contagious abortion. Immunization studies with vaccines of graded virulence, *Vet. Rec.*, 50, 1097, 1938.
58. **McEwen, A. D.,** Experiments on contagious abortion. The immunity of cattle inoculated with vaccines of graded virulence, *Vet. Rec.*, 52, 815, 1940.
59. **Edwards, S. J., de Ropp, R. S., and McLeod, D. H.,** A study of the immunological properties and infectivity of *Brucella abortus* strain 45/20, McEwen, in cattle, *Vet. Rec.*, 57, 259, 1945.
60. **Corbel, M. J.,** The immune response to *Brucella abortus* 45/20 adjuvant vaccine in terms of immunoglobulin class, *Dev. Biol. Stand.*, 31, 141, 1975.
61. **Beh, K. J.,** Immunoglobulin class specificity of non-agglutinating antibody produced in cattle following *Brucella abortus* 45/20 vaccination, *Aust. Vet. J.*, 51, 481, 1975.
62. **Roerink, J. H. G.,** Investigation of the usefulness of the non-agglutinogenic *Brucella abortus* adjuvant vaccine Duphavac N. A. in the control of bovine brucellosis, *Vet. Rec.*, 80, 727, 1967.
63. **Cunningham, B.,** Vaccination of cattle with killed 45/20 adjuvant vaccine. Effects on serological and milk ring tests when used in cattle previously exposed to infection or vaccinated with S. 19, *Vet. Rec.*, 86, 2, 1970.
64. **Chukwu, C. C.,** Serological investigation on cattle vaccinated with a killed *Brucella abortus* strain 45/20 adjuvant vaccine, *Int. J. Zoon.*, 12, 14, 1985.
65. **Hall, W. T. K., Ludford, C. G., and Ward, W. H.,** Infection and serological responses in cattle given 45/20 vaccine and later challenged with *Brucella abortus, Aust, Vet. J.*, 52, 409, 1976.
66. **Ray, W. C.,** An assessment of investigations conducted in the USA on *Brucella abortus* strain 45/20 bacterins, *Dev. Biol. Stand.*, 31, 335, 1976.
67. **McKeon, F. W.,** A recent trial comparing two 45/20 adjuvant brucella vaccines, *Dev. Biol. Stand.*, 31, 343, 1976.
68. **Cunningham, B.,** Some variations in properties of strain 19 and killed strain 45/20 adjuvant "Brucella Abortus" vaccines, *Dev. Biol. Stand.*, 12, 11, 1970.

69. **Hendricks, J. B. and Ray, W. C.,** Studies on cattle vaccinated with live strain 19 and/or 45/20 adjuvant *Brucella abortus* vaccines, in, *Proc. 74th Annu. Meet. U.S. Anim. Health Assoc.*, 1970, 147.
70. **Brinley Morgan, W. J. and McDiarmid, A.,** Adjuvant vaccines prepared from killed *Brucella abortus* strain 45/20, *Vet. Rec.*, 83, 184, 1968.
71. **Alton, G. G., Corner, L. A., and Plackett, P.,** Vaccination of cattle against brucellosis using either a reduced dose of strain 19 or one or two doses of 45/20 vaccine, *Aust. Vet. J.*, 60, 175, 1983.
72. **Chukwu, C.,** Comparison of cell-mediated immune responses to *Brucella abortus* strain 19 vaccine and *Brucella abortus* killed 45/20 adjuvant vaccine in cattle, *Microbios Lett.*, 34, 93, 1987.
73. **Penderleith, R. W. J.,** Some observations in brucellosis in a Jersey herd, 1965-1969, *Vet. Rec.*, 87, 404, 1970.
74. **Tacken, P. H. W.,** De enting van jongvee als hulpmiddel fer onderkenning van een jeugdinfectie met abortus Bang, *Tijdschr. Diergeneeskd.*, 89, 1703, 1964.
75. **Cunningham, B. and O'Connor, M.,** The use of killed 45/20 adjuvant vaccine as a diagnostic agent in the final stages of the eradication of brucellosis: the clearance of brucellosis from problem herds by interpretation of anamnestic serological responses, *Vet. Rec.*, 89, 680, 1971.
76. **Reid, M. A., and Harvey, P. R.,** The use of *Brucella abortus* 45/20 adjuvant vaccine as a diagnostic aid in the brucellosis eradication campaign in Papua, New Guinea, *Aust. Vet. J.*, 48, 495, 1972.
77. **Nicoletti, P.,** Use of 45/20 bacterin to detect latent infection in brucellosis, in *Bovine Brucellosis,* Crawford, R. P. and Hidalgo, R. J., Eds., Texas A & M University Press, College Station, 1977, 72.
78. **Sutherland, S. S.,** The anamnestic response to *Brucella abortus*-infected and vaccinated cattle, *Vet. Microbiol.*, 8, 405, 1983.
79. **Van Drimmelen, C. and Horwell, F. D.,** Preliminary findings with the use of *Brucella melitensis* strain Rev. 1 as a vaccine against brucellosis in cattle, *Bull. Off. Int. Epiz.*, 62, 987, 1964.
80. **Horwell, F. D. and Van Drimmelen, G. C.,** *Brucella melitensis* strain Rev 1 as a vaccine for cattle, *J. S. Afr. Vet. Med. Assoc.*, 42, 233, 1971.
81. **Garcia-Carrillo, C.,** Comparison of *B. melitensis* Rev. 1 and *B. abortus* strain 19 as a vaccine against brucellosis in cattle, *Zentralbl. Veterinaermed. B.*, 27, 131, 1980.
82. **King, N. B., Edgington, B. H., and Frank, N. A.,** The use of Huddleson's Brucella M vaccine under field conditions, *Am. J. Vet. Res.*, 19, 93, 1958.
83. **Shumilov, K. V., Kas'yanov, A. N., Romakhov, V. A., Al'bertyan, M. P., Klochkov, A. A., Akulov, A. V., Kosilov, I. A., and Novitskii, A. A.,** Properties of vaccine strain 104-M of *Brucella abortus, Trudy Vses. Inst. Eksp. Vet.*, 61, 10, 1986; Abstr. 5815 in *Vet. Bull.*, 1984.
84. **Al'bertyan, M. P.,** Comparison of pathological changes in calves vaccinated with *Brucella abortus* strains 104-M and 19, and challenged with virulent *B. abortus* strain 54, *Byull. Vses. Inst. Eksp. Vet.*, 42, 64, 1981; Abstr. 6911 in *Vet. Bull.*, 1983.
85. **Xie Xin,** Orally administrable brucellosis vaccine: *Brucella suis* strain 2 vaccine, *Vaccine,* 4, 212, 1986.
86. **Live, I. and Danks, A. G.,** The use of ether-killed *Brucella abortus* in saline-in-oil emulsion in a herd of cattle with brucellosis, *Am. J. Vet. Res.*, 12, 175, 1951.
87. **Gwatkin, R., Connel, R., and Duthie, R. C.,** Brucellosis. Vaccination of pregnant heifers one month prior to infection with ether-killed saline-oil emulsion of *Brucella abortus* and strain 19, *Can. J. Comp. Med.*, 18, 102, 1954.
88. **Renoux, G., Nicholas, J. A., Imbert, R., Quechon, M.,** Calf vaccination against *Brucella abortus:* comparison of four vaccines, *Bull. Acad. Vet. Fr.*, 37, 139, 1964; Abstr. 4381 in *Vet. Bull.*, 1964.
89. **Plommet, M., Renoux, G., Philippon, A., Lorentz, C., and Gestin, J.,** Brucellose bovine experimentale. I. comparaison de l'efficacite des vaccins B 19 et H 38, *Ann. Rech. Vet.*, 1, 189, 1970.
90. **Meyer, M. E. and Gibbons, R. W.,** Results of trial use of H-38 vaccine for immunizing beef heifers against experimental exposure to *Brucella abortus,* strain 2308, in *Proc 82nd Annu. Meet. U.S. Anim. Health Assoc.*, 1978, 106.
91. **Bonneau, M., Valette, L., and Pilet, Ch.,** Etude preliminaire et comparee du vaccin antibrucellique P.B. chez les bovins, *Dev. Biol. Stand.*, 12, 63, 1970.
92. **Kolar, J.,** Some observations on antigenic properties and immunogenic effects of Brucella vaccine PB 19, *Br. Vet. J.*, 134, 384, 1978.
93. **Sterne, M., Trim, G., and Broughton, E. S.,** Immunisation of laboratory animals and cattle with non-agglutinogenic extracts of *Brucella abortus* strain 45/20, *J. Med. Microbiol.*, 4, 185, 1971.
94. **Confer, A. W., Tabatabai, L. B., Deyoe, B. L., Oltjen, S. L., Hall, S. M., Oltjen, J. W., Morton, R. J., Fulnechek, D. L., Smith, R. E., and Smith, R. A.,** Vaccination of cattle with chemically modified and unmodified salt-extractable proteins from *Brucella abortus, Vet. Microbiol.*, 15, 325, 1987.

Chapter 12

THE PATHOGENESIS AND PATHOBIOLOGY OF *BRUCELLA* INFECTION IN DOMESTIC ANIMALS

Fred M. Enright

TABLE OF CONTENTS

I. INTRODUCTION

The pathogenesis of brucellosis in various domestic animals is remarkably similar despite the fact that five different species of bacteria in the genus *Brucella* are responsible for the infections. These similarities are due to common events surrounding the interaction of the bacteria with susceptible cells in their preferred hosts. The brucellae are facultative intracellular Gram-negative bacteria which have adapted to survival within host cells.

Typically, brucellosis in domestic animals is viewed as a chronic infectious disease which usually persists for the life of the animal. Brucella-infected animals generally develop granulomatous inflammatory responses which are often located within lymphoid tissues and organs with a prominent reticuloendothelial component. The localization and persistence of brucellae within these tissues and organs follow in the wake of widespread distribution of the bacteria during a generalized stage of infection. Localization of brucellae within the female and male reproductive tracts accounts for the principal clinical symptoms of infection — abortion and male infertility.[1] Much research effort, both past and present, has been directed toward an understanding of the events associated with genital brucella infections and abortion. This scientific preoccupation with only one aspect of the disease has, in some ways, contributed to a general lack of knowledge and understanding of the acute and subacute generalized phases of the infection.[2] A more complete understanding of the initial events involved in the establishment of brucella infections will be of great importance in the development and application of technology to better control brucellosis.

As with any infectious disease, the initiation of brucella infections is dependent on the dose and virulence of the bacteria and upon the relative resistance and susceptibility of the host as determined by both innate and acquired immunity. Susceptibility of cattle to *B. abortus* infection is influenced by the age, sex, and reproductive status of the individual animal. Sexually mature, pregnant cattle are more susceptible to *B. abortus* infection than sexually immature cattle of either sex. Even among pregnant cattle, susceptibility to infection may increase with the length of gestation (see Chapter 7).

Experimental studies involving challenge infections with a variety of strains and dose levels of *B. abortus* have clearly indicated that within the general population of cattle a significant proportion of individual animals are innately resistant to infection. The host defense mechanism responsible for such resistance and the mode of its inheritance are not fully understood.[3]

Experimental infection studies with a variety of *B. abortus* strains, in a variety of hosts, have demonstrated substantial differences in the pathogenic potential among different strains of *B. abortus*. Their pathogenic potential is related to well-documented biochemical and structural differences between *B. abortus* strains.[1,20] The considerable body of knowledge on these characteristics far exceeds our understanding of the interactions of these strains with the host's cells responsible for defense and immunity.[1] Resistance to brucella infections depends on host factors which prevent the establishment of mucosal infections, or upon host factors which facilitate the destruction of brucellae in tissues adjacent to the sites of initial bacterial invasion. In either case, the generalized phase of infection characterized by spread of brucellae to distant locations is prevented.

II. PATHOGENESIS

A. MUCOSAL INFECTION

Regardless of the primary route of natural transmission of the various brucellae infecting domestic animals, the organisms must possess the ability to attach to mucosal surfaces; they must be able to establish a niche and penetrate the epithelial lining of the mucous membrane. While numerous studies have documented the ability of many brucella species to breech the conjunctival, oral, pharyngeal, intestinal, respiratory, vaginal, and preputial mucosa, none has demonstrated how this, a critical event in the pathogenesis of brucella infections, occurs.[4] A better understanding of the mechanisms which allow brucellae to penetrate the mucosa may lead to the development of more effective vaccination procedures.

Brucella invasion of the mucosal epithelium of the conjunctiva and lachrymal ducts elicits an acute inflammatory reaction in the submucosal tissues. Both quantitative and qualitative differences in the submucosal inflammatory responses at 6 and 8 d postinoculation have been noted in strain 19-vaccinated and nonvaccinated cattle challenged with *B. abortus* strain 2308. As might be expected, the submucosal inflammatory reactions in vaccinated cattle are more intense and consist of increased numbers of mononuclear, polymorphonuclear, and eosinophilic leukocytes (Figures 1, 2, and 3). In addition, dramatic increases in large metachromatic-staining mononuclear cells are observed in challenged vaccinated cattle. In both vaccinated and nonvaccinated cattle, scattered plasma cells and focal lymphoid aggregates containing lymphoid follicles were noted. Despite the differences noted in the submucosal inflammatory reactions, brucellae were recovered from these sites in both vaccinated and nonvaccinated animals from 2 to 8 d postchallenge.[5] The effectiveness of such inflammatory reactions in destroying or limiting the spread of brucellae is not known. It is tempting to speculate that the more intense inflammatory reactions observed in vaccinated cattle may result in increased killing of brucellae by both cellular and humoral bactericidal systems. Clarification of the role of the submucosal inflammatory responses in brucellosis resistance must await much more detailed studies in both laboratory animals and cattle.

B. EXTENSION OF INFECTION TO LOCAL LYMPH NODES

Invading brucellae which have escaped the submucosal defenses are distributed to regional lymph nodes by lymphatic drainage. It is not known whether the brucellae which colonize the regional lymph nodes are carried there within phagocytic cells or arrive as free organisms. Challenge strains of *B. abortus* were isolated from the parotid lymph nodes 2 to 4 d postconjunctival challenge.[5] In an earlier study on the pathogenesis of *B. abortus* infections of cattle, 11 pregnant cattle (stage of gestation not indicated), varying in age from 4.5 to 13.5 years of age, were inoculated in the right conjunctival sac with 1.5×10^8 colony-forming units (CFU) of *B. abortus* strain 544.[6] These cattle were slaughtered at approximately weekly intervals postchallenge, and a variety of tissues were collected, cultured, and examined histologically. The parotid lymph node draining the right eye contained 10^5 CFU of *B. abortus* per gram of tissue, 8 d postinoculation. The number of *B. abortus* recovered from this lymph node varied with individual cows through 45 d postchallenge. Brucellae were, however, consistently isolated from this lymph node. The mandibular lymph nodes were infected in 10 of the 11 cattle. The number of brucellae present in these nodes varied from 10^2 CFU/g at 8 d to a peak number of 10^5 CFU/g at 15 and 22 d PI., after which time lower numbers of brucella were isolated. Despite the small number of cattle examined, this study suggested that larger numbers of bacteria (CFU per gram of tissue) and greater persistence of infection are found in the lymph nodes draining the site of inoculation. The lower initial number of brucellae infecting the lymph nodes draining the oral cavity suggests that fewer *Brucellae* gained access to the nasal and posterior oral mucosa by lachrymal duct drainage.

It further demonstrates that up to 4 weeks may be required for the bacteria colonizing lymph nodes draining the oral mucosa to achieve the level of infection present at 1 week in the parotid node. Failure to maintain high levels of infection in lymph nodes more distal to the inoculation site suggests that the greater the number of bacteria which initially colonize the node, the greater the chance of prolonged persistence of the infection in that node. Brucellae were not recovered from the suprapharyngeal lymph node in 4 of the 11 cows, and infection was not detected in the intestinal lymph nodes of any of the cattle. If one

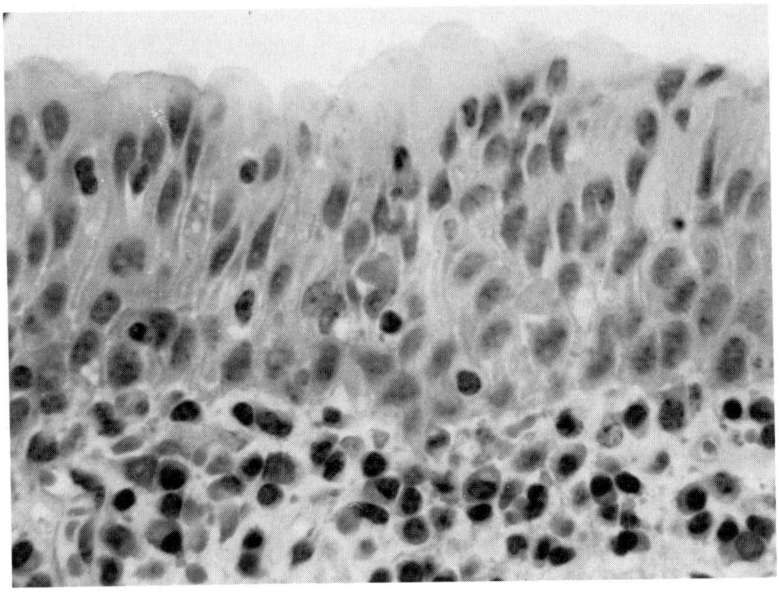

A

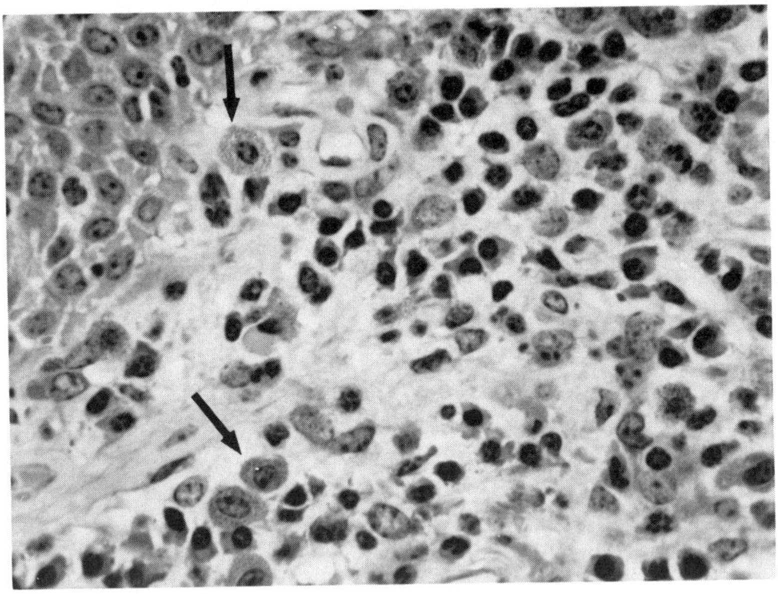

B

FIGURE 1. Submucosal inflammatory reaction around the lachrymal duct of cows 6 d after challenge with *B. abortus* strain 2308. (A) Nonvaccinated cow with an accumulation of mononuclear leukocytes within the submucosa and mucosa. Moderate numbers of plasma cells are present. (H & E; magnification × 400.) (B) Strain 19-vaccinated cow with a moderately heavy infiltration of the submucosa by a variety of inflammatory cells. Note the frequent occurrence of large granular mononuclear cells (solid arrows). (H & E; magnification × 400.)

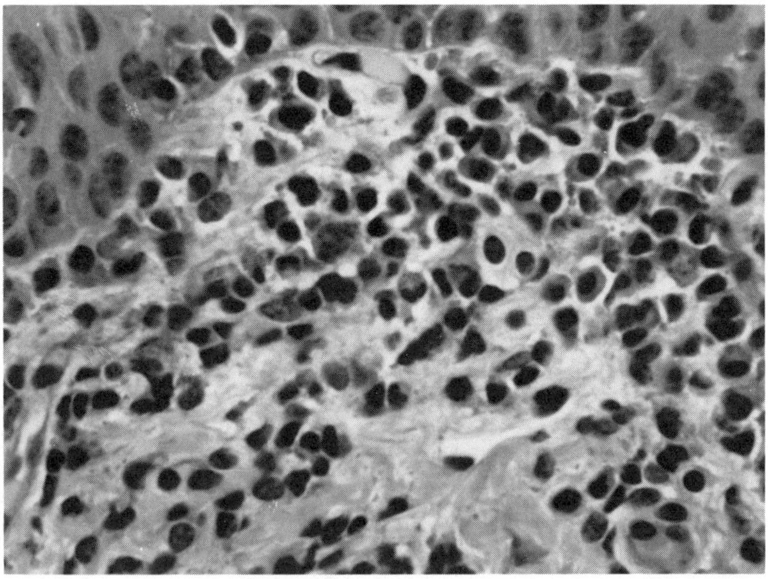

A

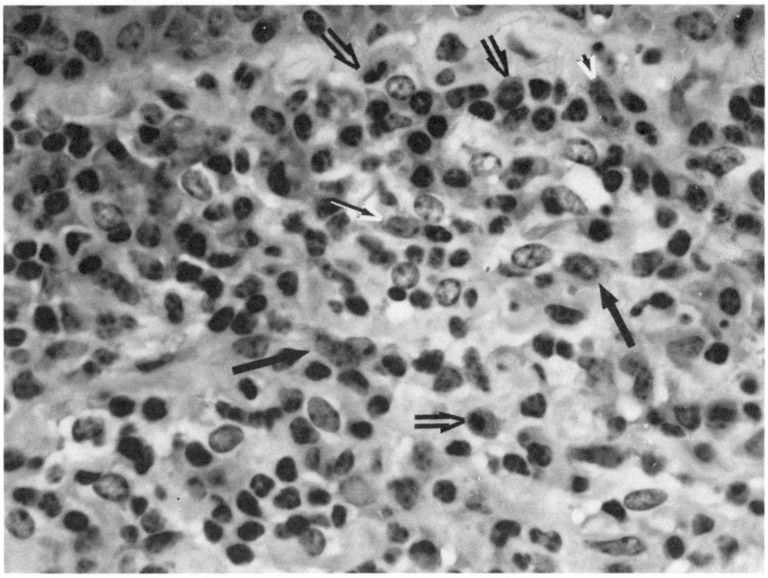

B

FIGURE 2. Submucosal inflammatory reactions around the lachrymal duct of cows 8 d after challenge with *Brucella abortus* strain 2308. (A) Nonvaccinated cow with a moderate number of plasma cells, lymphocytes, and macrophages in submucosa. (H & E; magnification × 400.) (B) Strain 19-vaccinated cow with a large number of inflammatory cells comprised of macrophages (solid arrows), mast cells (open arrows) and numerous eosinophils (black and white arrows.) (H & E, magnification × 400.)

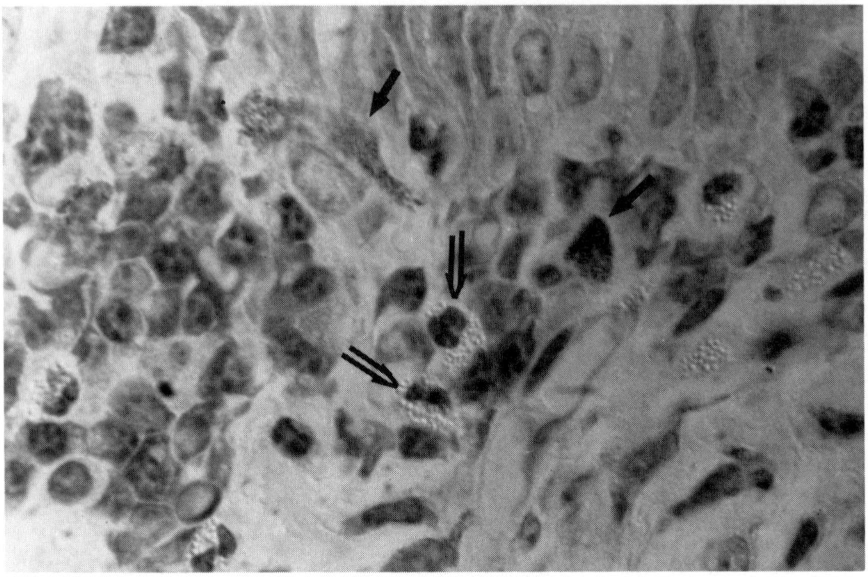

FIGURE 3. Submucosal inflammatory reaction in a vaccinated cow 8 d after challenge with *B. abortus* strain 2308. Note the large number of mast cells with metachromatic granules (solid arrows) and eosinophils which are unstained (open arrows). (Toluidine blue/DIC; magnification × 720.)

assumes that colonization of lymph nodes draining mucosal sites more distal to the site of initial infection is due to secondary invasion of the mucosa, then such studies suggest that fewer and fewer brucellae are available "downstream" to cross the mucosal barrier. Such observations may be critical in determining the optimal route of administration of brucella vaccines.

Infected lymph nodes are enlarged due to both lymphoid and reticuloendothelial hyperplasia, and to the infiltration of inflammatory cells. In addition, multiple areas of hemorrhage are present within infected nodes by 1 to 2 weeks postinoculation. Multiple foci of extramedullary hematopoiesis adjacent to lymphoid follicles are frequently observed during the first week postinfection. Both neutrophilic and eosinophilic precursor cells are observed in these foci (Figure 4). These hematopoietic foci may have been misinterpreted as inflammatory infiltrates in earlier reports. *B. abortus* infection in several different mouse strains resulted in increased levels of hematopoietic colony stimulating factors.[7] As early as 4 d postinoculation, varying numbers of granulomatous inflammatory foci are noted, particularly within the cortex of the draining lymph node. These reactions are composed of monocytes and macrophages and epithelioid cells (Figure 5).[8] Moderate to large of numbers of mature neutrophils and eosinophils are noted packed within small blood vessels within the lymph node and in the capsule. The tissue surrounding these vessels is infiltrated with these cells.

C. BACTEREMIA AND SECONDARY LOCALIZATION

The mucosal membrane barrier represents the host's first line of defense against brucella infections. The inflammatory reactions occurring in the submucosal tissues and within the lymph node draining the invasion site are the host's second and third lines of defense. Failure to destroy *B. abortus* within the draining lymph node results in persistence of infection and the eventual escape of the agent via the blood. Brucellae are able to survive within host leukocytes and may utilize both neutrophils and macrophages for protection from humoral and cellular bactericidal mechanisms during the periods of hematogenous spread. This, in fact, has been demonstrated in *B. canis* infections.[9] Prolonged periods of bacteremia are not

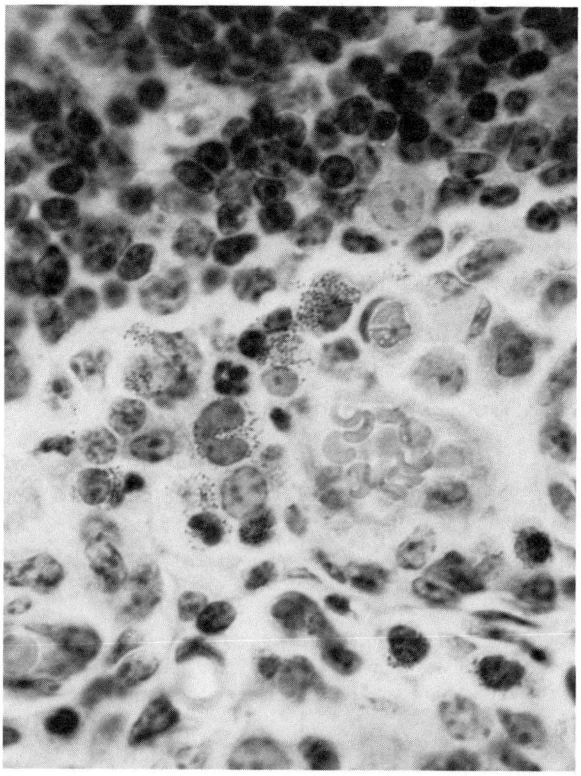

FIGURE 4. Fetal thymus from a late gestational bovine fetus infected with *B. suis* 14 d earlier. The interlobular connective tissue contains nests of eosinophils in varying developmental stages. (H & E; magnification × 720.)

uncommon in brucella infections. Following conjunctival exposure brucellae were isolated from the spleen as early as 15 d PI. Distal lymph nodes, the mammary gland, and the pregnant uterus were colonized by 22 to 29 d PI.[6] During the bacteremic phase, brucellae may localize in a wide variety of tissues. However, the bacteria are most frequently isolated from lymphoid tissues, the mammary gland, and the reproductive tract. Infections may also become established in bones, joints, the eye, and, occasionally, the brain. In males, secondary localization occurs most frequently in lymph nodes, the testes, epididymis, and accessory sex organs. In general, the species of brucellae which are commonly associated with prolonged periods of bacteremia, *B. suis* and *B. canis*, are also noted for their tendency to localize more frequently in a wider variety of secondary sites.[1]

D. REPRODUCTIVE TRACT LOCALIZATION

Brucella infections of both the male and female reproductive tract, at one time, were thought to be the result of a specific tropism of the bacteria to tissues with a high concentration of erythritol.[10] High concentrations of erythritol are found in tissues of both the male and female reproductive tracts of ruminants and swine. Erythritol stimulates the growth of some strains of *B. abortus*. Other strains, including *B. abortus* strain 19, are not stimulated by erythritol, yet these strains are capable of causing genital infections and abortions.[2] Placental localization and abortions are also observed in experimental *B. abortus* infections of a variety of laboratory rodents which do not have detectable levels of erythritol in their tissues. Following intravenous inoculation of *B. abortus* into pregnant mice, the initial localization

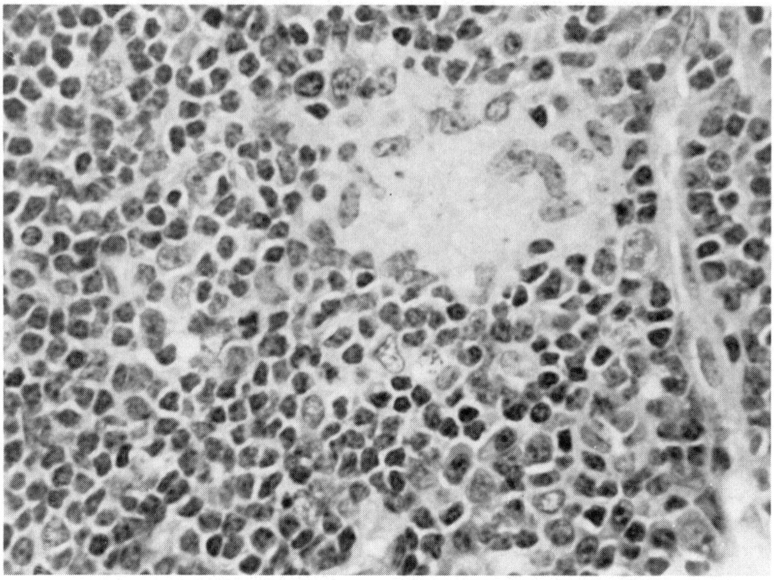

FIGURE 5. Bovine lymph node draining a subcutaneous inoculation site. A stable rough mutant of *B. abortus* 2308 was given 7 d earlier. The granulomatous loci are composed of histocytic and epithelioid cells. (H & E; magnification × 400.)

of bacteria within the placenta is much less than the number of bacteria localized in the spleen. However, by 72 h postinoculation, extremely large numbers of brucellae were present within placental tissues while relatively small numbers of brucellae were present within the spleen.[11] The fact that the placental environment offers a privileged site for brucella replication should come as no surprise. In the earliest descriptive reports of bovine brucellosis, massive numbers of bacteria were observed within chorionic trophoblasts.[12] Both ultrastructural and immunohistochemical studies have demonstrated very large numbers of brucellae within the rough endoplasmic reticulum of the chorionic trophoblasts of intravenously infected pregnant goats.[13] Mesenchymal, mesothelial, yolk endodermal, and hepatic cells in *B. abortus*-infected chick embryos also support the growth of bacteria within their rough endoplasmic reticulum.[14] Both the location and the number of brucellae within these cells differ markedly from the relatively small number of brucellae located within the phagosomes of macrophages. The chorionic trophoblasts are metabolically active cells capable of producing a variety of hormones and secretory proteins which may stimulate the growth of brucellae.

Following experimental intravenous infection of pregnant goats, *B. abortus* was first localized within the erythrophagocytic trophoblasts of the placentome. Subsequently, very large number of brucellae were present within periplacentomal chorioallantonic trophoblasts. Necrosis of trophoblasts with ulceration of the chorioallantonic membrane was accompanied by an inflammatory exudate composed of necrotic debris, macrophages, and polymorphonuclear leukocytes. Both placentomal and fetal infection and inflammation were preceded by infection of the chorioallantoic membrane. Despite the presence of a severe placentitis, only mild inflammatory changes were noted within the endometrium. The endometrial epithelium remained intact, and *B. abortus* was not demonstrated in either the endometrium or lamina propria. Severe ulcerative endometritis and metritis were not observed until abortions has occurred and were probably due to postparturient infections with a variety of bacteria other than *B. abortus*.[15]

E. PLACENTITIS AND ABORTIONS

It may come as a surprise to many that the mechanisms of brucella-induced abortions are poorly understood. The earliest descriptions of the disease in cattle suggested that placentitis, which is commonly observed, resulted in impaired placental function which prevented the delivery of oxygen and nutrients to the fetus and/or the removal of waste products from the fetus. The extent and severity of placentitis in *B. abortus* infection is highly variable. At one extreme, widespread destruction of placentomes and fetal membranes is found, while at the other extreme only focal damage with minimal placentitis is noted. Brucella abortions may occur in both of these situations. In cases with minimal placentitis (placental damage), it is particularly difficult to embrace a theory based on interference with placental transfer of nutrients and failure to remove waste products due to the remarkable compensatory capacity of the placenta.[1]

Several studies have suggested that brucella endotoxins may be responsible for abortions.[6,13] Crude and purified lipopolysaccharides (LPS) and lipid A from a variety of Gram-negative bacteria can induce abortion in a number of species. Placental changes in endotoxin-inoculated laboratory rodents ranged from widespread vascular damage and ischemia to profound alteration in the levels of various prostaglandins which play a role in pregnancy maintenance.[16,17] *B. abortus* endoantigen (crude endotoxin) caused abortions in guinea pigs 24 to 36 h after inoculation.[18] The same material caused one pregnant heifer to abort 3 d after inoculation. *B. abortus* strain 2308 LPS was inoculated into bovine fetuses between 120 and 160 d of gestation. The LPS was given twice at 24-h spaced intervals. Two of three inoculated fetuses were aborted 72 to 96 h following the second inoculation. Age-matched fetuses inoculated with either saline or with the LPS complexed with anti-brucella LPS bovine antibody remained normal.[19] Purified LPS from both smooth and rough strains of *B. abortus* demonstrated a wide range of biological activities including abortions in mice.[20]

The maintenance of pregnancy and initiation of delivery vary considerably among different species. In cattle and sheep, progesterone produced by the corpus luteum during the first half of gestation is essential to maintain pregnancy. During the second half of gestation, progesterone produced by both the placenta and the fetus maintains pregnancy. Thus, diseases which affect either the placenta or the fetus during the second half of gestation may alter their ability to maintain pregnancy and result in either abortions or premature delivery.

F. FETAL STRESS DUE TO BRUCELLA INFECTIONS

The hormonal interactions in the termination of pregnancy are complex, but shifts in the production of progesterone in favor of estrogen result in the endometrial production of $PGF2_{\alpha}$ which is capable of inducing the normal or premature delivery process. Brucella infections of both bovine and sheep fetuses result in rapid elevations of cortisol levels, presumably due to an increase in the fetal adrenal cortical production of steroids. High cortisol levels may then result in decreased progesterone production and increase estrogen production by the placenta. This hormonal shift is associated with increased $PGF2_{\alpha}$ production by the endometrium and results in premature delivery.[21] In *B. abortus* strain 2308-infected bovine fetuses, substantial elevations of cortisol were present as early as 6 d postinfection.[22] Increased levels of fetal cortisol were paralleled by histological evidence of adrenal cortical hyperplasia following infection. Fetal sheep between 120 to 126 d of gestation were infected directly with *B. abortus* strain 2308. Marked increases in cortisol levels were detected in infected fetuses between 24 to 120 h after infection. Histological evidence of fetal adrenal cortical hyperplasia was not obvious during the 24- to 120-h interval. The fetal cortisol levels were elevated 20-fold 24 h postinfection; at 48 h postinoculation, a tenfold increase in fetal cortisol was noted; and 30 times normal levels of cortisol occurred at 72 and 120 h PI. Cortisol levels in infected fetuses were as high or higher than cortisol levels in fetuses at the initiation of labor. In parallel studies, brucella-infected sheep fetuses were generally delivered dead between 5 to 7 d PI.[23] The intervals between infection and delivery closely parallel the intervals to delivery in similarly aged fetuses infused with exogenous cortisol.[24]

These and other studies clearly suggest that fetal infections, and the resultant stress associated with infections, result in elevation of steroids and that these changes are associated with premature delivery and/or abortions. It is also equally clear that no single event in *in utero* brucella infection has been identified which is responsible for the abortions.

Experimental infections of bovine, sheep, and goat fetuses in the last third of gestation result in abortions of nonautolyzed fetuses at about 12 d postinoculation in cattle and between 6 to 9 d PI. in sheep and goats.[22,23,25] Late gestational bovine fetuses inoculated with 6.5 $\times$ 10^3 CFUs of *B. suis* were delivered dead or weak 11 to 14 d PI. Interestingly, both cattle and goat fetuses inoculated directly with less pathogenic strains of *B. abortus* (strain 19 and a rough mutant of strain 2308) were carried *in utero* for longer intervals PI. Two bovine fetuses at 230 and 240 d of gestation were inoculated subcutaneously with 1.0 $\times$ 10^8 CFUs of strain 19 and were delivered dead in excellent postmortem condition on day 270 of gestation (term). Both fetuses had gross and histological lesions. Fetal goats inoculated with strain 19 were aborted 11 d PI. with lesions. Fetal goats inoculated with a stable rough mutant of strain 2308 *B. abortus* were carried to term and delivered alive. Both gross and histological lesions typical of congenital brucella infections were present, and the rough mutant strain was recovered from multiple fetal tissues.[25]

The significance of fetal infection, inflammation, and stress in the abortion process requires reconsideration, as both strain 19 and the stable rough mutant resulted in fetal infections, with typical inflammatory reactions, but failed to induce premature delivery. These studies suggest that interactions of more pathogenic strains of *B. abortus* with placental tissues may be more important than their interactions with the fetus in initiating premature delivery.

G. *IN VITRO* BRUCELLA INFECTION OF CHORIOALLANTOIC MEMBRANE EXPLANTS

To understand the differences in the pathogenic potential of various strains of *B. abortus* better, midgestational bovine chorioallantoic membrane explants were established and infected with three different strains of *B. abortus*. The surface areas of each membrane explant were equal to 1 cm^2. *B. abortus* strain 2308, strain 19, and the rough strain 2308 (RB 51) were adjusted to 1.0 $\times$ 10^6 CFUs in serum-free tissue culture media. Groups of six explants were inoculated with one of the three strains of *B. abortus* by flooding the chorionic surface of the explants. The inoculated explants were incubated for 2 h at 37°C in a 10% CO_2 atmosphere. Following incubation, each explant surface was washed two times with physiological saline and then completely covered with tissue culture media without serum supplementation. At 12 h postinoculation, groups of three membranes, each inoculated with one of the three strains of *B. abortus,* were homogenized and cultured. The remaining three explant membranes from each group were harvested at this time, fixed in buffered formalin, sectioned, stained, and histologically examined. Large numbers of the rough strain (7.4 $\times$ 10^5 CFUs) were recovered from the homogenized explants. Intermediate numbers of strain 19 (1.4 $\times$ 10^5 CFUs) were recovered. Surprisingly, fewer smooth strain 2308 brucellae (7.5 $\times$ 10^4 CFUs) were associated with the explants. Histological examination of the smooth 2308-infected explants revealed near total destruction of the chorionic trophoblasts. Both strain 19 and the rough strain resulted in slight necrosis of trophoblastic cells (Figure 6). This and other *in vitro* culture systems may be utilized to understand better the early events involved in the infection of trophoblasts from varying stages of gestation. The direct effects of brucella infection on the secretory functions of these cells may also be determined.[26]

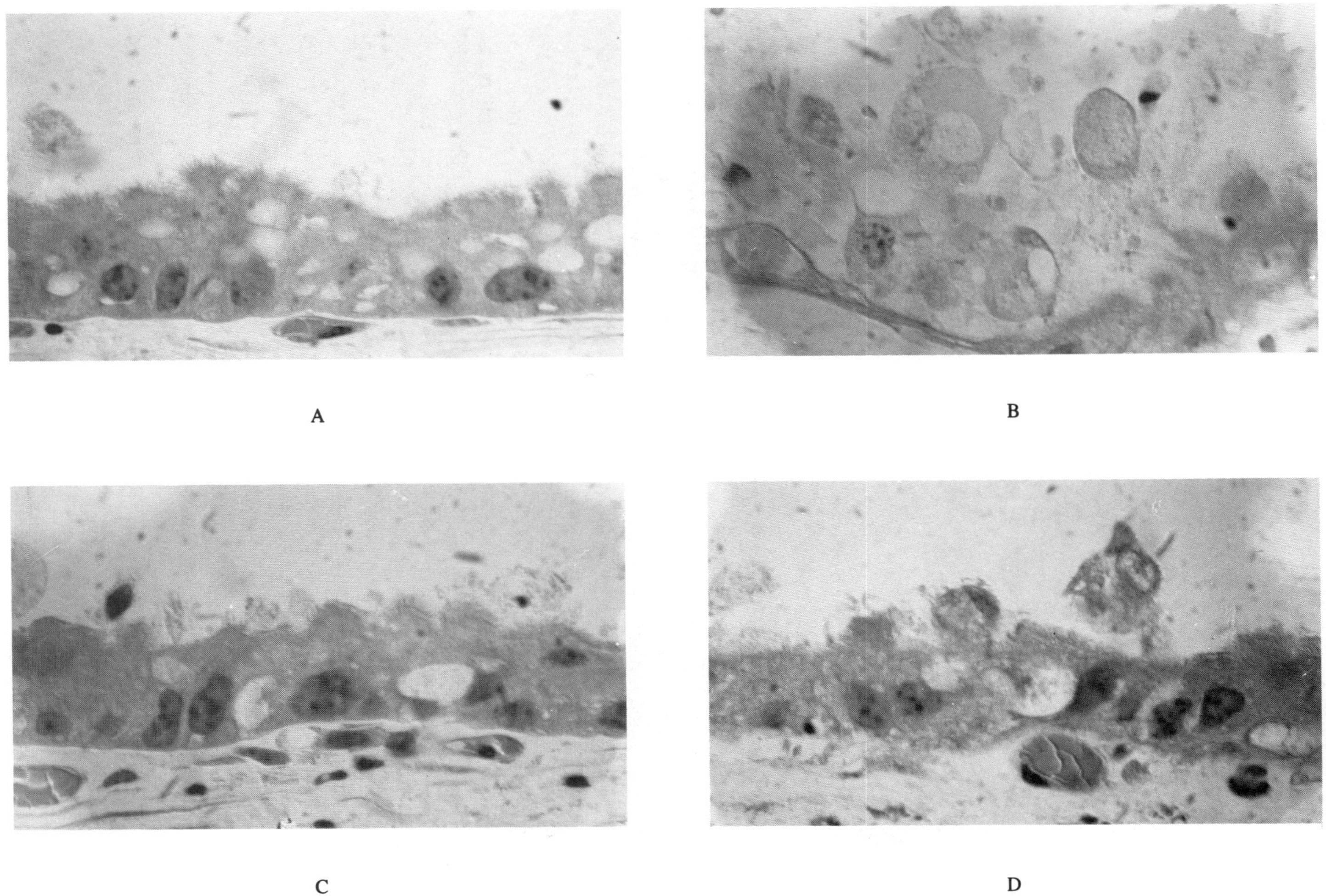

FIGURE 6. Sections of chorioallantoic membrane explants inoculated with various strains of *B. abortus* 12 h earlier. (H & E; magnification × 720.) (A) Control membrane. The trophoblasts are intact and mildly vaculated; (B) Strain 2308. The membrane is covered by very thin trophoblasts, the majority of which are necrotic; (C) RB51-inoculated explants revealed no significant cytopathology; (D) Strain 19-inoculated explant demonstrates mild cellular degeneration.

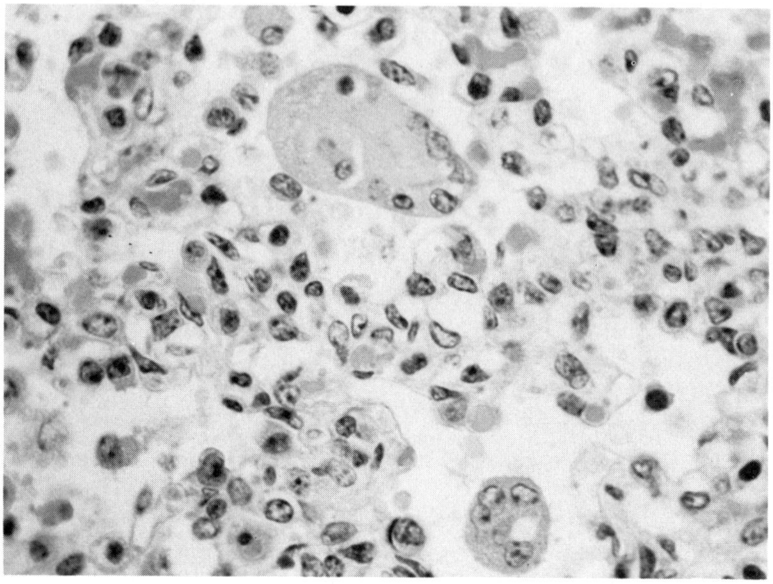

FIGURE 7. Fetal lung from a bovine fetus infected with *B. suis*. The alveoli are filled with macrophages and multiple giant cells. (H & E; magnification × 720.)

H. HISTOLOGICAL, IMMUNOLOGICAL AND HEMATOLOGICAL RESPONSES IN BRUCELLA-INFECTED FETUSES

Ruminant fetuses of different species respond to infection by similar brucella species in a similar manner. Infected fetuses demonstrate lymphoreticular hyperplasia of multiple lymphoid tissues. Granulomatous foci composed of macrophages, epithelioid cells, giant cells, and variable numbers of granulocytic cells are observed in multiple lymphoid tissues and in the lung (Figure 7). These reactions are occasionally observed in other organ systems, such as the kidney and liver.[22,27,28] Inflammatory lesions were apparent as early as 4 d PI. in infected sheep fetuses. The lesions were most obvious in the fetal lungs and consisted of interstitial infiltrates of mononuclear and polymorphonuclear leukocytes. Both cell populations were components of an exudate within the terminal airways and within potential alveolar spaces.[27] Giant cells are infrequently observed in the granulomatous reaction in both lymphoid tissues and the lung, except in *B. suis*-infected bovine fetuses which developed granulomatous reactions with moderate numbers of giant cells. As early as 6 d PI in infected sheep fetuses and by 9 d PI in infected bovine fetuses, lymphoid hyperplasia with primary follicle formation is obvious. *B. abortus* strain 2308-infected fetal sheep seldom survive more than 6 d PI and developed only slight elevations of IgM and IgG_1 and IgG_2. No brucella antibodies were detected by the ELISA test in these infected sheep fetuses. Antibodies to brucella were, however, detected by 10 d PI in strain 2308-infected bovine fetuses.[22,27]

Following brucella infection, bovine fetuses developed a leukocytosis with significant increases in the number of mature neutrophils and significant decreases in the numbers of lymphocytes and eosinophils. A slight increase in numbers of both band neutrophils and monocytes was also observed. Similar hematologic changes were observed in brucella-infected sheep fetuses. Marked increases in mature neutrophils and decreases in lymphocytes and eosinophils were noted by 24 h PI. These changes persisted through day 6 PI. Increased numbers of immature neutrophils and mature blood monocytes were noted in sheep fetuses on day 4 PI. These changes coincide with the profound elevation of fetal cortisol observed in infected fetuses and were presumably a reflection of fetal stress. Increases in immature neutrophils on day 4 PI also coincide with increases in extramedullary hematopoiesis observed in a variety of fetal tissues.[23] Following brucella infection, bovine fetuses develop striking increases in the staining intensity of nonspecific esterase (NSE) positive blood monocytes. Six of 9 noninfected control bovine fetuses had negative NSE-stained circulating monocytes, while 14 of 15 infected fetuses had intensely stained monocytes. One infected fetus had weakly stained monocytes. These studies demonstrate that late gestational bovine, sheep, and goat fetuses respond to brucella infections in a fashion similar to the responses observed in infected adults. Unlike brucella-infected adults, infected fetuses are unable to slow the growth of brucellae in their tissues. Ruminant fetuses remain one of the most susceptible hosts for brucella infections. Only the prohibitive costs associated with these experimental animals prevent them from being utilized more often in pathogenesis studies.

III. PATHOBIOLOGY

The interactions of brucellae with the cells of the host's immune and defense systems are of primary importance in determining resistance or susceptibility to brucellosis. The balance between host factors, which serve to eliminate the bacteria, and virulence factors of the bacteria, which allow the agent to escape host defense mechanisms, ultimately determine the outcome of the infection.

Host factors which destroy brucellae are varied and include lytic proteins in serum, specific antibodies directed against surface components of the bacteria, and phagocytic leukocytes. In mice, immunity to brucellae can be successfully transferred to genetically identical naive individuals with a specific subpopulation of T-lymphocytes. These lymphocytes recognize specific antigenic determinants on the brucella surface and produce a number of soluble mediators (lymphokines), which are capable of expanding clones of other lymphoid cells responsible for immunity, of attracting leukocytes, and of activating their bactericidal capabilities.[29] In the absence of adequate numbers of specific T-lymphocytes in a brucella-naive animal, the bacteria themselves and their products can directly interact with a wide variety of nonlymphoid cells to generate a number of biologically active substances (cytokines), which are capable of stimulating the production of leukocytes and of increasing their functional status. Nonspecific immune enhancement of defense mechanisms are common in the acute stages of brucella infections. The inherent ability of macrophages, and perhaps neutrophils, to respond to these nonspecific factors probably determines the degree of innate resistance to brucellae in the individual animal. This natural resistance to brucella infection is under genetic control in swine, rabbits, mice, and, probably, cattle. Resistance to a number of intracellular pathogens including brucellae is mediated by gene products expressed in macrophages.[3]

Serum components in nonimmune cattle are able to kill *B. abortus*. In some instances, both complement components and specific anti-*Brucella* antibodies in the sera of immune cattle enhance the levels of ''brucellacidal'' activity. There is also conflicting evidence that specific antibodies may actually decrease the brucellacidal activity of sera. Bovine complement can kill *B. abortus*. During the reaction, hemolytic complement is consumed.[30] *B. abortus* LPS and intact brucella do not activate complement by the alternate pathway.[31]

The importance of the brucellacidal activity of serum *in vivo* is not known. The ability to kill brucella in the blood is generally considered a minor component in the host's defense system. If, in fact, the factors responsible for this activity are present within the exudates associated with the mucosal inflammatory reactions at the initial sites of infection, then they may be of considerable value in killing the relatively small number of brucellae which breech the mucosal epithelium.

The ability of phagocytic leukocytes to bind to, ingest, and kill brucella is the principal mechanism of host defense. Phagocytic leukocytes participate in both the acute and chronic phases of brucella infections. In most chronic diseases caused by obligate or facultative intracellular bacteria, the macrophages are considered to be the phagocytic cell which is most responsible for immunity. A number of *in vitro* studies have demonstrated that the neutrophil is, in fact, better equipped to phagocytize and kill brucella than are macrophages.[32] Neutrophils are a prominent component in the initial acute inflammatory reactions associated with brucella infections. Extracts of neutrophil granules, in particular the myeloperoxidase-hydrogen peroxide-halide (MPO-H_2O_2-halide) complex, is most efficient in killing a variety of brucella strains.[33]

Studies comparing the phagocytic and killing abilities of human and bovine neutrophils against rough and smooth strains of *B. abortus* noted that both strains were readily ingested, but were resistant to killing. The smooth strain proved to be more resistant to killing than the rough strain.[32,33] Other studies with bovine neutrophils have demonstrated equal killing of ingested smooth and rough strains of brucella, but noted that significantly fewer rough bacteria were ingested. Cytological studies strongly suggest that the diminished ingestion was due to a failure of the rough brucella to attach to the phagocyte's membrane.[34] The inability of neutrophils to kill ingested *B. abortus* was due either to a failure to degranulate or to undergo the respiratory burst. Recent data suggest that opsonized *B. abortus* does stimulate the respiratory burst, but may interfere with degranulation (phagosome-lysosome fusion).[35] Water extracts of *B. abortus* strain 2308 also inhibited the fusion of phagosomes and lysosomes in mouse peritoneal macrophages.[36] Both viable and nonviable *B. abortus* were able to inhibit degranulation in neutrophils. These findings suggested that surface components of brucellae were responsible for the inhibitory action.

Small molecular weight (<1000 Da) components of *B. abortus* strain 2308 inhibited the MPO-H_2O_2-halide bactericidal activity of neutrophils. These substances specifically inhibited the degranulation of primary granules. The substances demonstrated little or no interference with the ability of neutrophils to ingest and kill *Staphylococcus aureus* or to generate an oxidative burst. This inhibitory activity was due to 5′ guanosine monophosphate (GMP) and to adenine.[37,38] Other bacteria which resist neutrophil killing also release large amounts of both compounds from their surfaces. Guanosine also inhibited lysosomal degranulation in mouse peritoneal macrophages.[39] Whether or not the concentrations of these substances *in vivo* are sufficient to inhibit degranulation is not known. Utilizing this novel approach, the potential intracellular pathogen is able to mimic the release of these compounds by dead bacteria undergoing degradation. Presumably, the release of these compounds from the degraded organism would prevent continued and unnecessary lysosomal degranulation. In the case of the pathogenic bacteria, failure to degranulate would aid in intracellular survival.

The participation of significant numbers of neutrophils in the inflammatory responses in both the submucosa and lymph nodes of brucella-infected adults and their participation in a variety of tissue reactions in infected fetuses and experimentally infected laboratory rodents prompted a comparative study of the phagocytic and bactericidal ability of fetal sheep and adult sheep neutrophils. The acridine orange uptake assay was used to determine both the phagocytic and killing capabilities of neutrophils against *B. abortus* strain 2308. The tests were conducted in autologous serum in control and brucella-infected fetal sheep and mature ewes. Control fetuses demonstrated a low background level of phagocytic activity directed toward both *B. abortus* (17.0%) and *S. aureus* (28.5%), as compared to 31.7% phagocytic activity against *B. abortus*, and 65.0% activity against *S. aureus* in control ewes. At 24 to 144 h postinfection, fetal phagocytic activity against *B. abortus* increased to 50% and to a 57% level of activity against *S. aureus*. The adult activity increased to 58.7% against *B. abortus* and to 75.3% against *S. aureus*.

Neutrophils from both control fetuses and adults demonstrated a low level of killing activity against both *B. abortus*, 10.0 and 18.2%, and *S. aureus*, 12.0 and 22.0%, respectively. Following infection the ability of fetal neutrophils to kill brucellae increased to 49%

while that in the adults increased to 42%. The ability of infected fetal neutrophils to kill *S. aureus* increased to 34%, and no increases were observed in the killing ability of infected ewes' neutrophils against *S. aureus*. The phagocytic and killing activity of control fetal neutrophils was consistently lower than that observed in the adult; however, following infection with *B. abortus* no differences in these activities were noted for both the fetus and the adult.[23] These studies once again demonstrate the ability of later-term ruminant fetuses to respond as well as the adult to *B. abortus* infections. Even though *in vitro* tests such as these need to be cautiously evaluated, similar results have been reported by others after vaccination or infection of adult cattle with either strain 19 or strain 2308 *B. abortus*. It is most important to realize that these observed increases in neutrophil activity occurred in the absence of detectable *B. abortus*-specific antibody. These observations, combined with more recent findings, suggest that the functional activities of mature neutrophils are subject to regulation and that under the appropriate stimulation, up-regulation of functional activities essential to host defense mechanisms occurs quite rapidly.

It goes without saying that while the above data on the interaction of leukocytes with *B. abortus* is exciting, it must be interpreted with caution. In such studies, seldom are all the brucellae ingested, and rarely are they all killed. At this point, not enough is known to suggest that the ability of phagocytic leukocytes to kill 70, 80, or 90% of the *B. abortus in vitro* will prevent the remaining 10, 20, or 30% from establishing infection in the host.

IV. PATHOLOGY

A. *BRUCELLA ABORTUS*

Brucella abortus is the principal cause of brucellosis in domestic cattle. Many of the pathological changes in lymph nodes and in the female reproductive tract have been previously discussed. During the bacteremic phase, the bacteria may localize in a variety of tissues. Localization most frequently occurs in the female and male reproductive tract, the mammary gland, and the lymphoid tissues. Less frequently, localization of bacteria occurs within bone marrow, bones, renal cortex, and synovial membranes. In general, these localized brucella infections result in focal granulomatous lesions composed primarily of macrophages, lymphocytes, plasma cells, and variable numbers of neutrophils. There is generally little necrosis, and fibrosis is rarely observed.[1] Carpal bursitis in cattle may be caused by *B. abortus*. Typically, the lesions are granulomatous with villus proliferation of the synovium. Occasionally, focal accumulations of lymphoid tissue with germinal centers are noted in the subsynovial tissues.[40,41]

A focal nonsuppurative interstitial nephritis, similar to white spotted kidney disease of calves, has been associated with *B. abortus* infections. The lesions are composed of infiltrates of macrophages, large numbers of lymphocytes, and plasma cells. They are most often limited to the renal cortex.[40]

The mammary gland and the supramammary lymph nodes are frequently infected in bovine brucellosis. Focally distributed throughout the mammary gland, interstitial infiltrates of plasma cells, lymphocytes, and macrophages with variable numbers of neutrophils are present. Focal granulomatous lesions are present with the supramammary lymph nodes. The chronic interstitial inflammatory lesions within the mammary gland are not specific for *B. abortus*.[40] The pathogenic mechanisms accounting for the frequency of mammary gland infection, persistence of the infections, and the triggering of shedding in mammary gland secretions are not well understood.[40]

A variety of strains of *B. abortus*, including strain 19, is responsible for a necrotizing orchitis in bulls. Most cases are acute, and the damage is irreversible. One or both testicles may be affected. There is swelling of the scrotum and considerable inflammation of the tunics. Spaces between the tunics become distended, with a hemorrhagic and fibrinopurulent exudate. Initially, necrotic foci are present within the testicular parenchyma. The foci become progressively larger, and coalesce to produce total testicular necrosis, which may either become sequestered and surrounded by fibrous tissue or undergo liquefaction and become an exudate-filled cavity surrounded by fibrous wall.[42]

It is not uncommon to observe a focal necrotizing epididymitis in *B. abortus*-infected bulls. Spermatic granulomas are frequently observed secondarily to the brucella-induced inflammatory reactions in the epididymis. *B. abortus* is a major cause of seminal vesiculitis and prostatitis in bulls. *B. abortus* infection of the seminal vesicles may result in a fibrinopurulent exudate with variable degrees of necrosis and dystrophic calcification.[43] *Brucella*-caused prostatitis, on the other hand, is less suppurative. Degenerative changes are common in the prostatic epithelium, and the exudate tends to be catarrhal in character. Both accessory sex organs are enlarged as a result of the inflammation.

Brucella infections of the male reproductive tract usually result in either permanent or temporary sterility. The mixing of normal sperm, produced by an unaffected testicle, with semen containing bacteria, necrotic debris, and inflammatory exudates from the infected testicle, the epididymis, the seminal vesicle, or the prostate, will result in poor quality semen.[40]

B. *BRUCELLA MELITENSIS*

Brucella melitensis is primarily responsible for brucellosis in sheep and goats. The disease in goats closely resembles the disease produced in cattle infected with *B. abortus*. During the initial bacteremic phase of the infection in goats, individuals may become severely ill and die suddenly, while other individuals are asymptomatic. Localization of *B. melitensis* in the mammary gland may result in acute mastitis with palpable nodules of inflamed tissue within the gland and the production of clotted and watery milk. Abortions in the latter stages of pregnancy remain the primary clinical sign of infection. Repeated abortions in successive pregnancies may occur.[44] Experimental infections of pregnant sheep with *B. melitensis* result in placental changes which are more necrotizing and less exudative than those observed in placentitis caused by *B. abortus*.[45,46]

C. *BRUCELLA OVIS*

B. ovis is considered the least pathogenic of the five species of *Brucella* which infect domestic animals. *B. ovis* is an important cause of epididymitis in rams and is less frequently associated with abortions. Venereal transmission is the most important route of exposure.[47] As with other *brucella* infections there is a period of generalized infection. The bacteria can be recovered from multiple tissues, but their presence elicits no detectable host cellular responses. *B. ovis* localizes in the epididymis, where it generally causes an epididymitis which is usually limited to the tail of epididymis. Following localization in the epididymis, there is perivascular edema and infiltration of lymphocytes and monocytes into the peritubular tissue. Neutrophils become a component of the inflammatory exudates, and the tubular epithelium becomes hyperplastic and undergoes hydropic degeneration. The epithelial cells are eventually destroyed, either by bacterial products or by the underlying inflammation and extravasation of spermatozoa. The tissue responses to extravasated sperm lead to the formation of large spermatic granulomas, which may result in blockage of the epididymis.[48]

Experimental placental infections with *B. ovis* result in plaque-like thickenings of the intercotyledonary placenta. The plaques may coalesce to resemble yellow-white chamois leather. The placental reaction consists of edema and mild to moderate perarteritis and arteritis. Placental necrosis and cellular exudation tend to be minimal in *B. ovis* infections.[49]

Many lambs are alive at the initiation of parturition. Dead fetuses may be edematous, and their tissues are hemoglobin stained. Strands of fibrin may coat both the organs and walls of the thoracic and peritoneal cavities. Calcified plaques on the walls and soles of the hooves are characteristic.[44] The immune response and granulomatous inflammatory responses of *B. ovis*-infected ovine fetuses are similar to those observed in *B. abortus*-infected bovine fetuses. However, the inflammatory reaction in *B. ovis*-infected fetuses tend to be milder than those observed in fetuses infected with other species of *Brucella*. An acute interstitial nephritis and hepatic portal inflammatory reactions are frequently seen in *B. ovis*-infected fetuses. Chronic fetal infections with *B. ovis* have been reported. *In utero* infections of fetal sheep with *B. ovis* resulted in abortions which ranged from 23 to 80 d postinoculation.[50] This extended period of infection is in marked contrast to an average abortion interval of 12 d for *B. abortus*-infected bovine fetuses, and to as few as 6 to 9 d for *B. abortus*-infected sheep and goat fetuses.

D. *BRUCELLA SUIS*

B. suis is responsible for most brucellosis in swine. The disease in swine differs somewhat from brucellosis in cattle, sheep, and goats. *B. suis* tends to produce focal granulomatous reactions with a considerable degree of coagulative necrosis. These lesions tend to be more widespread and more frequently localized in bones and joints. In addition, the nonpregnant uterus is frequently infected.[51]

The initial events in *B. suis* infections are similar to those observed in *B. abortus* infections. The infection may be transmitted by ingestion and by coitus. Venereal transmission may be the more common of the two routes of transmission. Boars are as susceptible as sows and frequently develop orchitis and epididymitis and shed the bacteria in semen for extended periods of time. Sows with genital infections may also shed *B. suis* from the uterus for up to $2^1/_2$ years.

Following infection,[51,52] extensive lymphadenitis develops in lymph nodes draining the initial sites of entry. Following oral transmission, the lymph nodes of the head and neck are most frequently involved. If transmission is via the venereal route, then the inguinal and iliac lymph nodes are most frequently involved. The initial bacteremic phase of the infection may persist for years and may be responsible for the wide range of tissues secondarily infected in swine brucellosis. Localization occurs frequently in both the male and female reproductive tracts and in the skeletal system, joints, mammary glands, lymph nodes, spleen, kidney, bladder, and brain.[52]

The typical lesions are granulomatous with sheets of histocytic and epithelioid cells with a central zone of caseous or coagulative necrosis. The lesions are usually partially or completely encapsulated by a fibrous capsule. The necrotic portions of the granulomas are heavily infiltrated with neutrophils. Liquefaction or mineralization of the necrotic zones can also occur. Giant cells are infrequently observed within the granulomas.[52]

Localization of *B. suis* within synovial membranes often leads to a severe exudative purulent or fibrinopurulent arthritis. *B. suis* typically localizes in the epiphysis of lumbar vertebrae. The osteomyelitits produced is composed of granulomatous foci with necrotic centers. The reactions commonly contain numerous neutrophils and frequently undergo liquefaction. Intervertebral disks are often affected and may be destroyed by the inflammatory reaction. Damage to the adjacent spinal cord and major nerves may result in hind limb lameness and/or paralysis.[52]

Unlike other brucella infections, *B. suis* may localize in the nonpregnant uterus and uterine tubes. Variable numbers of small 2- to 3-mm white-to-yellow nodules are seeded throughout the uterine mucosa. The nodules are composed of macrophages, lymphocytes, and plasma cells. They frequently have necrotic centers and are surrounded by a variable amount of fibrous tissue. Well-developed hyperplastic lymphocytic nodules are also present. Neutrophils and large numbers of mononuclear cells infiltrate the uterine epithelium and collect within the uterine lumen. Uterine glands are dilated with leukocytes and mucus. The chronic endometritis which results interferes with conception and may account for the prolonged shedding of brucella from the female reproductive tract.[52]

Abortions due to *B. suis* generally occur in late gestation. Infected fetuses develop typical lesions; however, they may be difficult to distinguish due to autolysis. The incidence of stillborn and weak piglets may be very high in *B. suis*-infected sows.

E. *BRUCELLA CANIS*

Brucella canis causes infectious abortions in bitches and epididymitis, testicular atrophy, and sterility in infected males. Both oral and venereal transmission are important in *B. canis* infections. As with the other brucella infections, there is initial localization of the bacteria in lymph nodes draining the site of infection. The retropharyngeal lymph nodes are most often affected after oral infection, while the superficial inguinal and iliac lymph nodes are affected following venereal exposure. The lymph nodes are enlarged and edematous due to both lymphoid hyperplasia and granulomatous inflammation.[9]

Susceptible dogs generally develop a bacteremic phase of infection 1 to 3 weeks following exposure. The bacteremia may persist for several months to several years. Localization of *B. canis* in many tissues may occur. As with *B. suis* infections of swine, localization within lumbar vertebrae is not uncommon.[53] Infection of the pregnant uterus results in a typical placentitis, fetal infection, and a chronic granulomatous endometritis. Infected fetuses develop a granulomatous pneumonia and may demonstrate endocarditis, hepatitis, and nephritis. Aborted fetuses and the placenta are seldom available for examination, as they are most often eaten by the bitch.[9]

Scrotal swelling is apparent 3 to 5 weeks following oral inoculation of *B. canis*. It is caused by a fibrinopurulent exudate in the tunica vaginalis. Scrotal ulceration may occur. For the most part, the inflammatory reactions in the epididymis and prostate are nonsuppurative. The reactions are composed of macrophages and lymphocytes which infiltrate the interstitial tissues of both organs. There is degeneration and destruction of adjacent epithelial cells and tissue fibrosis. Complete obliteration of the epididymal ducts, however, is rare. *Brucella canis* can persist in the prostate for many months.[54]

Spermatic granulomas may form following extravasation of sperm, and antisperm antibodies are frequently detected. An autoantibody of the IgA class has been demonstrated on head-to-head agglutinated spermatozoa within accessory sex organs. Immune responses to sperm antigens may explain the infertility observed in infected male dogs.[55]

REFERENCES

1. **Thoen, C. O. and Enright, F. M.,** Brucella, in *Pathogenesis of Bacterial Infections in Animals,* Gyles, C. L. and Thoen, C. O., Eds., Iowa State University Press, Ames, 1986, chap. 20.
2. **Berman, D. T.,** *Brucellosis Research: An Evaluation,* National Research Council, National Academy of Sciences, Washington, D.C., 1977, chap. 7.
3. **Harmon, B. G., Templeton, J. W., Crawford, R. P., Heck, F. C., Williams, J. D., and Adams, L. G.,** Macrophage function and immune response of naturally resistant and susceptible cattle to *Brucella abortus,* in *Genetic Control of Host Resistance to Infection and Malignancy,* Skamene, E., Ed., Alan R. Liss, New York, 1985, 345.
4. **Olitzki, A.,** *Immunological Methods in Brucellosis Research,* Part II: *In Vivo Procedures,* S. Karger, Basel, 1970, chap. 1.
5. **Enright, F. M.,** unpublished data, 1985.
6. **Payne, J. M.,** The pathogenesis of experimental brucellosis in the pregnant cow, *J. Pathol. Bacteriol.,* 78, 447, 1959.
7. **Cheers, C. and Young, A. M.,** Serum colony stimulating activity and colony forming cells in murine brucellosis: relationship to immunopathology, *Microb. Pathogenesis,* 3, 185, 1987.
8. **Enright, F. M.,** unpublished data, 1986.
9. **Carmichael, L. E. and Kenney, R. M.,** Canine brucellosis: the clinical disease, pathogenesis and immune response, *J. Am. Vet. Med. Assoc.,* 156, 1726, 1970.

10. **Keppie, J., Williams, A. E., Witt, K., and Smith, H.,** The role of erythritol in tissue localization of the *Brucellae, Br. J. Exp. Pathol.,* 46, 104, 1965.
11. **Bosseray, N.,** Colonization of mouse placentas by *Brucella abortus* inoculated during pregnancy, *Br. J. Exp. Pathol.,* 61, 361, 1980.
12. **Smith, T.,** A characteristic localization of *Bacillus abortus* in the bovine fetal membranes, *J. Exp. Med.,* 29, 451, 1919.
13. **Anderson, T. D., Cheville, N. F., and Meador, V. P.,** Pathogenesis of placentitis in the goat inoculated with *Brucella abortus.* II. Ultrastructural studies, *Vet. Pathol.,* 23, 227, 1986.
14. **Detilleux, P. G., Cheville, N. F., and Deyoe, B. L.,** Pathogenesis of *Brucella abortus* in chicken embryos, *Vet. Pathol.,* 25, 138, 1988.
15. **Anderson, T. D., Cheville, N. F., and Meador, V. P.,** Pathogenesis of placentitis in the goat inoculated with *Brucella abortus.* I. Gross and histologic lesions, *Vet. Pathol.,* 23, 219, 1986.
16. **Morrison, D. C. and Ulevitch, R.,** The effects of bacterial endotoxins on host mediation systems. A review, *Am. J. Pathol.,* 93, 527, 1978.
17. **Harper, M. J. K.,** Effect of endotoxin treatment on prostaglandin metabolism by the rabbit uterus, oviduct, and lung, *J. Reprod. Fertil.,* 58, 101, 1980.
18. **Pennell, R. B. and Huddleson, I. F.,** Chemical constitution and biological properties of the endo-antigen of the *Brucella* group of microorganisms, *Mich. State Coll. Agric. Exp. Stn. Tech. Bull.,* No. 156, 1937.
19. **Adams, L. G.,** personal communication, 1986.
20. **Moreno, E., Berman, D. T., and Boettcher, L. A.,** Biological activities of *Brucella abortus* lipopolysaccharides, *Infect. Immun.,* 31, 362, 1981.
21. **Liggins, G. L.,** Endocrinology of parturition, in *Fetal Endocrinology,* Novy, M. J. and Resko, J. A., Eds., Academic Press, New York, 1981, 211.
22. **Enright, F. M., Walker, J. V., Jeffers, G. W., and Deyoe, B. L.,** Cellular and humoral responses of *Brucella abortus* infected fetuses, *Am. J. Vet. Res.,* 45, 424, 1984.
23. **Gorham, S. L.,** Morphologic Lesions, Hematological and Immunological Responses and Neutrophil Functions in Fetal Neonatal and Adult Sheep Infected with *Brucella abortus,* Ph.D. thesis, Louisiana State University, Baton Rouge, 1983.
24. **Jack, P. M. B., Nathanielsz, P. W., Thomas, A. L., and Stevens, D. H.,** Ultrastructural changes in the placenta of the ewe following foetal infusion of cortisol, *Q. J. Exp. Physiol.,* 60, 171, 1975.
25. **Enright, F. M., Roop, M., III, and Schurig, G.,** unpublished data, 1985.
26. **Enright, F. M.,** unpublished data, 1987.
27. **Gorham, S. L., Enright, F. M., Snider, T. G., III, and Roberts, E. D.,** Morphologic lesions in *Brucella abortus* infected ovine fetuses, *Vet. Pathol.,* 23, 331, 1986.
28. **Meador, V. P., Hagemoser, W. A., and Deyoe, B. L.,** Histopathologic findings in *Brucella abortus*-infected, pregnant goats, *Am. J. Vet. Res.,* 49, 274, 1988.
29. **Cheers, C.,** Pathogenesis and cellular immunity in experimental murine brucellosis, in *Biological Standardization,* Vol. 56, 3rd Int. Symp. on Brucellosis, Vallette, L. and Heunessen, W., Eds., Algiers, S. Karger, Basel, 1984, 237.
30. **Irwin, M. R. and Berman, D. T.,** *Bactericidal Tests in Brucellosis in Brucellosis,* Larson, C. T. and Soule, M. H., Eds., Waverly Press, Baltimore, 1950, 85.
31. **Hoffmann, E. M., and Houle, J. J.,** Failure of *Brucella abortus* lipopolysaccharide (LPS) to activate the alternative pathway of complement, *Vet. Immunol. Immunopathol.,* 5, 65, 1984.
32. **Kreutzer, D. L., Dreyfus, L. A., and Robertson, D. C.,** Interactions of polymorphonuclear leukocytes with smooth and rough strains of *Brucella abortus, Infect. Immun.,* 23, 737, 1979.
33. **Riley, L. K.,** Interactions of *Brucella abortus* with Polymorphonuclear Leukocytes, Ph.D. dissertation, University of Kansas, Lawrence, 1979.
34. **Enright, F. M.,** unpublished data, 1984.
35. **Canning, P. C., Deyoe, B. L., and Roth, J. A.,** Opsonin-dependent stimulation of bovine neutrophil oxidative metabolism by *Brucella abortus, Am. J. Vet. Res.,* 49, 160, 1988.
36. **Frenchick, P. J., Markham, R. J. F., and Cochrane, A. H.,** Inhibition of phagosome-lysosome fusion in macrophages by soluble extracts of virulent *Brucella abortus, Am. J. Vet. Res.,* 45, 332, 1985.
37. **Canning, P. C., Roth, J. A., Tabatabai, L. B., and Deyoe, B. L.,** Isolation of components of *Brucella abortus* responsible for inhibition of function in bovine neutrophils, *J. Infect. Dis.,* 152, 913, 1985.
38. **Bertram, T. A., Canning, P. C., and Roth, J. A.,** Preferential inhibition of primary granule release from bovine neutrophils by an extract from *Brucella abortus, Infect. Immun.,* 52, 285, 1986.
39. **Riches, D. W. H., Watkins, J. L., Henson, P. M., and Stanworth, D. R.,** Regulation of macrophage lysosomal secretion by adenosine, adenosine phosphate esters, and related structural analogues of Adenosine, *J. Leu. Biol.,* 37, 545, 1985.
40. **Jubb, K. V. F., Kennedy, P. C., and Palmer, N.,** *Pathology of Domestic Animals,* Vol. 3, Academic Press, Orlando, FL, 1985, 345.

41. **Bracewell, C. D. and Corbel, M. J.,** An association between arthritis and persistent serological reactions to *Brucella abortus* in cattle from apparently brucellosis-free herds, *Vet. Rec.*, 106, 99, 1980.
42. **Jubb, K. V. F., Kennedy, P. C., and Palmer, N.,** *Pathology of Domestic Animals,* Vol. 3, Academic Press, Orlando, FL, 1985, 432.
43. **Jubb, K. V. F., Kennedy, P. C., and Palmer, N.,** *Pathology of Domestic Animals,* Vol. 3, Academic Press, Orlando, FL, 1985, 448.
44. **Jubb, K. V. F., Kennedy, P. C., and Palmer, N.,** *Pathology of Domestic Animals,* Vol. 3, Academic Press, Orlando, FL, 1985, 350.
45. **Mollelo, J. A., Flint, J. C., Collier, J. R., and Jensen, R.,** Placental pathology. III. Placental lesions of sheep experimentally infected with *Brucella melitensis, Am. J. Vet. Res.*, 24, 905, 1963.
46. **Mollelo, J. A., Jensen, R., Collier, J. R., and Flint, J. C.,** Placental pathology. III. Placental lesions of sheep experimentally infected with *Brucella abortus, Am. J. Vet. Res.*, 24, 915, 1963.
47. **Jubb, K. V. F., Kennedy, P. C., and Palmer, N.,** *Pathology of Domestic Animals,* Vol. 3, Academic Press, Orlando, FL, 1985, 349.
48. **Jubb, K. V. F., Kennedy, P. C., and Palmer, N.,** *Pathology of Domestic Animals,* Vol. 3, Academic Press, Orlando, FL, 1985, 444.
49. **Mollelo, J. A., Jensen, R., Flint, J. C., and Collier, J. R.,** Placenta pathology. I. Placental lesions of sheep experimentally infected with *Brucella ovis, Am. J. Vet. Res.*, 24, 897, 1963.
50. **Osburn, B. I. and Kennedy, P. C.,** Pathologic and immunologic responses of the fetal lamb to *Brucella ovis, Pathol. Vet.*, 3, 110, 1966.
51. **Jubb, K. V. F., Kennedy, P. C., and Palmer, N.,** *Pathology of Domestic Animals,* Vol. 3, Academic Press, Orlando, FL, 1985, 347.
52. **Jubb, K. V. F., Kennedy, P. C., and Palmer, N.,** *Pathology of Domestic Animals,* Vol. 3, Academic Press, Orlando, FL, 1985, 349.
53. **Jubb, K. V. F., Kennedy, P. C., and Palmer, N.,** *Pathology of Domestic Animals,* Vol. 1, Academic Press, Orlando, FL, 1985, 119.
54. **Jubb, K. V. F., Kennedy, P. C., and Palmer, N.,** *Pathology of Domestic Animals,* Vol. 3, Academic Press, Orlando, FL, 1985, 446.
55. **Serikawa, T., Takada, H., Kondo, Y., Muraguchi, T., and Yamada, J.,** Multiplication of *Brucella suis* in male reproductive organs and detection of autoantibody to spermatozoa in canine brucellosis, in *Biological Standardization,* Vol. 56, 3rd Int. Symp. on Brucellosis, Vallette, L. and Heunessen, W., Eds., S. Karger, Basel, 1984, 295.

Chapter 13

BRUCELLOSIS IN WILDLIFE

Donald S. Davis

TABLE OF CONTENTS

I. INTRODUCTION

Most of the historical investigations and, hence, the scientific literature have predominantly dealt with diseases in domestic animals or humans caused by various species of the genus *Brucella*. There are, however, numerous examples of widespread brucellosis in free-ranging, semidomesticated, and captive wildlife species. As brucellosis is being actively eradicated or, at least, successfully controlled in most domestic species in the more industrialized or advanced countries of the world, the presence of *Brucella* species in wildlife populations increases in importance as a potential future health threat to sympatric livestock industries and human populations. This pattern will, in every likelihood, be repeated in Third-World countries as improved vaccines and diagnostic techniques are eventually applied to the problem of brucellosis in their domestic livestock. At present, even in the face of the powerful tools of modern medical technology, there is no known example of a successful eradication of brucellosis from a free-ranging wildlife population without the accompanying eradication of that infected wildlife population. It is, therefore, highly probable that the last remaining reservoir of brucellosis on this earth will be in a wildlife population. This increasing epidemiologic significance of brucellosis in wildlife populations dictates that investigations should be conducted to understand better the disease as it exists in nature and that innovative disease prevention and control methods be tested for their efficacy in wildlife applications.

II. LITERATURE REVIEW

Brucellosis in wildlife is not a new phenomenon. It has a historical basis almost equal in magnitude to that of the genus *Brucella*. Within three decades of Bruce's discovery[1,2] of *B. melitensis* as the cause of Malta Fever in 1887 and the isolation of *B. abortus* from aborted cattle by Bang[3] in 1897, serologic evidence of brucellosis was reported by Mohler[4] in 1917 as positive agglutination reactions in sera from three bison (*Bison bison*) cows from Yellowstone National Park. The following portion of this chapter will, in large part, be a review of the literature organized by the wildlife host species (or taxonomically similar groups) in which the various species or strains (biovars) of the genus *Brucella* are known to occur or cause disease. The review will be followed by a discussion of the significance of brucellosis in wildlife with regard to future eradication efforts and potential threat to livestock, exotic animals, and human health.

A. WILD RUMINANTS

1. Bison (American Buffalo, *Bison bison*)

Brucella abortus was the first species of the genus to be isolated from a wild or nondomesticated host when Creech[5] in 1930 isolated the bacterium from the testicle of a bison bull killed on the National Bison Range, Moiese, MT. Earlier, in 1917, Mohler[4] reported serologic evidence of the presence of brucellosis in bison from three bison (more commonly called buffalo) cows from Yellowstone National Park after abortions were observed in two of the three bison cows. In 1930, Rush[6] observed *Brucella*-positive reactions in three of five serum samples from bison from the Yellowstone National Park herd. Then later, Rush[6] collected 110 bison sera from the Yellowstone and Moiese herds and found 58 positive and 25 suspicious by an agglutination test. Subsequent and more extensive testing by Tunnicliff and Marsh[7] in 1935 of bison sera collected over several years from the National Bison Range and Yellowstone National park indicated a greater than 60% (305/484) reactor rate tested. Moore[8] in 1947 tested 37 sera collected from bison in Elk Island National Park, Alberta, Canada in the winter of 1946—1947. Of these, 16.2% were found to be positive and 13.5% questionable. Sera from 343 bison collected in 1956 by Corner and Connell[9] from Elk Island National Park reacted to a *Brucella* agglutination test at a rate of 42.3% positive. Six of six bison from Utah were found to possess serum antibodies which reacted with *B. abortus* tube agglutination antigens at titers of 20 or greater.[10] More recently, serum samples collected from 2365 free-ranging bison in Wood Buffalo National Park in Alberta and Northwest Territories, Canada during the period from 1959 to 1974 were tested for brucellosis, and 31.2% (739) reacted positively to the tube agglutination test.[11] Isolates of *B. abortus* collected from bison from Yellowstone National Park have been shown to be infective and pathogenic in susceptible cattle.[12] Brucellosis in bison (presumably limited to *B. abortus*) continues to be a problem in both Canada and the U.S. in private herds as well as in some herds on public lands.

2. Elk (Wapiti, *Cervus elaphus canadensis*)

Rush[6] in 1932 found the first serologic evidence of *Brucella* infections in elk when 13 of 67 serum samples from Yellowstone National park were positive or suspicious on the plate agglutination test. Further serologic evidence of brucellosis in elk was documented in 1935 by Tunnicliff and Marsh[7] when they classified 6 of 32 elk from Yellowstone National Park as reactors to the agglutination test. They also speculated that the elk population had become infected with brucellosis from an association with infected bison in Yellowstone National Park. Honess and Winter[13] in 1956 agreed with this theory and further stated that brucellosis had been spread from domestic cattle to bison and from the infected bison to the elk. They based this belief solely on the fact that while bison were infected at a greater than 50% rate, elk on the same range were found to be infected with brucellosis at about a 10% rate. In Canada, Corner and Connell[9] in 1958 reported that of 221 elk sera collected from 1956 to 1957 in Elk Island National Park, 4 were suspicious and 25 were positive to the brucellosis agglutination test. In a serologic survey of wild ruminants in Colorado by Adrian and Keiss,[14] sera collected from 1967 to 1976 from 3833 elk were tested by the rapid screening agglutination test for brucellosis. Ten were positive at a 1/25 dilution, and one was positive at a 1/50 dilution. Vaughn et al.[15] in 1973 found none of 54 elk from Idaho to have detectable levels of *Brucella* antibodies. Three of 113 elk from Utah in 1978 tested by the slide agglutination test were found to be positive.[16] The results of the most comprehensive serologic and bacteriologic survey of brucellosis in elk were reported by Thorne et al.[17] in 1978. The incidence of brucellosis in 1165 elk from Wyoming examined over a 5-year period by testing serum samples by the standard plate agglutination, buffered *Brucella* antigen, Rivanol, and complement fixation techniques, was found to be 31% in this study. *Brucella abortus* biotype 1 was isolated from 17 of 45 elk. The results of this investigation indicated that "no single serologic test should be relied upon to diagnosis brucellosis in elk." In a separate report, Thorne et al.[18] discussed the clinical effects of experimental infections of elk with *B. abortus* strain 2308 and the methods of transmission from experimentally infected to susceptible elk. In 1985, McCorquadale and DiGiacomo[19] reported a summary of serologic surveys for *B. abortus* in populations of elk in North America. Out of a total of 7267 elk tested, 446 (6.1%) were classified as reactors. Elk, like the bison in certain regions of the western U.S., are obviously infected with *B. abortus* and as such pose difficult problems for the states involved as well as the Federal Brucellosis Eradication Program.

3. Reindeer and Caribou (*Rangifer* spp.)

Serologic evidence of brucellosis in Alaskan Eskimo natives has been recorded since 1939. The infective agent was originally thought to be *B. abortus* and the human cases attributable to contact with domestic cattle or swine.[20] In 1958 and 1959, *Brucella* spp. organisms were isolated from blood and bone marrow from two Eskimos in Alaska. The organisms resembled, but were not identical to a strain of *B. suis*.[21] Epidemiologic evidence based on the occurrence of this nontypical strain of *Brucella* suggested that caribou or wild reindeer (*Rangifer tarandus*) were the animal reservoirs and the source of the human infections in Alaska.[22] Huntley et al.[20] in 1963 first isolated the *Brucella* organism from the Arctic caribou herd and stated that this organism was pathogenic to humans, and brucellosis was prevalent among the Alaskan native populations that were dependent on caribou for food. There has been some speculation that brucellosis may have been endemic in the prehistoric caribou of North America, while others believe that the disease was introduced into Alaska at the turn of the century by the reindeer brought from Russia. While there is some evidence to support each of these views, it is interesting to note that all of the isolates from North American reindeer and caribou and those in Siberia are indistinguishable from one another and are now classified as *B. suis* biotype 4.[23] In any case, *B. suis* biotype 4 is now recognized as a circumpolar etiologic agent of brucellosis with caribou and reindeer as its primary host but capable of infecting humans, dogs, and a variety of carnivorous hosts.[24] More recently, serologic surveys in Alaska, Canada, and Siberia have continued to document the presence and distribution of brucellosis in caribou populations in North America and Russia.[25-30] Tessaro and Forbes[31] in 1986 reported *B. suis* biotype 4 as isolated from a barren ground caribou (*R.t. groenlandicus*) and presented an excellent review of the distribution of human and animal cases of brucellosis in northern Canada. Attempts to vaccinate semi-domesticated herds of reindeer in Alaska with killed *B. abortus* strain 45/20 have met with mixed results[32] as have attempts to utilize strain 19 vaccine in Russian reindeer.[33] The documented prevalence of *B. suis* biotype 4 in a widespread and migratory species such as caribou in Alaska, Canada, and Siberia puts severe strains on the imagination when the realistic chances of the total success of governmental brucellosis eradication programs in the future are considered.

4. Mule Deer and White-Tailed Deer (*Odocoileus hemionus* and *O. virginianus*)

In terms of numbers, white-tailed deer and mule deer have been the most extensively studied wild ruminant species with respect to their possible involvement with the disease of brucellosis. Most of this attention is the apparent result of two facts, the first being the large numbers and wide geographic distribution of the genus *Odocoileus* in North America and, second, the commonly observed close association of these deer and domestic livestock on shared pastures and ranges. Over 30 serologic and bacteriologic surveys in the U.S. for brucellosis in white-tailed and mule deer have been published.[10,34-63] Relatively few have found any positive results. From a total of 25,000+ deer from 29 states tested for brucellosis over the last 40 years by a variety of serologic techniques in 50+ published studies, only 0.29% (72) of the deer sera reacted at any level, and only 46 (0.18%) reacted at diagnostic levels.[64] In only one instance has *B. abortus* been isolated from a free-ranging white-tailed deer.[65] From the results of this overwhelming accumulation of data, it is fairly safe to state that white-tailed deer and mule deer have not in the past 40 years had any large involvement in the spread of brucellosis in the U.S. This does not mean that they may never become involved in the future because both species have been shown to be susceptible to brucellosis under experimental conditions.[66-68] The fact that these two deer species are largely browsers instead of grazers (like elk and bison) may spatially isolate them for exposure, in most cases, from contaminated forage.

5. Moose (*Alces alces*)

Little is actually known about brucellosis in moose. There have only been four cases reported in the literature. Fenstermacher and Olsen,[69] described the first case of moose brucellosis in 1942, when *B. abortus* was isolated from the tissues of a young free-ranging male moose in Minnesota. The young male was found in an apparently weakened state in the wild and was observed for some time before it was killed and necropsied. A fibrinous peritonitis was observed on the liver, and the left lobe of the lungs was adherent to the thoracic cavity. The base of the heart "was covered with a straw-colored gelatinous exudate." Histopathologic examination revealed a fibrinous myocarditis and pericarditis. Serum from this moose was found to be positive to the agglutination test at a dilution of 1:50,000. The second case of brucellosis in a moose was also in the U.S. and was described by Jellison et al.[70] in 1953. The animal was a young female moose also observed in the wild in an obviously ill condition. After several days of observation, the moose died at the Red Rocks Lakes Wildlife Refuge in Montana. The only significant pathological findings seen at necropsy were "a marked pericarditis and myocarditis". The pericardinal sac was distended with fluid, and a fibrinous film was observed over the anterior portion of the heart. Serum from this moose was tested several times and complete "agglutination was observed in one test at a serum dilution of 1:20,000". Pleural fluid and pericardial fluid from this female moose were cultured, and *B. abortus* was isolated from both. In a subsequent serologic survey for brucellosis in the same year on 44 hunter-killed moose in Montana, Jellison et al.[70] reported 35 negative, 3 positive at a 1:20 dilution, and 6 positive at a 1:60 dilution. The other two cases of brucellosis in moose were reported from Canada in 1958 by Corner and Connell.[9] During a depopulation program in Elk Island National Park in Alberta, on two separate occasions emaciated and weak male moose were observed. Both were shot and necropsied. Severe fibrinous pericarditis and fibrinous peritonitis were observed in both. *B. abortus* was recovered from the tissues of the 4-year-old male, and the serum from the 1 1/2-year-old moose was positive on the brucellosis agglutination test to 1:12,800. Serum from 124 apparently normal moose collected during the same study were serologically negative for brucellosis. Thorne et al.[17] have postulated that the lack of serologically positive moose found in subsequent studies is the result of the usually terminal fate of moose infected with *Brucella*. No reactors were found in 104 moose sera collected in British Columbia in 1977—1978 during an outbreak of brucellosis in cattle.[71] Zarnke and Yuill[28] reported no positive serologic reactors in 44 moose from Alberta. Zarnke[30] found 1 of 39 moose positive for brucellosis in a serologic survey for microbial pathogens in Alaska wildlife. None of 208 moose tested in 1984 from Quebec, had antibodies to *B. abortus*.[72] Further controlled studies on brucellosis in the moose will be necessary before the entire effects of the disease in this species are understood.

6. Pronghorn Antelope (*Antilocarpa americana*)

A number of studies have surveyed this North American species for the presence of brucellosis, probably again mainly on the basis of the close and commonly observed contact between pronghorns and domestic livestock on the ranges of the West. Ryff[73] in 1951 found no *Brucella* reactors in five pronghorns tested in Wyoming. Bolin and Eveleth[49] (also in 1951) found no *Brucella* seroreactivity in the 65 antelope from North Dakota that were tested. Erling,[36] in testing wild animals in Arizona, could document no serologic evidence of brucellosis in 215 pronghorn antelope. Stauber[74] et al. also found no evidence of *B. abortus* in 146 pronghorn sera in Idaho. The largest number of pronghorns to be tested for brucellosis was in Colorado, when 5272 were evaluated by the plate agglutination test.[14] In that study, only one pronghorn serum sample was positive at the 1:50 dilution. Pronghorns from Alberta and Saskatchewan were tested, and no reactors were found in the 110 sera examined.[75] In a more recent study, there were no serologic reactors to *Brucella* in 233 pronghorn serum samples collected by hunters in Nebraska.[76] In the light of the above-known evidence, it does not appear that pronghorn antelope (in spite of their close association with domestic livestock) are at present an important wildlife reservoir for brucellosis.

7. Other Deer Species

There have been quite a few studies in which other cervid or deer-like species were involved. The axis or chital deer (*Axis axis*) is a native of India and now is a common free-ranging exotic in Texas and some other states. The axis deer has been shown to be susceptible to *B. abortus* infections under experimental conditions. The prevalence of brucellosis in free-ranging populations of axis deer in Texas is unknown.[77] In California 42 axis deer were found to be serologically negative for brucellosis.[78] Sika or Japanese deer (*Cervus nippon*) in England (4 of 5)[79] and New Zealand (1 of 42)[80] have shown some *Brucella* seroreactivity. A similar situation with regard to brucellosis exists with fallow deer (*Dama dama*). Serologic surveys for brucellosis in fallow deer have had mixed results. In England, McDiarmid[79] found 7 of 58 (12%) fallow deer reactive to the tube agglutination test. In a later study, McDiarmid and Matthews[81] found no reactors to the tube agglutination test in 383 English fallow deer. Reimann et al.[78] found no *Brucella* reactors in 42 fallow deer from California, and Daniel[80] reported the same for 57 fallow deer tested in New Zealand. Red deer (*Cervus elaphus*), even though it is a congeneric of elk or wapiti, does not seem to be involved in brucellosis. No reactors were found in the 156 red deer tested in England[81] or the 109 New Zealand red deer evaluated by Daniel.[80] In a geographic distributional study of wildlife diseases in Switzerland, Bouvier[82] reported the inability to isolate *Brucella* from the tissues of 72 red deer. Even less has been reported on some of the other species of deer. No evidence of *Brucella* antibody activity could be found in 8 Muntjac deer (*Muntiacus reevesi*) or 20 Chinese water deer (*Hydropotes inermis*).[81] Roe deer (*Capreolus capreolus*) have been serologically surveyed in England where a total of 70 were found to be negative.[79,81] Bouvier,[82] however, reported that *B. abortus* was isolated from the tissues of 2 of 126 roe deer cultured in Switzerland. The question of the distribution and importance of brucellosis in most species of wild deer is still largely unknown, and at best, the few known facts could be characterized as conflicting and confusing.

8. African Buffalo (*Syncerus caffer*)

In the last 20 years there have been several studies in relation to brucellosis and the African buffalo. The earliest reported testing of African buffalo was in 1966 by Sachs and Staak[83] during their investigations of brucellosis in antelopes of the Serengeti when they tested four buffalo and found them to be negative. Roth[84] reported a positive agglutination titer in one buffalo from Rhodesia and a suspicious reaction in another from a total of 43 tested. Sachs and Staak[85] reported that 4 of 23 African buffalo from northern Tanzania reacted to *Brucella* antigens. DeVos and Van Niekerk[86] in 1969 found 36 of 253 buffalo from Kruger National Park, South Africa to be serologically reactive. Serologic evidence of brucellosis in 15 of 102 buffalo tested in Rhodesia was reported in 1972 by Condy and Vickers.[87] Kaliner and Staak[88] first isolated *B. abortus* biotype 3 in 1973 from a buffalo with orchitis from Tanzania. Gradwell et al.[89] in 1977 isolated *B. abortus* biotype 1 from 4 of 68 buffalo from Kruger National Park. Further serologic evidence of brucellosis in free-ranging buffalo was reported in Rhodesia by Condy and Vickers[90] in 1976 and in South Africa by Herr and Marshall[91] in 1981. The latest report of brucellosis in African buffalo was by Waghela and Karstad,[92] who reported that 30% of the buffalo tested from the Masai Mara area of Kenya had detectable *Brucella* antibody levels. Waghela and Karstad[92] also stated that an increase in shared grazing and watering areas between wildlife and domestic cattle ''makes the presence of infections by *Brucella* spp. in wildlife an important consideration in any program for control of brucellosis''.

9. African Ungulates

There have been five major serologic surveys for brucellosis in African ungulates or plains herbivores. Rollinson[93] in Uganda, Sacks and Staak[83] in Tanzania, De Vos and Van Niekerk[86] in South Africa, Condy and Vickers[87] in Rhodesia, and Waghela and Karstad[92] in Kenya reported on the occurrence of brucellosis in the animals that they tested. Antibodies to *Brucella* antigens were found in the sera of the following species: 7 of 237 kudu (*Tra-*

gelaphus strepsiceros) and 1 of 33 bushbuck (*T. scriptus*) from Rhodesia; 12 of 71 eland (*Taurotragus oryx*) from Rhodesia; 3 of 11 eland from Uganda; 3 of 111 Rhodesian duiker (*Sylvicapra grimmia*); 1 of 6 Rhodesian waterbuck (*Kobus ellipsiprymnus*) and 1 of 1 waterbuck from South Africa; 3 of 58 Rhodesian sable antelope (*Hippotragus niger*); 2 of 33 topi (*Damaliscus korrigum*) from Tanzania; 1 of 25 Rhodesian grysbok (*Raphicerus sharpei*); 101 of 1036 impala (*Aepyceras melampus*) from Rhodesia, 4 of 32 Tanzanian impala, and 1 of 120 South African impala; 2 of 7 Grant's gazelle (*Gazella granti*) and 2 of 7 Thompson's gazelle (*G. thomsonii*) from Tanzania; 13 of 50 zebra (*Equus burchellia*) from Rhodesia and 1 of 1 South African zebra; 1 of 9 Rhodesian hippopotamuses (*Hippopotamus amphibius*), 7 of 51 South Aftican hippos, and 67 of 145 hippos from Uganda; and 31 of 175 blue wildebeest (*Connochaetes taurinus*) from Kenya, 2 of 38 Rhodesian wildebeest, and 8 of 22 wildebeest from Tanzania. In these same studies, no serologic evidence of brucellosis was found in the following species: giraffe (*Giraffe camelopardalis*), nyala (*Tragelaphus angasi*), reedbuck (*Redunca arundinum*), roan antelope (*Hippotragus equinus*), tsessby (*Damaliscus lunatus*), klipspringer (*Oreotragus oreotragus*), steenbok (*Raphicerus campestris*), suni (*Nesotragus moschatus*), elephant (*Loxodonta africana*), rhinoceros (*Ceratotherium simun*), bushpig (*Potamochoerus porcus*), or warthog (*Phacochoerus aethiopicus*). *Brucella melitensis* was isolated from an impala (*Aepyceros melampus*) by Schiemann and Staak[94] in 1971, and Condy and Vickers[95] in 1969 isolated *B. abortus* from a Rhodesian waterbuck.

10. Other Wild Ruminants

Brucellosis is a problem in some populations of Asian buffalo (*Bubalis bubalis*). Zaki,[96] for instance, reported that of 200 Asian buffalo tested in Egypt, 75 were found to be reactive. Recently, *Brucella suis* biovar 4 (previously thought to be limited to reindeer and caribou and those animals which feed on them) was for the first time reported in a muskox (*Ovibus moschatus moschatus*).[97] Bouvier[82] recovered *B. abortus* from the tissues of 6 of 489 Swiss chamois (*Rupicapra rupicapra*), but 14 chamois tested in New Zealand were negative for brucellosis, as were 93 Himalayan tahr (*Hemitragus jemlahicus*).[80] Forty-three bighorn sheep (*Ovis canadensis*) from Arizona[36] and 9 bighorns from Canada[28] showed no *Brucella* seroreactivity. Four percent (3 of 73) of Dall's sheep (*Ovis dalli dalli*) in Alaska were positive to the plate agglutination test for brucellosis.[98]

B. NONRUMINANT WILDLIFE

1. Wild Canids

There are several studies in the literature that document the presence of at least serologic evidence of brucellosis in wild canids from most continents. Neiland[99] in 1970 reported serologic evidence that timber wolves (*Canis lupus*) in Alaska can become naturally infected with rangiferine brucellosis. Wolves have also been experimentally infected by intraperitoneal inoculations of 1×10^8 colony forming units (CFU) of *Brucella suis* biovar 4.[100] Rangiferine brucellosis has also been reported in Arctic foxes (*Alopex lagopus*) and red foxes (*Vulpes vulpes*) in Russia[101] and Alaska.[102] Rementsova[103] also reports the natural occurrence of *B. suis* infections in Russian red foxes, and Pavlov[104] found the same to be true in Bulgarian red foxes. Positive reactions were detected in a red fox from New York and two coyotes (*Canis latrans*) from Texas during a serologic survey in the U.S. conducted on seven species of wild carnivores from five states.[105] Serologic evidence of *B. abortus* infections was found by McCaughey[106] in red foxes in Northern Ireland. Seven of 87 red foxes from an area in Wales where cattle brucellosis is endemic were positive at low titers to the standard agglutination tube test (SAT) and the complement fixation test (CFT). Another Welsh fox in the same study had a 320 titer on the SAT and a 200 on the CFT. *Brucella abortus* biovar 1 was recovered from the same red fox.[107] Randhawa et al.[108] in 1977 reported agglutinin prevalence rates (which they attributed to *B. canis*) in 198 coyotes from Texas as 7.1, 8.9, and 6.7% as determined by the card test, the rapid slide agglutination test, and the salt 2-mercaptoethanol tube agglutination test, respectively. Hoq[59] in 1978 reported *Brucella* agglutins in 9 of 148 coyotes from California tested by the rapid plate agglutination technique. Davis et al.[109] in 1979 found 18% of 51 coyotes from eastern Texas to be serologically positive for brucellosis by several techniques, and *B. abortus* biovar 1 was isolated from 7 of 43 of the same coyotes. Congenital transmission was also documented in coyotes. Two species of Argentine wild foxes (*Dusicyon gymnocercus* and *D. griseus*) were shown to be naturally infected with *B. abortus* biovar 1 by both serologic and bacteriologic methods by Szyfres and Tome in 1966.[110] In Tanzania, Sachs and Staak[85] in 1968 reported antibodies against *B. abortus* in sera from the black-backed jackal (*Canis mesomelas*), spotted hyena (*Crocuta crocuta*), and wild hunting dog (*Lycaon pictus*). Scanlan et al.[111] in 1984 successfully infected 8 gray foxes (*Urocyon cinereoargenteus*) by feeding them 4.4×10^{10} CFU of *B. abortus* strain 2308 mixed in commercial dog food. Davis et al.[112] in 1988 experimentally infected coyotes by feeding them macerated bovine tissue which had been infused with 1×10^{8} CFU *B. abortus* strain 2308. These infected coyotes were capable of secondarily transmitting the *B. abortus* to susceptible cattle and causing subsequent abortions in the cattle under confined conditions. The prevailing theory on brucellosis in wild canids is that they generally become exposed by scavenging on aborted fetuses and placental membranes, and that while they may become infected, the wild canids do not under most circumstances play a preeminent role in the distribution of *Brucella* species in nature. These same canids are, however, excellent sentinel animals for the presence of brucellosis within an area and probably could be utilized more effectively as such.

2. Feral Swine (*Sus scofa*)

Although not an indigenous wild species in North America, populations of feral swine (domesticated pigs that have returned to the wild state) are of sufficient number, ecological impact, and distribution to consider them wildlife in many areas, particularly in the southeastern U.S. Brucellosis in feral swine in the continental U.S. was first reported by Wood et al.[113] in 1976 in South Carolina. Of 255 of the feral hogs tested, 18% were reactors as determined by the card, complement fixation, and the Rivanol precipitation tests. *Brucella suis* biovar 1 was isolated from the lymph nodes of one boar. Of feral swine in Hawaii, 24% (10/42) had been previously found to have low agglutination titers to *Brucella* antigens by Nichols[114] in 1962. Becker et al.[115] in 1978 tested 95 sera from feral swine trapped in Florida by the card, standard tube, complement fixation, and Rivanol precipitation test and found 50 (53%) to be positive to at least one test. Tissues were collected from 9 swine, and *B. suis* biovar 1 was isolated from the lymph nodes of all 9. The standard tube test was the only serologic technique utilized that detected *Brucella* antibodies in all culture-positive swine. Bigler et al.[116] in 1977, while investigating epidemiologic trends of brucellosis in Florida, reported 22% (6/27) human cases of brucellosis during 1974—1975 as attributable to hunter contact with swine. Lawhorn[117] in 1984 reported that 8 of 76 feral swine from southeast Texas were positive for brucellosis. National Veterinary Services Laboratory, Ames, IA has isolated *B. suis* from several cattle herds in Florida,[128] and *B. suis* was also isolated from a cow in Texas.[129] The wide geographic distribution of feral swine and the fact that they are known to be capable of transmission of brucellosis to humans and domestic livestock should be of some serious concern to governmental agencies responsible from the control and/or eradication of this disease.

3. Capybara (*Hydrochaeris hydrochaeris*)

An interesting situation in wild capybara in Venezuela was reported by Lord and Flores[118] in 1983. In a bacteriologic and serologic study of 201 capybara, 23 isolations were made from spleen and lymph node tissues. Eight of the isolations were found to be *B. abortus* biovars 2, 3, 4, and 5, and the other 15 isolates were determined to be *B. suis* biovars 2 and 3. Fifty-eight percent of the capybara sera were determined by several serologic techniques to be positive. Plata[119] first found antibodies to *Brucella* in the sera of capybara from Venezuela, and in later studies Bello et al.[120-122] also found seroreactive capybara and isolated *B. abortus* biovars 1 and 2 from their tissues. The epidemiologic significance of these mixed *Brucella* infections in an alternative host such as the capybara with regard to disease prevention and control measures is not well understood.

4. Miscellaneous Large and Medium Sized Mammals

There are a few other mammalian species (not covered above) from which *Brucella* species have been isolated and a larger number of animals which have been shown to possess antibodies for *Brucella*. In large part, these mammals are carnivorous and probably have become infected by eating infected prey or scavenging on aborted fetuses or placental membranes. *Brucella abortus* biovar 1 was isolated from a raccoon (*Procyon lotor*) and four opossums (*Didelphis virginiana*) trapped on a farm in Alabama on which an infected herd of cattle was present.[123] Serologically positive raccoons were also found in California.[59] Moore and Schnurrenberger[64] in 1981 experimentally infected opossums with *B. abortus* and subsequently isolated the organism from tissues. In other investigations, the following species were found by various serologic techniques to have detectable *Brucella* antibody levels: grizzly bear (*Ursus arctos*);[102] black bear (*U. americanus*);[28,124] and the striped skunk (*Mephitis mephitis*), spotted skunk (*Spilogale gracilis*), badger (*Taxidea taxus*), and bobcat (*Lynx rufus*).[59] No serologic evidence of brucellosis was reported in the following species: weasel (*Mustela nivalis*);[125]lion (*Panthera leo*);[86] javelina (*Tayassu tajacu*);[36] and the genet cat (*Genetta genetta*), chacma baboon (*Papio urinus*), and the ant bear (*Orycteropus afer*).[87]

5. Lagomorphs and Rodents

While *Brucella abortus* has been isolated from rodents on occasion,[126,127] most rodent infections seem to be from areas where there are a large number of infected cattle. Literally scores of small mammalian species have been surveyed extensively by serologic and bacteriologic methods for the presence of brucellosis[10,52,64,103] and almost without exception no concentrations of *Brucella* capable of infecting humans or livestock were found. Thorpe et al.[10] reported 15 *Brucella neotomae* isolations in wood rats (*Neotoma lepida*) and one isolation from their fleas (*Orchopeus sexdentatus*). An isolation each of *Brucella suis* and possibly *B. melitensis* was found in black tailed jackrabbits in the same 10-year-long study in which over 29,100 wildlife serum samples and tissue specimens and 120,000 ectoparasites were examined.

III. EPIDEMIOLOGICAL SIGNIFICANCE OF BRUCELLOSIS IN WILDLIFE POPULATIONS

McCorquedale and DiGiacomo[19] (1985) in their review of brucellosis in wild ungulates and cattle stated that there "have been no published reports of wild ungulates hindering eradication of bovine brucellosis in any part of the world". While at face value that statement may be true, one may also ask, in what country or area has brucellosis been eradicated where large numbers of infected ungulates or wildlife are present? A synonym for eradication is extermination, and both literally mean to wipe out or destroy completely. The overwhelming existing evidence is that no species of *Brucella* is in any imminent danger whatsoever of extinction in spite of governmental pronouncements that areas or entire countries are "free of brucellosis". Rangiferine brucellosis in reindeer and caribou is circumpolar in distribution, well established in large populations of animals that migrate thousands of miles each year, and clearly is the source of disease in humans, their domestic animals, and other wildlife species in the Holarctic regions of the U.S., Canada, Europe, and Russia. Bovine brucellosis is being effectively combated in North America, but infected populations of bison, elk, moose in Canada and in the U.S. are of sufficient size and geographic distribution to create a real and serious threat to the livestock producers in certain areas of both countries. In the southeastern portion of the U.S., feral swine have shown their ability to transmit brucellosis to cattle and humans. The exact distribution and prevalence of *Brucella*- infected feral swine in the U.S. are not known. The large number of African ungulates with brucellosis, such as the African buffalo, the eland, the hippopotomas, impala, wildebeest, zebra, and kudu concerns those familiar with wildlife disease and the ease of exchange of disease between grazing wildlife and domestic livestock in Africa.[92] Mixed infections of brucellosis in capybara in South America may present problems for humans in those areas which utilize the capybara as a food source or for domestic livestock producers that are trying to eliminate brucellosis from their herds.

One should not overstate the issue of wildlife brucellosis. The preponderance of known *Brucella* infections is still in domestic livestock. The picture in industrialized countries is, however, changing as brucellosis in domestic animals is being brought under some control, while wildlife brucellosis in these same countries is either fairly constant or increasing. The picture is not completely bleak. With the powerful modern tools of molecular genetics and biotechnology, new vaccines are being developed, and improved diagnostic techniques are being presently evaluated. If and when these new methods of detecting and preventing brucellosis are successfully developed and available, they must be combined with innovative strategies and delivery systems for use in free-ranging wildlife species, for it is in the prevention of brucellosis in wildlife populations (as well as in domestic animals) that any realistic hope for the eradication of the disease lies. The only other alternative, which is the elimination of wildlife reservoirs of brucellosis, is economically, ecologically, and ethically unacceptable.

REFERENCES

1. **Bruce, D.,** Note on the discovery of a microorganism in Malta Fever, *Practitioner,* 40, 241, 1887.
2. **Bruce, D.,** The micrococcus of Malta Fever, *Practitioner,* 40, 241, 249, 1888.
3. **Bang, B.,** The etiology of epizootic abortion, *J. Comp. Pathol. Ther.,* 10, 125, 1897.
4. **Mohler, J. F.,** Annual Report of the United States Bureau of Animal Industry, 1917, 106.
5. **Creech, G. T.,** *Brucella abortus* infection in a male bison, *North Am. Vet.,* 11, 35, 1930.
6. **Rush, W. M.,** Bang's disease in Yellowstone National Park buffalo and elk herds, *J. Mammal.,* 13, 371, 1932.
7. **Tunnicliff, E. A. and Marsh, H.,** Bang's disease in bison and elk in the Yellowstone National Park and the National Bison Range, *J. Am. Vet. Med. Assoc.,* 86, 745, 1935.
8. **Moore, T. A.,** A survey of buffalo and elk herds to determine the extent of *Brucella* infection, *Can. J. Comp. Med.,* 11, 131, 1947.
9. **Corner, A. H. and Connell, R.,** Brucellosis in bison, elk and moose in Elk Island National Park, Alberta, Canada, *Can. J. Comp. Med.,* 22, 9, 1958.
10. **Thorpe, B. D., Sidwell, R. W., Bushman, J. B., Smart, K. L., and Noyes, R.,** Brucellosis in wildlife and livestock in west central Utah, *J. Am. Vet. Med. Assoc.,* 146, 225, 1965.
11. **Choquette, L. P. E., Broughton, E., Cousineau, J. G., and Novakowski, N. S.,** Parasites and diseases of bison in Canada. IV. Serological survey for brucellosis in northern Canada, *J. Wildl. Dis.,* 14, 329, 1978.
12. **Colgrove, C. S., Williams, E. Y., Morgan, R. L., and Kopec, J. D.,** *Brucella abortus* isolated from Yellowstone Park bison. Pathogenicity for cattle, Development Project Report, National Veterinary Services, 1985, 1.
13. **Honess, R. F. and Winter, K. B.,** Diseases of wildlife in Wyoming, *Wyo. Game Fish Comm.,* 9, 279, 1956.

14. **Adrian, W. J. and Keiss, R. E.,** Survey of Colorado's wild ruminants for serologic titers to brucellosis and leptospirosis, *J. Wildl. Dis.*, 13, 429, 1977.
15. **Vaughn, H. W., Knight, R. R., and Frank, R. W.,** A study of reproduction, disease and physiological blood and serum levels in Idaho elk, *J. Wildl. Dis.*, 9, 296, 1973.
16. **Merrell, C. L. and Wright, D. N.,** A serologic survey of mule deer and elk in Utah, *J. Wildl. Dis.*, 14, 471, 1978.
17. **Thorne, E. T., Morton, J. K., and Thomas, G. M.,** Brucellosis in elk. I. Serologic and bacteriologic survey in Wyoming, *J. Wildl. Dis.*, 14, 74, 1978.
18. **Thorne, E. T., Morton, J. K., Blunt, F. M. et al.,** Brucellosis in elk. II. Clinical effects and means of transmission as determined through artificial infection, *J. Wildl. Dis.*, 14, 280, 1978.
19. **McCorquadale, S. M. and DiGiacomo, R. F.,** The role of wild North American ungulates in the epidemiology of bovine burcellosis: a review, *J. Wildl. Dis.*, 21(4), 251, 1985.
20. **Huntley, B. E., Philip, E. R. N., and Maynard, J. E.,** Survey of brucellosis in Alaska, *J. Infect. Dis.*, 112, 100, 1963.
21. **Meyer, M. E.,** Species identity and epidemiology of Brucella strains isolated from Alaskan Eskimos, *J. Infect. Dis.*, 114, 169, 1964.
22. **Toshach, S. R.,** *Brucella melitensis* in the Northwest Territories, *Can. J. Public Health*, 46, 155, 1955.
23. **Meyer, M. E.,** Identification and virulence studies of *Brucella* strains isolated from Eskimos and reindeer in Alaska, Canada, and Russia, *Am. J. Vet. Res.*, 27, 253, 1966.
24. **Neiland, R. A., King, J. A., Huntley, B. E., et al.,** The diseases and parasites of Alaskan wildlife populations. I, *Bull. Wildl. Dis. Assoc.*, 4, 27, 1968.
25. **Broughton, E., Choquette, L. P. E., Cousineau, J. G., and Miller, F. L.,** Brucellosis in reindeer, *Rangifer tarandus* (L.), and the migratory barren ground caribou, *Rangifer tarandus groenlandicus* (L.), in Canada, *Can. J. Zool.*, 48, 1023, 1970.
26. **Davidov, N. N.,** On the epizootiology of brucellosis in domestic reindeer in Yakutia, *Proc. 5th All-Union Symp. Biological Problems in the North,* Magadan, U.S.S.R., 1974, 229.
27. **Gorban, L. V.,** Natural foci of brucellosis in the far north, *Microb. Epidemiol. Immunobiol.*, vol. 8, 1977.
28. **Zarnke, R. and Yuill, T. M.,** Serologic survey for selected microbial agents in mammals from Alberta, *J. Wildl. Dis.*, 17, 453, 1981.
29. **Dieterich, R. A.,** Bacterial diseases, in *Alaskan Wildlife Diseases,* Dieterich, R. A., ed., University of Alaska Press, Fairbanks, 1981, 45.
30. **Zarnke, R. L.,** Serologic survey for selected microbial pathogens in Alaskan wildlife, *J. Wildl. Dis.*, 114, 169, 1983.
31. **Tessaro, S. V. and Forbes, L. B.,** *Brucella suis* biotype 4: a case of granulomatous nephritis in a barren ground caribou *(Rangifer tarandus groenlandicus L.)* with a review of the distribution of rangerifine brucellosis in Canada, *J. Wildl. Dis.*, 22, 479, 1986.
32. **Dieterich, R. A., Deyoe, B. L., and Morton, J. K.,** Effects of killed *Brucella abortus* strain 45/20 vaccine on reindeer later challenge-exposed with *B. suis* type 4, *Am. J. Vet. Res.*, 42, 131, 1981.
33. **Orloff, E. S.,** Brucellosis in reindeer, in *Proc. 17th World Vet. Congr., Hannover, F.R.G.*, 1963, 1, 585.
32. **Dieterich, R. A., Deyoe, B. L., and Morton, J. K.,** Effects of killed *Brucella abortus* strain 45/20 vaccine on reindeer later challenge-exposed with *B. suis* type 4 *Am. J. Vet. Res.*, 42, 131, 1981.
33. **Orloff, E. S.,** Brucellosis in reindeer, in *Proc. 17th World Vet. Congr., Hannover 1963,* 1, 585.
34. **Hayes, F. A., Gerard, W. T., Shotts, E. B., et al.,** Brucellosis in white-tailed deer, *J. Am. Vet. Med. Assoc.*, 137, 190, 1960.
35. **Shotts, E. B., Greer, W. E., and Hayes, F. A.,** A preliminary survey of the incidence of brucellosis and leptospirosis among white-tailed deer *(Odocileus virginanus)* of the Southeast, *J. Am. Vet. Med. Assoc.*, 133, 359, 1958.
36. **Erling, H. G.,** *Results of Brucellosis Testing of Wild Animals of Arizona,* Spec. Rep., Project W-52-R-6, Arizona Game and Fish Dept., Wildlife Restoration Division, 1956.
37. **Woolsey, J. J., cited by Fay, L. D.,** The current status of brucellosis in white-tailed and mule deer in the United States, *Trans. 26th Annu. Meet. North Am. Wildl. Nat. Res. Conf.*, 1961, 203.
38. **Ferris, D. H., Hanson, L. E., Rhoades, H. E., et al.,** Bacteriologic and serologic investigations of brucellosis and leptospirosis in Illinois deer, *J. Am. Vet. Med. Assoc.*, 139, 892, 1961.
39. **Haugen, A. O., cited by Fay, L. D.,** The current status of brucellosis in white-tailed and mule deer in the United States, *Trans. 26th Annu. Meet. North Am. Wildl. Nat. Res. Conf.*, 1961, 203.
40. **Witter, J. F., cited by Fay, L. D.,** The current status of brucellosis in white-tailed and mule deer in the United States, *Trans. 26th Annu. Meet. North Am. Wildl. Nat. Res. Conf.*, 1961, 203.
41. **Fay, L. D. and Youatt, W. G., cited by Fay, L. D.,** The current status of brucellosis in whitetailed and mule deer in the United States, *Trans. 26th Annu. Meet. North Am. Wildl. Nat. Res. Conf.*, 1961, 203.
42. **Jones, R. L., Tamayo, R. I., Porath, W., Giessman, N., Selby, L. S., and Buening, G. M.,** A serologic survey of brucellosis in white-tailed deer *(Odocoileus virginianus)* in Missouri, *J. Wildl. Dis.*, 19, 321, 1983.

43. **Fernstermacher, R., Olsen, O. W., and Pomeroy, B. S.,** Some diseases of white-tailed deer in Minnesota, *Cornell Vet.*, 33, 323, 1943.
44. **Wedman, E. E. and Driver, F. C.,** Leptospirosis and brucellosis titers in deer blood, *J. Am. Vet. Med. Assoc.*, 130, 513, 1957.
45. **Steen, M. O., Broh, H., and Robb, D.,** A survey of brucellosis in white-tailed deer in Missouri, *J. Wildl. Manage.*, 19, 320, 1955.
46. **Wildlife Disease Association, cited by Fay, L. D.,** The current status of brucellosis in white-tailed and mule deer in the United States, *Trans. 26th Annu. Meet. North Am. Wildl. Nat. Res. Conf.*, 1961, 203.
47. **Critcher, T. S., cited by Fay, L. D.,** The current status of brucellosis in white-tailed and mule deer in the United States, *Trans. 26th Annu. Meet. North Am. Wildl. Nat. Res. Conf.*, 1961, 203.
48. **Bolin, F. M., Goldsby, A. I., Rheault, P., et al.,** Brucellosis in North Dakota deer — a survey, *N. D. Agric. Exp. Stn. Bull.*, 11,102,1949
49. **Bolin, F. M. and Eveleth, D. F.,** Where deer and antelope play — North Dakota's bit game animals are not reservoirs of brucellosis, *N. D. Agric. Exp. Stn. Bull.*, 14, 46, 1951.
50. **Fleming, W. J. and Nusbaum, S. R.,** A survey of deer on the Seneca Army Depot for evidence of leptospirosis, brucellosis and toxoplasmosis, *N. Y. Fish Game J.*, 26, 198, 1979.
51. **Friend, M. and Halterman, L. G.,** Serologic survey of two deer herds in New York State, *Bull. Wildl. Dis. Assoc.*, 3, 32, 1967.
52. **Boeer, W. J., Crawford, R. P., Hildalgo, R. J., et al.,** Small mammals and white-tailed deer as possible reservoir hosts of *Brucella abortus* in Texas, *J. Wildl. Dis.*, 16, 19, 1980.
53. **Trainer, D. O., Jr. and Hanson, R. P.,** Leptospirosis and brucellosis serologic reactors in Wisconsin deer, *J. Wildl. Manage.*, 24, 44, 1960.
54. **Stauber, E. H., Nellis, C. H., Magonigle, T. A., et al.,** Prevalence of reactors to selected livestock pathogens in Idaho mule deer, *J. Wildl. Manage.*, 41, 515, 1977.
55. **Greenley, J. C., cited by Fay, L. D.,** The current status of brucellosis in white-tailed and mule deer in the United States, *Trans. 26th Annu. Meet. North Am. Wildl. Nat. Res. Conf.*, 1961, 203.
56. **Lee, L., cited by Fay, L. D.,** The current status of brucellosis in white-tailed and mule deer in the United States, *Trans. 26th Annu. Meet. North Am. Wildl. Nat. Res. Conf.*, 1961, 203.
57. **Low, J. B., cited by Fay, L. D.,** The current status of brucellosis in white-tailed and mule deer in the United States, *Trans. 26th Annu. Meet. North Am. Wildl. Nat. Res. Conf.*, 1961, 203.
58. **Hepworth, W. G., cited by Fay, L. D.,** The current status of brucellosis in white-tailed and mule deer in the United States, *Trans. 26th Annu. Meet. North Am. Wildl. Nat. Res. Conf.*, 1961, 203.
59. **Hoq, M. A.,** A serologic survey of Brucella agglutinins in wildlife and sheep, *Calif. Vet.*, 35, 15, 1978.
60. **Davis, R. W., cited by Fay, L. D.,** The current status of brucellosis in white-tailed and mule deer in the United States, *Trans. 26th Annu. Meet. North Am. Wildl. Nat. Res. Conf.*, 1961, 203.
61. **Shaw, W. M., cited by Fay, L. D.,** The current status of brucellosis in white-tailed and mule deer in the United States, *Trans. 26th Annu. Meet. North Am. Wildl. Nat. Res. Conf.*, 1961, 203.
62. **Newby, F. E., cited by Fay, L. D.,** The current status of brucellosis in white-tailed and mule deer in the United States, *Trans. 26th Annu. Meet. North Am. Wildl. Nat. Res. Conf.*, 1961, 203.
63. **Bailey, B. and Cuming, R. B., cited by Fay, L. D.,** The current status of brucellosis in white-tailed and mule deer in the United States, *Trans. 26th Annu. Meet. North Am. Wildl. Natl. Res. Conf.*, 1961, 203.
64. **Moore, C. G. and Schnurrenberger, P. R.,** Experimental infection of opossums with *Brucella abortus*, *J. Am. Vet. Med. Assoc.*, 179, 1113, 1981.
65. **Corey, R. R., Paulissen, L. J., and Swartz, D.,** Prevalence of *Brucellae* in the wildlife of Arkansas, *J. Wildl. Dis.*, 39 (WD-63-2), 1964.
66. **Youatt, W. G. and Fay, L. D.,** Experimental brucellosis in white-tailed, *Am. J. Vet. Res.*, 20, 925, 1959.
67. **Thorpe, B. D., Sidwell, R. W., and Lundgren, D. L.,** Experimental studies with four species of *Brucella* in selected wildlife, laboratory, and domestic animals, *Am. J. Trop. Med. Hyg.*, 16, 665, 1967.
68. **Baker, M. F., Dills, G. J., and Hayes, F. A.,** Further experimental studies on brucellosis in white-tailed deer, *J. Wildl. Manage.*, 26, 27, 1962.
69. **Fernstermacher, R. and Olsen, D. W.,** Further studies of disease affecting moose. III, *Cornell Vet.*, 32, 241, 1942.
70. **Jellison, W. L., Fishel, C. W., and Cheatum, E. L.,** Brucellosis in a moose, *Alces americana*, *J. Wildl. Manage.*, 17, 217, 1953.
71. **Hudson, M., Child, K. N., Halter, D. F., Fujino, K. K., and Hudson, K. A.,** Brucellosis in moose *(Alces alces)*. A serologic survey in an open range cattle area of north central British Columbia recently infected with bovine brucellosis, *Can. Vet. J.*, 21, 47, 1980.
72. **Bourque, M. and Higgins, R.,** Serologic studies on brucellosis, leptospirosis and tularemia in moose. *(Alces alces)* in Quebec, *J. Wildl. Dis.*, 20, 95, 1984.
73. **Ryff, J. E.,** Brucellosis absent in Wyoming game, *Wyo. Wildl.*, 15, 13, 1951.
74. **Stauber, E. K., Autenreith, R., Markham, O. D., and Whitbeck, U.,** A serioepidemiologic survey of three pronghorn *(Antilocarpa americana)* populations in southeastern Idaho 1975—1977, *J. Wildl. Dis.*, 16, 109, 1980.

75. **Barrett, M. W. and Chalmers, G. A.,** A serologic survey of pronghorns in Alberta and Saskatchewan, 1970—1972, *J. Wildl. Dis.*, 11(2), 157, 1975.
76. **Johnson, J. L., Barber, T. L., Frey, M. L., and Nason, G.,** Serosurvey for selected pathogens in hunter-killed pronghorns in western Nebraska, *J. Wildl. Dis.*, 22, 87, 1986.
77. **Davis, D. S., Heck, F. C., and Adams, L. G.,** Experimental infection of captive axis deer with *Brucella abortus, J. Wildl. Dis.*, 20(3), 177, 1984.
78. **Reimann, H. P., Ruppanner, R., Willeberg, P., et al.,** Serologic profiles of exotic deer at Point Reyes National Seashore, *J. Am. Vet. Med. Assoc.*, 1175, 911, 1979.
79. **McDiarmid, A.,** The occurrence of agglutinins for *Brucella abortus* in the blood of wild deer in the south of England, *Vet. Rec.*, 63, 469, 1951.
80. **Daniel, M. J.,** A survey of diseases in Fallow, Virginia and Japanese deer, chamois, tahr and feral goats and pigs in New Zealand, *N.Z. J. Sci.*, 10, 949, 1967.
81. **McDiarmid, A. and Matthews, P. R. J.,** Brucellosis in wildlife, *Vet Rec.*, 95, 559, 1974.
82. **Bouvier, G.,** Distribution geographique de quelques maladies du gibier et des animaux sauvages de la Suisse, *Bull. Off. Int. Epizoot.*, 61, 67, 1964.
83. **Sachs, R. and Staak, C.,** Evidence of brucellosis in antelopes of the Serengeti, *Vet. Rec.*, 79, 857, 1966.
84. **Roth, H. H.,** A survey of brucellosis in game animals in Rhodesia, *Bull Epizoot. Dis. Afr.*, 15, 133, 1967.
85. **Sachs, R. and Staak, C.,** Serological investigation of brucellosis in game animals in Tanzania, *Bull. Epizoot. Dis. Afr.*, 16, 91, 1968.
86. **DeVos, V. and Van Niekerk, C. A. W. J.,** Brucellosis in the Kruger National Park, *J. S. Afr. Vet. Med. Asso.*, 40, 331, 1969.
87. **Condy, J. B. and Vickers, D. B.,** Brucellosis in Rhodesian wildlife, *J. S. Afr. Vet. Med. Assoc.*, 43, 175, 1972.
88. **Kaliner, G. and Staak, C.,** A case of orchitis caused by *Brucella abortus* in the African buffalo, *J. Wildl. Dis.*, 9, 251, 1973.
89. **Gradwell, D. V., Schuttee, A. P., Van Niekerk, C. A. W. J., et al.,** The isolation of *Brucella abortus* biotype 1 from African buffalo in the Kruger National Park, *J. S. Afr. Vet. Med. Assoc.*, 48, 41, 1977.
90. **Condy, J. B. and Vickers, D. B.,** Brucellosis in buffalo *(Syncerus caffer)* in Wankie National Park, *Rhod. Vet. J.*, 7, 58, 1976.
91. **Herr, S. and Marshall, C.,** Brucellosis in free-living African buffalo *(Syncerus caffer)*, a serological survey, Onderstepoort, *J. Vet. Res.*, 48, 133, 1981.
92. **Waghela, S. and Karstad, L.,** Antibodies to *Brucella* spp. among blue wildebeest and African buffalo in Kenya, *J. Wildl. Dis.*, 22, 189, 1986.
93. **Rollinson, D. H. L.,** *Brucella* agglutinins in East African game animals, *Vet. Rec.*, 74, 904, 1962.
94. **Schiemann, B. and Staak, C.,** *Brucella melitensis* in impala *(Aepyceros melampus), Vet. Rec.*, 88, 344, 1971.
95. **Condy, J. B. and Vickers, D. B.,** The isolation of *Brucella abortus* from a waterbuck, *Vet. Rec.*, 85, 200, 1969.
96. **Zaki, R.,** *Brucella abortus* infection among buffaloes in Egypt, *J. Comp. Pathol.*, 58, 73, 1948.
97. **Gates, C. C, Wobeser, G., and Forbes, L. B.,** Rangiferine brucellosis in a muskox, *Ovibos moshatus moschatus* (Zimmermann), *J. Wildl. Dis.*, 20(3), 177, 1984.
98. **Foreyt, W. J., Smith, T. C., Evermann, J. F., and Heimer, W. E.,** Hematologic, serum chemistry, and serologic values of Dall's sheep *(Ovis dalli dalli)* in Alaska, *J. Wildl. Dis.*, 19, 136, 1983.
99. **Neiland, K. A.,** Rangiferine brucellosis in Alasakan canids, *J. Wildl. Dis.*, 6, 136, 1970.
100. **Neiland, K. A. and Miller, L. G.,** Experimental *Brucella suis* type 4 infections in domestic and wild Alaskan carnivores, *J. Wildl. Dis.*, 17, 183, 1981.
101. **Pinigin, A. F. and Zabrodin, V. A.,** On the nidality of brucellosis, *Vestni. Skh. Nauki (Moscow)*, 7, 96, 1970.
102. **Neiland, K. A.,** Further observations on rangiferine brucellosis in Alaskan carnivores, *J. Wildl. Dis.*, 11, 45, 1975.
103. **Rementsova, M. M.,** La brucellose des animoux sauvages, *Bull. Off. Int. Epizoot.*, 61, 99, 1964.
104. **Pavlov, P. I.,** (T) Report on the sanitary state of domestic animals in Bulgaria on May 1, 1960 (T) Chap. 4-donors and recipients, *Bull. Off. Int. Epizoot.*, 54, 272, 1960.
105. **Hoff, G. L., Bigler, W. J., Trainer, D. O., Debbie, J. E., Brown, G. M., Winkler, W. G., et al.,** Survey of selected carnivore and opossum serums for agglutinins to *Brucella canis, J. Am. Vet. Med. Assoc.*, 165(9), 830, 1974.
106. **McCaughey, W. J.,** Brucellosis in wildlife, *Symp. Zool. Soc. London*, 24, 99, 1968.
107. **Davies, G., Ockey, J. H., and Lloyd, H. G.,** Isolation of *Brucella abortus* from a fox *(Vulpes vulpes), State Vet. J.*, 28(84), 250, 1973.
108. **Randhawa, A. S., Dieterich, W. H., Hunter, C. C., Kelly, V. P., Johnson, T. C., et al.,** Prevalence of seropositive reactions to *Brucella canis* in a limited survey of domestic cats, *J. Am. Vet. Med. Assoc.*, 171(3), 267, 1977.

109. **Davis, D. S., Boeer, W. J., Mims, J. P., Heck, F. C., and Adams, L. G.,** *Brucella abortus* in coyotes. I. A serological and bacteriologic survey in eastern Texas, *J. Wildl. Dis.*, 15(3), 367, 1979.
110. **Szyfres, B. and Tome, J. G.,** Natural *Brucella* infection in Argentine Wild Foxes, *Bull. W. H. O.*, 34, 919, 1966.
111. **Scanlan, C. M., Pidgeon, G. L., Sawngo, L. J., Hannon, S. S., and Galik, P. A.,** Experimental infection of gray foxes Urocyon cinereoargenteus with *Brucella abortus, J. Wildl. Dis.*, 20(1), 27, 1984.
112. **Davis, D. S., Heck, F. C., Williams, J. D., Simpson, T. R., and Adams, L. G.,** Interspecific transmission of *Brucella abortus* from experimentally infected coyotes *(Canis latrans)* to parturient cattle, *J. Wildl. Dis.*, 24(3), 533, 1988.
113. **Wood, G. W., Hendricks, J. B., and Goodman, D. E.,** Brucellosis in feral swine, *J. Wildl. Dis.*, 12, 579, 1976.
114. **Nichols, L., Jr.,** Ecology of the wild pig, PR Project W5-R-13, Hawaii Division of Fish and Games, 1962, 20.
115. **Becker, H. N., Belden, R. C., Breault, T., et al.,** Brucellosis in feral swine in Florida, *J. Am. Vet. Med. Assoc.*, 173, 1181, 1978.
116. **Bigler, W. J., Hoff, G. L., Hemmert, W. H., Tomas, J. A., and Janowski, H. T.,** Trends of brucellosis in Florida: an epidemiologic review, *Am. J. Epidemiol.*, 105(3), 245, 1977.
117. **Lawhorn, B.,** Pseudorabies and brucellosis in East Texas piney wood rooter hogs, *Park Producers*, July-August, 1984, 4.
118. **Lord, V. R. and Flores, R. C.,** *Brucella* spp. from capybara *(Hydrochoerus hydrochaeris)* in Venezuela: serologic studies and metabolic characterization of isolates, *J. Wildl. Dis.*, 19, 308, 1983.
119. **Plata, G. V.,** Muestreo serológico en chiguires *(Hydrochoerus hydrochaeris)* Estado Apure, Proyecto, *Conicit Df. Df.*, 030 Sl. Maracay, Venezuela, p. 6, 1975.
120. **Bello, A., Mogollon, P., Villegas, M., et al.,** La brucelosis en los animales salvages. I. El chigüire *(Hydrochoerus hydrochaeris), Vet. Trop.*, 1, 117, 1976.
121. **Bello, A., Mogollon, P., Ramirez, M., Rodriquez, V., Laserna, R. D., Perez, M., Moreno, J., and Lord, R. D.,** Brucelosis en chiguires del Estado Apure *(Hydrochoerus hydrochaeris), Acta Cient. Venez.*, 29, Suppl. 1, 178, 1978.
122. **Bello, A., Lord, V. D., Mogollon, P., Laserna, R. D., De Salmeron, C., Ramirez, M., Moreno, J., Toro, M., and Ramos, J.,** Estudio epidemiológico de la brucelosis en chiguires *(Hydrochoerus hydrochaeris)* del Estado Apure, *Acta Cient. Venez.*, 30(1), 31, 1979.
123. **Swann, A. I., Schnurrenberger, P. R., Brown, R. R., et al.,** *Brucella abortus* isolations from wild animals, *Vet. Rec.*, 106, 57, 1980.
124. **Binninger, C. E., Beecham, J. J., Thomas, L. A., et al.,** A serologic survey for selected infectious diseases of black bears in Idaho, *J. Wildl. Dis.*, 16, 423, 1980.
125. **Verger, J. M. and Dufrenoy, J.,** Isolement de *Brucella abortus* biotype 1 chez le mulot *(Apodemus sylvaticus), Ouest. Med.*, 25, 989, 1972.
126. **Fitch, C. P. and Bishop, L. M.,** The wild rat as a host of *Brucella abortus, Cornell Vet.*, 28, 304, 1938.
127. **Bosworth, T. J.,** Further observations on the wild rat as a carrier of *Brucella abortus, J. Comp. Pathol. Ther.*, 53, 42, 1940.
128. **Ewalt, D.,** National Veterinary Services Laboratory, Ames, IA, personal communication.
129. **Alexander, J.,** Texas Animal Health Commission, personal communication.

Chapter 14

BRUCELLA CANIS

Leland E. Carmichael

TABLE OF CONTENTS

I. INTRODUCTION

Brucellosis in dogs caused by *Brucella canis* is a contagious bacterial disease that is characterized by abortions in females and epididymitis, testicular atrophy, and infertility in males. The disease is insidious, and many dogs do not have prominent signs. It is of particular importance to dog breeders since infection with *B. canis* usually ends a dog's reproductive career.

In the late 1960s several investigators reported increased incidence of abortion, epididymitis, and reproductive failure in dogs in the U.S. Outbreaks were reported mainly in large commercial breeding kennels and in packs of field dogs, mainly beagles. A Gram-negative bacterium morphologically similar to members of the genus *Brucella* was isolated from placental and fetal tissues of aborted pups,[1-6] and the organism was later named *Brucella canis*[7-9] after initial consideration as a biotype of *B. suis.*[10] The disease was initially observed in beagles, but it subsequently was identified as the cause of widespread abortions and reproductive failures in various breeds, and its occurrence has since been reported in several countries.[11-21] Although cases were not documented prior to 1966, it is likely that the illness had been overlooked for many years. The origin of *B canis* is unknown, but its biochemical properties are similar to *B. suis.*[10] Like *B. ovis,* it has never been observed in the smooth (S) phase and is designated rough (R) because it lacks the somatic lipopolysaccharide (LPS) side chain antigens of S-brucella.[9] Unlike the other brucellae, however, it is always naturally mucoid.[7]

Natural infections with the three "classical" *Brucella* species (*B. abortus, B. suis,* and *B. melitensis*) also may occur in dogs. However, such infections are sporadic, and epizootics do not occur.[22,23] The infection in dogs caused by those species is self-limiting, and the organisms do not appear to persist or shed into the environment. However, the theory that the dog is a terminal host has been challenged.[10,24] Most canine infections are acquired through drinking infected milk or ingesting aborted material from livestock or infected offal.

Within a few years of its initial identification, *B. canis* was recognized in countries as distant as Japan[13,16,19] and Madagascar,[11] but highest seroprevalence rates appear to occur in Mexico, Central America, and South America,[12,17,21] especially where stray dogs are allowed to roam freely. In the U.S., the prevalence is highest in the South and Southwest, and its impact is greatest in large breeding kennels, packs of field dogs, or wherever large numbers of dogs are kept. The disease also has caused significant losses in private kennels.

II. CHARACTERISTICS OF *BRUCELLA CANIS*

Brucella canis is a small, Gram-negative coccobacillus that is morphologically indistinguishable from other brucellae.[3-8] It is aerobic and grows luxuriantly on enriched media such as brucella broth (Albimi). Other media that support abundant growth are 5% sheep blood agar, trypticase-soy agar, tryptose medium, and Thayer-Martin medium.[25,26] Growth is somewhat slow, requiring 2 to 3 d for colonies to mature, and it is inhibited by CO_2. After several days of incubation, the translucent colonies become very mucoid (ropy in broth). Mature colonies are generally 1 to 1.5 mm in diameter. The natural mucoidness of *B. canis* is a unique property. Therefore, the standard brucellosis test antigens for *B. abortus* are not useful for the diagnosis of *B. canis* infections. The LPS endotoxin of S-brucella organisms associated with agglutination is reduced, or lacking, in *B. canis.*[8,9]

Internal protein antigens extracted from whole cells are common to *B. canis* and the other *Brucella* species, but shared protein antigens have not been observed with other Gram-negative bacteria.[27,28] *B. canis* cross-reacts extensively with the other R-brucella such as *B. ovis* or *B. abortus* 45/20, and with LPS moieties of certain additional bacterial species such as mucoid strains of *Pseudomonas* sp., *Bordetella bronchiseptica, Staphylococcus epider-*

midis, and *Actinobacillus equuli.*[7,26] The close antigenic similarity between *B. canis* and *B. ovis* has been used to advantage in serologic tests for *B. canis.*[29-32] However, private *B. canis* and *B. ovis* determinants have been revealed by agarose gel immunodiffusion tests using both polyclonal[9,30] and monoclonal antibodies.[33]

Biochemically, *B. canis* is similar to *B. suis.*[7,8,10] It is a strong producer of urease, is H_2S negative, is weakly oxidase positive, reduces nitrate, and is similar to *B. suis* in its growth patterns on thionine and basic fuchsin and in its oxidative metabolism of urea cycle amino acids. Erythritol is only weakly metabolized oxidatively, and the organism is not lysed by phage Tb. Its taxonomic position as a member of the genus *Brucella* was decided by DNA homology studies where similar polynucleotide homologies were found between *B. canis* and the established members of the genus *Brucella.*[34,35] Acceptance of *B. canis* as a separate species by the Subcommittee on Taxonomy of *Brucella*[36] was based primarily on its singular pathogenicity for dogs and wild canids; low pathogenicity for domestic animal species, rodents, and man; and its natural transmissibility only among canines.

III. CLINICAL DISEASE IN DOGS

The disease has been observed naturally only in dogs and in wild canids. *B. canis* has a limited host range, unlike the S-brucellae that infect several domestic animal species. Host range susceptibility and pathogenicity of *B. canis* for various animal species has been reviewed in several publications.[37-40]

The pathogenesis of *B. canis* has not been studied extensively, but it most likely follows the pattern of brucella infections in other animals. Infection can begin on any mucous membrane, provided that sufficient numbers of organisms are introduced. The incubation period following oronasal infection is variable, but a leukocyte-associated bacteremia can usually be demonstrated after postinfection (PI) week 1 to 2. The minimal oral infectious dose for dogs is about 10^6 bacteria,[41] with somewhat fewer numbers required for infection by way of the vagina. The conjunctival infectious dose is less, being about 10^4 to 10^5 organisms.[42]

Brucellae localize in the lymph nodes, spleen, bone marrow, and reproductive tract of infected dogs. A common, and diagnostically significant, feature of the disease is the long period of sustained bacteremia where greater than 10^3 bacteria per milliliter of blood are commonly found after month 1 PI. The bacteremia may be as short as 6 months, but it commonly persists longer than 1 to 2 years and has been observed as long as 5 years.[43] Spontaneous recovery from infection may occur 1 to more than 5 years PI. Following cessation of the bacteremia, organisms may persist for many months in the spleen, lymph nodes, bone marrow, prostate gland, and epididymides.

Clinical manifestations of canine brucellosis have been extensively reviewed[37-40] and only salient features are noted here. Signs are generally associated with the reproductive tracts of both sexes. Despite systemic infection, dogs are rarely observed ill. Fever is uncommon, probably due to the relative lack of endotoxin. Most infections are not diagnosed by routine history or physical examination, but dogs usually have enlarged lymph nodes; prostatitis and epididymitis are common in males.

In females the principal sign is abortion. Abortion may occur at any time, but generally after 45 to 55 d of gestation. Early embryonic deaths and abortion 10 to 20 d after mating have been observed, but they are often unnoticed and regarded as conception failures. As with brucellosis in other species, infections with *B. canis* do not interfere with normal estrous cycles. Up to 85% of bitches that abort may subsequently have normal litters; however, some infected females experience intermittent reproductive failures. Rarely, bitches may abort up to four times in succession and have as many as three consecutive unsuccessful matings. Infected dams may carry infected pups to term and whelp both live and dead puppies within a single litter. Most pups that are born alive will die within a few hours or days, but those that survive have generalized enlargement of lymph nodes as the primary sign until they reach sexual maturity.

In males, the disease is characterized by poor reproductive performance and/or epididymitis. Physical examination usually reveals epididymitis, atrophy of one or both testes, and scrotal dermatitis. The volume of ejaculate is diminished, and semen from infected males usually contain large numbers of abnormal sperm and inflammatory cells, especially during the initial 3 months of infection. Abnormalities include immature sperm, swollen midpieces, bent tails, retained protoplasmic droplets, and detached heads. Spermagglutination is noted in the seminal fluid after PI week 15. Also seen are clumps of inflammatory cells that consist of neutrophils that surround macrophages and adherent spermatozoa, some containing phagocytized sperm. Chronically infected males may be azoospermic. The seminal plasma of infected dogs contains IgA spermagglutinins and IgG cytophilic antibodies which persist as long as 10 months PI. However, antibodies to *B. canis* are not detected in the seminal plasma. Spermagglutinins also occur in the sera of males infected for more than 4 months, and cutaneous delayed-type hypersensitivity reactions are observed after testing with saline extracts of canine testicular tissue.[44] Immune responses produced against spermatozoa are believed to contribute to the infertility and eventual spermatogenic arrest.[44,45]

Less common manifestations of *B. canis* infection are diskospondylitis,[46] polyarthritis, glomerulonephritis, or recurrent anterior uveitis with corneal edema.[38,47,48]

Pathological changes consist of lymphoid and reticular cell hyperplasia of all lymphoid tissues, including Peyer's patches, the gall bladder, and eyelids. There is inflammatory cell infiltration of all genitourinary organs. A necrotizing vasculitis occurs in the target tissues of the gonadal steroids, including the prostate, epididymides, and vulva; milder changes are observed in the testes, ductus deferens, urinary bladder, and ureters.[38,40,49] Chronically infected dogs may have subacute to chronic endometritis, granulomatous prostatitis, testicular atrophy and fibrosis, hyaline thickening of the basement membrane of glomeruli with minimal cellular infiltration or proliferation, or meningitis. Diskospondylitis is not an uncommon feature of chronic infection, and anterior uveitis has been seen in some chronically infected dogs. Ocular pathology includes granulomatous iridocyclitis, exudative retinitis, and corneal edema.

IV. DISEASE IN MAN

Brucella canis is infectious for humans,[50-56] but man is considered relatively resistant, and the disease is mild in contrast to infections caused by the three classical species. It apparently requires massive exposure to produce clinical signs. Asymptomatic cases are suspected based on serosurveys,[57,58] but the total number of reported cases worldwide appears to be less than 40. The most common manner of infection has been through laboratory accidents or direct contact with aborting bitches; however, it is suspected that male dogs also may transmit the disease.[50] The source of many human cases was not determined, and the diagnoses in some instances can be questioned since the sole laboratory criterion for infection was a positive agglutination test result.[55]

Nonhuman primates *(Macaca arctioides)* were found susceptible to experimental infection with large numbers of organisms ($\sim 10^{10}$ bacteria), and focal granulomatous lesions that varied in magnitude were observed in liver, spleen, and lymphoid tissues.[59] The bacteremia was transient, and organisms could not be demonstrated after 7 weeks PI. Genital lesions were not observed.

The incubation period in man is unknown, and pathologic lesions in culture-positive humans are not described. The reported symptoms in man include malaise, sore throat, headaches, chills, and nausea. Weight loss and lymphadenopathy also have been reported in symptomatic patients. In one instance,[51] *B. canis* was isolated at surgery from an aortic blood clot of an individual who died 3 years later of unrelated causes. All patients with confirmed infection had agglutinin titers of 1:200 or greater, but titers declined within a few weeks after a 2- to 3-week course of tetracycline treatment.

V. EPIZOOTIOLOGY

A. NATURAL HOST RANGE

Unlike the S-brucellae that infect several domestic animal species, *B. canis* has a limited host range (reviewed in References 4, 23, 37—39, and 60—62). Only dogs and wild canids have been found naturally susceptible to infection. Cattle, sheep, and swine are highly resistant to infection. Cats appear moderately susceptible, and with development of a bacteremia and low agglutinin titers following oral exposures to *B. canis*. Guinea pigs, mice, rats, and nonhuman primates also are susceptible to experimental infection, but, as in man, the infection is relatively mild. The rabbit may be more susceptible than the other laboratory animal species; rabbits developed orchitis and peritoneal abscesses after inoculation with large doses given intraperitoneally.[7] Tests of 770 wild carnivore and opossum sera indicated a negligible seroprevalence when agglutinin titers of 1:200 or greater were considered significant.[63] Antibodies were not detected in the sera of striped skunks, gray foxes, wolves, or opposums. Significant titers were observed, however, in one bobcat, one red fox, and two coyotes.

B. DISTRIBUTION

Prevalence rates in the U.S. vary from approximately 1 to nearly 19%, with highest frequencies (~8%) observed in rural areas of the southern states.[37-39] Seroprevalence rates ranging from 1.6 to 2.9% have been reported in Japan.[16,64] As noted above, canine brucellosis also has been documented in Argentina, Brazil, Czechoslovakia, Germany, the Malagasy Republic, Peru, and Tunisia. Recently, *B. canis* was recovered from purebred dogs originating in Spain and tested on their arrival in Mexico.[65] In Czechoslovakia and Japan, outbreaks were traced to the importation of beagle dogs from the U.S.[19,66] Highest prevalence rates have been reported in Mexico and Peru.[17] In a serological and bacteriologic study of canine brucellosis in Mexico, 28% of 500 stray-dog sera had agglutinin titers greater than 1:100.[12] An additional study of stray dogs in the Mexico City area that included autopsy of each animal revealed a similar seroprevalence, and *B. canis* was isolated from 7 of 59 (~12%) dogs.[21] A study of 203 sera from hospital patients in Mexico City with undiagnosed febrile illnesses revealed a 13.3% rate of agglutination titers greater than 1:100.[65] Virtually all patients had received antibiotic treatment, and *B. canis* was not isolated from blood samples.

The determination of seroprevalence is strongly influenced by the mode of testing and interpretation since *B. canis* shares antigenic determinants with all R-brucellae as well as with several unrelated organisms, some of which are cited above. False positive reactions are common with all methods that measure antibodies to R-envelope LPS antigens, e.g., agglutination or immunodiffusion tests.

C. TRANSMISSION

Canine brucellosis is most frequent in strays, or where dogs are housed in large numbers or maintained as packs. Factors affecting natural endemicity have not been determined, but the seroprevalence is higher in stray dogs than in household pets. Pet dogs in economically depressed environments appear to have a higher seroprevalence rate than do those in suburban areas. When the disease is introduced into a kennel, it spreads rapidly.

Natural transmission of canine brucellosis occurs by several routes, but it is most commonly spread by placental and fetal tissue and vaginal discharges that follow an abortion. Vaginal discharges may contain up to 10^{10} organisms per milliliter,[41] and shedding by this route may continue for several weeks following an abortion. Other body secretions contain lower concentrations of bacteria and are less important in disseminating infection. Milk of infected bitches is a potential source of infection for surviving pups,[4,23] but most are infected *in utero*. *Brucella canis* also has been isolated from ticks *(Rhipicephalus sanguinis)* feeding on an infected dog.[67]

Seminal fluid and urine from dogs that harbor the organisms in the prostate glands and epididymides have been determined as important sources of infection.[42,68,69] The rate of *B. canis* isolation from the semen of infected males is high for the initial 6 to 8 weeks of infection, but intermittent shedding of the organism in low numbers has been observed for up to 60 weeks PI.[68] Brucellae have been cultured from the prostate gland and epididymides for longer than 2 months after the bacteremia ceased, and venereal transmission can continue for at least 2 years in dogs that appear clinically normal.

Urinary excretion begins a few weeks after the onset of bacteremia and continues for at least 3 months. Concentrations of 10^3 to 10^6 organisms per milliliter of urine have been observed in males, with lesser numbers of bacteria in females.[42] The tendency of males to shed organisms in the urine probably relates to localization in the prostate and epididymis, which, in the dog, are in close association with the bladder. It has been demonstrated that *B. canis* can be transmitted from infected to noninfected sexually mature males after several weeks or months of cohabitation.[41,42] Brucellosis was not transmitted between sexually immature males or females during 10 months of close contact in isolation units; however, transmission was observed in both sexes following estrus or a mating.

Infected females transmit *B. canis* through oronasal contact with infectious placental, fetal, or vaginal discharges following an abortion; at estrus, or during a mating. Because a low number of organisms is shed in the urine,[41,42,69] it is unlikely that urine is a significant mechanism of spread by females.

VI. LABORATORY DIAGNOSIS

A. PHYSICAL EXAMINATION

Clinical signs are usually not adequate to diagnose canine brucellosis. The disease is insidious in its onset and variable in clinical manifestations. It should be considered in the differential diagnosis whenever there is a history of abortion in females or poor reproductive performance in either sex. Epididymitis, testicular atrophy, scrotal dermatitis, poor semen quality, and, occasionally, sterility are the salient signs in males. The possibility of brucellosis also should be considered in cases of diskospondylitis.

There are no characteristic hematologic changes, but examination of semen from infected males is helpful since sperm abnormalities are evident by 5 weeks PI and become pronounced by 8 weeks.[68]

B. SEROLOGICAL DIAGNOSIS

Several methods have been used by laboratories in the serodiagnosis of canine brucellosis, but none are ideal. Detailed methods for the bacteriologic and serological diagnoses have been published elsewhere.[21,25,26,30,43]*

The serodiagnosis of *B. canis* infection presents difficulties for several reasons. Antigen preparation with wildtype organisms has presented obstacles because colonial growth is always mucoid (M+) and, as with other M+ brucellae,[70] growth is viscous in saline below pH 6.8; M+ cells are hydrophobic and adhere tenaciously to agar or form tenacious, ropy

* A recently revised third edition of *Laboratory Techniques in Brucellosis*,[25] has been published and should be consulted for recent developments.[88]

masses in broth media at a pH below 7.0.[71,72] Ropy masses are obviated by adjusting growth media and suspending buffers to pH 7.4. Hot saline (HS) or sodium deoxycholate (SDC) antigenic extracts which aggregate below pH 6.8 also do not clump at alkaline pH. In addition, there is the important issue of broad cross-reactivity between cell envelope LPS antigens of *B. canis,* R-phase brucella, and certain other bacterial species.

All serologic tests are unreliable during the initial 4 weeks of infection, some 2 to 3 weeks after the onset of bacteremia. Thereafter, dogs have sustained antibody responses against bacterial LPS and protein antigens. Antibodies can be detected by a variety of methods until several months after the bacteremia has ceased.[1,30,32,39,40,43] Following a 1-year to more than 5-year period of bacteremia (in untreated dogs), antibody titers decline after a 4- to 6-month period and then become negative or equivocal with most tests. Antibodies may be detected in the sera of chronically infected dogs for more than 16 months after the bacteremia has ceased; however, results of agglutination and agarose immunodiffusion (AGID) test utilizing cell envelope antigenic extracts then become indefinite. This is especially true in chronically infected, abacteremic males with localized infections of the prostate or epididymides. Animals that have received a course of antibiotic therapy also may have sustained low (1:25 to 1:50) agglutinin titers.

Three serological methods for canine brucellosis are most commonly used by veterinary diagnostic laboratories.[25,32,43,73,74] These include the rapid slide agglutination test (SAT), the tube agglutination test (TAT), and the modified 2-mercaptoethanol tube agglutination test (ME-TAT). In addition, some laboratories employ AGID tests that use HS or SDC extracts of *B. canis*[30,73] or *B. ovis*[29,31] as antigen. More recently, an AGID test that utilizes protein antigens extracted from the cytoplasm (cytoplasmic protein [CP] antigen, soluble antigens, internal antigens) by gentle agitation in acidified buffer or by sonication followed by high-speed centrifugation (100,000 × *g* for 2 h) has proved highly accurate in confirming the serodiagnosis of canine brucellosis by agglutination or other tests that employed cell envelope extracts.[30] In addition, a recent modification of the SAT that uses a less-mucoid (M−) variant of *B. canis* as antigen has been reported[75] to reduce the percentage of false positive reactors.

Other serological methods that have been employed occasionally include counterimmunoelectrophoresis,[76] indirect immunofluorescence,[77] and complement-fixation tests.[29,78]

1. Slide Agglutination Test

An SAT similar to the *B. abortus* plate agglutinin test, but employing *B. ovis* as antigen, was developed as a rapid procedure for the presumptive serodiagnosis of canine brucellosis.[32] The test is quick and simple, and antibody can be detected early in the course of infection. For this reason, it is the most common screening procedure used by veterinarians in the U.S. The results obtained with the SAT in healthy and in infected dogs correlate well (99% agreement) with the TAT. The test is highly accurate in identifying *noninfected* animals since false negative reactions are rare. However, positive test results in field dogs must be interpreted with caution since a high proportion (~50%) of positive, but not negative, results are false positive reactions.

In an effort to reduce the rate of false positive reactions, the SAT was modified[74] to include brief treatment of serum with 0.2 *M* 2-mercaptoethanol prior to adding the stained antigen. Because of the high correlation between *negative* SAT results and absence of infection and significant (~50%) reduction in the rate of false positive reactions, the 2-mercaptoethanol SAT (ME-SAT) is now widely used as a screening test.

Antigen is prepared from heat-inactivated and Rose Bengal-stained *B. ovis.* Bacterial growth from solid media is washed several times in phosphate-buffered saline (0.15 *M*, pH 7.4), heat inactivated for 2 h at 60°C, and stained with 1% Rose Bengal dye (at least 90% dye content, 5 cc of 1% dye per 100 ml 6 to 8% cells) overnight at 4°C with constant stirring. Organisms are washed at least twice in tris-malic acid (TM) buffer (0.4 *M*, pH 7.0—7.2) and then suspended to a final packed cell volume of 6%. A commercial kit for this test is available (Pitman Moore, Inc., Washington Crossing, NJ 80560).

In the SAT procedure, one drop (~50 μl) of suspect or positive control serum, and the same amount of 0.2 *M* 2-ME, is added to separated portions of a slide or scored plastic plate. The solution is briefly (at least 10 s) mixed, and 50 μl of the antigen are then added and mixed. Results are read and compared with the positive control serum within 2 min. Alteratively, 0.1-ml amounts of serum and 2-ME may be mixed in a small glass tube and allowed to stand for a brief period. One drop of the mixture is then added to a glass slide and stirred with the same amount of antigen. Since 2-ME is rapidly inactivated when not properly stored, it should be kept in a dark, tightly stoppered bottle at 4°C when not in use.

Since a high percentage of the positive results obtained with field sera in the ME-SAT still can be false reactions,[30,43,75] additional serologic and bacteriologic studies must be done before a diagnosis can be established.

The ME-SAT is often positive in infected dogs as early as 3 weeks PI, but not usually until 4 to 12 weeks, and tests remain positive until several months after the bacteremia has ceased. Since there is little difference in the sensitivity and specificity between the ME-TAT and slide test results,[43,79] the ME-SAT is considered the preferred test for initial screening. It is simple and provides information that is similar to the ME-TAT.

A recent modification of the ME-SAT has further reduced the rate of false positive reactions.[75] The modified procedure is essentially the same as described for the ME-SAT that uses *B. ovis,* but it employs the M− variant of *B. canis* as antigen. It has the advantage of reducing the rate of false positives from 50 to 12% without forfeit of sensitivity.

The variant strain is not only less mucoid, but bacterial cells are less hydrophobic, have a higher net negative charge, and are less virulent than wild-type (M+) organisms.[71,72] Polyclonal antibodies from dogs experimentally infected with the M− strain react strongly with *B. canis* cells, but only weakly with the R-*B. ovis* or *B. abortus* 45/20. Also, monoclonal antibodies highly specific for *B. canis* determinants (Figure 1) were readily obtained in mice immunized with LPS extracts (HS) from M− cells. In contrast, monoclonal antibodies obtained with M+ *B. canis* antigens all cross-reacted in AGID tests with SDC extracts of R-*Brucella, Streptococcus fecalis, Staphylococcus epidermidis* (mucoid isolate), and *Pseudomonas aeruginosa* (mucoid isolate).[33]

2. 2-Mercaptoethanol Tube Agglutination Test

Several tube agglutination procedures have been described,[25,26,43,80] but the ME-TAT has been the one most widely used by veterinary diagnostic laboratories in the U.S. The test is commonly used for confirmation of infection in ME-SAT-positive dogs, but it is subject to considerable variation in interpretation and suffers the same problems of false positive reactions as does the ME-SAT. The test was developed at the National Veterinary Services Laboratories of the U.S. Department of Agriculture, Ames, IA, and provided free of charge to qualified laboratories. Detailed methods of antigen preparation and test procedure have been published elsewhere.[25]

Infected dogs generally become seropositive 3 to 6 weeks after oronasal exposure, and they remain positive (ME-TAT titers > 1:200) for several months after the bacteremia has abated (Figure 2). Field experience indicates that neither the ME-SAT nor the ME-TAT are accurate during the initial 12 weeks of infection.[43,73]

Dogs are judged "negative" if there is no agglutination at the 1:50 serum dilution; "suspicious" if incomplete or complete reaction at the 1:50 or 1:100 dilution, or incomplete at the 1:200 dilution; and "positive" when complete agglutination occurs at the 1:200 dilution or higher.

If a result is negative, the animal may be healthy or in the early stage of infection; if

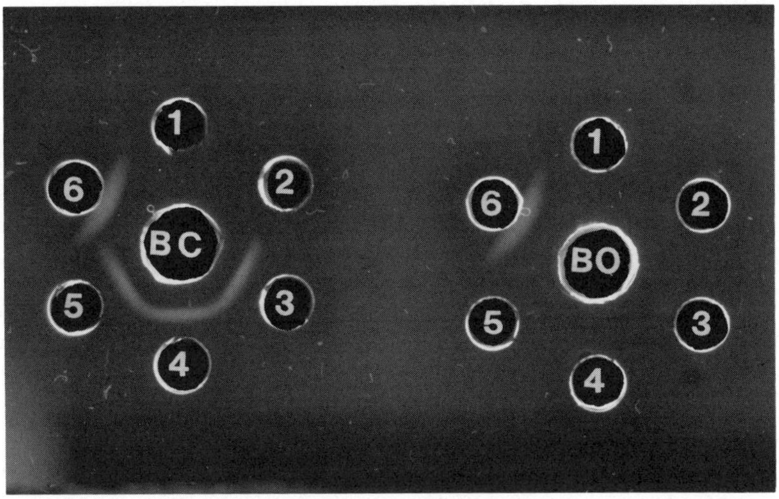

FIGURE 1. Agarose-gel immunodiffusion reactions between *B. canis* (BC) or *B. ovis* (BO) antigens (center wells) and (1) canine parvovirus monoclonal antibodies (MAb), mouse ascitic fluid; (2) canine parvovirus mouse, MAb hybridoma fluid (controls); (3) and (5) different *B. canis*-specific MAb mouse hybridoma fluids; (4) *B. canis*-specific mouse ascitic fluid MAb; (6) cross-reactive *B. canis* Mab. Most (33 of 35) MAbs produced were cross-reactive in AGID tests with other R-*Brucella* as well as certain other bacterial species.

''suspicious'', it may indicate early infection or a false positive reaction. A second serum sample should, therefore, be obtained 30 d later and retested. If the second test is positive, the animal should be presumptively diagnosed as infected and hemocultures should be done. Blood should be cultured from all suspect or seropositive dogs before a diagnosis is rendered. Antibiotic treatment prior to testing obscures serological as well as bacteriological results and should not be instituted until diagnostic tests have been completed.

The ME-TAT is most useful to monitor the responses of dogs in kennels where canine brucellosis has been diagnosed. It also is a valuable test to evaluate responses to antibiotic therapy.[81]

3. Agarose Gel Immunodiffusion Tests

a. Cell Envelope Antigens (LPS Antigens)

Two AGID methods are described for the serodiagnosis of *B. canis* infection.[30,31,43] One is similar to that described for the diagnosis of *B. ovis*[29] but employs SDC or HS extracts of *B. canis*[30,43] or *B. ovis*.[31] In both instances, the antigenic activity resides in the LPS fractions. Somatic cell extracts employed in AGID tests, therefore, suffer from the same problems of nonspecific reactions as do agglutination tests, and caution must be exercised in their interpretation. Sera from noninfected dogs may give strong precipitin reactions that are difficult to differentiate from those of infected dogs and lead to erroneous positive interpretations.[30,43] The use of a positive control serum, preferably prepared in specific pathogen-free (SPF) dogs, should be included in all AGID tests.

AGID tests employing cell envelope antigens reveal precipitins in the sera of infected dogs within 6 to 12 weeks PI, and antibodies may persist for several years (see Figure 2). With the somatic antigens, three precipitin lines usually develop, but only one is considered specific for *B. canis* (Figure 3). The precipitin reaction highly specific for *B. canis* is not seen when *B. ovis* or *B. abortus* 45/20 is used as the source of antigen.[30] Although specific and nonspecific reactions can be distinguished by experienced personnel, interpretation is particularly difficult in sera from chronically infected dogs.

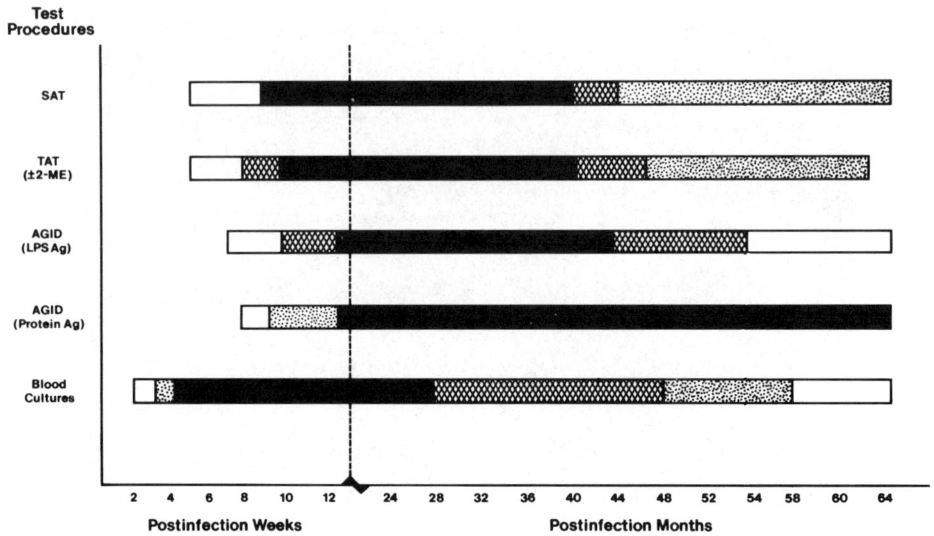

FIGURE 2. Antibody responses and hemoculture results of experimentally infected dogs (n = 8) over a period of 64 months. Empty bars indicate less than 25% positive; dotted bars mean 25 to 50% positive; cross-hatched bars signify 50% positive; black bars mean all dogs positive. See text for test procedure abbreviations. (From Carmichael, L. E., Zoha, S. J., and Fores-Castro, R., in *Developments in Biological Standardization,* Vol. 56, S. Karger, Basel, 1984, 376. With permission.)

b. Protein Antigens

The second type of AGID test employs protein-rich extracts from the bacterial cell cytoplasm. The extracts are chiefly protein, but they also contain small, but variable, amounts of cross-reactive LPS and polysaccharide. Complete removal of cross-reacting material has been difficult, but it may be substantially reduced by treating the extracts with lipid solvents such as xylene or methylene chloride, or by affinity chromatography using Detoxi-Gel (Pierce Chemical Co., Rockford, IL 61105) or wheat germ lectin.[33] Use of cytoplasmic protein (CP) antigens in AGID tests, but not latex agglutination or enzyme-linked immunosorbent tests,[33] has been highly accurate in detecting antibodies in the sera of dogs when other methods gave equivocal or even negative results.[30,43] We have used sonication or disruption of cells by high pressure in a Ribi-Sorvall Cell Fractionator, but extraction of methanol-fixed cells by vigorous agitation in acidified buffer in a ''Bead-Beater'' (Biospec Products, Bartlesville, OK 74005) also gives satisfactory preparations. Detailed procedures for preparing protein-rich extracts are given in several publications (see References 27, 28, 30, 43, and 82).

Cytoplasmic salt-extractable protein (CSP) antigens prepared from *B. abortus*[82] may be used with similar results. The cytoplasmic proteins are highly conserved within the genus *Brucella.*[27] We have not observed qualitative differences in SDS-PAGE polypeptide profiles between protein extracts of the various species of *Brucella,* but quantitative differences exist and may reflect the method of extraction. When CP antigens are used, canine infections caused by other *Brucella* species cannot be differentiated from those caused by *B. canis.*

Cytoplasmic antigens typically give two to four precipitin lines with sera from infected dogs, and reactions may be observed for as long as 68 months. Reactions do not occur, however, with sera from healthy dogs that give false positive reactions by agglutination or AGID tests using cell wall extracts.[30,43] Positive reactions with CP antigens are highly specific for antibodies to *Brucella;* however, sensitivity is low during the initial 8 to 12 weeks of infection. Precipitins persist in the sera of chronically infected dogs and have been demonstrated 12 months after the bacteremia has ceased, a time when other test results were equivocal or negative.

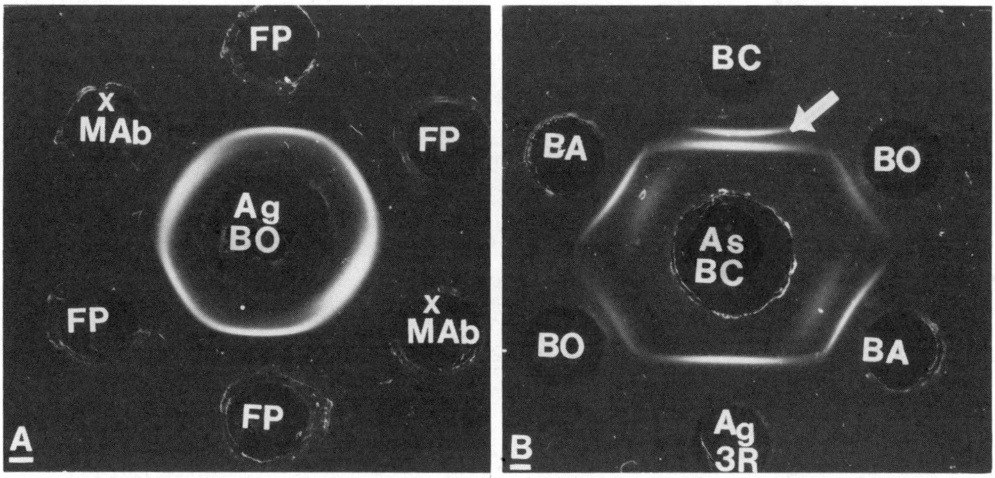

FIGURE 3. Agarose immunodiffusion reactions of (A) *B. ovis* antigen (Ag/BO) with cross-reactive *B. canis* monoclonal antibodies (x/MAb) or false positive (FP) sera from healthy dogs; (B) *B. canis* antisera (As/BC) with hot buffered saline extracts of *B. canis* (BC), *B. ovis* (BO), and the cross-reactive antigen 3-R.[30] Arrow indicates the *B. canis*-specific antigen (antigen 2-R). The cross-reactive MAb shown in (A) gives precipitin lines with all R-brucellae tested and with cell wall extracts from several other bacterial species. (Modified from Zoha, S. J. and Carmichael, L. E., *Veterinary Microbiology,* 7, 43, 1982. With permission.)

Antibody response patterns of an infected dog to CP antigenic polypeptides from the time of oronasal exposure over a 36-week period are shown in Figure 4.

4. Bacteriologic Methods

Detailed methods for the culture of *B. canis* have been described in several publications (see References 7, 23, 25, and 26). Most media that support growth of the other brucellae may be used for the cultivation of *B. canis,* but we routinely use Albimi Brucella Broth (BBL, Becton, Dickinson & Co., Cockeysville, MD 21030). Tryptose media is satisfactory, with or without added serum. Trypticase soy agar also has been used, but 5% serum is necessary for good growth. Potato infusion agar, useful for growing many *brucella* does not support satisfactory growth of *B. canis.*[23] Antibiotics such as bacitracin, polymyxin, and cyclohexamide may be added for potentially contaminated material or for subculturing from broth media.[25] Thayer-Martin Media (Difco Laboratories, Detroit, MI 48201) also supports excellent growth of *B. canis* from contaminated specimens. To prepare solid media, 1.5 to 2% agar is added to brucella broth base. If the pH is adjusted to pH 7.4, growth is less mucoid, and organisms to be used for antigen are easier to harvest.

Like other *brucellae,* growth is slow. Turbidity is first detected in broth after 2 to 3 d, and a ropy sediment forms after several days of growth. Incubation should be done aerobically at 35 to 37°C for CO_2 is inhibitory. Growth on solid media may not be visible until 48 h after subculture from broth, and mature colonies (1 to 1.5 mm in size) develop after 3 to 4 d. Colonies become very mucoid after several days of incubation, especially when the media pH is below 6.8.[71] Although *B. canis* is always isolated in the R form, colonies do not appear greatly different from those of the S-brucella.

a. Culture from Blood and Urine

Hemoculture is the most practical means of isolating *B. canis* from an infected dog, provided the animal had not received antibiotic therapy. A leukocyte-associated bacteremia begins as early as 7 d after oronasal exposure, and it lasts for several months or years. Blood for isolation (about 5 ml) should be collected in heparin or 1% sodium citrate. After 5 to 7

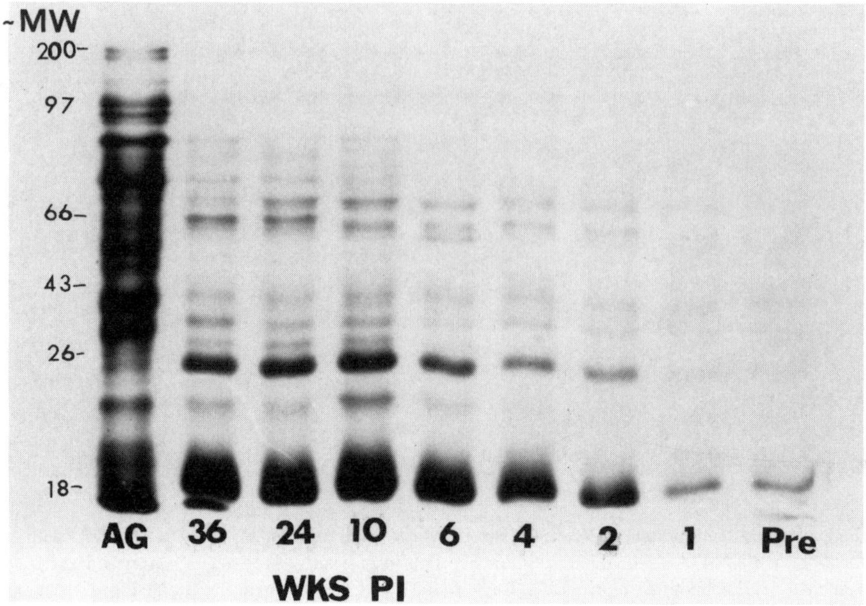

FIGURE 4. Immunoprecipitation patterns of serial serum samples from a dog experimentally infected with *B. canis*. ^{35}S methionine-labeled cytoplasmic protein (CP) antigens precipitated by pre- and postinfection sera and separated by electrophoresis in 10% SDS-PAGE. Lanes are designated: AG (antigen); Pre (reacted with preinfection sera); numbers, weeks postinfection (PI). MW = molecular weight standards.

d of incubation in 10 ml of liquid medium, broth is spread or streaked onto solid medium and examined after 3 to 5 d for typical growth. Isolation rates may be increased by freezing blood-broth mixtures at −70°C and rapidly thawing them before inoculation onto solid medium.[23]

In some instances, isolation of organisms from the urine is successful in dogs that have become abacteremic.[42] Unless cystocentesis is performed, isolation from urine is generally unrewarding due to overgrowth by contaminants, even with use of antibiotic-containing media. The frequency of urine isolation from male dogs is much higher than from females.[41,42,69]

b. Culture from Tissue Specimens

Brucella canis can be isolated with ease from hemoculture-positive dogs. Organisms are especially abundant in the lymph nodes, spleen, liver, and male reproductive organs. In chronically infected, abacteremic dogs the sites of most frequent isolation are the prostate and epididymides of males, and the spleen, lymph nodes, and, sometimes bone marrow of both sexes.[33,37,69] The direct method of smearing portions of tissue onto solid medium containing selected antibiotics has proved satisfactory in most instances.

VII. PREVENTION AND CONTROL

Prevention is particularly important in large breeding kennels. Unlike brucellosis in domestic livestock, the canine disease is not reportable in the U.S., and knowledge of prevalence is based on limited serosurveys. The incidence of the disease in dogs is probably little different from when it was initially recognized, with the exception of certain large commercial kennels and organizations of dog breeders or hunt clubs which have privately instituted preventative measures. When canine brucellosis is diagnosed in a kennel, control becomes a dilemma. Elimination of the infection is not only time consuming and costly, but the disease also presents emotional crises, especially when dogs with exceptional genetic traits are affected.

Prevention in breeding kennels involves a closely monitored plan of hygiene and animal care.[38,62] Serologic testing should be done at least twice at monthly intervals on all dogs to be introduced into a breeding kennel. Bitches in a kennel should be tested several weeks before their expected estrus. If tests other than the ME-SAT are required to certify freedom from brucellosis, there will be time to carry out those procedures before the heat period. No new animals should be introduced into a breeding colony until they have been shown to be negative on two brucellosis tests taken 1 month apart. Tests on all animals in a kennel should be done at least yearly, or whenever reproductive problems arise such as abortions or failures to conceive.

If a bitch aborts, canine brucellosis should be suspected until it is proved otherwise. Dogs with positive or questionable screening test (ME-SAT or ME-TAT) results should be isolated and evaluated by hemocultures and serologic testing using more specific tests, e.g., the AGID employing CP antigens. Any dog found to be infected should be removed from the kennel and destroyed. In some instances, owners refuse to destroy an animal, and spaying or castration plus antibiotic therapy and monitoring for at least 3 months is minimally indicated. Results of at least three monthly tests of all dogs in a kennel with proved infection should be negative before the kennel can be considered free of infection.

Since the principal routes of spread are by way of contaminated discharges after an abortion or during a mating by infected dogs of either sex, all persons in contact with suspect or infected dogs should thoroughly disinfect their hands as well as contaminated areas. Commercial disinfectants such as quaternary ammonium compounds and halogens (iodine or chlorine solutions) rapidly destroy *B. canis* organisms.

When a positive case is reported in a household pet, an owner may request antibiotic therapy. In breeding kennels infected animals, especially males, should be destroyed unless there are compelling reasons to attempt treatment, e.g., to save a valuable blood line. The decision to treat should be based on informed consultation between a client and the veterinarian because of the expense and possibility of failure. The request should be discouraged, or declined, if posttreatment monitoring is not agreed on. Recrudescence of active infection may recur weeks or months after antibiotics are discontinued and treated animals should be monitored for at least three posttreatment months by the ME-SAT or ME-TAT. In addition, infected pets should be castrated or spayed. In one reported case,[83] *B. canis* was isolated from the uterine stump 3 years after ovariohysterectomy.

In order for treatment to be considered successful, *B. canis* must be completely eliminated from the body. Several treatment regimens have been effective,[38,39,73,79,84-87] and the references cited should be referred to for details. Large doses of tetracycline combined with streptomycin, administered over a period of 3 to 4 weeks, have proved most effective. Treatment has been successful in approximately 80% of early cases,[81,87] i.e., in dogs infected less than 2 months. Success has been greater with infected females, probably because of the greater difficulty in eliminating foci of infection from the male genital tract, especially the prostate gland. Owners should be informed of the potential health hazard in keeping infected pets and of the uncertain outcome, and expense, of treatment.

There is no acceptable vaccine for canine brucellosis.[40] The desirability of a vaccine is doubtful, especially where serological monitoring is accessible, since an effective vaccine would be required to provide durable immunity, but not obscure the serodiagnosis.

REFERENCES

1. **Carmichael, L. E. and Kenney, R. M.,** Canine abortion caused by *Brucella canis, J. Am. Vet. Med. Assoc.,* 152, 605, 1968.
2. **Kimberling, C. V., Luchsinger, D. W., and Anderson, R. K.,** Three cases of canine brucellosis, *J. Am. Vet. Med. Assoc.,* 148, 900, 1966.
3. **Moore, J. A. and Bennett, M.,** A previously undescribed organism associated with canine abortion, *Vet. Rec.,* 80, 604, 1967.
4. **Morisset, R. and Spink, W. W.,** Epidemic canine brucellosis due to a new species, *Brucella canis, Lancet,* 2, 1000, 1969.
5. **Taul, L. K., Powell, H. S., and Baker, O. E.,** Canine abortion due to an unclassified gram-negative bacterium, *Vet. Rec. Small Anim. Clin.,* 62, 543, 1967.
6. **Hill, W. A., Van-Hoosier, G. L., and McCormick, N.,** Epidemic abortion in a canine production colony. I. Epizootiology, clinical features, and control procedures, *Lab. Anim. Care,* 20, 295, 1970.
7. **Carmichael, L. E. and Bruner, D. W.,** Characteristics of a newly-recognized species of *Brucella* responsible for infectious canine abortions, *Cornell Vet.,* 58, 579, 1968.
8. **Jones, L. M., Zanardi, M., Leong, D., and Wilson, J. B.,** Taxonomic position in the genus *Brucella* of the causative agent of canine abortion, *J. Bacteriol.,* 95, 625, 1968.
9. **Diaz, R., Jones, L., and Wilson, J. B.,** Antigenic relationship of the gram-negative organism causing canine abortion to smooth and rough brucellae, *J. Bacteriol.,* 95, 618, 1968.
10. **Meyer, M. E.,** Brucella organisms isolated from dogs: comparison of characteristics of members of the genus *Brucella, Am. J. Vet. Res.,* 30, 1751, 1969.
11. **Verger, J. M., Gate, M., Piechaud, M., Chatelain, R., Ramisse, J., and Blancou, J.,** Isolement de *Brucella suis* biotype 5 a Madagascar, chez une chienne, validite du nom d'espece *Brucella canis, Ann. Microbiol. Paris,* 126A, 57, 1975.
12. **Flores-Castro, R. and Segura, R.,** A serological and bacteriological survey of canine brucellosis in Mexico, *Cornell Vet.,* 66, 347, 1975.
13. **Yamauchi, C., Suzuki, T., Nomura, T., Kukita, Y., Iwaki, T., Kazuno, Y., and Ghoda, A.,** Canine brucellosis in a beagle breeding colony, *Jpn. J. Vet. Sci.,* 36, 175, 1974.
14. **von Kreudener, R. B.,** Isolierung und Bestimmung von *Brucella canis* aus einem Beaglebestand, *Zentralbl. Veterinaermed. Reihe B,* 21, 307, 1974.
15. **Godoy, A. M.,** Brucella canis in Brasil, *Proc. Natl. Meet. Microbiol. (Brasil),* 5, 77, 1974.
16. **Ueda, K., Saegusa, J., Fujiwara, K., Muto, S., Okada, K., Hasegawa, A., Saegusa, S., and Usui, K.,** Detection of *Brucella canis* in dogs in the Tokyo area, *Jpn. J. Vet. Sci.,* 36, 539, 1974.
17. **Englehardt, C. E. M.,** Incidencia de *Brucella canis* en perros en el destrito de Chiclayo, Master's thesis (English summary), University Nacional Pedro Ruiz Gallo, Peru, 1974.
18. **El Bahri, L., Ben Osman, F., and Chadli, A.,** Serological survey for canine brucellosis in Tunis, *Arch. Inst. Pasteur Tunis,* p. 315, 1970.
19. **Azuma, R. and Isayama, Y.,** Canine brucellosis, *J. Jpn. Vet. Med. Assoc.,* 26, 111, 1973.
20. **Ramacciotti, F.,** Aislamiento de Brucella canis efectuado por primera vez en la Republica Argeentina con transmision humana, *Rev. Med. Vet. (Buenos Aires),* 59, 69, 1978.
21. **Flores-Castro, R., Suarez, F., Ramirez-Pfeiffer, C., and Carmichael, L. E.,** Canine brucellosis: bacteriological and serological investigation of naturally infected dogs in Mexico City, *J. Clin. Microbiol.,* 6, 591, 1977.
22. **Morse, E. V.,** Canine Brucellosis — a review of the literature, *J. Am. Vet. Med. Assoc.,* 119, 304, 1951.
23. **Deyoe, B. L.,** Studies on the pathogenesis of a canine abortion agent *(Brucella canis)* in dogs and other domestic animals, Ph.D. thesis, Iowa State University, Ames, 1970.
24. **Kormendy, B. and Nagy, G. Y.,** The supposed involvement of dogs carrying *Brucella suis* in the spread of swine brucellosis, *Acta Vet. Acad. Sci. Hung.,* 30, 3, 1982.
25. **Alton, G. G., Jones, L. M., and Pietz, D. E.** *Laboratory Techniques in Brucellosis,* 2nd ed., World Health Organization, Geneva, 1975, 149.
26. **Carmichael, L. E.,** Canine brucellosis: isolation, diagnosis, transmission, in *Proc. 71st Annu. Meet. U.S. Livestock Sanitary Assoc.,* U.S. Livestock Sanitary Association, Richmond, VA, 1968, 517.
27. **Berman, D. T., Wilson, B. L., Moreno, E., Angus, R. D., and Jones, L. M.,** Characterization of *Brucella abortus* soluble antigen employed in immunoassay, *J. Clin. Microbiol.,* 11, 355, 1980.
28. **Stemshorn, B. and Nielsen, K.,** The bovine immune response to *B. abortus:* water soluble antigen precipitated by sera of some naturally infected cattle, *Can. J. Comp. Med.,* 41, 152, 1977.
29. **Myers, D. M., Jones, L. M., and Varela-Diaz, V. M.,** Studies of antigens for complement fixation and gel diffusion tests in the diagnosis of infections caused by *Brucella ovis* and other *Brucella, Appl. Microbiol.,* 23, 894, 1972.
30. **Zoha, S. J. and Carmichael, L. E.,** Serological responses of dogs to cell wall and internal antigens of *Brucella canis (B. canis), Vet. Microbiol.,* 7, 35, 1982.

31. **Myers, D. M., Varela-Diaz, V. M., and Coltori, E. A.,** Comparative sensitivity of gel diffusion and tube agglutination tests for detection of *Brucella canis* antibodies in experimentally infected dogs, *Appl. Microbiol.*, 28, 1, 1974.
32. **George, L. W. and Carmichael, L. E.,** Development of a rose bengal-stained plate-test antigen for the rapid diagnosis of *Brucella canis* infection, *Cornell Vet.*, 68, 530, 1978.
33. **Carmichael, L. E. and Joubert, J. C.,** unpublished data, 1985.
34. **Hoyer, B. H. and McCullough, N. B.,** Homologies of deoxyribonucleic acids from *Brucella ovis,* canine abortion organisms, and other *Brucella* species, *J. Bacteriol.*, 96, 1783, 1968.
35. **Verger, J.-M., Grimont, F., Grimont, P. A. D., and Grayon, M.,** *Brucella,* a monospecific genus as shown by deoxyribonucleic acid hybridization, *Int. J. Syst. Bacteriol.*, 35, 292, 1985.
36. **Corbell, M. J. and Brinley-Morgan, W. J.,** Genus *Brucella* (Meyer and Shaw 1920, 173[AL]), in *Bergey's Manual of Systematic Bacteriology,* Vol. 1, Kreig, N. R. and Holt, S. D., Eds., Williams & Wilkins, Baltimore, 1984, 377.
37. **Carmichael, L. E.,** Brucellosis *(Brucella canis),* in *CRC Handbook Series in Zoonoses,* Vol. 2, Steele, J. H., Ed., CRC Press, Boca Raton, FL, 1979, 185.
38. **Greene, C. E. and George, L. W.,** Canine brucellosis, in *Clinical Microbiology and Infectious Diseases of the Dog and Cat,* Greene, C. E., Ed., W. B. Saunders, Philadelphia, 1984, 646.
39. **Pollock, R. H. V.,** Canine brucellosis: current status, *Compend. Cont. Educ. Pract. Vet.*, 1, 255, 1979.
40. **Carmichael, L. E. and Kenney, R. M.,** Canine brucellosis: the clinical disease, pathogenesis, and immune response, *J. Am. Vet. Med. Assoc.*, 156, 1726, 1970.
41. **Carmichael, L. E. and Joubert, J. C.,** Transmission *of Brucella canis* by contact exposure, *Cornell Vet.*, 78, 63, 1988.
42. **Serikawa, T. and Muraguchi, T.,** Significance of urine in transmission of canine brucellosis, *Jpn. J. Vet. Sci.*, 41, 607, 1979.
43. **Carmichael, L. E., Zoha, S. J., and Flores-Castro, R.,** Problems in the serodiagnosis of canine brucellosis: dog responses to cell wall and internal antigens of *Brucella canis,* in *Developments in Biological Standardization,* Vol. 56, S. Karger, Basel, 1984, 371.
44. **George, L. W. and Carmichael, L. E.,** Antisperm responses in male dogs with chronic *Brucella canis* infections, *Am. J. Vet. Res.*, 45, 274, 1984.
45. **Serikawa, T., Takada, H., Kondo, Y., Muraguchi, T., and Yamada, J.,** Multiplication of *Brucella canis* in male reproductive organs and detection of autoantibody to spermatozoa in canine brucellosis, in *Developments in Biological Standardization,* Vol. 56, S. Karger, Basel, 1984, 295.
46. **Henderson, R. A., Horlein, B. F., Kramer, T. T., and Meyer, M. E.,** Discospondylitis in three dogs infected with *Brucella canis,* J. Am. Vet. Med. Assoc., 165, 451, 1974.
47. **Saegusa, J., Ueda, K., Goto, Y., and Fujiwara, K.,** Ocular lesions in experimental canine brucellosis, *Jpn. J. Vet. Sci.*, 39, 181, 1978.
48. **Reike, J. A. and Rhodes, H. E.,** Brucella canis isolated from the eye of a dog, *J. Am. Vet. Med. Assoc.*, 165, 1734, 1975.
49. **Gleiser, C. A., Sheldon, W. G., Van Hoosier, G. L., and Hill, W. A.,** Pathologic changes in dogs infected with *Brucella* organism, *Lab. Anim. Sci.*, 21, 540, 1971.
50. **Swenson, R. M., Carmichael, L. E., and Cundy, K. P.,** Human infection with *Brucella canis, Ann. Intern. Med.*, 76, 435, 1972.
51. Center for Disease Control, Human *Brucella canis* infection, in Brucellosis Surveillance Annual Summary, U.S. Department of Health, Education, and Welfare, Washington, D. C., 1966 through 1972, 1974.
52. **Munford, R. S., Weaver, R. E., Patton, C., Feeley, J. C., and Feldman, R. A.,** Human disease caused by *Brucella canis* — a clinical and epidemiologic study of two cases, *JAMA*, 231, 1267, 1975.
53. **Blankenship, R. M. and Sanford, J. P.,** *Brucella canis* — a cause of undulant fever, *Am. J. Med.*, 59, 424, 1975.
54. **Spink, W. W. and Morisset, R.,** Epidemic canine brucellosis due to a new species: *Brucella canis, Trans. Am. Clin. Climatol. Assoc.*, 81, 43, 1970.
55. **Polt, S. S., Dismukes, W. E., Flint, A., and Shaefer, J.,** Human brucellosis caused by *Brucella canis, Ann. Intern. Med.*, 97, 717, 1982.
56. **Rifkin, G. D., Supena, R. B., and Axelson, J. A.,** Case report: *Brucella canis* bacteremia: a case with negative *B. canis* agglutinins, *Am. J. Med. Sci.*, 276, 113, 1978.
57. **Lewis, G. E. and Anderson, J. K.,** The incidence of *Brucella canis* antibodies in sera of military recruits, *Am. J. Public Health,* 63, 204, 1973.
58. **Weber, A.,** Seroepidemiologic investigations on the incidence of *Brucella canis* antibodies in man, *Zentralbl. Bakteriol. Parasitenkd. Infektionskr. Hyg. Abt. 1 Orig. Reihe A,* 238, 237, 1977.
59. **Percy, D. H., Egwu, I. N., and Johas, A. M.,** Experimental *Brucella canis* infection in the monkey *(Macaca arctoides), Can. J. Comp. Med.*, 36, 221, 1972.
60. **Pickerill, P. A.,** Canine Brucellosis: Serological Host Range, and Epidemiological Studies, Ph.D. thesis, Cornell University, Ithaca, New York, 1970.

61. **Moore, J. A. and Gupta, B. N.,** Epizootiology, diagnosis, and control of *Brucella canis, J. Am. Vet. Med. Assoc.,* 156, 1737, 1970.
62. **Pickerill, P. A.,** Comments on epizootiology and control of canine brucellosis, *J. Am. Vet. Med. Assoc.,* 156, 1741, 1970.
63. **Hoff, G. L., Bigler, W. J., Trainer, D. O., Debbie, J. G., Brown, G. M., Winkler, W. G., Richards, S. H., and Reardon, M.,** Survey of selected carnivore and opossum serums for agglutinins to *Brucella canis, J. Am. Vet. Med. Assoc.,* 165, 830, 1974.
64. **Wada, T., Handa, S., and Mohri, S.,** Serological survey on agglutinins to *Brucella canis* in dogs of the Kyushu district, *Jpn. J. Vet. Sci.,* 41, 339, 1979.
65. **Flores-Castro, R. and Carmichael, L. E.,** unpublished data, 1983—1986.
66. **Kolar, J.,** General discussion, in *Developments in Biological Standardization,* Vol. 31, S. Karger, Basel, 1976, 247.
67. **Peres, J. N., Godoy, A. M., Barg, L., and Costa, J. O.,** Isolamento de *Brucella canis* de carrapatos *(Rhipicephalus sanguineus), Arg. Esc. Vet. Univ. Fed. Minas Gerais Belo Horizonte,* 83, 51, 1981.
68. **George, L. W., Duncan, J. R., and Carmichael, L. E.,** Semen examination in dogs with canine brucellosis, *Am. J. Vet. Res.,* 40, 1589, 1979.
69. **Moore, J. A. and Kakuk, T. J.,** Male dogs naturally infected with *Brucella canis, J. Am. Vet. Med. Assoc.,* 155, 2034, 1969.
70. **Huddleson, I. F.,** The mucoid phases of the genus *Brucella, Am. J. Vet. Res.,* 7, 5, 1946.
71. **Zoha, S. J. and Carmichael, L. E.,** Properties of *Brucella canis* surface antigens associated with colonial mucoidness, *Cornell Vet.,* 71, 428, 1981.
72. **Zoha, S. J. and Carmichael, L. E.,** Properties of cell wall antigens of virulent *Brucella canis* and a less mucoid variant of reduced pathogenicity, *Am. J. Vet. Res.,* 43, 171, 1982.
73. **Flores-Castro, R. and Carmichael, L. E.,** Canine brucellosis: current status of methods for diagnosis and treatment, in *Proc. 27th Gaines Symposium,* White Plains, N.Y., 1977, 17.
74. **Badakhsh, F. F., Carmichael, L. E., and Douglass, J. A.,** Improved rapid slide agglutination test for presumptive diagnosis of canine brucellosis, *J. Clin. Microbiol.,* 15, 286, 1982.
75. **Carmichael, L. E. and Joubert, J. C.,** A rapid slide agglutination test for the serodiagnosis of *Brucella canis* infection that employs a variant (M−) organism as antigen, *Cornell Vet.,* 77, 3, 1987.
76. **Myers, D. M. and Varela-Diaz, V. M.,** Serological and bacteriological detection of *Brucella canis* infection of stray dogs in Moreno, Argentina, *Cornell Vet.,* 258, 1980.
77. **Weber, A. and Hussein, N. A.,** The indirect immunofluorescence test for detection of *Brucella canis* infection in beagles, *Zentralbl. Veterinaermed. Reihe B,* 24, 151, 1976.
78. **Weber, A. and Krauss, H.,** The complement-fixation test for the detection of *Brucella canis* infections in dogs, *Zentralbl. Veterinaermed. Reihe B,* 24, 746, 1977.
79. **Nicoletti, P. and Chase, A.,** An evaluation of methods to diagnose *Brucella canis* infection in dogs, *Compend. Cont. Educ. Pract. Vet.,* 9, 1071, 1987.
80. **Weber, A.,** Serological demonstration of Brucella canis infections by comparative slow agglutination at 37°C and 50°C, *Zentralbl. Veterinaermed. Reihe B,* 27, 412, 1980.
81. **Nicoletti, P. and Chase, A.,** The use of antibiotics to control canine brucellosis, *Compend. Cont. Educ. Pract. Vet.,* 9, 1063, 1987.
82. **Tabatabai, L. B., Deyoe, B. L., and Ritchie, A. E.,** Isolation and characterization of toxic fractions from *Brucella abortus, Infect. Immun.,* 26, 668, 1979.
83. **Dillon, A. R. and Henderson, R. A.,** *Brucella canis* in a uterine stump abscess in a bitch, *J. Am. Vet. Med. Assoc.,* 178, 987, 1981.
84. **Jennings, P. B., Crumrine, M. H., Lewis, G. E., Jr., and Fariss, B. L.,** The effect of a two-state antibiotic regimen on dogs infected with *Brucella canis, J. Am. Vet. Med. Assoc.,* 164, 513, 1974.
85. **Flores-Castro, R. and Carmichael, L. E.,** *Brucella canis* infection in dogs: treatment trials, *Rev. Latinoam. Microbiol.,* 23, 75, 1981.
86. **Terakado, N., Ueda, H., Sugawara, H., and Isayama, Y.,** Drug susceptibility of *Brucella canis* isolated from dogs, *Jpn. J. Vet. Sci.,* 40, 291, 1978.
87. **Zoha, S. J. and Walsh, R.,** Effect of a two-stage antibiotic treatment regimen on dogs naturally infected with *Brucella canis, J. Am. Vet. Med. Assoc.,* 180, 1474, 1982.
88. **Alton, G. G., Jones, L. M., Angus, R. O., and Verger, J. M.,** INRA, Paris, 1988, 169.

Chapter 15

BRUCELLA OVIS

Jose M. Blasco

TABLE OF CONTENTS

I. INTRODUCTION

Brucella ovis causes a chronic disease of sheep characterized by testicular alterations and subsequent low fertility of rams and occasional abortions in ewes. In addition, increased perinatal mortality is frequently reported in infected flocks. Although there are no exact data it has been estimated by simulation models that the disease causes substantial economic losses to the sheep industry in infected flocks without control programs.[1]

The disease has been reported in New Zealand, Australia, the U.S., Argentina, the U.S.S.R., Czechoslovakia, Romania, Hungary, France, Germany, Spain, Canada, Mexico, Uruguay, Peru, Chile, Brazil, and South Africa, and probably occurs in other sheep-raising countries.

II. ETIOLOGY

The causative microorganism was first isolated in New Zealand by McFarlane et al.[2] and in Australia as a brucella-like organism by Simmons and Hall.[3] It was initially considered to be a stable mutant of *Brucella melitensis*.[4] Because of the presence of surface antigens common with rough strains of *B. abortus* and *B. melitensis,* Buddle[5,6] proposed the names of *Brucella ovis* as a new species, or *Brucella brucei* var. *ovis australasiae* as a new variety. The same author named the new disease ''infectious ram epididymitis''. Although the microorganism resembled *Neisseria*[7] and *Haemophilus,*[8] it was soon shown that there was an extensive cross-reactivity with the soluble antigens of *Brucella melitensis*.[9] In addition, 94% similarity in polynucleotide sequences of DNA from *B. ovis* and *B. suis* was observed.[10] Thus, the micoorganism was included in the genus *Brucella.*

On the basis of DNA homology, it has recently been proposed that only one species *(B. melitensis)* be recognized in the genus *Brucella,*[11] using specific epithets in a vernacular form to refer to biovars (e.g., *B. melitensis* biovar *ovis*). This proposal has yet to be accepted by the taxonomy subcommittee.

B. ovis are non-sporeforming and nonencapsulated Gram-negative bacilli or coccobacilli, 0.7 to 1.2 μm long by 0.5 to 0.7 μm wide, that are also stained[12] by Stamp's method. *B. ovis* grows on basal media (trypticase soy agar, blood agar base, or Columbia agar) enriched with 5 to 10% sera or blood. In addition, it can be selectively isolated on Thayer-Martin's modified medium.[13,14] (See Appendix for preparation.) It requires an atmosphere of 10 to 20% CO_2 for primary isolation, although CO_2-independent strains can be isolated. Colonies are generally visible after 3 to 5 d of incubation at 34 to 37°C. It also grows at 27°C, but colonies appear in 6 to 10 d.[15]

The organism does not cause hemolysis on blood-containing media. Colonies are circular, convex, with entire edge, being always of nonsmooth type when examined by oblique illumination, and positive in the acriflavine test.[14] In contrast to most *Brucellae, B. ovis* lacks urease activity and fails to reduce nitrate to nitrite. The organism is catalase positive, oxidase negative, does not produce H_2S, and, although it does not grow in the presence of methyl violet, it usually grows in the presence of standard concentrations of basic fuchsin and thionin. *B. ovis* oxidizes L-alanine, D-alanine, L-asparagine, D-asparagine, L-glutamic acid, DL-serine, and adonitol. It does not oxidize L-arabinose, D-galactose, D-glucose, D-ribose, *meso*-erythritol, D-xylose, L-arginine, DL-citruline, DL-ornithine, or L-lysine. The cultures are not lysed by *Brucella*-phages of the Tb, Wb, M51-S708, Fz, BK, MC/75, D, or R groups at the routine test dilution (RTD) or $10^4 \times$ RTD. However, bacteria are lysed by phage R/O at both RTD or $10^4 \times$ RTD. No biotypes are recognized.[12]

Most laboratories are not prepared for complete identification, and a practical schedule for presumptive identification is needed. In our laboratory, the vast majority of *B. ovis* isolates were correctly identified on the basis of colonial morphology by direct observation, using obliquely transmitted light, Gram staining, catalase, oxidase, urease, and acriflavine tests.

III. EPIDEMIOLOGY

A. INFECTION IN SHEEP

Although the incidence varies, *B. ovis* infection is a worldwide disease of sheep-producing countries. When the disease is first reported in a country, the incidence is uniformly high, with 20 to 60% of rams infected, and 45 to 75% of the flocks affected.[15-17] In countries with advanced control programs, the incidence is much lower but complete eradication is extremely difficult to achieve.

From the literature concerning natural transmission of the disease, most authors agree that passive venereal transmission via the ewe is the main route.[5,18] Ram-to-ram transmission can also occur by direct contact or after housing healthy rams in pens previously occupied by infected ones.[19] Sodomy could be another means of transmission,[15,19] and shearing wounds may also be implicated.[19]

In contrast to rams, ewes appear to be relatively resistant. After being mated by infected rams, only a few ewes develop an active infection leading to abortion and dead or weak lambs.[5,18,19] A similar picture has been described in experimental infection of pregnant ewes.[20,21] Carryover of infection from one pregnancy to the next is quite infrequent.[18,19] In addition, lambs born to infected ewes seldom develop active infection, and they do not do so even after nursing on contaminated milk.[15] Altogether, these findings demonstrate that the role of ewes in the active transmission of infection is less important.

Most outbreaks of the disease occur immediately after the mating season. Undoubtedly, complex and not fully understood mechanisms of transmission exist, but passive ram-to-ram venereal transmission via the ewe appear to be the most important way for maintaining and spreading the disease. The probability of infection depends mainly on the route of infection, the infecting dose, and some intrinsic characteristics of the animals, such as age and breed.

Under experimental conditions, successful infection of rams can be achieved by a variety of routes: orally, intravenously, intratesticularly, conjunctivally, intrapreputially, subcutaneously, through skin lesions, intrarectally, and intranasally.[15,22,23] Although no extensive comparative studies have been carried out, the best results have been obtained conjunctivally, intrapreputially, or by using both routes simultaneously.[24,25] Doses of 5×10^8 to 10^9 colony forming units (CFU) of *B. ovis* are sufficient to achieve rates of infection close to 100%.[24-27]

Whether age affects the susceptibility to infection is a matter of controversy. The infection has been demonstrated in 4-month-old rams, suggesting that animals at, or soon after puberty, are susceptible to *B. ovis*.[28] However, since venereal transmission seems to be the chief method of spreading, adult animals are clearly more liable to be naturally infected. Moreover, it has been repeatedly observed that the incidence of both testicular alterations and brucellosis increases with age[29-33] (Figure 1) and relates to the sexual experience of the animals.[34] There is no published information on the effect of age on susceptibility under experimental conditions.

There are many references suggesting that susceptibility may vary among breeds of rams. The Merino breed appears to be less frequently infected than British breeds in the same environment.[31,35-40] The same observation has been made in comparing imported with native breeds. On a statistical basis, Spanish native and Merino-derived breeds are more resistant to ovine brucellosis than other imported European breeds[41] (Figure 2). Although genetic resistance to the disease could be important, it has been suggested that susceptibility could also relate to differences in growth rates and sexual precocity and activity.[31]

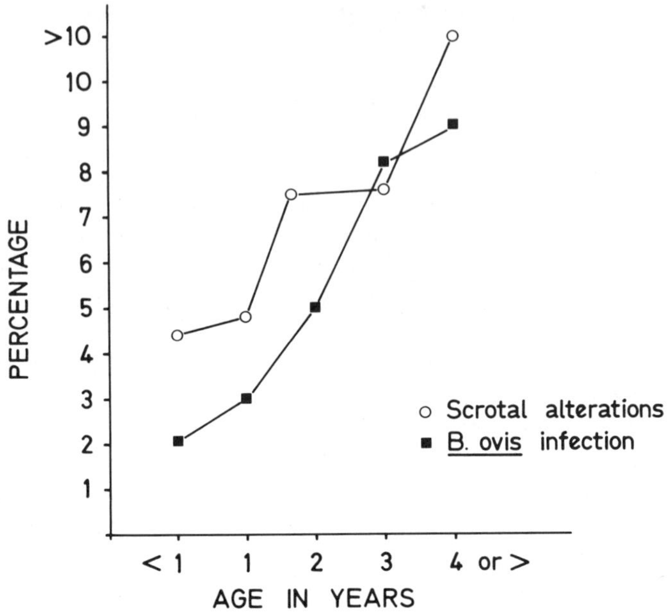

FIGURE 1. Influence of age on the incidence of palpable scrotal abnormalities and *B. ovis* infection in 2398 rams examined. (Adapted from Blasco.[33])

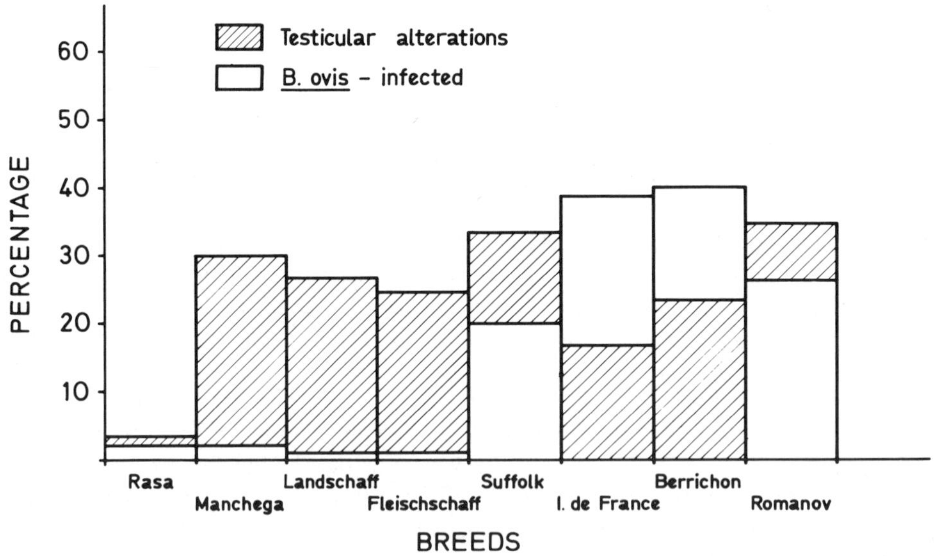

FIGURE 2. Percentage of rams from different breeds showing testicular alterations and brucellosis under similar management conditions in Spain. (Rasa and Manchega are native breeds of Spain and the other breeds are imported from European countries; adapted from Marin et al.[41])

B. INFECTION IN SPECIES OTHER THAN SHEEP

B. ovis infection appears to be exclusively an infection of sheep and affects essentially rams. Although some serological surveys have suggested human contagion exposure,[15] there are no reports on the isolation of *B. ovis* from man.

While there are references on experimental infections of animals other than sheep, few clear-cut results are available. Experimental inoculation of *B. ovis* in male goats[42-44] leads to genital and extragenital colonization in some animals, and subsequent development of lesions similar to those observed in rams. In extensive management systems, goats and sheep frequently cohabit and, therefore, transmission from sheep to goats and vice versa could happen. However, isolation of *B. ovis* from natural cases of infection in goats has never been reported. The infection has been also reproduced in wild deer.[45]

Small laboratory animals have been infected by a variety of routes with doses varying from 10^4 to 10^{11} CFU with variable success.[15,46] Although rabbits, rats, gerbils, hamsters, mice, and guinea pigs have been used, a satisfactory laboratory animal model has not been found for *B. ovis* infection.

IV. PATHOGENESIS

As in infections due to other *Brucella,* there is a relatively long latency period before clinical signs appear.[47] Under experimental conditions, the bacteria remain confined to the lymph nodes close to entry sites for 2 to 3 weeks, and then bacteremia develops, followed by infection of reticuloendothelial organs, lymph nodes distant from entry sites, genital organs, and accessory sexual glands. In experimentally infected rams, bacteria have been isolated from the liver, kidney, spleen, testicles, epididymis, vesicular glands, bulbourethral glands, ampullae and iliac, prescapular, precrural, submaxillary, parotid, and retropharyngeal lymph nodes.[25-27] However, the target organs are the epididymis and accessory sexual glands. For this reason, semen excretion of the microorganism occurs in most (but not all) infected rams. In seropositive animals giving consistently negative semen cultures, the bacteria might be located in organs other than those that are a source of semen components. This hypothesis is consistent with the isolation of *B. ovis* from the spleen and iliac lymph nodes of some experimentally infected animals in which the bacteria could not be isolated from genital organs and accessory sexual glands.[25,27]

Initial localization in the epididymis is accompanied by perivascular edema and infiltration of lymphocytes, monocytes, and neutrophils into peritubular tissue. Subsequently, the inflamed tubular epithelium develops papillary hyperplasia and local hydropic degeneration with subsequent formation of intraepithelial cysts. Eventual destruction of epithelium, either by bacteria or by inflammatory reaction, leads to an extravasation of spermatozoa. Host response to extravasated spermatozoa leads to formation of large spermatic granulomas (Figure 3), which may cause complete blockage of the epididymis with subsequent testicular degeneration and fibrosis.[47]

The pathogenesis of *B. ovis* in ewes is not fully explained. Ewes experimentally exposed to *B. ovis* either before mating or in late pregnancy do not abort. Only ewes exposed at early or midpregnancy develop infection and could eventually abort[20,47] As the infection progresses, the bacteria become localized in the placenta and reach the fetus through chorion vessels.[15,21] Although abortion is not frequent, infected ewes develop placentitis causing fetal malnutrition which results in weak lambs.[15,20] The immunological and granulomatous inflammatory responses of *B. ovis*-infected fetuses are similar to those observed in *B. abortus*-infected bovine fetuses.[47]

V. PATHOLOGY

A. MACROSCOPIC LESIONS IN RAMS

Although *B. ovis* causes epididymitis, there are infected rams that do not develop palpable epididymitis. In a survey conducted on 267 seropositive rams, only 125 (46.8%) showed palpable testicular alterations.[33] Even though the number of animals showing actual testicular alterations is increased after histological examination of testes and epididymis, an important proportion of *B. ovis*-infected rams do not show scrotal lesions. In most cases, testicular

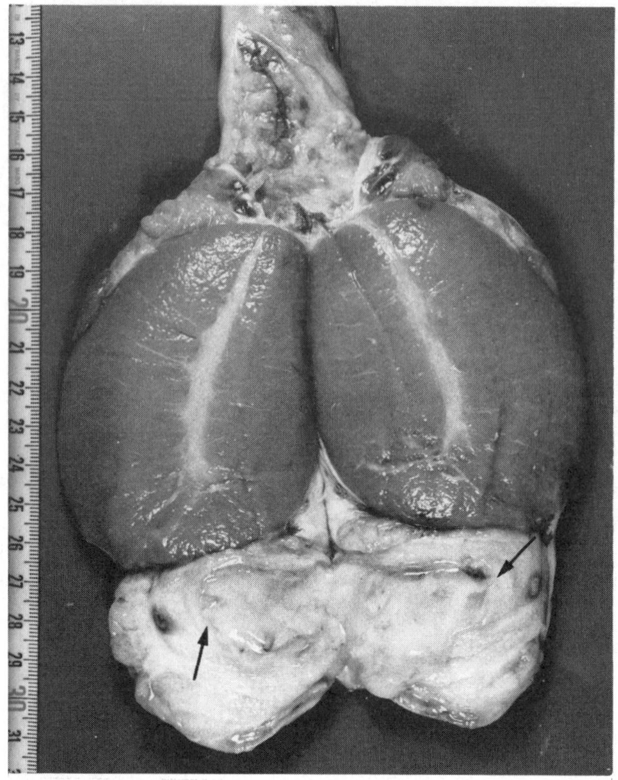

FIGURE 3. Testicle and epididymis of a *B. ovis*-infected ram. Note the large spermatic granuloma in tail epididymis (arrow).

localization is unilateral (Figure 4), with the tail of epididymis being the most common site. The head and body of the epididymis are also frequently affected (Figure 5).

Testicular atrophy and a variable degree of enlargement of the tail of the epididymis are characteristics of the chronic phase of the disease. The macroscopic appearance of the testes is usually normal, but granulomas and calcification may be apparent on the cut surface.[48]

The affected epididymis appears firm, showing a white cut surface as a consequence of connective tissue proliferation. One or more abscesses like spermatoceles filled with creamy or caseous substances are frequently found in the thick connective tissue.

Hemorrhage and exudative inflammation in the tunica vaginalis are frequent findings, as a result of rupture of the basic lesion (spermatocele) of the epididymis. The organization of this exudate leads to the formation of adhesions between the two layers of the tunica vaginalis.

The seminal vesicles are frequently enlarged, with dilated or fluid-filled ducts observed on the cut surface. No macroscopic changes are usually observed in the bulbourethral glands, prostate, and ampullae.

B. HISTOLOGICAL LESIONS OF RAMS

Microscopic examination of affected epididymides shows interstitial edema, fibrosis, and perivascular infiltrates of lymphocytes and plasma cells. Epididymal ducts show epithelial hyperplasia with intraepithelial cysts containing neutrophils and cellular debris (Figure 6). Spermatic granulomas made up of extravasated sperm surrounded by lymphocytes, epitheliod, and giant cells are also frequent findings.[15,25,27,48-50]

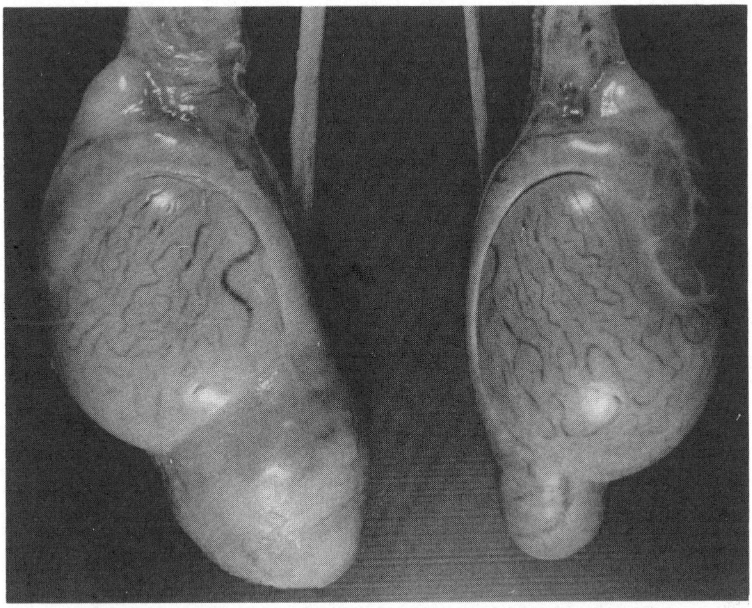

FIGURE 4. Characteristic unilateral tail epididymitis of *B. ovis*-infected rams.

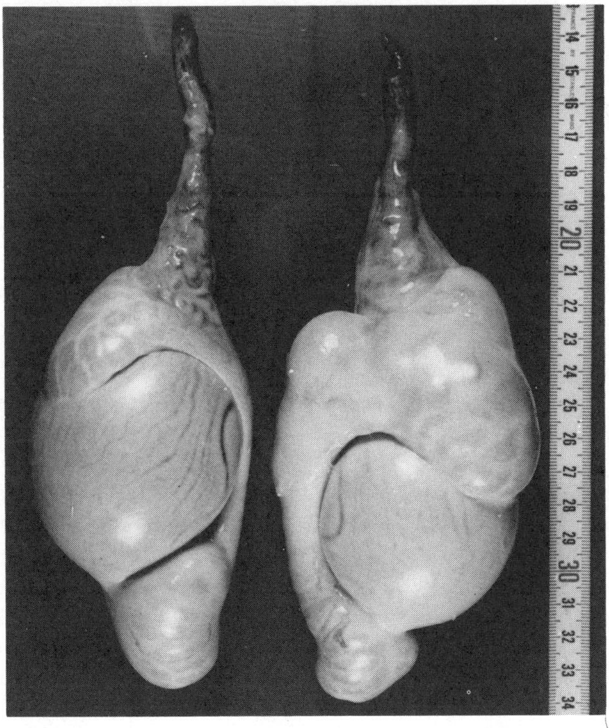

FIGURE 5. Bilateral epididymitis in a *B. ovis* — infected ram. Note the head and body epididymitis in the right testicle.

Testicular atrophy derives from regressive processes in the testicular epithelium and cessation of spermatogenesis. Proliferation of intertubular connective tissue, small extratubular spermatic granulomas, and necrosis and calcification of seminiferous ducts may also occur.[27,48]

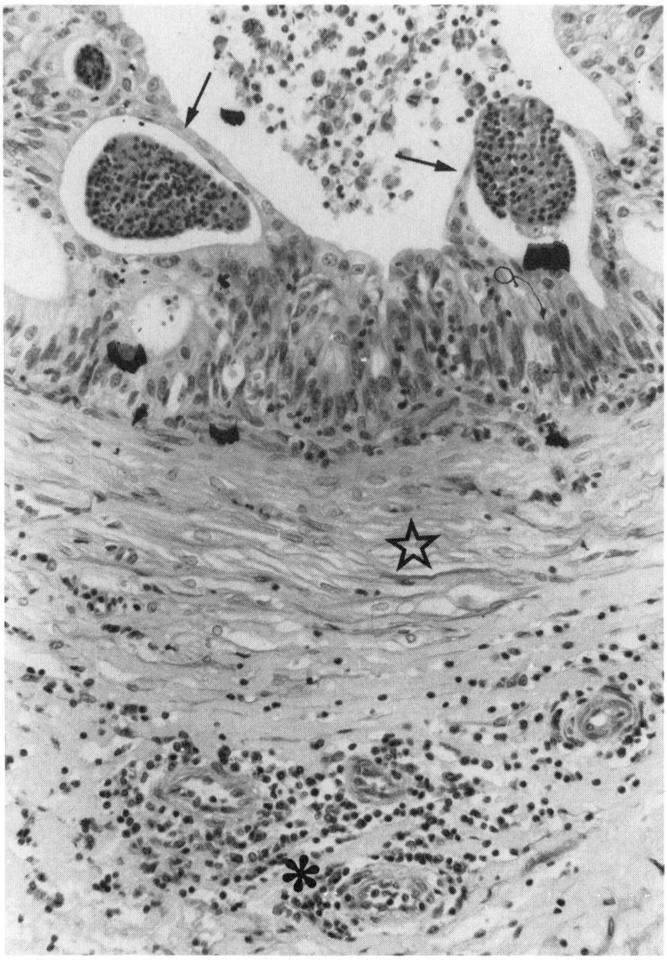

FIGURE 6. Characteristic microscopic lesions in the epididymis of *B. ovis*-infected rams. Note the epithelial hyperplasia with intraepithelial cysts (↓), connective hyperplasia (☆), and perivascular infiltrates of round cells (*). (Hematoxylin-eosin stain, × 250. From Blasco, J. M., Marin, C. M., Barberán, M., Moriyón, I., and Diaz, R., *Vet. Microbiol.*, 14, 381, 1987. With permission.)

Infiltration by both plasma cells and lymphocytes, fibrosis, and widespread epithelial hyperplasia (Figure 7) with intraepithelial cysts containing neutrophils, are the main findings of the altered seminal vesicles.[27,49,51]

Inflammation of ampullae is frequently observed in association with epididymitis. Focal areas of hyperplasia with empty intraepithelial cysts and accumulation of neutrophils in the lumen of the sack-like dilatations of epithelium, are frequently seen (Figure 8). Fibrosis and accumulation of round cells are observed in the lamina propia of the septa.[51,52]

Microscopic examination of the prostate and bulbourethral glands reveals slight infiltration of round cells and focal glandular hypertrophy. Papilliform formations and concretions *(corpora amylacea)* are also frequent in both accessory glands.[51,52]

C. PATHOLOGY IN EWES

Infected ewes show a yellowish purulent exudate, varying from a small amount on the surface of an intact chorioallantoic membrane to an abundant amount that fills interplacentomae areas. The exudate contains bacteria, macrophages, neutrophils, and desquamated

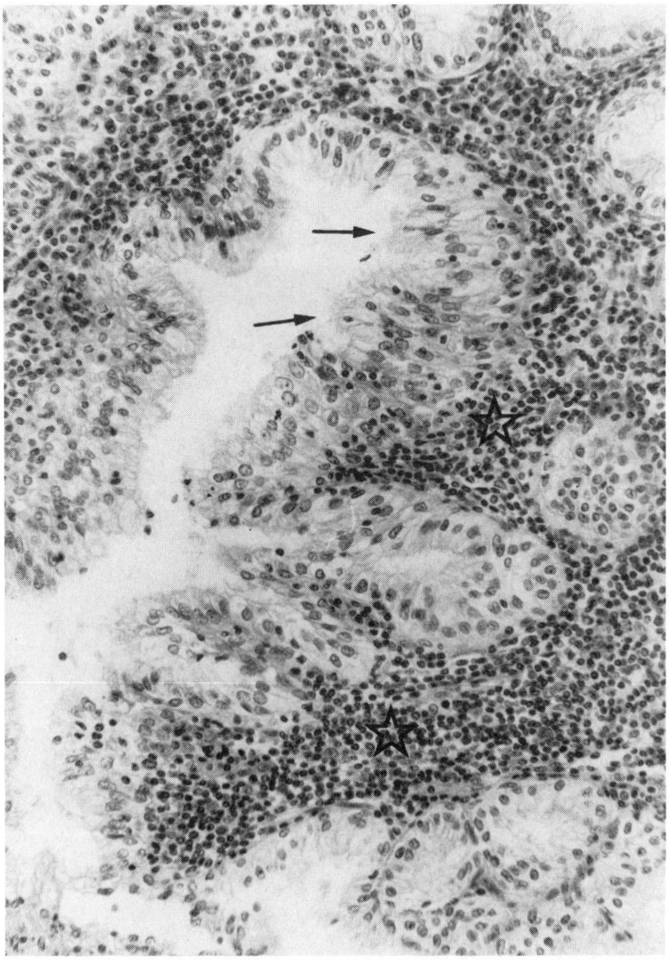

FIGURE 7. Seminal vesicle of a *B. ovis* - infected ram showing epithelial hyperplasia (↓) and widespread infiltrates of round cells (☆). (Hematoxylin-eosin; magnification × 250.)

epithelial cells. Fibrosis, thickening and edema of the chorioallantoic membrane are frequently observed in severe cases. The cotyledons may show various stages of necrosis. The chorionic epithelial cells may be enlarged and contain bacteria. Focal necrosis of intercotyledonary and chorionic epithelium are frequently observed. Arterial lesions are also quite common, with vessels filled with fibrinous thrombi, enlarged endothelial cells, and proliferative cells of endothelium of the tunica intima.[15]

No pathognomonic lesions have been recognized for *B. ovis* infection.

VI. DIAGNOSIS

A. CLINICAL DIAGNOSIS

The existence of other bacteria causing clinical epididymitis makes it extremely difficult to diagnose *B. ovis* infection on the basis of testicular palpation exclusively. The most frequently reported microbial causes of epididymitis in rams include *Actinobacillus seminis, Actinobacillus actinomycetem-comitans, Histophilus ovis, Haemophilus* spp., *Corynebacterium pseudotuberculosis ovis, Chlamydia psitacii,* and others.[34,53-59]

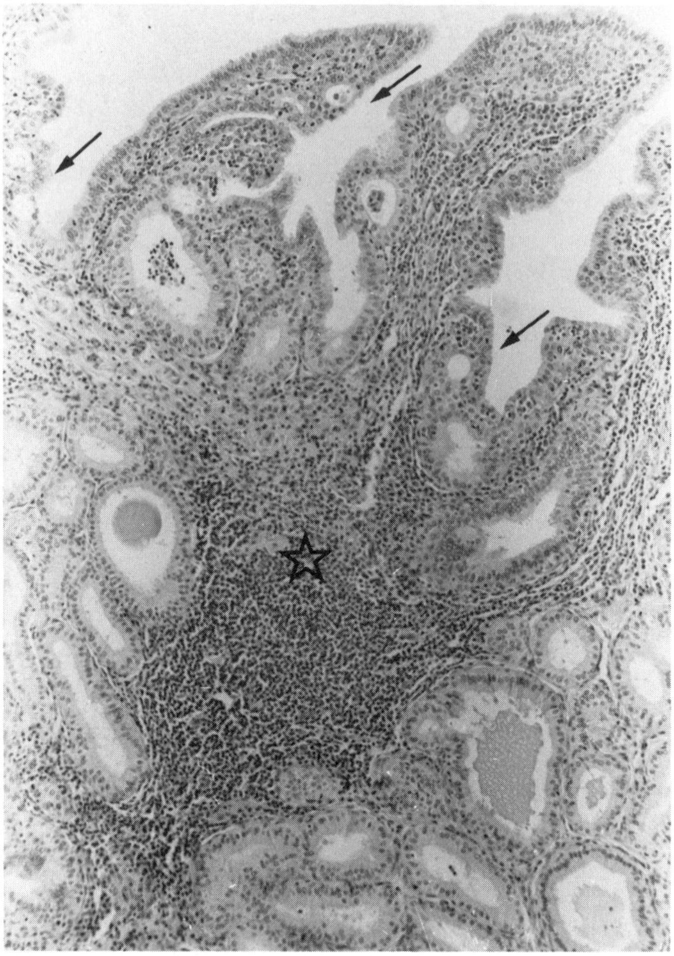

FIGURE 8. Characteristic histological lesions in the ampullae of *B. ovis*-infected rams. Note the focal areas of epithelial hyperplasia (↓) and the infiltration of plasma cells (☆). (Hematoxylin-eosin; magnification × 100.)

In addition, infected animals showing palpable epididymitis can often return to normal clinical palpations in a few weeks, although histological examination would demonstrate lesions in the epididymis or in accessory sexual glands.[25,33]

Infected rams show poor semen quality and lowered fertility.[60-62] Both the total concentration and the proportion of normal living spermatozoa are spectacularly reduced in rams with testicular alterations[60,62] (Figure 9). Even in the absence of palpable epididymitis, infected animals may have poor semen quality. Since clinical symptoms are not always clear and infected rams frequently maintain a residual degree of fertility[61] (Figure 10), it is not possible to reach a definite diagnosis only by palpation of the scrotal contents and examination of fertility scores of rams.

B. BACTERIOLOGICAL EXAMINATION OF SEMEN SAMPLES

The semen of rams can be obtained easily by means of electroejaculation. For bacteriological examination, semen can be collected in plastic bags or by means of swabs taken from the preputial cavity after electroejaculation.

Semen smears can be examined after Gram or Stamp staining and characteristic coccobacilli demonstrated in many infected animals.[63] However, other bacteria with similar mor-

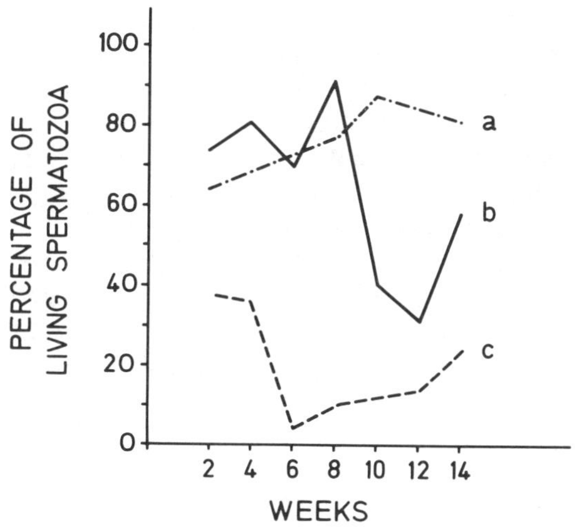

FIGURE 9. Normal living spermatozoa yields of healthy rams (a); *B. ovis*-infected rams showing palpable epididymitis at week 12 (b); and *B. ovis*-infected rams showing epididymitis from the beginning of the experiment (c). (Adapted from Folch et al.[62])

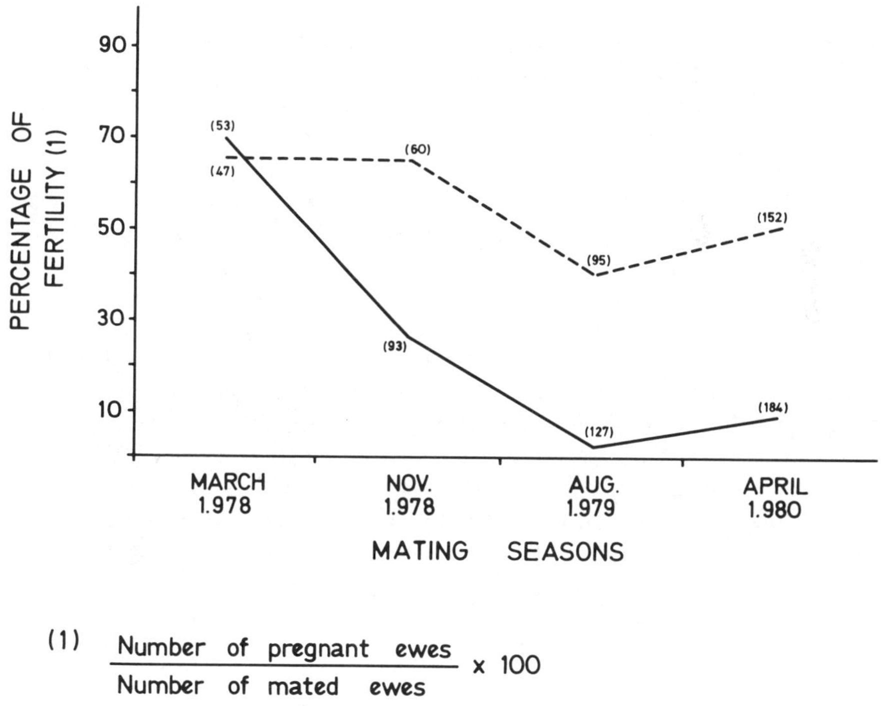

FIGURE 10. Evolution of fertility of four healthy (discontinuous line) and three *B. ovis*-infected rams (continuous line) during four subsequent mating seasons. The infection was serologically detected at the end of the first mating season (March 1978). The numbers in brackets are the total number of mated ewes. (Adapted from Blasco et al.[61])

phology or staining characteristics (*B. melitensis* and *Chlamydia* spp.) can also be present in such samples. Both direct and indirect immunofluorescent methods have been successfully applied to the specific identification of *B. ovis* in semen smears.[64,65] As a consequence of inflammation, neutrophils, and desquamated epithelial cells may also be observed.[63]

For isolation of *B. ovis,* semen samples are smeared directly onto plates of appropriate culture medium (see Etiology or Appendix) and incubated in 10% CO_2. General purpose media are often impractical because of contaminating microorganisms and as a result, the use of selective media, such as freshly prepared Thayer-Martin's medium, is preferred.

Because sampling and transport are cumbersome and because some infected animals show intermittent or no excretion of *B. ovis,*[27,63,66,67] bacteriological examination of semen samples is not a practical and reliable means of diagnosis in large-scale programs.

C. SEROLOGICAL DIAGNOSIS

1. Antigens of *B. ovis*

The cell surface lipopolysaccharide (LPS) is the main surface antigen of most Gram-negative bacteria, and this is also true for smooth brucellae. However, *B. ovis* is permanently rough and, therefore, its cell surface is different from that of smooth *B. abortus, B. melitensis,* or *B. suis.*[9,68-70] The major differences in cell surface structure between smooth and rough bacteria are derived from differences in the corresponding LPSs.

Like the LPS of rough *B. abortus* and *B. melitensis* mutants, the LPS of *B. ovis* can be extracted from whole cells using organic solvents.[71-74] The petroleum ether-chloroform-phenol (PCP) method, originally developed for rough mutants of enterobacteria,[71] is the method of choice to obtain *B. ovis* LPS since it yields preparations essentially free from protein or nucleic acids.[73-75] However, the high hydrophobicity of this purified LPS represents a problem for its use in many serological tests.

The structure of the LPS of *B. ovis* has not been elucidated. Although it contains sugars present also in the core-lipid A of the LPS of other *Brucellae,*[76] purified *B. ovis* LPS gives only a reaction of partial identity with the LPS of rough mutants of *B. abortus* and *B. melitensis* or with the rough LPS present in hot saline extracts (see below) of *B. canis.*[69,76] These observations strongly suggest the presence of species-specific antigenic determinants on the LPS of *B. ovis.*[69]

The cross-reactivity with the core-lipid A determinants of other brucellae and the different state of aggregation of the rough LPSs (disperse molecules) and smooth LPSs (micelles) in aqueous solutions account for some discrepancies observed in the serological tests. For example, in the smooth LPS (micelles) the core-lipid A antigenic determinants are not exposed and, therefore, anti-*B. ovis* LPS antibodies do not precipitate smooth LPS. On the other hand, the cross-reactivity becomes apparent when rough or smooth LPSs are adsorbed to a solid phase, as in many enzyme immunoassays.[25,73] The same parameters may explain why some animals vaccinated with *B. melitensis* Rev. 1 or infected with *B. melitensis* develop precipitin lines with *B. ovis* LPS in gel immunodiffusion[27] (Figure 11). *In vivo* smooth to rough dissociation of *B. melitensis*[77] could also increase the amount of antibody to rough antigenic determinants in some animals.

In addition to the LPS, the outer membrane of *B. ovis* contains three groups of major proteins also present in the outer membrane of other *Brucella* species.[75,78] Preliminary results indicate that the outer membrane proteins are more exposed on *B. ovis* than in *B. melitensis* smooth cells.[75] This observation could explain why, in contrast to *B. melitensis* infection, some outer membrane proteins of *B. ovis* play an important antigenic role in ram epididymitis.[73]

When rough *Brucella* cells are heat extracted with saline (hot saline method, HS), they yield water-soluble antigenic extracts whose major component precipitates with sera to rough brucellae.[69,70] For this reason, the HS extracts have been referred to as the "rough-specific antigen". The chemical characterization of the HS extracts has shown that they are enriched

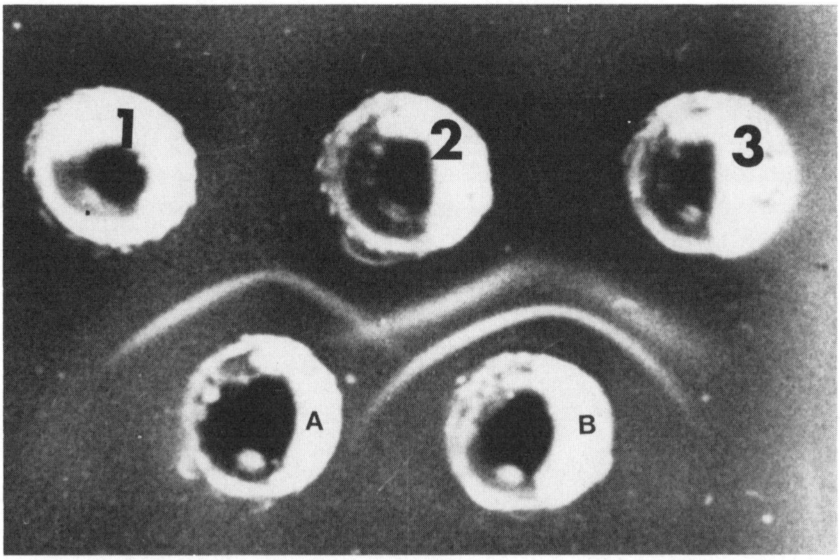

FIGURE 11. Humoral immune response measured by agar gel immunodiffusion to both rough HS (A) and smooth LPS (B) antigens. The serum of a Rev. 1-vaccinated animal (2) gives precipitin lines of exactly like those of *B. ovis* (1) and *B. melitensis* (3) antisera.

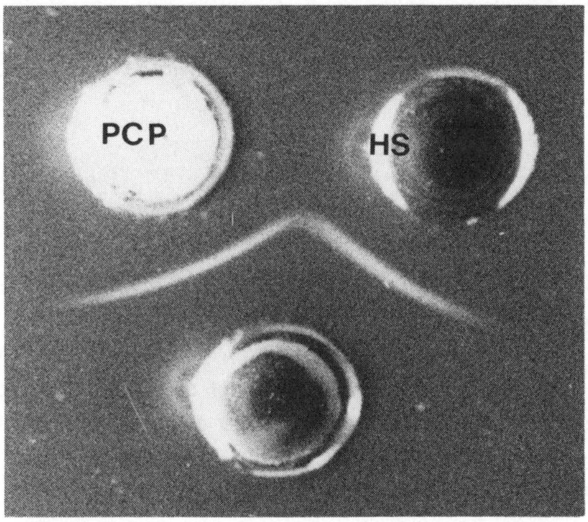

FIGURE 12. Immunological identity of purified rough LPS (PCP) and HS antigens when examined against serum of a ram infected with *B. ovis*.

in rough LPS (Figure 12), group 3 outer membrane proteins, and other outer membrane components.[33,73,75] Thus, HS extracts contain LPS determinants specific for *B. ovis*, but also additional antigenic components (Figure 13), some of them shared with rough and smooth *B. melitensis* and other brucellae.[75,78] Such components account for the cross-reactivity observed with the HS and sera of sheep infected with *B. melitensis* in enzyme immunoassay.[73,75]

The HS extract, or preparations obtained by similar means, has been widely used for the serological diagnosis of *B. ovis* infection. Its water solubility and high content of relevant cell surface antigens explain its activity in serological tests. However, in areas where *B. melitensis* also exists, the specificity of the diagnosis with regard to *B. ovis* has to be interpreted taking into account the results of serological tests for smooth brucellae.[73]

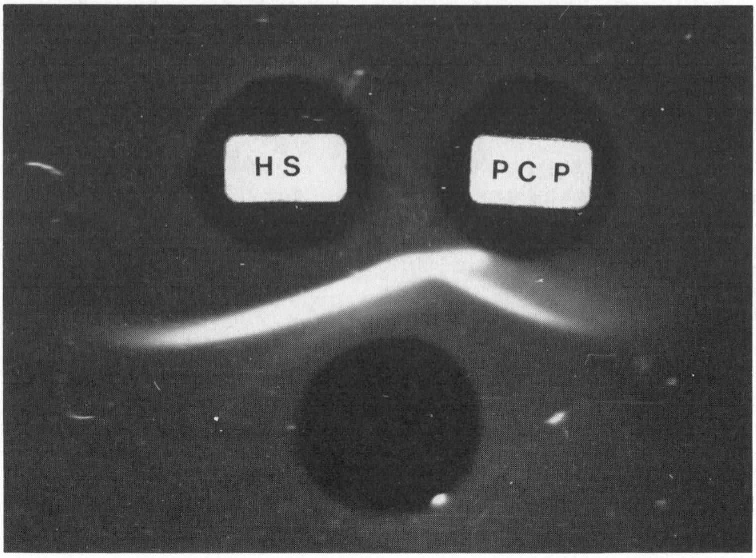

FIGURE 13. Reaction of partial identity between the HS and purified rough LPS (PCP). This reaction is observed only with some sera from *B. ovis*-infected rams.

There is little information available on *B. ovis* antigens not exposed on the cell surface. As far as is known, cytoplasmic antigens are common to all *Brucella* species.[9,68] Moreover, the soluble protein antigens extracted from whole *B. melitensis* 115 cells by the cold saline method[79] are indistinguishable from those obtained from *B. ovis,* at least by double gel diffusion or immunoelectrophoresis with sera to *B. ovis*.[33]

The procedures for extraction of *B. ovis* antigens are presented in the Appendix.

2. Serological Tests

A large variety of techniques have been applied to detect antibodies to *B. ovis,* including agglutination, indirect hemagglutination, indirect immunofluorescence, gel precipitation, complement fixation, and enzyme immunoassay.[63-68,73,74,80-95] These tests differ in practical aspects as well as in sensitivity and specificity. The most efficient and widely used are the double gel diffusion, the complement fixation, and the enzyme-linked immunosorbent assay.

a. Complement Fixation Test (CF)

This is the most widely used test for the serodiagnosis of *B. ovis* infection. Both the sensitivity and the specificity of CF test depend on the antigen employed, the temperature (warm or cold) of the incubation in the first step of the method, and the time of incubation.

B. ovis antigens frequently contain a significant amount of anticomplementary activity which has been associated with the presence of whole bacteria or cellular debris in the preparations.[96] Thus, the antigens used in CF are mostly HS extracts, or extracts of similar characteristics, and sonicated cell-free extracts. The HS extract possesses better sensitivity than sonicated cell-free extracts[90] or than protein-free rough LPS obtained by the petroleum ether-chloroform-phenol method.[25] In addition, HS extracts are very specific since intense cross-reactions are given mostly by other rough-*Brucellae* which do not infect sheep under natural conditions. However, some cross-reactivity in the CF can be seen with sera of sheep naturally infected with *B. melitensis,* or after Rev. 1 vaccination.[25]

With respect to the choice of incubation temperature of the antigen-antibody-complement mixture, in our experience, differences in sensitivity between warm (37°C, 30 to 40 min) or cold (4°C, overnight) fixation are not significant. However, the cold method has been reported to give better sensitivity than the warm method.[89,90] Some reports indicate that the sensitivity of the warm method is improved if longer (60 min) incubation times are used.[88]

In comparative studies, the sensitivity of CF with HS extracts has been found to be similar to that of the double gel diffusion test (see below),[25,80,83,84] and lower than that of the ELISA with the same antigen.[91,93,94]

From a practical standpoint, the CF has some disadvantages that have to be considered before choosing it as the main diagnostic test in eradication programs: complexity, the necessity for serum heat inactivation, anticomplementary activity of some sera, difficulty in performing with hemolyzed sera, and the prozone phenomena.

The methodology of a warm CF test for antibody to *B. ovis* is described in the Appendix.

b. Gel Precipitation Tests

Different antigenic extracts have been used to detect precipitating antibodies to *B. ovis* by double gel diffusion (GD). The test used more often is a micro-slide adaptation of the method of Ouchterlony[97] using HS extract as antigen.[83] This test is easy to perform, and has a sensitivity similar to that of CF, provided the two following precautions are taken.[25,33,80,84] First, every batch of HS extract, if possible freeze-dried, has to be titrated against several positive sera (optimal concentration varies usually from 2.5 to 20 mg/ml).[33] Second, hypertonic NaCl gels have to be used. Gels with 5% NaCl improve the precipitation reaction as compared to gels containing 0.85% NaCl,[80] and with 10% NaCl the sensitivity of the gel diffusion test is optimal for ovine sera.[98]

In general, the antibodies involved in the gel diffusion test are mostly of the IgG class.[70] This could explain the high correlation between CF and gel diffusion results.

Because of its sensitivity, simplicity, and easy interpretation, the gel diffusion test with HS extract can be performed in nonspecialized laboratories. The main disadvantages of this test are that it has not been standardized and that the results can only be qualitatively interpreted.

When used in gel diffusion, soluble protein (SP) extracts prepared by the cold saline method[79] detect only about 54% of rams infected with *B. ovis*. In addition, with this test alone it is not possible to distinguish whether a seropositive animal is infected by *B. ovis* or by *B. melitensis*.

Protein-free rough LPS, such as that extracted by the PCP method from *B. ovis,* is highly insoluble. For double gel diffusion, this insolubility poses a problem and it is not resolved by sonication, addition of triethylamine, or by coupling to BSA.[25,33,73]

Counter-immunoelectrophoresis (CIE) with HS extracts has also been applied to the serological diagnosis of *B. ovis* infection, and the results are similar to those of gel diffusion and CF with the same antigen.[85] The main advantage of CIE is the reduction in the time required for the precipitation to occur. However, the test is more expensive and special equipment and trained personnel are needed. CIE has not been widely employed for diagnostic purposes.

A description of the gel diffusion method is presented in the Appendix.

c. Enzyme-Linked Immunosorbent Assay (ELISA)

Indirect ELISA on polystyrene plates has been applied to the serological diagnosis of *B. ovis* infection by several authors.[73,74,91-95,99] The main differences in the method used are the type of substrate and antigen selected.

The substrates used have been ABTS (2-2′-azino-di-3-ethylbenzthiazoline sulfonate), 5-amino-2-hydroxybenzoic acid, and O-phenylene-diamine hydrochloride. In a comparative study,[25] these three substrates gave satisfactory results, although with ABTS the differences in optical density between the sera from infected and uninfected rams were more pronounced.

As in other tests, both sensitivity and specificity depend mostly on the antigen used. Antigenic extracts obtained after disruption by sonication, HS extracts, and LPS extracted

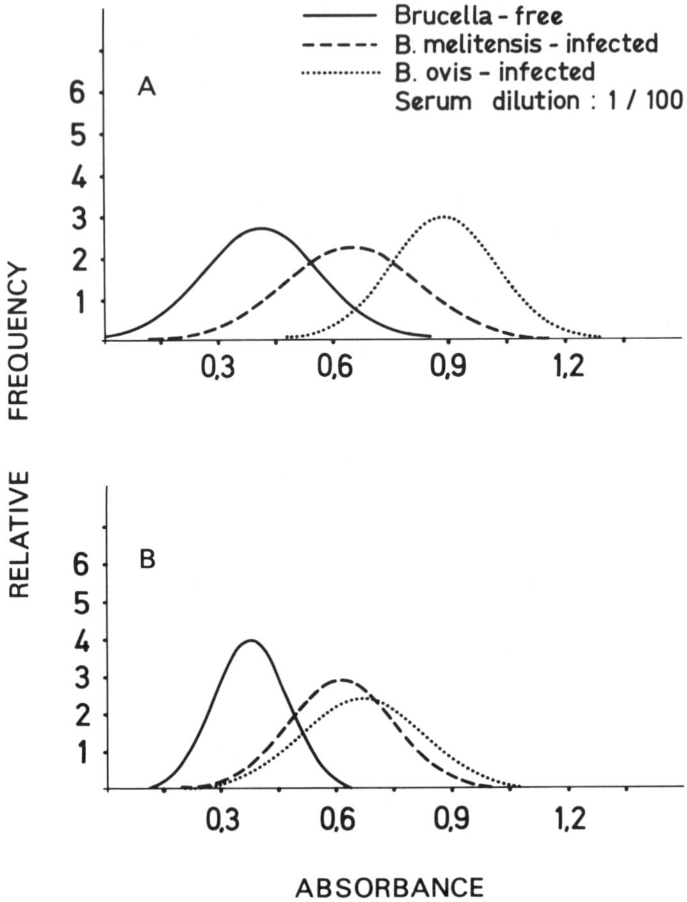

FIGURE 14. Distribution of ELISA OD, with purified rough LPS (PCP) (panel A) and HS (panel B) antigens in *Brucella*-free rams, *B. melitensis*-infected ewes, and *B. ovis*-infected rams. (From Riezu-Boj, J. I., Moriyón, I., Blasco, J. M., Marin, C. M., and Diaz, R., *J. Clin. Microbiol.*, 23, 938, 1986. With permission.)

with petroleum ether-chloroform-phenol have been used. Among them, extracts of the HS type are the most common and they are, in our opinion, the best antigen for ELISA.

Comparative studies have shown that ELISA with HS extracts has better sensitivity than either gel diffusion or CF.[67,92-94] Although rough LPS extracted with petroleum ether-chloroform-phenol also gives good results,[74] it has been shown that ELISA with HS extracts is more sensitive.[73] However, the same study showed that cross-reactivity of sera of *B. melitensis*-infected sheep in ELISA is stronger (Figure 14) with HS extract than with purified rough LPS[73] (see also VI.C.1).

In a study comparing gel diffusion, CF, and ELISA (IgG conjugates) with characterized HS extracts (Table 1), the best individual results were obtained with the ELISA test. However, there were some sera which precipitated HS in gel diffusion but did not give positive readings in the ELISA. Thus, the combination of gel diffusion and ELISA resulted in optimal sensitivity. On the other hand, the combination of CF and ELISA did not improve upon the sensitivity of ELISA alone[25] (Table 2).

Protein extract obtained by the cold saline method can also be used in the ELISA test. With this antigen, 49 of 83 rams infected with *B. ovis* were detected.[25]

The methodology of the ELISA test used in our laboratory is detailed in the Appendix.

TABLE 1
Chemical Composition of *B. ovis* Antigenic Extracts[a]

Antigen	% Protein	% KDO[b]	% RNA	Yield[c]
HS	44—50[a]	0.43	8.6	0.9—2.7
PCP	0.5—0.8	2.75	10.6	1.4—2.2

[a] When included, ranges represent the highest and lowest values obtained in an analysis of five different batches of extracts.
[b] KDO = 2-3-Deoxyoctulosonic acid.
[c] Percentage of cell dry weight.

Adapted from Riezu-Boj et al.[73]

TABLE 2
Comparative Results of Double Gel Diffusion (GD), Complement Fixation (CF), and ELISA Tests with HS Antigen for the Diagnosis of *B. ovis* Infection of Rams

Test	Sensitivity(%)[a]	Specificity(%)[b]
GD	96.4	100
CF	92.7	100
ELISA	97.6	100
GD + ELISA	100	100
CF + ELISA	97.6	100

[a] Calculated using sera from 83 *B. ovis*-culture positive rams.
[b] Calculated using sera from 83 *Brucella*-free rams.

Data from Reference 25.

D. ALLERGIC DIAGNOSIS

As with *B. melitensis* infection, cellular immunity is stimulated in rams by *B. ovis* infection, and the ensuing immune response may be detected by delayed type hypersensitivity (DTH) skin test reaction.[80] Evidence from studies with guinea pigs indicate that this response is not species specific[100] and that, as far as is known, LPS does not elicit DTH in brucellosis. The main antigens involved in DTH are soluble proteins.[80,101]

The cutaneous DTH reaction has been used for the diagnosis of *B. melitensis* infection of sheep,[102,103] with cold saline protein extracts of the rough *B. melitensis* 115 strain as allergen. This sort of extracts is, by a variety of criteria, free of LPS which could cause antibody-mediated reactions overlapping with characteristic DTH reactions.[80,100,101] The *B. melitensis* 115 allergen also causes cutaneous DTH reactions in rams infected with *B. ovis* but, as is the case with brucellosis caused by *B. melitensis,*[103] there is no correlation between allergic results and serological or bacterological data.[80]

A skin hypersensitivity reaction in rams has also been described during *B. ovis* infection after intradermal inoculation with the HS antigen.[24] The presence of LPS and proteins in these extracts precludes, however, a clear interpretation of these results.

Finally, it has to be pointed out that allergic reactions cannot distinguish whether a ram is infected by *B. ovis* or *B. melitensis,* and that Rev. 1 vaccination also elicits DTH reactions.[104]

VII. TREATMENT

In spite of successful antibiotic treatment of human brucellosis, this therapy has seldom been employed in animal brucellosis. The existence of eradication programs and high costs of antibiotics are the main reasons precluding this treatment. There are, however, some reports on the successful use of antibiotic therapy in *B. ovis* ram epididymitis.

The combined use of aureomycin and streptomycin cured epididymitis and stopped seminal excretion of *B. ovis* in infected rams but semen quality after treatment was poor.[105] On the other hand, streptomycin alone or combined with sulfamethazine, failed to cure animals.[105] In a different study, 27 of 32 infected rams treated with streptomycin plus tetracycline were serologically negative and showed no epididymitis symptoms 8 months after the treatment.[106] Likewise, a combination of long-acting oxytetracycline with dihydrostreptomycin eliminated *B. ovis* from 11 of 12 infected rams. In contrast, only 4 of 12 animals were bacteriologically negative when treated with oxytetracycline alone. Many animals bacteriologically negative after antibiotic therapy showed gross histological lesions in the epididymis and ampullae, limiting the efficacy of the treatment.[52]

Although additional studies should be carried out to improve the efficacy of treatments and to determine the recovery in fertility, at this moment antibiotic therapy is not recommended for treatment of *B. ovis* infection, except for selected animals of high genetic value.

VIII. CONTROL

A. TEST AND SLAUGHTER POLICY

The easiest way to decrease the incidence of *B. ovis* infection in an isolated flock is by culling rams showing palpable testicular alterations. However, even though the disease can be reduced to low levels of incidence, it cannot be eradicated by this method because of some infected animals showing no testicular alterations. As pointed out above (see sections VIA and B), this problem is not solved by the use of semen cultures, although they could be useful if applied in combination with serological tests.[67]

The most practical way to detect infected animals is serology. The combined use of scrotal palpation and CF has been successfully applied to the eradication of *B. ovis*.[107-110] Also, eradication has been achieved in two years by the combined use of scrotal palpation and gel diffusion with HS antigen, both tests performed every six months[33] (Table 3). The simpler gel diffusion test appears to be more practical than the CF test and of similar diagnostic value.

Although there is limited information available on the application of the ELISA, this assay would be expected to improve the efficacy of control programs based on a single test. There is one report in which it was shown that by means of scrotal palpation, ELISA and subsequent culling, the infection was eliminated from a flock in two years.[95] As mentioned above (see VI.C.2.c), the combination of ELISA and the gel diffusion test gives better results than the use of either test individually.

B. VACCINATION

Vaccination is the most economical and practical means for midterm control in areas with high incidence of *B. ovis* infection. In these areas, eradication by test and slaughter alone is often economically impossible, and vaccines should be used first to decrease the incidence of infection.

Although formalin-killed bacteria (bacterins) in adjuvant have been widely used, subcutaneous injection of saline-in-oil bacterins confers only a relatively low degree protection,[111-120] and aluminum hydroxide-adsorbed bacterins have even less efficacy.[115] The vaccination with saline-in-oil bacterin and revaccination 24 weeks later with the same vaccine has the same efficacy as the simultaneous inoculation of bacterin and live *B. abortus* strain

TABLE 3
Evolution of the Percentage of Culled Rams in a *B. ovis*-Infected Flock on the Basis of Scrotal Palpation and Double Gel Diffusion Test Performed every 6 Months

Year	Control no.	Rams tested	Rams culled
1983	1	190	37
	2	140	11
1984	3	107	2
	4	99	1
1985	5	120	0
	6	140	0

Data from Reference 33.

19 vaccine[115] (see below). In addition to the relatively low efficacy of bacterins, this method has disadvantages such as the high costs of annual revaccination, the long lasting serological response which interferes with serological tests, and interferences with management and mating practices.

The simultaneous immunization at different subcutaneous sites with *B. abortus* strain 19 and *B. ovis* bacterin in either aluminum hydroxide or saline-in-oil adjuvants has been considered very efficient in protecting rams from infection.[111-120] However, this method has important disadvantages. First, strain 19 can colonize the genitalia of rams and be excreted in the semen, a situation that can persist for a long period of time.[115] Second, strain 19 can cause epididymitis after vaccination.[121] Third, epiphysitis and lameness caused by *B. abortus* strain 19 has been shown to occur in rams vaccinated as yearlings.[122] Finally, this immunization elicits a long-lasting immune response which interferes with the serological tests for both *B. ovis* and *B. melitensis*,[123] and prevents exports to countries requiring negative blood tests.

Live *B. melitensis* strain Rev. 1 is probably the best available vaccine for *B. ovis* ram epididymitis. Under experimental conditions, the protection afforded in 3- to 5-month-old rams by a single subcutaneous dose of 10^9 CFU ranges between 74 and 100%,[25-27,121,124,125] depending on whether extragenital or only genital organs were cultured bacteriologically. In a comparative study performed in young rams, Rev. 1 provided better immunity than the attenuated *B. abortus* strain 19 commonly used in cattle.[121] It is also unclear whether Rev. 1 vaccination of adult rams produces serviceable immunity.[121,125] In our experience, the subcutaneous or conjunctival vaccination of 13-month-old rams protects rams against a *B. ovis* challenge infecting all the unvaccinated controls.[25]

Strain Rev. 1 seems to be avirulent for rams. Repeated inoculation in the same yearling animals has no pathogenic effects[126] and Rev. 1 seems also to be innocuous in adult rams.[25] However, there is one reported case in which Rev. 1 was isolated from a vaccinated ram that developed epididymitis[127] and, therefore, the pathogenicity of this strain should be reevaluated in large-scale vaccination programs.

Vaccination with Rev. 1 interferes with the serological tests used for *B. melitensis* infection. In addition, although it has been reported that Rev. 1 vaccination does not elicit antibodies to *B. ovis*, low CF titers to *B. ovis* after Rev. 1 vaccination have been demonstrated by others.[25,125,126] Also precipitating antibodies to rough antigens have been observed in a low proportion of Rev. 1-vaccinated rams[27] (see VI.C.a). It is unclear whether Rev. 1 revaccination improves the immunity conferred by a single dose.[27,125] In addition, revaccination causes increased titers to *B. melitensis* antigens and would often be difficult to perform.

Strain Rev. 1 is also effective in preventing *B. melitensis* infection of sheep.[104,128-130] Thus, in countries where both *B. melitensis* and *B. ovis* are prevalent, Rev. 1 should be the vaccine of choice in programs against sheep brucellosis. However, the interference of Rev. 1 vaccination with serological tests for *B. melitensis* antibody should be kept in mind. Up to now, no diagnostic test solving this problem has been developed,[98] but alternative methods of vaccination could prove to be useful. Conjunctival vaccination with Rev. 1 protects ewes against *B. melitensis,* and elicits only a low humoral immune response to smooth antigens.[104,130] In our experience, this vaccination procedure applied to young rams results in a degree of protection against *B. ovis* similar to that achieved by subcutaneous vaccination. Moreover, while 10% of subcutaneously vaccinated rams were still positive in the CF for *B. melitensis* 1 year after vaccination, all conjunctivally vaccinated animals were negative in the same test 5 months after vaccination.[25]

In countries where the use of the Rev. 1 strain is not recommended, only bacterins or subcellular vaccines can be recommended. The development of subcellular vaccines and of adjuvants with improved efficacy[131] should be the objective of research for the prophylaxis of *B. ovis* in these countries.

ACKNOWLEDGMENTS

The author is grateful to Clara M. Marin for unpublished data on serological diagnosis and vaccination. Special thanks to Montserrat Barberán for histological data and Ignacio Moriyón for helpful discussion and preparation of the manuscript. Research at the author's laboratory is supported in part by CAYCIT (grant 2910), CONAI (grant 2085), and the U.S.-Spanish Joint Committee.

APPENDIX

PREPARATION OF THE CULTURE MEDIUM FOR THE SELECTIVE ISOLATION OF *B. OVIS*

B. ovis can be isolated in nonselective media such as blood agar base enriched with 10% sterile equine, ovine, or bovine sera or in the same medium with 7 to 10% sterile ovine blood. However, the inoculum may contain other bacteria which often overgrow *B. ovis*. This is especially true when semen is cultured. For the preparation of the *B. ovis* selective medium, the procedure of Brown et al.,[13] modified as follows, is recommended:

For 1000 ml medium, 36 g of GC Medium Base (Gibco, BBL, etc.) is dissolved in 920 ml distilled water by heating with continuous stirring. After autoclaving, the medium is cooled to 50 to 60°C in a water bath, and 70 ml of aseptically collected ovine blood is carefully added and mixed, avoiding bubble formation. The mixture is then heated with slow stirring in a water bath to 80°C to lyse the red cells. After cooling to 45 to 50°C, the following inhibitors are added: 10 mg Nitrofurantoin (Sigma) dissolved in 1 ml of 0.1 *M* NaOH, and 10 ml commercial VCN inhibitor (Gibco, BBL, etc.). The VCN mixture can be replaced by 3 mg Vancomycin (Sigma), 7.5 mg Colistin Methanesulfonate (Sigma) and 12,500 IU Nystatin (Sigma).

The ovine blood is aseptically collected from the jugular vein of healthy rams (after local disinfection) with blood-collecting tubes (Venoject or Vacutainer) containing EDTA or sodium citrate.

If bacterial contamination appears, the concentration of agar in the basal medium can be increased to 3%.[12] If contamination by fungi is the problem, the quality of Nystatin should be checked. Once prepared, the plates should not be stored for long periods of time, and freshly prepared medium is always recommended.

PREPARATION OF CELLULAR MASS FOR ANTIGENIC EXTRACTION

Because of its innocuousness and CO_2 independence, the use of strain 198 REO of *B. ovis* is recommended.

Slants of trypticase soy agar enriched with 0.5% yeast extract or of blood agar base are inoculated and incubated at 37°C. After 48 h, the bacteria are washed off the slant with sterile saline soln. and the suspension is inoculated into 2000-ml flasks containing 800 ml trypticase soy broth with 0.5% yeast extract. After 48 to 72 h incubation at 37°C in an orbital incubator at 150 r.p.m., the cells are separated by centrifugation (7,500 × *g*; 20 min) and washed twice with saline. The bacteria can be stored frozen, freeze-dried, or dried with acetone, depending on the ultimate use.

HOT SALINE EXTRACTION[68,69]

Cells are resuspended in 0.85% saline (12 g of dried cells or 30 g of wet packed cells in 150 ml) and steamed in the autoclave for 15 to 30 min. After cooling, the suspension is centrifuged (15,000 × *g*, 4°C, 15 min) and the supernatant fluid is filtered and dialyzed against distilled water (3 × 100 volumes) at 4°C for 2 days. The dialyzed fluid is ultracentrifuged (100,000 × *g*, 4°C, 6 h), and the sediment resuspended in a small amount of distilled water and freeze-dried. This material is the HS antigen. For serological tests, the HS antigen is resuspended in distilled water at a concentration varying from 2.5 mg/ml to 20 mg/ml depending on the batch, and is titrated against a panel of positive sera. The resuspended HS antigen is kept at 4°C with 0.5% phenol as preservative. Freezing and thawing should be avoided.

LPS EXTRACTION (PETROLEUM ETHER-CHLOROFORM-PHENOL METHOD)

Acetone-dried cells are resuspended in a mixture (8:5:2) of petroleum ether-chloroform-phenol (10 g in 40 ml; this mixture is explosive and adequate aeration and protection are recommended). The suspension is homogenized with a blender and centrifuged (5800 × g, 15 min) in solvent-resistant tubes. The supernatant solution is collected and the precipitate resuspended again in the same amount of solvent mixture, and the process repeated twice. The three supernatant solutions are mixed, filtered (Whatman #1), and the chloroform and petroleum ether are evaporated off in a flash-evaporator. After cooling the remaining phenol phase to room temperature, the rough LPS is precipitated by dropwise saturation of the phenol with water (slight stirring). This is a critical step and the formation of a water phase should be avoided. The precipitated LPS is collected by centrifugation (9500 × *g*, 15 min, room temperature), and the viscous sediment washed three times (same centrifuge conditions) with 80% phenol and twice with ethyl ether. Finally, the extract is resuspended in distilled water and ultracentrifuged (100,000 × g, 4°C, 6 h). The sediment, which contains the rough LPS, is resuspended in a small amount of distilled water and freeze-dried.

EXTRACTION OF PROTEINS BY THE COLD SALINE METHOD[79]

Acetone-dried *B. ovis* cells are resuspended (5/100 w/v) in 2.5% NaCl) and the mixture held at 4°C for 48 to 72 h with continuous stirring. The suspension is centrifuged (13,000 × *g*, 30 min, 4°C), and the supernatant fluid precipitated with cold ethanol (24 h at 4°C). The resulting precipitate is collected by centrifugation (5000 × *g*, 15 min, 4°C), resuspended in distilled water, and dialized against distilled water (24 h, 4°C, 1 × 100 volumes). After ultracentrifugation (100,000 × *g*, 6 h), the supernatant fluid, containing a mixture of soluble proteins but lacking rough LPS, is collected and freeze-dried.

COMPLEMENT FIXATION (CF) TEST FOR *B. OVIS*

Sheep red blood cells collected in Alsever's solution (proportion of 1/1) are stabilized at 4°C for a week. Before using, red blood cells are washed twice in Veronal-buffered saline (VBS) (pH 7.3) and resuspended in the same buffer at a concentration of 2% (v/v). The indicator mixture is prepared daily by incubating at 37°C, with continuous shaking equal amounts of 2% red blood suspension and anti-red cell serum in VBS (the use of commercial anti-red cell serum of known titer simplifies this step; in our laboratory, the Pasteur Institute serum is used routinely at 1/800 in VBS).

Fresh guinea pig serum is used as a complement source. After titration in the presence of antigen, two 100% hemolytic units in a volume of 25 μl are used.

The test sera are diluted in VBS and inactivated at 60 to 63°C for 30 min in a water bath.

The stock solution of HS antigen (2.5 to 20 mg/ml) is diluted in VBS and titrated (checkerboard titration) against sera from infected rams.

The test is performed in microplates. Double dilutions of serum samples are made with the use of a micropipette or 25 μl microdiluter (50 μl of 1/5 dilution in wells of row A, and 25 μl of VBS in rows B to row G) and equal volumes (25 μl each) of titrated complement and antigen solutions dispensed in rows B to G. Wells of row A are the controls for anticomplementary activity of sera (only the complement is added), and a 25 μl volume of VBS is added. The plates are covered to avoid evaporation and, after incubation at 37°C for 30 to 40 min, the indicator mixture is dispensed (50 μl/well) in all wells from rows A to G. The plates are then carefully shaken and incubated again at 37°C with slow continuous shaking (or manual shaking at 10-min intervals). The test is interpreted after centrifugation of the plates or sedimentation of red blood cells. The results are expressed as titer (highest dilution of sera giving 50% or less of hemolysis).

Wells of row H contain the controls of the indicator mixture (indicator mixture: 50 μl), antigen (25 μl antigen, 25 μl complement, and 50 μl indicator mixture) and complement (25 μl complement and 50 μl indicator mixture). The volume of all the controls is completed brought up to 100 μl with VBS. Two positive sera of known titer and two negative sera are included as controls for each test.

DOUBLE GEL DIFFUSION TEST (HYPERTONIC GELS)

Reagents

- Borate buffer:
  - boric acid .... 12.4 g
  - potassium chloride .... 14.5 g
  - sodium azide solution (1/100 w/v) in distilled water .... 1 ml
  - distilled water .... 1600 ml
  - adjust to pH 8,3 with 0.2 M NaOH and complete to 2000 ml with distilled water.
- Agarose
- NaCl

Preparation of gels:

Add 0.8 to 1 g agarose and 10 g of NaCl to 100 ml of borate buffer and dissolve by boiling with stirring. On a level surface, cover glass slides with the necessary amount of molten gel to form a bed of 2.5 mm (3.5 ml approximately for standard microslides). After the gel has solidified, holes of 3 mm diameter at 3 mm spacing forming a hexagonal pattern are cut with a gel puncher. A 3 mm diameter center hole is also made. Problem sera are placed in alternate wells separated by a control positive serum (infection proved by bacteriological culture of *B. ovis*) with the antigen in the central well. Interpretation of results is made after incubation for 24, 48, and 72 h at room temperature in a humid chamber. A positive reaction is a precipitin line between the wells of the problem sera and the antigen that gives total or partial identity with the precipitin line of the positive controls. Precipitin lines not giving total identity may appear and probably correspond to minor antigenic components of HS antigen extracts (antibodies to these components can be elicited by infections due to brucellae other than *B. ovis)*.

The HS antigen is used diluted in distilled water.

ENZYME-LINKED IMMUNOSORBENT ASSAY (ELISA)

Materials

- Immunoplates. Good quality plates are essential to obtain consistent results. There are differences in adsorption among different brands and even among batches of the same brand.
- Multichannel micropipettes (50—200 μl)
- Adsorption buffer (0.06 *M* carbonate-bicarbonate buffer, pH 9.6)
- Washing buffer (0.01 *M* phosphate-buffered saline, containing 0.05% Tween 20)
- Antigen (stock of HS antigen at 1 mg/ml in water)
- Conjugate (rabbit anti-sheep IgG conjugated with horseradish peroxidase, H + L specific from Nordic Labs, is used in our laboratory, but other conjugates against sheep immunoglobulins are also available).
- Substrate: 0.1% solution (w/v) ABTS (2-2′ azinobis 3-ethylbenzthiazoline sulfonic acid, diamonium salt) in citrate Buffer (0.05 *M*, pH 4), containing 0.004% H_2O_2.
- Stopping solution (0.1 *M* HF)
- Spectrophotometer (with filter of 412 nm) if OD readings are to be made.

Method

Titrations should be performed for every batch of antigen and lot of conjugate by checking different amounts of antigen against serial dilutions of several positive and negative control sera, with different dilutions of conjugate.

The indirect ELISA is performed as follows:

1. Antigen

Most HS antigen batches work well at concentrations of 5 to 15 μg/ml in adsorption buffer. The antigen solution is dispensed (100 μl/well) into the polystyrene plates, and the plates are sealed or placed in a humid chamber to avoid evaporation and incubated at 37°C overnight. Unbound antigen is removed by three washes of PBS/Tween and the plates are dried at room temperature. The coated plates can be used immediately or stored at 4°C (the stability in these conditions is good for at least 3 months).

Purified rough LPS can be adsorbed in the same fashion. For extracts containing soluble protein, PBS rather than carbonate is used and adsorption is achieved by evaporation at 45°C in a desiccator.

2. Sera

An amount of 200 μl of an initial serum dilution ranging from 1/25 to 1/200 in PBS/Tween is dispensed in wells of row A, except in those of columns 1 and 12. Wells from rows B to H contain 100 μl PBS/Tween. Double serial dilutions are made from rows A to H. The final volume in each well is 100 μl.

The plates are covered, incubated at 37°C for 40 min, and washed three times with PBS/Tween.

3. Conjugate

The titrated conjugate is diluted in PBS/Tween, added (100 μl) to the wells, and the plate is covered and incubated for 40 min at 37°C. After incubation, the plates are washed again three times with PBS/Tween.

4. Substrate

The solution of ABTS H_2O_2 in citrate buffer is added (100 μl/well) and the plates are incubated 15 min at room temperature with continuous shaking. The reaction is stopped by addition of 0.1 *M* HFl (100 μl/well).

5. Reading and Interpretation

The optical density is read automatically in a spectrophotometer (Titertek Multiskan) at 412 nm. One plate is tested with several sera from both infected and negative rams as controls, and the mean and standard error of the OD of the positive and negative control sera are calculated for each dilution. Each test plate contains a blank (wells of column 1) and a control reaction with four dilutions of a positive serum and of a negative one (wells of column 12). A reaction is considered positive when its OD is significantly higher than the mean OD of the corresponding dilution of the negative control sera tested the same day.

Although statistical analysis based on OD from a single dilution cannot be made, visual assessment of titers is often satisfactory for diagnostic purposes.

REFERENCES

1. **Carpenter, T. E., Berry, S. L., and Glenn, J. S.,** Economics of *Brucella ovis* control in sheep: computerized decision-tree analysis, *J. Am. Vet. Med. Assoc.*, 8, 983, 1987.
2. **McFarlane, D., Salisbury, R. M., Osborne, H. G., and Jebson, J. L.,** Investigations into sheep abortion in New Zealand during the 1950 lambing season, *Aust. Vet. J.*, 28, 221, 1952.
3. **Simmons, G. C. and Hall, W. T. K.,** Epididymitis of rams. Preliminary studies on the occurrence and pathogenicity of a *Brucella*-like organism, *Aust. Vet. J.*, 29, 33, 1953.
4. **Buddle, M. B. and Boyes, B. W.,** A *Brucella* mutant causing genital disease of sheep in New Zealand, *Aust. Vet. J.*, 29, 145, 1953.
5. **Buddle, M. B.,** Observations on the transmission of *Brucella* infection in sheep, *N. Z. Vet. J.*, 3, 10, 1955.
6. **Buddle, M. B.,** Studies on *Brucella ovis* (n.sp.), a cause of genital disease of sheep in New Zealand and Australia, *J. Hyg.*, 54, 351, 1956.
7. **Meyer, M. E. and Cameron, H. S.,** Studies on the etiological agent of epididymitis in rams, *Am. J. Vet. Res.*, 17, 495, 1956.
8. **Kennedy, P. C., Frazier, L. M., and McGowan, B.,** Epididymitis in rams: pathology and bacteriology, *Cornell Vet.*, 46, 303, 1956.
9. **Diaz, R., Jones, L. M., and Wilson, J. B.,** Antigenic relationship of *Brucella ovis* and *Brucella melitensis*, *J. Bacteriol.*, 93, 1262, 1967.
10. **Hoyer, B. H. and McCullough, N. B.,** Homologies of deoxyribonucleic acids from *Brucella ovis*, canine abortion organism and other *Brucella* species, *J. Bacteriol.*, 96, 1783, 1968.
11. **Verger, J. M., Grimont, F., Grimont, P. A. D., and Grayon, M.,** *Brucella*, a monospecific genus as shown by deoxyribonucleic acid hybridization, *Int. J. Syst. Bacteriol.*, 35, 292, 1985.
12. **Corbel, M. J., Gill, K. P. W., and Thomas, E. L.,** Methods for the identification of *Brucella*, MAFF, U.K., ADAS, RCV 22, 1978.
13. **Brown, G. M., Ranger, C. R., and Kelley, D. J.,** Selective media for the isolation of *Brucella ovis*, *Cornell Vet.*, 61, 265, 1971.
14. **Alton, G. G., Jones, L. M., and Pietz, D. E.,** Las técnicas de laboratorio en la Brucelosis, *WHO Monogr. Ser.*, 55, 1976.
15. **Meyer, M. E.,** *Brucella ovis*, in *Handbach der bakteriellen Infektionen bei Tiernen*, Blobel and Scheleba, Eds., Jena, Germany, 1982, chap. 6.
16. **Blasco, J. M., Buen, L., Estrada, A., Gracia, J., Llena, J., and Ortilles, A.,** Alteraciones testiculares y brucelosis en moruecos de la region aragonesa, *Noticias Neosan*, 211, 147, 1983.

17. **Sancho, F., Marin, C. M., and Blasco, J. M.,** Evolucion de la brucelosis ovina en una agrupación de defensa sanitaria, *Inf. Tecn. Econ. Agraria,* 5, 431, 1985.
18. **Hartley, W. J., Jebson, J. L., and McFarlane, D.,** Some observations on natural transmission of ovine brucellosis, *N. Z. Vet. J.,* 3, 5, 1955.
19. **Clapp, K. H., Keogh, J., and Richards, M. H.,** Epidemiology of ovine brucellosis in South Australia, *Aust. Vet. J.,* 38, 482, 1962.
20. **Ris, D. R.,** The bacteriology and serology of ewes inoculated with viable *Brucella ovis* organisms, *N. Z. Vet. J.,* 18, 2, 1970.
21. **Collier, J. R. and Molello, J. A.,** Comparative distribution of *Brucella abortus, Brucella melitensis* and *Brucella ovis* in experimentally infected pregnant sheep, *Am. J. Vet. Res.,* 25, 930, 1964.
22. **Lawrence, W. E.,** Ovine brucellosis: a review of the disease in sheep manifested by epididymitis and abortion, *Br. Vet. J.,* 117, 435, 1961.
23. **Plant, J. W., Eamens, G. J., and Seaman, J. T.,** Serological, bacteriological and pathological changes in rams following different routes of exposure to *Brucella ovis, Aust. Vet. J.,* 63, 409, 1986.
24. **Afzal, M., Tengerdy, R. P., Brodie, S. J., Demartini, J. C., Ellis, R. P., Jones, R. L., and Kimberling, C. V.,** Immune response in rams experimentally infected with *Brucella ovis, Res. Vet. Sci.,* 41, 85, 1986.
25. **Marin, C. M.,** Contribución al diagnóstico serologico, profilaxis vacunal y tratamiento de la infección por *B. ovis* en moruecos, Doctoral thesis, Universidad de Zaragoza, 1989, 147 pp.
26. **Fensterbank, R., Pardon, P., and Marly, J.,** Efficacy of *Brucella melitensis* Rev 1 vaccine against *Brucella ovis* infection of rams, *Ann. Rech. Vet.,* 13, 185, 1982.
27. **Blasco, J. M., Marin, C. M., Barberán, M., Moriyón, I., and Diaz, R.,** Immunization with *Brucella melitensis* Rev 1 against *Brucella ovis* infection of rams, *Vet. Microbiol.,* 14, 381, 1987.
28. **Burguess, G. W., McDonald, J. W., and Norris, M. J.,** Epidemiological studies on ovine brucellosis in selected ram flocks, *Aust. Vet. J.,* 59, 45, 1982.
29. **Murray, R. M.,** Scrotal abnormalities in rams in tropical Queensland with particular reference to ovine brucellosis and its control, *Aust. Vet. J.,* 45, 63, 1969.
30. **Rival, M. D.,** Clinical reproductive abnormalities of rams, in *Animal Production in Australia. Proc. Australian Society of Animal Production,* 14, 176, 1983.
31. **De Wet, J. A. L. and Erasmus, J. A.,** Epididymitis of rams in the central and southern districts of the Orange free state, *J. S. Afr. Vet. Assoc.,* 55, 173, 1984.
32. **Walker, R. L., LeaMaster, B. R., Stellflug, J. N., and Biberstein, E. L.,** Association of age of ram with distribution of epididymal lesions and etiologic agent, *J. Am. Vet. Med. Assoc.,* 188, 393, 1986.
33. **Blasco, J. M.,** Contribucion al estudio de la brucelosis ovina por *Brucella ovis,* Doctoral thesis, Fac. Veterinaria Zaragoza, Spain, 1986, 101 pp.
34. **Bulgin, M. S. and Anderson, B. C.,** Association of sexual experience with isolation of various bacteria in cases of ovine epididymitis, *J. Am. Vet. Med. Assoc.,* 182, 372, 1983.
35. **Hadju, S.,** Serological investigation and control of infectious epididymitis and ovine brucellosis in Slovakia. *Vet. Bull.,* 3, 664, 1962.
36. **Clapp, K. H., Keogh, Y., and Richards, M. H.,** Epidemiology of ovine brucellosis in South Australia, *Aust. Vet. J.,* 38, 482, 1962.
37. **Watt, D. A.,** Investigations of ovine brucellosis in merino rams of Western Australia, *Aust. Vet. J.,* 46, 506, 1970.
38. **Cameron, R. D., Carles, A. B., and Lauerman, L. H.,** The incidence of *Brucella ovis* in some Kenya flocks and its relationship to clinical lesions and semen quality, *Vet. Rec.,* 89, 552, 1971.
39. **Brown, G.M., Pietz, D. E., and Price, D. A.,** Studies on the transmission of *Brucella ovis* infection in rams, *Cornell Vet.,* 63, 29, 1973.
40. **Mercy, A. R., Robertson, G. M., Goulder, R. K., and McKenzie, A. P.,** The prevalence of ovine brucellosis in cull merino rams in Western Australia, *Aust. Vet. J.,* 62, 137, 1985.
41. **Marin, C. M., J. de Bagués, M. P., Blasco, J. M., and Gabiña, D.,** Incidence of reproductive abnormalities in different breeds of rams, in *U.S. — Spain Joint Seminar on sheep breeding,* Zaragoza, Spain, 1986, 303.
42. **Garcia-Carrillo, C., Cuba-Caparó, A., and Myers, D. M.,** Susceptibilidad comparada de cabritos y carneros a la infeccion causada por *Brucella ovis:* estudios serológicos, bacteriológicos y patológicos, *Gac. Vet. Arg.,* 36, 355, 1974.
43. **Garcia-Carrillo, C., Casas-Olascoaga, R., Cuba-Caparó, A., Lucero, N., and Szyfres, B.,** Infección experimental de caprinos machos con *Brucella ovis.* Estudio bacteriológico, serológico e histopatológico, *Rev. Asoc. Arg. Microbiol.,* 9, 101, 1977.
44. **Burguess, G. W., Spencer, T. L., and Norris, M. J.,** Experimental infection of goats with *Brucella ovis, Aust. Vet. J.,* 62, 262, 1985.
45. **Barron, S.J., Kocan, A. A., Morton, R. J., Thedford, T. R., and McCain, C. S.,** Susceptibility of male white-tailed deer *(Odocoileus virginianus)* to *Brucella ovis* infection, *Am. J. Vet. Res.,* 46, 1762, 1985.

46. **Cuba-Caparó, A. and Myers, D. M.,** Pathogenesis of epididymitis caused by *Brucella ovis* in laboratory animals. *Am. J. Vet. Res.*, 34, 1077, 1973.
47. **Thoen, C. O. and Enright, F.,** *Brucella,* in *Pathogenesis of bacterial infections in animals,* Gyles, G. L. and Thoen, C. O., Eds., Iowa State University Press, Ames, 1986, chap. 20.
48. **Badiola, J. J., Barberán, M., Garcia de Jalón, J. A., and Bascuas, J. A.,** Caracteristicas reproductivas de moruecos Romanov afectados de epididimitis infecciosa. III. Alteraciones anatomopatológicas, *Anal. Inst. Nac. Invest. Agrar. Ser. Gan.*, 17, 69, 1982.
49. **Biberstein, E. L., McGowan, B., Olander, H., and Kennedy, P. C.,** Epididymitis in rams. Studies on pathogenesis, *Cornell Vet.*, 54, 27, 1964.
50. **Rahaley, R. S. and Dennis, S. M.,** Histopathology of experimental brucellosis in rams following vaccination with *Brucella ovis, Aust. Vet. J.*, 61, 353, 1984.
51. **Foster, R. A., Ladds, P. V., Briggs, G. D., and Hoffmann, D.,** Pathology of the accessory sex glands of rams infected with *Brucella ovis, Aust. Vet. J.*, 64, 248, 1987.
52. **Marin, C. M. J., de Bagués, M. P., Barberan, M., and Blasco, J. M.,** Efficacy of long acting oxytetracycline combined or not with strepotomycin for treating *Brucella ovis* infection of rams, *Am. J. Vet. Res.*, 50, 560, 1989.
53. **Livingstone, C. W. and Hardy, W. T.,** Isolation of *Actinobacillus seminis* from ovine epididymitis, *Am. J. Vet. Res.*, 25, 660, 1964.
54. **Ekdahl, M. O., Money, D. F., and Martin, C. A.,** Some aspects of epididymitis of rams in New Zealand, *N. Z. Vet. J.*, 16, 81, 1968.
55. **Rodolakis, A. and Bernard, K.,** Isolement de *Chlamydia* des organes genitaux de beliers atteints d'epididymite, *Bull. Acad. Vet. France,* 50, 65, 1977.
56. **De Long, W. J., Waldhalm, D. G., and Hall, R. F.,** Bacterial isolates associated with epididymitis in rams from Idaho and eastern Oregon flocks, *Am. J. Vet. Res.*, 40, 101, 1979.
57. **Williamson, P. and Nairn, M. E.,** Lesions caused by *Corynebacterium pseudotuberculosis* in the scrotum of rams, *Aust. Vet. J.*, 56, 496, 1980.
58. **Burguess, G. W. and McDowell, J. W.,** *Escherichia coli* epididymitis and seminal vesiculitis in a ram, *Aust. Vet. J.*, 57, 479, 1981.
59. **Webb, R. F.,** Clinical findings and pathological changes in *Histophilus ovis* infections of sheep, *Res. Vet. Sci.*, 35, 30, 1983.
60. **Cameron, R. D. A. and Lauerman, L. H.,** Characteristics of semen changes during *Brucella ovis* infection in rams, *Vet. Rec.*, 99, 231, 1976.
61. **Blasco, J. M., Folch, J., and Barberan, M.,** Caracteristicas reproductivas de moruecos Romanov afectados de epididimitis infecciosa. I. Observaciones clinicas y alteraciones de la fertilidad, *Anal. Inst. Nac. Invest. Agrar. Ser. Gan.*, 15, 67, 1982.
62. **Folch, J., Blasco, J. M., and Uriarte, J.,** Caracteristicas reproductivas de moruecos Romanov afectados de epididimitis infecciosa. II. Alteraciones seminales, *Anal. Inst. Nac. Invest. Agrar. Ser. Gan.*, 15, 77, 1982.
63. **Webb, R. F., Quinn, C. A., Cockram, F. A., and Husband, A. J.,** Evaluation of procedures for the diagnosis of *Brucella ovis* infection in rams, *Aust. Vet. J.*, 56, 172, 1980.
64. **Van Drimmelen, G. C., Botes, H. J., Claassen, Ross, W. F., and Viljoen, C. C.,** The use of fluorescent antibody in the diagnosis of *Brucella ovigenitalium* infection in sheep semen smears, *J. S. Afr. Vet. Med. Assoc.*, 34, 265, 1963.
65. **Ajai, C. O., Cook, J. E., and Dennis, S. M.,** Diagnosing ovine epididymitis by immunofluorescence, *Vet. Rec.*, 107, 421, 1980.
66. **Hughes, K. L. and Claxton, P. D.,** *Brucella ovis* infection. I. An evaluation of microbiological, serological and clinical methods of diagnosis in the ram, *Aust. Vet. J.*, 44, 41, 1968.
67. **Worthington, R. W., Stevenson, B. J., and de Lisle, G. W.,** Serology and semen culture for the diagnosis of *Brucella ovis* infection in chronically infected rams, *N. Z. Vet. J.*, 33, 84, 1985.
68. **Diaz, R., Jones, L. M., and Wilson, J. B.,** Antigenic relationship of the gram-negative canine organism causing canine abortion to smooth and rough *Brucellae, J. Bacteriology,* 95, 618, 1968.
69. **Diaz, R. and Bosseray, N.,** Identification d'un composé antigénique spécifique de la phase rugueuse (R) des *Brucella, Ann. Rech. Vet.*, 4, 283, 1973.
70. **Myers, D. M., Jones, L. M., and Varela-Diaz, V.,** Studies of antigens for complement fixation and gel diffusion tests in the diagnosis of infections caused by *Brucella ovis* and other *Brucella, Appl. Microbiol.*, 23, 894, 1972.
71. **Galanos, C., Luderitz, O., and Westphal, O.,** A new method for the extraction of R lipopolysaccharides, *Eur. J. Biochem.*, 9, 245, 1969.
72. **Moreno, E.,** Purification, chemical characterization and biological activity of *Brucella* lipopolysaccharide, Ph.D. thesis, University of Wisconsin, Madison, 1980, 169 pp.
73. **Riezu-Boj, J. I., Moriyón, I., Blasco, J. M., Marin, C. M., and Diaz, R.,** Comparison of lipopolysaccharide and outer membrane protein-lipopolysaccharide extracts in an enzyme-linked immunosorbent assay for the diagnosis of *Brucella ovis* infection, *J. Clin. Microbiol.*, 23, 938, 1986.

74. **Afzal, M., Tengerdy, R., Squire, P. G., and Ellis, R. P.,** Characterization of *Brucella ovis* lipopolysaccharide and its use for diagnosis of ram epididymitis by enzyme-linked immunosorbent assay, *J. Clin. Microbiol.*, 20, 1159, 1984.
75. **Gamazo, C., Winter, A. J., Moriyon, I., Riezu-Boj, J. I., Blasco, J. M., and Diaz, R.,** Comparative analyses of proteins extracted by hot saline or released spontaneously into outer membrane blebs from field strains of *Brucella ovis* and *Brucella melitensis, Infect. Immun.*, 57, 1419, 1989.
76. **Moreno, E., Jones, L. M., and Berman, D. T.,** Immunochemical characterization of rough *Brucella* lipopolysaccharides, *Infect. Immun.*, 43, 779, 1984.
77. **Alton, G. G.,** The epidemiology of *Brucella melitensis* infection in sheep and goats, in *Brucella melitensis,* Plommet, M. and Verger, J. M., Eds., Martinus Nijhoff, Dordrecht, 1985, 187.
78. **Santos, J. M., Verstreate, D. R., Perera, V. Y., and Winter, A. J.,** Outer membrane proteins from rough strains of four *Brucella* species, *Infect. Immun.*, 46, 188, 1984.
79. **Bhongbhibhat, N., Elberg, S. S., and Chen, T. H.,** Characterization of *Brucella* skin test antigens, *J. Infect. Dis.*, 122, 70, 1970.
80. **Jones, L. M., Dubray, G., and Marly, J.,** Comparison of methods of diagnosis of *Brucella ovis* infection of rams, *Ann. Rech. Vet.*, 6, 11, 1975.
81. **Ris, D. R. and Te Punga, W. A.,** An indirect hemagglutination test for the detection of *Brucella ovis* antibodies. I. Development of the test, *N. Z. Vet. J.*, 11, 94, 1963.
82. **Cox, J. C., Gorrie, C. J., Nairn, R. C., and Ward, H. A.,** A comparison of methods for the serological diagnosis of *Brucella ovis* infection, *Br. Vet. J.*, 133, 442, 1977.
83. **Myers, D. M. and Siniuk, A.,** Preliminary report on the development of a diffusion in gel method for the diagnosis of ram epididymitis, *Appl. Microbiol.*, 19, 335, 1970.
84. **Myers, D. M.,** Field evaluation of the gel diffusion test for the diagnosis of ram epididymitis caused by *Brucella ovis, Appl. Microbiol.*, 26, 855, 1973.
85. **Myers, D. M. and Varela-Diaz, V.,** Serodiagnosis of ram epididymitis by counter immunoelectrophoresis using *Brucella ovis* surface R antigen, *J. Clin. Microbiol.*, 10, 451, 1979.
86. **Biberstein, E. L., McGowan, B., Robinson, E. A., and Harrold, D. R.,** Epididymitis in rams. Studies on laboratory diagnosis, *Cornell Vet.*, 48, 31, 1958.
87. **Sanchis, R. and Giauffret, A.,** Mise au point d'une microtechnique de fixation du complement appliquée au serodiagnostic de l'epididymite contagieuse du bélier, *Recl. Med. Vet.*, 152, 305, 1976.
88. **Searson, J. E.,** Sensitivity and specificity of two microtitre complement fixation tests for the diagnosis of *Brucella ovis* infection in rams, *Aust. Vet. J.*, 58, 5, 1982.
89. **Burguess, G. W. and Norris, M. J.,** Evaluation of the cold complement fixation test for diagnosis of ovine brucellosis, *Aust. Vet. J.*, 59, 23, 1982.
90. **Anon.,** The complement fixation test for the diagnosis of *Brucella ovis* infection of rams. An Animal Health Division Report, *N. Z. Vet. J.*, 31, 157, 1983.
91. **Rahaley, R. S., Dennis, S. M., and Smeltzer, M. S.,** Comparison of the enzyme-linked immunosorbent assay and complement fixation test for detecting *Brucella ovis* antibodies in sheep, *Vet. Rec.*, 113, 467, 1983.
92. **Worthington, R. W., Weddell, W., and Penrose, M. E.,** A comparison of three serological tests for the diagnosis of *B. ovis* infection in rams, *N. Z. Vet. J.*, 32, 58, 1984.
93. **Ris, D. R., Hamel, K. L., and Long, D. L.,** Comparison of an enzyme-linked immunospecific assay (ELISA) with the cold complement fixation test for the serodiagnosis of *Brucella ovis* infection, *N. Z. Vet. J.*, 32, 18, 1984.
94. **Spencer, T. L. and Burguess, G. W.,** Enzyme-linked immunosorbent assay for *Brucella ovis* specific antibody in ram sera, *Res. Vet. Sci.*, 36, 194, 1984.
95. **Walker, R. L., LeaMaster, B. R., Stellflug, J. N., and Biberstein, E. L.,** Use of enzyme-linked immunosorbent assay for detection of antibodies to *Brucella ovis* in sheep: field trial, *Am. J. Vet. Res.*, 46, 1642, 1985.
96. **Clapp, K. H.,** A comparison of various antigens used in the complement fixation test for ovine brucellosis, *Aust. Vet. J.*, 37, 188, 1961.
97. **Ouchterlony, O.,** Antigen-antibody reactions in gels, *Acta Path. Microbiol. Scand.*, 26, 507, 1949.
98. **Blasco, J. M., Moriyón, I., Marin, C. M., and Diaz, R.,** Evaluation of a radial immunodiffusion test for differentiating infected from Rev 1 vaccinated sheep, in *Brucella melitensis,* Plommet, M. and Verger, J. M., Eds., Martinus Nijhoff, Dordrecht, 1985, 147.
99. **Chin, J. C.,** Comparison of different antigenic preparations for the detection of ovine serum antibodies against *Brucella ovis* by ELISA, *Aust. Vet. J.*, 60, 261, 1983.
100. **Jones, L. M.,** Specificity of *Brucella* protein antigens and role of lipopolysaccharide antigens in eliciting delayed hypersensitivity reactions in sensitized guinea pigs, *Ann. Rech. Vet.*, 5, 189, 1974.
101. **Jones, L. M., Diaz, R., and Taylor, A. G.,** Characterization of allergens prepared from smooth and rough strains of *Brucella melitensis, Br. J. Exp. Path.*, 54, 492, 1973.
102. **Jones, L. M. and Marly, J.,** Titration of a *Brucella* protein antigen in sheep sensitized with *Brucella melitensis, Ann. Rech. Vet.*, 6, 11, 1975.

103. **Fensterbank, R.,** Allergic diagnosis of brucellosis, in *Brucella melitensis,* Plommet, M. and Verger, J. M., Eds., Martinus Nijhoff, Dordrecht, 1985, 167.
104. **Fensterbank, R., Pardon, P., and Marly, J.,** Vaccination of ewes by single conjunctival administration of *Brucella melitensis* Rev. 1 vaccine, *Ann. Rech. Vet.,* 16, 351, 1985.
105. **Kuppuswamy, P. B.,** Chemotherapy of brucellosis in rams, *N. Z. Vet. J.,* 2, 110, 1954.
106. **Giauffret, A. and Sanchis, R.,** Etude d'un foyer d'epididymite contagieuse du bélier. Eradication de la maladie, *Bull. Off. Int. Epizooties,* 82, 581, 1974.
107. **West, D. M.,** The control of ovine brucellosis, in *Proc. 8th Seminar of the NZV Society,* 1978, 45.
108. **West, D. M. and Bruere, A. N.,** Accreditation for freedom from ovine brucellosis, *N. Z. Vet. J.,* 27, 263, 1979.
109. **West, D. M. and Bruere, A. N.,** The *Brucella ovis* complement fixation test, *N. Z. Vet. J.,* 31, 124, 1983.
110. **Bruere, A. N. and West, D. M.,** Effectiveness of the New Zealand brucellosis control scheme for sheep, *Proc. N. Z. Soc. Animal Production,* 47, 49, 1987.
111. **Buddle, M. B.,** Production of immunity against ovine brucellosis, *N. Z. Vet. J.,* 2, 99, 1954.
112. **Buddle, M. B.,** Vaccination against *Brucella ovis* infection in sheep, *N. Z. Vet. J.,* 5, 43, 1957.
113. **Buddle, M. B.,** Vaccination in the control of *Brucella ovis* infection in sheep, *N. Z. Vet. J.,* 6, 41, 1958.
114. **Buddle, M. B.,** Production of immunity in rams against *Brucella ovis* infection, *N. Z. Vet. J.,* 10, 111, 1962.
115. **Buddle, M. B., Calverley, F. K., and Boyes, B. W.,** *Brucella ovis* vaccination of rams, *N. Z. Vet. J.,* 11, 90, 1963.
116. **Claxton, P. D.,** *Brucella ovis* vaccination of rams, *Aust. Vet. J.,* 44, 48, 1968.
117. **Swift, B. L. and Maki, L. R.,** Immunological studies on three ram epididymitis bacterins, *Cornell Vet.,* 58, 659, 1968.
118. **Haughey, K. G. and Hughes, K. L.,** The apparent failure of a vaccine to protect rams against naturally occuring *Brucella ovis* infection, *Aust. Vet. J.,* 45, 10, 1969.
119. **Biberstein, E. L., McGowan, B., Robinson, E. A., and Harrold, D. R.,** Epididymitis in rams. Studies on immunity, *Cornell Vet.,* 52, 214, 1962.
120. **McGowan, B. and Harrold, D. R.,** Epididymitis in rams: studies on vaccine efficacy, *Cornell Vet.,* 69, 73, 1979.
121. **Van Heerden, K. M. and Van Rensburg, S. W.,** The immunization of rams against ovine brucellosis, *J. S. Afr. Vet. Med. Assoc.,* 33, 143, 1962.
122. **West, D. M., Johnstone, A. C., Bruere, A. N., and Chapman, H. M.,** Epiphysitis in rams following vaccination against *Brucella ovis* infection, *N. Z. Vet. J.,* 26, 133, 1978.
123. **Ris, D. R.,** The persistance of antibodies against *Brucella ovis* and *Brucella abortus* in rams following vaccination: a field study, *N. Z. Vet. J.,* 15, 94, 1967.
124. **Garcia-Carrillo, C.,** Protection of rams against *Brucella ovis* infection by *Brucella melitensis* Rev 1 vaccine, *Zentralbl. Vet. Med.,* 28B, 425, 1981.
125. **Gradwell, D. V. and Van Zyl, F. E.,** Effectivity of Rev 1 vaccine in rams against *Brucella ovis* infection, *J. S. Afr. Vet. Assoc.,* 46, 349, 1975.
126. **Erasmus, J. A. and Bergh, E. C.,** Ovine brucellosis: repeated vaccination with Rev 1 vaccine and the prevalence of the disease in the Winburg district, *J. S. Afr. Vet. Assoc.,* 56, 205, 1985.
127. **Lantier, F. and Fensterbank, R.,** Kinetics of Rev 1 infection in sheep, in *Brucella melitensis,* Plommet, M. and Verger, J. M., Eds., Martinus Nijhoff, Dordrecht, 1985, 247.
128. **Alton, G. G. and Elberg, S. S.,** Rev 1 *Brucella melitensis* vaccine. A review of ten years of study, *Vet. Bull.,* 37, 793, 1967.
129. **Elberg, S. S.,** Rev 1 *Brucella melitensis* vaccine. II. 1968—1980, *Vet. Bull.,* 51, 67, 1981.
130. **Fensterbank, R., Pardon, P., and Marly, J.,** Comparison between subcutaneous and conjunctival route of vaccination with Rev 1 strain against *Brucella melitensis* infection in ewes, *Ann. Rech. Vet.,* 13, 295, 1982.
131. **Afzal, M., Tengerdy, R. P., Ellis, R. P., Kimberling, C. V., and Morris, C. J.,** Protection of rams against epididymitis by a *Brucella ovis* — Vitamin E adjuvant vaccine, *Vet. Immunol. Immunopath.,* 7, 293, 1984.

Chapter 16

BRUCELLA MELITENSIS, 1887 TO 1987

Godfrey G. Alton

The early history of brucellosis, then known under a variety of names such as Mediterranean or Malta fever, was particularly associated with the island of Malta. It was here that three crucial events took place: (1) the description of the disease as an entity distinct from other similar fevers by Marston;[1] (2) the isolation and description of the organism by Bruce;[2] and (3) the discovery of its animal reservoir by Zammit,[3] a member of the Mediterranean Fever Commission. All this has been accurately and interestingly described by Spink in the book *The Nature of Brucellosis*.[4]

Bruce submitted his article describing the isolation of *B. melitensis* from the Station Hospital, Valetta. This remarkable building was constructed in 1574 by the Knights of Malta. It was known as the Sacred Infirmary and dedicated to the welfare of the sick of all nations and creeds. At one time it was claimed to be the longest building in Europe; it has an undivided interior forming a single, immensely long hospital ward. In 1676 a school of anatomy and surgery was established here by the ruling Grandmaster of the knights, and this became part of the origins of the medical school, now part of the University of Malta. The Sacred Infirmary was taken over by the British military in about 1800, renamed, somewhat less picturesquely, the Station Hospital, and used as a "fever ward" housing mostly typhoid and Mediterranean fever cases. Here in 1887 Bruce and the Maltese Doctor Caruana-Scicluna, by growing the organism in culture, "confirmed the existence of the microbe of the subtle fever" as stated in Maltese on the plaque on the wall of the building commemorating the event. "Confirmed" is an apt description since Bruce and his helpers had already detected the organism microscopically in previous cases. The hospital was destroyed by bombing in World War II, and the building remained a ruin for many years but has now been elegantly restored as a conference center and would be an excellent location for a conference on brucellosis.

The workers in Malta were aware that this particular fever was by no means confined to that island. To quote from Bruce,[2] "It is identical with the Rock Fever of Gibraltar, the Neapolitan Fever of Naples, the Country Fever of Constantinople and the New Fever of Crete . . . I find that it occurs also at Cagliari in Sardinia, Catania in Sicily, Smyrna and Tunis." There are references in the Bible to sheep and goats casting their young and a description in early Greek medical literature of a disease resembling human brucellosis. It would not be unreasonable to speculate that *B. melitensis* infection was spread throughout the Mediterranean littoral and its islands by Phoenician traders during the first millenium B.C. from their bases on the east coast. In 1897, Hughes published his monograph *Mediterranean, Malta or Undulant Fever*.[5] This gives an excellent description of the human disease and contains a comprehensive bibliography of the disease up to that time.

After the discovery of the organism, the ravages of the disease continued unabated. Bruce in his second article[6] stated that in the preceding year 109 soldiers were affected out of the 2200 that depended on him for medical care. This represents an incidence of 495 cases per 10,000 at risk. Their average stay in hospital was 83 d.

It had become evident to Bruce and the others involved that the disease was not transmitted person to person. Their frustration, caused by the lack of knowledge of the transmission of the infection, led to the setting up of the Mediterranean Fever Commission. On January 25, 1904, Mr. Lyttelton, the Colonial Secretary in London, wrote a letter to the War Office proposing the establishment of a commission to investigate the problem. Eventually, the Government of Malta was involved, and the Royal Society of London asked to provide scientific management. The best talent available was recruited, and the Commission under the leadership of Colonel Bruce began work in Malta on June 13th, 1907. The first article by a member of the Commission was received for publication on 14th July 1904.

The success of the commission was greatly aided by the recent development of the agglutination test for the differential diagnosis of typhoid and Mediterranean fevers. This development is generally attributed to Wright[7] who was professor of pathology at the British Army Medical School and Hospital at Netley in England, where persistent cases of Malta and typhoid fever were repatriated for treatment. Bruce had passed through this school before taking up his appointment in Malta.

The story is well known of how the Commission, having experienced difficulty with the monkeys they had been using as an animal model, looked for a cheaper and more readily available alternative and in Malta the goat seemed an obvious choice. At that time the goats were walked through the towns once a day and milked on the customer's doorstep. It may be thought that once this animal had been chosen, its role in providing the source of infection was bound to be discovered. Credit must, however, be given to the researchers for their scientific approach in applying the agglutination test to the blood of the goats they had purchased before trying to infect them.

The building where the laboratory work was done, once the Palace of Justice in Valetta, is now, as then, the headquarters of the Public Health Department, now called Department of Health and Environment. The laboratory in which the Mediterranean Fever Commission worked, including the roof where the experimental monkeys, guinea pigs, rabbits, and mice were kept, has been converted into a museum, with much of the original equipment on view.

The prohibition of drinking untreated milk went far toward solving the problem of Mediterranean fever among the military in Malta, and their investigations were soon terminated.

The interest now moved to France where in the Department of Gard near the Mediterranean coast an epidemic of Mediterranean fever was diagnosed in the people of a group of mountain villages and its origin traced to goats.[8] This gave rise to the publication of a monograph on the human disease by Cantaloube.[9] The famous Medical School of the University of Montpellier became interested in the disease and, supported by the Health Organization of the League of Nations, established a research center where, among other investigations, work commenced on developing a vaccine.

In Russia, brucellosis was first diagnosed in 1913 in goats at Tashkent. Nine years later, six cases of human brucellosis were diagnosed in Azerbaijan and the source of infection traced to sheep and goats. However, *B. melitensis* was not isolated in Russia until 1930, after it had become the Soviet Union. A major investigation into *B. melitensis* infection in sheep was soon initiated in the Soviet Union, where the development of large-scale sheep farming had provided suitable conditions for *B. melitensis* infection to prosper. All aspects of the disease were investigated, including measures for its diagnosis and control. The results of these investigations were published in Russian[10] and detailed summaries of all the papers in French.[11]

In the mid-1930s Japanese veterinarians working in Manchuria had the opportunity to observe acute outbreaks of *B. melitensis* infection in two large flocks of sheep over successive lambing periods, and in their attendants. Extensive bacteriological and epidemiological investigations were undertaken, and a lanolin-based vaccine was developed.[12,13] Again, detailed summaries are available in French.[12,13]

To return to Malta, little was done to protect the civilian population during the first four decades of this century. Measures were taken forbidding the entry of goats into the towns for milking, and 1939 the government established a dairy to collect, pasteurize, and market milk. These efforts were nullified when Malta was fiercely attacked during World War II, and the highest number of human cases ever recorded there was in 1946 when 2400 were reported, about 80 per 10,000. After the war, the prohibition against the sale of nonpasteurized milk was gradually extended to the whole of Malta and its sister island Gozo, and by 1959 the annual human incidence had fallen to 220 cases.

Beginning in 1956, investigations were undertaken in Malta to establish an effective method of immunizing goats and to investigate diagnostic methods with a view to developing an eradication program.[14] Extensive investigations led to the selection of Rev. 1 vaccine, and vaccination of young goats began in 1959. Despite only about 50% of female kids being vaccinated, the campaign seemed to be causing a drop in the number of cases of human brucellosis, and by 1968 the annual incidence had fallen to 26 cases, or 0.86 per 10,000. At this level complacency set in, the control measures were not actively pursued, and the level of human brucellosis remained about the same for the next 20 years: there were 29 cases in 1985 and 22 in 1986. This supposedly low incidence is about double that of the worst recorded year for human brucellosis in the U.S.

B. melitensis infection was, most likely, introduced into the Americas by the Spanish conquistadors with infected goats. Peru and Mexico, the earliest parts colonized, remain to this day the countries with the highest incidence. This infection must have penetrated the U.S. from the south and retreated again as effective control measures were applied. A consignment of goats from Malta landed in New York in 1905 and contained some infected animals, but the whole batch was slaughtered, and so far as we know there was no spread of infection.

During and soon after the end of World War II there was much movement of people and traffic in food animals in Germany, Eastern Europe, and the U.S.S.R. This resulted in a large resurgence of infection with *B. melitensis,* especially in sheep. In many of these areas in the years following the reestablishment of peace, the disease was eliminated by depopulating infected flocks.

In the 1960s a high incidence of human infection with *B. melitensis* was reported from both Mongolias. In each case the disease was brought under control by vaccination of small ruminants. In the Mongolian Republic, Rev. 1 was used in a widespread campaign supported by the World Health Organization (WHO), while in Inner Mongolia, (including the region previously administered by Japan), Chinese scientists developed their own vaccines and, starting about 1968, used them to mount a successful campaign to control *B. melitensis* infection in sheep.[15]

A most dramatic increase in the incidence of *B. melitensis* infection occurred in the countries bordering the Persian Gulf, where, during the 1980s, a large and spreading epidemic developed. Long prevalent in Iran, *B. melitensis* has only recently become a serious problem in other oil-rich states where increasing affluence has stimulated the introduction of intensive sheep farming. In Saudi Arabia, Radwan et al.[16] described an outbreak in a flock of 38,000 sheep, kept under intensive conditions, where the serological positivity had reached 24% and where some of the sheep handlers had contracted brucellosis. In nearby Kuwait, the available figures show 32 cases of human brucellosis in 1980, 77 in 1982 and an estimated average of 2000 in both 1985 and 1986.[17]

Attempts to produce a vaccine against *B. melitensis* go back to the days of the Mediterranean Fever Commission. Those of the Japanese in Manchuria have been mentioned, and a considerable amount of work was done in France between the wars where, among other approaches, a combined vaccine comprising an attenuated strain of *B. abortus* in combination with various nonliving preparations of *B. melitensis* found some favor. After World War II *B. abortus* strain 19 was widely used in the U.S.S.R. to immunize sheep, and some success was reported. WHO and the Food and Agriculture Organization (FAO) jointly supported research into various aspects of brucellosis, and in 1952 a meeting of the

Expert Committee on Brucellosis appointed by these organizations recommended, and financially supported, a program of work to evaluate existing vaccines for sheep and goats and to develop and evaluate a killed *B. melitensis* oil-adjuvant vaccine. An extensive series of experiments was undertaken at the Pasteur Institute in Tunis where the adjuvant vaccine, subsequently known as H38, was shown to immunize adequately but also to possess formidable side effects. It was about this time that Rev. 1 first made its appearance[18] and proved its potential as an immunizing agent albeit with some drawbacks. In spite of these drawbacks, the living attenuated Rev. 1 vaccine has been used successfully to immunize many millions of sheep and goats throughout the world.[19]

So far, eradication has only been achieved on a national scale by locating infected flocks and depopulating them. However, widespread vaccination with Rev. 1 has already greatly reduced the prevalence of *B. melitensis* infection in sheep and goats in many areas, and there are good prospects that further research into diagnostic methods will produce effective and economic programs that will make the option of eradication more feasible. This work has great significance because human infection with *B. melitensis* is now perhaps the most important zoonosis in the world.

REFERENCES

1. **Marston, J. A.,** Report on fever, (Malta), *Army Med. Dept. Rep.*, 3, 486, 1861.
2. **Bruce, D.,** Note on the discovery of a micro-organism in Malta fever, *Practitioner,* 39, 161, 1887.
3. **Zammit, T.,** A preliminary note on the examination of the blood of goats suffering from Mediterranean fever, in *Reports of the Commission on Mediterranean Fever, Part III,* Harrison and Sons, London, 1905, 83.
4. **Spink, W. W.,** *The Nature of Brucellosis,* University of Minnesota Press, Minneapolis, 1956.
5. **Hughes, M. L.,** *Mediterranean, Malta or Undulant Fever,* Macmillan, London, 1897.
6. **Bruce, D.,** The micrococcus of Malta fever, *Practitioner,* 40, 241, 1888.
7. **Wright, A. E.** Note on the technique of serum diagnosis of acute specific fevers, *Br. Med. J.*, January 16, 139, 1897.
8. **Aubert, P., Cantaloube, P., and Thibault, E.,** Une epidemie de fievre de Malte dans le Department du Gard, *Ann. Inst. Pasteur,* 24, 376, 1910.
9. **Cantaloube, P.,** *La fievre de Malta en France,* Maloine, A., Paris, 1911.
10. **Sdrodowski, P. F.,** Ed., *Reports of the Commission to Investigate Sheep Brucellosis,* All-Union Institute for Experimental Medicine A. M. Gorki, Moscow, 1937.
11. **Russeff, Chr.,** Travaux de la mission d'etudes de la brucellose ovine en U.R.S.S. (1933—1936), *Off. Int. Epizoot. Bull.*, 17, 553, 1938.
12. **Itabashi, K., Watanabe, S., Ito, S., Tajima, Y., and Otaki, K.,** Studies on contagious abortion in sheep. II, *J. Jpn. Soc. Vet. Sci.*, 16, 63, 1937; summary in *Off. Int. Epizoot. Bull.*, 15, 1000, 1938.
13. **Watanabe, S., Tajima, Y., Ito, S., Itabishi, K., and Yoda, H.,** Studies on contagious abortion in sheep. III, *J. Jpn. Soc. Vet. Sci.*, 16, 66, 1937; summary in *Off. Int. Epizoot. Bull.*, 15, 1002, 1938.
14. **Alton, G. G.,** The control of animal brucellosis in Malta, *St. Luke's Hosp. Gaz. Malta,* 3, 74, 1968.
15. **Xie Xin,** Orally administrable brucellosis vaccine: *Brucella suis* strain 2 vaccine, *Vaccine,* 4, 212, 1986.
16. **Radwan, A. I., Bekairi, S. I., Al-Mukayel, A. A., and Abu Agla, O. S.,** Control of ovine brucellosis in Najdi sheep in Saudi Arabia, *Trop. Anim. Health Prod.*, 16, 213, 1984.
17. **Lulu, A. R., Araj, Khateeb, M. I., Mustafa, M. Y., Yusuf, A. R., and Fenech, F. F.,** Human brucellosis in Kuwait: a prospective study of 400 cases, *Q. J. Med.*, 66, 39, 1988.
18. **Elberg, S. S. and Faunce, O. K.,** Immunization against *Brucella* infection. VI. Immunity conferred on goats by a non-dependent mutant from a streptomycin-dependent mutant strain of *Brucella melitensis, J. Bacteriol.*, 73, 211, 1957.
19. **Elberg, S. S.,** Rev. 1 *Brucella melitensis* vaccine. II. 1968—1980 (review) *Vet. Bull.*, 51, 67, 1981.

Chapter 17

BRUCELLA MELITENSIS

Godfrey G. Alton

TABLE OF CONTENTS

I. INTRODUCTION

The natural hosts of *Brucella melitensis* are adult goats and certain breeds of sheep, among which it spreads readily, causing abortion. It is the least species specific of the brucellae and from the reservoir of infection in goats and sheep transmits to many other species. These, except for cattle,[1] among which it spreads to some extent, are almost always end hosts. *B. melitensis* is highly pathogenic for man, causing one of the most serious zoonoses in the world.

B. melitensis was the first species in the genus *Brucella* to be described. It was first isolated by Bruce in 1887[2] from the spleens of soldiers dying of Mediterranean fever on the island of Malta. Bruce called it *Micrococcus melitensis*. Mediterranean fever did not appear to spread among people, and the origins of the disease remained a mystery for nearly 20 years until it was discovered[3,4] that goats were affected and were providing the source of infection for the human population.

II. GEOGRAPHIC DISTRIBUTION

The natural home of *B. melitensis* appears to be the Mediterranean, especially along its northern and eastern shores, stretching through Central Asia as far south as the Arabian peninsula and as far east as the two Mongolias. Parts of Latin America are also seriously affected, especially Mexico, Peru, and northern Argentina, presumably following the migration of people from the Mediterranean region. Sporadic occurrence is common to Africa and India. Here the available data derive mainly from serological surveys and are imprecise. In the U.S. in recent years, *B. melitensis* caused sporadic outbreaks in goats near the Mexican border.[5] The country is now believed to be free, as are Canada, Northern Europe (except for sporadic incursions from the south), Southeast Asia, Australia, and New Zealand.

The goat was originally considered the principal host of *B. melitensis,* and this continues to be the case in certain areas, notably Malta and Latin America, where sheep are not significantly infected even when kept in close contact with goats.[6] In many other areas the disease is more important in sheep, despite the fact that goats are always more susceptible. There are two principal reasons for this difference. First, there is great variation in the susceptibility of different breeds of sheep: the Maltese sheep, and presumably those of Latin America are almost totally resistant, while in areas bordering the Mediterranean and through West and Central Asia, the principal breeds of sheep are almost as susceptible to *B. melitensis* infection as goats. Second, in these latter areas sheep are the predominant species of small ruminant, being kept for the production of much-favored dairy products and meat, often in large flocks, in conditions that encourage the spread of infection.

In recent years some of the countries of West Asia have sought to increase supplies of meat and milk by encouraging intensive sheep farming. In some of the very large flocks established for this purpose severe outbreaks of *B. melitensis* infection have occurred, leading to epidemics of human brucellosis.[7]

Of the three different biovars of *B. melitensis* (see below) biovar 1 predominates almost exclusively in Latin America[8] and Malta,[9] where goats are the only species involved. Biovar 2 is reported, often along with biovar 1, from areas where sheep play a major role and predominates in Italy[10] and Greece.[11] Biovar 3 cultures have been described in small proportion from most areas where biovar 2 is also present. There appears to be no difference in the nature of the disease caused by these three biovars.

III. THE BACTERIUM *BRUCELLA MELITENSIS*

The taxonomy of the genus *Brucella* is described in Chapter 1, as is the proposal that since DNA hybridization studies indicate that the various members of the genus show more than 90% similarity in their DNA base sequences, the genus *Brucella* should be regarded as consisting of a single species.[12] As this proposal has not yet been adopted, the existing nomenclature is used in this chapter. *B. melitensis* has been described in detail by Corbel and Morgan[13] along with the other presently recognized species.

B. melitensis has the same morphology and staining characteristics as other members of the genus. It grows on the media routinely used for other brucellae, without needing added CO_2, but somewhat less luxuriantly. The addition of serum is not essential, but growth is often improved by the presence of serum in the medium. *B. melitensis* colonies on solid medium resemble those of the other members of the genus. It does not produce H_2S and grows in the presence of both dyes thionine and basic fuchsin. Smooth strains of *B. melitensis* are lysed only by the Berkeley phage, Bk_2, Iz, and not by the phages Tb, Wb, Fz, or the rough phages at routine test dilutions.[14] It should be remembered that *B. melitensis* is very prone to S-R changes in colony type and that phage typing is very dependent on the dissociation status of the culture. The species can also be distinguished from other brucellae by its oxidative metabolic pattern. It oxidizes the usual amino acid substrates, except those of the urea cycle, but of the carbohydrate substrates normally included only glucose and i-erythritol are utilized.[15]

The three biovars of *B. melitensis* are distinguishable one from the other only by their immunochemical reactions. As with *B. abortus,* the immunodominant surface antigen is the lipopolysaccharide (LPS). The LPS of the type species, biovar 1, of *B. melitensis* possesses the M type of ‘O’-chain polysaccharide, whereas biovar 2 possesses the A chain. The chemistry of these LPS molecules is described in Chapter 2. The differences are recognized in routine typing by agglutination in monospecific (cross-absorbed) anti-A and anti-M sera. The chemistry of the LPS 'O' chain of biovar 3, which shows partial agglutination with both monospecific sera, has not yet been worked out. At present, polyclonal monospecific sera are generally used in typing tests, but monoclonal sera have been prepared against the A[16,17] and M[18] antigens, and the quality of sera used in these tests should improve greatly when A- M-chain-specific monoclonal sera become more generally available.

The precise recognition of biovar 3, especially its differentiation from biovar 2, sometimes appears equivocal. For example, laboratory A classified 95% of more than 300 strains of *B. melitensis* isolated from humans as biovar 2, while laboratory B in the same area found none of 61 strains isolated from animals to be biovar 2, but 55 of them to be biovar 3. These discrepancies may be due to problems associated with the production of sufficiently discriminatory monospecific sera or the common occurrence of dissociated cultures of *B. melitensis,* some of which autoagglutinate.

B. melitensis is urease positive, and various strains differ markedly in their degree of positivity. This feature may be of use as an epidemiological marker.

IV. NATURE OF THE DISEASE IN FEMALE SHEEP AND GOATS

B. melitensis infection in female sheep and goats causes disease in sexually mature animals. Usually the only symptom is abortion, followed in some cases by retention of the fetal membranes. Manifestations of the infection in male sheep and goats and in cattle will be dealt with later in the chapter.

A. INFECTION

The 50% infective dose for virulent *B. melitensis,* administered via the conjunctiva, has been calculated as 2×10^4 for Swedish goats and 4×10^5 for Swedish sheep.[19] In both cases the animals were adult and not pregnant. It needs to be borne in mind that the sheep may not have been representative of the more susceptible breeds.

The infection in female sheep and goats follows a course very similar to *B. abortus* infection in cattle, which is described in Chapter 12. Briefly, *B. melitensis* usually enters the animal's body through the mouth, nose, or eye, though invasion through the skin, especially if abraded, has been shown to be possible.[20] This latter route is of particular interest because opportunities for infection via the teats must occur when numerous animals are being milked by hand, and because the excretion of *B. melitensis* in the milk often persists for a long time, especially in goats.

Brucellae are facultative intracellular parasites of the reticuloendothelial system. After gaining entrance to the body, the organisms encounter the cellular and humoral defenses of the body but generally succeed in arriving via the lymph channels at the nearest lymph node. The fate of the invading bacteria is mainly determined by the cellular defenses of the host, chiefly macrophages and T lymphocytes, though specific antibody undoubtedly plays a part (immune mechanisms in brucellosis are discussed in Chapters 3 and 4). The outcome depends on the age, immune status of the host, whether or not it is pregnant, and the virulence of the invading bacteria and their numbers. When the bacteria prevail over the body defenses, a bacteremia is generally established.

B. THE BACTEREMIC PHASE

Biggi[21] found that after experimental infection, via the conjunctiva and subcutaneously, 83% of pregnant and 73% of nonpregnant Italian sheep became positive on hemoculture. The bacteremia was detectable after 10 to 20 d and persisted for 30 to 45 d. Shimi and Tabatabayi[22] infected ten fat-tailed Iranian sheep via the conjunctiva and subsequently isolated

B. melitensis from the blood of seven of them. In most of these cases the bacteremia lasted more than 2 months and in one sheep as long as 302 d. Information regarding the frequency of bacteremia in goats is more difficult to find. One study[23] showed that after natural exposure the numbers positive on regular blood culture corresponded approximately with the numbers found later to have had a generalized infection. During the bacteremic phase the brucellae may be found in almost any organ or tissue, but this does not necessarily indicate that the bacteria are established in these sites.

If the animal is pregnant, bacteremia often leads to the invasion of the uterus where, in sheep and goats, the synthesis of the sugar erythritol in the developing membranes[24] greatly aids the multiplication of brucellae, as it does in cattle. At the same time, infection generally becomes established in various lymph nodes and organs, often in the udder and sometimes in the spleen.

C. THE SEROLOGICAL RESPONSE

Sheep and goats infected parenterally with large doses of *B. melitensis,* whether virulent or attenuated and whether they are pregnant or not, respond within 1 week with a sharp rise in serum antibody. This reaches a peak at about 2 weeks, often reaching titers of 2500 in the agglutination and complement fixation tests and then, depending on the course of the infection, begins to decline. Following infection by natural exposure, a serological response can be expected within 2 to 4 weeks,[25] but the response is variable and may be absent altogether. Invasion of the pregnant uterus can be expected to produce a large and persistent rise, but this may be delayed until after the act of abortion or parturition at the normal time. Invasion of the lactating udder causes a lesser serological response, and localization confined to a small number of lymph nodes may fail to stimulate any response at all, or only a minimal one. This last situation is dangerous since such infections can escape serological detection, then, at least in goats, flare up at a later date.

The pattern of the serological response in terms of immunoglobulin production has been little studied in small ruminants, but available information[26] suggests close similarity with that in cattle, i.e., production of IgM followed within a week or two by a switch to IgG, both isotypes falling to a low level in the more chronic stage of infection, with IgG predominating after the 27th week.

D. LOCALIZATION IN THE PREGNANT UTERUS

Infection of the pregnant uterus often ends in abortion. Following the parenteral inoculation of large doses of *B. melitensis,* pregnant goats are almost certain to abort, after a delay of 3 to 4 weeks. However, one study[27] showed that out of 40 pregnant goats that became massively infected in the uterus following natural exposure, 24 aborted and 16 gave birth at the normal time, i.e., only 60% aborted. Goats that have aborted at one pregnancy are less likely to do so at the next, but are likely to shed following parturition.

Pregnant sheep, likewise, almost always abort 4 to 12 weeks after experimental parenteral[28] or conjunctival[22] infection. However, when pregnant sheep become infected naturally, they also are less likely to abort. When Entessar et al.[29] exposed 25 pregnant nonvaccinated Iranian sheep to natural infection, most became infected, but only 8 suffered detectable uterine invasion, and of these only 2 aborted. It is said that new infections in nonpregnant sheep never result in abortion during the next pregnancy. Instead, a state of immunity develops.[30,31] This is not so in goats where existing infection may invade the uterus when the animal becomes pregnant. Sheep seldom abort a second time. Watanabe et al.[32] found that 84% of those that had aborted at one pregnancy lambed normally at the next.

Retention of the fetal membranes often follows involvement of the uterus, but highly satisfactory levels of fertility in infected goats have been reported.[33] Evidence supporting the widely held opinion that brucellosis causes infertility is hard to find.

Abortion usually occurs during the last 2 months of pregnancy. The young may be born alive but weak. Death or lack of full development of the fetus appears to be due to interruption of blood supply rather than toxicity of the invading organisms, since heavily infected kids born to term often appear healthy and some aborted fetuses are found to have escaped infection. Some aborted fetuses may have an excess of blood-stained fluids in the body cavities, with enlarged liver and spleen. Others appear normal. Lesions when present are not pathognomic. In aborted fetuses brucellae are most numerous in the lungs, stomach contents, and spleen.

Infected fetal membranes show changes affecting part or all of the membrane. The cotyledons lose their blood-red appearance, becoming thickened, necrotic, and dull gray in color. There may also be thickening and sometimes edema of the membrane between the cotyledons. These lesions are packed with brucellae, and, therefore, infected or potentially infected membranes should be handled with care.

Excretion of brucellae from the infected uterus invariably follows abortion or an infected birth; in goats it is copious and prolonged, generally lasting for 2 to 3 months, sometimes longer. In many cases it will recommence after the succeeding pregnancy.[33] In sheep, excretion is less prolonged, usually ceasing within 3 weeks after abortion or a full-term parturition but in exceptional cases persisting for up to 3 months,[22,34,35] and even on rare occasions recommencing in the succeeding pregnancy after a normal birth.[36] Since the completely involuted uterus is not attractive for brucellae, the maintenance of infection in the animal from one breeding season to the next must depend on the persistence of the infection in other sites.

E. *B. MELITENSIS* IN KIDS AND LAMBS

Viable kids or lambs from an infected dam may themselves be infected with *B. melitensis*[37] but generally lose the infection by the time they are 2 months old.[38] Meador and Deyoe[39] observed that kids from infected dams lacked anti-*Brucella* antibody before receiving colostrum. Specific antibody, presumably of maternal origin, is often present soon after birth. It gradually diminishes, disappearing by the time of weaning even when the kid is still harboring brucellae. Goats appear unable to produce antibody before they are at least 2 months old. These observations need to be reconciled with the findings of Richardson et al.[40] and Hugh et al.[41] who found that sheep fetuses were able to mount both a cellular and a humoral response *in utero* but were not able to produce antibody during the first month of life. One practical implication is that serological tests during the first 2 months of life have no diagnostic significance. Whether or not a small proportion of infected lambs and kids retains inapparent infection into adult life, as happens with calves, remains a matter of speculation, but is likely, since in order to establish brucellosis-free flocks of sheep by rearing young lambs in isolation, it is necessary to take lambs only from serologically and allergically negative mothers.[42]

F. LOCALIZATION IN THE UDDER

The udder is a most important predilection site for *B. melitensis*. Infection in lactating nonpregnant goats is likely to lead to colonization of the udder with excretion of *B. melitensis* in the milk.[43] In goats, about two thirds of acute infections acquired naturally during pregnancy lead to infection of the udder and excretion of the bacteria in the milk during the subsequent lactation.[27] In some goats excretion may cease during this lactation, but in many it persists and often continues copiously during the next.[33] Greatly reduced milk yield follows abortion, and infection of the udder following a normal birth also leads to a considerable reduction in yield. In spite of this, physical signs of mastitis are seldom detectable in naturally infected goats.

Sheep that abort often excrete the bacteria in the milk, but generally for not more than

2 months. However, exceptionally, excretion may continue for 140[44] and even 180[45] d. We have been unable to find any record of excretion being continued into a second lactation.

Brucellae found in goats milk may be in the rough phase. So far this phenomenon has only been reported from an area where biovar 1 was almost the only biovar present.[46] All the colonies may be smooth, all may be rough, or there may be a mixture. Animals shedding only rough bacteria may be negative to serological tests carried out with smooth antigens. Rough *B. melitensis* cultures agglutinate in anti-rough *Brucella* serum, no matter what the species, but not with anti-smooth *Brucella* sera or monospecific anti-M or anti-A sera. Autoagglutination of rough suspensions may occur in agglutination tests, and, therefore, negative control sera are essential when cultures of *B. melitensis* are tested against monospecific or other sera.

G. OTHER CHANNELS OF EXCRETION

It is sometimes said that *B. melitensis* is frequently excreted in the urine of goats and sheep.[47] Two factors cast doubt on the veracity of this finding. First, it is often very difficult to exclude the possibility of contamination from other sites when obtaining samples of urine from the female urogenital tract. Second, *B. melitensis* is rarely found in the kidney at autopsy. Nevertheless, it seems likely that *B. melitensis* can be excreted on rare occasions in the urine.

H. SYMPTOMS

Symptoms, other than abortion, sometimes followed by retention of the fetal membranes are not usually noticed in naturally infected sheep and goats. However, when animals are artificially infected with large doses of bacteria given parenterally, symptoms including loss of condition, mastitis, and lameness have been reported in goats[48] and in sheep.[49] These, no doubt, occur on rare occasions in natural infection.

I. DURATION OF INFECTION AND IMMUNITY

In goats the infection may vary in duration from a very short period in slight infections rapidly overcome (especially in vaccinated animals) to persistence for years, where excretion of the organism in the milk may continue for at least two lactations and perhaps many more. The situation in sheep is very different, though this varies with the susceptibility of the breed. Striedter et al.[50] infected 101 sheep subcutaneously, and after the injection of small doses they were only able to recover the organism from local lymph nodes between 15 and 30 d post inoculation. Larger doses produced abortion and more prolonged infection (45 to 90 d), but in 54 sheep experimentally and 35 naturally infected, all were found to be bacteriologically negative between 4 and 13 months after infection. Taylor et al.[35] reported persistence in exceptional cases up to 17 months. Following recovery from infection with *B. melitensis,* sheep are very resistant to reinfection. Werschilova and Striedter[51] demonstrated powerful resistance to reinfection, even during pregnancy, up to 8 to 9 months, after which it began to decline, although some resistance was still demonstrable after 2 years.

J. PATHOLOGY

The macroscopic changes in the fetus and fetal membranes have already been described. In other organs and tissues lesions are rarely seen. In the acute phase there may be some swelling of lymph nodes and perhaps of the spleen and liver caused by infiltration of plasma cells and polymorphs with swelling and proliferation of the reticuloendothelial cells. Microscopic granulomas are formed.

The brucellar placentitis that develops in the infected ruminant uterus has been studied in considerable detail using *B. abortus*-infected goats as a model.[52,53] Histological studies suggest that bacteremic brucellae are endocytosed by erythrophagocytic trophoblasts and that placental vasculitis is important in producing abortion. Ultrastructural studies using colloidal gold-labeled anti-*B. abortus* IgG_1 showed that within the cells the bacteria were multiplying in the cisternae of rough endothelial reticulum (RER) with hypertrophy of affected RER.

K. INTERACTION BETWEEN SHEEP POX AND *BRUCELLA MELITENSIS* INFECTION

There are reports in the literature suggesting that *B. melitensis* infection can be excerbated by concurrent infection with the virus of sheep pox, and furthermore, that vaccination against these two diseases with a short interval between may also be dangerous.

Tarassow[54] described an outbreak of sheep pox which appeared to activate latent *Brucella* infections producing arthritis as a prominent symptom. *B. melitensis* was isolated from a number of sheep that had been infected previously but which were thought to have rid themselves of the infection. After the sheep pox epidemic was over, some of the sheep that had been affected with the reactivated brucellosis were slaughtered and examined, but no *B. melitensis* could be found.

Crowther et al.[55] vaccinated 49 pregnant sheep with reduced doses of Rev. 1, and 4 d later the sheep were vaccinated against sheep pox. Twenty-one abortions occurred during the 3 weeks that followed the Rev. 1 vaccination, an incubation period shorter than that seen in pure *B. melitensis* infection. Nevertheless, Rev. 1 was isolated from 8 of them. Later, another 16 sheep and 15 goats were vaccinated with reduced doses of Rev. 1, but this time not followed by sheep pox vaccination. There were no abortions. In the latter series of vaccinations the average dose of Rev. 1 was somewhat lower than in the former, but the suspicion remains that it was the sheep pox vaccination following so shortly after Rev. 1 that was responsible for the abortions.

V. *B. MELITENSIS* INFECTION IN MALE SHEEP AND GOATS

Male sheep and goats in an infected herd not infrequently become infected with *B. melitensis* themselves. Male sheep, like females are more resistant than goats. Andreani[56] exposed eight rams to a flock of artificially infected ewes, three of the rams being prevented from having sexual contact with the ewes. Four of the rams developed positive blood cultures, including one that had been denied sexual contact. A fifth ram was positive serologically, yet none developed any symptoms, and all eight were negative pathologically and bacteriologically when sacrificed 7 to 8 months after exposure commenced. This suggests that in sheep the infection may be short lived and symptomless. When symptoms do develop, orchitis is normally the only manifestation, usually appearing as a large firm swelling of one testicle. Reports of hygromas and lesions in joints have been quoted by Renoux.[57] It is likely that such lesions are very rare.

In male goats orchitis, generally unilateral, is quite common. It has the outward appearance of the same condition in bulls infected by *B. abortus* but appears to have been little studied.

VI. EPIDEMIOLOGY

Only when they are excreting the bacteria do sheep and goats present any danger to other animals or to man. Abortion, accompanied as it is by the emission of huge numbers of brucellae, is the chief source of infection for the onward transmission of the disease. If it is accepted that kids and lambs are resistant, excretion in the milk is likely to play a lesser part in transmission to other animals, and then only where the animals are milked; it is, however, a major source of infection for humans.

The chief danger of introducing the disease into a previously clean country or region is by the purchase of infected animals. In France there is a strong correlation between the prevalence of brucellosis in small ruminants and the practice of transhumance, i.e., the mingling of animals from different flocks and their transport to mountain pastures during summer for grazing.[35,58] Intermingling of flocks may occur under nomadic and seminomadic conditions of husbandry and also in static village flocks where animals are taken daily for grazing on common pasture.

A. SHEEP

The self-limiting nature of the disease in sheep, which is seldom accompanied by prolonged excretion of the bacteria, means that the disease does not easily maintain itself in small flocks except where replacement animals are purchased, where there is contact with other flocks, and where milking is practiced.[35] The infection is more evident in large flocks even when milking is not practiced,[7] and abortion rates of 10 to 16% have been reported.[7,32] Such flocks are often crowded together at night and at lambing, resulting in heavily infected pens which must play an important part in maintaining the infection between lambing seasons.[44] A number of abortions at one lambing period is often followed by fewer at the next, one or more years of more or less normal reproduction, after which the cycle recurs.

B. GOATS

Infection is generally introduced in the same way as in sheep.[59] The greater susceptibility of the goat, with more copious and prolonged excretion of the bacteria, results more often in abortion storms in goat herds. Nevertheless, a chronic state may soon become established within the herd, with only rare abortions, but the disease is liable to flare up if susceptible animals are introduced.

It is not thought that male sheep or goats play any significant role in spreading the infection to other animals, except, perhaps, mechanically. Neither are wild animals nor insect vectors thought to play any significant part in the epidemiology of sheep and goat brucellosis.

The occurrence of brucellosis among humans may be the first indication of the presence of *B. melitensis* in sheep and goats.

C. SURVIVAL OF *B. MELITENSIS* IN THE ENVIRONMENT

Horrocks[60] reported that *B. melitensis* survived in dried soil at 18°C for 69 to 72 d, but when the soil was wet brucellae could not be recovered after 7 d. Poliakov[61] reported that at low temperatures brucellae survived on wooden walls and the floors of pens for 4 months. Sunlight kills brucellae in hours.

VII. DIAGNOSIS

Isolation and identification of the bacterium is the only certain way of diagnosing *B. melitensis* infection, and this is often a practical method, especially in goats. However, there are many circumstances where culture methods are not practical or even possible, and it becomes necessary to rely on serological or allergic tests.

A. BACTERIOLOGICAL METHODS

When brucellae are present in tissues in large numbers, they may be identified in stained smears. This method depends on the fact that smooth brucellae are mildly acid fast. It is applicable to smears made from fetal membranes, fetal stomach contents, or fetal lung, vaginal discharge collected on a swab, or similar material. Smears are stained either by a modified Ziehl Nielsen method or by the Koster method.[15] The usefulness of this method depends, to some extent, on the disease situation in the area. The method does not distinguish between *B. melitensis* and *B. ovis,* and *Chlamydia* can only be distinguished from *B. melitensis* by an experienced operator. "Staining" of brucellae in smears by immunofluorescent methods has been described.[62] Background fluorescence tends to be a problem here, and the method has not been much used. Results obtained by tinctorial or immunofluorescent methods need to be verified by culture. The abundance of antigenic material generally present in infected fetuses and especially fetal membranes invites the development of more specific methods for antigen detection, such as enzyme immunoassay, to produce more effective differential diagnosis.

Culture methods suitable for the isolation and identification of *B. melitensis* are very similar to those recommended for the other *Brucella* species.[15,63] *B. melitensis* grows on the basal media recommended for other species of *Brucella,* but growth is somewhat less luxuriant than is the case with either *B. abortus* or *B. suis*. Growth of *B. melitensis* is not dependent on serum in the medium or an atmosphere enhanced by the addition of CO_2, but is somewhat improved by the presence of either or both of these. Selective media are usually required for the culture of specimens from the field, especially fetal membranes which are usually grossly contaminated. The selective medium of Kuzdas and Morse,[64] containing the antibiotics cycloheximide, bacitracin, and polymyxin B has been used extensively for the culture of *B. melitensis* and is adequate in most situations. Newer selective media such as those of Farrell[65] and Brodie and Sinton[66] are likely to be even more effective in suppressing contamination, but they have not been extensively proved for culture of *B. melitensis.* The appearance of colonies of *B. melitensis* is the same as those of other members of the genus. Isolates may be identified as brucellae by agglutination in specific anti-*Brucella* serum and as *B. melitensis* by routine typing tests. The characteristics of *B. melitensis* are described in section III.

Culture methods are most effective in diagnosis when applied to birth products such as fetal membranes, fetal lung, and stomach contents. Vaginal discharge taken on a swab is valuable for culture if taken immediately after parturition in sheep and during the first 2 months in goats. Culture of milk is more likely to yield positive results in goats than in sheep; chronic *Brucella* infection in goats is particularly associated with the presence of *B. melitensis* in the milk.[33] Culture of lymph nodes and tissues taken at the abattoir can be carried out without spoiling the carcass. Favored nodes and tissues for small ruminants are the same as for cattle.[15]

Animal inoculation, usually of guinea pigs or mice,[15] is sometimes resorted to for isolating *B. melitensis* from grossly contaminated materials such as cheeses, or where antibiotics are not available. Culture on selective medium is generally preferred.

B. SEROLOGICAL TESTS

The serological response of sheep and goats to *B. melitensis* infection has already been discussed in Section VI.A.

The serological tests generally used for the diagnosis of *B. melitensis* infection — Rose Bengal, agglutination and complement fixation — employ the same particulate antigens that are used for the diagnosis of *B. abortus* infection in cattle. The versatility of these antigens depends on their ability to react equally with anti-*B. abortus* sera (with the A antigen) and anti-*B. melitensis* sera. This cross-reaction is complete in sera against *B. melitensis* biovar 2 which also has the A antigen, and generally, but not always, with sera against *B. melitensis* biovar 1, which has the M antigen. There is evidence to suggest that in those regions, such as Latin America, where infection with *B. melitensis* biovar 1 predominates, some increase in diagnostic efficiency could be gained by employing antigens prepared from *B. melitensis* biovar 1.[67,68] This would not, of course, apply to the many regions where biovar 2 predominates.

It is often stated that serological tests in sheep and goats should be regarded as herd tests, since none of them is adequate for diagnosis in individual animals.[69,70] The buffered brucella antigen tests (card and Rose Bengal) appear promising, especially as herd tests, but have not been adequately evaluated, and the agglutination tests have serious deficiencies, as will be described. The complement fixation test appears to be as effective in sheep and goats as it is in cattle but is too complex for general use in some of the areas where *B. melitensis* is a serious problem.

The milk ring test, so useful with cow's milk, is not effective with the milk of sheep and goats. The fat globules in the milk of sheep and goats are smaller, and the antigen when agglutinated may rise to the surface to form a ring but generally falls to the base of the tube or remains suspended in the milk column.

Research in enzyme immunoassay holds promise for improved efficiency in detecting antibody to *B. melitensis,* especially antibody in milk. There is no apparent reason why this method should be less effective in small ruminants than it is in cattle, but to date published information is scarce.

The effectiveness of diagnostic methods as eradication tools varies with the conditions of husbandry under which the animals are kept. Serological tests have been used under favorable conditions to establish brucellosis-free herds of sheep and goats by removal of reacting animals,[9] sometimes in combination with the allergic reaction.[71] It follows that serological tests as eradication tools need to be evaluated in the conditions under which they are to be used.

All the currently available vaccines for sheep and goats stimulate the production of antibody which persists for varying periods of time after vaccination. The interference of vaccinal antibody with serological diagnosis varies with the different tests, and the different vaccines, and is discussed later under vaccines.

1. The Rose Bengal Test

The Rose Bengal test (RBT) and the card test are essentially the same and are known collectively as the buffered brucella antigen tests.[15] Because of its simplicity, the RBT is of great potential usefulness for routine diagnosis, but so far there is no body of evidence on its routine field use in small ruminants. There are, however, numerous favorable reports on the effectiveness of the RBT in sheep and goats.[72-75] The published evidence casting doubt on the efficiency of the test in these species is not convincing. The sensitivity and specificity appear to be at least as high as the serum agglutination test, and the RBT can be recommended at least as a herd test. Trap and Gaumont[76] found that the sensitivity of the test was increased by reducing the cell concentration from the usual 8 to 5%, and Fensterbank and Maquere[77] were able to eradicate brucellosis from a flock of sheep using the RBT in combination with an allergic test. The RBT is influenced by the temperature at which it is performed, and the antigen needs protection against heat during transport and storage. The above discussion relates to nonvaccinated animals; the RBT is very sensitive to vaccinal antibody, seriously limiting its use in vaccinated animals.

2. The Agglutination Test

This test may be done as a tube or a plate test.[15] The latter is very convenient for testing the sera of sheep and goats as a laboratory is not required. It is known that the ability of some immunoglobulin isotypes to agglutinate *Brucella* cells in suspension is affected by differences in the electrolyte concentration of the solution in which they are suspended.[78] Tube agglutination tests with sheep and goat sera are usually done with antigen and sera diluted in 5% saline solution with an added 0.5% of phenol. This should assure the detection of IgG_1 and IgG_2 in addition to IgM.

The agglutination test may be adequate as a herd test, but there is considerable doubt as to its value as a test for individual animals. Titers of 160 international units (IU) and above generally indicate that the animal is infected, but titers below this are very difficult to interpret since infected animals often have very low titers and not infrequently are completely negative, even in established infections.[22,79-81] This applies particularly to animals in an area where brucellosis is endemic, to those that are not pregnant, and to those that have been vaccinated. However, successful use of the plate agglutination test in controlling *B. melitensis* infection in sheep has been reported.[7]

Like the RBT, the agglutination test is very sensitive to antibody resulting from vaccination.

3. The Complement Fixation Test

Despite its complexity and the diversity of techniques employed in different laboratories, there is practically unanimous agreement that the complement fixation test (CFT) is the most effective of those commonly used for the diagnosis of brucellosis in individual sheep and goats.[82] The method is complex, and complement, which is very labile, needs to be accurately titrated. This requires good laboratory facilities and well trained personnel. In new infections the CFT becomes positive 1 to 2 weeks after the agglutination test, when the latter does become positive. This disadvantage seems to be without practical significance. Anticomplementary reactions are common, especially with sheep sera. This problem is reduced by using a warm rather than cold fixation technique and/or by inactivating the sera to be tested at 62°C.[15] This temperature inactivates early antibody and thus retards somewhat further the appearance of a positive reaction in new infections. The CFT is outstanding among the available diagnostic tests in its relative lack of sensitivity to antibody resulting from vaccination, particularly vaccination with Rev. 1.

4. Other Tests

There is evidence to suggest that the antiglobulin (Coombs) test is effective in diagnosing brucellosis in sheep[83,84] and also in goats, but because of the complexity of the technique its use may be restricted to special situations.

Counterimmunoelectrophoresis (CIE) has been proposed for the diagnosis of *B. melitensis* infection in sheep.[75,85] In both publications cited, the system was evaluated against the serum agglutination test, itself an imperfect criterion. Given the necessary apparatus, this is a simple and rapid procedure, but the evidence of its efficiency is by no means conclusive. Depending on a protein antigen the test will also be positive in the presence of antibody against *B. ovis*. However, it will not react with antibody against other bacterial species such as *Yersinia*. Cross-reactions are discussed later.

C. ALLERGIC TESTS

The delayed-type hypersensitivity (DTH) demonstrable in brucella-infected animals has been exploited for diagnosis to a greater extent in sheep and goats than in cattle. Early allergens were filtrates from old broth cultures. They tended to stimulate the production of antibody and to sensitize the animal and were generally unsatisfactory. An allergen produced by acid hydrolysis of brucella cells (Brucellysate, Brucellin fraction F) and lacking these inconveniences has been in use for many years in several countries.[62,71,86] More recently, a protein antigen (Brucellin INRA) free from smooth lipopolysaccharide (LPS) has been developed.[86,87] It is prepared from a rough strain of *B. melitensis* by extraction in cold hypertonic saline as described by Bhongbhibhat et al.[88] The preparation and standardization of this allergen has been described in detail.[15,89] It is now available commercially.

In sheep the reactions to the intradermic injection of allergen on the side of the neck or in the caudal fold are sometimes not very distinct, and the most suitable method of applying the allergen is to inject 0.5 ml subcutaneously in the lower eyelid. The usual dose of brucellin INRA by this route is 50 μg. The reaction is observed from above and in front; at 48 h positive reactions are seen as various degrees of swelling of the eyelid, from a very slight filling, only discernible by comparison with the other eye, to a large swelling, which closes the eye, produces tears, and spreads down the face. In goats the intradermal reactions are somewhat more pronounced, and the skin at the side of the neck or the caudal fold is satisfactory for intradermic injection.[90] Unfortunately, a long-lasting allergic state (for 2 years or more) is provoked by vaccination with either Rev. 1 or H38 vaccines.[91] The subject of cross-reactions with the allergic test is discussed in the next section. There should be no cross-reactions with *Yersinia enterocolytica,* but cross-reactions with *B. ovis* are to be expected.

Positive reactions are not seen in nonvaccinated, brucellosis-free flocks. In infected flocks, when reactions to the allergic test are compared with those to serological tests, a number of animals are seen to react to both tests, with smaller numbers reacting to either one or the other, but not to both.[92] Each of these latter groups may contain animals infected with *B. melitensis*. In general, the allergic test is considered not suitable for diagnosis in individual sheep and goats, but is very practical for screening to detect infected flocks.[93]

D. CROSS-REACTIONS IN SEROLOGICAL AND ALLERGIC TESTS

Where *B. melitensis* and *B. ovis* infections coexist in sheep populations, the question of cross-reactions between them assumes considerable importance. The immunochemistry of smooth and rough *Brucella* antigens is discussed in Chapter 4 but, unfortunately, we have little information on how these factors affect the routine diagnosis of *B. melitensis* infection. The smooth *Brucella* antigens used in the Rose Bengal, agglutination, complement fixation, and antiglobulin (Coombs) tests depend on smooth LPS, and, therefore, they should not react with antibody against the rough *B. ovis* which lacks smooth LPS. The experience of Erasmus and Bergh[94] using the smooth and rough *B. ovis* CF tests in Rev. 1-vaccinated rams suggests that this is so. However, in the case of enzyme immunoassay, preliminary reports[95] suggest that extensive cross-reaction occurs. The allergic and gel diffusion tests (including CIE) are mediated by protein antigens which are common to rough and smooth brucellae of whatever species, and, therefore, these tests are positive in sheep infected with *B. melitensis* and with *B. ovis*. As stated above, the allergic reaction is very useful in detecting *Brucella* infection in sheep flocks, but in view of the cross-reaction with *B. ovis*, it would, in certain epidemiological situations, be necessary to confirm positive allergic reactions with one of the more specific serological tests. These problems should not arise in testing goats since they are not affected by *B. ovis* in nature. However, *B. melitensis* vaccines dissociate readily, and the use of vaccine containing a significant number of dissociated cells is known to stimulate antibody in both sheep and goats indistinguishable from that seen in *B. ovis* infection.

In both sheep and goats, infection with *Y. enterocolytica* type 0.9 is likely to cause a cross-reaction in serological tests with smooth *Brucella* antigens in the same way as in cattle. Mittal and Tizzard[96] reported that *Y. enterocolytica* 0.9 infections were uncommon in goats in Canada, but Krogstad[97] found that while *Y. enterocolytica* was common in goats in Sweden, the infections encountered there were all with types other than 0.9, and they did not cross-react with *Brucella* antigens.

So far little has been reported concerning cross-reactions between brucellae and other bacterial infections in sheep and goats, but further investigations may well reveal a situation similar to that already discussed for cattle (see Chapter 8 and Alton et al.).[15] Allergic and gel diffusion tests, being specific for the genus, could be useful in distinguishing reactions caused by bacteria of the genus *Brucella* from those of other genera.

VIII. CONTROLLING *B. MELITENSIS* INFECTION IN SHEEP AND GOATS

Conditions in many regions where *B. melitensis* infection is prevalent are quite primitive, and control measures difficult to apply. There are often a lack of facilities, water, disinfectants, etc., and added to this, a widespread lack of knowledge on the part of animal owners and attendants. In many cases information campaigns by agriculture and health authorities are required as a first step toward any worthwhile measures of control. The Joint FAO/WHO Expert Committee on Brucellosis[82] offers advice on this subject in the context of the WHO campaign "Health for all by the year 2000".

Hygienic measures, to be discussed later, are important in limiting the spread of infection between animals and for protecting human beings from a dangerous zoonosis, but even when properly applied cannot be expected to do more than assist in controlling outbreaks.

So far, *B. melitensis* infection has only been eradicated on a national scale by identifying infected flocks and slaughtering the whole flock. Such a strategy will be limited to countries with a low prevalence or where the disease has recently been introduced and where highly developed infrastructure facilities and sufficient finance are available.

Because of the doubtful reliability of serological tests in individual animals the feasibility of eradication by test and slaughter seems to vary with the conditions under which the animals are kept. For example, in a pilot project on the island of Gozo (Malta),[9] 50 herds of goats were freed from infection by testing with the CFT and slaughtering reactors; many of the goats had been vaccinated with Rev. 1 vaccine when they were kids. Here the goats were kept in small herds and tied individually, always in the same place, when not grazing. These conditions are unusual. Generally, throughout the eastern mediterranean and central Asian regions large flocks, usually of sheep, are grazed during the day and penned at night, often loose in crowded enclosures where lambing may take place. Here eradication by test and slaughter is very much more difficult, but not impossible. Trilenko et al.[98] succeeded in eliminating the disease from several large flocks of sheep by combining vaccination with test and slaughter. In Cyprus, also, a national eradication campaign based on periodic testing without vaccination has had some success.[99]

In spite of these local successes, the vast majority of affected countries will need to rely on vaccination for the control of *B. melitensis* infection in sheep and goats, at least for the forseeable future. This has been recognized by the Commission of the European Communities.[100] It could be regarded as a measure for reducing the prevalence of the disease eventually to a level where eradication by test and slaughter could be considered.

A. VACCINATION

B. abortus strain 19 vaccine has been used for vaccinating sheep and goats but is no longer recommended for general use. However, in the U.S.S.R. brucellosis has been eradicated from large flocks of sheep by eliminating reactors, vaccinating adult ewes with strain 19, and ewe lambs with Rev. 1, and repeating serological tests annually.[101]

Of the vaccines now used for immunizing sheep and goats against *B. melitensis,* Rev. 1 vaccine is generally preferred.[82,100] Another living, attenuated vaccine, *B. suis* strain S2, which is administered orally, has been developed and used successfully on a large scale in China.[102] The nonliving *B. melitensis* H38 adjuvant vaccine is also used occasionally.

B. *B. MELITENSIS* REV. 1 VACCINE

The literature on Rev. 1 vaccine up to 1985 has been reviewed in some detail (see References 103 to 105).

Rev. 1 vaccine is a living attenuated vaccine prepared from the Rev. 1 strain of *B. melitensis*. Rev. 1 is a nondependent reverse mutant derived from a streptomycin-dependent strain of *B. melitensis,* which itself originated from a virulent strain.[106] The strain has several characteristics that allow its identification and differentiation from other strains of *B. melitensis,* such as colony size on solid medium, dye sensitivity, resistance to antibiotics, and persistence in guinea pigs.[15] The Rev. 1 strain retains some virulence. The normal vaccinal dose (10^9 cells) injected subcutaneously during pregnancy will regularly produce abortion with excretion from the vagina, and sometimes in the milk in both sheep[107,29] and goats.[108] However, the stability of the level of attenuation and its inability to transmit from vaccinated to nonvaccinated animals has been amply demonstrated for sheep[107] and goats.[108]

In goats, whether they are pregnant or not, vaccination is likely to give rise to a bacteremia and generalized infection, but it has been found that the infection is practically eliminated by 14 weeks post vaccination.[106,109] Sheep are believed to rid themselves of infection somewhat more quickly, but persistence may vary from breed to breed. In young male sheep of four French breeds, Rev. 1 infection was practically cleared in 11 weeks.[110]

In a number of experiments in which kids (3 to 8 months old) were vaccinated with Rev. 1 and then bred, excretion of the vaccine strain was never detected after parturition. However, if adult goats are vaccinated during lactation, a small proportion will excrete Rev. 1 in the milk. This is usually transitory, involving small numbers of bacteria, but has been known to persist into a second lactation.[27] Sheep vaccinated during lactation rarely excrete Rev. 1 in the milk; Entessar et al.[29] vaccinated 22 Iranian sheep, and Jones and Marley[111] 53 French ewes during lactation without subsequently being able to detect the vaccine strain in the milk. Ne'eman[112] examined 437 samples of milk from Awassi sheep vaccinated during lactation and isolated Rev. 1 from four of them.

The ability of the vaccine to produce a high level of immunity against both artificial and natural challenge has been convincingly demonstrated both for sheep[29,113-116] and goats.[117-121]

The duration of immunity conferred by vaccination with Rev. 1 was investigated by vaccinating Maltese goats when they were 4 to 12 months of age and challenging some at $2^1/_2$,[117] and others at $4^1/_2$[118] years after vaccination. Those challenged at $4^1/_2$ years were as resistant as those challenged at shorter intervals after vaccination, and it was concluded that immunity could be considered lifelong. Similar results were observed in Iranian sheep challenged $2^1/_2$ years after vaccination.[29]

1. The Serological Response to Rev. 1 Vaccination

Vaccination with Rev. 1 at the usual dose of 10^9 stimulates a serological response which begins 2 to 3 d after vaccination and rises to a peak at about 2 weeks, usually of 1000 IU or more in the agglutination and CF tests. The titers fall rapidly during the next month and then more slowly, depending on the particular test used.

Varela Diaz et al.[122] studied the levels of different immunoglobulin isotypes present in serum during 13 months following vaccination with Rev. 1 in a group of 30 goats. The response closely resembled that seen in cattle following strain 19 vaccination, i.e., a rapid development during the month following vaccination, first, of IgM followed later by IgG. During the next 3 months, the IgM fell to a low level where it persisted. The level of IgG immunoglobulins also fell, but by 4 months after vaccination they were undetectable in all except two of the goats, while the two exceptions continued to exhibit low levels of activity attributable to IgG throughout the rest of the period of observation.

Rev. 1 vaccination of goats that are already infected with *B. melitensis* had no effect on existing serological titers.[33]

Of the routine serological tests, the complement fixation test is least affected by Rev. 1 vaccination; this test is the only one likely to be effective in the diagnosis of *B. melitensis* infection in sheep and goats that have been vaccinated with full doses of Rev. 1 vaccine. In the vast majority of vaccinated goats, the CFT has returned to negative by 1 year after vaccination, especially when the serum samples are inactivated at 62°C.[123] More persistent titers have been reported in vaccinated sheep,[124] but the differences may be due to differences in the techniques used in the test.

The agglutination tests (tube and plate) remain positive much longer in vaccinated sheep and goats than the CFT. Young goats retained titers of up to 160 IU 4 years after vaccination, even though they had been maintained in a brucellosis-free environment.[123] In fact, agglutination titers reached a minimum 1 year after vaccination, then showed a slight but progressive increase in each of the next 3 years. In vaccinated lambs, titers seem to disappear more rapidly, but low titers may persist for at least 2 years.[29]

Rev. 1 vaccination also gives rise to long-term positivity in the Rose Bengal and allergic intrapalpebral tests. Durand et al.[91] reported that 50% of sheep and goats remain positive to the RBT after 1 year and a significant proportion after 2 years. They found that the intrapalpebral allergic test returned to negative sooner than the RBT, but even then 2.5 to 4% of vaccinated animals were still positive after 1 to 2 years compared to 5 to 15% for the RBT.

2. Recommendations for Using Rev. 1 Vaccine

In routine use the recommended dose of Rev. 1 is 10^9 viable organisms in a volume of 1 to 2 ml injected subcutaneously. Revaccination is not usually recommended in view of the duration of the immunity conferred, but Absatirov et al.[125] reported that an annual booster is of value in ewes. Because of the ability of Rev. 1 vaccine to cause abortion, pregnant animals should not be vaccinated, and at least 1 month should be allowed between vaccination and mating. For this reason and particularly because the antibody stimulated by Rev. 1 vaccination interferes for a time with serological diagnosis, it is generally recommended that vaccination be restricted to young animals aged between 3 and 8 months. This allows time for vaccinal antibody to subside before the vaccinates become adult and productive when they may need to be tested. However, in many situations, restricting vaccination to young animals seriously reduces its impact; acute outbreaks almost always occur when most of the flock is pregnant and live vaccines, such as Rev. 1, cannot be used in full doses. What is needed is a method that can be used when required to raise the level of resistance of the whole flock quickly. Also, and especially where there is an extended lambing season, it is often difficult and expensive to find enough kids and lambs at the right age for vaccination to provide effective cover.

In areas where sheep have a restricted breeding season there will be several months in the year when none of the flock is pregnant, and here the adults may be vaccinated with the routine dose of Rev. 1 as soon as practical after lambing, provided the rams can be kept away from the ewes during this period and for 1 month after the vaccinations have been completed. Excretion of Rev. 1 in the milk is likely to occur only rarely, to be short-lasting, and to be of no importance when the ewes are not milked. Serological testing is feasible 1 year later, provided the complement fixation test is used. In view of the varying susceptibility of different breeds of sheep to *B. melitensis* infection and to vaccination, a preliminary trial should be carried out in all cases where there is any doubt as to how the particular breed of sheep is going to react to the vaccine. In a self-contained flock the vaccination of adults is only necessary in the first year of a vaccination campaign; subsequently, all the female progeny can be vaccinated when they are between the ages of 3 and 8 months.

With goats the opportunities for vaccinating adults are limited. The breeding season is more prolonged, and adult goats are more likely to excrete Rev. 1 in the milk and for a longer time than sheep. They are almost always milked. These difficulties have prompted research into alternative methods of vaccinating adult goats (and sheep), including those that may be pregnant, e.g., vaccination with reduced doses and conjunctival vaccination. These developments are discussed later.

3. Field Experience with Rev. 1 Vaccine

Rev. 1 vaccine has now been in field use for more than a quarter of a century. The primitive conditions under which vaccination of sheep and goats is often done make evaluation of the results difficult. In many regions evaluation has to be based on the effect of animal vaccination on the prevalence of human brucellosis. Rev. 1 was introduced into routine use in Malta in 1959, and during the next 7 years the number of human cases registered each year fell from 260 to 29. Much of this fall must have been due to vaccination in spite of its being restricted to kids and less than half of the eligible animals being vaccinated each year.

Since 1959 many millions of sheep and goats have been vaccinated with Rev. 1 in France, Greece, Iran, Israel, Italy, Mongolia, Peru, Portugal, Spain, and the U.S.S.R. among others. From the U.S.S.R. Ulasevich et al.[116] reported that Rev. 1 vaccination of lambs along with periodic serological testing of adults and elimination of reactors succeeded in clearing 12 large flocks of *B. melitensis* infection. Kolar[69] reported that in Mongolia the widespread use of Rev. 1 in sheep and goats was accompanied by a fall in the incidence of human brucellosis from 4.8 per 10,000 population in 1974 to 0.23 in 1981. Good results have also been reported from Italy,[126] where, in one district, widespread vaccination of lambs was accompanied by a marked fall in the occurrence of human brucellosis. In contrast, Landau[127] reported that in Israel "there has been no clear-cut decline in the incidence of human brucellosis" resulting from the use of Rev. 1.

Two approaches are presently being investigated in an attempt to make Rev. 1 vaccination more universally applicable.

4. Vaccination with Reduced Doses of Rev. 1

A reduced dose of 5×10^4 Rev. 1 cells appears to protect goats almost as well as the full dose without causing abortion or excretion of the vaccine strain and with a much reduced serological response, particularly noticeable with the Rose Bengal (card) test.[120,128-130] In the latter study, all of 18 adult female goats vaccinated subcutaneously with 5×10^4 Rev. 1 cells were negative serologically by 105 d after vaccination. The tests used were agglutination, complement fixation, and card. In spite of these favorable reports, reduced dose vaccination does not appear to have been much used in the field.

5. Conjunctival Vaccination with Rev. 1

Conjunctival vaccination, like reduced-dose vaccination, offers two potential advantages: first, a reduction in the serological and/or allergic response, and second, a safe method for vaccinating adult animals that may be pregnant. When young kids[131] or lambs[132] are given the normal dose of Rev. 1 vaccine (10^9) conjunctivally, the serological, but not the allergic, response is greatly curtailed without significant reduction in the immunity conferred. The serological tests, including the RBT, return to negative by 4 months after vaccination. Preliminary results in adult sheep confirm that conjunctival vaccination curtails the serological response,[133] but it appears that some reduction in dose will be needed to remove the risk of abortion.[134]

6. Production of Rev. 1 Vaccine

Specific requirements for *B. melitensis* Rev. 1 vaccine have been formulated by the World Health Organization.[135] The requirements are very similar to those for *B. abortus* strain 19 vaccine. Either solid[15] or liquid[136] medium may be used. *B. melitensis* is more liable to S-R changes than other brucellae, and in some laboratories dissociation has been troublesome in the production of Rev. 1 vaccine. Dissociation is discouraged by limiting the heat applied during sterilization of the medium (filtering rather than autoclaving, where possible), otherwise by keeping the medium for at least a week after autoclaving before seeding. Some workers add serum to the medium to discourage dissociation.[15] It has already been pointed out that Rev. 1 vaccine containing numerous dissociated cells will stimulate antibody which reacts with *B. ovis* antigens.

7. Safety in Handling Rev. 1 Vaccine

Though greatly attenuated, the Rev. 1 strain is capable, under certain circumstances, of causing local hypersensitivity reactions or generalized brucellosis. Workers have contracted brucellosis from aerosols produced in the manufacturing process,[137] and accidental self-inoculation with vaccine may occur in the field. Many veterinarians and vaccinators have become sensitized to *Brucella* infection in the course of their work, and reactions to accidental inoculation of vaccine occurring within 1 or 2 h are generally due to hypersensitivity rather than infection and need treating as such.

In outbreaks of brucellosis, the work of handling sheep and goats for vaccination exposes the personnel involved to considerable risk of contracting brucellosis by contact with infected animals; this leads sometimes to unjustified accusations against the vaccine. Those involved should be aware of the risks, and proper medical care should be available to them.

C. ORALLY ADMINISTRABLE *B. SUIS* STRAIN 2 VACCINE

This vaccine, which can be given orally, was developed in China.[102] It is of special interest for the vaccination of sheep and goats in difficult and arid terrain where it can be given in drinking water.

The vaccine consists of a smooth attenuated strain of *B. suis* biovar 1, apparently indistinguishable, except in its virulence, from field strains. It has about the same virulence for guinea pigs as strain 19. The stability of its level of attenuation has been established by passage in pregnant sheep, goats, and pigs. It causes abortion in pregnant sheep and goats if injected in full doses but not when given orally. Good protection has been demonstrated in sheep and goats, persisting for up to 4 years. The antibody produced in response to oral administration of S2 vaccine, measured by the complement fixation and agglutination tests, disappears by 9 months postvaccination. This vaccine has been used in Inner Mongolia for almost 20 years to vaccinate many millions of sheep and goats. Here the vaccine is usually administered in drinking water. No problems have emerged, and it is claimed that vaccination has reduced the incidence of brucellosis in the region to a very low level.

The FAO/WHO Expert Committee on brucellosis[82] has shown interest in the S2 vaccine, and evaluations are in progress in a number of countries.

D. *B. MELITENSIS* H38 ADJUVANT VACCINE

This is a killed vaccine prepared from a virulent smooth strain of *B. melitensis* biovar 1, called H38, emulsified in an oily adjuvant.[113] It is available commercially. Other smooth strains of *B. melitensis* biovar 1 can be used to produce a very similar vaccine.[23]

The immunity produced is similar to that given by Rev. 1 vaccination, but, according to Gaumont et al.,[121] is somewhat less solid. These workers gave a single dose; others[7] prefer to use two doses. The advantage of H38 vaccine over Rev. 1 is that it can be given to animals of any age whether they are pregnant or not. Against this, H38 vaccine has two considerable disadvantages. First, it gives rise to a prolonged serological and allergic response which prevents diagnosis or surveillance by these means for several years after vaccination. Second, it produces a local reaction at the site of inoculation that sometimes leads to suppuration and often antagonizes owners.

Good results have been reported following the vaccination of small ruminants with H38 vaccine in the field,[138] but there have been disappointments. Because of the disadvantages mentioned, H38 vaccine has not received wide acceptance in field use.

E. HYGIENIC MEASURES FOR THE CONTROL OF *B. MELITENSIS* INFECTION

Hygenic measures help to limit the spread of infection in the flock. Furthermore, the resistance provided by vaccination against *B. melitensis* infection is by no means absolute, and measures that will lessen the natural challenge presented to vaccinated sheep and goats will improve the chances of success. Briefly, there are five main areas needing attention.

1. Owners need to be educated in the basic principles and to know whether their flocks are infected or not.
2. The provision of clean accommodation at parturition that can be disinfected is very advantageous. Individual pens, where possible, are advantageous. In contrast, the

repeated use of fouled, crowded pens for lambing and kidding aids the spread and the persistence of infection.

3. In an infected herd, all parturitions should be regarded as potential sources of infection, and the nonliving birth products (membranes and aborted fetuses) should be incinerated where possible, otherwise buried deeply. Heavy, leak-proof, agricultural plastic sacks are useful for transporting potentially contaminated materials. Contaminated areas should be disinfected before and after sweeping or hosing down. Animals that abort should be removed, but not sold, except for immediate slaughter.
4. Unneccessary visitors, especially children, should be kept away during the lambing period.
5. Hygiene at milking should be adequate to prevent the spread of infection from udder to udder or to the milker.

F. ERADICATION BY TEST AND SLAUGHTER

The most certain way to eradicate *B. melitensis* infection is to identify infected herds and to slaughter the whole herd.

It has already been pointed out that few countries with a *B. melitensis* problem will have the necessary means to eradicate the infection on a national scale by test and slaughter of reactors and will need to rely on vaccination to control the disease. Nevertheless, where the overall individual prevalence of the disease is below 2% and where flocks are maintained under closely controlled conditions and can be protected against reentry of infection, eradication may be feasible, perhaps on a farm or regional basis, provided (1) an adequate veterinary organization for surveillance and laboratory testing is available and (2) the administrative means and will to implement the program, including the control of movements of animals into clean herds, is present.

The basic requirements for an eradication program are listed below, but any widespread campaign should be preceded by a pilot project which, if successful, can be expanded into a full-scale eradication program. It is essential that the willing cooperation of animal owners be assured before such a program is introduced.

1. Flocks not already known to be infected need to be surveyed to establish the presence or absence of infection. In nonvaccinated flocks the allergic palpebral test described in Section VII.C is recommended for this purpose. It may be advisable to remove all reactors to this test in addition to those detected by subsequent serological investigations.[71]
2. In infected flocks, all animals of breeding age need to be individually identified by tattoo or ear tag, then bled for serological testing at intervals of 30 to 60 d until a clean test is achieved. After this, two clean tests at 6-month intervals suffice to allow the flock to be considered as brucellosis-free. The sera should be tested by the complement fixation test. In the absence of vaccination, the RBT has been used successfully in place of the CFT.[77]
3. In flocks where the Rev. 1 vaccine has been used, the CFT will be required for serological diagnosis. Various combinations of vaccination and testing have been successful in bringing about eradication.[101]
4. Reacting animals must be removed for slaughter as quickly as possible and, where necessary, adequate compensation paid to owners.
5. New animals must not be introduced into brucellosis-free flocks from sources other than those proved to be brucellosis-free.
6. At intervals of 1 or 2 years, certified brucellosis-free flocks should be subjected to the same flock test that is used in surveillance.

IX. *BRUCELLA MELITENSIS* INFECTION IN CATTLE

In regions where *B. melitensis* is prevalent in sheep and goats, cattle are liable to catch the infection from them. *B. melitensis* was isolated from cow's milk in Malta as early as 1905 by Shaw,[1] and 30% of the cows he examined had positive agglutination titers. In 1980, Feinhaken and Dafni[139] reported that 40% of *Brucella* strains isolated from cattle in Israel were *B. melitensis.*

It has not yet been established whether *B. melitensis* can maintain itself indefinitely in a cattle population in the absence of infected sheep or goats, but an extensive outbreak has been reported in a herd of cows where no sheep or goats were present.[140] *B. melitensis* infection in cattle sometimes causes them to abort, but less frequently than *B. abortus,* and a focus of infection in cattle may die out spontaneously. Unfortunately, colonization of the udder is frequent, and excretion of the organism in the milk may be prolonged for months or years and has frequently led to epidemics of brucellosis in people working with the cattle or drinking the milk.

The serological response is similar to that seen in *B. abortus* infection, and eradication by conventional methods of test and slaughter is said to be relatively easy. If circumstances should require control by vaccination, then available evidence suggests that vaccination of calves and also young yaks with Rev. 1 would be effective in protecting against both *B. melitensis* and *B. abortus.*[141,142]

B. melitensis infection in cattle has been reviewed in detail by Verger,[143] who provides an extensive bibliography.

X. *BRUCELLA MELITENSIS* IN OTHER ANIMALS

A. DOGS

Sheep dogs working with infected flocks frequently become infected with *B. melitensis* but rid themselves of the infection relatively quickly. Islamov, quoted by Hellman,[144] killed and examined 36 serologically positive sheep dogs from infected flocks. During the 6 months following the start of lambing, 9 of 25 dogs examined yielded *B. melitensis,* while during the 7 months after lambing had ended, all of the 11 dogs examined were negative bacteriologically. This suggests that dogs are not a factor in perpetuating the disease in a flock, and we have no evidence to contradict this. Nevertheless, the eradication system in West Germany requires that when flocks of sheep are depopulated, sheep dogs also must be eliminated.

Dogs and wolves may be more important as mechanical disseminators of infection by carrying away infected materials such as fetuses or fetal membranes.

B. CAMELS

There is serological evidence suggesting a high level of *Brucella* infection in camels in a number of countries, but the rather high susceptibility of this species to *B. abortus* infection[145] tends to obscure the true position of *B. melitensis*. The isolation of *B. melitensis* from camels has been reported only rarely.[146]

C. SAIGAS

Saigas are sheep-like antelopes with strange inflatable pouch-like nostrils. *Saiga tatarica* is encountered on marginal grazing lands in the Kazarkistan and Stavropol districts of the U.S.S.R. Saigas often share watering places with domestic sheep. *B. melitensis* is said to spread naturally among Saigas and to cause abortion.[147] The position seems somewhat obscure, but they are considered a possible source of *Brucella* infection for domestic animals.

D. OTHER ANIMALS

Any of the small mammals that come in contact directly or indirectly with infected sheep or goats seem liable to temporary infection with *B. melitensis*. Rats, mice, rabbits, hares, wild guinea pigs, and even domestic poultry have all been found infected, but there is no evidence of significant disease being caused or of any ability to transmit infection. Wild birds, especially vultures, have the ability to transmit infection mechanically, but we have no information on whether they play a part in the epidemiology of the disease.

Outbreaks of brucellosis in fur-bearing animals such as mink have been traced to the feeding of infected meat, e.g., dead lambs containing *B. melitensis*.[148] This poses a threat to workers killing and skinning these animals.

Ticks and biting insects have been found to harbor *B. melitensis*, but none of these creatures have been incriminated in the epidemiology of the disease.

XI. ZOONOTIC ASPECTS OF *B. MELITENSIS* INFECTION

In the more difficult agricultural regions of the world, owners of sheep and goats are often hard pressed to earn a living for themselves and their families. The economic wastage resulting from loss of young replacement animals, reduction of milk yield, etc. is keenly felt, even when the cause is not clearly understood, but this is greatly exceeded in terms of human misery by the impact of *B. melitensis* infection on human health. *B. melitensis* is transmitted more readily from animals to man than other members of the genus. The resulting disease is severe, often long lasting, and rural populations may be unable to support the expense of extended antibiotic treatment. Since every human case has an animal origin, the benefits of controlling or, better still, eradicating the disease from the animal population are obvious.

Communication from animal to man may be direct or indirect.

Direct contact with infected animals is dangerous for animal attendants, owners of infected animals, and their families. Acute outbreaks often occur among people working in or living near an infected premise. Ezilondo[149] described an outbreak in which 113 persons in a village of 351 inhabitants became ill. The maximum danger is during the lambing or kidding period when infected animals, whether they have aborted or not, may discharge billions of bacteria from the uterus in birth products and the discharges that follow. In goats especially, this emission may continue for 2 to 3 months, causing goat pens to become saturated with infection. Under dusty conditions aerosols containing dangerous numbers of brucellae may also be responsible for human infection even on public roads.

Dairy products are the main source of infection for people who do not have direct contact with animals. The catastrophic effect of drinking untreated, infected goat milk was dramatically demonstrated among military personnel in Malta. In 1905 they experienced 913 cases of brucellosis (Mediterranean fever), but in 1907, after the drinking of goat's milk had been prohibited, the number had fallen to 21.[150] In some countries of West Asia, human brucellosis caused by *B. melitensis* appears to be increasing rapidly; e.g., available figures for Kuwait show 32 cases in 1980, 77 in 1982, and an estimated average of 2000 for each of the 2 years 1985 and 1986.[151] Much of the milk drunk is now rendered safe by pasteurization or boiling, but cheeses made from sheep or goat milk are very popular in most countries where sheep or goats are the main source of milk. These cheeses are generally prepared from untreated milk and by the use of rennet that may have come from *Brucella*-infected stomachs of lambs or kids. During the course of cheese manufacture any brucellae present in the milk become trapped in the clot and, so, concentrated in the cheese. This type of cheese is usually consumed within 1 to 2 weeks of preparation. Such cheeses are a major vehicle for transmission of *B. melitensis* to the human population, sometimes in areas remote from where the cheese was prepared. Escalente and Held[152] recorded one such outbreak in which more than 6000 people became ill. In Iran, Sabbaghian[153] examined 1220 specimens from fresh sheep cheese and found that 87 (7%) of them contained *B. melitensis*. The average number of brucellae per gram in the contaminated cheeses was 2120, but 40% of them had become negative after 1 week, and all except one were free from brucellae after 9 weeks. In this series the pH of the positive samples varied between 5 and 7.6 at the time the cultures were made. Alton[154] calculated the half-life of *B. melitensis* in the fresh cheeses of Malta to be about 1 day. Cream and ice cream prepared from goat milk has also been incriminated.[6] The incidence of human brucellosis acquired from dairy products is also seasonal, reaching a peak soon after kidding and lambing times.

Abattoir workers handling infected sheep or goats are also at risk, especially from the contents of uteri and udders. In the U.S.S.R. the handling of raw wool has been identified as a frequent source of infection among the workers involved. *B. melitensis* is one of the most easily acquired laboratory infections. Detailed suggestions for protecting human populations have been published.[82]

The zoonotic aspects of *B. melitensis* infection were discussed in some detail in a previous CRC publication.[6]

REFERENCES

1. **Shaw, E. A.,** Mediterranean fever in goats, cows and other animals, in *Reports of the Commission on Mediterranean Fever, Part IV,* Harison and Sons, London, 1906, 16.
2. **Bruce, D.,** Note on the discovery of a micro-organism in Malta fever, *Practioner,* 39, 161, 1887.
3. **Zammit, T.,** A preliminary note on the examination of the blood of goats suffering from Mediterranean fever, in *Reports of the Commission on Mediterranean Fever, Part III,* Harrison and Sons, London, 1905, 83.
4. **Horrocks, W. H.,** A preliminary note on goats as a means of propagation of Mediterranean fever, in *Reports of the Commission on Mediterranean Fever, Part III,* Harrison and Sons, London, 1905, 84.
5. **Whiting, R. D., White, B. M., and Stiles, F. C.,** An epizootic of *Brucella melitensis* infection in Texas, *J. Am. Vet. Med. Assoc.,* 157, 1860, 1970.
6. **Flores-Castro, R. and Baer, G. M.,** Brucellosis *(Brucella melitensis),* Zoonotic implications, in *CRC Handbook Series in Zoonoses, Section A: Bacterial, Rickettsial, and Mycotic Diseases,* CRC Press, Boca Raton, FL, 1980, 195.
7. **Radwan, A. I., Bekairi, S. I., Al-Mukayel, A. A., and Abu Agla, O. S.,** Control of ovine brucellosis in Najdi sheep in Saudi Arabia, *Trop. Anim. Health Prod.,* 16, 213, 1984.
8. **Garcia-Carrillo, C., Szyfres, B., and Gonzales-Tome, J.,** Tipificacion de brucelas aisladas del hombre y los animales en America Latina, *Rev. Latin Am. Microbiol.,* 14, 117, 1972.
9. **Alton, G. G.,** The control of animal brucellosis in Malta, *St. Luke's Hosp. Gaz. Malta,* 3(2), 74, 1968.
10. **Caporale, V., Donatella Nannini, Petracca, G., and Paganico, G.,** Epidemiology and control of *Brucella melitensis* in Italy, in *Brucella melitensis,* Verger, J. M. and Plommet, M., Eds., Martinus Nijhoff, Dordrecht, 1985, 51.
11. **Danielides, V.,** Brucellosis in Greece, *J. Infect.,* 3, 392, 1981.
12. **Verger, J. M., Grimont, F., Grimont, P. A. D., and Grayon, M.,** *Brucella,* a monospecific genus as shown by deoxyribonucleic acid hybridization, *Int. J. Syst. Bacteriol.,* 35, 292, 1985.
13. **Corbel, M. J. and Morgan, J. W. B.,** Genus Brucella, Meyer and Shaw 1920, in *Bergey's Manual of Systemic Bacteriology, Volume 1,* Krieg, N. R., Ed., Williams & Wilkins, Balitmore, 1984, 377.
14. **Corbel, M. J.,** Recent advances in Brucella-phage research, *Vet. Bull.,* 54, 65, 1984.
15. **Alton, G. G., Jones, L. M., Angus, R. D., and Verger, J. M.,** *Techniques for the Brucellosis Laboratory,* I.N.R.A., Paris, 1988.
16. **Quin, R., Campbell, A. M., and Phillips, A. P.,** A monoclonal antibody specific for the A antigen of *Brucella* spp., *J. Gen. Microbiol.,* 130, 2285, 1984.
17. **Cappuccinelli, P., Fiori, P. L., Gargani, G., and Pacetti, A. M.,** Antigenic differences between *Brucella abortus* and *Brucella melitensis* recognized by a monoclonal antibody, *Microbiologica,* 9, 179, 1986.
18. **Greiser-Wilke, I., Moennig, V., Thon, D., and Rauter, K.,** Characterization of monoclonal antibodies against *Brucella melitensis, Zentralbl. Veterinaermed. Reihe,* B, 32, 616, 1985.

19. **Renoux, G., Alton, G. G., and Mahaffey, L. W.,** Etudes sur la Brucellose ovine et caprine, II, Etablissemment de la D. 50 d'une souche de *Brucella melitensis* pour la chevre et la brebis, *Arch. Inst. Pasteur Tunis,* 32, 65, 1955.
20. **Tarassow, I.,** Infection brucellique experimentale des moutons par le peau et les muqueuses, *Off. Int. Epizoot., Bull.,* 17, 555, 1938.
21. **Biggi, P.,** Indagini sulla batteriemia brucellare in pecore sperimentalmente infettate, *Ann. Fac. Med. Vet. Pisa,* 9, 264, 1956.
22. **Shimi, A. and Tabatabayi, A. H.,** Pathological, bacteriological and serological responses of ewes experimentally infected with *Brucella melitensis, Bull. Off. Int. Epizoot.* 93, 1411, 1981.
23. **Alton, G. G.,** A comparison under natural conditions of two vaccines for the immunization of goats against *Brucella melitensis* infection, *Res. Vet. Sci.,* 2, 232, 1961.
24. **Keppie, J., Williams, A. E., Witt, K., and Smith, H.,** The role of erythritol in the tissue localization of the brucellae, *Br. J. Exp. Pathol.,* 46, 104, 1965.
25. **Alton, G. G.,** Serological tests for the diagnosis in goats, in *Report of the FAO/WHO Meeting on the Control of Brucellosis in the Mediterranean Region,* Food and Agriculture Organization, Rome, 1964, 64.
26. **Madad, A.,** *A Study of Specific Serum Immunoglobulins in Ovine Brucellosis, Summary of thesis,* Aristotle University, Thessaloniki, Greece, undated.
27. **Alton, G. G.,** The epidemiology of *Brucella melitensis* infection in sheep and goats, in *Brucella melitensis,* Verger, J. M. and Plommet, M., Eds., Martinus Nijhoff, Dordrecht, 1985, 187.
28. **Paltrinieri, S., Biggi, P., and Farina, R.,** Brucellosi ovina ed autosterlizzazione spontanea, *Ann. Fac. Vet. Med. Pisa,* 9, 251, 1956.
29. **Entessar, F., Ardalan, A., Ebadi, A., and Jones, L. M.,** Effect of living Rev. 1 vaccine in producing long term immunity against *Brucella melitensis* infection in sheep in Iran, *J. Comp. Pathol.,* 77, 367, 1967.
30. **Werschilowa, P. and Tarassow, I.,** Pathogenie des avortements brucelliques des moutons, *Off. Int. Epizoot. Bull.,* 17, 556, 1938.
31. **Biggi, P.,** L'aborto brucellare negli ovini e la sua patogenesi, *Ann. Fac. Med. Vet. Pisa,* 10, 97, 1957 (publ. 1958).
32. **Watanable, S., Tajima, Y., Ito, S., Itabashi, K., and Yoda, H.,** Etudes sur l'avortement epizootique du mouton, *Off. Int. Epizoot. Bull.,* 15, 1002, 1938.
33. **Alton, G. G.,** The reaction of goats naturally infected with *Brucella melitensis* to vaccination with a living attenuated vaccine, *Res. Vet. Sci.,* 3, 326, 1962.
34. **Farina, R. and Biggi, P.,** Sulla eliminazione delle brucelle per via vaginale in pecore sperimentalmente infettate, *Ann. Fac. Med. Vet. Pisa,* 9, 278, 1956.
35. **Taylor, R. M., Lisbonne, M., Vidal, L., and Hazemann, R. H.,** Quelques notes epidemiologiques sur l'infection des chevres et des brebis par *Br. melitensis, Rev. Med. Vet.,* 90, 185, 1938.
36. **Tazima, Y., Ito, S., Okuno, Y., Oyama, T., and Itabashi, K. J.,** Excretion apres le part chez les brebis naturellement infectées par *Br. melitensis* (summary), *Rec. Med. Vet.,* 116, 335, 1940.
37. **Pisanu, S.,** Presenza di *Brucella melitensis* in capretti regolarmente macellati, *Atti. Soc. Ital. Sci. Vet.,* 39, 622, 1985 (publ 1986).
38. **Alton, G. G.,** *Brucella melitensis* infection in female kids born into a heavily infected environment, *Br. Vet. J.,* 126, 61, 1970.
39. **Meador, V. P. and Deyoe, B. L.,** Experimentally induced *Brucella abortus* infection in pregnant goats, *Am. J. Vet. Res.,* 47, 2337, 1986.
40. **Richardson, M., Beck, C. C., and Clark, D. T.,** Prenatal immunization of the lamb to Brucella, dissociation of immunocompetence and reactivity, *J. Immunol.,* 101, 1363, 1968.
41. **Hugh, A. R., Trevella, W., Simpson-Morgan, M. W., and Morris, B.,** The lymph-borne response of fetal lamb lymph nodes to challenge with *Brucella abortus* in utero, *Aust. J. Exp. Biol. Med. Sci.,* 63, 381, 1985.
42. **Didovets, S. R.,** Programme d'eradication de la brucellose ovine en Ukraine, *Off. Int. Epizoot. Bull.,* 63, 1225, 1965.
43. **Renoux, G., Alton, G. G., and Sacquet, E.,** Etudes sur la brucellose ovine et caprine. X. Elimination de *B. melitensis* par le lait de chevres suedoises artificiellement infectees, *Arch. Inst. Pasteur Tunis,* 33, 413, 1956.
44. **Itabashi, K., Watanabe, S., Ito, Y., Tajima, Y., and Otaki, K.,** Etudes sur l'avortment epizootique du mouton (summary), *Off. Int. Epizoot. Bull.,* 15, 1000, 1938.
45. **Biggi, P.,** Eliminazione delle brucelle con il latte in pecore sperimentalment infettate, *Ann. Fac. Med. Vet. Pisa,* 9, 271, 1956.
46. **Alton, G. G.,** The occurrence of dissociated strains of *Brucella melitensis* in the milk of goats in Malta, *J. Comp. Pathol.,* 70, 10, 1960.
47. **Werschilowa, P.,** L'excretion de bacilles par le lait, l'urine, la secretion vaginale chez les moutons naturellement et experimentalement infectes, *Off. Int. Epizoot. Bull.,* 17, 557, 1938.
48. **Polding J. B.,** in *Brucellosis in Man and Animals,* Huddleson, I. F., The Commonwealth Fund, New York, 1943, 197.

49. **Tarassow, I.,** Contribution a l'aspect clinique de la brucellose ovine, *Off. Int. Epizoot. Bull.*, 17, 557, 1938.
50. **Striedter, W., Woskressenski, B., and Kotlarowa, H.,** Sur la bacterioscopie des moutons naturellement et artificiellement infectes par des Brucella, *Off. Int. Epizoot. Bull.*, 17, 554, 1938.
51. **Werschilowa, P. and Striedter, W.,** Etudes sur l'immunite dans la brucellose ovine par la methode de la surinfection, *Off. Int. Epizoot. Bull.*, 17, 563, 1938.
52. **Anderson, T. D., Meador, V. P., and Cheville, N. F.,** Pathogenesis of placentitis in the goat inoculated with *Brucella abortus.* I. Gross and histologic lesions, *Vet. Pathol.*, 23, 219, 1986.
53. **Anderson, T. D., Cheville, N. F., and Meador, V. P.,** Pathogenesis of placentitis in the goat inoculated with *Brucella abortus.* II. Ultrastructural studies, *Vet. Pathol.*, 23, 227, 1986.
54. **Tarassow, I.,** La brucellose activee sous l'influence de la clavelee, *Off. Int. Epizoot. Bull.*, 17, 564, 1938.
55. **Crowther, R. W., Orphanides, A., and Polydorou, K.,** Vaccinaton of adult sheep with reduced doses of *Brucella melitensis* strain Rev. 1, *Trop. Anim. Health Prod.*, 9, 85, 1977.
56. **Andreani, E.,** Decorso dell'infezione brucellare in montoni sotoposti a contagio naturale, *Ann. Fac. Med. Vet. Pisa,* 19, 331, 1967.
57. **Renouz, G.,** Brucellosis in goats and sheep, in *Advances in Veterinary Science,* Vol. 3, Brandley, C. A. and Jungherr, E. L., Eds., 1957, 242.
58. **Gilles, G.,** Quelques considerations sur la brucellose des petits ruminants a *Brucella melitensis, Rec. Med. Vet.*, 153, 207, 1977.
59. **Zapatel, J. and Malaga, H.,** Epidemiologia de la brucelosis caprina in el Peru, *Bol. Of. Sanit. Panam.*, 71, 121, 1971.
60. **Horrocks, W. H.,** On the duration of the life of the *Micrococcus melitensis* outside the human body, in *Reports of the Commission on Mediterranean Fever, Part I,* 5, 1905.
61. **Poliakov, A. A.,** Les mesures zoo-sanitaires dans les foyers d'infection brucellique, *Off. Int. Epizoot. Bull.*, 63, 1077, 1965.
62. **Alton, G. G., Jones, L. M., and Pietz, D. E.,** *Laboratory Techniques in Brucellosis, 2nd ed., WHO Monogr. Ser. No. 55,* World Health Organization, Geneva, 1975, 26.
63. **Corbel, M. J.,** Bacteriological procedures in the diagnosis of *Brucella melitensis* infection, in *Brucella melitensis,* Verger, J. M. and Plommet, M., Eds., Martinus Nijhoff, Dordrecht, 1985, 105.
64. **Kuzdas, C. D. and Morse, E. V.,** A selective medium for the isolation of brucellae from contaminated materials, *J. Bacteriol.*, 66, 502, 1953.
65. **Farrell, I. D.,** The development of a new selective medium for the isolation of *Brucella abortus* strains from contaminated sources, *Res. Vet. Sci.*, 16, 280, 1974.
66. **Brodie, J. and Sinton, G. T.,** Fluid and solid media for isolation of *Brucella abortus, J. Hyg.*, 74, 359, 1975.
67. **Alton, G. G.,** Standardization of agglutinating antigens for the diagnosis of brucellosis, *Res. Vet. Sci.*, 12, 330, 1971.
68. **Corbel, M. J.,** Comparison of *Brucella abortus* and *B. melitensis* antigens for the Rose Bengal plate test on sera from cattle infected with *B. abortus* biovar-5, *Vet. Rec.*, 117, 385, 1985.
69. **Kolar, J.,** Diagnosis and control of brucellosis in small ruminants, *Prev. Vet. Med.*, 2, 215, 1984.
70. **Farina, R.,** Current serological methods in *B. melitensis* diagnosis, in *B. melitensis,* Verger, J. M. and Plommet, M., Eds., Martinus Nijhoff, Dordrecht, 1985, 139.
71. **Anon.,** La lutte contre la brucellose ovine et caprine en URSS, *Off. Int. Epizoot. Bull.*, 82, 83, 1974.
72. **Contini, A., Coni, V., and Casu, A.,** Diagnosi sierologica di brucellosi negli ovini e caprini con l'antigene tamponato, *Atti. Soc. Ital. Sci.*, Vet. 27, 640, 1973.
73. **Gyantzis, D. O.,** Adaptation of the Rose Bengal test for the diagnosis of brucellosis, *Bull. Helenic Vet. Med.*, 32, 341, 1981.
74. **Waghela, S., Wandera, J. G., and Wagner, G. G.,** Comparison of four serological tests in the diagnosis of caprine brucellosis, *Res. Vet. Sci.*, 28, 168, 1980.
75. **Muhammed, S. I., Mohammadi, H., and Saadi-nam, H.,** A comparison of counter-immunoelectrophoresis with the Rose Bengal and serum tube agglutination tests in the diagnosis of brucellosis in sheep, *Vet. Microbiol.*, 5, 223, 1980.
76. **Trap, D. and Gaumont, A. J. R.,** Le diagnostic serologique de la brucellose bovine et ovine par l'epreuve a l'antigene tamponne, *Bull. Mens. Soc. Vet. Prat. Fr.*, 60, 301, 1976.
77. **Fensterbank, R. and Maquere, M.,** Assainissement d'un troupeau ovin atteint de brucellose par les moyens de la prophylaxie sanitaire en utilisant l'epreuve au Rose Bengale, *Rec. Med. Vet.*, 154, 657, 1978.
78. **Diaz, R. and Levieux, D.,** Role respectif en serologie de la brucellose bovine des antigenes et des immunoglobulines G_1 et G_2 dans les tests d'agglutination, de Coombs et au Rose Bengale ainsi que dans le phenomene de zone, *C. R. Acad. Sci. Paris, Ser. D,* 274, 1593, 1972.
79. **Renoux, G. and Alton, G.,** Etudes sur la brucellose ovine et caprine. IV. Reactions serologique dans le sang et le lait de chevres recemment infectees, par *Br. Melitensis, Arch. Inst. Pasteur Tunis,* 32, 523, 1955.

80. **Renoux, G., Alton, G., and Mahaffey, L. W.,** Etudes sur la brucellose ovine et caprine. IV. Reactions serologiques dans le sang de brebis infectees par *Br. melitensis, Arch. Inst. Pasteur Tunis,* 33, 33, 1956.
81. **Biggi, P.,** Il comportamento della sieroagglutinazione in ovini sperimentalmente infettati con *Br. Melitensis, Ann. Fac. Vet. Med. Pisa,* 9, 287, 1956.
82. Joint FAO/WHO Expert Committee on Brucellosis 6th Rep., *Tech. Rep. Ser. No. 740,* World Health Organization, Geneva, 1986.
83. **Unel, S., Williams, C. F., and Stableforth, A. W.,** Relative value of the agglutination test, complement fixation test and Coombs (antiglobulin) test in the detection of *Brucella melitensis* infection in sheep, *J. Comp. Pathol.,* 79, 155, 1969.
84. **Le Pennec, J.,** Serologie antibrucellique ovine. I. Des anticorps vaccinaux et infectieux des serums negative a la sero-agglutination de Wright et positifs a la deviation du complement, *Bull. Soc. Vet. Prat. Fr.,* 51, 94, 1967.
85. **Trap, D. and Gaumont, R.,** Comparaison entre electrosynerese et epreuves serologiques classiques dans le diagnostic de la brucellose ovine, *Ann. Rech. Vet.,* 13, 33, 1982.
86. Joint FAO/WHO Expert Committee on Brucellosis, 6th Rep., *Tech. Rep. Ser. 740, Brucellins, Annex 4,* World Health Organization, Geneva, 1986, 118.
87. **Jones, L. M., Diaz, R., and Tayler, A. G.,** Characterization of allergens prepared from smooth and rough strains of *Brucella melitensis, Br. J. Exp. Pathol.,* 54, 492, 1973.
88. **Bhongbhibhat, N., Elberg, S., and Chen, T. H.,** Characterization of *Brucella* skin-test antigens, *J. Infect. Dis.,* 122, 70, 1970.
89. **Jones, L. M. and Marley, J.,** Titration of a Brucella protein allergen in sheep sensitized with *Brucella melitensis, Ann. Rec. Vet.,* 6, 173, 1975.
90. **Fensterbank, R.,** Allergic diagnosis of brucellosis, in *Brucella melitensis,* Verger, J. M. and Plommet, M., Eds., Martinus Nijhoff, Dordrecht, 1985, 167.
91. **Durand, M. P., Loquerie, R., Fabre, F., Molinie, M., and Lingelser, J.,** Application d'un test allergique de detection de la brucellose en elevages ovins vaccines contre cette infection, *Bull. Acad. Vet. Fr.,* 57, 231, 1984.
92. **Loquerie, R. and Durand, M. P.,** Applications cliniques d'un test cutane d'hypersensibilite retardee pour depistage de la brucellose ovine et caprine au moyen d'un allergene commercial, *Dev. Biol. Stand.,* 56, 407, 1984.
93. **Ebadi, A. and Zowghi, E.,** The use of allergic test in the diagnosis of *Brucella melitensis* infection in sheep, *Br. Vet. J.,* 139, 456, 1983.
94. **Erasmus, J. A. and Bergh, E. C.,** Ovine brucellosis: repeated vaccination with Rev. 1 vaccine and the prevalence of the disease in the Winburgh district, *J. S. Afr. Med. Assoc.,* 56, 205, 1985.
95. **Riezu-Boj, J. I., Moriyon, I., Blasco, J. M., Marin, C. M., and Diaz, R.,** Comparison of lipopolysaccharide and outer membrane protein-lipopolysaccharide extracts in an enzyme-linked immunosorbant assay for the diagnosis of *B. ovis* infections, *J. Clin. Microbiol.,* 23, 938, 1986.
96. **Mittal, K. R. and Tizzard, I. R.,** Studies on the relationship between *Yersinia enterocolitica* IX and *Brucella abortus* agglutinins in naturally infected animals, *Res. Vet. Sci.,* 28, 311, 1980.
97. **Krogstad, O.,** *Yersinia enterocolytica* infection in goat. A serological and bacteriological investigation, *Acta Vet. Scand.,* 15, 597, 1974.
98. **Trilenko, P. A., Fishbein, V. Y., Balandi, F. G., Gukov, A. S., and Fomenko, T. V.,** Eradication of brucellosis from a sheep farm, *Veterinaria (Moscow),* 11, 54, 1971.
99. **Polydorou, K.,** Caprine Brucellosis in Cyprus, in *Les Maladies de la Chevre: Goat Diseases,* Yvore, P. and Perrin, G., Eds., Les Colloques de l'INRA, No 28, INRA, Paris, 1984, 93.
100. **Anon.,** in *Brucella melitensis,* Verger, J. M. and Plommet, M., Eds., Martinus Nijhoff, Dordrecht, 1985, 262.
101. **Bazalei, F. K., Evstafiadi, K. P., Taran, I. F., and Tsybin, B. P.,** Use of *Brucella abortus* strain 19 *and Brucella melitensis* Rev. 1 vaccines for the prophylaxis of ovine brucellosis, *Veterinariya (Moscow),* 4, 32, 1985.
102. **Xie Xin,** Orally administrable brucellosis vaccine: *Brucella suis* strain 2 vaccine, *Vaccine,* 4, 212, 1986.
103. **Alton, G. G. and Elberg, S. S.,** Rev. 1 *Brucella melitensis* vaccine, a review of 10 years of study, *Vet. Bull.,* 37, 793, 1967.
104. **Elberg, S. S.,** Rev. 1 *Brucella melitensis* vaccine. II. 1968—1980 (review), *Vet. Bull.,* 51, 67, 1981.
105. **Alton, G. G.,** Rev. 1 and H38 *Brucella melitensis* vaccines, in *Brucella melitensis,* Verger, J. M. and Plommet, M., Eds., Martinus Nijhoff, Dordrecht, 1985, 215.
106. **Elberg, S. S. and Faunce, K.,** Immunization against *Brucella* infection. VI. Immunity conferred on goats by a non-dependent mutant from a streptomycin-dependent mutant strain of *Brucella melitensis, J. Bacteriol.,* 73, 211, 1957.
107. **Ne'eman, L.,** Virulence stability of the Rev. 1 strain of *Brucella melitensis* on passage in pregnant sheep: the safety of the Rev. 1 strain of *Brucella melitensis* for pregnant sheep by natural contact, *Refu. Vet.,* 25, 202 and 265, 1968.

108. **Alton, G. G., Elberg, S. S., and Crouch, D.**, Rev. 1 *Brucella melitensis* vaccine, the stability of the degree of attenuation, *J. Comp. Pathol.*, 77, 293, 1967.
109. **Elberg, S. S.**, Immunization against *Brucella* infection. VII. Immunological and epidemiological studies in Cordoba, Spain, *Bull. W. H. O.*, 20, 1033, 1959.
110. **Lantier, F. and Fensterbank, R.**, Kinetics of Rev. 1 infection in sheep, in *Brucella melitensis*, Verger, J. M. and Plommet, M., Eds., Martinus Nijhoff, Dordrecht, 1985, 247.
111. **Jones, L. M. and Marley, J.**, Serological and bacteriological studies of ewes vaccinated with *Brucella melitensis* strain Rev. 1 during lactation, *Ann. Rech. Vet.*, 6, 67, 1975.
112. **Ne'eman, L.**, personal communication, 1966.
113. **Renoux, G.**, Etudes sur la brucellose ovine et caprine. XX. Vaccination des brebis contre l'infection a *Brucella melitensis*, comparaison de trois vaccins, *Arch. Inst. Pasteur Tunis*, 35, 251, 1958.
114. **Jones, L. M., Entessar, F., and Ardalan, A.**, Comparison of living vaccines in producing immunity against natural *Brucella melitensis* infection in sheep and goats in Iran, *J. Comp. Pathol.*, 74, 17, 1964.
115. **Paltrinieri, S., Farina, R., and Andreani, E.**, Esperimento di vaccinazione contro la brucellosi ovina da *Br. Melitensis* mediante impiego dello stipite Rev. 1, *Zooprofilassi*, 23, 241, 1968.
116. **Ulasevich, P. S., Aliverdiev, A. A., and Jusupov, O. Y.**, Immunization of sheep with *Brucella melitensis* strain Rev. 1 vaccine, *Veterinariya, (Moscow)*, 12, 57, 1975.
117. **Alton, G. G.**, Duration of the immunity produced in goats by the Rev. 1 *Brucella melitensis* vaccine, *J. Comp. Pathol.*, 76, 241, 1966.
118. **Alton, G. G.**, Further studies on the duration of immunity produced in goats by the Rev. 1 *Brucella melitensis* vaccine, *J. Comp. Pathol.*, 78, 173, 1968.
119. **Morgan, J. B. W., Littlejohn, A. I., Makinnon, D. J., and Lawson, J. R.**, The degree of protection given by living vaccines against experimental infection with *Brucella melitensis* in goats, *Bull. W. H. O.*, 34, 33, 1966.
120. **Alton, G. G., Jones, L. M., Garcia Carrillo, C., and Trenchi, A.**, *Brucella melitensis* Rev. 1 and *Brucella abortus* 45/20 vaccines in goats: immunity, *Am. J. Vet. Res.*, 33, 1747, 1972.
121. **Gaumont, R., Trap, D., and Dhennin, L.**, Immunisation de la brebis et de la chevre contre l'infection experimentale a *Brucella melitensis*. Comparison des vaccins Rev. 1 et H38, *Dev. Biol. Stand.*, 56, 629, 1984.
122. **Varela-Diaz, V. M., Jones, L. M., and Perez-Esandi, M. V.**, *Brucella melitensis* Rev. 1 and *Brucella abortus* 45/20 vaccines in goats: pattern of immunoglobulin production after vaccination and challenge, *Am. J. Vet. Res.*, 34, 203, 1973.
123. **Alton, G. G.**, Rev. 1 *Brucella melitensis* vaccine, serological reactions in Maltese goats, *J. Comp. Pathol.*, 77, 327, 1967.
124. **Fensterbank, R., Pardon, P., and Marley, J.**, Comparison between subcutaneous and conjunctival route of vaccination of Rev. 1 strain against *Brucella melitensis*, infection in ewes, *Ann. Rech. Vet.*, 13, 295, 1982.
125. **Absatirov, G. G., Abutalipov, A. A., and Amanzholov, B. A.**, Revaccinating sheep with *Brucella melitensis* Rev. 1 vaccine, *Veterinariya (Moscow)*, 5, 16, 1987.
126. **Farina, R., Agrimi, P., Andreani, E., and Casanovi, E.**, Sulla profilassi della brucellosi ovina mediante l'impiego dello stipite Rev. 1. Valutazione dei resultati ottenuti nel territorio di alcuni comuni della provincia di Lucca, *Ann. Fac. Med. Vet. Pisa*, 34, 39, 1982.
127. **Landau, M.**, personal communication, 1986.
128. **Jones, L. M., Garcia Carrillo, C., and Alton, G. G.**, *Brucella melitensis* Rev. 1 and *Brucella abortus* 45/20 vaccines in goats: serologic tests, *Am. J. Vet. Res.*, 34, 199, 1973.
129. **Alton, G. G.**, Vaccination of goats with reduced doses of Rev. 1 *Brucella melitensis* vaccine, *Res. Vet. Sci.*, 11, 54, 1970.
130. **Diaz Aparicio, E., Prado, A. F. J., Ontiveros, C. L., and Batalla, C. D.**, Evaluacion serologica de anticuerpos postvacunales en cabras adultas, vaceunadas con una dosis reducida (5×10^4) de Rev. 1 en una zona enzootica de brucelosis, *Tec. Pecu. Mex.*, 47, 137, 1984.
131. **Fensterbank, R., Verger, J. M., and Grayon, M.**, Conjunctival vaccination of young goats with *Brucella melitensis* strain Rev. 1, *Ann. Rech. Vet.*, 18, 397, 1987.
132. **Fensterbank, R., Pardon, R., and Marley, J.**, Vaccination of ewes by a single conjunctival administration of *Brucella melitensis* Rev. 1 vaccine, *Ann. Rech. Vet.*, 16, 351, 1985.
133. **Ciuchini, F., Fischetti, R., and Pistoia, C.**, Revizzazione per via congiunctivale degli ovine. I. Incidenza e dinamica della risposta sierologica, *Clin. Vet.*, 109, 172, 1986.
134. **Jimenez de Bagues, M. P., Marin, C. M., Baberan, M., and Blasco, J. M.**, Responses of ewes to *B. melitensis* Rev. 1 vaccine administered by subcutaneous or conjunctival routes at different stages of pregnancy, *Ann. Rech. Vet.*, 20, 205, 1989.
135. Twenty-eighth Rep. WHO Expert Committee on Biological Standardization (annex 4), *WHO Tech. Rep. Ser. No. 610*, World Health Organization, Geneva, 1977, 85.
136. **Garcia Carrillo, C. and Arce, C.**, Produccion de vacuna *Brucella melitensis* Rev. 1 en tangue de fermentacion, *Rev. Vet. Med. Argentina*, 54, 1, 1973.

137. **Olle-Goig, J. E. and Canela-Soler, J.,** An outbreak of *Brucella melitensis* infection by airborne transmission among laboratory workers, *Am. J. Public Health,* 77, 335, 1987.
138. **Gilles, G. and Letroteur, P.,** Les avortements infectieux de la brebis dans le departement des Bouches-du-Rhone, *Rev. Med. Vet.,* 129, 591, 1978.
139. **Feinhaken, D. and Dafni, I.,** Identification of Brucella isolates in Israel, 1970—1979, *Refu. Vet.,* 37, 117, 1980.
140. **Ademollo, A., Paltrinieri, S., Pellegrini, D., and Farina, R.,** Relazione sulle brucellosi, bovina, ovina e caprina in Italia, *Zooprofilassi,* 20, 41, 1965.
141. **Garcia Carrillo, C.,** Comparison of *B. melitensis* Rev. 1 and *B. abortus* strain 19 as a vaccine against brucellosis in cattle, *Zentrabl. Veterinaermed.,* 27B, 131, 1980.
142. **Tserendash, Ch.,** Use of Rev. 1 *Brucella melitensis* vaccine in cattle in Mongolia, *Tr. Vses. Inst. Eksp. Vet.,* 61, 21, 1984.
143. **Verger, J. M.,** *B. melitensis* infection in cattle, in *Brucella melitensis,* Verger, J. M. and Plommet, M., Eds., Martinus Nijhoff, Dordrecht, 1985, 197.
144. **Hellman, E.,** *Brucella melitensis,* in *Handbuch der bakteriellen Infektionen bei Tieren, Band IV,* Blobel, H. and Schliesser, T., Eds., Gustav Fischer, Stuttgart, 1982, 214.
145. **Kasyanov, A. N. and Aslanyan, R. G.,** Epizootology and clinical appearance of animal brucellosis, in *Zoonosis Control,* Lysenko, A., Ed., Centre of International Projects GKNT, Moscow, 1982, 100.
146. **Meyer, M. E.,** The epizootology of brucellosis and its relationship to the identification of *Brucella* organisms, *Am. J. Vet. Res.,* 25, 553, 1964.
147. **Taran, I. F.,** Spontaneous and experimental brucellosis in saiga antelopes, *Zh. Mikrobiol. Moscow,* 5, 99, 1960.
148. **Minkov, G. B., Chvatseva, S. S., Rudenko, B. S., Kaledina, R. F., Kunicyn, O. V., and Svidcenko, G. A.,** An outbreak of brucellosis in an animal breeding establishment, *Zh. Mikrobiol. Moscow,* 46, 97, 1969.
149. **Ezilondo, A.,** Brucellosis epidemica en la Comarca Lagunera, *Bol. Epidemiol.,* 19, 51, 1955.
150. **Eyre, J. W. H.,** The Milroy lectures on melitensis septicaemia (Malta or Mediterranean Fever), *Lancet,* 1, 1677, 1908.
151. **Lulu, A. R., Araj., Khateeb, M. I., Mustafa, M. Y., Yusuf, A. R., and Fenech, F. F.,** Human brucellosis in kuwait: a prospective study of 400 cases, *Q. J. Med.,* 66, 39, 1988.
152. **Escalante, J. A. and Held, J. R.,** Brucellosis in Peru, *J. Am. Vet. Med. Assoc.,* 155, 2146, 1969.
153. **Sabbaghian, H.,** Fresh white cheese as a source of *Brucella* infection, *Public Health London,* 89, 165, 1975.
154. **Alton, G. G.,** Caprine brucellosis in Malta, Thesis, University of Edinburgh, 1971.

Chapter 18

BRUCELLA SUIS

Godfrey G. Alton

TABLE OF CONTENTS

I. INTRODUCTION

Brucellosis in swine is caused by one or another of the first three biovars of *Brucella suis*. Biovar 4 causes disease in reindeer, and this infection is described in Chapter 13 as is murine brucellosis caused by the remaining biovar, 5. Infection of swine with *B. abortus* or *B. melitensis* occurs rarely and is generally symptomless.

B. suis biovar 2, however, causes a disease that is different in some respects from that caused by the other two biovars that affect the pig.[1] The European hare *(Lepus capensis)* is also a principal host of biovar 2 and can form a reservoir of infection from which pigs, both domestic and wild, may become infected.[2] Nevertheless, infection with biovar 2 spreads as readily among pigs both within and between herds as the other two biovars. In pigs, miliary brucellosis of the uterus is a feature of the disease caused by biovar 2. Otherwise, it closely resembles that caused by biovars 1 and 3. *B. suis* biovar 2 also differs in not being a human pathogen. Its geographic distribution is confined to Europe in a broad band between Scandinavia and the Balkans. The disease in hares is characterized by the formation of nodules, varying in size from a millet seed to a cherry or even larger; these often become purulent. Such nodules may occur in almost any location, sometimes subcutaneously or intramuscularly, but most often in the reproductive organs. The bodily condition of the hare may be surprisingly unaffected. In the rest of this chapter we shall be mainly concerned with biovars 1 and 3 and the disease they cause.

Brucellosis in swine was first described by Traum[3] in 1914 in swine herds in Indiana, and for several years was thought to be a specially pathogenic type of *B. abortus*. The credit for naming *B. suis* as a separate species goes to Huddleson.[4]

II. BACTERIOLOGY

B. suis biovars 1, 2, and 3 occur in nature in the smooth form. With regard to morphological staining characteristics and colony appearance on solid media, they are indistinguishable from other smooth *Brucella* species. They grow well on the usual *Brucella* media without the addition of serum or enrichment of the atmosphere with carbon dioxide. They are resistant to the antibiotics generally used to prepare selective media for the culture of *Brucella*. Isolations from pigs can be presumptively identified as brucellae by agglutination in anti-smooth *Brucella* serum or monospecific serum. All three biovars have the A-type lipopolysaccharide and, therefore, agglutinate with the monospecific A serum, polyclonal or monoclonal. Species and biovar identification is accomplished by routine typing tests such as production of hydrogen sulfide, growth in the presence of dyes, agglutination in monospecific serum, phage typing (see Table 1), and, where necessary, oxidative metabolic tests. Additional tests such as the urease reaction and inhibition by safranin may be useful. More details are given in Chapter 1 on taxonomy and by Alton et al.[5] It should be noted that biovar 1 may be atypical in being resistant to basic fuchsin and safranin.[6]

In the days before the development of phage typing in the U.S.S.R. (reviewed by Drozevkina)[7] and oxidative metabololic tests in the U.S.,[8] it was difficult if not impossible to distinguish *B. suis* biovar 3 from *B. melitensis* biovar 2. Therefore, the frequent reports of *B. melitensis* in pigs in the U.S. before the early 1960s referred almost certainly to *B. suis* biovar 3. For instance, McCullough et al.[9] reported isolating 11 strains of *B. melitensis* and 14 strains of *B. suis* in an abattoir in Chicago. In reality they must have been dealing with 11 strains of biovar 3 and 14 strains of biovar 1 of *B. suis*. Phage typing can be done easily in small laboratories, but oxidative metabolic tests are best done at laboratories specializing in such work. A list of reference laboratories prepared to undertake this work and their addresses is given by Alton et al.[5] Isolates from pigs identified in local laboratories as *B. melitensis,* or strains from sheep or goats identified locally as *B. suis,* should be checked by phage and oxidative metabolic tests.

TABLE 1
Differential Characteristics of *B. suis* Biovars 1, 2, and 3

			Dye tests		Agglutination with monospecific sera		Lysis by phage		
							Tb		
Biovar	Requirement for CO_2	H_2S production	Thionine (20 μg/ml)	Basic fuchsin (20 μg/ml)	A	M	RTD	$10^4 \times$ RTD	R/C RTD
B. suis 1	−	+	+	−[a]	+	−	−	+	−
B. suis 2	−	−	+	−	+	−	−	+	−
B. suis 3	−	−	+	+	+	−	−	+	−

[a] Some strains grow.

III. HOST RANGE OF *B. SUIS* BIOVARS 1 AND 3

The natural host of *B. suis* biovars 1 and 3 is the pig, mainly domesticated pigs but in some areas wild pigs. In others feral pigs are significantly involved. Extensive infection in feral swine has been reported from Queensland in Australia[10] and the southern states of the U.S.[11,12]

B. suis causes a most important zoonosis. This will be discussed in more detail later in this chapter.

B. suis infection in bovine species also has considerable importance. While cattle are resistant to generalized infection with *B. suis,*[13] they sometimes acquire inapparent infections which cause them to react to routine tests in bovine eradication campaigns. Cooke and Noble,[14] working in Queensland, Australia, recorded in the isolation of *B. suis* from single bovine reactors in several routine herd tests. Feral swine were assumed to have provided the source of the infection. The infections were confined to the lymph nodes of the head. Research is in progress aimed at developing a method of distinguishing these reactions from those caused by *B. abortus.* Cattle also are susceptible to infection of the udder with *B. suis,* resulting in copious and persistent excretion which may give rise to extensive epidemics of brucellosis in humans.[15,16] It may be worth pointing out here that the ruminant udder is capable of providing a fertile medium for all three main species of *Brucella.*

Garcia-Carrillo[17] quotes reports from Colombia of *B. suis* infection in sheep and goats.

There are several reports of *B. suis* infection in dogs.[18] The infection results generally from feeding raw pig meat. It is usually symptomless, but there may be orchitis or epididiymitis.

IV. GEOGRAPHIC DISTRIBUTION

Canada appears to be free from swine brucellosis, and in the U.S. measures for its eradication are far advanced. In Latin America the infection, chiefly biovar I, is widespread. Infection rates up to 20% and more in individual swine have been reported from some areas in Argentina[19] and Brazil.[20] In other areas of these countries a much lower prevalence has been recorded. Garcia-Carrillo[17] provides detailed information for the whole American continent.

In Europe, Britain has never experienced *B. suis* infection, and apart from the special case of biovar 2, Scandinavia also appears to be free. But in practically every other European country there is a low prevalence of biovar 1. *B. suis* appears to be absent from the largely Muslim western Asia.

In Africa, swine brucellosis occurs sporadically in a number of countries. The population of domestic swine in the continent is very small, and there is a general lack of information on the disease.

On the Indian subcontinent the pig is considered unclean by both Hindus and Muslims. Nevertheless, there are reports from India of the occurrence of *B. suis.* These reports are generally based on serological investigations,[21] and it is difficult to estimate the likely overall prevalence.

There is no doubt that swine brucellosis is a major problem in Southeast Asia and the Pacific islands. In Singapore and south China, the prevailing biovar seems to be biovar 3, whereas offshore it is biovar 1, sometimes atypical. *B. suis* was isolated from 15% of pigs in a random sample taken at a large Indonesian abattoir,[22] and a high incidence of serological reactors was reported from pigs in the Philippines and Taiwan. In Australia, the infection appears confined to Queensland and mostly to feral swine. Pacific islands from which swine brucellosis has been reported include Papua New Guinea, Fiji, Tonga, Samoa, and Tahiti. No doubt other islands are affected.

V. DEVELOPMENTS IN THE PIG INDUSTRY

Second only to poultry, pig farming is the quickest and most practical way for countries, particularly newly industrialized countries, to increase their supply of animal protein. This is taking place where religious taboos against the pig do not apply as in east Asia. China, for example, has an estimated 340 million pigs. In many countries traditional backyard pig pens are being supplemented by very large commercial units, and artificial insemination is being introduced. Many of these developments are taking place in regions where *B. suis* is endemic in the native pig population, and attention will need to be paid to avoiding the introduction of brucellosis, especially into artificial insemination units.

VI. EPIDEMIOLOGY

Entry of infection into previously free herds occurs generally by the purchase of infected animals, male or female, or by the communal use of a boar, who may transmit the disease mechanically or who may be excreting *B. suis* in his semen even when appearing healthy. The potential for spreading *B. suis* infection through an ineffectively controlled artificial insemination service is enormous. Another potentially important means could be the feeding of kitchen waste containing uncooked meat to pigs. The fact that many Pacific islands appear to be affected supports the idea that this type of waste from ships could be a means of international spread.

A proportion of piglets will acquire infection from their mothers, either by being born infected or by ingesting the bacteria in their mother's milk. Spread of infection within the herd seems most likely to occur in the same way as it does in ruminants, by contamination of foodstuffs and surroundings by infected discharges, usually of uterine origin. The mouth must be the main portal of entry. Pigs will readily consume aborted fetuses and membranes, but conjunctival infection is possible as is penetration of the skin. The boar also can spread the infection at mating both within and outside the herd. *B. suis* has been shown to persist in some arthropods such as ticks, but this method seems unlikely to be a common means of spread.

B. suis appears to have the same viability under natural conditions and susceptibility to disinfectants as other *Brucella* species.

VII. IMMUNITY

The extent to which recovered pigs can be reinfected has been investigated by a number of workers. Orlov[23] was able to reinfect 1 of 29 pigs and Manthei,[24] 3 of 30 sows. Hutchins et al.[25] showed that although about half of recovered sows could be reinfected during later pregnancies, there was no interference with the course of the pregnancy. Infection with *B. suis* evidently produces a degree of resistance, but it is difficult to find evidence as to how long this may last.

VIII. PATHOGENICITY

Deyoe[26] considered *B. suis* biotypes 1 and 3 to be equally pathogenic for swine, but a substantial proportion of individual pigs, even of the more susceptible breeds, are resistant to infection. Genetic control of susceptibility was demonstrated by Cameron et al.[27] However, Cedro et al.[28] were able to infect all of a group of pigs with 10^5 cells of *B. suis* administered either intramuscularly or subcutaneously.

There seems little relation between the age of the animal and its susceptibility to infection. Both sexes appear to be more or less equally susceptible and even the castrated male pig is quite likely to have the infection.

In those pigs that do become infected, the sequence of events is similar to that seen in brucellosis in most other animal species. After invasion of the lymph nodes near the point of entry of infection, a bacteremia develops, lasting on average about 5 weeks but occasionally prolonged to 34 months or more.[29] Bacteremia leads to widespread involvement of lymph nodes, but the number of affected nodes becomes less with time. The bacteria may infect a number of organs, particularly those concerned with reproduction, but occasionally spleen, liver, bones, joints, tendons, and bursae. The majority of pigs free themselves from infection in a matter of months rather than years, but pathological changes in boars, once established, are less reversible than those in sows and may persist for years.

IX. CLINICAL SIGNS

Infection with *B. suis* may be symptomless, and it is not unusual for the owner to be unaware that his herd is infected. The bodily condition is seldom affected in the absence of specific lesions.

In the female, abortion is the principal symptom. This may occur at any time during pregnancy. Early abortions may be missed, and the owner may think he has an infertility problem. On the other hand, infected piglets may be carried to term and may appear normal. Dead and weak piglets may also be produced at or near term. The mother may suffer a discharge from the vagina. Mastitis sometimes occurs and may lead to abscess formation.

In the boar there may be no symptoms, even when brucellae are present in his semen. Or there may be swelling and later atrophy of the epididymes and testes, generally unilaterally. His sexual performance may or may not be affected.

Lameness with swollen joints, bursae, and tendons may occur in pigs of all ages. Symptoms involving the spine, leading sometimes to paralysis, are fairly frequent. There is a general tendency for abscesses to form.

X. PATHOLOGY

Piglets may be aborted at any stage of development. They may have swollen organs with edematous, often blood-stained, fluid, or they may look normal. The fetus may be expelled with the membrane intact. The mother may suffer catarrhal inflammation of the uterus and mastitis.

In the male it is often possible to isolate the organisms in the absence of visible pathological changes, and vice versa. The seminal vesicles are the genital organs most often involved. In the early stages they may be accutely inflamed, and this may give way to chronic infection with sclerosis and minute abscesses or a large single abscess developing sometimes. The prostate may show similar changes. The testes may be swollen. Sometimes there is abscess formation and later sclerosis and atrophy. Occasionally, nodules occur in the spleen or liver and abscesses in bones, joints, and bursae in both sexes.

The lesion is essentially a granuloma, with hyperplasia of connective tissues, necrosis, and minute abscesses which sometimes coalesce to form one large purulent abscess that may develop caseation. Lymph nodes often show no visible changes, but in the vicinity of other lesions may become yellowish, small, and hard with minute abscesses.

Histopathologically, there is an inflammatory lesion characterized by accumulations of polymorphs, macrophages, and giant cells with hyperplasia of the reticular tissue. Necrosis may be seen leading to the development of minute abscesses embedded in fibrous tissue.

XI. DIAGNOSIS

We have already seen that the effect of this disease is very variable in the affected animal. During their first 2 to 3 months, pigs seem to be able to harbor *B. suis* infection but not to be capable of producing antibody. In an infected herd there will be recently recovered animals giving false positive results. There will also be animals with mature infections that are likely to be positive to all diagnostic tests, although some of these may be negative or give indefinite reactions to serological and/or allergic tests. Thus, serological and allergic diagnosis is not likely to be easy, and the effectiveness of serological and allergic tests can be expected to vary under different epidemiological circumstances.

A. BACTERIOLOGICAL DIAGNOSIS

The microscopic methods described for *B. melitensis* in Chapter 5 and Chapter 17, Section VII.A can also be used for detecting *B. suis*. However, it is generally accepted that isolation and identification of the organism provides the only certain diagnosis, and there are usually plenty of opportunities for this in swine brucellosis. *B. suis* grows readily without additional CO_2 on the common *Brucella* media. In infections with biovar 1, it has been shown that routine culture of a small sample of lymph nodes from carcasses will yield as many positive results as serological examination.[22] This is important, as virtually all the produce of the industry passes through abattoirs, and lymph nodes can be removed without damage to the carcass. Preferred lymph nodes are mandibular, retropharyngeal, and internal iliac. Nodes can be dipped in alcohol, the alcohol burned off, and the node diced with scissors and processed in a stomacher before being spread on solid medium. This is a very practical survey strategy. Specimens that may be cultured from live animals include blood, often positive in the more acute phases of the disease, products of abortion or suspicious live births from sows, and vaginal swabs. Semen samples may be taken from boars, and sometimes castrated testes become available. The contents of swollen joints, abscesses, etc. should be taken for culture. The characteristics used to identify *B. suis* have already been described, and further details are available.[5]

B. SEROLOGICAL DIAGNOSIS

Antigens containing smooth whole cells of *B. abortus* are generally used for the diagnosis of *B. suis* infections. Since these organisms have the same surface lipopolysaccharide, nothing would be gained by employing a *B. suis* antigen.

The agglutination test, whether done in tubes or on a plate, is the test most ofen used, but deficiencies in both sensitivity and specificity prevent this test from being reliable in the diagnosis of brucellosis in individual pigs. Nevertheless, it is considered adequate as a herd test,[30] based on the assumption that an infected herd will contain at least one animal with 100 IU of serum agglutinin. *B. suis* has been isolated on numerous occasions from pigs that are negative to the agglutination test.[31] On the other hand, false positive reactions are not infrequent. Hoerlein[32] found titers up to 400 IUs in pigs in noninfected herds. These higher titers were generally transient, but low titers could persist for several months. Some of these false positive reactions are attributed by Deyoe[33] to heterospecific agglutinins which the agglutination test cannot distinguish from anti-*Brucella* IgM. Introduction of procedures that reduce the sensitivity of these tests to IgM, as in the heat inactivation, mercaptoethanol, acidified antigen, and Rivanol tests, may improve the specificity of the test but may reduce its sensitivity, especially in recent infections. The complement fixation test is less effective in pigs than it is in ruminants, but has been used in some Balkan countries as the preferred test for swine brucellosis. The card or Rose Bengal test (RBT) also operates at low pH and appears to be at least as effective as any of those mentioned so far for the diagnosis of swine brucellosis. But even this test can only be relied on as a herd test rather than a test for individual animals.

To date, only two reports[34,35] on the use of enzyme immunoassay (EIA) for diagnosing porcine brucellosis have been noted. The test appears more sensitive than the standard procedures. The second of the cited papers gave a sensitivity of 85% for EIA, compared with 62% for the RBT, 42% for the (European) tube agglutination test, and only 38% for the CFT in 66 pigs that were positive on culture. The specificity of these tests in the larger number of culture-negative pigs from the same sample was 78, 88, 95, and 92%, respectively. These figures need to be viewed in light of the fact that only a small number of lymph nodes were cultured from each animal, and some infected animals may well have been negative on culture. The figures would, however, seem to justify a trial of the EIA as an eradication test.

C. SEROLOGICAL CROSS-REACTIONS

At least some of the false positive serological reactions already mentioned must be cross-reactions caused by antibody produced in response to infections with bacteria of genera other than *Brucella*. The most complete cross-reaction is produced by infection with *Yersinia enterocolytica* serotype 0:9, due to a shared A chain on the surface lipopolysaccharide. This serotype occurs less commonly in pigs than other serotypes of *Y. enterocolytica*. Nevertheless, it does occur and does give rise to IgG.[36,37] Other bacterial infections in pigs such as *Salmonella urbana* and *Francisella tularensis* may produce cross-reactions when the sera are tested by the agglutination test.[38] These reactions are generally IgM mediated, and such sera are often negative when tested by complement fixation, Rose Bengal, or 2-mercapto-ethanol tests. These cross-reactions often interfere needlessly with international traffic in breeding swine when agglutination tests are specified. They also emphasize the usefulness of culture and allergic methods of diagnosis.

D. LYMPHOCYTE STIMULATION

Kaneene et al.[39] studied cell-mediated responses (CMI) by *in vitro* stimulation of lymphocytes in groups of infected and noninfected swine. On a group basis, there was good correlation between the presence of antibody and CMI but not in individuals. However, there was good correlation between the isolation of *B. suis* and CMI. This complex technique is likely to be of practical value only in research work and may be of use in assessing levels of resistance to infection.

E. ALLERGIC TESTS

The allergic skin test has been reported[40] to be as accurate as any of the traditional serological tests, and for use as a herd test it is more convenient. For many years an allergen (sometimes called Brucellin fraction F), prepared by acid hydrolysis of a smooth culture of *Brucella*,[41] has been used in the U.S.S.R., China, and a number of other countries. A newer protein allergen (Brucellin INRA) prepared from a rough strain of *B. melitensis*,[42] being free from smooth lipopolysaccharide, may offer advantages. There should be no tendency for the skin test to cross-react in bacterial infections with genera other than *Brucella*.

XII. PORCINE BRUCELLOSIS AS A ZOONOSIS

Biovars 1 and 3 of *B. suis* are both very pathogenic for man, and numerous reports of human brucellosis caused by them have come from the U.S. In some Latin American countries also *B. suis* is a major cause of human brucellosis. For example, in Argentina, Garcia-Carrillo et al.[43] report the numbers of different species of *Brucella* isolated from humans during the years 1965 to 1983 as *B. abortus* 49, *B. melitensis* 19, and *B. suis* 79 (this includes the so-called *B. atypica* which is regarded as an aberrant *B. suis* biovar 1). Reports from Brazil suggest a similar, if not greater, incidence of *B. suis* in humans in that country.

As pigs do not produce dairy products, transmission to humans must take place by other means which result in this form of human brucellosis being almost entirely occupational. Possible exceptions to this rule include those occasions when *B. suis* finds its way into the bovine udder and the milk is drunk unpasteurized. For example, Beattie and Rice[15] described a milk-borne epidemic caused by *B. suis* involving 30 people, and Borts et al.[16] described another similar incident involving 50 families, some with multiple cases. In each case the epidemic was traced to milk from one dairy farm. There is also the possibility of transmission via meat products.

During the bacteremic phase of the infection in pigs, *B. suis* is present in all tissues, so the question of its survival in pig products has considerable importance. Normal cooking temperatures can be relied on to render the meat safe. The process of smoking ham is also said to make it safe.[44] Pickling, on the other hand, reduces the numbers of viable brucellae, but some survive for up to 126 d. There is doubt about the safety of certain types of sausage, e.g., salami, depending on the dimensions of the sausage and the amount of heat applied in its preparation.

Many *Brucella* infections among occupationally exposed people are subclinical. This presents problems in diagnosis and emphasizes the usefulness of periodic serological testing of people at risk to establish background titers.

Most at risk to infection with *B. suis* are abattoir workers and farmers. This seems to be a disease whose spread is favored by modern equipment in abattoirs. Electric saws scatter minute fragments of bone marrow, pressure hoses produce aerosols from contaminated surfaces, and air conditioning systems can complete the distribution of brucellae throughout the plant. Hendricks[45] described an outbreak in a pig-processing plant in Iowa where 128 cases occurred during a period of 9 months.

Brucellosis among pig farmers has received much less attention. Perhaps it does not occur frequently. Or more likely, as cases tend to be sporadic, many may be missed, or if diagnosed, not reported. This may explain why little is heard of human brucellosis caused by *B. suis* in Asia, even when a high prevalence has been demonstrated in the pig population.

The huge modernization and commercialization of the pig industry that is taking place in Asia, and probably elsewhere, may present an opportunity for human epidemics of *B. suis* to occur in the same way as epidemics of brucellosis caused by *B. melitensis* have followed the development of intensive sheep farming in the Middle East. This possibility needs to be brought to the attention of authorities national and regional, so that preventive measures can be undertaken.

XIII. CONTROL METHODS

A. ERADICATION

Given the nature of the industry in most countries, with its rapid turnover of stock, control by test and slaughter ought to be easy. Unfortunately, the means for efficient testing are not available at the present time. The U.S. Department of Agriculture has operated a successful eradication scheme for a number of years, and its approach has been summed up by Deyoe and Manthei:[29]

> There are 3 acceptable plans: . . . Plan 1 consists of depopulation of the entire herd, which is by far the most successful and economical in the long run. Plan 2 is a procedure designed to salvage irreplaceable bloodlines and basically consists of marketing the adult swine for slaughter and retaining weanling pigs for breeding stock; a plan that is not always successful and necessitates considerable isolation and retesting requirements. Plan 3 consists of removing only serological reactors and retesting the herd as many times as necessary. This latter procedure is rarely successful if the herd is actually infected, but is the plan of choice if the herd contains only a single reactor or only a very low proportion of the herd are reactors and there is a reasonable doubt that brucellosis exists in the herd.

In other countries also, where herds have been freed from brucellosis, this has usually been done by identifying infected herds, slaughtering the whole herd, and replacing with clean animals.

B. VACCINATION

In the absence of eradication procedures other than depopulation, control of the disease would be more readily achieved if an effective vaccine were available.

Numerous attempts at producing a vaccine against porcine brucellosis can be cited. Many of these experimental vaccines were shown to be capable of inducing a degree of resistance, but for various reasons none found any general acceptance in field use.

It has been demonstrated that *B. abortus* strain 19 is not capable of producing a worthwhile degree of immunity in pigs.[31,46]

A number of living attenuated strains of *B. suis* have been shown to produce some degree of immunity, e.g., King 8[31] and Dyulevo.[47] In the U.S.S.R., Orlov[23] used an attenuated strain, designated 61, which seemed to produce a good resistance, and more recently Chinese investigators[48] have reported on their *B. suis* S2 vaccine strain. The Chinese report that the S2 vaccine protects pigs against conjunctival challenge but not against infection by mating with a boar that is excreting *B. suis* in his semen. This raises an interesting point which will be addressed later.

A nonliving vaccine, if effective, would be preferable, and Foster and Edens[49] found that endotoxin-containing preparations induced definite resistance to a considerable conjunctival challenge. They also found that conjugation of the antigen to bovine gamma globulin appeared to enhance the effectiveness of the vaccine.

In Argentina, Cedro and Cisale[50] reported success in an extensive field trial with a vaccine composed of a living attenuated strain of *B. abortus* combined with a heat-extracted *B. suis* antigen, the latter being repeated after 1 month.

More recently, Garcia-Carrillo et al.[51] gave details of a trial in guinea pigs in which Cedro's vaccine was compared with two killed preparations, one consisting of a killed culture of *B. abortus* and the other a killed culture of *B. suis* (both in adjuvant) and the living attenuated *B. melitensis* Rev. 1 vaccine. They found that all the vaccines produced resistance to small challenge doses, but only Rev. 1 produced resistance against the higher challenge dose, a modest 10^3 cells of virulent *B. suis*.

XIV. RESEARCH REQUIREMENTS IN PORCINE BRUCELLOSIS

A. DIAGNOSIS

An improved serological test, capable of diagnosing the disease in individual pigs, is a prime need, e.g., for purchased animals, for boars entering artificial insemination units, and for making feasible eradication by test and slaughter, etc. The enzyme immunoassay may well be adequate for this. Work is in hand to produce a standard system and to define the limits of its effectiveness.

A method for detecting the presence of antigen in aborted materials, lymph nodes, etc. would be useful in diagnosis, in surveys, and for monitoring the disease in individual farms.

B. VACCINES

There is an urgent need for a vaccine for use in areas of high prevalence in the third world. An attenuated living *B. suis* strain seems to offer the best hope of a quick result. The stimulation of long-lasting antibody by an effective vaccine need not inhibit its use since the disease situation in a herd could be monitored by examining the nonbreeding progeny which would not need to be vaccinated. The progress being made in research into vaccination against *B. abortus* with nonliving agents could also be applicable to *B. suis,* given the similarity of the surface antigens.

Reports from China suggest that immunity conferred by vaccination may not operate against infection introduced directly into the uterus with semen, and this is a definite requirement in vaccination against porcine brucellosis. It may be important to know whether *B. suis* introduced with semen can remain viable long enough in the uterus to infect a developing fetus and/or its membranes that may not have developed immune mechanisms. If this is so, the provision of an artificial insemination service using known uninfected boars may be a necessary means of control.

Results obtained with *Brucella* vaccines in laboratory animals have often not been reproducible in the target domestic species. A method for assessing resistance in domestic animals without the need for challenge would greatly enhance the scope for developing new vaccines and for evaluating improvements to existing products, estimating optimal doses, etc. It is possible that *in vitro* methods for measuring lymphocyte responses to *Brucella* antigens, such as those reported by Kaneene et al.,[39] may prove adequate to fulfill this role.

REFERENCES

1. **Thomsen, A.,** *Brucella Infection in Swine. Studies from an Epizootic in Denmark 1929—1932, Acta Pathol. Microbiol. Scand Suppl.*, 21, 1934.
2. **Scheibner, E.,** Evolution de la brucellose porcine et la lutte contre cette maladie dans la Republique Federal Allemagne, *Bull. Off. Int. Epizoot.*, 82, 113, 1974.
3. **Traum, J.,** *Report of the Chief of the Bureau of Animal Industry*, U.S Department of Agriculture, Washington D. C., 1914, 30.
4. **Huddleson, I. F.,** The differentiation of the species of the genus *Brucella, Bull. Mich. Agric. Exp. Stn.*, No. 100, 1929.
5. **Alton, G. G., Jones, L. M., Angus, R. D., and Verger, J.-M.,** *Techniques for the Brucellosis Laboratory*, I.N.R.A., Paris, 1988.
6. **Corbel, M. J., Thomas, E. L., and Garcia-Carrillo, C.,** Taxonomic studies on some atypical strains of *Brucella suis, Br. Vet. J.*, 140, 34, 1984.
7. **Drozevkina, M. S.,** The present position in *Brucella* phage research. A review of the literature, *Bull. W. H. O.*, 29, 43, 1963.
8. **Meyer, M. E. and Cameron, H. S.,** Metabolic characterization of the genus *Brucella*. II. Oxidative metabolic patterns of the described biotypes, *J. Bacteriol.*, 82, 396, 1961.
9. **McCullough, N. B., Eisle, C. W., and Pavelchek, E.,** Survey of brucellosis in slaughtered hogs, *Public Health Rep.*, 66, 205, 1951.
10. **Norton, J. H. and Thomas, A. D.,** *Brucella suis* in feral pigs, *Aust. Vet. J.*, 52, 293, 1976.
11. **Becker, H. N., Belden, R. C., Breault, T., Burridge, M. J., Frankenberger, W. B., and Nicoletti, P.,** Brucellosis in feral swine in Florida, *J. Am. Vet. Med. Assoc.*, 173, 1181, 1978.
12. **Zygmont, S. M., Nettles, V. F., Shotts, E. B., Jr., Carmen, W. A., and Blackburn, B. O.,** Brucellosis in wild swine; a serologic and bacteriologic survey in the South Eastern United States and Hawaii, *J. Am. Vet. Med. Assoc.*, 181, 1285, 1982.
13. **Norton, J. H. and Thomas, A. D.,** *Brucella suis* infection in pregnant cattle, *Aust. Vet. J.*, 55, 525, 1979.
14. **Cook, D. R. and Noble, J. W.,** Isolation of *B. suis* from cattle, *Aust. Vet. J.*, 61, 263, 1984.
15. **Beattie, C. P. and Rice, R. M.,** Undulant fever due to *Brucella* of the porcine type- *Brucella suis, JAMA*, 102, 1670, 1934.
16. **Borts, I. H., Harris, D. M., Joynt, M. F., Jennings, J. R., and Jordan, C. F.,** A milk borne epidemic of brucellosis caused by the porcine type of *Brucella, (Brucella suis)* in a raw milk supply, *JAMA*, 121, 319, 1943.
17. **Garcia-Carrillo, C.,** *La Brucelosis de los Animales en America y su Relacion con la Infeccion Humana*, O. I. C., Paris, 1987.
18. **Hellmann, E. and Sprenger, H.-U.,** Ein Betrag zur *Brucella suis*-Infektion des Hundes, *Berl. und Muench. Tiererztl. Wochenschr.*, 91, 385, 1978.
19. **Cedro, V. C. F., Cisale, H. O., and Cacchione, R. A.,** Algunos aspectos de la brucellosis porcina, *Rev. Invest. Ganad.*, 3, 39, 1958.
20. **Schlogel, F.,** Elementos para a cauistica da brucelose suina no Parana, *Rev. Esc. Agric. Vet.*, 1, 25, 1965.
21. **Choudhary, S. P., Singh, S. N., Narayan, K. G., and Kalimuddin, M.,** Sero-prevalence of porcine brucellosis in Bihar, *Indian Vet. J.*, 60, 336, 1983.
22. **Priadi, A., Hirst, R. G., Chasanah, U., Nurrhadi, A., Emmins, J. J., Darodjat, M., and Soeroso, M.,** Animal brucellosis in Indonesia — *Brucella suis* infection detected by an enzyme-linked immunosorbent assay, in *Efficient Animal Production for Asian Welfare*, Proc. 3rd AAAP Animal Science Congress, Vol. 1, Seoul, Korea, 1985, 507.

23. **Orlov, E. S.,** Studies on pathogenesis, diagnosis and specific prophylaxis of swine brucellosis, *Off. Int. Epizoot. Bull.*, 63, 993, 1965.
24. **Manthei, C. A.,** Brucellosis, in *Diseases of Swine,* 2nd ed., Dunne, H. W., Ed., Iowa State University Press, Ames, 1964.
25. **Hutchins, L. M., Bonham, C. R., and Delen, A. L.,** Studies on brucellosis of swine. II. Exposure and re-exposure experiments with *Brucella suis, Am. J. Vet. Res.,* 7, 11, 1946.
26. **Deyoe, B. L.,** Pathogenesis of three strains of *Brucella suis* in swine, *Am. J. Vet. Res.,* 28, 951, 1967.
27. **Cameron, H. S., Gregory, P. W., and Hughes, E. H.,** Studies in genetic resistance in swine to *Brucella* infection. II. A bacteriological examination of resistant stock, *Cornell Vet.,* 31, 21, 1941.
28. **Cedro, V. C. F., De Benedetti, L. M. E., and Carcia-Carrillo, C.,** Brucelosis experimental en porcinos. I. Determinacion de dosis infectante, *Rev. Invest. Agropecu. Ser. 4,* 8, 91, 1971.
29. **Deyoe, B. L. and Manthei, C. A.,** Brucellosis, in *Diseases of Swine,* 4th ed., Dunne, H. W. and Leman, A. D., Eds., Iowa State University Press, Ames, 1975.
30. Joint FAO/WHO Expert Committee on Brucellosis, 6th Rep., *Tech. Rep. Ser. 740,* World Health Organization, Geneva, 1986.
31. **Manthei, C. A.,** Research on swine brucellosis by the Bureau of Animal Industry 1941—1947, *Am. J. Vet. Res.,* 9, 40, 1948.
32. **Hoerlein, A. B.,** Studies on swine brucellosis. III. The differentiation of specific and non-specific agglutination titers, *Cornell Vet.,* 43, 28, 1953.
33. **Deyoe, B. L.,** Diagnostic tests for swine brucellosis, in *Proc. 53rd Annu. Meet. Livestock Conservation,* 1969, 20.
34. **Thoen, C. O., Hopkins, M. P., Armbrust, A. L., Angus, R. D., and Pietz, D. E.,** Development of an enzyme-linked immunosorbent assay for detecting antibodies in sera of *Brucella suis*-infected swine, *Can. J. Comp. Med.,* 44, 294, 1980.
35. **Priadi, A., Chasanah, U., Hirst, R. G., Emmins, J. J., van der Giessen, J., and Soeroso, M.,** Development of an enzyme-linked immunosorbent assay (ELISA) for detecting antibody to *Brucella suis* in porcine sera, *Penyakit Hewan,* 17, 66, 1985; *Vet. Bull.,* 56 (Abstr. 7528).
36. **Bockemuhl, J. and Roth, J.,** *Brucella*-Titer bei subklinischen Infectionen mit *Yersinia enterocolytica* Serotyp 0:9 in einem Schweinezuchtbetrieb, *Zentralbl. Bakteriol. Parasitenkd. Infektionskr. Hyg. Abt. 1 Orig. Reihe A,* 240, 94, 1978.
37. **Nattermann, H., Horsch, F., and Zimmermann, H.,** *Yersinia enterocolytica*-Infektionen in Sweinebestanden und ihre differentialdiagnostische Bedeutung., *Monatsch. Veterinaermed.,* 35, 225, 1980.
38. **Karpinski, T. M., Skwarek, P., and Zorawski, C.,** Agglutinins cross-reacting with *Brucella, Salmonella, Yersinia,* and *Francisella* antigens in the serum of swine, *Med. Weter.,* 36, 394, 1980; *Vet. Bull.,* 51 (Abstr. 3864).
39. **Kaneene, J. M., Anderson, R. K., Johnson, D. J., Angus, R. D., Muscoplat, C. C., Pietz, D. E., Vanderwagon, L. C., and Sloane, E. E.,** Cell mediated immune responses in swine from a herd infected with *Brucella suis, Am. J. Vet. Res.,* 39, 1607, 1978.
40. **Sen, G. P., Joshi, T. P., and Pundit, S. V.,** Comparison of allergic reaction with serological tests for diagnosis of swine brucellosis, *Indian Vet. J.,* 53, 241, 1976.
41. **Ottosen, H. E. and Plum, N.,** A nonantigenic allergic agent for intradermal brucellosis tests, *Am. J. Vet. Res.,* 10, 5, 1949.
42. **Jones, L. M., Diaz, R., and Taylor A. G.,** Characterization of allergens prepared from smooth and rough strains of *Brucella melitensis, Br. J. Exp. Pathol.,* 54, 492, 1973.
43. **Garcia-Carrillo, C., Turovetzky, A., and Lucero, N.,** Especies y biotipos de *Brucella* aislados del hombre en la Argentina, *Medicina (Buenos Aires),* 45, 20, 1985.
44. **Hutchins, L. M., Bunnel, D. E., Donham, C. R., Eisele, W. C., and McCullough, N. B.,** The viability of *B. melitensis* (sic) in naturally infected cured hams, *Public Health Rep.,* 66, 1402, 1951.
45. **Hendricks, St. L.,** Public health aspects of swine brucellosis, *Southwest. Vet.,* 15, 199, 1962.
46. **Lindley, D. C. and Lander, J. R.,** Observations on the use of *Brucella abortus* strain 19 in swine including multiple intradermal doses, *J. Am. Vet. Med. Assoc.,* 115, 359, 1949.
47. **Genev, H., Hoistoforov, L., and Todorov, T.,** Versuche zur Immunoprophylaxe bei Schweinebrucellose, *Vet. Med. Nauki (Sofija),* 3, 387, 1966.
48. **Xie Xin,** Orally administrable brucellosis vaccine: *Brucella suis* strain 2 vaccine, *Vaccine,* 4, 212, 1986.
49. **Foster, J. W. and Edens, J. D.,** Immunization of swine against *Brucella* infection, *Symp. Ser. Immunobiol. Stand.,* Vol. 12, S. Karger, Basel, 1970, 135.
50. **Cedro, V. C. F. and Cisale, H. O.,** Una vacuna eficaz contra la brucelosis porcina, *Rev. Invest. Granad.,* 12, 221, 1961.
51. **Garcia-Carrillo, C., Cedro, V. C. F., Casas Olascoago, R., and Bagnat, E.,** Evaluacion en cobayos vacunas contra *B. suis, Rev. Med. Vet. Buenos Aires,* 58, 177, 1977.

Chapter 19

LABORATORY ANIMAL MODELS FOR BRUCELLOSIS STUDIES

Casimiro García-Carrillo

TABLE OF CONTENTS

I. INTRODUCTION

The selection of a laboratory animal as an experimental model for brucellosis that is capable of developing lesions similar to those encountered in man and large animals is a difficult task. The ideal model is one that can become infected through the same natural routes, have equal sensitivity, show similar clinical signs, and display the same pathological changes as humans and large animals.

There are many factors involved in the pathogenesis of the disease of brucellosis. Inflammatory reactions and degenerative lesions are found in certain organs, although the microorganism is frequently recovered from apparently normal tissues. Findings in laboratory animals do not always apply to man, cattle, goats, etc., but in certain circumstances, and with some limitations, corrrelations can be established. Guinea pigs and mice have been the principal animals used in laboratory experiments. Some studies have been performed using rabbits, rats, hamsters, and gerbils. In this chapter, the information available on brucellosis studies in these animals is summarized.

II. GUINEA PIG (*CAVIA PORCELLUS*)

The most commonly used types of guinea pigs are the Hartley, Dunkan-Hartley, and short hair (English and American), which are kept as outbred stocks, and the inbred strains 2 and 13. There are some 13 or more inbred strains and a number of mutants, each of which has its own particular characteristics.

General characteristics[1,2]

Gestation period	63 (59—72) d
Litter size	2 to 6
Birth weight	60—100 g
Breeding age and weight	male: 90—120 d (600—700 g)
	female 60—90 d (350—450 g)
Adult weight	male 900—1200 g
	female 700—900 g

A. PATHOGENESIS

Studies using guinea pigs to determine virulence of recently isolated brucella strains are almost as old as records of documenting brucella itself. Guinea pigs are probably the animals that are most susceptible to brucella infection. As few as 11 cells of some brucella strains are sufficient to cause infection.[3,4]

Schierholz and Kaufman[5] studied 28 strains of brucella trying to establish differences in virulence according to origin. After intramuscular inoculation with 10^9, 10^7, 10^5, and 10^3 cells of several field strains, all animals became infected, and no differences related to dose could be observed. Guinea pigs infected with *B. abortus* were susceptible to superinfection with *B. suis*.[6] Those infected with *B. neotomae* resisted about 100 ID_{50} of *B. suis*, but guinea pigs infected with *B. suis* were very resistant to *B. neotomae*.[7]

In nature, *Brucella suis* biovar 2 has a very specific pathogenicity for pigs and hares. In guinea pigs, it caused macroscopic and microscopic lesions similar to those produced by other brucellae.[10] Guinea pig infections produced by strains of *B. suis* isolated from reindeer were quite similar to those infections produced by other biovars of *B. suis*.[12]

All brucellae, with exceptions such as *B. ovis*, were pathogenic for guinea pigs. Lesions were consistently recorded in the liver, spleen, lungs and lymph nodes a few hours after subcutaneous inoculation. The organism could be isolated from blood, lymph nodes and internal organs.[8,9] Initially, lesions were observed in lymph nodes at the site of inoculation; later, hepatic granulomas and splenomegaly were found. The granuloma, as reported by Fabyan in 1912, is the basic histologic lesion in guinea pigs, and the same type of lesion has been reported in humans, mice, rats, rabbits, dogs, and monkeys.[8]

An important property of brucella is its ability to survive and even replicate within phagocytic cells. Most of the brucellae injected intravenously were removed by the liver and spleen and concentrated in those organs. Chronic infections were established in spite of reticuloendothelial activity.[11]

Brucella infection has a great influence on body weight gain in guinea pigs. Some authors consider it possible to construct a virulence scale for brucella strains according to their impact on animal growth, although any strain, including vaccine strain 19, has a negative influence on body growth.

Some authors have observed differences in the pathogenicity of different species of *Brucella* for guinea pigs. Godgluck[13] studied the pathological changes produced by *B. abortus, B. suis,* and *B. melitensis*. *Brucella suis* produced much more pathology in the spleen, liver, and testicles, with a tendency to suppuration and giant-cell accumulation, than did *B. melitensis* and *B. abortus*.[14]

B. ROUTE OF INFECTION

Guinea pigs are generally known to be very sensitive to infection by virtually any route: subcutaneous, conjunctival, intraperitoneal, intranasal, intravenous, vaginal, oral, or cutaneous scarification. The conjunctival route, considered the best for large animals, is not routinely used in guinea pigs for many reasons. A dose that is 50 to 100 times greater than that required to induce infection by the subcutaneous route is needed; some authors indicate that it must be 20,000 times greater.[4] Either conjunctival or respiratory routes are more representative than others in simulating natural conditions for an experimental challenge, but safety reasons prevent their routine use.[15,16] A large number of cells is required for percutaneous infection.[17]

C. INFLUENCE OF SEX

Female guinea pigs have generally been preferred for immunogenic studies, even though sensitivity to brucella infection is practically the same for both sexes. Experiments using 200 males and 200 females showed that there was no significant difference between the number of brucella cells necessary to infect 50% of the animals of either sex.[18]

Some sex-related differences in immunogenic capacity have been observed. For example, the protection afforded by strain 19 vaccine in females tended to be higher than in males, and the serological response also tended to be higher in females.[18]

High concentrations of brucella were recovered from the semen of guinea pigs experimentally infected by the subcutaneous, intraperitoneal, intravenous, and oral routes.[19,20] Guinea pig semen had little anti-brucella activity: the microorganism remained viable for

49 days in infected semen. Nevertheless, Hillaert et al.[19] were unable to demonstrate natural transmission from infected males to females placed with them for several weeks. Lesions in infected males were in the spleen, liver, testis, and epididymis.[20]

D. CONGENITAL TRANSMISSION

According to Bosseray and Díaz,[21] the guinea pig may be used as an experimental model to study the congenital or neonatal transmission of brucellosis. When 13 pregnant guinea pigs were challenged with 5×10^4 *B. abortus* strain 544, 5 aborted, 7 had normal parturitions, and one had 2 live and 2 dead offspring. At 2 months of age, most of the offspring were found to be infected.

E. SPLENIC INDEX

Splenic lesions are almost invariably found, and it is easy to observe macroscopic alterations in infected guinea pigs. Sometimes the alteration is merely a discrete hypertrophy or a slight increase in spleen size relative to the expected normal size.

The splenic weight index is a ratio of the spleen weight to the body weight of the animal. This ratio follows a curvilinear regression.[22] The ratio spleen weight/body weight increases from birth to a body weight of 300 to 400 g, with different curves for males and females. Thereafter, the ratio continuously decreases for the rest of the animal's life (Figure 1).

Many researchers use alterations of splenic index as indicators of infection. This simple method is very useful if the expected values take into account the animals' body weight at the time of sacrifice. Comparisons of indices, therefore, are valid between groups only when the animals have equal body weights and are of the same sex.

F. ANTIGENIC STUDIES

Guinea pigs, mice, and rabbits have been used to differentiate brucellosis from infections involving cross-reacting *Yersinia enterocolitica*. Only those animals inoculated with *Y. enterocolitica* developed antibodies to enterobacterial common antigen (ECA), whereas all animals inoculated with *B. abortus* or *Y. enterocolitica* 09 produced both direct agglutinins to *B. abortus* and indirect agglutinins to *Y. enterocolitica* 09 somatic antigen.[23]

G. IMMUNIZATION

Experiments to study the effects of immunization of guinea pigs against brucella infection have been made since 1910. According to Huddleson,[24] Ascoli was the first who tried and failed to immunize guinea pigs against brucellosis by injecting heat-killed cultures of *B. abortus*. Negative results were also obtained by Stafseth (1920), Hagan (1922), Schroeder and Cotton (1924) and others. McEwen and Roberts (1936) demonstrated that the killed vaccine did not produce appreciable immunity, but in some experiments conducted using live vaccines, 90% of the animals developed immunity, whereas all of the controls became infected.[24]

Bendixen and Jorgensen,[25] in 1940, conducted several experiments in guinea pigs to determine the virulence of strain G (Copenhagen vaccine strain) as compared with strain 19 (U.S. vaccine strain), strain W, and strain K. It was found that both strain 19 and strain G were of considerably lower virulence than the others.

The Huddleson strain 85 I was used during the 1930s in the U.S., but it had little protective effect for cattle. When this strain was tested in guinea pigs, it did not produce any demonstrable degree of immunity.[25]

According to Huddleson, cell suspensions of live mucoid *B. suis* and *B. melitensis* were capable of producing a high degree of immunity against experimental infection with *B. suis* and *B. melitensis*. Furthermore, vaccinated guinea pigs did not produce agglutinins, an important consideration for its use in cattle.[26] These results were partially confirmed in cattle.[27]

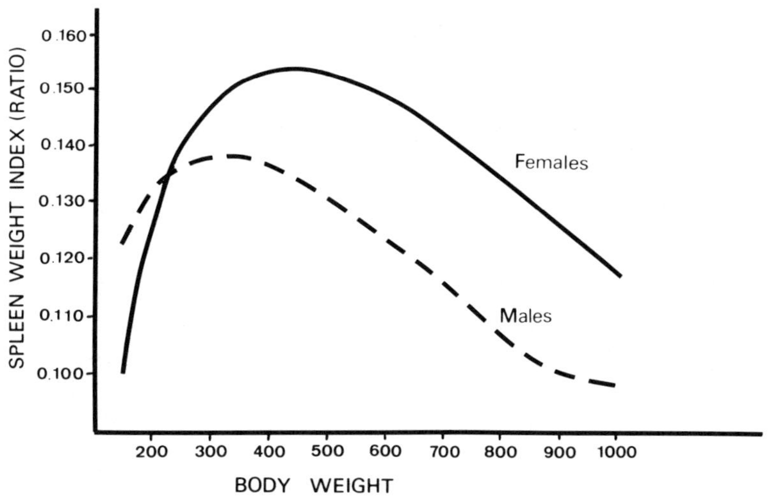

FIGURE 1. Changes of the splenic weight index related to the body weight in normal American short hair guinea pigs.

Live et al.[28] in 1945 demonstrated that sonic extract of *B. abortus,* injected subcutaneously or intracutaneously, produced an active immunity against moderate infecting doses of virulent *B. abortus*. Ether-killed whole brucella cells also produced some protection against a moderate challenge.

Many different types of soluble antigens extracted from brucella have been designed for use in humans to protect persons at risks against occupational disease. The immunogenicity of those products was tested in guinea pigs.[29-32] Woodard et al.[31,32] observed some immunogenic capacity in a purified protein derivative-like preparation of *B. abortus* strain 45/20, either alone or in association with defined adjuvants. Eight weeks after immunization, all guinea pigs were intramuscularly challenged with 1.04×10^4 CFU of *B. abortus* 2308, and 2 weeks later, their spleens were cultured for brucella. The challenge strain was recovered from most of the vaccinated animals. The evaluation criterion used for this study was lenient because the degree of protection was expressed in terms of the reduction in number of brucella organisms recovered from vaccinated animals as compared to the number recovered from the controls rather than the complete elimination of brucella from the vaccinates.

Keppie et al.[30] compared guinea pigs and mice as models in his studies of the immunogenic properties of killed vaccine with adjuvants. Oily adjuvant did not appreciably increase the immunogenicity for mice of either a 544 or 45/20 extract vaccine, as it did for guinea pigs. With the results obtained from the guinea pig, Keppie et al.[30] concluded that a whole-extract vaccine from strain 544 would be suitable for a trial in man, but would be inappropriate for cattle because the agglutinin response would interfere with the subsequent diagnosis of infected animals.

H. DURATION OF IMMUNITY

Results of experiments on the duration of immunity have been variable and contradictory. Manthei et al.[33] concluded that resistance induced in cattle immunized with strain 19 did not decrease with time, but this observation is inconsistent with results obtained from guinea pigs. In the latter species, the protection of animals vaccinated with either strain 19 or Rev. 1 was higher 1 month after vaccination than 1 year later.[34]

I. CROSS-IMMUNIZATION

Culture filtrates from *B. abortus, B. melitensis,* and *B. suis* were inoculated into guinea pigs to determine whether there was any cross-protection against challenge with different

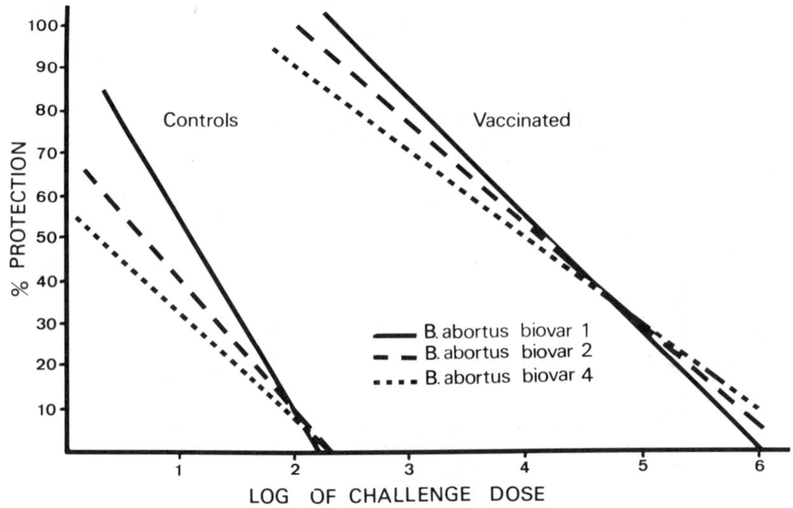

FIGURE 2. Immunity in guinea pigs vaccinated with strain 19 and challenged with graduated doses of field strains of *B. abortus* biovar 1, biovar 2, and biovar 4.

species of virulent strains. Crude cell wall preparations of *B. abortus, B. melitensis,* and *B. suis* protected guinea pigs against infection with all three *Brucella* species.[29]

A practical question was whether strain 19 protected cattle against natural infection with different biovars of *B. abortus*. Guinea pigs vaccinated with strain 19 were challenged with field virulent strains of biovars 1, 2, and 4. Some differences were observed between the challenge strains. Pathogenicity for guinea pigs of the strain represented by biovar 2 was higher than that of the two other strains. Nevertheless, all vaccinated guinea pigs showed a significant immunity against the three biovars in comparison to the controls, and there was no significant difference in the degree of protection afforded against each biovar[35,36] (Figure 2).

J. ADJUVANTS

Mineral oil, trehalose dimycolate (TDM), muramyl dipeptide (MDP), Freund's complete and incomplete adjuvants, and many other products have been tested and found to increase immunity.[37,38] Results obtained by many authors have demonstrated that most adjuvants enhance antibody production and that some of them increase immunity for a relatively short time. It is necessary to emphasize, however, that a beneficial effect of adjuvants has been obtained only with brucella extracts and inactivated brucella vaccines.

All known adjuvants have a deleterious effect on the protection afforded by living vaccines such as strain 19 or Rev. 1. This adverse effect is produced not only when the adjuvant is mixed with the vaccine, but also when inoculated simultaneously with the vaccine but in a different site.[39,40] Curiously, the serological response in animals that received strain 19 plus an adjuvant was as high as or higher than in animals which were better protected with strain 19 alone.[39]

K. POTENCY TESTS

Three methods are currently used for evaluating protection tests: (1) graduated vaccinal doses given against a fixed challenge; (2) graduated challenge doses given against a constant vaccinal dose; and (3) constant vaccine doses given against constant challenge doses.

The first method is probably the most extensively used for all types of vaccines including killed *Brucella* vaccines. With live vaccines, the dose response obtained by this method is very poor. At low vaccine doses, a linear relationship may be observed between the number

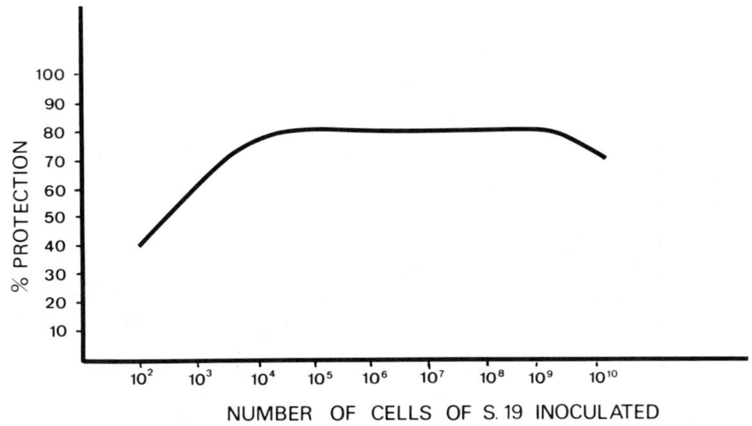

FIGURE 3. Protection afforded in guinea pigs by *B. abortus* strain 19 against an intramuscular challenge of 10^3 cells of virulent *B. abortus* strain 2308.

of strain 19 microorganisms administered and the protection afforded against challenge,[41] but as the dose of vaccine increases there is a tendency for the response to plateau (Figure 3).

The differences in immunity produced in guinea pigs by the administration of 10^7 and 10^{10} viable cells of strain 19 were not statistically significant, nor were the levels of protection afforded by *B. melitensis* Rev. 1 in doses of 6.2×10^5 and 1.08×10^9.[34] However, the immunity produced in guinea pigs by the administration of 100 cells of Rev. 1 was insignificantly lower than that produced by giving 10^8 cells of the same vaccine.[42,43]

Thornton and Muskett[41] observed that, when the vaccine dose was increased from $7 \times 10^{4.56}$ to 7×10^9 organisms, the protection against 5000 organisms of strain 544 was practically unchanged. In addition, the vaccine dose normally used in guinea pigs (which is 1/15 of that used in cattle) can be diluted over 10,000-fold before protection against the fixed challenge will decrease.

A graduated challenge dose produces a linear response curve in controls and vaccinated animals (Figures 2 and 4). Results obtained by different authors using different challenge strains produced the same type of linear regression relative to the challenge dose.[44,45] Consequently, this should be the method of choice for potency test assays. For the practical testing of living vaccines, a sufficiently critical procedure could, in general, be as follows:

Animals — Sixty guinea pigs weighing from 450 to 500 g, preferably of one sex or with the sexes equally represented in each group. Forty animals are vaccinated and 20 left as controls.

Vaccination — (a) *Brucella abortus* strain 19 in dose of 10^7 colony-forming units (cfu) subcutaneously; (b) *Brucella melitensis* Rev. 1 in dose of 10^6 cfu subcutaneously.

Challenge — Eight weeks after vaccination using *B. abortus* 2308, which is prepared as follows: *B. abortus* 2308, maintained lyophilized or in liquid nitrogen, is cultured in solid medium and a 48-h subculture is suspended in buffered 0.1% peptone in 0.85% saline. Dilution of this stock suspension is made to obtain concentrations of 10^4, 10^3, and 10^2, cfu. Vaccinated animals are divided into three unequal groups: 10 animals are challenged with 10^4 cfu; 20 are challenged with 10^3; and 10 are challenged with 10^2 cfu. One group of ten controls is given 10^3 cfu, and the other, 10^2 cfu. One (1) ml of the challenge suspension is given intramuscularly in each case.

Necropsy — Five to six weeks after challenge, all sixty animals are weighed and killed. Their spleens are observed for macroscopic alterations (mainly hypertrophy), weighed, and related to the body weight to calculate splenic indices. Spleens are examined bacteriologically by inoculation on suitable culture medium and incubation with 10% CO_2.

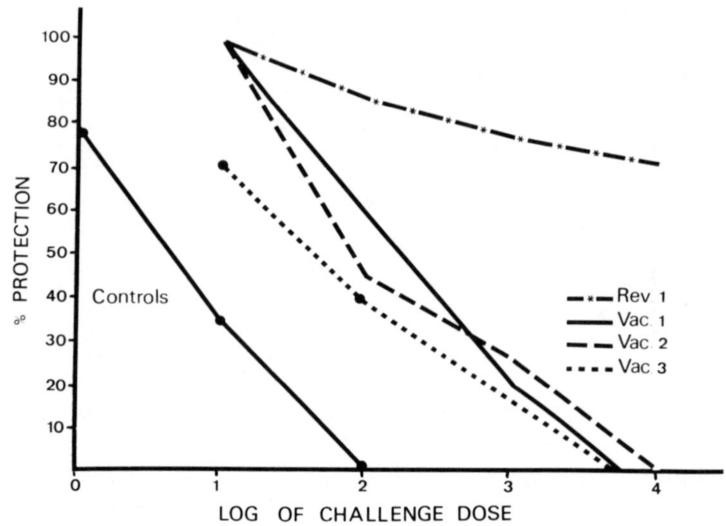

FIGURE 4. Protection afforded in guinea pigs by three killed vaccines in oily adjuvants and living Rev. 1 vaccine against graduated challenge doses of *Brucella abortus* strain 2308.

Interpretation of results — All animals with positive bacteriological cultures are considered infected regardless of the number of brucella organisms recovered. All animals with splenic indices that are more than three standard deviations higher than expected are considered positive.

Expected values of acceptance — (a) All controls at the challenge level of 10^3 should be infected; at the 10^2 level, two animals may fail to be infected (normally all are infected). (b) Of the animals vaccinated and subsequently challenged with 10^3 cfu, 70% are expected to be bacteriologically negative, and 50% should have splenic indices within the normal range. The level of protection expected in the group challenged with 10^2 should be almost 100%, and protection of 40% or more of the animals in the group challenged with 10^4 cfu is acceptable.

L. TESTS FOR NEW VACCINES

The general scheme designed for living vaccines (strain 19 and Rev. 1) which is described above could be used for any vaccine.

The guinea pig model is considerably more valuable than the mouse model for the evaluation of new vaccines. Occasionally, results that were promising in mice failed to materialize in guinea pigs, whereas the results obtained using guinea pigs were usually correlated with those obtained in other animals.[46,47]

The virulence of *B. suis* strain 2 vaccine (Chinese origin) was tested in guinea pigs to determine whether passages in susceptible animals increased its pathogenicity. The behavior of this strain was approximately the same as that of strain 19; after passages in pregnant sheep, goats, sows, and guinea pigs, its virulence remained practically unchanged.[48] The immunity provoked in guinea pigs by strain 2 is very promising, and this strain may prove useful for immunizing large animals.

The guinea pig is an acceptable model for screening and for the routine assay of vaccines[49] as well as for new products. In the author's judgment, those products that are poorly immunogenic in guinea pigs are probably not very immunogenic for other animals or for man.

In Figure 4, the protection afforded by three killed vaccines in oily adjuvants is compared with living Rev. 1 vaccine. The higher the challenge dose, the greater is the difference between inactivated and living vaccines.

M. CHALLENGE STRAIN

Brucella abortus 544 and *B. abortus* 2308 are the strains most extensively used for experimentally challenging animals vaccinated with living or killed *B. abortus* vaccines. In the author's experience, more variation was observed in the results of repeated experiments using strain 544 than 2308. Possibly, as pointed out by Hopkins et al.[50], these variations were related to the use of strain 544 variants. Strain 2308 apparently has maintained a higher level of virulence than 544, and the results obtained using this agent have been more reproducible.[3] Consequently, *B. abortus* 2308 is recommended for challenging cattle and guinea pigs.

N. HYPERSENSITIVITY

Lu Shi-Liang[51] used guinea pigs to demonstrate delayed hypersensitivity in the knee and liver. The same animal model has been used to study the desensitization by brucellysin. Jones and Berman[16] observed that, with few exceptions, guinea pigs given killed vaccines without adjuvant did not develop hypersensitivity. Those which received killed vaccines with Freund's incomplete adjuvant developed larger reactions than those given living organisms. Less variability in the size of reactions has been observed in animals inoculated with adjuvant.

O. POTENCY TEST FOR ALLERGENS

Delayed hypersensitivity is a common manifestation of human brucellosis, and the phenomenon plays an important role in the pathogenesis of this disease. Allergic tests are used for diagnosis of brucellosis in humans and animals, mainly in goats and sheep. The allergens, methods of sensitization, and concentration of allergens are usually tested in guinea pigs.[54-55]

The biological potency of each batch of brucellin must be compared to a standard allergen. De Reviers and Fensterbank[56] have proposed a methodology to control new batches of brucellin using guinea pigs, which consists of sensitizing eight females by the subcutaneous inoculation of the reference brucellin in Freund's adjuvant. Each animal is inoculated intradermally 4 to 5 weeks later with four different dilutions of the test batch and four dilutions of the standard. The sites for inoculation of the first animal are selected at random and the other seven by circular permutation. A table showing the diameters of the reactions produced by both allergens is then analyzed statistically.

P. TREATMENT

Brucella is an intracellular parasite, well protected from the effects of antibodies and antibiotics. Consequently, *in vitro* antimicrobial activity may not correlate with the efficacy of these agents in treatment regimens. Many antibiotics and chemotherapeutic agents have been used to treat brucellosis in man. Tetracycline is effective when infection is detected early, but chronic infection in man takes a long time to be cured, and relapses sometime occur.

The guinea pig has been used experimentally to evaluate the therapeutic effects of tetracycline and other chemotherapeutic agents. Certain new drugs, such as Minocycline, are theoretically superior because they have good *in vitro* activity and show good intracellular penetration. Sometimes the effectiveness of new drugs is confirmed by experiments in guinea pigs, and in many other cases, the drug is ruled out because its *in vivo* activity does not correlate with its *in vitro* action or because it is poorly tolerated.

A stable plurilamellar vesicle (SPLV) was used by Fountain et al.[57] as a carrier of antibiotics to treat brucella infections in experimentally infected animals. Guinea pigs infected with *B. abortus* were treated with SPLV-entrapped streptomycin on days 7 and 10 after infection, and all detectable bacteria were eliminated. Comparisons in infected guinea pigs have also demonstrated the superiority of SPLV-entrapped aminoglycosides over free aminoglycosides in eliminating intracellular bacteria.

The antibiotic rifampicin is active against brucella. The spleens of experimentally infected guinea pigs are cleared of bacteria after 20 days of oral rifampicin treatment. Contradictory results regarding combinations of rifampicin and trimethoprim have been reported. According to some authors, the combination had a synergistic effect, while other workers reported either no increase in effectiveness or an antagonistic effect.[58,59]

III. MOUSE (*MUS MUSCULUS*)[60]

Gestation period	20 (19—21) d
Litter size	4—12
Birth weight	0.5—1.5 g
Breeding age and weight	male 50 days (20—25 g)
	female 50—60 days (15—23 g)
Adult weight	male 20—40 g
	female 18—40 g

Strains and stocks — There are several hundred inbred strains, numerous congenic strains and mutants, and more than 50 outbred stocks of mice. All have their specific characteristics and there is considerable variability in many of their physiologic parameters. Their susceptibility to infections also varies according to their genetic constitution.

A. SUSCEPTIBILITY

The potential role of wild rodents as reservoirs of brucella has been reported by many authors. Thorpe et al. conducted experimental studies on the susceptibility of wildlife and laboratory animals to experimental infection with *B. abortus, B. suis, B. melitensis,* and *B. neotomae*. Eight species of mice were generally found to be relatively susceptible to all strains of brucella, while rats, lagomorphs, and squirrels were more resistant. Guinea pigs were also susceptible, except to *B. neotomae*. A few studies are available on the pathogenicity of *B. neotomae*. Thorpe et al.,[61] for example, reported extreme splenic hyperplasia in deer mice and white mice infected with *B. neotomae*. Some of their spleens were found to weigh as much as 4 g.

Taran and Rybasov,[62] in a comparative study of susceptibility, inoculated guinea pigs, mice, rats, and sheep. Guinea pigs inoculated with 10^1 to 10^2 cells of *B. melitensis,* and *B. suis* developed severe granulomatous lesions. In mice, it was necessary to administer 10^4 to 10^6 cells to produce lesions. Rats were infected with 10^9 cells, while sheep were resistant to the three strains used.

B. PATHOGENESIS

The etiologic agent of brucellosis enters the bloodstream and causes bacteremia regardless of the route of infection. Mice develop chronic infections, not only when challenged with virulent brucella strains, but also with attenuated vaccine strains.[63]

A steady increase in the number of strain 19 bacteria was found in the spleens of CBA mice over the first week after inoculation. Similar results were obtained by Birmingham and Jeska working with C57BL/6 mice inoculated with *B. abortus* strain 19. Analysis of the number of bacteria colony-forming units indicated that the infection has three phases: bacterial multiplication during the first 7 d; a period of high bactericidal activity until days 15 to 20; and a latent phase, characterized by a slow decline of bacterial numbers.[64] During the subsequent 5 to 6 weeks, the bacterial number was stable, and brucella could be isolated up to 3 months after infection.[63] A marked splenomegaly was observed.

Resistance to infection could be transferred to normal mice by T lymphocytes from the spleens of mice that had been infected for 3 months.[63] Mice have been used to demonstrate that B lymphocytes are not required to transfer immunity. The cell-mediated immunity is efficiently produced by living vaccines, while immunity produced by killed vaccines depends mainly on antibody levels.[63,65] Blood clearance of *B. abortus* is increased by preopsonization with specific antibodies, indicating that opsonins are limiting factors in the development of resistance to brucella.[66]

Genetic factors are important in mice. Strain 19 caused a chronic infection in CBA/H, while BALB/c, C57BL/10, and B10Br were more resistant to infection.[65,66] Genetic factors in combination with different brucella species have been studied. Sex was also important, females being more resistant. This resistance factor was inherited by F_1 and backcrosses as a partially dominant character.[67-69]

Pathogen-free female mice of the C57BL/6 and DBA/2 strains have been used to study autoimmunity and rheumatoid factor induced after inoculation with phenol insoluble (PI) fraction and virulent *Brucella suis*.[70] Results obtained by Cannat et al.[70] suggest that the PI fraction is active in inducing autoimmune phenomena.

C. ENDOTOXINS

Abernathy et al.[71] explored the susceptibility of infected mice to endotoxin. Mice infected with *B. melitensis* and *B. abortus* strain 19 were challenged with endotoxin 2 weeks after infection. Mortality was 90% in infected animals, while uninfected mice survived at the same endotoxin dose. This increased susceptibility of mice was interpreted as being the result of reactivation of a chronic brucella infection. However, the results of Abernathy et al. did not reveal an increase in the number of brucellae from the spleens of infected mice, and it was observed that the degree of infection, as measured by spleen cultures, had no influence upon susceptibility to endotoxin.

D. MACROPHAGE FUNCTION

Resistance to infection depends upon the successful collaboration between specifically sensitized T lymphocytes and macrophages. Birmingham and Jeska[64] demonstrated an enhancement of a variety of macrophage functions, including surface receptor activity, phagocytic capacity, and locomotor functions that may play a direct role in mediating resistance to brucella. The enhanced macrophage activity was detected very early during infection, before the onset of the bactericidal phase, and persisted 60 days post-infection.

E. ANTIBODY-DEPENDENT CYTOTOXICITY

Cytotoxic activity mediated by antibody dependent cells has been studied in C57BL/6 mice. K cells from experimentally infected mice produced lysis of brucella in the presence of specific antibodies.[72] Cytotoxic activity increased 5 to 7 d after inoculation of *B. abortus*, reaching a maximum on day 10.

F. ROUTE

Jacotot and Vallée,[73] in 1953, studied the susceptibility of mice and rats according to the route of infection. Intraperitoneal inoculation of strain 19 and virulent *B. abortus* gave variable results according to dosage; 18 of 20 rats inoculated intracardially with pathogenic strains died 2 to 23 d later. Variable results were obtained in mice.

G. REINFECTION

Virulent brucella multiplies in the spleen and other organs of mice, and resistance is acquired. One month after inoculation, mice are very resistant to intravenous challenge with other virulent brucella strains.[74]

H. L-FORMS

L-forms have been isolated from infected mice and treated with penicillin. Schmitt-Slomska et al.[75] were unable to obtain vigorously growing L-forms from *B. suis* on artificial media containing penicillin, ampicillin, carbenicillin, or glycine. From mice killed 1 or 2 weeks after infection with *Brucella,* and 5 d after the last dose of penicillin, only L-form-type growth appeared in primary culture.[75]

I. CELL AND SERUM-MEDIATED PROTECTION

Resistance to placental colonization for 3 d was produced by administering immune serum 24 h before or just after challenging 15-d pregnant mice with *B. abortus* 544.[76,77] Experiments in mice have demonstrated that immunity is the consequence of the interaction of T lymphocytes and immune serum.[78] The major mechanism for controlling the infection is cell-mediated immunity (CMI) and the cells responsible for transferring the resistance have been shown to be Ly 1+ 2+ T lymphocytes.[63] Monoclonal anti-Ly-1 antibodies injected before infection produced a 10-fold increase in bacterial numbers 25 d after infection and removed the cell population which transfer resistance.[63]

There is no doubt that living vaccines are able to stimulate CMI very efficiently. Protection afforded by killed vaccines, on the other hand, will depend on the capability of an adjuvant to stimulate T lymphocyte population.

It was demonstrated in mice, guinea pigs, and large animals that immunity is produced most efficiently as a result of exposure to a living vaccine or natural infection. Plommet and Plommet felt that this phenomenon was not so simple, and using the mouse model, demonstrated that vaccinal immunity depended partially on a serum factor,[78] and that protection is the consequence of the interaction of T lymphocytes and immune serum.

J. BRUCELLOSIS IN PREGNANT MICE

Abortion is a common sequella of brucella infection in cows, goats, and many other animals. Nevertheless, Bosseray[79] challenged pregnant mice at day 3, 7, 11, or 15 of pregnancy with *B. abortus* strain 544, using several routes, and observed no abortions or fetal deaths. Brucellosis did not modify pregnancy in mice, although placental colonization occurred as early as 5 min postchallenge.[80] The pathogenesis of placental colonization found in mice was similar to that observed in rats.[79] Mice immunized 1 month before pregnancy with a killed H38 vaccine and living strain 19 were resistant to challenge on the 7th day of pregnancy.[77,79]

The sensitivity to brucellosis of calves increases with age until they reach puberty, but tests conducted by Kawashima et al. in mice to study the influence of gonadotrophic hormones on this phenomenon were inconclusive. No relationship could be established in mice between brucella infection and sex hormones.[81]

K. VERTICAL TRANSMISSION

Pregnant mice were challenged with *B. abortus* 544, and 60% of the newborn were infected.[82]

L. VACCINES

Many authors use mice for the initial testing of proposed vaccines for their immunogenic properties.[83] Their main reason is that mice are less costly than other animals.

Mice immunized with H-38 *B. melitensis* killed vaccine 36 d before pregnancy were protected against an intraperitoneal challenge with 2×10^5 *B. abortus* 544.[79]

Protein-bound cell wall peptidoglycan fractions extracted from brucella protected mice as well as H38 killed and B-19 attenuated living vaccines.[76,77]

Monoclonal antibodies specific for the 0 polysaccharide of *B. abortus* inoculated before challenge decreased the number of viable counts in spleens and livers. Monoclonal antibodies specific to porin of *B. abortus* failed to confer protection. This finding of Montaraz et al.[84] may be a consequence of the mouse inability to fix complement on whole cells. The hypothesis that anti-O antibodies play an important role in protective immunity is consistent with results obtained in mice.[84,85] The protection given by living vaccines is poorly related to the dose. A standard dose of 1×10^5 colony-forming is reasonable for testing living vaccines in mice.[86]

1. Combined Vaccines

Three living, attenuated strains of *Brucella, Salmonella* and *Chlamydia* species were studied to determine their capacity to immunize mice. Plommet et al. found that the combination of living vaccines without adjuvants has no deleterious effect on the protection they respectively afforded.[87] The adverse effect of simultaneous inoculation demonstrated in guinea pigs was due to adjuvant effects.[39]

2. Methods for Vaccine Evaluation

The ideal method for vaccine testing should be simple, quick and inexpensive, and should also give clear-cut results. So far, the only existing method of vaccine assessment utilizes animals which are first vaccinated and then submitted to challenge. In practice, guinea pigs, and mice are the laboratory animals used for this purpose; however, the species used is not consistent from one laboratory to another.

Bosseray et al.[88] propose mice as the animal model for titration of biological activity and quality control. Immunity is evaluated relative to the number of brucella organisms recovered from the spleen.

A reference vaccine should be available. With killed vaccines, a dose response curve is obtained,[88] while with live vaccines, maximum protection is sometimes afforded by a very small dose.

Besides some differences that are related to the animal species used, factors such as challenge dose, route of challenge, and other parameters have to be considered. Pilet and Le Garrec, in a comparative study, concluded that mice yield the most consistent results, and with rats or guinea pigs significant differences were observed.[89-91] Geissler, in 1959, found a high level of resistance in mice 72 h after intraperitoneal inoculation of strain 19, while immunity reached its peak on the 12th day following immunization by the subcutaneous route.[92] Pilet et al.[93] reported a good immune response in mice vaccinated with heat-killed strain 19 vaccine at 1 day of age; it was similar to that produced by vaccination at 1 month of age. No information is available to confirm these results.

Plommet and Bosseray[86] proposed a general method to control anti-*Brucella* vaccine using mice and a reference vaccine. A system of units was used to compare different vaccines, either live or killed, and the potency of given vaccines was evaluated relative to the level of splenic infection in mice inoculated under standardized conditions. Killed vaccines showed a correlation between dose and response, while the level of immunity afforded by living vaccine was poorly correlated with the dose. According to Plommet and Bosseray, living vaccine should be tested in a dose of 1×10^5 living cells. In the case of strain 19, this would be equivalent to 20 U. The dose for cattle should be 5000 U.[86]

3. Treatment with Vaccine

For the last 30 years, many patients with brucellosis have been treated with brucella extracts, although there were many doubts regarding the beneficial effect of this treatment. Mice chronically infected with *B. abortus* were treated by subcutaneous injections of a *B. abortus* cell-wall vaccine suspended in oil adjuvant. Vaccinations administered four times at 15 to 20-d intervals did not impair the evolution of chronic brucellosis.[94]

Some studies have claimed that levamisole is helpful in the treatment of brucellosis in humans. Experiments in mice, however, demonstrated that levamisole treatment did not affect the clearance of brucella from the spleen. Results obtained by Young and McMillan cast further doubts on the usefulness of levamisole in the treatment of brucella infections.[95]

M. TREATMENT

Tetracyclines have been used for many years as the standard treatment of human brucellosis. Mice have been used to study the *in vivo* action of antibiotics. Trials showed that, in these animals, rifampicin was as good as or even better than tetracycline.[58,96] Ampicillin and the trimethoprim-sulfametoxazole combination showed slight activity.[97]

Fountain et al. observed that the intraperitoneal administration of stable plurilamellar vesicle-encapsulated aminoglycosides was effective against *B. canis* infection in mice.[57] Infected mice were treated with dihydrostreptomycin, gentamicin, kanamycin, or streptomycin-free or entrapped in SPLVs. In all cases treated twice with SPLV-entrapped aminoglycoside (10 mg/kg) on days 7 and 10 after infection, no bacteria could be detected in the spleen on the 17th day after infection.

IV. RABBIT (*ORYCTOLAGUS CUNICULUS*)[98]

Gestation period	31 (28—34) d
Litter size	4 to 10
Birth weight	50 (30—100) g
Breeding age and weight*	male 6—10 months
	female 5—9 months
Adult weight	male 2—5 kg
	female 2—6 kg

Breeds — There are more than 35 breeds of rabbits. Those most commonly used in research are the New Zealand White and California (medium-sized breeds, weighing 2 to 5 kg). The small Dutch and Polish breeds, weighing less than 2 kg, are sometimes used, whereas the large breeds (more than 5 kg) such as Flemish and Checkered Giants are rarely used. There are few inbred strains, but more than 70 outbred stocks of rabbits exist, which are derived from several breeds.

Normal rabbits are partially susceptible to brucella infection. In one experiment, New Zealand rabbits weighing from 3.5 to 4.2 kg were inoculated intramuscularly with 10^2 to 10^6 cfu of *B. abortus* 2308. A poor relationship was observed between the doses given and the numbers of rabbits infected. The maximum of 50% having positive splenic cultures was obtained at levels 10^4 and 10^6 cfu. The spleens of culture-positive infected rabbits had an average size (1.860 g) only a little larger than those of normal animals (1.519 g). Their serological response, in terms of IgG agglutinins, was very good and dose related. Antibody titers were as high 10 days after inoculation as on day 50, when animals were killed.

The same levels of infection and serological responses were obtained with Dutch rabbits weighing 1.7 to 2.3 kg following i.m. inoculation of *B. abortus* 2308. No differences in spleen size, however, could be demonstrated between infected and noninfected rabbits of this breed (in one experiment).

Local inflammatory reactions were produced in rabbits by the intradermal injection of living brucella, and no differences were observed in the degree of dermal pathogenicity of the 3 species of brucella.[99]

Pregnancy increases susceptibility to generalize brucella infection. Kniazeff et al.[100] demonstrated that both pseudopregnancy and progesterone treatment during estrus also increased susceptibility to infection. Despite the increase in systemic susceptibility to brucella infection, the infecting organism was not recovered from the uterus of pseudopregnant or progesterone-treated rabbits.

* Weight varies considerably depending on breed and strain/stock.

V. RAT (*RATTUS NORVEGICUS*)[60]

Gestation period	21 (20—22) d
Litter size	7 to 14
Birth weight	5—6 g
Breeding age and weight*	male 65—110 d (250—400 g)
	female 65—110 days (200—300 g)
Adult weight*	male 450—520 g
	female 250—300 g

Strains and stocks — There are more than 110 inbred strains of rats such as the ACI, BN (Brown-Norway), BUF (Buffalo), F344 (Fischer 344), LEW (Lewis), and WF (Wistar-Furth), various congenic strains and mutants as well as a number of outbred stocks which include the Wistar, Sprague-Dawley, Holtzman, and Long-Evans.

Rats are relatively resistant to *Brucella* infection. It was necessary to administer a high dose (several million cells) to produce generalized infection, and the spleen was usually free of *Brucella* a month after experimental infection. As a laboratory model for testing immunizing agents, the rat gave some contradictory results. Jacotot and Vallée, for example, reported a better immunity in rats inoculated with a phenol-killed strain 19 than with living organisms.[101,102] A serological response to concentrated suspension of killed *B. abortus* was early detected between 4 and 5 days postinoculation.[103]

The urine of rats contained *B. abortus* 24 to 48 h after feeding them with infected materials.[104]

Some reports show that gray rats *(Rattus norvegicus)* are more susceptible to *B. abortus* than to *B. melitensis* and *B. suis*.

VI. HAMSTER (*MESOCRICETUS AURATUS* — SYRIAN OR GOLDEN HAMSTER)[60]

Gestation period	16 (15—18) d
Litter size	5 to 9
Birth weight	2 g
Breeding age and weight**	male 10—14 weeks
	female 6—10 weeks
Adult weight	male 85—130 g
	female 95—150 g

Strains and stocks — There are at least 40 inbred strains and 15 outbred stocks of Syrian hamsters.

Susceptibility — Graded doses of 10^1 to 10^4 of *B. abortus* 2308 were inoculated i.m. into hamsters. In two consecutive experiments conducted under the same conditions, the infecting agent was recovered 56 d after inoculation from animals at each dose level. However, no real correlation could be established between dose and number of infected animals. At all levels, some bacteriologically and serologically negative animals were found. Infected hamsters showed an average of spleen weight of 0.175 g, while the spleens of noninfected (negative culture) and control animals (not inoculated) weighed 0.100 g.

Based on results reported to date, the hamster does not appear to be a good animal model in which to study *B. abortus* infection, because great individual differences in susceptibility exist. Other experimental conditions, such as routes of inoculation, lapsed time from challenge to sacrifice, etc., need to be studied, however, before the usefulness of this species in brucellosis research can be completely evaluated.

* Varies according to strain and stock.

** Too much variability in weight depending on stock or strain to define this parameter in general.

VII. GERBIL (*MERIONES UNGUICULATUS*)[60]

Gestation period	25 (24—26 d)
Litter size	4 to 6
Birth weight	2.5—3.0 g
Breeding age and weight	male 70—85 d (50—55 g)
	female 65—85 d (45—50 g)
Adult weight	male 80—110 g
	female 70—100 g

Strains and stocks — The majority of gerbils (Mongolian gerbil) are maintained as outbred stocks; there is at least one inbred strain.

Very limited data have been accumulated on brucellosis research in gerbils. This species is susceptible to infection by the i.m. inoculation of *B. abortus,* as few as 20 brucella cells being sufficient to produce disease in some individuals. However, the variation in susceptibility from one individual to another is significant.

Laboratory animal experiments with *B. ovis* — *B. ovis* is the only brucella with an exclusive host, epididymitis and placental inflammation being the natural manifestations of infection. No animals other than sheep are known to be naturally susceptible. Cuba-Caparó and Myers, searching for a laboratory animal model in which to study *B. ovis* infection, inoculated mice, guinea pigs, gerbils, hamsters, and both Lewis and Wistar rats intratesticularly. *Brucella ovis* was repeatedly recovered from the testicles, livers, and spleens of gerbils for 5 months.[105] In the same study, *B. ovis* was occasionally isolated from mice, hamsters, and guinea pigs. No information is available on *B. ovis* infection by other routes in laboratory animals.

ACKNOWLEDGMENTS

The author would like to acknowledge S. Albertelli, S. Potkay, and A. Gutiérrez-Reto for their assistance in the preparation of this manuscript.

REFERENCES

1. **Sisk, D. B.,** Physiology, in *The Biology of Guinea Pigs,* Wagner, J. E. and Manning, P. J., Eds., Academic Press, New York, 1976, chap. 7.
2. **Schilling, P. W.,** Selected topics on medicine of laboratory animals. The guinea pig, *Scient. Techn. Monogr. Ser.,* 13, Centro Panamericano de Fiebre Aftosa, Río de Janeiro, 1984.
3. **García Carrillo, C. and Casas Olascoaga, R.,** Comparison of two challenge strains used to test the potency of *Brucella abortus* vaccines, *Br. Vet. J.,* 133, 554, 1977.
4. **Ne'eman, L. and Jones, L. M.,** The evaluation of the virulence of a recently isolated strain of *Brucella melitensis* (Y 85) for guinea pigs and pregnant sheep, *Refu. Vet.,* 20, 169, 1963.
5. **Schierholz, H. and Kaufmann, R.,** Experimente zur Virulenzbestimmung von Bangbakterien im Kleintierversuch, *Zentralbl. Bakteriol. Parasitenkd. Infektionskr. Hyg. Abt. Orig.,* 179, 509, 1960.
6. **Pollack, A. D., Kelly, E. H., Gorelick, A. N., Braun, W., and Victor, J.,** Superinfection in experimental brucellosis of guinea pigs, *J. Infect. Dis.,* 90, 267, 1952.
7. **Stoenner, H. G.,** The behavior of *Brucella neotomae* and *Brucella suis* in reciprocal superinfection experiments in mice and guinea pigs, *Am. J. Vet. Res.,* 24, 376, 1963.

8. **Watanabe, M., Yamano, T., Ohchi, T., and Inui, S.,** Studies on experimental infection of guinea pigs with *Brucella abortus*. I. Correlation among distribution of *B. abortus* in the body, appearance of agglutinin in serum and formation of hepatic focal lesions in early stage of infection, *Bull. Nat. Inst. Anim. Health,* 38, 75, 1960.
9. **Rolle, M.,** Organotropismus der Brucellen bei nicht immunen und immunisierten Meerschweinchen, *Zentralbl. Veterinaermed. Reihe B,* 6, 714, 1959.
10. **Sterba, F.,** Morphological changes in brucellosis in guinea pigs due to experimental infection with *Brucella suis, Vet. Med. (Praha),* 29, 353, 1984.
11. **Guerra, H., Deter, R. L., and Williams, R. P.,** Infection at the subcellular level. II. Distribution and fate of intravenously injected brucellae within phagocytic cells of guinea pigs, *Infect. Immun.,* 8, 694, 1973.
12. **Grekova, N. S. and Gubina, E. A.,** (The course of brucellosis in guinea pigs infected with *Brucella* isolated from reindeer), *Zh. Mikrobiol. Epidemiol. Immunobiol.,* 42, 92, 1965; abstracted in *Vet. Bull. (London),* 35, 486, 1965.
13. **Godgluck, G.,** Die pathologisch-anatomischen und-histologischen Veranderungen bei mit Brucellosebakterien (Abortus Bang, Suis, Melitensis) infizierten Meerschweinchen und ihre diagnostische Bedeutung, *Zentralbl. Bakteriol. Parasitenkd. Infektionskr. Hyg. Abt. Orig.,* 159, 63, 1952.
14. **Moulton, J. E. and Meyer, M. E.,** The pathogenesis of *Brucella suis* infection in guinea pigs. Lesions of the spleen, liver, testis, and articulations, *Cornell Vet.,* 48, 165, 1958.
15. **Harper, G. J.,** *Brucella suis* infection of guinea pigs by the respiratory route, *Br. J. Exp. Pathol.,* 36, 60, 1955.
16. **Jones, L. M. and Berman, D. T.,** Antibody response, delayed hypersensitivity and immunity in guinea pigs induced by smooth and rough strains of *Brucella abortus, J. Infect. Dis.,* 124, 47, 1971.
17. **Maestrone, G.,** Limiti al passaggio di *B. abortus* nella cavia per via cutanea, *Atti Soc. Ital. Sci. Vet.,* 10, 641, 1957.
18. **García-Carrillo, C.,** Relación del sexo del cobayo con la sensibilidad a la infección por *Brucella abortus* y la respuesta immunológica, *Rev. Asoc. Argent. Microbiol.,* 10, 8, 1978.
19. **Hillaert, E. L., Hutchings, L. M., and Andrews, F. N.,** Brucellosis in male guinea pigs, *Am. J. Vet. Res.,* 11, 84, 1950.
20. **Ueda, K. and Imaizumi, K.,** Studies on experimental brucellosis. *Brucella abortus* infection in the male guinea pig and effect of sensitization on various organs, *Jpn. J. Vet. Sci.,* 31, 287, 1969.
21. **Bosseray, N. and Díaz, R.,** Brucellose congenitale du cobaye, *Ann. Rech. Vet.,* 5, 147, 1974.
22. **García-Carrillo, C.,** Relationship between body weight and spleen size in guinea pigs, *Lab. Animals,* 11, 175, 1977.
23. **Mittal, K. R., Ricciardi, I. D., and Tizard, L. R.,** Indirect hemagglutination employing enterobacterial common antigen and *Yersinia* somatic antigen: a technique to differentiate brucellosis from infections involving cross-reacting *Yersinia enterocolitica, J. Clin. Microbiol.,* 11, 149, 1980.
24. **Huddleson, I. F.,** Immunity in brucellosis, *Bacteriol. Rev.,* 6, 111, 1942.
25. **Bendixen, H. C. and Jorgensen, J.,** Investigations on the virulence and immunizing power of some *Brucella* strains with a special view to judging of strains employed for vaccination of cattle, in *The Royal Veterinary Agriculture College Yearbook,* Royal Veterinary and Agriculture College, Copenhagen, 1940, 80.
26. **Huddleson, I. F.,** The immunization of guinea pigs with mucoid phases of *Brucella, Am. J. Vet. Res.,* 8, 374, 1947.
27. **Huddleson, I. F. and Bennett, G. R.,** The vaccinal immunizing value of a mucoid-growth phase of *Brucella suis* against brucellosis in cattle, *Mich. Agric. Exp. Stn. Q. Bull.,* 31, 139, 1948.
28. **Live, I., Sperling, F. G., and Stubbs, E. L.,** Studies on the immunization of guinea pigs against *Brucella* infection, *J. Infect. Dis.,* 77, 16, 1945.
29. **Keppie, J., Witt, K., and Smith, H.,** Cross immunization of guinea pigs with products of *Brucella abortus, melitensis and suis, Br. J. Exp. Pathol.,* 44, 84, 1963.
30. **Keppie, J., Witt, K., and Smith, H.,** The immunization of guinea pigs and mice with a whole-culture extract of a smooth and a rough strain of *Brucella abortus, Br. J. Exp. Pathol.,* 53, 518, 1972.
31. **Woodard, L. F., Toone, N. M., and McLaughlin, C. A.,** Immunogenic properties of soluble antigens on whole cells of *Brucella abortus* strain 45/20 associated with immunoadjuvants. I. Soluble antigens, *Can. J. Comp. Med.,* 44, 453, 1980.
32. **Woodard, L. F., Toone, N. M., and McLaughlin, C. A.,** Immunogenic properties of soluble antigens on whole cells of *Brucella abortus* strain 45/20 associated with immunoadjuvants. II. Whole cells, *Can. J. Comp. Med.,* 44, 456, 1980.
33. **Manthei, C. A., Mingle, C. K., and Carter, R. W.,** Duration of immunity to brucellosis induced in cattle with strain 19 vaccine, *Proc. 88th Annu. Meet. Am. Vet. Med. Assoc.,* American Veterinary Medical Association, Milwaukee, 1951, 128.
34. **García-Carrillo, C.,** Comparación de la inmunidad en cobayos vacunados con cepa 19 y con Rev. 1 desafiados con *B. abortus* un mes y un año después, *J. Vet. Med.,* 33, 231, 1986.

35. **García-Carrillo, C., Casas Olascoaga, R., Szyfres, B., and Trenchi, A.,** Protección de la vacuna cepa 19 de *B. abortus* biotipos 1, 2 y 4. Ensayos en cobayos, *Rev. Med. Vet. (Buenos Aires),* 58, 1, 1977.
36. **García-Carrillo, C., Casas Olascoaga, R., Szyfres, B., and Trenchi, A.,** Patogenicidad para el cobayo de algunas cepas de 3 biotipos de *Brucella abortus, Rev. Med. Vet. (Buenos Aires),* 58, 15, 1977.
37. **Kulshrestha, R. C., Atal, P. R., and Wahi, P. N.,** A comparison of immunizing capacity of *Brucella* vaccines in guinea pigs, *Indian J. Anim. Sci.,* 42, 586, 1972.
38. **Woodard, L. F., Toone, N. M., and Jasman, R. L.,** *Brucella abortus* vaccines: comparison of protection provided by immunopotentiated 45/20 bacterins and live strain 19 vaccine in guinea pigs, *Am. J. Vet. Res.,* 42, 1959, 1981.
39. **García-Carrillo, C.,** Efecto de la aplicación simultánea de vacuna *B. abortus* cepa 19 y vacuna antiaftosa sobre la inmunidad contra brucelosis en cobayos, *Rev. Asoc. Argent. Microbiol.,* 10, 1, 1978.
40. **García-Carrillo, C.,** Efecto de la vacuna antiaftosa en adyuvante oleoso aplicada un mes antes o un mes después que la vacuna *B. abortus* cepa 19 sobre la inmunidad contra la brucelosis en cobayos, *Rev. Med. Vet. (Buenos Aires),* 68, 240, 1987.
41. **Thornton, D. H. and Muskett, J. C.,** The use of laboratory animals in the potency test of *Brucella abortus* S.19 vaccine, *J. Comp. Pathol.,* 82, 201, 1972.
42. **Alton, G.,** Vaccination of guinea pigs with Rev. 1 *Brucella melitensis* vaccine. The response to very large and very small doses, *Res. Vet. Sci.,* 10, 326, 1969.
43. **Alton, G.,** Trial in guinea pigs of four methods of vaccination against *Brucella melitensis* infection, *Res. Vet. Sci.,* 10, 329, 1969.
44. **Jones, L. M., Thomson, P. D., and Alton, G. G.,** Immunity against *Brucella* infection in guinea pigs produced by three different vaccines, *J. Comp. Pathol.,* 68, 416, 1958.
45. **García-Carrillo, C., Cedro V. C. F., Casas Olascoaga, R., and Bagnat, E.,** Evaluación en cobayos de vacunas contra *Brucella suis, Rev. Med. Vet. (Buenos Aires),* 58, 177, 1977.
46. **García-Carrillo, C. and Trenchi, A.,** Estudio comparativo de la inmunidad conferida a cobayos por varios tipos de vacunas antibrucélicas, *Rev. Med. Vet. (Buenos Aires),* 55, 381, 1974.
47. **Foster, T. L. and Brewer, J. H.,** A lysed, charcoal-adsorbed vaccine for brucellosis, *Proc. 88th Annu. Meet. U.S. Animal Health Association,* U.S. Animal Health Association, Fort Worth, Texas, 1984, 152.
48. **Xie Xin,** Orally administrable brucellosis vaccine: *Brucella suis* strain 2 vaccine, *Vaccine,* 4, 212, 1986.
49. **Cameron, C. M.,** Comparative efficacy of six *Brucella* vaccines in guinea pigs, *Onderstepoort J. Vet. Res.,* 46, 229, 1979.
50. **Hopkins, I. G., Thorton, D. H., and Hebert, N.,** Comparison of *Brucella abortus* (strain 544) variants, *J. Biol. Stand.,* 9, 421, 1981.
51. **Lu Shi-liang,** A study of delayed hypersensitivity and desensitization in guinea pigs after *Brucella* infection, in *Developments in Biological Standardization,* Vol. 56, S. Karger, Basel, 1984, 317.
52. **Jones, L. M.,** A potency assay for *Brucella* allergens in guinea pigs, in *Developments in Biological Standardization,* Vol. 31, S. Karger, Basel, 1976, 165.
53. **Jones, L. M. and Berman, D. T.,** Antibody-mediated and delayed-type hypersensitivity reactions to *Brucella* skin test antigens in guinea pigs, *Infect. Immun.,* 11, 360, 1975.
54. **Jones, L. M., Berman, D. T., and Díaz, R.,** Factors influencing biometric potency assays of *Brucella* allergens to guinea-pigs, *Br. J. Exp. Pathol.,* 54, 368, 1973.
55. **Jones, L. M., Díaz, R., and Taylor, A. G.,** Characterization of allergens prepared from smooth and rough strains of *Brucella melitensis, Br. J. Exp. Pathol.,* 54, 492, 1973.
56. **De Reviers, M. and Fensterbank, R.,** Controle biologique de la brucelline INRA sur cobayes sensibilisés: guide pour l'analyse statistique des resultats, *Ann. Rech. Vet.,* 15, 395, 1984.
57. **Fountain, M. W., Weiss, S. J., Fountain, A. G., Shen, A., and Lenk, R. P.,** Treatment of *Brucella canis* and *Brucella abortus* in vitro and in vivo by stable plurilamellar vesicle-encapsulated aminoglycosides, *J. Infec. Dis.,* 152, 529, 1985.
58. **Philippon, A., Kazmierczak, A., Plommet, M., Marly, J., and Nevot, R.,** Traitement de la brucellose expérimentale de la souris et du cobaye par la rifampicine, in *Developments in Biological Standardization,* Vol. 31, S. Karger, Basel, 1976, 279.
59. **Stuart, F. A.,** Comparison of rifampicin and tetracycline based regimens in the treatment of experimental brucellosis, *J. Infect.,* 5, 27, 1982.
60. **Peplow, A. M., Peplow, P. V., and Hafez, E. S. E.,** Parameters of reproduction, in *Handbook of Laboratory Animal Science,* Melby, E. C., Jr. and Altman, N. H., Eds., CRC Press, Cleveland, 1974, 107.
61. **Thorpe, B. D., Sidwell, R. W., and Lundgren, D. L.,** Experimental studies with four species of *Brucella* in selected wildlife, laboratory, and domestic animals, *Am. J. Trop. Med. Hyg.,* 16, 665, 1967.
62. **Taran, I. F. and Rybasov, N. A.,** (Comparative study of susceptibility and sensitivity to infection of laboratory animals and sheep with various species of *Brucella*), *Zh. Mikrobiol. Epidemid. Immunobiol.,* No. 10, 97, 1971; abstracted in *Vet. Bull. (London),* 42, 578, 1972.
63. **Cheers, C.,** Pathogenesis and cellular immunity in experimental murine brucellosis, in *Developments in Biological Standardization,* Vol. 56, S. Karger, Basel, 1984, 237.

64. **Birmingham, J. R. and Jeska, E.,** Characterization of macrophage functions in mice infected with *Brucella abortus, Infect. Immun.,* 32, 1079, 1981.
65. **Ho, M. and Cheers, C.,** Resistance and susceptibility of mice to bacterial infection. IV. Genetic and cellular basis of resistance to chronic infection with *Brucella abortus, J. Infect. Dis.,* 146, 381, 1982.
66. **Cheers, C. and Ho, M.,** Resistance and susceptibility of mice to bacterial infection. VI. Functional specificity in natural resistance to facultative intracellular bacteria, *J. Reticuloendothel. Soc.,* 34, 299, 1983.
67. **Cannat, A. and Serre, A.,** Genetic factors involved in murine susceptibility to experimental brucellosis, *Ann. Immunol. (Paris),* 132D, 15, 1981.
68. **Plommet, M. and Plommet, A. M.,** Evolution de l'infection splénique de souris de quatre lignees, inoculées par voie veineuse, par trois doses de *Brucella abortus, Ann. Rech. Vet.,* 12, 345, 1981.
69. **Cannat, A. and Serre, A.,** Genetic factors involved in murine resistance to experimental brucellosis, in *Developments in Biological Standardization,* Vol. 56, S. Karger, Basel, 1984, 357.
70. **Cannat, A., Escande, A., Peraldi, F., and Serre, A.,** Induction of autoantibodies and circulating immune complexes in mice after injection of *Brucella* fraction "Pi" or inoculation with live *Brucella suis, Ann. Immunol. (Paris),* 134, 43, 1983.
71. **Abernathy, R. S., Bradley, G. M., and Spink, W. W.,** Increased susceptibility of mice with brucellosis to bacterial endotoxins, *J. Immunol.,* 81, 271, 1958.
72. **Galdiero, F., Romano Carratelli, C., Nuzzo, I., and Folgore, A.,** Cytotoxic antibody dependent cells in mice experimentally infected with *Brucella abortus, Microbiologica,* 8, 217, 1985.
73. **Jacotot, H. and Vallée, A.,** Brucellose expérimentale du rat blanc apres inoculation intracardiaque, *Ann. Inst. Pasteur (Paris),* 84, 964, 1953.
74. **Pardon, P. and Marly, J. R.,** Resistance of *Brucella abortus* infected mice to intravenous or intraperitoneal *Brucella* infection, *Ann. Immunol. (Paris),* 127, 57, 1976.
75. **Schmitt-Slomska, J., Caravano, R., Anoal, M., Gay, B., and Roux, J.,** Isolation of L-forms from the spleens of *Brucella suis*-infected penicillin-treated mice, *Ann. Microbiol. (Paris),* 132, 253, 1981.
76. **Bosseray, N.,** Vaccine and serum-mediated protection against *Brucella* infection of mouse placenta, *Br. J. Exp. Pathol.,* 64, 617, 1983.
77. **Bosseray, N.,** Infection du placenta de la souris par *Brucella.* Pathogénie et immunite, in *Developments in Biological Standardization,* Vol. 56, S. Karger, Basel, 1984, 283.
78. **Plommet, M. and Plommet, A. M.,** Effect des sérums immuns sur l'évolution de la brucellose experimentale de la souris, in *Developments in Biological Standardization,* Vol. 56, S. Karger, Basel, 1984, 271.
79. **Bosseray, N.,** Colonization of mouse placentas by *Brucella abortus* inoculated during pregnancy, *Br. J. Exp. Pathol.,* 61, 361, 1980.
80. **Bosseray, N.,** Kinetics of placental colonization of mice inoculated intra-venously with *Brucella abortus* at day 15 of pregnancy, *Br. J. Exp. Pathol.,* 64, 612, 1983.
81. **Kawashima, H., Yanagawa, R., Isayama, Y., and Hirota, E.,** Experimental studies on *Brucella* infection in mice under the influence of sex hormones, *Bull. Nat. Inst. Anim. Health,* 34, 40, 1958.
82. **Bosseray, N.,** Mother to young transmission of *Brucella abortus* infection in mouse model, *Ann. Rech. Vet.,* 13, 341, 1982.
83. **Suire, A.,** Les vaccinations anti-*Brucella* par la souche vivante B.112, *Ann. Inst. Pasteur (Paris),* 99, 241, 1960.
84. **Montaraz, J. A., Winter, A. J., Hunter, D. M., Sowa, B. A., Wu, M. M., and Adams, L. G.,** Protection against *Brucella abortus* in mice with O-poly-saccharide-specific monoclonal antibodies, *Infect. Immun.,* 51, 961, 1986.
85. **Montaraz, J. A. and Winter, A. J.,** Comparison of living and nonliving vaccines for *Brucella abortus* in BALB/C mice, *Infect. Immun.,* 53, 245, 1986.
86. **Plommet, M. and Bosseray, N.,** Propositions pour une méthode générale de controle d'activité des vaccins antibrucelliques, in *Developments in Biological Standardization,* Vol. 56, S. Karger, Basel, 1984, 247.
87. **Plommet, M., Bosseray, N., Lantier, F., Bernard, F., Pardon, P., and Rodolakis, A.,** Simultaneous vaccination by three living attenuated strains of *Brucella, Salmonella,* and *Chlamydia* in mice, *Vaccine,* 5, 27, 1987.
88. **Bosseray, N., Plommet, A. M., and Plommet, M.,** Theoretical, practical and statistical basis for a general control method of activity for anti-*Brucella* vaccines, in *Developments in Biological Standardization,* Vol. 56, S. Karger, Basel, 1984, 257.
89. **Pilet, C. and Le Garrec, Y.,** Controle, sur petits animaux de laboratorie, du pouvoir immunogene des vaccins antibrucelliques inactivés. I. Sur le choix des especes animales et de l'époque du sacrifice, lors du controle du pouvoir immunogene par le test de l'infection, *Ann. Inst. Pasteur (Paris),* 110, 755, 1966.
90. **Pilet, C. and Le Garrec, Y.,** Controle sur petits animaux de laboratoire du pouvoir immunogene des vaccins antibrucelliques inactivés. II. Sur les modalités de traitement avant ensemencement des organes infectés. Résultats et conséquences, *Ann. Inst. Pasteur (Paris),* 112, 384, 1967.
91. **Pilet, C. and Le Garrec, Y.,** Testing the immunogenicity of inactivated brucellosis vaccines in small laboratory animals, *Symp. Ser. Immunobiol. Stand.,* 10, 125, 1969.

92. **Geissler, A.,** Die Immunreaktionen der weissen Maus nach der Impfung mit Abortus-Bang-Impfstoffen, *Zentralbl. Veterinaermed. Reihe B,* 6, 981, 1959.
93. **Pilet, C., Mallick, B. B., and Le Garrec, Y.,** Immunisation contre la brucellose du nouveau-né issu de mere saine et non immunisée, *Bull. Acad. Vet. Fr.,* 40, 287, 1967.
94. **Plommet, M., Plommet, A. M., and Bosseray, N.,** L'évolution de la brucellose chronique de la souris n'est pas ameliorée par l'administration d'un antigene vaccinal, *Ann. Rech. Vet.,* 13, 127, 1982.
95. **Young, E. J. and McMillan, K.,** Effects of levamisole in murine brucellosis, *J. Infect. Dis.,* 156, 530, 1987.
96. **Philippon, A. and Kazmierczak, A.,** Rifampicine et tetracycline dans le traitement de la brucellose expérimentale de la souris, *Ann. Inst. Pasteur (Paris),* 123, 459, 1972.
97. **Philippon, A., Kazmierczak, A., and Nevot, P.,** Traitement de la brucellose de la souris par les antibiotiques (tetracycline, ampicilline, trimethoprime-sulfamethoxazole) en association avec *Corynebacterium parvum, Ann. Inst. Pasteur (Paris),* 123, 349, 1972.
98. **Thompson, H. V. and Worden, A. N.,** *The rabbit,* Collins Clear-type Press, London, 1956.
99. **Davoli, R. and Gargani, G.,** Infezione intradermica sperimentale da brucelle nel coniglio, *Vet. Ital.,* 6, 3, 1955.
100. **Kniazeff, A. J., Elberg, S. S., and Traum, J.,** Pathogenesis of experimental brucellosis of rabbits in pseudopregnancy, *J. Comp. Pathol.,* 74, 141, 1964.
101. **Jacotot, H. and Vallée, A.,** Essais comparatifs d'immunisation du rat blanc contre la brucellose experimentale au moyen de vaccin vivant et de vaccin tué, *Ann. Inst. Pasteur (Paris),* 87, 152, 1954.
102. **Jacotot, H. and Vallée, A.,** Essai d'immunisation du rat blanc contre la brucellose, *Ann. Inst. Pasteur (Paris),* 94, 372, 1958.
103. **Halliday, R.,** The relationship between the occurrence of mortality and the development of active immunity in the young rat, *Proc. R. Soc. London Ser. B,* 161, 208, 1964.
104. **Harms, R. and Wegener, K. H.,** Übertragung der Rinderbrucellose durch Ratten, *Berl. Münch. Tieraerztl. Wochenschr.,* 66, 242, 1953.
105. **Cuba Caparo, A. and Myers, D. M.,** Pathogenesis of epididymitis caused by *Brucella ovis* in laboratory animals, *Am. J. Vet. Res.,* 34, 1077, 1973.

INDEX

A

B

C

D

E

F

G

H

I

Q

R

S

T

U

V

W

X

Z